Springer Series in Solid-State Sciences

Volume 181

The Springer Series in Solid-State Sciences consists of fundamental scientific books prepared by leading researchers in the field. They strive to communicate, in a systematic and comprehensive way, the basic principles as well as new developments in theoretical and experimental solid-state physics.

More information about this series at http://www.springer.com/series/682

Friedhelm Bechstedt

Many-Body Approach to Electronic Excitations

Concepts and Applications

Springer

Friedhelm Bechstedt
Department of Physics and Astronomy
Friedrich-Schiller University
Jena
Germany

ISSN 0171-1873 ISSN 2197-4179 (electronic)
ISBN 978-3-662-44592-1 ISBN 978-3-662-44593-8 (eBook)
DOI 10.1007/978-3-662-44593-8

Library of Congress Control Number: 2014947656

Springer Heidelberg New York Dordrecht London

Printed on acid-free paper

Springer is part of Springer Science+Business Media (www.springer.com)

To Andreas, Susanne and Uta

Preface

In recent decades, the modern first-principles theory of real materials and its practical implementation in computer codes has made enormous progress. It is illustrated by rapid advances in basic theory, computational methods, and applications. The electronic structure theory has advanced to the point where not only an accurate description of properties of condensed matter such as solids, nanosystems, and molecules is possible but bold predictions of yet unmade materials and of unsuspected physical properties are being made.

A subdiscipline of the electronic structure field is the investigation of excited states of matter to learn more about materials properties or to find novel physical effects, sometimes called 'theoretical spectroscopy'. It combines quantum-mechanics-based many-body theories and computer simulations to understand the interaction of radiation and matter. The understanding of both the interaction of matter and radiation – visible or ultraviolet light, X-rays, and electron beams – and our capability to analyze and predict the materials reaction enhances our ability to design new materials, improve devices, and understand our environment. Beyond the interpretation of results of experimental spectroscopies from an atomistic quantum-mechanical point of view, the theoretical spectroscopy has reached predictive power for properties of complex materials critical to the development of new technologies. It includes to predict atomic arrangements to design materials with a desired spectroscopic property.

The literature, including the number of good books, on the modern electronic structure theory can hardly be overlooked today. By contrast, the rapidly developing field of electronic excitations of matter has been documented only in very few books, mainly as an appendix to the conventional electronic structure theory with a stronger focus to ground state properties and a reduced relationship to the theory and calculation of spectral properties, which can be directly compared with measurements. The theoretical spectroscopy is based on the many-body perturbation theory (MBPT) developed by Kadanoff, Baym, Hedin, Lundqvist, and many other colleagues already in the 1960s of the last century. However, the first numerical simulation of single-particle excitations by Hybertsen and Louie became possible in the mid-1980s after the implementation of the density functional theory, developed originally for the ground states. Based on these '*ab initio*' (called in physics but not in chemistry) methods the first simulation of electron-hole pair excitations and the computation of optical spectra from first principles have been demonstrated 1998 by Reining, Del Sole and others. Meanwhile, the solution of Dyson and Bethe-Salpeter equations has been developed to standard methods which supplement many available electronic structure codes. However, there is an increasing opening of a gap between the knowledge of the theoretical basics and the frequent use of such codes to study real problems.

The purpose of this book is therefore to provide a unified exposition of the many-body theory and methods of electronic structure calculations, together with instructive examples for computational methods and actual applications or comparisons with measured data. The theoretical and numerical methods are developed toward the calculation of charged and neutral electronic excitations as well as complete electron or optical spectra. The presentation is focused on the many-body perturbation theory based on Green functions. Other approaches to electronic excitations as the time-dependent density functional theory, the dynamical mean-field theory or the quantum Monte Carlo method are only mentioned for the benefit to follow a clear 'red line' from the basic theory to numerical calculation of spectra. The author apologizes for this and other subjective decisions, for instance the selection of examples. The aim of the book is to serve graduate students as well as researchers in the field. Consequently, it not only provides a text for courses on electronic structure or to serve as supplementary material for courses on condensed matter physics, quantum chemistry and materials science but an advanced text for Ph.D. students and scientists working in the field of theoretical spectroscopy. Even the second half of each part of the book, in particular the final results and the application of the theory to real problems, should be interesting for experimentalists. The book is also intended to improve communication between the two communities in physics and chemistry, despite the fact that the theoretical methods mainly originate from the physics of the inhomogeneous electron gas. All readers are encouraged to provide feedback to the author suggesting updates, corrections, additions, etc.

The text is divided into four parts. Part I describes condensed matter in terms of the many-body quantum mechanics and quantum field theory. Special care is taken to characterize not only the motion of the electrons in the field of nuclei but also the electron-electron interaction. Besides the (longitudinal) Coulomb interaction of the electrons also their (transverse) interaction via the entire electromagnetic field generated by the moving electrons and their spins are described. The terms exchange and correlation are introduced. Electron exchange is described within the Hartree-Fock theory. The description of electronic excitations asks for starting electronic structures. Therefore, Part II is devoted to the density functional theory, in particular to the use of the Kohn-Sham ansatz and to widely used exchange-correlation functionals. Generalizations to spatially non-local functionals and the inclusion of dispersion forces are also presented. This part does not compete with specialized books about density functional theory. Rather, it only serves to illustrate how starting atomic geometries and electronic structures can be made available on a first-principles basis for the subsequent studies of electronic excitations. The concept of thermodynamic Green functions within the framework of the grand canonical statistics is applied in Part III. It addresses the derivation of the set of fundamental equations of the many-body theory based on Matsubara Green functions. The quasiparticle concept is introduced. The understanding and the explicit use of the developed scheme are mainly illustrated in the framework of the Hedin GW approximation. The success of the approach is demonstrated for all kinds of condensed matter including a comparison with experimental data as, e.g., obtained by means of photoemission spectroscopy. The not fully understood problem of satellite structures to single-particle excitations is discussed in the last chapter. Part IV describes electron-hole-pair and collective excitations. The Bethe-Salpeter equations for the polarization and density correlation functions based on the two-particle Green function are derived. For the description of optical spectra the influence of the spin structure and the inclusion of local-field effects are discussed. The Bethe-Salpeter equation for the macroscopic polarization function is solved within the GW approximation. The relationship to excitons of different kinds and consequences for optical spectroscopies are illustrated. As a culmination of the present-day treatment, it is clearly demonstrated that an optical or energy-loss spectrum can be only computed in agreement with experiment if excitonic and quasiparticle effects are included. Finally, the inclusion of dynamical effects and free carriers and their consequences are described.

Jena, July 2014 Friedhelm Bechstedt

Acknowledgments

Many people and several institutions have played an important role in shaping the author and his work related to electronic excitations and their many-body treatment. For the first time, in the beginning of the 1970s of the last century at the Humboldt University in Berlin H. Stolz made the many-body theory of electron gases accessible to me. Later in 1976–1982, during my postdoc time, R. Enderlein (Humboldt University Berlin) and V.L. Bonch-Bruevich (Moscow State University) stimulated my interest in this field and the preparation of my habilitation thesis 'On the theory of core electron excitations in semiconductors'. Already in 1986 R. Del Sole (University II Rome) suggested to start joint work directed to the application of the GW approximation to calculate the electronic structure of solids. This was the begin of a fruitful collaboration for decades. It converged in a first institutional network EPSI (Electronic Properties of Semiconductors and Insulators) of eight European groups. In addition, to R. Del Sole also the groups of C.-O. Almbladh/U. von Barth (University Lund), R. Godby (University of Cambridge and York), L. Hedin/ O. Gunnarson (MPI Solid-State Physics Stuttgart), and L. Reining (Ecole Polytechnique Palaiseau) took part. Three years later the network blew up to EXCAM (Electronic Exchange and Correlation in Advanced Materials) with a further group, that of A. Rubio (University Valladolid and San Sebastian).

In the year 2000 there was a continuation with another network NANOPHASE (Nanoscale Photon Absorption and Spectroscopy with Electrons) and a new member M. Scheffler (FHI MPG Berlin). Five years later a further extension resulted in the Network of Excellence NANOQUANTA (Nanoscale Quantum Simulation for Nanostructures and Advanced Materials) with the additional groups of E.K.U. Gross (Free University Berlin and MPG Halle), X. Gonze (University Louvain), and G. Onida (University of Milan). All these activities culminated 2009 in the foundation of ETSF (European Theoretical Spectroscopy Facility) consisting currently of 17 European and US groups, among them G. Kresse (University of Vienna), C. Draxl (Humboldt University Berlin) and J. Rehr (University of Washington). The scientific discussions and collaborations within these networks were not only a great pleasure and benefit but also sharpened our scientific topics at the University of Jena toward the *ab initio* many-body description of electronic excitations in electron and optical spectra of condensed matter including surfaces, nanostructures and molecules. Moreover, former students and postdocs O. Pulci (Rome), A. Schleife (Urbana-Champaign) and W.G. Schmidt (Paderborn) grew up in the network communities and now run their own research groups. My personal scientific background concerning the electronic structure methods, the many-body Green function theory and their applications has been deepened during longer research stays at the Stanford University (W.A. Harrison), Fritz-Haber Institute Berlin (M. Scheffler), University II Rome (R. Del Sole), University of California San Diego (L.J. Sham), Ecole Polytechnique Palaiseau (L. Reining), and University of California Santa Barbara (C.G. Van de Walle). In addition to the European Community, for all these activities funding from the Deutsche Forschungsgemeinschaft, the Volkswagen Foundation, the German Academic Exchange Service (DAAD), the Carl-Zeiss Foundation, and the Austrian Fonds zur Förderung der wissenschaftlichen Forschung (FFW) has to be acknowledged.

The book is based on lectures on 'Density Functional Theory', 'Elementary Excitations in Solids', 'Green Function Theory', 'Exchange and Correlation', 'Many-Body Theory', etc., given in the last twenty years at the Friedrich-Schiller-Universität Jena for Diploma and Ph.D. students as well as postdocs. The actual writing of the book has been influenced by many discussions with colleagues around the world. Several of them who provided figures are specifically acknowledged in the text. I would also like to thank colleagues for a critical reading of parts of the manuscript: G. Cappellini, J. Furthmüller, D. Ködderitsch, L. Kühl-Teles, M. Marsili, J. Paier, C. Rödl, M. Rohlfing, A. Schleife, and W.G. Schmidt. The typing of the LaTeX manuscript was achieved with competence and infinite patience by my secretary Sylvia Hennig. This also holds for the preparation or modification of many of the figures.

Contents

Acronyms

Density Functional Theory

B3LYP	Becke three-parameter Lee-Yang-Parr
EOS	Equation of state
DFT	Density functional theory
DFPT	Density functional perturbation theory
EXX	Exact exchange
GGA	Generalized gradient approximation
gKS	Generalized Kohn-Sham
HSE	Heyd-Scuseria-Ernzerhof
KS	Kohn-Sham
LDA	Local density approximation
LSDA	Local spin density approximation
OEP	Optimized effective potential
SIC	Self-interaction correction
sX	Screened exchange
TDDFT	Time-dependent density functional theory
TFD	Thomas-Fermi-Dirac
vdW	Van der Waals
PBE	Perdew-Burke-Ernzerhof
PW91	Perdew-Wang 1991
PZ	Perdew-Zunger

Electronic Structure Methods

APW	Augmented plane wave
BZ	Brillouin zone
CBM	Conduction band minimum
DOS	Density of states
EEX	Exact exchange

HOMO	Highest occupied molecular orbital
IP	Ionization potential
KKR	Korringa-Kohn-Rostocker
MP	Monkhorst-Pack
LAPW	Linearized augmented plane wave
LMTO	Linear muffin-tin orbital
LUMO	Lowest unoccupied molecular orbital
PAW	Projector-augmented wave
PP	Pseudopotential
PW	Plane wave
TB	Tight-binding
VASP	Vienna ab initio simulation package
VBM	Valence band maximum

General Terms

FWHM	Full width at half maximum
HWHM	Half width at half maximum
MAE	Mean absolute error
MARE	Mean absolute relative error
rh	Rhombohedral
rs	Rocksalt
rt	Rutile
wz	Wurtzite
zb	Zinc blende

Many-body Perturbation Theory

AC	Adiabatic connection
BSE	Bethe-Salpeter equation
COH	Coulomb-hole
DE	Dyson equation
DMFT	Dynamical mean field theory
FDT	Fluctuation-dissipation theorem
GW	Green function-screened potential
GWA	GW approximation
LF	Local-field
MBPT	Many-body perturbation theory
QP	Quasiparticle
RPA	Random-phase approximation
SEX	Screened exchange

SPP	Single-plasmon pole
TDA	Tamm-Dancoff approximation
XC	Exchange-correlation

Quantum Chemistry

CI	Configuration interaction
CC	Coupled cluster
HF	Hartree-Fock
HFA	Hartree-Fock approximation
MP2	Møller-Plesset perturbation theory
QMC	Quantum Monte Carlo
SCF	Self-consistent field
SE	Single excitation
SOSEX	Screened-second order exchange
UHF	Unrestricted Hartree-Fock

Spectroscopies and Other Experimental Techniques

AFM	Atomic force microscopy
ARPES	Angle-resolved photoemission spectroscopy
ARUPS	Angle-resolved ultraviolet photoemission spectroscopy
EELS	Electron energy loss spectroscopy
IPES	Inverse photoemission spectroscopy
IXS	Inelastic X-ray scattering
KRIPES	k-resolved inverse photoemission spectroscopy
MBE	Molecular beam epitaxy
MOVPE	Metal organic vapor phase epitaxy
PES	Photoelectron spectroscopy
RAS	Reflectance anisotropy spectroscopy
SDR	Surface differential reflectance
STM	Scanning tunneling microscopy
STS	Scanning tunneling spectroscopy
XAS	X-ray absorption spectroscopy

Symbols

Constants

a_B	Bohr radius
c	Speed of light
$\hbar$	Planck constant
k_B	Boltzmann constant
m	Free electron mass
R_∞	Hydrogen Rydberg
α_s	Sommerfeld finestructure constant
ε_0	Permittivity of vacuum
μ_0	Permeability of free space

Crystal Quantities

$\boldsymbol{a}_i$	Basis vector of Bravais lattice
$\boldsymbol{b}_j$	Basis vector of reciprocal lattice
E_{cut}	Plane wave energy cutoff
g	Charge asymmetry coefficient
$\mathbf{G}$	Vector of reciprocal lattice
$\mathbf{k}$	Bloch wave vector
M_l	Mass of nucleus at position $\mathbf{R}_l$
$\mathbf{r}_l$	Vector of atomic basis
$\mathbf{R}_l$	Atomic position
$\mathbf{R}$	Bravais lattice vecto
Z_l	Nuclear number at $\mathbf{R}_l$
Ω	Sample volume
Ω_0	Volume of unit cell
ω_p	Plasma frequency of valence electrons
ν	Bloch band index

Energy Contributions

E_{C}	Correlation energy
E_{coh}	Cohesive energy
E_{H}	Hartree (direct Coulomb) energy
E_{kin}	Kinetic energy
E_{KS}	Kohn-Sham energy
E_{tot}	Total energy
E_{X}	Exchange energy
E_{XC}	Exchange-correlation energy
E_{nn}	Repulsion energy of nuclei
F_{HK}	Universal Hohenberg-Kohn functional
T_s	kinetic energy of non-interacting electrons
$\epsilon_{\mathrm{XC}}(\mathbf{x};[n])$	Exchange-correlation energy per particle

Green and Response Functions

$A_{ss'}(\mathbf{xx}',\omega)$	Spectral function
$A_{\nu\nu'}(\mathbf{k},\omega)$	Spectral function in Bloch representation
$G^{>}_{ss'}(\mathbf{xx}',t-t')$	Electron propagator
$G^{<}_{ss'}(\mathbf{xx}',t-t')$	Hole propagator
$G_{ss'}(\mathbf{xx}',t-t')$	Single-particle Green function
$G_{ss'}(11')$	Single-particle Green function
$G_{s_1s_2,s_{1'}s_{2'}}(12,1'2')$	Two-particle Green function
$L_{ss'}(11')$	Density correlation function
$L^{\gtrless}(11')$	Spin-averaged electron-hole-pair propagator
$\hat{L}(\mathbf{xx}',\omega)$	Spectral function of density correlation function
$P(11')$	Two-point polarization function
$P(11',22')$	Four-point polarization function
$\hat{P}(\mathbf{xx}',\omega)$	Spectral function of polarization function
$P^M(11',22')$	Macroscopic polarization function
$S(\mathbf{q},\omega)$	Dynamic structure factor
$\mathcal{S}(t,t_0)$	Scattering matrix
z_n	Poles of Fermi function
$\tilde{z}_m$	Poles of Bose function
$\Gamma_{s_1s_2,s_3s_{3'}}(12,3)$	Vertex function
$\epsilon(\mathbf{xx}',t-t')$	Dielectric function
$\epsilon(\mathbf{q}+\mathbf{G},\mathbf{q}+\mathbf{G}';\omega)$	Fourier transformed dielectric function (tensor)
$\epsilon^M(\mathbf{q},\omega)$	Macroscopic dielectric function
$\tilde{\Sigma}_{ss'}(11')$	Self-energy due to longitudinal electron-electron interaction
$\Sigma_{ss'}(11')$	Exchange-correlation self-energy
$\Sigma^{\mathrm{C}}_{ss'}(11')$	Correlation self-energy
$\Sigma^{\mathrm{H}}_{ss'}(11')$	Hartree self-energy

$\Sigma^{X}_{ss'}(11')$	Exchange self-energy
$\Xi_{s_1 s_{1'}, s_2 s_{2'}}(11', 22')$	Kernel of BSE for polarization function
$\tilde{\Xi}_{s_1 s_{1'}, s_2 s_{2'}}(11', 22')$	Kernel of BSE for density correlation function
$\Xi^{M}_{s_1 s_{1'}, s_2 s_{2'}}(11', 22')$	Kernel of BSE for macroscopic polarization function
Δ_{scissors}	Scissors shift

Operators

$\hat{a}_{\lambda m_s}$	Annihilation operator of electron in state λm_s
$\hat{a}^{+}_{\lambda m_s}$	Creation operator of electron in state λm_s
$\mathscr{H}$	Hamiltonian of interacting electrons with perturbation
$\mathscr{H}_0$	Hamiltonian without perturbation
$\mathscr{H}_{\text{ext}}$	External perturbation
$\hat{H}$	Single-particle Hamiltonian
$H(c\mathbf{k}m_{s_c} v\mathbf{k}m_{s_v}, c'\mathbf{k}'m_{s_{c'}} v'\mathbf{k}'m_{s_{v'}})$	Electron-hole-pair Hamiltonian
$\hat{\mathbf{j}}_p(\mathbf{x})$	Paramagnetic current density operator
$\hat{\mathbf{m}}(\mathbf{x})$	Magnetization density operator
$\hat{n}(\mathbf{x})$	Electron density operator
$\hat{n}_s(\mathbf{x})$	Spin density operator
$\hat{N}$	Particle number operator
$\mathbf{p}$	Momentum operator of an electron
$\hat{\mathbf{s}}(\mathbf{x})$	Spin density operator
$\hat{\mathbf{S}}$	Total spin operator
$\hat{T}$	Electronic kinetic energy operator
$\mathcal{T}$	Wick time-ordering operator
$\hat{U}$	Operator of electron-electron interaction
$\hat{V}$	Operator of electron-nuclei interaction
$\hat{W}_0$	Grand canonical statistical operator
$\mathbf{x}$	Space operator (or coordinate)
$1 = \mathbf{x}t$	Space-time variable
σ	Vector of Pauli spin matrices
$\psi^{+}_s(\mathbf{x}, t)$	Creation operator of an electron
$\psi^{+}_s(1)$	Creation operator of an electron
$\psi_s(\mathbf{x}, t)$	Annihilation operator of an electron
$\psi_s(1)$	Annihilation operator of an electron

Potentials and Fields

$\mathbf{A}(\mathbf{x})$	Vector potential
$\mathbf{B}(\mathbf{x})$	Magnetic field

$\mathbf{E}(\mathbf{x})$	Electric field
U	On-site Coulomb interaction
$\mathrm{v}(\mathbf{x})$	Bare Coulomb potential
$V_{\mathrm{n}}(\mathbf{x})$	Potential energy of electron at **x** in the field of nuclei
$V(\mathbf{x})$	Potential energy of an electron in a mean-field approximation
$V_{\mathrm{H}}(\mathbf{x})$	Hartree potential
$V_{\mathrm{KS}}(\mathbf{x})$	Kohn-Sham potential
$V_{\mathrm{XC}}(\mathbf{x})$	Exchange-correlation potential
$\varphi(\mathbf{x})$	Scalar potential
$\phi_{ss'}(\mathbf{x})$	Perturbation potential
$W(\mathbf{xx}', t-t')$	Screened Coulomb potential
$\hat{W}(\mathbf{xx}', \omega)$	Spectral function of screened Coulomb potential

Single-, Two- and Many-particle Quantities

A	Electron affinity
$A_\Lambda(c\mathbf{k}m_{sc}, v\mathbf{k}m_{sv})$	Electron-hole-pair eigenvector
$B^{\mathbf{kk}'}_{\substack{m_s m_s \\ vv'}}(\mathbf{q}+\mathbf{G})$	Bloch integral
$D(\hbar\omega)$	Single-particle density of states
E_B	Exciton binding energy
E_g	Fundamental quasiparticle gap
E_g^{KS}	Fundamental Kohn-Sham gap
E_Λ	Electron-hole-pair excitation energy
E_Ψ	Energy of many-electron system in state Ψ
I	Ionization energy
m_v^*	Effective mass
$M_{\lambda\lambda'}^{m_s}(\hat{\mathbf{q}})$	Optical transition matrix element
m_s	Quantum number of (z-component of) spin
$n(\mathbf{x})$	Electron density
r_s	Average distance of electrons in units of a_B
$\bar{\mathrm{v}}^{\mathbf{kk}'}_{\substack{cv \\ c'v'}}$	Bare electron-hole-exchange matrix element
$W^{\mathbf{kk}'}_{\substack{cc' \\ vv'}}$	Matrix element of screened interaction
$\varepsilon^{\mathrm{QP}}_{\lambda m_s}$	Quasiparticle energy
$\varepsilon_{\lambda m_s}$	Single-electron energy
$\varsigma_\alpha(ss')$	Electron-hole-pair spin function
$\varphi_{\lambda m_s}(\mathbf{x})$	Electron wave function in spin channel m_s
$\phi_\Lambda(\mathbf{x}, s)$	Pauli spinor
$\Phi_\Lambda(\mathbf{x})$	Excitonic wave function

$\chi_{\frac{1}{2}m_s}(s)$	Single-particle spin function
$\lvert\Psi\rangle$	Many-electron wave function in Hilbert or Fock space

Thermodynamics

B	Isothermal bulk modulus
F	Helmholtz free energy
$f(\hbar\omega)$	Fermi function
$g(\hbar\omega)$	Bose function
G	Gibbs free enthalpy
N	Particle number
p	pressure
S	Entropy
T	Absolute temperature
U	Internal energy
Ω_g	Grand thermodynamic potential
μ	Chemical potential of electrons
ε_F	Fermi energy: chemical potential of electrons at 0 K
ε_F^0	Fermi energy of non-interacting electrons

Part I
Electron-Electron Interaction

Chapter 1
Born-Oppenheimer Approximation

Abstract In condensed matter the motion of the electrons is determined by the electric field generated by the nuclei and their mutual interaction. The conditions for neglecting the vibrations of the nuclei around their fixed positions are discussed. The arrangement of the nuclei rules the symmetry and classification of electronic states. The strength of the electron-nucleus interaction is used to distinguish valence and core electrons.

1.1 Solids and Molecules as Many-Body Systems

Systems of condensed matter such as molecules and solids consist of atomic nuclei l of mass M_l and charge Z_l at positions $\mathbf{R}_l$ and electrons i with mass m and charge $-e$ ($e > 0$) at positions $\mathbf{x}_i$ and spin variables $\mathbf{s}_i = \frac{\hbar}{2}\boldsymbol{\sigma}$ with $\boldsymbol{\sigma}$ as the vector of Pauli spin matrices

$$\sigma_x = \begin{pmatrix} 0 & 1 \\ 1 & 0 \end{pmatrix}, \quad \sigma_y = \begin{pmatrix} 0 & -i \\ i & 0 \end{pmatrix}, \quad \text{and} \quad \sigma_z = \begin{pmatrix} 1 & 0 \\ 0 & -1 \end{pmatrix}. \tag{1.1}$$

The particles can move in the systems, i.e., they possess momentum operators $\mathbf{P}_l = -i\hbar\nabla_{\mathbf{R}_l}$ (nuclei) or $\mathbf{p}_i = -i\hbar\nabla_{\mathbf{x}_i}$ (electrons). In the spirit of the 'first quantization' $\mathbf{R}_l$ and $\mathbf{x}_i$ represent canonical position operators. A possible spin of the nuclei is not considered in this book. Each system with all its transport, optical, magnetic, mechanical and thermal properties represents a quantum-mechanical many-body system. For example, in a cube of 1 cm^3 of a silicon crystal one finds 5×10^{22} nuclei and 7×10^{23} core and valence electrons. For N_n nuclei the total number of electrons N in a neutral system is given by $N = \sum_{l=1}^{N_\text{n}} Z_l$. The Hamiltonian for such a system consists of a sum of five terms: the kinetic energies of the nuclei and electrons, the interactions between nuclei, between electrons and nuclei, and between electrons. The first two types of interactions can be approximately described by Coulomb potentials

$$\text{v}(\mathbf{x}) = \frac{e^2}{4\pi\varepsilon_0|\mathbf{x}|} \tag{1.2}$$

F. Bechstedt, *Many-Body Approach to Electronic Excitations*,
Springer Series in Solid-State Sciences 181, DOI 10.1007/978-3-662-44593-8_1

with the permittivity of vacuum $\varepsilon_0 = 8.854188 \times 10^{-12}$ As/Vm using consistently SI units in this book. In the non-relativistic limit the Hamiltonian of the system has the form

$$\mathscr{H}_{\text{sys}} = \frac{1}{2}\sum_{l=1}^{N_{\text{n}}} \frac{1}{M_l}\mathbf{P}_l^2 + \frac{1}{2}\sum_{i=1}^{N} \frac{1}{m}\mathbf{p}_i^2 + \frac{1}{2}\sum_{\substack{l,l'=1 \\ (l \neq l')}}^{N_{\text{n}}} Z_l Z_{l'} \text{v}(\mathbf{R}_l - \mathbf{R}_{l'}) - \sum_{l=1}^{N_{\text{n}}}\sum_{i=1}^{N} Z_l \text{v}(\mathbf{x}_i - \mathbf{R}_l) + \mathscr{H}_{ee}. \tag{1.3}$$

In this limit it holds $\mathscr{H}_{ee} = \frac{1}{2}\sum_{\substack{i,i'=1 \\ (i \neq i')}}^{N} \text{v}(\mathbf{x}_i - \mathbf{x}_{i'})$ for the electron-electron interaction. It is also described by the Coulomb potential (1.2). Relativistic corrections will be discussed in the next chapter.

If $\mathscr{H}_{ee}$ is known one can solve the Schrödinger equation of the many-body system with the Hamiltonian (1.3) and, in principle, predict all its properties, e.g. its geometric structure including the arrangement of all the nuclei, its thermodynamic behavior, its electrical or thermal transport properties, its excitation spectra, etc. However, such a complete solution of the many-body problem is certainly impossible. Due to the nucleus-nucleus, electron-nucleus, and electron-electron interactions, the Hamiltonian cannot be separated into a sum over Hamiltonians of independent particles. One must therefore solve a coupled system of (differential) equations with a huge number of variables of the order of 10^{23}. This is not feasible numerically and also makes less sense from the viewpoint of physics.

1.2 Decoupling of Electron and Nucleus Motion

To make any progress in the theoretical treatment, one is compelled to make certain approximations, mainly motivated by physical considerations. The first obvious fact is that the protons and neutrons in the nuclei are by a factor 1836 more massive than the electrons. For that reason, in general, the nuclei will have a much slower dynamics than the electrons. For instance, the frequencies of atomic vibrations in a typical solid, e.g. a semiconductor, are less than 10^{13} s^{-1}. The energy required to excite electrons is given by its fundamental band gap of the order of 1 eV. The frequencies of electronic motion in such a semiconductor are therefore of the order of 10^{15} s^{-1}. Consequently, electrons can respond to atomic vibrations almost instantaneously. They follow the atomic motion adiabatically. On the other hand, one may say, nuclei cannot follow the motion of the electrons and only see a time-averaged electronic potential. In the Born-Oppenheimer or adiabatic approximation [1, 2], studying the motion of electrons, the kinetic energy of the nuclei (first term in (1.3)) can be neglected to a first approximation. Experimental observations confirm this idea. Crystallographic studies show that the 'atoms' make up a static, i.e., time-independent, arrangement $\{\mathbf{R}_l\}$ with small oscillations about their equilibrium positions.

In the older formulation of the adiabatic approximation Born and Oppenheimer [1] expanded the Hamiltonian (1.3) in terms of $\left(\frac{m}{M_l}\right)^{1/4}$, the fourth root of the mass ratio between an electron and any of the nuclear masses. The criterion $\left(\frac{m}{M_l}\right)^{1/4} \ll 1$ has been later reformulated by Born and Huang [2]. In any case it indicates the validity of the adiabatic approximation for not too light elements. Consequently, the treatment of the motion of electrons and nuclei can be decoupled. The neglected interaction can be included in a later step as electron-phonon interaction.

The criterion for semiconductors, insulators, and molecules, small vibrational frequencies compared to the fundamental energy gap, is seemingly violated for metals, while the mass criterion is still fulfilled for sufficiently heavy elements. Nevertheless, more recent investigations [3] showed that in general the adiabatic approximation is also valid for metals. In metals non-adiabaticity is governed by the ratio of a characteristic phonon frequency and the Fermi energy. For a wide range of temperatures the thermal energy $k_B T$ is usually lower than the Fermi energy ε_F of the electron gas in a metal. Therefore, electronic excitations are confined to a narrow region around the Fermi surface, and most of the properties of a metal are little affected by neglecting non-adiabatic contributions due to the motion of the metal electrons.

As a result of the Born-Oppenheimer approximation the motion of the electrons will be investigated for fixed positions of the nuclei $\{\mathbf{R}_l\}$. Then the third term in (1.3), the energy of repulsion of the nuclei, does not depend on electronic coordinates and, hence, only shifts the total energy of the electrons by the fixed value $E_{\mathrm{nn}} = \frac{1}{2}\sum_{l,l'=1}^{N_{\mathrm{n}}} Z_l Z_{l'} \mathrm{v}(\mathbf{R}_l - \mathbf{R}_{l'})$. Electronic energies and wave functions will be obtained as functions of the nuclei positions. The residual impact of the charged nuclei on the electrons can be described by a potential energy

$$V_{\mathrm{n}}(\mathbf{x}) = -\sum_{l=1}^{N_{\mathrm{n}}} Z_l \mathrm{v}(\mathbf{x} - \mathbf{R}_l) \tag{1.4}$$

of an electron at position $\mathbf{x}$. The interaction of the spins of nuclei and electrons is not taken into account. Consequently, hyperfine splittings in localized electron systems will not be discussed. We have to point out that in the framework of the description of the pure electronic problem the potential energy (1.4) can be considered as an 'external' (from the point of view of the electrons) potential which depends parametrically on the atomic positions.

1.3 Atomic Arrangements

The atomic positions in condensed matter can be in general arbitrary, particularly in amorphous systems. However, there are many examples, e.g. molecules and crystals, where the positions of the nuclei underly certain point and/or translational and space symmetries. In addition, for the (numerical) modeling of non-crystalline solids, such

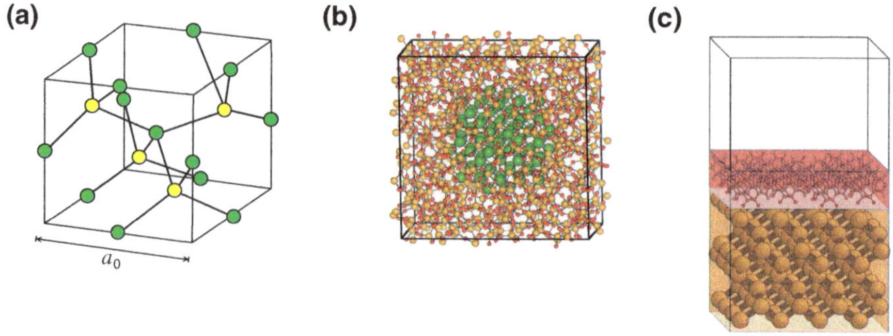

Fig. 1.1 Illustration of primitive and non-primitive unit cells of three-dimensional periodic atomic arrangements: (**a**) Cube with edge length a_0 of a zinc-blende crystal with cations (*yellow dots*) and anions (*green dots*), (**b**) Si nanocrystal with 99 atoms (*green spheres*) in an amorphous SiO_2 matrix with Si (*yellow dots*) and oxygen (*red dots*) atoms arranged in a simple cubic lattice (Reprinted with permission from [5]. Copyright 2012 by the American Physical Society.), and (**c**) orthorhombic material slab to simulate a Si(001) surface covered by organic molecules together with a separating vacuum region [6]

as nanostructures, molecules, clusters, surfaces, and interfaces, frequently periodic arrangements of such objects in supercells are used, i.e., artificial translationally invariant structures. In addition to supercell arrangements repeated slab descriptions are applied [4]. Illustrations of characteristic unit cells and atomic arrangements for a crystal, embedded nanocrystals, and an isolated surface with adsorbate film are displayed in Fig. 1.1. The regularity of the atomic arrangements of such objects in one, two or three dimensions can be mathematically described by a space group of symmetry operations which transforms the arrangement into itself while leaving one space point fixed. Modeling high-symmetric nanoobjects such as a benzene molecule as illustrated in Fig. 1.2, the shape of the supercell and the orientation of the nanoobject in the supercell should be chosen in such away that its point symmetry is not broken. One subgroup of the space group could be a translational symmetry group or simply a translational group of spatial translations $\{\mathbf{R}\}$, which can be represented by primitive basis vectors $\boldsymbol{a}_1$, $\boldsymbol{a}_2$, and $\boldsymbol{a}_3$ according to

$$\mathbf{R} = \sum_{i=1}^{3} n_i \boldsymbol{a}_i \qquad (n_i \; - \; \text{integer}). \tag{1.5}$$

The set $\{\mathbf{R}\}$ defines a Bravais lattice [8]. In three dimensions 14 Bravais lattice types exist. The parallelepiped spanned by the basis vectors is a primitive unit cell with the volume

$$\Omega_0 = \boldsymbol{a}_1(\boldsymbol{a}_2 \times \boldsymbol{a}_3). \tag{1.6}$$

The parallelepiped is not imperative. Other unit cell shapes are possible. The primitive unit cell which represents the point-group symmetry by visual inspection is the so-called Wigner-Seitz cell. The center of this cell lies on a point $\mathbf{R}$ and its surface is

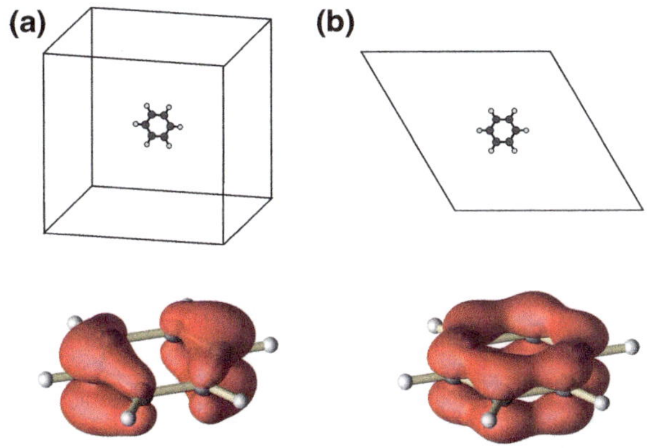

Fig. 1.2 Arrangement of a benzene molecule in a (100) plane of an arrangement of cubic supercells (**a**) or in the (0001) plane of a hexagonal supercell (**b**). The point groups of the three-dimensional repeated supercell systems D_{2h} (**a**) and D_{6h} (**b**) strongly influence the computed π-electron density (*red clouds*) but not the total one. In the case (**a**) an unphysical symmetry break is clearly visible. Courtesy of M. Preuss, Universität Jena

formed by the perpendicular bisector planes which divide in half the line segments joining the center $\mathbf{R}$ to adjacent lattice points $\mathbf{R}'$.

The arrangements of the nuclei $\{\mathbf{R}_l\}$ in the natural or artificial translationally invariant structures can be related to Bravais lattice vectors according to

$$\mathbf{R}_l = \mathbf{R} + \mathbf{r}_l. \tag{1.7}$$

The set $\{\mathbf{r}_l\}$ of vectors $\mathbf{r}_l$ describes the atomic positions in one unit cell. It is therefore called atomic basis. Together with the Bravais lattice vectors $\{\mathbf{R}\}$ they describe all atomic positions in the natural or artificial crystal.

The set of all wave vectors $\mathbf{G}$, which yield plane waves with the periodicity of a given Bravais lattice, is known as its reciprocal lattice $\{\mathbf{G}\}$. These vectors are defined as

$$\mathbf{G} = \sum_{j=1}^{3} m_j \mathbf{b}_j \qquad (m_j \;-\; \text{integer}) \tag{1.8}$$

with basis vectors $\mathbf{b}_1$, $\mathbf{b}_2$, and $\mathbf{b}_3$ which satisfy the relation

$$\boldsymbol{a}_i \cdot \mathbf{b}_j = 2\pi\delta_{ij}. \tag{1.9}$$

The construction procedure, which gives in real space the Wigner-Seitz cell, leads in reciprocal space to the Brillouin zone (BZ) with the volume

$$\mathbf{b}_1(\mathbf{b}_2 \times \mathbf{b}_3) = \frac{(2\pi)^3}{\Omega_0}. \tag{1.10}$$

Three-dimensional examples are illustrated in Fig. 1.3. It is common to denote the high-symmetry points in the BZ by capital letters. Greek letters are assigned to symmetry points (and lines) in the interior of the BZ but latin letters to symmetry points (and lines) on its surface. The irreducible part of the BZ is also indicated for the (highest) point-group symmetry of the corresponding Bravais lattice.

Each Bravais lattice point $\mathbf{R}$ is related to a translational operator $T_{\mathbf{R}} = e^{\mathbf{R}\nabla_{\mathbf{x}}}$ which, when operating on any function $f(\mathbf{x})$, shifts the argument according to

$$T_{\mathbf{R}} f(\mathbf{x}) = f(\mathbf{x} + \mathbf{R}). \tag{1.11}$$

The translational operators form an Abelian group with a complete system of eigenfunctions $\varphi_{\nu\mathbf{k}}(\mathbf{x})$, which can be classified by means of the eigenvalues $\mathbf{k}$ of the operator $-i\nabla_{\mathbf{x}}$. It holds the Bloch theorem [7]

$$T_{\mathbf{R}}\varphi_{\nu\mathbf{k}}(\mathbf{x}) = e^{i\mathbf{k}\mathbf{R}}\varphi_{\nu\mathbf{k}}(\mathbf{x}) = \varphi_{\nu\mathbf{k}}(\mathbf{x} + \mathbf{R}). \tag{1.12}$$

Since $e^{i(\mathbf{k}+\mathbf{G})\mathbf{R}} = e^{i\mathbf{k}\mathbf{R}}$ all eigenvalues of $T_{\mathbf{R}}$ are obtained with $\mathbf{k} \in \text{BZ}$. For any given $\mathbf{k}$, a countably infinite set of eigenfunctions exists. They can be labeled with an additional index ν, following for instance the increasing value of energy. The operator $T_{\mathbf{R}}$ commutes with the single-particle Hamilton operator $\hat{H}$ of the electronic system taken in a certain approximation. Since the two operators have a simultaneous system of eigenfunctions, and relation (1.12) represents the Bloch theorem [8], functions $\{\varphi_{\nu\mathbf{k}}(\mathbf{x})\}$ can be identified with Bloch functions

$$\varphi_{\nu\mathbf{k}}(\mathbf{x}) = \frac{1}{\sqrt{\Omega}} e^{i\mathbf{k}\mathbf{x}} u_{\nu\mathbf{k}}(\mathbf{x}) \tag{1.13}$$

with the periodic Bloch factor $u_{\nu\mathbf{k}}(\mathbf{x}) = u_{\nu\mathbf{k}}(\mathbf{x} + \mathbf{R})$, the total volume Ω of the condensed-matter system, and the Bloch wave vector $\mathbf{k} \in \text{BZ}$. The additional index ν may be directly related to the Bloch energy eigenvalues $\varepsilon_\nu(\mathbf{k})$, i.e., to the Bloch band index.

In many analytical and numerical studies the infinite periodic systems are described by macroscopic (but finite) ones with volume Ω. Such a volume can be identified with a parallelepiped with edges $G_1\boldsymbol{a}_1$, $G_2\boldsymbol{a}_2$, and $G_3\boldsymbol{a}_3$, where G_1, G_2, and G_3 are large (but finite) integer numbers. Then the Born-von Karman periodic boundary conditions [9] on the Bloch eigenfunctions (1.13) require

$$\varphi_{\nu\mathbf{k}}(\mathbf{x}) = \varphi_{\nu\mathbf{k}}(\mathbf{x} + G_i\boldsymbol{a}_i) \qquad (i = 1, 2, 3). \tag{1.14}$$

Together with (1.12) they lead to the conditions $G_i\mathbf{k}\boldsymbol{a}_i = 2\pi m_i$ (m_i – integer) with the solution

$$\mathbf{k} = \sum_{i=1}^{3} \frac{m_i}{G_i}\mathbf{b}_i. \tag{1.15}$$

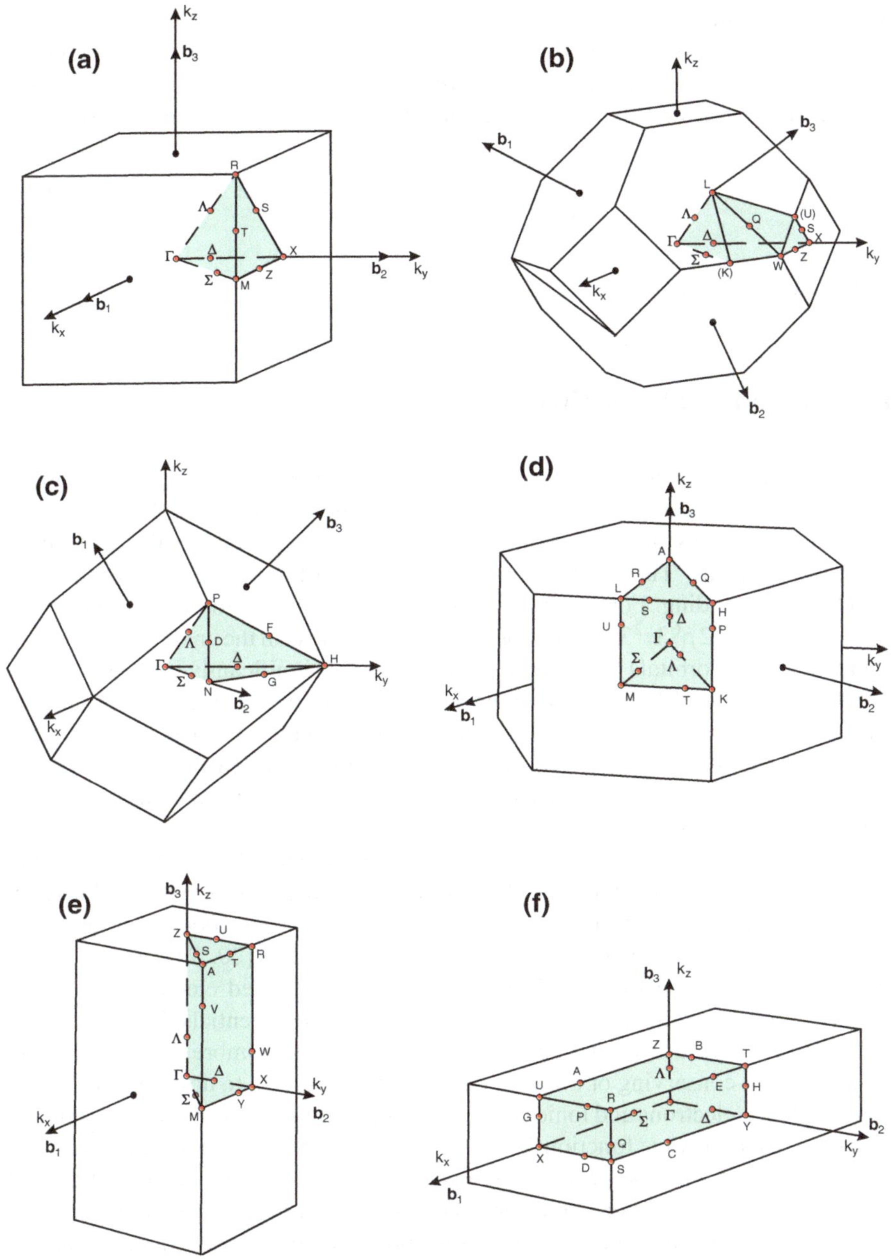

Fig. 1.3 Brillouin zone and high-symmetry lines/points for the simple cubic (sc) lattice (*cube*) (**a**), the face-centered cubic (fcc) lattice (*truncated octahedron*) (**b**), the body-centered cubic (bcc) lattice (*rhombic dodecahedron*) (**c**), the hexagonal (h) Bravais lattice (**d**), the simple tetragonal (st) Bravais lattice (**e**), and the simple orthorhombic (sor) Bravais lattice (**f**). The *green* region shows the irreducible part of a BZ

Since the volume of the BZ $\mathbf{b}_1(\mathbf{b}_2 \times \mathbf{b}_3) = \frac{(2\pi)^3}{\Omega_0}$ is given by (1.10), (1.15) asserts that the number of possible wave vectors $\mathbf{k}$ in such a primitive cell of the reciprocal lattice is equal to the number N_c of the lattice points in the volume $\Omega = N_c\Omega_0$. The numerical advantage of the use of periodic boundary conditions is that it holds

$$\frac{1}{\Omega}\sum_{\mathbf{k}} \to \frac{1}{(2\pi)^3}\int_{\mathrm{BZ}} d^3\mathbf{k} \tag{1.16}$$

in the limit $\Omega \to \infty$.

1.4 Core and Valence Electrons

Obviously, despite fixed nuclei, the eigenvalue problem of the resulting many-particle Hamiltonian (1.3) can generally not be solved without further simplifications. For solids and molecules the electrons can be separated into two groups, valence electrons and core electrons. They can be distinguished according to their contribution to the chemical bonding. For instance, the electronic configuration of a Si atom with $Z = 14$ is $1s^2 2s^2 2p^6 3s^2 3p^2$. The core electrons are those in the energetically lowest, completely filled orbitals $1s$, $2s$, and $2p$. They are not significantly influenced by the chemical bonding of Si atoms to other ones. They are mostly localized around the nucleus, so they can be 'lumped' together with the nucleus to form an ion core with a valence $Z_{\mathrm{val}} = 4$. The outer electrons in the incompletely filled shell, such in the $3s$ and $3p$ states, are called valence electrons because of their substantial contribution to the chemical bonding.

Several modern electronic-structure codes, e.g. WIEN2k [10], EXCITING [11], CRYSTAL09 [12], and FHI-aims [13, 14], are still all-electron codes and take the motion of the core electrons into account. However, other codes such as VASP [15–17], ABINIT [18, 19], and QUANTUM ESPRESSO [20, 21] are (or can be [13, 14]) restricted to valence electrons. The more localized electrons are frozen into the core. Consequently, instead of bare Coulomb potentials $Z_l v(\mathbf{x} - \mathbf{R}_l)$ in (1.3), these codes use pseudopotentials (see Sect. 8.3.2), more precisely atomic *ab initio* norm-conserving or ultrasoft pseudopotentials [22], to describe the interaction of valence electrons and ionic cores. The disadvantage of such pseudopotentials, to lead to 'pseudo-wave functions' which are too smooth in the core regions, is currently overcome with the projector augmented wave (PAW) formulation [23, 24]. The resulting wave functions of the valence electrons are all-electron wave functions with the correct nodal structure around the cores. There are also examples for efficient total-energy codes such as SIESTA [25, 26] which are based on a description of valence electrons by a few localized orbitals.

References

1. M. Born, R. Oppenheimer, Quantum theory of the molecules. Ann. d. Physik **84**, 457–484 (1927)
2. M. Born, K. Huang, *Dynamic Theory of Crystal Lattices* (Oxford University Press, Oxford, 1954)
3. E.G. Brovman, Yu.M. Kagan, Phonons in non-transition metals, in *Dynamical Properties of Solids*, vol. I, ed. by G.K. Horton, A.A. Maradudin (North-Holland, Amsterdam, 1974), pp. 191–301
4. F. Bechstedt, *Principles of Surface Physics* (Springer, Berlin, 2003)
5. K. Seino, F. Bechstedt, P. Kroll, Tunneling of electrons between Si nanocrystals embedded in a SiO_2 matrix. Phys. Rev. B **86**, 075312 (2012)
6. A. Hermann, Ab initio Untersuchung eines molekularen π-Elektronensystems auf der Si(001)-Oberfläche. Diploma thesis, Friedrich-Schiller-Universität Jena (2004)
7. F. Bloch, Über die Quantenmechanik der Elektronen in Kristallgittern. Z. Phys. **52**, 555–560 (1928)
8. Ch. Kittel, *Introduction to Solid State Physics* (Wiley, Hoboken, 2005)
9. M. Born, Th. von Kármán, Über Schwingungen in Raumgittern. Z. Physik **13**, 297–309 (1912)
10. http://www.wien2k.at/
11. http://exciting-code.org/
12. http://www.cse.clrc.ac.uk/cmg/CRYSTAL/
13. https://aimsclub.fhi-berlin.mpg.de/aims
14. M. Fuchs, M. Scheffer, Ab initio pseudopotentials for electronic structure calculations of poly-atomic systems using density functional theory. Comput. Phys. Commun. **119**, 67–98 (1999)
15. www.vasp.at/
16. G. Kresse, J. Furthmüller, Efficient iterative schemes for *ab initio* total-energy calculations using a plane-wave basis set. Phys. Rev. B **54**, 11169–11186 (1996)
17. G. Kresse, J. Furthmüller, Efficiency of ab-initio total energy calculations for metals and semi-conductors using a plane-wave basis set. Comput. Mater. Sci. **6**, 11–50 (1996)
18. http://www.abinit.org/
19. X. Xonze, J.-M. Beuken, R. Caracas, F. Detraux, M. Fuchs, G.-M. Riganese, L. Sindic, M. Verstraete, G. Zerah, F. Jollet, M. Torrent, A. Roay, M. Mikami, Ph Ghosez, J.-Y. Raty, D.C. Allan, First-principles computation of material properties: the ABINIT software project. Comput. Mater. Sci. **25**, 478–492 (2002)
20. http://www.quantum-espresso.org/
21. P. Gianozzi, S. Baroni, N. Bonini, M. Calandra, R. Car, C. Cavazzoni, D. Ceresoli, G.L. Chiarotti, M. Cococcioni, I. Dabo, A. Dal Corso, S. de Gironcoli, S. Fabris, G. Fratesi, R. Gebauer, U. Gerstmann, C. Gougoussis, A. Kokalj, M. Lazzeri, L. Martin-Samos, N. Marzari, F. Mauri, R. Mazzarello, S. Paolini, A. Pasquarello, L. Paulatto, S. Scandolo, G. Sclauzero, A.P. Seitsonen, A. Smogunov, P. Umari, R.M. Wentzcovitch, QUANTUM ESPRESSO: a modular and open-source software project for quantum simulations of materials. J. Phys. Condens. Matter **21**, 395502 (2009)
22. R.M. Martin, *Electronic Structure. Basic Theory and Practical Methods* (Cambridge University Press, Cambridge, 2004)
23. P.E. Blöchl, Projector augmented-wave method. Phys. Rev. B **50**, 17953–17979 (1994)
24. G. Kresse, D. Joubert, From ultrasoft pseudopotentials to the projector augmented-wave method. Phys. Rev. B **59**, 1758–1775 (1999)
25. http://www.icmab.es/siesta/
26. J.M. Soler, E. Artacho, J.D. Gale, A. García, J. Junquera, P. Ordejón, D. Sánchez-Portal, The SIESTA method for ab initio order-N materials simulation. J. Phys. Condens. Matter **14**, 2745–2780 (2002)

Chapter 2
Hamiltonian of Interacting Electrons

Abstract Relativistic effects also influence the motion of electrons. They are described in the framework of the Pauli equation but with a velocity operator from the more general Dirac theory. The moving electrons generate an electromagnetic field beyond the electric field due to the nuclei. It describes the electron-electron interaction and depends on the position, momentum and spin operator of each individual electron in a self-consistent manner. The electromagnetic field is calculated up to the second order in the ratio of electron velocity and speed of light. Besides the well-known scalar-relativistic corrections, the Darwin and mass-correction terms, and the spin-orbit interaction known for isolated atoms, an additional relativistic effect, the Breit interaction, is described by the coupling of the vector potential to the mechanical momentum and of the magnetic field to the electron spin. In addition to the non-relativistic mutual Coulomb interaction of the electrons, the longitudinal one, a relativistic transverse interaction appears, which, however, can be neglected in non-magnetic systems or systems where the spin-orbit coupling predominates the magnetic dipole-dipole interaction.

2.1 Quasi-relativistic Electrons

Relativistic effects are essential for heavy atoms and carried over to molecules and solids essentially unchanged. For their description on the level of independent electrons the Schrödinger equation of an almost independent electron has to be replaced by the famous equation proposed by Dirac in 1928 [1, 2]. As spin has been shown to be one of the most important consequences of the relativistic nature of an electron, the Dirac equation has, in general, to use as starting point for its description. Thereby the relativistic motion of an electron or its antiparticle, the positron, is characterized by means of a 'wave function' that has four components, the so-called Dirac spinor. For a many-electron system with all possible interactions between the particles the many-body equivalent of the Dirac equation is too complex for an accurate numerical treatment. Rather, in the physics of condensed matter one is interested in a perturbative treatment of relativistic effects but for interacting electrons. In the non-relativistic limit the electron and positron parts of the Dirac equation may be separated by means of the Foldy-Wouthuysen transformation [3] to give an equation

F. Bechstedt, *Many-Body Approach to Electronic Excitations*,
Springer Series in Solid-State Sciences 181, DOI 10.1007/978-3-662-44593-8_2

for the two-component wave function describing the electron alone. We follow this idea to describe the electron motion by a two-component spinor but leave this straightforward mathematical way. Instead, we try to describe such effects following more the physical intuition but take the electron-electron interaction beyond the electrostatic repulsion into account.

The interaction of two relativistic Dirac electrons is not only described by the Coulomb potential $\mathrm{v}(\mathbf{x})$ as indicated in (1.3). Rather, it has to be modified by the Breit interaction potential [4–7]

$$\mathrm{v}(\mathbf{x}_i - \mathbf{x}_j) \to \mathrm{v}(\mathbf{x}_i - \mathbf{x}_j)\left\{1 - \boldsymbol{\alpha}_i \cdot \boldsymbol{\alpha}_j + \frac{1}{2}\left(\boldsymbol{\alpha}_i \cdot \boldsymbol{\alpha}_j - \frac{\left[\boldsymbol{\alpha}_i \cdot (\mathbf{x}_i - \mathbf{x}_j)\right]\left[\boldsymbol{\alpha}_j \cdot (\mathbf{x}_i - \mathbf{x}_j)\right]}{|\mathbf{x}_i - \mathbf{x}_j|^2}\right)\right\} \tag{2.1}$$

with $\boldsymbol{\alpha}_j$ as the vector of Dirac α matrices, which act on the spinor wave function of an electron j [8]. Their off-diagonal elements are given by the Pauli spin matrices (1.1), while the diagonal elements are zero [9]. The Breit interaction modification in (2.1) corresponds to the lowest-order relativistic corrections. Besides the magnetic interaction of the electrons [second term in (2.1)], it contains also retardation effects (third term in parenthesis in (2.1)). The first correction term to the bare Coulomb interaction in (2.1), the magnetic interaction, is also known as Gaunt term. The sum of both corrections represents the actual (frequency-independent) Breit term [7]. Indeed, the Breit interaction can be taken via (2.1) into the electronic-structure calculations of molecules and solids [9–11].

In molecular and condensed-matter physics usually also the Dirac description of the electrons is approximated. This particularly holds for studying electronic excitations and not ground states as in a relativistic density functional theory [12, 13]. Only contributions up to the second order in the ratio of electron velocity and speed of light c are taken into account. The four-component spinors of the Dirac theory are decoupled. Only two-component spinors are used to describe the electrons. The Pauli Hamilton operator is derived from the Dirac Hamiltonian [8]. For vanishing motion of the nuclei and neglecting for a moment the repulsion of the nuclei, according to Pauli [14] the quasi-relativistic two-component Hamiltonian of an ensemble of electrons at $\mathbf{x}_j$ with spin $\mathbf{s}_j$ can be rewritten in second order in c^{-2} to [7, 8, 14, 15]

$$\begin{aligned}\mathcal{H} = \sum_{j=1}^{N} \Bigg\{ & \frac{1}{2m}\left[\mathbf{p}_j + e\mathbf{A}(\mathbf{x}_j)\right]^2 + V_{\mathrm{n}}(\mathbf{x}_j) - e\varphi(\mathbf{x}_j) \\ & - \frac{\hbar e}{2i(2mc)^2}\mathbf{p}_j\left[\mathbf{E}_{\mathrm{n}}(\mathbf{x}_j) + \mathbf{E}(\mathbf{x}_j)\right] - \frac{1}{2m(2mc)^2}\left[\mathbf{p}_j + e\mathbf{A}(\mathbf{x}_j)\right]^4 \\ & + \frac{e}{2(mc)^2}\mathbf{s}_j\left(\left[\mathbf{E}_{\mathrm{n}}(\mathbf{x}_j) + \mathbf{E}(\mathbf{x}_j)\right] \times \left[\mathbf{p}_j + e\mathbf{A}(\mathbf{x}_j)\right]\right) \\ & + \frac{e}{m}\mathbf{s}_j\mathbf{B}(\mathbf{x}_j)\Bigg\}.\end{aligned} \tag{2.2}$$

Here notations of the scalar product $\boldsymbol{ab}$ and the vector product $\boldsymbol{a} \times \boldsymbol{b}$ of two Cartesian vectors with three components have been introduced.

A more elegant way to derive the Hamiltonian (2.2) up contributions $\sim\frac{1}{c^2}$ could be the use of the results of the Foldy-Wouthuysen transformation [3] of the Dirac problem for a particle in an external electromagnetic field. Here the electromagnetic field is due to the nuclei and electrons. The fixed nuclei generate the electrostatic field

$$\mathbf{E}_{\mathrm{n}}(\mathbf{x}) = -\frac{1}{e} \text{ grad } V_{\mathrm{n}}(\mathbf{x}) \tag{2.3}$$

with the potential energy $V_{\mathrm{n}}(\mathbf{x})$ (1.4) of a quasi-relativistic electron. In addition, an electromagnetic field with electric and magnetic components $\mathbf{E}(\mathbf{x})$ and $\mathbf{B}(\mathbf{x})$, which are described by a scalar potential $\varphi(\mathbf{x})$ and a vector potential $\mathbf{A}(\mathbf{x})$ according to

$$\begin{aligned} \mathbf{E}(\mathbf{x}) &= -\text{ grad } \varphi(\mathbf{x}) - \frac{\partial}{\partial t}\mathbf{A}(\mathbf{x}), \\ \mathbf{B}(\mathbf{x}) &= \text{curl } \mathbf{A}(\mathbf{x}) \end{aligned} \tag{2.4}$$

may act on the electrons.

For vanishing electromagnetic field (usually assumed as a macroscopic external one) and neglecting the relativistic corrections, (2.2) changes over into a Hamiltonian of N non-interacting electrons which move in the potential $V_{\mathrm{n}}(\mathbf{x})$. It is given by the first line of (2.2) for $\varphi(\mathbf{x}) = 0$ and $\mathbf{A}(\mathbf{x}) = 0$. The second line in (2.2) describes scalar-relativistic corrections $\sim (2mc)^{-2}$. The first term in this line is the so-called Darwin term, which may be interpreted as the first relativistic correction to the potential energy of an electron since it is proportional to the electron density and the distribution of nuclear charges. The second term $\sim\mathbf{p}^4$ in this line corresponds to the first relativistic correction to the kinetic energy. It describes a mass enhancement and is sometimes called mass-velocity term. The term in the third line is also proportional to $(mc)^{-2}$. It couples the spin $\mathbf{s}_j$ with the spatial motion of an electron with momentum $\mathbf{p}_j$ (modified by the vector potential) at position $\mathbf{x}_j$ under the action of the total electric field. It is therefore called spin-orbit interaction. The term in the last line represents the interaction of the magnetic field with the magnetic moment related to the electron spin.

2.2 Electromagnetic Field Due to Electrons

Formally the Hamiltonian (2.2) does not contain any electron-electron interaction. It enters the many-body problem in a natural way if the electromagnetic field $\mathbf{E}(\mathbf{x})$, $\mathbf{B}(\mathbf{x})$ with the potentials $\varphi(\mathbf{x})$, $\mathbf{A}(\mathbf{x})$ is interpreted as the microscopic electromagnetic field generated by the moving electrons with their charges and spins. Of course, such a replacement has to be done with care, in order to avoid double counting of pair interactions. That means, for instance, that the term $-e\varphi(\mathbf{x})$ has to be replaced by $-\frac{e}{2}\varphi(\mathbf{x})$, if $\varphi(\mathbf{x})$ is basically given as the Coulomb interaction between two electrons (2.11). However, also other contributions to the Hamiltonian have to be weighted by $\frac{1}{2}$. The second complication is related to the fact that the field components now become operators in the electron coordinates.

The electrons at positions $\mathbf{x}_j$ give rise to an electron density operator

$$\hat{n}(\mathbf{x}) = \sum_{j=1}^{N} \delta(\mathbf{x} - \mathbf{x}_j). \tag{2.5}$$

Each electron possesses a velocity operator

$$\hat{\mathbf{v}}_j = \frac{i}{\hbar}[\mathscr{H}, \mathbf{x}_j]_- \tag{2.6}$$

which leads to a current density operator

$$\hat{\mathbf{j}}(\mathbf{x}) = \frac{1}{2}\sum_{j=1}^{N} \left[\hat{\mathbf{v}}_j, \delta(\mathbf{x} - \mathbf{x}_j)\right]_+, \tag{2.7}$$

where the brackets in (2.6) and (2.7)

$$\left[\hat{A}, \hat{B}\right]_{\mp} = \hat{A}\hat{B} \mp \hat{B}\hat{A} \tag{2.8}$$

denote the commutation ($-$) or anticommutation ($+$) relation of the operators $\hat{A}$ and $\hat{B}$.

The electromagnetic field due to the moving electrons, $\mathbf{E}(\mathbf{x})$ and $\mathbf{B}(\mathbf{x})$, is described by the Maxwell equations in vacuum with the sources $\hat{n}(\mathbf{x})$ and $\hat{\mathbf{j}}(\mathbf{x})$. The potentials (2.4) fulfill the condition of gauge invariance. We use the Coulomb gauge

$$\text{div}\,\mathbf{A}(\mathbf{x}) = 0. \tag{2.9}$$

Then the scalar potential is given by the Poisson equation

$$\Delta_{\mathbf{x}}\varphi(\mathbf{x}) = \frac{e}{\varepsilon_0}\hat{n}(\mathbf{x}), \tag{2.10}$$

and hence by

$$\varphi(\mathbf{x}) = -\frac{e}{4\pi\varepsilon_0}\int_{\Omega} d^3\mathbf{x}'\frac{\hat{n}(\mathbf{x}')}{|\mathbf{x} - \mathbf{x}'|} = -\frac{1}{e}\sum_{j=1}^{N} \mathrm{v}(\mathbf{x} - \mathbf{x}_j). \tag{2.11}$$

The vector potential is given by the wave equation

$$\Delta_{\mathbf{x}}\mathbf{A}(\mathbf{x}) - \frac{1}{c^2}\frac{\partial^2}{\partial t^2}\mathbf{A}(\mathbf{x}) = \mu_0 e\hat{\mathbf{j}}(\mathbf{x}) + \frac{1}{c^2}\frac{\partial}{\partial t}\,\mathrm{grad}\,\varphi(\mathbf{x}) \tag{2.12}$$

with the permeability of free space $\mu_0 = 4\pi \times 10^{-7}$ Vs/Am, which is related to the velocity of light in vacuum by $\varepsilon_0\mu_0 = 1/c^2$. According to (2.11) and (2.12), both potentials $\varphi(\mathbf{x})$ and $\mathbf{A}(\mathbf{x})$ become operators in the electron coordinates. As a formal generalization a time dependence of the quantities $\mathbf{A}(\mathbf{x})$, $\varphi(\mathbf{x})$, and $\hat{\mathbf{j}}(\mathbf{x})$ is taken into account.

According to (2.12) the vector potential $\mathbf{A}(\mathbf{x})$ is influenced by retardation effects. However, it is by itself a functional of the vector potential due to the Hamiltonian (2.2) that determines the particle velocities (2.6) appearing in the definition of the current density operator (2.7). We approximately treat this self-consistency by expanding the Hamiltonian (2.2) up to second-order relativistic corrections $\sim c^{-2}$. In lowest order the vector potential itself is proportional to c^{-1}. That means, in general, it can be neglected in the first relativistic corrections to the kinetic energy and the spin-orbit interaction in (2.2). For the particle velocity it holds (which can be better seen directly from the Dirac theory than from the Pauli description [7])

$$\hat{\mathbf{v}}_j = \frac{1}{m}\left\{\mathbf{p}_j + e\mathbf{A}(\mathbf{x}_j) + \frac{2i}{\hbar}(\mathbf{p}_j \times \mathbf{s}_j)\right\}. \tag{2.13}$$

The first two summands give the electromagnetic momentum of an electron, while the third contribution is due to the spin motion. Consequently the current density operator can be divided in three contributions

$$\hat{\mathbf{j}}(\mathbf{x}) = \hat{\mathbf{j}}_p(\mathbf{x}) + \hat{\mathbf{j}}_d(\mathbf{x}) + \hat{\mathbf{j}}_s(\mathbf{x}), \tag{2.14}$$

where

$$\hat{\mathbf{j}}_p(\mathbf{x}) = \frac{1}{2m}\sum_{j=1}^{N}\left[\mathbf{p}_j, \delta(\mathbf{x} - \mathbf{x}_j)\right]_+ \tag{2.15}$$

is the paramagnetic current density operator,

$$\hat{\mathbf{j}}_d(\mathbf{x}) = \frac{e}{m}\hat{n}(\mathbf{x})\mathbf{A}(\mathbf{x}) \tag{2.16}$$

is the diamagnetic current density operator, and

$$\hat{\mathbf{j}}_s(\mathbf{x}) = \frac{1}{m}\nabla_{\mathbf{x}} \times \hat{\mathbf{s}}(\mathbf{x}) \tag{2.17}$$

is the spin density current operator with the vector of the spin density operator

$$\hat{\mathbf{s}}(\mathbf{x}) = \frac{\hbar}{2} \sum_{j=1}^{N} \delta(\mathbf{x} - \mathbf{x}_j) \boldsymbol{\sigma} .$$

Because of the restriction to terms $\sim c^{-2}$ the diamagnetic contribution to the velocity operator does not play a role. Including retardation effects, in lowest order in c^{-1} the solution of (2.12) can be written as

$$\mathbf{A}^{(1)}(\mathbf{x}) = -\frac{\mu_0 e}{8\pi m} \sum_{j=1}^{N} \left\{ \frac{\boldsymbol{a}_j}{|\mathbf{x} - \mathbf{x}_j|} + \frac{[\boldsymbol{a}_j (\mathbf{x} - \mathbf{x}_j)](\mathbf{x} - \mathbf{x}_j)}{|\mathbf{x} - \mathbf{x}_j|^3} \right\} \tag{2.18}$$

with the generalized velocity operator

$$\boldsymbol{a}_j = \frac{1}{m} \mathbf{p}_j + \frac{2i}{m\hbar} (\mathbf{p}_j \times \mathbf{s}_j). \tag{2.19}$$

This operator enters the kinetic energy and the spin-magnetic field coupling in the Hamiltonian (2.2).

2.3 Relativistic and Non-relativistic Contributions

The resulting Hamiltonian (2.2) for quasi-relativistic electrons can be divided according to

$$\mathscr{H} = \mathscr{H}_0 + \mathscr{H}_{sr} + \mathscr{H}_{so} + \mathscr{H}_B . \tag{2.20}$$

The first contribution is the non-relativistic Hamiltonian of the electrons (without self-interaction)

$$\mathscr{H}_0 = \sum_{j=1}^{N} \left\{ \frac{1}{2m} \mathbf{p}_j^2 + V_{\mathrm{n}}(\mathbf{x}_j) \right\} + \frac{1}{2} \sum_{\substack{i,j=1 \\ (i \neq j)}}^{N} \mathrm{v}(\mathbf{x}_i - \mathbf{x}_j). \tag{2.21}$$

It contains the longitudinal electron-electron interaction mediated by the Coulomb potential (1.2). The scalar-relativistic corrections, i.e., Darwin term and mass correction, are

$$\mathscr{H}_{sr} = \frac{1}{2(2mc)^2} \sum_{j=1}^{N} \left\{ \hbar^2 \Delta_{\mathbf{x}} \left[V_{\mathrm{n}}(\mathbf{x}) - e\varphi(\mathbf{x}) \right]_{\mathbf{x}=\mathbf{x}_j} - \frac{1}{m} \mathbf{p}_j^4 \right\} . \tag{2.22}$$

The spin-orbit coupling follows as

$$\mathscr{H}_{so} = -\frac{1}{2(mc)^2} \sum_{j=1}^{N} \mathbf{s}_j \left\{ \nabla_{\mathbf{x}} \left[V_{\mathrm{n}}(\mathbf{x}) - e\varphi(\mathbf{x}) \right]_{\mathbf{x}=\mathbf{x}_j} \times \mathbf{p}_j \right\}. \tag{2.23}$$

The Breit Hamiltonian or the (transverse) electron-electron interaction, $\mathscr{H}_B$, that is not mediated by the pairwise Coulomb interaction $\mathrm{v}(\mathbf{x}_i - \mathbf{x}_l)$ in (2.21) is given by

$$\mathscr{H}_B = \frac{e}{m} \sum_{j=1}^{N} \left\{ \mathbf{A}^{(1)}(\mathbf{x}_j)\mathbf{p}_j + \frac{1}{2} e \left[\mathbf{A}^{(1)}(\mathbf{x}_j) \right]^2 + \mathbf{s}_j \mathbf{B}^{(1)}(\mathbf{x}_j) \right\}. \tag{2.24}$$

The higher order term $\sim \left[\mathbf{A}^{(1)}(\mathbf{x})\right]^2$ is omitted below. Consequently, diamagnetic contributions do not appear.

For nearly homogeneous magnetic fields and the Coulomb gauge (2.9) an approximate expression

$$\mathbf{A}^{(1)}(\mathbf{x}) = \frac{1}{2} \mathbf{B}^{(1)}(\mathbf{x}) \times \mathbf{x} \tag{2.25}$$

holds. Then, the Breit contribution to the Hamiltonian (2.24) takes the form

$$\mathscr{H}_B = \frac{e}{2m} \sum_{j=1}^{N} (\boldsymbol{\ell}_j + 2\mathbf{s}_j) \mathbf{B}^{(1)}(\mathbf{x}_j), \tag{2.26}$$

where the orbital angular momentum operator $\boldsymbol{\ell} = \mathbf{x} \times \mathbf{p}$ of an individual electron has been introduced. With the (operator of the) magnetic moment

$$\mathbf{m}_j = -\frac{1}{\hbar} \mu_B (\boldsymbol{\ell}_j + 2\mathbf{s}_j) \tag{2.27}$$

of an electron j ($j = 1, \ldots, N$) and the Bohr magneton $\mu_B = \frac{e\hbar}{2m} = 0.579 \times 10^{-4} \frac{\mathrm{eV}}{\mathrm{T}}$ one finds

$$\mathscr{H}_B = -\sum_{j=1}^{N} \mathbf{m}_j \mathbf{B}^{(1)}(\mathbf{x}_j), \tag{2.28}$$

i.e., the energy of the magnetic moments $\mathbf{m}_j$ of the electrons $j = 1, \ldots, N$ in their own magnetic field $\mathbf{B}^{(1)}$. Possible double counting is still included in (2.28). The factor 2 in front of the spin operator in the definition (2.27) approximately describes the Landé factor of an electron. Deviations of the value $g = 2$ remain small for electrons in condensed matter. For instance, for amorphous hydrogenated silicon the value $g = 2.0044$ has been measured [16]. Despite the operator character of all quantities $\mathbf{m}_j$ and $\mathbf{B}^{(1)}(\mathbf{x}_j)$, the approximation (2.28) allows an easy physical interpretation of the Breit contribution.

In general, the vector potential (2.18) and the resulting magnetic field (2.4) can be divided into orbital and spin contributions according to the paramagnetic and spin current action,

$$\begin{aligned}
\mathbf{A}_{\text{orbit}}(\mathbf{x}) &= -\frac{\mu_0 e}{8\pi m}\sum_{j=1}^{N}\left\{\frac{\mathbf{p}_j}{|\mathbf{x}-\mathbf{x}_j|} + \frac{[\mathbf{p}_j(\mathbf{x}-\mathbf{x}_j)](\mathbf{x}-\mathbf{x}_j)}{|\mathbf{x}-\mathbf{x}_j|^3}\right\}, \\
\mathbf{B}_{\text{orbit}}(\mathbf{x}) &= \frac{\mu_0 e}{4\pi m}\sum_{j=1}^{N}\frac{(\mathbf{x}-\mathbf{x}_j)}{|\mathbf{x}-\mathbf{x}_j|^3}\times\mathbf{p}_j, \\
\mathbf{A}_{\text{spin}}(\mathbf{x}) &= \frac{\mu_0 e}{4\pi m}\sum_{j=1}^{N}\frac{(\mathbf{x}-\mathbf{x}_j)}{|\mathbf{x}-\mathbf{x}_j|^3}\times\mathbf{s}_j, \\
\mathbf{B}_{\text{spin}}(\mathbf{x}) &= \frac{\mu_0 e}{4\pi m}\sum_{j=1}^{N}\left\{\frac{\mathbf{s}_j}{|\mathbf{x}-\mathbf{x}_j|^3} - 3\frac{\left[\mathbf{s}_j(\mathbf{x}-\mathbf{x}_j)\right](\mathbf{x}-\mathbf{x}_j)}{|\mathbf{x}-\mathbf{x}_j|^5}\right\}.
\end{aligned} \tag{2.29}$$

The fields (2.29) allow us to rewrite the Hamiltonian contribution (2.24) to the well-known Breit Hamiltonian (or adding the other relativistic corrections $\mathscr{H}_{sr} + \mathscr{H}_{so}$ to the Breit-Pauli Hamiltonian) [4, 6, 7, 17, 18]

$$\mathscr{H}_B = \frac{e}{2m}\sum_{j=1}^{N}\left\{\mathbf{A}_{\text{orbit}}(\mathbf{x}_j)\mathbf{p}_j + \mathbf{A}_{\text{spin}}(\mathbf{x}_j)\mathbf{p}_j + \mathbf{B}_{\text{orbit}}(\mathbf{x}_j)\mathbf{s}_j + \mathbf{B}_{\text{spin}}(\mathbf{x}_j)\mathbf{s}_j\right\}, \tag{2.30}$$

where we have introduced a factor $\frac{1}{2}$ to account for the double counting of pair interactions in the magnetic or spin-polarized systems. The first term in expression (2.30) is a correction that partly accounts for retardation and can be described as the interaction between the magnetic dipole moments of the electrons, which arise from the orbital motion of the charges (also called orbit-orbit interaction) [7]. The second and third terms represent spin-orbit and orbit-spin interactions in addition to the spin-orbit interaction in $\mathscr{H}_{so}$ (2.23) related to the internal electric fields. They describe the coupling between orbital magnetic moments (from the orbital motion of the charged particles) and spin magnetic moments. Both terms can be summarized. The last term in (2.30) represents a spin-spin interaction mediated by spin magnetic moments. Together with (2.29) (lowest line) one sees that it has the form of a classical dipole-dipole interaction. On this level of the description we do not discuss what happens for particles at the same position. Many of the derivations of the Breit Hamiltonian [4, 7, 17, 18] describe this situation by a contact spin interaction proportional to $-\frac{8\pi}{3}(\mathbf{s}_i \cdot \mathbf{s}_j)\delta(\mathbf{x}_i - \mathbf{x}_j)$ similar to the dipole contact interaction known in the electrodynamics [19].

More in detail, with (2.29) the Breit Hamiltonian can be rewritten to [7]

$$\begin{aligned}\mathscr{H}_B = \frac{e^2}{4\pi\varepsilon_0}\frac{1}{(2mc)^2}\sum_{\substack{i,j=1\\(i\neq j)}}\Bigg\{ & -\frac{1}{|\mathbf{x}_i-\mathbf{x}_j|}\left[\mathbf{p}_i\mathbf{p}_j+\frac{[\mathbf{p}_i(\mathbf{x}_i-\mathbf{x}_j)][\mathbf{p}_j(\mathbf{x}_i-\mathbf{x}_j)]}{|\mathbf{x}_i-\mathbf{x}_j|^2}\right] \\ & +\frac{4}{|\mathbf{x}_i-\mathbf{x}_j|^2}\mathbf{s}_i\left(\frac{\mathbf{x}_i-\mathbf{x}_j}{|\mathbf{x}_i-\mathbf{x}_j|}\times\mathbf{p}_j\right) \\ & +\frac{2}{|\mathbf{x}_i-\mathbf{x}_j|^3}\left[\mathbf{s}_i\mathbf{s}_j-3\frac{[\mathbf{s}_i(\mathbf{x}_i-\mathbf{x}_j)][\mathbf{s}_j(\mathbf{x}_i-\mathbf{x}_j)]}{|\mathbf{x}_i-\mathbf{x}_j|^2}\right]\Bigg\}. \end{aligned} \tag{2.31}$$

The spin-contact term for electrons at the same position [7, 18] needs a special treatment and is therefore not given here. In (2.31) any self-interaction of an electron is omitted. Obviously, all contributions to the Breit interaction (2.31) are quadratic in the ratio of electron velocity and speed of light c. Therefore and because of $\alpha_s = \frac{e^2}{4\pi\varepsilon_0}\frac{1}{\hbar c}$, their prefactor can be rewritten to be proportional to the square of the Sommerfeld finestructure constant α_s.

2.4 Explicit Treatment of Relativistic Corrections

2.4.1 Scalar-Relativistic Corrections

The most important Hamiltonian for the description of the interacting inhomogeneous gas of electrons in the field $V_n(\mathbf{x})$ of the nuclei is given by the non-relativistic part of (2.2), $\mathscr{H}_0$ (2.21). It contains the (longitudinal) electron-electron interaction mediated by Coulomb interaction between the electrons. The majority of current electronic structure codes takes the scalar-relativistic corrections $\mathscr{H}_{sr}$ (2.22) to the kinetic and potential energy explicitly into account, for instance, in the constructed pseudopotentials, without mentioning this fact in the description of the actually used single-electron equation of motion.

Because of the smallness of the scalar-relativistic corrections $\mathscr{H}_{sr}$ (2.22) $\sim c^{-2}$ but also of the spin-orbit interaction $\mathscr{H}_{so}$ (2.23), further approximations are meaningful. From the point of view of physics the common replacement of the operator of the total electric field $\mathbf{E}(\mathbf{x})$ in the electronic system has to be explained by the expectation value $\langle\mathbf{E}(\mathbf{x})\rangle = \langle\Psi|\mathbf{E}(\mathbf{x})|\Psi\rangle$ with $|\Psi\rangle$ as the Hilbert state of the electronic system in which one is interested. Field fluctuations $\Delta\mathbf{E}(\mathbf{x}) = \mathbf{E}(\mathbf{x}) - \langle\mathbf{E}(\mathbf{x})\rangle$ are neglected. In other words, the operator of the Coulomb interaction of the electrons $-e\varphi(\mathbf{x})$ has to be replaced by its expectation value $\langle\varphi(\mathbf{x})\rangle = \langle\Psi|\varphi(\mathbf{x})|\Psi\rangle$. Practically in all implementations of electronic-structure codes, in the relativistic corrections $V_n(\mathbf{x}) - e\langle\varphi(\mathbf{x})\rangle$ is replaced by the effective potential $V(\mathbf{x})$ of the electrons used in the code. Thereby, $V(\mathbf{x}) \approx V_n(\mathbf{x}) - e\langle\varphi(\mathbf{x})\rangle$ is computed in a certain mean-field approximation for the classical Hartree as well as the non-classical exchange and correlation contributions to the (longitudinal) electron-electron interaction. Instead of (2.22) a mean-field approximation

$$\mathscr{H}_{sr} = \frac{1}{2(2mc)^2} \sum_{j=1}^{N} \left\{ \hbar^2 \Delta_{\mathbf{x}} V(\mathbf{x})\big|_{\mathbf{x}=\mathbf{x}_j} - \frac{1}{m} \mathbf{p}_j^4 \right\} \tag{2.32}$$

is applied [20].

2.4.2 Spin-Orbit Interaction

The described mean-field approximation $V(\mathbf{x}) \approx V_{\mathrm{n}}(\mathbf{x}) - e\langle\varphi(\mathbf{x})\rangle$ yields

$$\mathscr{H}_{so} = -\frac{1}{2(mc)^2} \sum_{j=1}^{N} \mathbf{s}_j \left[\nabla_{\mathbf{x}} V(\mathbf{x})\big|_{\mathbf{x}=\mathbf{x}_j} \times \mathbf{p}_j \right] \tag{2.33}$$

instead of (2.23). The gradient of $V(\mathbf{x})$ indicates especially strong relativistic effects for electron motion near the nuclei. In order to treat them in space regions near the nuclei, the potential can be approximately divided into effective contributions from these regions. According to (1.4) for $V_{\mathrm{n}}(\mathbf{x})$ it can be decomposed as

$$V(\mathbf{x}) = \sum_{l=1}^{N_{\mathrm{n}}} \tilde{V}(|\mathbf{x} - \mathbf{R}_l|) \tag{2.34}$$

with contributions $\tilde{V}(|\mathbf{x}|)$ which can be assumed to be spherically symmetric near the nucleus at the position $\mathbf{R}_l$. With $\nabla_{\mathbf{x}} \tilde{V}(r) = \frac{1}{r}\left[\frac{d}{dr}\tilde{V}(r)\right]\mathbf{x}$ and the abbreviation $r = |\mathbf{x}|$ the spin-orbit Hamiltonian (2.33) becomes

$$\mathscr{H}_{so} = -\frac{1}{2(mc)^2} \sum_{l=1}^{N_{\mathrm{n}}} \sum_{j=1}^{N} \frac{1}{r}\frac{d}{dr}\tilde{V}(r)\big|_{r=|\mathbf{x}_j - \mathbf{R}_l|} \mathbf{s}_j \boldsymbol{\ell}_{jl} \tag{2.35}$$

with the orbital momentum operator $\boldsymbol{\ell}_{jl} = (\mathbf{x}_j - \mathbf{R}_l) \times \mathbf{p}_j$ of the electron j near the nucleus at site $\mathbf{R}_l$.

The representation (2.35) of the spin-orbit interaction suggests a unification with the potential energy of the electrons in the field of the nuclei in (1.3). By means of (2.35) the potential energy $V_{\mathrm{n}}(\mathbf{x})$ (1.4) felt by one electron at $\mathbf{x}$ can be generalized to a spin-dependent potential $\tilde{V}_{\mathrm{n}}(\mathbf{x}, \mathbf{s})$ which follows as summation over the positions of the nuclei. Nevertheless, the use of this potential is related to two difficulties: (i) The dependence on the spin variable $\mathbf{s}$ asks for a non-collinear treatment. (ii) The potential

$V(\mathbf{x})$ or $\langle\varphi(\mathbf{x})\rangle$ occurring in the spin-orbit interaction requires an approximate self-consistent treatment of the (longitudinal) electron-electron interaction.

For the implementation of the spin-orbit interaction in electronic-structure codes the local nature of the factor $\frac{1}{r}\frac{d}{dr}\tilde{V}(r)$ is frequently used. Exploiting the fact that only the region close to the nucleus contributes to it, the spin-orbit coupling is only treated within a sphere around the core [21, 22] or taken *a priori* into account in the construction of the pseudopotentials [23, 24]. The latter approach follows the standard procedure to generate scalar-relativistic pseudopotentials that include the kinematic relativistic effects, mass velocity and Darwin term, from the fully relativistic all-electron solution of the electronic-structure problem of the free atoms [25–29]. The advantage of such a treatment is that only collinear spins have to be considered in each environment of a nucleus. The non-collinearity of the spins in the entire system is described by rotating the local Cartesian coordinate system in an appropriate manner.

2.4.3 Breit Interaction

The smallness of the relativistic corrections $\mathcal{H}_B$ (2.31) also suggests a mean-field approximation. For its illustration we make use of the assumption of a nearly homogeneous magnetic field (2.25). This leads to a description of the Breit interaction by a Hamiltonian (2.28) that describes the energy of magnetic moment operators $\mathbf{m}_j$ (2.27) of the electrons in their own field. On the other hand, such magnetic dipoles can be also considered to generate the magnetic field $\mathbf{B}^{(1)}(\mathbf{x})$ by [19]

$$\mathbf{B}^{(1)} = -\frac{\mu_0}{4\pi}\sum_{j=1}^{N}\left[\frac{\mathbf{m}_j}{|\mathbf{x}-\mathbf{x}_j|^3} - 3\frac{[\mathbf{m}_j(\mathbf{x}-\mathbf{x}_j)](\mathbf{x}-\mathbf{x}_j)}{|\mathbf{x}-\mathbf{x}_j|^5}\right]. \tag{2.36}$$

Together with expression (2.28) and a factor $\frac{1}{2}$ to avoid double counting of pair interactions, instead of (2.31), one finds

$$\mathcal{H}_B = \frac{\mu_0}{8\pi}\sum_{\substack{i,j=1\\(i\neq j)}}^{N}\left[\frac{\mathbf{m}_i\mathbf{m}_j}{|\mathbf{x}_i-\mathbf{x}_j|^3} - 3\frac{[\mathbf{m}_i(\mathbf{x}_i-\mathbf{x}_j)][\mathbf{m}_j(\mathbf{x}_i-\mathbf{x}_j)]}{|\mathbf{x}_i-\mathbf{x}_j|^5}\right], \tag{2.37}$$

a Hamiltonian that corresponds to the classical magnetic dipole-dipole interaction energy [30]. A mean-field approximation is suggested as the replacement of a dipole operator $\mathbf{m}_j$ by its expectation value $\langle\mathbf{m}_j\rangle$. Then, the expectation value of the approximate Hamiltonian $\mathcal{H}_B$ is described by the classical interaction energy [19].

While the spin-orbit interaction is important for splittings of electronic levels in the electronic structure of condensed matter consisting of heavy elements, its contribution $\langle\mathcal{H}_{so}\rangle$ to the total energy of an electronic system is usually negligible.

Indeed, calculations have shown that for simple solids built by elements up to those of the sixth row of the periodic table, the effects induced by spin-orbit coupling on the structural and elastic properties are quite small [22]. The same holds for the contribution $\langle \mathscr{H}_B \rangle$. It can be nearly neglected. In other words the transverse electron-electron interaction seems indeed to be negligible as done in the majority of the condensed matter and molecule descriptions. With $\langle \mathbf{m}_i \rangle = \langle \mathbf{m}_j \rangle \approx \mu_B$ (2.27) one may find for the interaction of N electrons in a system, which carry a magnetic moment,

$$\langle \mathscr{H}_B \rangle \approx \frac{N\mu_0}{4\pi}\mu_B^2 \frac{1}{R^3} \approx N \frac{e^2}{4\pi\varepsilon_0} \frac{\hbar^2}{4m^2c^2} \frac{1}{R^3}$$

ignoring the angular dependence of the interaction and assuming R to be a characteristic distance of two magnetic moments, perhaps $R \approx 4a_B$. With the Bohr radius of the hydrogen atom $a_B = \frac{4\pi\varepsilon_0}{e^2}\frac{\hbar^2}{m} = 5.29 \times 10^{-11}$ nm, the Rydberg energy $R_\infty = \left(\frac{e^2}{4\pi\varepsilon_0}\right)^2 \frac{m}{2\hbar^2} = 13.605$ eV, and the Sommerfeld finestructure constant $\alpha_s = \frac{e^2}{4\pi\varepsilon_0\hbar c} = \frac{1}{137.036}$ it arises

$$\langle \mathscr{H}_B \rangle \approx \frac{N}{2}\alpha_s^2 \left(\frac{a_B}{R}\right)^3 R_\infty \approx \frac{N}{8}\left(\frac{\alpha_s}{4}\right)^2 R_\infty \approx N \times 5.6 \times 10^{-6}\ \text{eV}.$$

This is a marginal energy compared to the electrostatic energies per particle in the system.

Spin-orbit interaction $\langle \mathscr{H}_{so} \rangle$ can only play a role for a reasonable orbital angular momentum of the relevant electrons. This immediately follows from the scalar product of spin and orbital moments in (2.35). However, even in this case it may only give rise to a small total-energy modification, but slightly larger than $\langle \mathscr{H}_B \rangle$.

Only in systems with magnetic ordering the two interactions $\mathscr{H}_{so}$ and $\mathscr{H}_B$ can become important for the total energy and the derived properties. One example is the magnetocrystalline anisotropy in magnetic crystals [31]. The orientation of the global or local magnetization, i.e., the so-called easy magnetization axis is determined by the interplay of $\mathscr{H}_{so}$ and $\mathscr{H}_B$ with the crystal field. Examples are ferromagnetic Co [31] that crystallizes in a hexagonal close-packed structure and antiferromagnetic MnO and CoO which crystallize in distorted rocksalt structures. The latter cases show that, despite the smallness of the magnetic interactions, they determine the easy axis (or easy plane) [32]. Thereby, because of the half-filled 3d shell of Mn^{2+} in MnO the local orbital moment and hence the spin-orbit coupling vanish. The magnetic anisotropy is ruled by $\mathscr{H}_B$ (2.37). The opposite situation occurs for CoO with a partially filled minority-spin 3d shell. Its easy axis is determined by $\mathscr{H}_{so}$.

2.5 Transverse Interaction in General

The transverse contribution to the electron-electron interaction in (2.2) is dominated by the magnetic field $\mathbf{B}(\mathbf{x}) = \text{curl } \mathbf{A}(\mathbf{x})$ and the transverse electric field $\mathbf{E}_\perp(\mathbf{x}) = -\frac{\partial}{\partial t}\mathbf{A}(\mathbf{x})$ (2.3), i.e., finally by the vector potential $\mathbf{A}(\mathbf{x})$ generated by the moving electrons. According to wave equation (2.12) it mainly goes back to the current density $\hat{\mathbf{j}}(\mathbf{x})$ (2.7) of the electrons. For that reason we investigate this operator in more detail, however, restrict the studies to systems with time-reversal symmetry. In addition, to make the proof short we omit for a moment the non-collinearity of the electron spins.

The time-reversal symmetry operator $\hat{K}$ is antilinear and antiunitary. It takes the complex conjugation of the orbitals in a Pauli spinor and rotates the spin by 180° according to the Pauli spin matrix σ_y (1.1). It commutes with the Hamiltonian $\mathscr{H}$,

$$[\hat{K}, \mathscr{H}]_- = 0. \tag{2.38}$$

The non-degenerate ground state $|\Psi_0\rangle$ of the considered system obeys the stationary Schrödinger equation $\mathscr{H}|\Psi_0\rangle = E_0|\Psi_0\rangle$. Since (2.38) holds, $\hat{K}$ and $\mathscr{H}$ possess the same eigenvector system. Therefore, it holds

$$\hat{K}|\Psi_0\rangle = \lambda|\Psi_0\rangle \tag{2.39}$$

with the condition $|\lambda|^2 = 1$ for the eigenvalues. The time-reversal symmetry operator is defined by [33]

$$\begin{aligned} \hat{K}\mathbf{x}_j\hat{K}^{-1} &= \mathbf{x}_j, \\ \hat{K}\mathbf{p}_j\hat{K}^{-1} &= -\mathbf{p}_j, \\ \hat{K}\mathbf{s}_j\hat{K}^{-1} &= -\mathbf{s}_j. \end{aligned} \tag{2.40}$$

As a consequence the relations

$$\begin{aligned} \hat{K}\hat{\mathbf{j}}(\mathbf{x})\hat{K}^{-1} &= -\hat{\mathbf{j}}(\mathbf{x}), \\ \hat{K}\mathbf{A}(\mathbf{x})\hat{K}^{-1} &= -\mathbf{A}(\mathbf{x}) \end{aligned} \tag{2.41}$$

hold with (2.7) and (2.18).

Most interesting are the ground-state expectation values of an operator $\hat{\mathbf{O}}(\mathbf{x}) = \mathbf{A}(\mathbf{x})$ or $\hat{\mathbf{j}}(\mathbf{x})$,

$$\begin{aligned} \langle\Psi_0|\hat{\mathbf{O}}(\mathbf{x})|\Psi_0\rangle &= \langle\Psi_0|\lambda\hat{\mathbf{O}}(\mathbf{x})\lambda^*|\Psi_0\rangle \\ &= \langle\Psi_0|\hat{K}\hat{\mathbf{O}}(\mathbf{x})\hat{K}^{-1}|\Psi_0\rangle \\ &= -\langle\Psi_0|\hat{\mathbf{O}}(\mathbf{x})|\Psi_0\rangle. \end{aligned}$$

Such a relation can be only fulfilled for vanishing expectation values $\langle\Psi_0|\hat{\mathbf{j}}(\mathbf{x})|\Psi_0\rangle = 0$ and $\langle\Psi_0|\mathbf{A}(\mathbf{x})|\Psi_0\rangle = 0$.

For the majority of electronic systems one may conclude that in a mean-field approximation the contribution of the transverse electron-electron interaction to the total energy vanishes. Only fluctuations of the transverse fields due to the moving electrons can have an influence. In any case, in the following considerations this fact allows us to restrict ourselves to the longitudinal interaction of the electrons mediated by Coulomb potentials. Only in cases where the magnetic anisotropy of an electronic system with a certain magnetic ordering is investigated, the transverse interaction $\mathscr{H}_B$ and the effect of the spin-orbit interaction $\mathscr{H}_{so}$ have to be taken into account to calculate the total energy of the system. The neglect of the transverse electron-electron interaction is certainly a good approximation to excited electronic states, while the spin-orbit coupling will influence the electronic structure in general, particularly for systems consisting of heavy atoms. As a consequence we will not investigate the transverse interaction in the following chapters.

References

1. P.A.M. Dirac, The quantum theory of the electron. Proc. Roy. Soc. London Ser. A **117**, 610–624 (1928)
2. P.A.M. Dirac, The quantum theory of the electron. Part II. Proc. Roy. Soc. London Ser. A **118**, 351–361 (1928)
3. L.L. Foldy, S.A. Wouthuysen, On the Dirac theory of spin 1/2 particles and its non-relativistic limit. Phys. Rev. **78**, 29–36 (1958)
4. G. Breit, The effect of retardation on the interaction of two electrons. Phys. Rev. **34**, 553–573 (1929)
5. G. Breit, The fine structure of He as a test of the spin interactions of two electrons. Phys. Rev. **36**, 383–397 (1930)
6. G. Breit, Dirac's equation for the spin-spin interaction of two electrons. Phys. Rev. **39**, 616–624 (1932)
7. H.A. Bethe, E.E. Salpeter, *Quantum Mechanics of One- and Two-Electron Systems* (Springer, Berlin, 1957)
8. P. Strange, *Relativistic Quantum Mechanics* (Cambridge University Press, Cambridge, 1998)
9. M. Reiher, A. Wolf, *Relativistic Quantum Chemistry. The Fundamental Theory of Molecular Science* (Wiley-VCH Verlag GmbH & Co. KGaA, Weinheim, 2009)
10. P. Pyykkö, Relativistic quantum chemistry. Adv. Quantum Chem. **11**, 353–409 (1978)
11. P. Pyykkö, Relativistic effects on periodic trends, in *The Effects of Relativity in Atoms, Molecules and the Solid-State*, ed. by S. Wilson, I.P. Grant, B.L. Gyorffy (Plenum Press, New York and London, 1991), pp. 1–13
12. E. Engel, Relativistic density functional theory: foundations and basic formalism, in *Relativistic Electronic Structure Theory, Part 1*, ed. by P. Schwerdtfeger (Elsevier, Amsterdam, 2002), pp. 523–621
13. E. Engel, R.M. Dreizler, S. Varga, B. Fricke, Relativistic density functional theory, in *Relativistic Effects in Heavy-Element Chemistry and Physics*, ed. by B.A. Hess (Wiley, New York, 2003), pp. 123–164
14. W. Pauli, Zur Quantenmechanik des magnetischen Elektrons. Z. Physik **43**, 601–623 (1927)
15. P.L. Taylor, O. Heinonen, *A Quantum Approach to Condensed Matter Physics* (Cambridge University Press, Cambridge, 2002)

16. M.S. Brand, M. Stutzmann, Spin-dependent conductivity in amorphous hydrogenated silicon. Phys. Rev. B **43**, 518–5187 (1991)
17. A.I. Akhiezer, V.B. Berestetsky, *Quantum Electrodynamics* (Interscience, New York, 1965)
18. N. Scott, P. Burke, Electron scattering by atoms and ions using the Breit-Pauli Hamiltonian: an R-Matrix approach. J. Phys. B. Atom. Molec. Phys. **13**, 4299–4314 (1980)
19. J. Jackson, *Classical Electrodynamics* (Wiley, New York, 1975)
20. D.D. Koelling, B.N. Harmon, A technique for relativistic spin-polarized calculations. J. Phys. C **10**, 3107–3114 (1977)
21. D. Hobbs, G. Kresse, J. Hafner, Fully unconstrained noncollinear magnetism within the projector augmented-wave method. Phys. Rev. B **62**, 11556–11570 (2000)
22. A. Dal Corso, Projector augmented wave method with spin-orbit coupling: applications to simple solids and zincblende-type semiconductors. Phys. Rev. B **86**, 085135 (2012)
23. G. Theurich, N.A. Hill, Self-consistent treatment of spin-orbit coupling in solids using relativistic fully separable *ab initio* pseudopotentials. Phys. Rev. B **64**, 073106 (2001)
24. A. Dal Corso, A. Mosca Conte, Spin-orbit coupling with ultrasoft pseudopotentials: application to Au and Pt. Phys. Rev. B **71**, 115106 (2005)
25. G.B. Bachelet, M. Schlüter, Relativistic norm-conserving pseudopotentials. Phys. Rev. B **25**, 3103–2108 (1982)
26. G.B. Bachelet, D.R. Hamann, M. Schlüter, Pseudopotentials that work: from H to Pu. Phys. Rev. B **26**, 4199–4228 (1982)
27. A.M. Rappe, K.M. Rabe, E. Kaxiras, J.D. Joannopoulos, Optimized pseudopotentials. Phys. Rev. B **41**, 1227–1230 (1990)
28. N. Troullier, J.L. Martins, Efficient pseudopotentials for plane-wave calculations. Phys. Rev. B **43**, 1993–2006 (1991)
29. D. Vanderbilt, Soft self-consistent pseudopotentials in a generalized eigenvalue formalism. Phys. Rev. B **41**, 7892–7895 (1990)
30. H.J.F. Jansen, Magnetic anisotropy in density-functional theory. Phys. Rev. B **59**, 4699–4707 (1999)
31. H. Alloul, *Introduction to the Physics of Electrons in Solids* (Springer, Heidelberg, 2011)
32. A. Schrön, C. Rödl, F. Bechstedt, Crystal symmetry and magnetic anisotropy of $3d$-transition metal monoxides. Phys. Rev. B **86**, 115134 (2012)
33. M. Lax, *Symmetry Principles in Solid State and Molecular Physics* (Dover Publications Inc, Mineola, 2011)

Chapter 3
Exchange and Correlation

Abstract A convenient description of an interacting, inhomogeneous electron gas is derived within the second quantization. Electrons are described by fermionic field operators depending on space, spin, and time with either creation or annihilation character. The many-electron Hamiltonian including the longitudinal electron-electron interaction takes a form that supports the physical intuition. The field operators allow the construction of many-electron states in the Hilbert as well as the Fock space. Corresponding expectation values of the Hamiltonian can be expressed by one- and two-particle density matrices. The two-particle one obeys a sum rule that describes classical Hartree and quantum-mechanical exchange-correlation contributions to the electron-electron interaction. It is directly related to the pair correlation and van Hove correlation function.

3.1 Field-Theoretical Description of Electrons

3.1.1 Hamiltonian

According to the discussion in Chap. 2 we restrict the electron-electron interaction to the longitudinal contribution. Relativistic corrections are omitted from the considerations. We are however aware that scalar-relativistic corrections are included in all explicit calculations using a common electronic-structure code, e.g. via the pseudopotentials. If the electronic structure of the studied condensed matter asks for the inclusion of spin-orbit interaction, additional comments will be added. In this case the spin-orbit interaction will be considered at least on the single-particle level. The spin of the electrons **s** is however always taken into account. So spin-polarized or magnetic condensed matter will be coincidentally discussed.

To study a system of N electrons in the volume Ω we investigate the Hamiltonian (2.21)

$$\mathscr{H}_0 = \hat{T} + \hat{V} + \hat{U} \tag{3.1}$$

F. Bechstedt, *Many-Body Approach to Electronic Excitations*,
Springer Series in Solid-State Sciences 181, DOI 10.1007/978-3-662-44593-8_3

in 'first quantization' with contributions

$$\hat{T} = \frac{1}{2m} \sum_{j=1}^{N} \mathbf{p}_j^2 \tag{3.2}$$

of the kinetic energy,

$$\hat{V} = \sum_{j=1}^{N} V_{\mathrm{n}}(\mathbf{x}_j) \tag{3.3}$$

of the potential energy of the electrons in the field of the nuclei, and

$$\hat{U} = \frac{1}{2} \sum_{\substack{i,j=1 \\ (i \neq j)}}^{N} \mathrm{v}(\mathbf{x}_i - \mathbf{x}_j) \tag{3.4}$$

of the (longitudinal) electron-electron interaction mediated by the Coulomb potential (1.2).

3.1.2 Field Operators

The particle operators $\mathbf{x}_j$, $\mathbf{p}_j$, and $\mathbf{s}_j$ are not very convenient for a statistical description of the system of electrons being subject to the Pauli exclusion principle [1]. Therefore, we introduce field operators $\psi_s(\mathbf{x})$ and $\psi_s^+(\mathbf{x})$ which are related to probabilities to annihilate or create electrons at a space point $\mathbf{x}$ with the spin variable $\mathbf{s}$ indicated by the index $s = \uparrow, \downarrow$ for spin-up and spin-down electrons [2]. The creation operator $\psi_s^+(\mathbf{x})$ is the adjoint of the annihilation operator $\psi_s(\mathbf{x})$.

The introduction into the field-theoretical description is given in several text books, e.g. [3]. For that reason only some basic knowledge will be repeated. For $N = 0$ the unique state of zero particles is denoted the 'vacuum' and written as

$$|0\rangle \qquad \text{(with } \langle 0| \text{ the conjugate state).} \tag{3.5}$$

In such a state an electron of spin s can be injected at the space point $\mathbf{x}$ applying a Hermitian-conjugated field operator as

$$\psi_s^+(\mathbf{x})|0\rangle \qquad \text{(with } \langle 0|\psi_s(\mathbf{x}) \text{ the conjugate state).} \tag{3.6}$$

The requirement that this state is normalized and orthogonal to that of a particle having a different spin index or a different position can be formulated as

$$\langle 0|\psi_{s'}(\mathbf{x}')\psi_s^+(\mathbf{x})|0\rangle = \delta_{ss'}\delta(\mathbf{x} - \mathbf{x}') \tag{3.7}$$

using the Kronecker delta for the discrete spin index and the Dirac delta function for the continuous space coordinate.

The orthonormalization relation (3.7) can be also interpreted as the orthogonality formulation between the conjugate vacuum state $\langle 0|$ and a state

$$\psi_{s'}(\mathbf{x}')\psi_s^+(\mathbf{x})|0\rangle. \tag{3.8}$$

For $s = s'$ and $\mathbf{x} = \mathbf{x}'$ it must be the vacuum state $|0\rangle$ that does not contain any particle. Consequently, the successive applications of operators $\psi_s^+(\mathbf{x})$ raise the number of electrons stepwise by one, while the successive applications of $\psi_s(\mathbf{x})$ lower it by one particle. The vacuum state has the lowest possible occupancy with

$$\psi_s(\mathbf{x})|0\rangle \equiv 0 \qquad (\text{or } \langle 0|\psi_s^+(\mathbf{x}) \equiv 0). \tag{3.9}$$

The Pauli exclusion principle comes into play with two or more particles, e.g.

$$\psi_s^+(\mathbf{x})\psi_{s'}^+(\mathbf{x}')|0\rangle \tag{3.10}$$

for a two-electron state. The Pauli exclusion principle requires a change of the sign under interchange of the two electrons, i.e.,

$$\psi_{s'}^+(\mathbf{x}')\psi_s^+(\mathbf{x})|0\rangle = -\psi_s^+(\mathbf{x})\psi_{s'}^+(\mathbf{x}')|0\rangle. \tag{3.11}$$

Together with the orthonormalization condition (3.7) this interchange requirement can be conveniently formulated by anticommutation relations

$$\begin{aligned} \left[\psi_s(\mathbf{x}), \psi_{s'}^+(\mathbf{x}')\right]_+ &= \delta_{ss'}\delta(\mathbf{x}-\mathbf{x}'), \\ \left[\psi_s(\mathbf{x}), \psi_{s'}(\mathbf{x}')\right]_+ &= \left[\psi_s^+(\mathbf{x}), \psi_{s'}^+(\mathbf{x}')\right]_+ = 0. \end{aligned} \tag{3.12}$$

The use of the additional (anti)commutation relations (3.12) is therefore frequently denoted as the formulation of the 'second' quantization [3, 4].

In 'first' quantization the collective operators such as the electron density one (2.5) are defined by sums over all N particles in the volume Ω. Given a state (3.6) or (3.10) with a certain number of particles, e.g. $N = 1$ and $N = 2$, such a quantity may be formulated as

$$\hat{n}_s(\mathbf{x}) = \psi_s^+(\mathbf{x})\psi_s(\mathbf{x}) \tag{3.13}$$

as the density of electrons with equal spin s. The total electron density operator follows as

$$\hat{n}(\mathbf{x}) = \sum_s \psi_s^+(\mathbf{x})\psi_s(\mathbf{x}), \tag{3.14}$$

where s is running over the two possible spin orientations $s = \uparrow, \downarrow$. Consequently, the operator $\hat{N}$ of the total number of electrons

$$\hat{N} = \int_{\Omega} d^3\mathbf{x}\hat{n}(\mathbf{x}) \tag{3.15}$$

is given by a space integral over the entire volume Ω. Currently particle conservation

$$[\mathcal{H}_0, \hat{N}]_- = 0 \tag{3.16}$$

still holds.

3.1.3 Second Quantization

The Hamiltonian of the interacting electrons (3.1) can be reformulated in a similar manner as the operator of the electron number (3.15). This is easily seen for the one-body parts $\hat{T}$ and $\hat{V}$ with

$$\hat{T} = \sum_s \int d^3\mathbf{x}\psi_s^+(\mathbf{x}) \left\{ -\frac{\hbar^2}{2m}\Delta_{\mathbf{x}} \right\} \psi_s(\mathbf{x}) \tag{3.17}$$

and

$$\begin{aligned} \hat{V} &= \sum_s \int d^3\mathbf{x}\psi_s^+(\mathbf{x}) V_{\mathrm{n}}(\mathbf{x})\psi_s(\mathbf{x}) \\ &= \int d^3\mathbf{x} V_{\mathrm{n}}(\mathbf{x})\hat{n}(\mathbf{x}). \end{aligned} \tag{3.18}$$

Here and in the following we do not indicate anymore that the space integral is taken over the volume Ω.

The simplicity of reformulation ceases when two-body terms, e.g. the Coulomb interaction of the electrons, are introduced. Instead of the simple form (3.4) in 'first' quantization the double sums require four field operators

$$\hat{U} = \frac{1}{2}\sum_{s,s'} \int d^3\mathbf{x} \int d^3\mathbf{x}' \psi_{s'}^+(\mathbf{x}')\psi_s^+(\mathbf{x}) \mathrm{v}(\mathbf{x}-\mathbf{x}')\psi_s(\mathbf{x})\psi_{s'}(\mathbf{x}'). \tag{3.19}$$

A more detailed description of the translation of Hamiltonian (3.1) into the formulation of the 'second' quantization (3.17), (3.18) and (3.19) can be found in many monographs. The reader is referred e.g. to the text books of Schiff [3] and Mahan [4]. The translation usually starts from a Lagrange formulation and a Hamilton density and can be easily performed with the rules described above.

3.2 Many-Electron States

3.2.1 Hilbert and Fock Spaces

The creation and annihilation operators of the electrons, $\psi_s^+(\mathbf{x})$ and $\psi_s(\mathbf{x})$, allow to construct many-body states in a Fock space with varying number of electrons. Examples are given in (3.5) for $N = 0$ and (3.10) for $N = 2$. In general, for a given $N = 0, 1, 2, \ldots$ one may construct eigenvectors of a Hilbert space $\mathfrak{H}_N$ by successive application of the creation operator according to

$$|\mathbf{x}_1 s_1, \ldots, \mathbf{x}_N s_N) = \psi_{s_1}^+(\mathbf{x}_1) \ldots \psi_{s_N}^+(\mathbf{x}_N)|0\rangle. \tag{3.20}$$

Here we use the Dirac bra-ket notation [5] for the N-particle state. These states should be normalized according to

$$(\mathbf{x}_1' s_1', \ldots, \mathbf{x}_N' s_N' | \mathbf{x}_1 s_1, \ldots, \mathbf{x}_N s_N) = \prod_{j=1}^{N} \delta(\mathbf{x}_j' - \mathbf{x}_j)\delta_{s_j' s_j}. \tag{3.21}$$

Despite the relation (3.11) the state vectors (3.20) are in general not antisymmetric under the interchange of any two electrons $\mathbf{x}_j s_j$ and $\mathbf{x}_{j'} s_{j'}$, in contrast to the two-electron state (3.10). To indicate this fact we have introduced the denotation $|\ldots)$. In any case the antisymmetry can be achieved by linear combinations

$$|\mathbf{x}_1 s_1, \ldots, \mathbf{x}_N s_N\rangle = \frac{1}{\sqrt{N!}} \sum_{P_\alpha} (-1)^{\nu_{P_\alpha}} P_\alpha |\mathbf{x}_{\alpha_1} s_{\alpha_1}, \ldots, \mathbf{x}_{\alpha_N} s_{\alpha_N}), \tag{3.22}$$

where ν_{P_α} is the number of permutations P_α of pair interchanges of particle indices α_j. With (3.21) these states are also orthonormalized. Therefore, the vectors (3.22) represent a complete antisymmetric basis in the N-electron Hilbert space $\mathfrak{H}_N$. The true N-particle eigenstates $|\Psi\rangle$ of $\mathscr{H}_0$ in second quantization (3.17)–(3.19) are coherent superpositions of the basis vectors (3.22),

$$|\Psi\rangle = \sum_{s_1, \ldots, s_N} \int d^3\mathbf{x}_1 \ldots \int d^3\mathbf{x}_N \, \Psi(\mathbf{x}_1 s_1 \ldots \mathbf{x}_N s_N) |\mathbf{x}_1 s_1, \ldots, \mathbf{x}_N s_N\rangle, \tag{3.23}$$

with expansion coefficients $\Psi(\mathbf{x}_1 s_1 \ldots \mathbf{x}_N s_N)$, which represent the spatial (and spin-dependent) part of the many-body wave function. The representations (3.22) and (3.23) may be regarded in the following way. The field operators in (3.20) create electrons at the points $\mathbf{x}_1, \mathbf{x}_2, \ldots, \mathbf{x}_N$ with spin coordinates $s_1, s_2, \ldots, s_N$. The resulting states are then multiplied by the relative contribution $\Psi(\mathbf{x} s_1 \ldots \mathbf{x}_N s_N)$. The integrations and summations complete the many-electron state by defining it over all space and all spin adjustments. Here again, the correct symmetry is guaranteed.

The states (3.22) are not only complete. They also fulfill a closure relation. One can define a projection operator onto the N-particle space

$$\mathscr{P}_N = \int \sum_{s_1...s_N} |\mathbf{x}_1 s_1, ..., \mathbf{x}_N s_N\rangle \frac{d^3\mathbf{x}_1...d^3\mathbf{x}_N}{N!} \langle \mathbf{x}_1 s_1, ..., \mathbf{x}_N s_N| \tag{3.24}$$

with

$$\langle\Psi|\Psi'\rangle = \langle\Psi|\mathscr{P}_N|\Psi'\rangle$$

and orthonormalized states $|\Psi\rangle$ and $|\Psi'\rangle$ of the Hilbert space $\mathfrak{H}_N$.

The various Hilbert spaces for $N = 0, 1, 2, ...$ allow to construct a Fock space according to the generalized product $\mathfrak{H}_0 \bigotimes \mathfrak{H}_1 \bigotimes \mathfrak{H}_2 \bigotimes ....$ In the Fock space it holds the closure relation

$$1 = \sum_{N=0}^{\infty} \mathscr{P}_N, \tag{3.25}$$

where $\mathscr{P}_0 = |0\rangle\langle 0|$ is the projection operator onto the vacuum state. In the Fock space there exist states which do not belong to a defined particle number. States for defined but different particle numbers are orthogonal.

The annihilation and creation operators, $\psi_s(\mathbf{x})$ and $\psi_s^+(\mathbf{x})$, are linear operators in the Fock space. They connect Hilbert spaces with particle numbers N and $N \pm 1$. With (3.6), (3.9), (3.12) and (3.22) one finds [6]

$$\begin{aligned}
\psi_s(\mathbf{x})|\mathbf{x}_1 s_1, ..., \mathbf{x}_N s_N\rangle &= \sum_{i=1}^{N} (-1)^{N-i} \delta(\mathbf{x} - \mathbf{x}_i)\delta_{ss_i} \\
&\quad \times |\mathbf{x}_1 s_1, ..., \mathbf{x}_{i-1} s_{i-1}, \mathbf{x}_{i+1} s_{i+1}, ..., \mathbf{x}_N s_N\rangle, \\
\psi_s^+(\mathbf{x})|\mathbf{x}_1 s_1, ..., \mathbf{x}_{i-1} s_{i-1}, \mathbf{x}_{i+1} s_{i+1} ..., \mathbf{x}_N s_N\rangle \\
&= (-1)^{N-i} |\mathbf{x}_1 s_1, ..., \mathbf{x}_{i-1} s_{i-1}, \mathbf{x} s, \mathbf{x}_{i+1} s_{i+1}, ..., \mathbf{x}_N s_N\rangle.
\end{aligned} \tag{3.26}$$

3.2.2 Many-Body Schrödinger Equation

For a continuous system one can define operators in the same way as for the discrete case. For an n-particle operator $\hat{\mathscr{O}}_n = \sum_{i_1...i_n}^{N} \hat{O}_n(\mathbf{x}_{i_1} s_{i_1} ... \mathbf{x}_{i_n} s_{i_n})$ with $\hat{\mathscr{O}}_n(\mathbf{x}_{i_1} s_{i_1} ... \mathbf{x}_{i_n} s_{i_n})$ as a function of n sets of coordinates $\mathbf{x}_i s_i$ and, eventually, the canonical momentum $\mathbf{p}_i$, one has the field-theoretical representation

$$\hat{\mathscr{O}}_n = \sum_{s_1...s_n} \int d^3\mathbf{x}_1 ... \int d^3\mathbf{x}_n \psi_{s_1}^+(\mathbf{x}_1)...\psi_{s_n}^+(\mathbf{x}_n) \hat{O}_n(\mathbf{x}_1 s_1 ... \mathbf{x}_n s_n) \psi_{s_n}(\mathbf{x}_n)...\psi_{s_1}(\mathbf{x}_1) \tag{3.27}$$

as used in (3.17) and (3.18) for the $n = 1$ case and in (3.19) for an $n = 2$ example.

For simplicity we only demonstrate the action of $\hat{\mathscr{O}}_1$, for instance for $\hat{V} = \sum_{i=1}^{N} V_{\rm n}(\mathbf{x}_i)$ (3.3), in detail. It holds

$$\hat{\mathscr{O}}_1|\Psi\rangle = \sum_{s}\sum_{s_1...s_N}\int d^3\mathbf{x}\int d^3\mathbf{x}_1...\int d^3\mathbf{x}_N\psi_s^+(\mathbf{x})\hat{O}_1(\mathbf{x}s)\Psi(\mathbf{x}_1s_1...\mathbf{x}_Ns_N)\psi_s(\mathbf{x}) \times |\mathbf{x}_1s_1,...,\mathbf{x}_Ns_N\rangle.$$

Using (3.26) it follows

$$\hat{\mathscr{O}}_1|\Psi\rangle = \sum_{s_1...s_N}\int d^3\mathbf{x}_1...\int d^3\mathbf{x}_N\sum_{i=1}^{N}(-1)^{N-i}\psi_{s_i}^+(\mathbf{x}_i)\left\{\hat{O}_1(\mathbf{x}_is_i)\Psi(\mathbf{x}_1s_1...\mathbf{x}_Ns_N)\right\} \times |\mathbf{x}_1s_1,...,\mathbf{x}_{i-1}s_{i-1},\mathbf{x}_{i+1}s_{i+1},...,\mathbf{x}_Ns_N\rangle.$$

The creation operator does not act on {...}. Therefore, with (3.26) one finds

$$\hat{\mathscr{O}}_1|\Psi\rangle = \sum_{s_1...s_N}\int d^3\mathbf{x}_1...\int d^3\mathbf{x}_N\left\{\sum_{i=1}^{N}\hat{O}_1(\mathbf{x}_is_i)\Psi(\mathbf{x}_1s_1...\mathbf{x}_Ns_N)\right\}|\mathbf{x}_1s_1,...,\mathbf{x}_Ns_N\rangle.$$

Because of the orthonormalization of the basis functions (3.21) and the expansion (3.23) the last equation can be rewritten to

$$\sum_{s'_1...s'_N}\int d^3\mathbf{x}'_1...\int d^3\mathbf{x}'_N\langle\mathbf{x}_1s_1,...,\mathbf{x}_Ns_N|\hat{O}_1|\mathbf{x}'_1s'_1,...,\mathbf{x}'_Ns'_N\rangle\Psi(\mathbf{x}'_1s'_1...\mathbf{x}'_Ns'_N) = \sum_{i=1}^{N}\hat{O}_1(\mathbf{x}_is_i)\Psi(\mathbf{x}_1s_1...\mathbf{x}_Ns_N). \quad (3.28)$$

Since a similar representation also holds for the two-particle operator, the Hamiltonian $\mathscr{H}_0$ (3.1) in the second quantization (3.17), (3.18) and (3.19) obeys a stationary Schrödinger equation

$$\mathscr{H}_0|\Psi\rangle = E_\Psi|\Psi\rangle, \quad (3.29)$$

which is equivalent to the equation

$$\mathscr{H}_0\Psi(\mathbf{x}_1s_1...\mathbf{x}_Ns_N) = E_\Psi\Psi(\mathbf{x}_1s_1...\mathbf{x}_Ns_N), \quad (3.30)$$

where $\mathscr{H}_0$ (3.1) is given by the three contributions (3.2), (3.3) and (3.3) in the particle picture. A generalization from an N-electron Hilbert space to the Fock space, where all operators are expressed in second quantization by field operators, is easily feasible [6].

3.2.3 Other Operators in Second Quantization

The field-theoretical representation (3.27) can also be applied to other operators than the operator of the total number of particles (3.15) or the Hamiltonian of the electron-electron interaction (3.19). Thereby the global operators are related to some densities in operator form such as the density operator (3.13). The matrix elements $\langle \mathbf{x}_1 s_1, ..., \mathbf{x}_N s_N | \hat{O}_1 | \mathbf{x}'_1 s'_1, ..., \mathbf{x}'_N s'_N \rangle$ in (3.28) can be reduced to the case $N = 1$. For the single-particle operators momentum $\mathbf{p}$ and spin $\mathbf{s}$ the corresponding space-spin representations are

$$\begin{aligned} \langle \mathbf{x}s|\mathbf{p}|\mathbf{x}'s'\rangle &= \delta_{ss'}(-i\hbar)\nabla_{\mathbf{x}}\delta(\mathbf{x}-\mathbf{x}'), \\ \langle \mathbf{x}s|\mathbf{s}|\mathbf{x}'s'\rangle &= \delta(\mathbf{x}-\mathbf{x}')\frac{\hbar}{2}\boldsymbol{\sigma}_{ss'} \end{aligned} \tag{3.31}$$

with $\boldsymbol{\sigma}_{ss'}$ as elements of the vector of the Pauli spin matrices (1.1). They allow us to write the operators of the paramagnetic current density (2.15) and the spin density in (2.17) as

$$\begin{aligned} \hat{\mathbf{j}}_p(\mathbf{x}) &= \frac{\hbar}{2mi}\sum_s \left\{\psi_s^+(\mathbf{x})[\nabla_{\mathbf{x}}\psi_s(\mathbf{x})] - \left[\nabla_{\mathbf{x}}\psi_s^+(\mathbf{x})\right]\psi_s(\mathbf{x})\right\}, \\ \hat{\mathbf{s}}(\mathbf{x}) &= \frac{\hbar}{2}\sum_{s,s'}\psi_s^+(\mathbf{x})\boldsymbol{\sigma}_{ss'}\psi_{s'}(\mathbf{x}). \end{aligned} \tag{3.32}$$

The latter representation is helpful for the formulation of the spin-orbit interaction $\mathscr{H}_{so}$ (2.33) if a mean-field approximation $-\mathrm{grad}\, V(\mathbf{x})$ is applied to describe the electric field acting on the electrons.

With (2.27) the spin density operator is directly related to the operator of the magnetization density (without orbital contribution) according to

$$\begin{aligned} \hat{\mathbf{m}}(\mathbf{x}) &= -\frac{2}{\hbar}\mu_B\hat{\mathbf{s}}(\mathbf{x}) \\ &= -\mu_B\sum_{s,s'}\psi_s^+(\mathbf{x})\boldsymbol{\sigma}_{ss'}\psi_{s'}(\mathbf{x}). \end{aligned} \tag{3.33}$$

With the Pauli spin matrices (1.1) its z-component is

$$\hat{m}_z(\mathbf{x}) = -\mu_B\left[\hat{n}_\uparrow(\mathbf{x}) - \hat{n}_\downarrow(\mathbf{x})\right]. \tag{3.34}$$

In the limit of collinear spins where the axis of spin quantization is fixed at the z-axis, the system can be described by two spin densities $\hat{n}_s(\mathbf{x})$ (3.13) while off-diagonal elements in s and s' must not be considered.

3.3 Density Matrices and Pair Correlation Function

3.3.1 Expectation Values

The Hamiltonian $\mathscr{H}_0$ (3.1) of the system without external perturbations does not explicitly depend on time. Therefore, it fulfills a stationary Schrödinger equation of type (3.29). Fixing the number of electrons to N, the eigenstates $|\Psi\rangle$ are defined in the Hilbert space $\mathfrak{H}_N$. The Schrödinger equation itself leads to the definition of the total energy of the N electrons in the many-particle state $|\Psi\rangle$ as the expectation value

$$E_\Psi = \langle\Psi|\mathscr{H}_0|\Psi\rangle. \tag{3.35}$$

In a similar way the electron density in this state $|\Psi\rangle$ is defined as expectation value of the density operator (3.14)

$$n_\Psi(\mathbf{x}) = \langle\Psi|\hat{n}(\mathbf{x})|\Psi\rangle \tag{3.36}$$

with the particle conservation $N = \int d^3\mathbf{x} n_\Psi(\mathbf{x})$. All these definitions are also valid for the ground state of the system $|\Psi_0\rangle$.

With the use of the field-theoretical formulation (3.17), (3.18) and (3.19) the expectation value of the Hamiltonian (3.35) can be divided into three contributions according to

$$\begin{aligned} E_\Psi = &- \frac{\hbar^2}{2m}\int d^3\mathbf{x}\left[\Delta_\mathbf{x} n_\Psi(\mathbf{x},\mathbf{x}')\right]_{\mathbf{x}'=\mathbf{x}} + \int d^3\mathbf{x} V_\mathrm{n}(\mathbf{x}) n_\Psi(\mathbf{x}) \\ &+ \frac{1}{2}\int d^3\mathbf{x}\int d^3\mathbf{x}' \mathrm{v}(\mathbf{x}-\mathbf{x}') m_\Psi(\mathbf{x},\mathbf{x}') \end{aligned} \tag{3.37}$$

with the spin-summed two-particle density matrix

$$m_\Psi(\mathbf{x},\mathbf{x}') = \sum_{s,s'}\langle\Psi|\psi_{s'}^+(\mathbf{x}')\psi_s^+(\mathbf{x})\psi_s(\mathbf{x})\psi_{s'}(\mathbf{x}')|\Psi\rangle \tag{3.38}$$

as well as the spin-summed one-particle density matrix

$$n_\Psi(\mathbf{x},\mathbf{x}') = \sum_s\langle\Psi|\psi_s^+(\mathbf{x})\psi_s(\mathbf{x}')|\Psi\rangle. \tag{3.39}$$

The diagonal elements of (3.39) are directly related to the electron density (3.36) by $n_\Psi(\mathbf{x}) = n_\Psi(\mathbf{x},\mathbf{x})$. We state that the energy expectation value (3.37) can be easily traced back to space integrals over one- and two-particle density matrices.

3.3.2 Sum Rule

By means of the anticommutation relations of the field operators (3.12) the definition of the two-particle density matrix (3.38) can be rewritten in terms of expectation values of the density operator (3.14) as

$$m_\Psi(\mathbf{x}, \mathbf{x}') = \langle \Psi | \hat{n}(\mathbf{x}')\hat{n}(\mathbf{x}) - \delta(\mathbf{x} - \mathbf{x}')\hat{n}(\mathbf{x}) | \Psi \rangle. \tag{3.40}$$

Thereby the term $-\delta(\mathbf{x}-\mathbf{x}')\hat{n}(\mathbf{x})$ describes an infinite self-interaction because of the singular Coulomb potential in (3.37). This indicates that the reformulation (3.40) is rather formal [7].

Nevertheless (3.40) is helpful to prove an important sum rule for the two-particle density matrix (3.38). With the definition of the operator of the particle number (3.15) one finds

$$\int d^3\mathbf{x}' m_\Psi(\mathbf{x}, \mathbf{x}') = \langle \Psi | [\hat{N} - \hat{1}]\hat{n}(\mathbf{x}) | \Psi \rangle.$$

Because of (3.16) $\mathcal{H}_0$ and $\hat{N}$ have a simultaneous system of eigenvectors $\{|\Psi\rangle\}$ with $\hat{N}|\Psi\rangle = N|\Psi\rangle$ regardless of the N-electron state being studied. It follows

$$\int d^3\mathbf{x}' m_\Psi(\mathbf{x}, \mathbf{x}') = (N - 1)n_\Psi(\mathbf{x}) \tag{3.41}$$

or

$$\int d^3\mathbf{x} \int d^3\mathbf{x}' m_\Psi(\mathbf{x}, \mathbf{x}') = N(N - 1).$$

The sum rule (3.41) applies for ground and excited states.

3.3.3 Pair Correlation Function

The occurrence of a product of operators of the electron density suggests the introduction of a (spin-averaged) pair correlation function [3] or pair distribution function [4]

$$g_\Psi(\mathbf{x}, \mathbf{x}') = \frac{m_\Psi(\mathbf{x}, \mathbf{x}')}{n_\Psi(\mathbf{x})n_\Psi(\mathbf{x}')}. \tag{3.42}$$

The calculation of the pair correlation function beyond the classical limit $g_\Psi(\mathbf{x}, \mathbf{x}') \equiv 1$ is the key problem of the description of an inhomogeneous gas of interacting electrons. Since the mutual interactions involve pairs of electrons, the two-body function $g_\Psi(\mathbf{x}, \mathbf{x}')$ determines important properties of the system, in particular the (total) energy E_Ψ in the state $|\Psi\rangle$. Besides the classical electron-electron repulsion $g_\Psi(\mathbf{x}, \mathbf{x}')$ also contains the 'statistical' repulsion of electrons, in order to guarantee the Pauli exclusion principle. This effect is mainly described by exchange of particles. In addition, other quantum-mechanical effects are included. They are summarized as phenomenon of correlation. Summarizing, $g_\Psi(\mathbf{x}, \mathbf{x}')$ represents the probability of finding an electron at $\mathbf{x}$ given that there is another one at $\mathbf{x}'$. The presence of this electron discourages other electrons from approaching it because of the Coulomb repulsion. Consequently, $g_\Psi(\mathbf{x}, \mathbf{x}')$ interpolates from zero at $\mathbf{x} = \mathbf{x}'$ (what we have to enlight below including the spin structure) to the value $g_\Psi(\mathbf{x}, \mathbf{x}') = 1$ at infinite distance $|\mathbf{x} - \mathbf{x}'| \to \infty$.

In the classical limit $g_\Psi(\mathbf{x}, \mathbf{x}') \equiv 1$ of the pair correlation function (3.42) the two-particle density matrix is only characterized by uncorrelated electrons according to $m_\Psi(\mathbf{x}, \mathbf{x}') = n_\Psi(\mathbf{x})n_\Psi(\mathbf{x}')$. In this limit, the expectation value of the longitudinal electron-electron interaction (3.19) gives rise to the classical interaction energy, the Hartree energy [8–10],

$$\begin{aligned} E_{\mathrm{H}}^{\Psi} &= \frac{1}{2}\int d^3\mathbf{x}\int d^3\mathbf{x}' \mathrm{v}(\mathbf{x}-\mathbf{x}')n_\Psi(\mathbf{x})n_\Psi(\mathbf{x}') \\ &= \frac{1}{2}\int d^3\mathbf{x} V_{\mathrm{H}}(\mathbf{x})n_\Psi(\mathbf{x}) \end{aligned} \tag{3.43}$$

with the Hartree potential

$$V_{\mathrm{H}}(\mathbf{x}) = \int d^3\mathbf{x}' \mathrm{v}(\mathbf{x}-\mathbf{x}')n_\Psi(\mathbf{x}') \tag{3.44}$$

in the many-body state $|\Psi\rangle$. Nevertheless, in order to simplify the description we drop the index Ψ from the potential denotation and, consequently, its state dependence in (3.44).

With the Hartree energy E_{H}^{Ψ} (3.43) the total energy (3.37) can be exactly decomposed into

$$E_\Psi = E_{\mathrm{kin}}^{\Psi} + E_{\mathrm{pot}}^{\Psi} + E_{\mathrm{H}}^{\Psi} + E_{\mathrm{XC}}^{\Psi}, \tag{3.45}$$

with the kinetic energy

$$E_{\mathrm{kin}}^{\Psi} = -\frac{\hbar^2}{2m}\int d^3\mathbf{x}\left[\Delta_{\mathbf{x}} n_\Psi(\mathbf{x}, \mathbf{x}')\right]_{\mathbf{x}'=\mathbf{x}}, \tag{3.46}$$

the potential energy in the field of the nuclei

$$E_{\mathrm{pot}}^{\Psi} = \int d^3\mathbf{x} V_{\mathrm{n}}(\mathbf{x})n_\Psi(\mathbf{x}), \tag{3.47}$$

and an additional contribution [11–13]

$$E^{\Psi}_{XC} = \frac{1}{2}\int d^3\mathbf{x}\int d^3\mathbf{x}' \mathrm{v}(\mathbf{x}-\mathbf{x}')n_{\Psi}(\mathbf{x})n_{\Psi}(\mathbf{x}')\left[g_{\Psi}(\mathbf{x},\mathbf{x}')-1\right], \tag{3.48}$$

the so-called exchange (X) and correlation (C) energy that characterizes the effects of the non-classical electron-electron interaction on the total energy. If the true pair correlation function is known, expression (3.45) gives the exact total energy of an N-electron system in the state $|\Psi\rangle$. For large distances between an electron at $\mathbf{x}$ and another one at $\mathbf{x}'$ it holds $\lim_{|\mathbf{x}-\mathbf{x}'|\to\infty} g_{\Psi}(\mathbf{x},\mathbf{x}') = 1$. Therefore, the difference $[g_{\Psi}(\mathbf{x},\mathbf{x}') - 1]$ is short range and vanishes for large $|\mathbf{x}-\mathbf{x}'|$. Consequently, for electron gases of low density the simplest approximation for the electron-electron interaction is the Hartree approximation [8–10], which is equivalent to assume that $g_{\Psi}(\mathbf{x},\mathbf{x}') \equiv 1$ everywhere. This approach corresponds to a completely uncorrelated system where the interaction energy turns into the classical electrostatic form (3.43) for a continuous electron distribution $n_{\Psi}(\mathbf{x})$.

3.3.4 Exchange-Correlation Hole

The XC contribution (3.48) can be formally rewritten into a classical electrostatic interaction energy of the electron distribution $n_{\Psi}(\mathbf{x})$ with another (non-local) distribution [13, 14]

$$n^{\Psi}_{XC}(\mathbf{x},\mathbf{x}') = n_{\Psi}(\mathbf{x}')[g_{\Psi}(\mathbf{x},\mathbf{x}')-1] \tag{3.49}$$

called exchange and correlation hole (density) surrounding a particle at $\mathbf{x}$.

The denotation is obvious with the definition of the pair correlation function (3.42) and the sum rule (3.41). The sum rule holds in a modified form as

$$\int d^3\mathbf{x}' n^{\Psi}_{XC}(\mathbf{x},\mathbf{x}') = -1. \tag{3.50}$$

The sign of the right-hand side indeed indicates a distribution of a missing electron, i.e., a hole. Exactly one electron is missing in the surroundings of an electron at $\mathbf{x}$, irrespective of the system state $|\Psi\rangle$, due to exchange and correlation.

The spatial distributions of the exchange and correlation contributions to the XC hole density are displayed in Fig. 3.1 for an excess electron fixed at the interstitial site between two Si bond chains in a (110) plane of crystalline silicon. They are calculated by means of a variational Monte Carlo method [15]. A definition of exchange is given in Sect. 4.1.2. The relation of the two contributions is illustrated by the much smaller scale for the correlation hole. The strong spatial localization of the contributions is clearly visible.

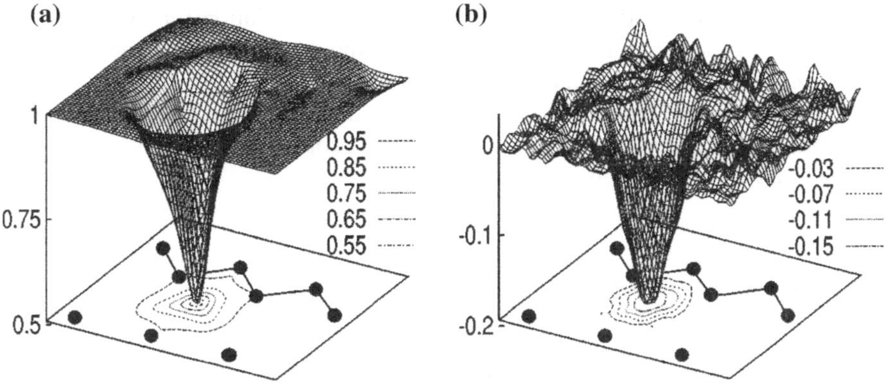

Fig. 3.1 Exchange (**a**) and (coupling-constant-integrated) correlation (**b**) hole for an electron fixed at a tetrahedral interstitial site of a Si crystal. The atoms and bonds in the (110) plane are schematically represented by dots and straight lines. Reprinted with permission from [15]. Copyright 1998 by the American Physical Society

3.4 Relation Between Correlation and Screening

3.4.1 Van Hove Correlation Function

The pair correlation function $g_\Psi(\mathbf{x}, \mathbf{x}')$ (3.42) and the two-particle density matrix $m_\Psi(\mathbf{x}, \mathbf{x}')$ (3.40) can be usefully generalized to describe time-dependent correlations. To see how to carry out this generalization, we start with the definition of the static structure factor of the inhomogeneous electron gas in the N-electron state $|\Psi\rangle$ per particle

$$S_\Psi(\mathbf{x}, \mathbf{x}') = \frac{1}{N}\left\{n_\Psi(\mathbf{x})\delta(\mathbf{x} - \mathbf{x}') + n_\Psi(\mathbf{x})n_\Psi(\mathbf{x}')\left[g_\Psi(\mathbf{x}, \mathbf{x}') - 1\right]\right\}. \tag{3.51}$$

The first term is related to the self-interaction while the second one describes exchange and correlation (3.48). The sum rule (3.50) implies that the space integral $\int d^3\mathbf{x}' S_\Psi(\mathbf{x}, \mathbf{x}') = 0$ vanishes.

With the definitions (3.40) and (3.42) the structure factor (3.51) takes the form

$$\begin{aligned} S_\Psi(\mathbf{x}, \mathbf{x}') &= \frac{1}{N}\left\{\langle\Psi|\hat{n}(\mathbf{x}')\hat{n}(\mathbf{x})|\Psi\rangle - \langle\Psi|\hat{n}(\mathbf{x})|\Psi\rangle\langle\Psi|\hat{n}(\mathbf{x}')|\Psi\rangle\right\} \\ &= \frac{1}{N}\langle\Psi|\Delta\hat{n}(\mathbf{x}')\Delta\hat{n}(\mathbf{x})|\Psi\rangle \end{aligned} \tag{3.52}$$

with the operator of spatial fluctuations of the electron density

$$\Delta\hat{n}(\mathbf{x}) = \hat{n}(\mathbf{x}) - \langle\Psi|\hat{n}(\mathbf{x})|\Psi\rangle. \tag{3.53}$$

The structure factor in the form (3.52) represents a correlation function of spatial fluctuations of the electron density at the same time. It has been first introduced by van Hove [11, 12, 14]. It contains contributions from self- and distinct-particle correlations. In the case of time-dependent density fluctuations $\Delta\hat{n}(\mathbf{x}, t)$, their correlation function can be related to the time-ordered microscopic inverse dielectric function [16], as we will discuss in more detail below.

3.4.2 Dynamic Structure Factor

The static structure factor (3.52) can be directly related to the dynamic structure factor $S(\mathbf{xx}', \omega)$ by a frequency integral $\frac{1}{N}\int_0^\infty \frac{d\hbar\omega}{2\pi} S(\mathbf{xx}', \omega)$ taking the fluctuation-dissipation theorem and the normalization to the particle number into account. For the sake of simplicity the dependence on the state is omitted from the denotation. Such a relation can be easily demonstrated for their space-Fourier transforms, but should not be further discussed here. Rather, corresponding sum rules will be shown in Sect. 13.1.2. For a more detailed discussion the reader is referred to Chap. 13. Here, we will only mention a possible dynamical generalization.

The dynamic structure factor is determined by a density-density response function. Compared to (3.52) a more generalized and, hence, time-dependent correlation function of the density fluctuations is introduced, e.g. by

$$\begin{aligned}\langle\Psi|\left[\Delta\hat{n}(\mathbf{x}, t), \Delta n(\mathbf{x}', t')\right]_- |\Psi\rangle &= \langle\Psi|\left[\hat{n}(\mathbf{x}, t), \hat{n}(\mathbf{x}', t')\right]_- |\Psi\rangle \\ &= \int_{-\infty}^{+\infty} \frac{d\hbar\omega}{2\pi} S(\mathbf{xx}', \omega) e^{-i\omega(t-t')}, \end{aligned} \tag{3.54}$$

with vanishing correlation for $|t - t'| \to \infty$. Usually, when calculating the response of the system to an external charge distribution such a response function appears. For $t > t'$ one encounters the retarded inverse dielectric function [16]

$$\epsilon_r^{-1}(\mathbf{xx}', t - t') = \delta(\mathbf{x} - \mathbf{x}')\delta(t - t') - 2\int d^3\mathbf{x}'' \mathrm{v}(\mathbf{x} - \mathbf{x}'') L^r(\mathbf{x}''\mathbf{x}', t - t') \tag{3.55}$$

with the corresponding spin-averaged density correlation function

$$L^r(\mathbf{xx}', t - t') = \frac{1}{2i\hbar}\theta(t - t')\langle\Psi|\left[\hat{n}(\mathbf{x}, t), \hat{n}(\mathbf{x}', t')\right]_- |\Psi\rangle. \tag{3.56}$$

The dynamic structure factor $S(\mathbf{xx}', \omega)$ is proportional to the spectral function $\hat{L}(\mathbf{xx}', \omega)$ (13.9) of the density correlation function. Therefore, it can be shown [6] that the dynamic structure factor multiplied with the bare Coulomb potential becomes the spectral function of the inverse dielectric function (see Sect. 13.1.3). Then $S(\mathbf{xx}', \omega)$ is proportional to the anti-Hermitian part of the Fourier-transformed inverse dielectric function [17, 18]. Within a more generalized view the direct relation of the (dynamic) structure factor and (the anti-Hermitian part of) the inverse dielectric function represents the fluctuation-dissipation theorem because fluctuations of the electron density in (3.52) are linked to effects of damping, energy loss, and dissipation in the anti-Hermitian part of (3.55). The relation of the structure factor to the density response of the system and, hence, to its inverse dielectric function indicates that the exchange-correlation effects $\sim[g_\Psi(\mathbf{x},\mathbf{x}')-1]$ in (3.49) are closely related to the screening reaction of the system. This idea has been well elaborated deriving the so-called adiabatic connection formula for XC effects (see Sect. 9.3.2) [11–13].

3.5 Spin Dependence

3.5.1 Spin Densities

The sum rule (3.50) indicates that in the entire volume Ω one electron is redistributed by exchange-correlation effects. This electron also possesses a spin variable s. The resulting spin dependence should also play a role in systems whose Hamiltonian $\mathscr{H}_0$ (3.1) does not explicitly depend on spin since spin-orbit interaction $\mathscr{H}_{so}$ (2.33) and the transverse electron-electron (Breit) interaction $\mathscr{H}_B$ (2.37) have been omitted. Nevertheless, spin polarization may occur in such a system if the numbers of spin-up and spin-down electrons are different.

Therefore, we introduce spin-dependent quantities instead of spin-averaged ones. The electron density (3.36) is generalized to the 2×2 Hermitian spin density matrix $\tilde{n}^\Psi(\mathbf{x})$ with the elements ($s, s' =\uparrow,\downarrow$)

$$n^\Psi_{ss'}(\mathbf{x}) = \langle\Psi|\psi^+_s(\mathbf{x})\psi_{s'}(\mathbf{x})|\Psi\rangle. \tag{3.57}$$

Its trace (Tr) and determinant (Det) are invariants with respect to spatial rotations. The trace of this matrix is the electron density

$$n_\Psi(\mathbf{x}) = \mathrm{Tr}\,\tilde{n}^\Psi(\mathbf{x}) = n^\Psi_{\uparrow\uparrow}(\mathbf{x}) + n^\Psi_{\downarrow\downarrow}(\mathbf{x}). \tag{3.58}$$

According to the definition (3.33) the absolute value $|\mathbf{m}_\Psi(\mathbf{x})|$ of the vector of the magnetization density

$$\mathbf{m}_\Psi(\mathbf{x}) = -\mu_B\sum_{s,s'}\boldsymbol{\sigma}_{ss'}n^\Psi_{ss'}(\mathbf{x}) \tag{3.59}$$

yields a second invariant because of $\mathbf{m}^2_\Psi(\mathbf{x}) = \mu_B^2\left\{[\mathrm{Tr}\,\tilde{n}^\Psi(\mathbf{x})]^2 - 4\,\mathrm{Det}\,\tilde{n}^\Psi(\mathbf{x})\right\}$. Its components are

$$m^\Psi_x(\mathbf{x}) = -2\mu_B\mathrm{Re}\,n^\Psi_{\uparrow\downarrow}(\mathbf{x}),$$

$$m_y^\Psi(\mathbf{x}) = -2\mu_B \mathrm{Im}\, n_{\uparrow\downarrow}^\Psi(\mathbf{x}),$$
$$m_z^\Psi(\mathbf{x}) = -\mu_B \left[n_{\uparrow\uparrow}^\Psi(\mathbf{x}) - n_{\downarrow\downarrow}^\Psi(\mathbf{x}) \right],$$

since $n_{\uparrow\downarrow}^\Psi = \left[n_{\downarrow\uparrow}^\Psi(\mathbf{x}) \right]^*$. This is a generalization of the collinear result (3.34). Likewise, the two independent invariants are $n_\Psi(\mathbf{x})$ and $|\mathbf{m}_\Psi(\mathbf{x})|$. The degree of spin polarization $\xi_\Psi(\mathbf{x})$ in the system and its direction $\mathbf{e}_\Psi^m(\mathbf{x})$ are defined as

$$\xi_\Psi(\mathbf{x}) = \frac{|\mathbf{m}_\Psi(\mathbf{x})|}{\mu_B n_\Psi(\mathbf{x})} \qquad (0 \le \xi_\Psi(\mathbf{x}) \le 1),$$
$$\mathbf{e}_\Psi^m(\mathbf{x}) = \frac{\mathbf{m}_\Psi(\mathbf{x})}{|\mathbf{m}_\Psi(\mathbf{x})|}. \tag{3.60}$$

Particle density $n_\Psi(\mathbf{x})$ and spin polarization $\xi_\Psi(\mathbf{x})$ form another pair of invariants derived from the spin density matrix.

3.5.2 Spin-Resolved Pair Correlation

Instead of (3.42) a spin-resolved pair correlation function, more precisely a corresponding Hermitian matrix, can be defined as

$$g_{ss'}^\Psi(\mathbf{x}, \mathbf{x}') = \langle \Psi | \hat{n}_{s'}(\mathbf{x}') \hat{n}_s(\mathbf{x}) - \delta_{ss'} \delta(\mathbf{x} - \mathbf{x}') \hat{n}_s(\mathbf{x}) | \Psi \rangle / n_{ss}^\Psi(\mathbf{x}) n_{s's'}^\Psi(\mathbf{x}') \tag{3.61}$$

with the diagonal elements of the spin density matrix (3.57) and the density operators (3.13). The exchange-correlation energy (3.48) takes the form

$$E_{\mathrm{XC}}^\Psi = \frac{1}{2} \sum_{s,s'} \int d^3\mathbf{x} \int d^3\mathbf{x}' \mathrm{v}(\mathbf{x} - \mathbf{x}') n_{ss}^\Psi(\mathbf{x}) n_{s's'}^\Psi(\mathbf{x}') \left[g_{ss'}^\Psi(\mathbf{x}, \mathbf{x}') - 1 \right]. \tag{3.62}$$

The four elements of the pair correlation function (3.61) are strongly dependent on spin. This is clearly indicated by their definition. The influence of the adjustment of the two spins on the variation of $g_{ss'}^\Psi(\mathbf{x}, \mathbf{x}')$ with the space coordinates can be best demonstrated in the case of a homogeneous electron gas in the ground state with uniform density n but without spin polarization. Because of the equal number of spin-up and spin-down electrons only two elements of the 2×2 matrix, e.g. $g_{\uparrow\uparrow}(\mathbf{x}, \mathbf{x}') = g_{\uparrow\uparrow}(|\mathbf{x} - \mathbf{x}'|)$ and $g_{\uparrow\downarrow}(\mathbf{x}, \mathbf{x}') = g_{\uparrow\downarrow}(|\mathbf{x} - \mathbf{x}'|)$, have to be studied. Corresponding results of quantum Monte-Carlo (QMC) simulations [15] are plotted in Fig. 3.2 and compared with results of model calculations [20–22]. Thereby, the homogeneous electron gas with the uniform density n is characterized by the dimensionless electron gas parameter $r_s = [3/(4\pi n a_B^3)]^{1/3}$.

For parallel spins the pair correlation function in Fig. 3.2 shows the limits discussed above in general for pair correlation functions independent of the electron density. It holds $g_{\uparrow\uparrow}(r \to \infty) = 1$ and $g_{\uparrow\uparrow}(r \to 0) = 0$. As a consequence, for

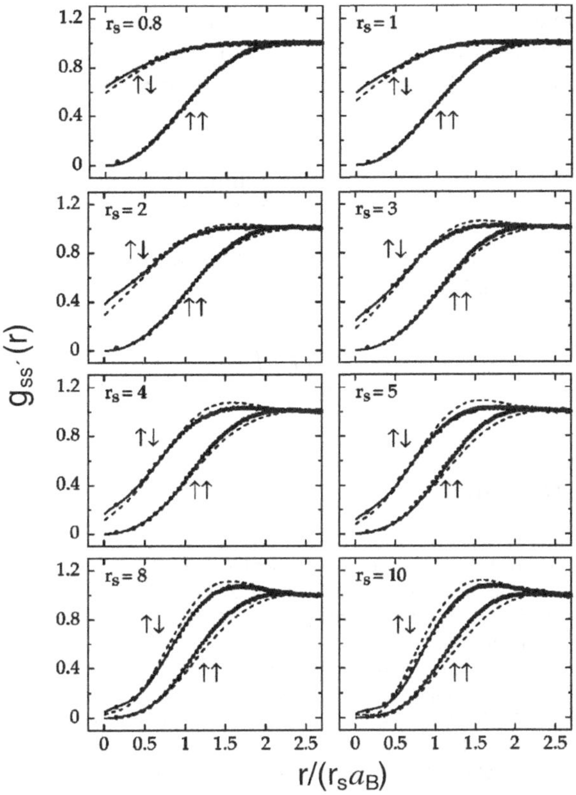

Fig. 3.2 Spin-resolved pair correlation functions $g_{\uparrow\uparrow}(r)$ and $g_{\uparrow\downarrow}(r)$ for an unpolarized homogeneous electron gas as a function of the particle distance r for various electron gas parameters r_s. *Dots*: QMC data [19], *solid line*: Perdew-Wang model [20, 21], and *dashed line*: Gori-Giorgi et al. [22]. Reprinted with permission from [22]. Copyright 2000 by the American Physical Society

large distances $r \to \infty$ exchange and correlation vanish while for vanishing distance $r \to 0$ a compensation of all contributions to the electron-electron interaction occurs. The function $g_{\uparrow\uparrow}(r)$ leads to an exchange-correlation hole. Because of the parallel spins [see (4.20)] the effects are dominated by exchange. Correlation is much more important for antiparallel spins, which are not kept apart by the exclusion principle. Correlation seems to tend to reduce the long-range part of the exchange hole, i.e., it tends to cause screening as discussed in Sect. 3.4. The hole related to $g_{\uparrow\downarrow}$ is therefore sometimes called depletion hole.

The hole behavior can be better understood introducing in the XC energy (3.62) the spin-dependent exchange-correlation density

$$n^{\Psi}_{\mathrm{XC}}(\mathbf{x}s, \mathbf{x}'s') = n^{\Psi}_{s's'}(\mathbf{x}') \left[g^{\Psi}_{ss'}(\mathbf{x}, \mathbf{x}') - 1 \right]. \tag{3.63}$$

Instead of (3.50) it fulfills the sum rule

$$\int d^3\mathbf{x}' n^{\Psi}_{\mathrm{XC}}(\mathbf{x}s, \mathbf{x}'s') = -\delta_{ss'}. \tag{3.64}$$

Indeed, globally only parallel spins contribute to the XC hole while the electron redistribution for antiparallel spins does globally not give a net hole. Since the sum rule (3.64) holds for arbitrary positions **x**, the XC hole density must vanish for large distances

$$\lim_{|\mathbf{x}-\mathbf{x}'|\to\infty} n^{\Psi}_{\mathrm{XC}}(\mathbf{x}s,\mathbf{x}'s') = 0. \tag{3.65}$$

Nevertheless the sum rule suggests a long-range XC potential. For large distances (3.64) yields

$$\lim_{|\mathbf{x}|\to\infty} \int d^3\mathbf{x}' \frac{n^{\Psi}_{\mathrm{XC}}(\mathbf{x}s,\mathbf{x}'s')}{|\mathbf{x}-\mathbf{x}'|} = -\frac{1}{|\mathbf{x}|}\delta_{ss'}. \tag{3.66}$$

References

1. W. Pauli, Über den Zusammenhang des Abschlusses der Elektronengruppen im Atom mit der Komplexstruktur der Spektren. Z. Phys. **31**, 765–785 (1925)
2. S.S. Schweber, *Introduction to Relativistic Quantum Field Theory* (Row Peterson and Co., Evanston, 1961)
3. L.I. Schiff, *Quantum Mechanics* (Mc-Graw Hill, New York, 1968)
4. G.D. Mahan, *Many-Particle Physics* (Plenum Press, New York, 1990)
5. P.A.M. Dirac, A new notation for quantum mechanics. Math. Proc. Cambr. Phil. Soc. **35**, 416–418 (1939)
6. H. Stolz, *Einführung in die Vielelektronentheorie der Kristalle* (Akademie-Verlag, Berlin, 1974)
7. H. Eschrig, *The Fundamentals of the Density Functional Theory* (Teubner-Verlagsgesellschaft, Stuttgart, 1996)
8. D.R. Hartree, The wave mechanics of an atom with a non-Coulomb central field. Part I. Theory and methods. Proc. Cambr. Phil. Soc. **24**, 89–110 (1928)
9. D.R. Hartree, The wave mechanics of an atom with a non-Coulomb central field. Part II. Some results and discussion. Proc. Cambr. Phil. Soc. **24**, 111–132 (1928)
10. D.R. Hartree, The wave mechanics of an atom with a non-Coulomb central field. Part III. Term values and intensities in series in optical spectra. Proc. Cambr. Phil. Soc. **24**, 426–437 (1928)
11. D.C. Langreth, J.P. Perdew, The exchange correlation energy of a metallic surface. Solid State Commun. **17**, 1425–1429 (1975)
12. D.C. Langreth, J.P. Perdew, Exchange-correlation energy of a metallic surface: wave-vector analysis. Phys. Rev. B **15**, 2884–2901 (1977)
13. O. Gunnarsson, B.I. Lundqvist, Exchange and correlation in atoms, molecules, and solids by the spin-density-functional formalism. Phys. Rev. B **13**, 4274–4298 (1976)
14. L. van Hove, Correlations in space and time and Born approximation scattering in systems of interacting particles. Phys. Rev. **95**, 249–262 (1954)
15. R.Q. Hood, M.Y. Chou, A.J. Williamson, G. Rajagopal, R.J. Needs, Exchange and correlation in silicon. Phys. Rev. B **57**, 8972–8982 (1998)
16. L. Hedin, S. Lundqvist, Effect of electron-electron and electron-phonon interactions on the one-electron states of solids. Solid State Phys. **23**, 1–181 (1969)
17. P. Nozieres, D. Pines, A dielectric formulation of the many body problem: application to the free electron gas. Nuovo Cimento **9**, 470–490 (1958)
18. W. Jones, N.H. March, *Theoretical Solid State Physics*, vol. 1 (Dover Publications Inc, New York, 1973)

19. G. Ortiz, M. Harris, P. Ballone, Correlation energy, structure factor, radial density distribution function, and momentum distribution of the spin-polarized electron gas. Phys. Rev. B **50**, 1391–1405 (1994)
20. J.P. Perdew, Y. Wang, Pair-distribution function and its coupling-constant average for the spin-polarized electron gas. Phys. Rev. B **46**, 12947–12954 (1992)
21. J.P. Perdew, Y. Wang, Erratum: pair-distribution function and its coupling-constant average for the spin-polarized electron gas. Phys. Rev. B **46**, 12947–12954 (1992) (Phys. Rev. B **56**, 7018–7018 (1997))
22. P. Gori-Giorgi, F. Sacchetti, G.B. Bachelet, Analytic structure factors and pair correlation functions for the unpolarized elecron gas. Phys. Rev. B **61**, 7353–7363 (2000)

Chapter 4
Hartree-Fock Approximation

Abstract The first quantum-mechanical effect of the electron-electron interaction beyond the Hartree approximation is the exchange of spin-parallel electrons. For the ground state of an electron gas the variation of the total energy according to the Rayleigh-Ritz variational principle leads to the Hartree-Fock approach. In the limit of collinear spins it is characterized by single-particle eigenfunctions and eigenvalues which can be computed by solving a Schrödinger-like equation for an electron. The electron-electron interaction is characterized by a local Hartree potential and a non-local and/or state-dependent exchange potential, the Fock operator. The quantities of the Hartree-Fock theory grant an easy interpretation of electronic excitations. In particular, the Hartree-Fock eigenvalues describe single-particle excitation energies in the framework of the frozen-orbital approximation. This result is the Koopmans theorem. For a homogeneous electron gas, more precisely the jellium model, the Hartree-Fock approximation gives an instructive description of the kinetic energy and exchange contributions to the total energy of the ground state, especially their density dependence. A physically intuitive variation of the exchange with the spin polarization of the electron gas is derived.

4.1 Exchange

4.1.1 Beyond Hartree Approximation

The Hartree approximation $g_{\Psi}(\mathbf{x}, \mathbf{x}') = 1$ (3.43) of (3.42) completely neglects any non-classical interaction in the inhomogeneous electron gas, i.e., exchange and correlation. N-particle wave functions $\Psi(\mathbf{x}_1 s_1 ... \mathbf{x}_N s_N)$ in (3.23) and (3.30) can be simply described by a product of single-particle wave functions. The next step of possible approximations could be that the electrons must obey the Pauli exclusion principle while they remain uncorrelated. Historically, in order to do so Fock [1, 2] recommended the use of advanced N-particle wave functions $\Psi(\mathbf{x}_1 s_1 ... \mathbf{x}_N s_N)$ based on appropriate linear combinations of single-particle wave functions which fulfill the antisymmetry condition and the exclusion principle. They are so-called Slater

F. Bechstedt, *Many-Body Approach to Electronic Excitations*,
Springer Series in Solid-State Sciences 181, DOI 10.1007/978-3-662-44593-8_4

determinants [3, 4]. As the result of their use the interaction of the electrons is described within the Hartree-Fock (HF) approximation [5]. A description of the many-body wave functions by linear combinations of different Slater determinants leads to mixing of configurations and, thus, describes configuration interactions (CIs) beyond the HF approach. Therefore, corrections to the energy in HF approximation are generally called configuration interactions by chemists but correlation by physicists.

4.1.2 Exchange Energy

In this chapter, we go another way to proceed and to introduce the exchange interaction. We consider an electron with a certain spin $s = \uparrow$ at the origin and look at the density of the other $(N-1)$ electrons. The exclusion principle forbids the presence of electrons with spin $\uparrow$ at the origin, but says nothing about electrons with spin $s = \downarrow$, which can well be located there. To include this behavior we apply the Wick theorem [6]. According to this theorem every matrix element $\langle \Psi | ... | \Psi \rangle$ of field operators (3.6) and (3.9) can be expressed as sums of expectation values of their products. To do so, one uses the contraction of two operators, more precisely of an annihilation and a creation operator. The contraction of two annihilation or creation operators vanishes in this approximation.

Following this idea we investigate the two-particle density matrix (3.38) and apply the Wick theorem. Thereby, according to (3.12) the number of commutations of two field operators determines the sign of a product of contractions. One finds

$$
\begin{aligned}
m_\Psi(\mathbf{x}, \mathbf{x}') &= \sum_{s,s'} \langle \Psi | \psi_{s'}^+(\mathbf{x}') \psi_s^+(\mathbf{x}) \psi_s(\mathbf{x}) \psi_{s'}(\mathbf{x}') | \Psi \rangle \\
&\approx \sum_{s,s'} \Big\{ \langle \Psi | \psi_{s'}^+(\mathbf{x}') \psi_{s'}(\mathbf{x}') | \Psi \rangle \langle \Psi | \psi_s^+(\mathbf{x}) \psi_s(\mathbf{x}) | \Psi \rangle \\
&\qquad - \langle \Psi | \psi_{s'}^+(\mathbf{x}') \psi_s(\mathbf{x}) | \Psi \rangle \langle \Psi | \psi_s^+(\mathbf{x}) \psi_{s'}(\mathbf{x}') | \Psi \rangle \Big\} \qquad (4.1) \\
&= n_\Psi(\mathbf{x}) n_\Psi(\mathbf{x}') - \sum_{s,s'} |n_{ss'}^\Psi(\mathbf{x}, \mathbf{x}')|^2
\end{aligned}
$$

with the definitions of the electron density (3.36) and the generalization of the spin density matrix (3.57), the exchange density $n_{ss'}^\Psi(\mathbf{x}, \mathbf{x}') = \langle \Psi | \psi_s^+(\mathbf{x}) \psi_{s'}(\mathbf{x}') | \Psi \rangle$.

Using this approach to the two-particle density matrix, the exchange-correlation energy E_{XC}^Ψ (3.48) in the total energy E_Ψ (3.45) reduces to the pure exchange energy

$$
\begin{aligned}
E_{\mathrm{X}}^{\Psi} &= -\frac{1}{2}\sum_{s,s'}\int d^3\mathbf{x}\int d^3\mathbf{x}'\mathrm{v}(\mathbf{x}-\mathbf{x}')|n_{ss'}^{\Psi}(\mathbf{x},\mathbf{x}')|^2 \\
&= \frac{1}{2}\sum_{s,s'}\int d^3\mathbf{x}\int d^3\mathbf{x}'\mathrm{v}(\mathbf{x}-\mathbf{x}')n_{ss}^{\Psi}(\mathbf{x})n_{s's'}^{\Psi}(\mathbf{x}')\left[g_{ss'}^{\Psi,\mathrm{X}}(\mathbf{x},\mathbf{x}')-1\right]
\end{aligned} \tag{4.2}
$$

with the pair correlation function in HF approximation

$$
g_{ss'}^{\Psi,\mathrm{X}}(\mathbf{x},\mathbf{x}') = 1 - \frac{|n_{ss'}^{\Psi}(\mathbf{x},\mathbf{x}')|^2}{n_{ss}^{\Psi}(\mathbf{x})n_{s's'}^{\Psi}(\mathbf{x}')}, \tag{4.3}
$$

where all densities are derived within this approximation. The exchange energy (4.2) is negative and, thus, reduces the total energy with respect to the Hartree approximation. It therefore stabilizes the electronic system against the classical electron-electron repulsion. The pair correlation function (4.3) describes the exchange depletion, or the exchange hole [7]. In the limit of collinear spins, it becomes immediately clear that only parallel spins $\sim\delta_{ss'}n_{ss}^{\Psi}(\mathbf{x},\mathbf{x}')$ contribute to the reduction of the pair correlation function in Hartree-Fock approximation.

The difference of the XC energy (3.62) and the exchange energy (4.2) leads to the correlation energy. Many studies suggest that the effect of correlation can be cast in terms of the remaining part of the pair correlation function beyond exchange. Then a decomposition of (3.63) according to

$$
n_{\mathrm{XC}}^{\Psi}(\mathbf{x}s,\mathbf{x}'s') = n_{\mathrm{X}}^{\Psi}(\mathbf{x}s,\mathbf{x}'s') + n_{\mathrm{C}}^{\Psi}(\mathbf{x}s,\mathbf{x}'s') \tag{4.4}
$$

is possible. Thereby the exchange hole density is defined by (3.63) but with the pair correlation function (4.3). Since the entire exchange-correlation hole density obeys the sum rule (3.64) and since within the HF approximation the exchange hole density $n_{\mathrm{X}}^{\Psi}(\mathbf{x}s,\mathbf{x}'s')$ (for parallel spins) obeys a similar sum rule (4.22), the correlation hole must approximately integrate to zero,

$$
\int d^3\mathbf{x}'n_{\mathrm{C}}^{\Psi}(\mathbf{x}s,\mathbf{x}'s) = 0, \tag{4.5}
$$

i.e., it merely redistributes the density of the hole. This result corresponds to the fact that correlation is most important for electrons of opposite spin, since electrons of the same spin are repelled as a consequence of the exclusion principle. This remaining correlation part, once exchange has been taken into account, is the major problem in many-body theory independent of investigating ground or excited states. We note that the terms exchange and correlation or exchange and correlation energies are frequently used but not very precise. They depend on the state $|\Psi\rangle$ of the N-electron system that is investigated and, of course, the approximations made for the various densities, density matrices or pair correlation functions.

4.2 Hartree-Fock Equations

4.2.1 Representation of Field Operators

We expand the field operators $\psi_s^+(\mathbf{x})$ and $\psi_s(\mathbf{x})$ in terms of single-particle spinors. Assuming no explicit spin dependence of the total Hamiltonian (3.1), the limit of collinear spins is valid. For the spin-half particles we factorize the two-component Pauli spinors from the very beginning. It holds

$$\begin{aligned}\psi_s^+(\mathbf{x}) &= \sum_{\lambda,m_s} \varphi_{\lambda m_s}^*(\mathbf{x})\chi_{\frac{1}{2}m_s}^+(s)\hat{a}_{\lambda m_s}^+,\\ \psi_s(\mathbf{x}) &= \sum_{\lambda,m_s} \varphi_{\lambda m_s}(\mathbf{x})\chi_{\frac{1}{2}m_s}(s)\hat{a}_{\lambda m_s}\end{aligned} \tag{4.6}$$

with the creation operator $\hat{a}_{\lambda m_s}^+$ (annihilation operator $\hat{a}_{\lambda m_s}$) of an electron in a single-particle state characterized by a set of orbital quantum numbers λ, e.g. ν and $\mathbf{k}$ in the case of translationally invariant systems (1.13). The unknown orbital parts $\varphi_{\lambda m_s}(\mathbf{x})$ of the single-electron spinors are functions of the space coordinate $\mathbf{x}$. They may depend on the quantum number $m_s = \pm\frac{1}{2}$ of the spin projection in spin-polarized systems. The spinors $\chi_{\frac{1}{2}m_s}(s)$ are functions of the spin variable s. In closed-shell systems without any spin polarization the functions $\varphi_{\lambda m_s}(\mathbf{x})$ are independent of m_s. In open-shell systems the neglect of m_s leads to the 'spin-restricted HF approximation', while the opposite case gives the 'unrestricted HF approximation'. On the other hand, in systems with spin polarization, the two possibilities $m_s = \pm\frac{1}{2}$ indicate the two spin channels. That means that despite the assumption of vanishing magnetic fields, a magnetization and, thus, a symmetry break occurs in the electron system. The projection axis of the spin operator is fixed in correspondence with the collinear limit.

The spin functions $\chi_{\frac{1}{2}m_s}(s)$ form an orthonormalized and complete set,

$$\begin{aligned}\sum_s \chi_{\frac{1}{2}m_s}^+(s)\chi_{\frac{1}{2}m_s'}(s) &= \delta_{m_s m_s'},\\ \sum_{m_s} \chi_{\frac{1}{2}m_s}(s)\chi_{\frac{1}{2}m_s}^+(s') &= \delta_{ss'}.\end{aligned} \tag{4.7}$$

They are eigenfunctions of the spin operator $\mathbf{s} = \frac{\hbar}{2}\boldsymbol{\sigma}$,

$$\begin{aligned}\mathbf{s}^2\chi_{\frac{1}{2}m_s}(s) &= \hbar^2\frac{1}{2}\left(1+\frac{1}{2}\right)\chi_{\frac{1}{2}m_s}(s),\\ s_z\chi_{\frac{1}{2}m_s}(s) &= \hbar m_s\chi_{\frac{1}{2}m_s}(s)\end{aligned} \tag{4.8}$$

with $m_s = \pm\frac{1}{2}$ as the quantum number of its z-component. The Pauli spin matrices (1.1) lead to the matrices $\mathbf{s}^2 = \frac{3}{4}\hbar^2 \begin{pmatrix} 1 & 0 \\ 0 & 1 \end{pmatrix}$ and $s_z = \frac{\hbar}{2} \begin{pmatrix} 1 & 0 \\ 0 & -1 \end{pmatrix}$ which have the simultaneous eigenvectors

$$\chi_{\frac{1}{2}\frac{1}{2}}(\uparrow) = \begin{pmatrix} 1 \\ 0 \end{pmatrix} \quad \text{and} \quad \chi_{\frac{1}{2}-\frac{1}{2}}(\downarrow) = \begin{pmatrix} 0 \\ 1 \end{pmatrix}. \tag{4.9}$$

4.2.2 Total Energy

We determine the unknown single-particle orbitals $\{\varphi_{\lambda m_s}(\mathbf{x})\}$ by two conditions: First, the electrons in these states $|\lambda m_s\rangle$ are non-interacting in the many-body state $|\Psi\rangle$, i.e., it holds

$$\langle \Psi | \hat{a}^+_{\lambda m_s} \hat{a}_{\lambda' m'_s} | \Psi \rangle = \delta_{\lambda\lambda'} \delta_{m_s m'_s} n^{\Psi}_{\lambda m_s} \tag{4.10}$$

with the occupation numbers $n^{\Psi}_{\lambda m_s}$ of the corresponding one-electron states of the N-electron system in state $|\Psi\rangle$. Neglecting temperature effects the single-particle states are occupied, $n^{\Psi}_{\lambda m_s} = 1$, or empty, $n^{\Psi}_{\lambda m_s} = 0$. The state dependence of the space-dependent orbitals $\varphi_{\lambda m_s}(\mathbf{x})$ is not explicitly considered in (4.6). With the explicit representation of the spin eigenvectors (4.9) the spin density matrix (3.57) becomes diagonal

$$n^{\Psi}_{ss'}(\mathbf{x}) = \delta_{ss'} \sum_{\lambda, m_s} n^{\Psi}_{\lambda m_s} \left|\varphi_{\lambda m_s}(\mathbf{x})\right|^2 \tag{4.11}$$

because of the collinearity, and the total electron density (3.58) is

$$n_{\Psi}(\mathbf{x}) = \sum_{\lambda, m_s} n^{\Psi}_{\lambda m_s} \left|\varphi_{\lambda m_s}(\mathbf{x})\right|^2 . \tag{4.12}$$

With (4.6), (4.10)–(4.12), and $E^{\Psi}_{\text{XC}} = E^{\Psi}_{\text{X}}$ the total energy (3.45) can be rewritten as

$$E_{\Psi} = E^{\Psi}_{\text{kin}} + E^{\Psi}_{\text{pot}} + E^{\Psi}_{\text{H}} + E^{\Psi}_{\text{X}} \tag{4.13}$$

with the kinetic energy (3.46) of these non-interacting electrons

$$E^{\Psi}_{\text{kin}} = \sum_{\lambda, m_s} n^{\Psi}_{\lambda m_s} \int d^3\mathbf{x} \varphi^*_{\lambda m_s}(\mathbf{x}) \left\{ -\frac{\hbar^2}{2m} \Delta_{\mathbf{x}} \right\} \varphi_{\lambda m_s}(\mathbf{x}), \tag{4.14}$$

the potential energy (3.47) of the electrons in the field of the nuclei and the Hartree energy (3.43), but both formulated with the electron density $n_{\Psi}(\mathbf{x})$ (4.12), as well as the exchange energy

$$E_{\mathrm{X}}^{\Psi} = -\frac{1}{2}\sum_{m_s}\int d^3\mathbf{x}\int d^3\mathbf{x}' \mathrm{v}(\mathbf{x}-\mathbf{x}')\left|\sum_{\lambda} n_{\lambda m_s}^{\Psi}\varphi_{\lambda m_s}^{*}(\mathbf{x}')\varphi_{\lambda m_s}(\mathbf{x})\right|^2 \tag{4.15}$$

to which only electrons with parallel spins contribute.

For the purpose of comparison we rewrite the Hartree energy (3.43) in terms of the unknown one-electron orbitals with (4.12) to

$$E_{\mathrm{H}}^{\Psi} = \frac{1}{2}\sum_{m_s,m_s'}\sum_{\lambda,\lambda'}\int d^3\mathbf{x}\int d^3\mathbf{x}' \mathrm{v}(\mathbf{x}-\mathbf{x}') n_{\lambda m_s}^{\Psi}\left|\varphi_{\lambda m_s}(\mathbf{x})\right|^2 n_{\lambda' m_s'}^{\Psi}\left|\varphi_{\lambda' m_s'}(\mathbf{x}')\right|^2 . \tag{4.16}$$

The comparison of (4.15) and (4.16) show that the self-interaction

$$\sim \frac{1}{2}\left[n_{\lambda m_s}^{\Psi}\right]^2\int d^3\mathbf{x}\int d^3\mathbf{x}' \mathrm{v}(\mathbf{x}-\mathbf{x}')\left|\varphi_{\lambda m_s}(\mathbf{x})\right|^2\left|\varphi_{\lambda m_s}(\mathbf{x}')\right|^2 \tag{4.17}$$

appears in the Hartree and the exchange energy. The two contributions cancel each other, so that the approach (4.10) is free of self-interaction. Therefore, the inclusion of exchange corrects an important failure of the Hartree theory, especially for localized states, the spurious influence of the self-interaction. The cancellation effect again suggests some uncertainties in the description of the (longitudinal) electron-electron interaction. It would be better to denote the energies (4.15) and (4.16) without the self-interaction (4.17) as direct and exchange interaction energies of the electrons.

For a better physical interpretation the exchange energy (4.15) is written according to (3.62) and (3.63) as an electrostatic interaction energy

$$E_{\mathrm{X}}^{\Psi} = \frac{1}{2}\sum_{m_s,m_s'}\int d^3\mathbf{x}\int d^3\mathbf{x}' n_{m_s}^{\Psi}(\mathbf{x})\mathrm{v}(\mathbf{x}-\mathbf{x}') n_{X}^{\Psi}(\mathbf{x}m_s,\mathbf{x}'m_s') \tag{4.18}$$

with the spin density

$$n_{m_s}^{\Psi}(\mathbf{x}) = \sum_{\lambda} n_{\lambda m_s}^{\Psi}\left|\varphi_{\lambda m_s}(\mathbf{x})\right|^2 \tag{4.19}$$

and the exchange density

$$n_{X}^{\Psi}(\mathbf{x}m_s,\mathbf{x}'m_s') = -\delta_{m_s m_s'}\frac{\left|\sum_{\lambda} n_{\lambda m_s}^{\Psi}\varphi_{\lambda m_s}^{*}(\mathbf{x}')\varphi_{\lambda m_s}(\mathbf{x})\right|^2}{\sum_{\lambda'} n_{\lambda' m_s}^{\Psi}\left|\varphi_{\lambda' m_s}(\mathbf{x})\right|^2}. \tag{4.20}$$

The latter density is spatially inhomogeneous but spin diagonal and negative. Its space dependence is dominated by the unknown orbitals $\varphi_{\lambda m_s}(\mathbf{x})$. If their orthonormalization in one spin channel m_s

$$\int d^3\mathbf{x}\varphi_{\lambda m_s}^{*}(\mathbf{x})\varphi_{\lambda' m_s}(\mathbf{x}) = \delta_{\lambda\lambda'} \tag{4.21}$$

is used together with the relation $\left[n^{\Psi}_{\lambda m_s}\right]^2 = n^{\Psi}_{\lambda m_s}$ one finds the well-known sum rule (3.64) in the form

$$\int d^3\mathbf{x}' n^{\Psi}_{\mathrm{X}}(\mathbf{x}m_s, \mathbf{x}'m_{s'}) = -\delta_{m_s m'_s}. \tag{4.22}$$

4.2.3 Ground State: Hartree-Fock Equations

In (4.14), (4.15), and (4.16) the total energy (4.13) of the collinear N-electron system in the state $|\Psi\rangle$ has been explicitly represented in terms of the orbitals $\{\varphi_{\lambda m_s}(\mathbf{x})\}$ which describe independent particles. The explicit energy expressions clearly demonstrate that the total energy E_Ψ is a functional of all (occupied) orbitals $\varphi_{\lambda m_s}(\mathbf{x})$. However, they are still unknown. For their determination we restrict the studies to the ground state $|\Psi_0\rangle$ of the system. The ground state should be subject to an optimization (more physically: minimization) by varying it with respect to all orbitals $\varphi_{\lambda m_s}(\mathbf{x})$ in agreement with the Rayleigh-Ritz variational principle. This is also the basic idea in text books of quantum chemistry (see e.g. [8]) despite the use of Slater determinants there.

Consequently, the second condition that completes the so-called Hartree-Fock approach is to minimize the total energy $E_{\Psi_0} = E_{\Psi_0}[\{\varphi_{\lambda m_s}(\mathbf{x})\}]$ with respect to all degrees of freedom in the orbitals $\varphi_{\lambda m_s}(\mathbf{x})$ with the restriction to the first condition (4.1) and the fact that the spins are quantized along an axis leading to the representations (4.6) and (4.7). In addition, the normalization condition of the many-body state $\langle\Psi_0|\Psi_0\rangle = 1$ and the orthonormalization of the orbitals $\varphi_{\lambda m_s}(\mathbf{x})$ in each spin channel (4.21) have to be taken into account. Together with the Langrange multipliers $\varepsilon^{m_s}_{\lambda\lambda'}$ the variation of $\varphi^*_{\lambda m_s}(\mathbf{x})$ leads to the functional derivative in each spin channel

$$\frac{\delta}{\delta\varphi_{\lambda m_s}(\mathbf{x})}\left\{E_{\Psi_0}\left[\{\varphi_{\lambda m_s}(\mathbf{x})\}\right] - \sum_{\lambda,\lambda'} n^{\Psi_0}_{\lambda m_s} n^{\Psi_0}_{\lambda' m_s} \varepsilon^{m_s}_{\lambda\lambda'} \int d^3\mathbf{x}\varphi^*_{\lambda m_s}(\mathbf{x})\varphi_{\lambda' m_s}(\mathbf{x})\right\} = 0. \tag{4.23}$$

With the explicit representations (4.14), (4.15), (4.16), and (3.47) with (4.12) of the energy contributions to the total ground-state energy (4.13) the functional derivative (4.23) leads to the HF equations

$$\left[-\frac{\hbar^2}{2m}\Delta_{\mathbf{x}} + V_{\mathrm{n}}(\mathbf{x}) + \int d^3\mathbf{x}' \mathrm{v}(\mathbf{x}-\mathbf{x}') n_{\Psi_0}(\mathbf{x}')\right]\varphi_{\lambda m_s}(\mathbf{x}) \tag{4.24}$$
$$-\sum_{\lambda'} n^{\Psi_0}_{\lambda' m_s}\int d^3\mathbf{x}' \mathrm{v}(\mathbf{x}-\mathbf{x}')\varphi^*_{\lambda' m_s}(\mathbf{x}')\varphi_{\lambda m_s}(\mathbf{x}')\varphi_{\lambda' m_s}(\mathbf{x}) = \sum_{\lambda'}\varepsilon^{m_s}_{\lambda\lambda'}\varphi_{\lambda' m_s}(\mathbf{x}).$$

In principle, they are only valid for occupied states $|\lambda m_s\rangle$ with $n^{\Psi_0}_{\lambda m_s} = 1$ but this restriction is usually not longer mentioned (and also not used).

The matrix $\varepsilon^{m_s}_{\lambda\lambda'}$ should be self-adjoint. Consequently a unitary transformation $\varphi'_{\lambda m_s}(\mathbf{x}) = \sum_{\lambda'} U_{\lambda\lambda'}\varphi_{\lambda' m_s}(\mathbf{x})$ diagonalizes the matrix with diagonal elements $\varepsilon_{\lambda m_s}$ [4]. The set $\{\varphi'_{\lambda m_s}(\mathbf{x})\}$ is called the canonical orbital representation. In each spin channel the HF equations for $\varphi'_{\lambda m_s}(\mathbf{x}) \to \varphi_{\lambda m_s}(\mathbf{x})$ can be formulated with a single-particle Hamiltonian $\hat{H}^{m_s}_{\mathrm{HF}}(\mathbf{x}, \mathbf{x}')$ being off-diagonal in the space coordinates as

$$\int d^3\mathbf{x}'\, \hat{H}^{m_s}_{\mathrm{HF}}(\mathbf{x}, \mathbf{x}')\varphi_{\lambda m_s}(\mathbf{x}') = \varepsilon_{\lambda m_s}\varphi_{\lambda m_s}(\mathbf{x}) \tag{4.25}$$

with the Hamilton density

$$\hat{H}^{m_s}_{\mathrm{HF}}(\mathbf{x}, \mathbf{x}') = \left[-\frac{\hbar^2}{2m}\Delta_{\mathbf{x}} + V_{\mathrm{n}}(\mathbf{x}) + V_{\mathrm{H}}(\mathbf{x})\right]\delta(\mathbf{x} - \mathbf{x}') + V^{m_s}_{\mathrm{X}}(\mathbf{x}, \mathbf{x}') \tag{4.26}$$

with the Hartree potential (3.44) and a spatially non-local and spin-channel-dependent potential, the Fock Hamilton operator,

$$V^{m_s}_{\mathrm{X}}(\mathbf{x}, \mathbf{x}') = -\sum_{\lambda'} n^{\Psi_0}_{\lambda' m_s}\mathrm{v}(\mathbf{x} - \mathbf{x}')\varphi^*_{\lambda' m_s}(\mathbf{x}')\varphi_{\lambda' m_s}(\mathbf{x}). \tag{4.27}$$

The HF equations can be rewritten with a fully state-dependent local Hamilton operator

$$\hat{H}^{\lambda m_s}_{\mathrm{HF}}(\mathbf{x})\varphi_{\lambda m_s}(\mathbf{x}) = \varepsilon_{\lambda m_s}\varphi_{\lambda m_s}(\mathbf{x}) \tag{4.28}$$

with

$$\hat{H}^{\lambda m_s}_{\mathrm{HF}}(\mathbf{x}) = -\frac{\hbar^2}{2m}\Delta_{\mathbf{x}} + V_{\mathrm{n}}(\mathbf{x}) + V_{\mathrm{H}}(\mathbf{x}) + V^{\lambda m_s}_{\mathrm{X}}(\mathbf{x}) \tag{4.29}$$

and a state-dependent exchange potential

$$V^{\lambda m_s}_{\mathrm{X}}(\mathbf{x}) = -\sum_{\lambda'} n^{\Psi_0}_{\lambda' m_s}\mathrm{v}(\mathbf{x} - \mathbf{x}')\varphi^*_{\lambda' m_s}(\mathbf{x}')\varphi_{\lambda m_s}(\mathbf{x}')\frac{\varphi_{\lambda' m_s}(\mathbf{x})}{\varphi_{\lambda m_s}(\mathbf{x})}. \tag{4.30}$$

Note that the equations (4.25) or (4.28) represent a set of integro-differential equations for each orbital $\varphi_{\lambda m_s}(\mathbf{x})$. The differential character is related to the Laplace operator in the kinetic energy, while the Hartree and exchange potentials include integrals over orbital combinations. The latter fact indicates the involved self-consistency via the self-consistent potential $V_{\mathrm{H}}(\mathbf{x}) + V^{\lambda m_s}_{\mathrm{X}}(\mathbf{x})$. Consequently the solution of (4.25) or (4.28) asks for a self-consistent procedure. The numerical difficulties to do so grow with the number of electrons and the required accuracy. For instance, one has to compute a usually large number of Coulomb integrals of the order of N^4.

4.3 Koopmans Theorem

4.3.1 HF Total Energy

The kinetic energy (4.14) of the N electrons in the HF ground state $|\Psi_0\rangle$ can be reformulated by using the HF equations (4.28) and the Lagrange multipliers $\varepsilon_{\lambda m}$. One finds

$$\begin{aligned} E_{\text{kin}}^{\Psi_0} &= \sum_{\lambda,m_s} n_{\lambda m_s}^{\Psi_0} \left\{ \varepsilon_{\lambda m_s} - \int d^3\mathbf{x} V_{\text{n}}(\mathbf{x}) \left|\varphi_{\lambda m_s}(\mathbf{x})\right|^2 \right\} \\ &\quad - \sum_{\lambda,m_s} \sum_{\lambda',m_s'} n_{\lambda m_s}^{\Psi_0} n_{\lambda' m_s'}^{\Psi_0} \int d^3\mathbf{x} \int d^3\mathbf{x}' \text{v}(\mathbf{x}-\mathbf{x}') \\ &\quad \times \left\{ \left|\varphi_{\lambda m_s}(\mathbf{x})\right|^2 \left|\varphi_{\lambda' m_s'}(\mathbf{x}')\right|^2 - \delta_{m_s m_s'} \varphi_{\lambda m_s}^*(\mathbf{x}) \varphi_{\lambda' m_s}^*(\mathbf{x}') \varphi_{\lambda m_s}(\mathbf{x}') \varphi_{\lambda' m_s}(\mathbf{x}) \right\}. \end{aligned} \tag{4.31}$$

Together with the three potential terms in (4.13) the total energy of the HF ground state $|\Psi_0\rangle$ of the N-electron system is

$$\begin{aligned} E_{\Psi_0} &= \sum_{\lambda,m_s} n_{\lambda m_s}^{\Psi_0} \varepsilon_{\lambda m_s} - \frac{1}{2} \sum_{\lambda,m_s} \sum_{\lambda',m_s'} n_{\lambda m_s}^{\Psi_0} n_{\lambda' m_s'}^{\Psi_0} \int d^3\mathbf{x} \int d^3\mathbf{x}' \text{v}(\mathbf{x}-\mathbf{x}') \\ &\quad \times \left\{ \left|\varphi_{\lambda m_s}(\mathbf{x})\right|^2 \left|\varphi_{\lambda' m_s'}(\mathbf{x}')\right|^2 - \delta_{m_s m_s'} \varphi_{\lambda m_s}^*(\mathbf{x}) \varphi_{\lambda' m_s}^*(\mathbf{x}') \varphi_{\lambda m_s}(\mathbf{x}') \varphi_{\lambda' m_s}(\mathbf{x}) \right\}. \end{aligned} \tag{4.32}$$

Introducing abbreviations for the Coulomb integrals

$$U_{\lambda\lambda'}^{m_s m_s'} = \int d^3\mathbf{x} \int d^3\mathbf{x}' \text{v}(\mathbf{x}-\mathbf{x}') \left|\varphi_{\lambda m_s}(\mathbf{x})\right|^2 \left|\varphi_{\lambda' m_s'}(\mathbf{x}')\right|^2 \tag{4.33}$$

and exchange integrals

$$J_{\lambda\lambda'}^{m_s m_s'} = \int d^3\mathbf{x} \int d^3\mathbf{x}' \text{v}(\mathbf{x}-\mathbf{x}') \varphi_{\lambda m_s}^*(\mathbf{x}) \varphi_{\lambda' m_s'}^*(\mathbf{x}') \varphi_{\lambda m_s}(\mathbf{x}') \varphi_{\lambda' m_s'}(\mathbf{x}), \tag{4.34}$$

the total energy can be rewritten in a more compact expression

$$E_{\Psi_0} = \sum_{\lambda,m_s} n_{\lambda m_s}^{\Psi_0} \varepsilon_{\lambda m_s} - \frac{1}{2} \sum_{\lambda,m_s} \sum_{\lambda',m_s'} n_{\lambda m_s}^{\Psi_0} n_{\lambda' m_s'}^{\Psi_0} \left\{ U_{\lambda\lambda'}^{m_s m_s'} - \delta_{m_s m_s'} J_{\lambda\lambda'}^{m_s m_s} \right\}. \tag{4.35}$$

The first term in (4.32) and (4.35), the major contribution, is the so-called band structure energy. The second term, which is free of self-interaction, corrects the double counting of the mutual Coulomb interaction of the electrons in the single-particle HF equations (4.28) as indicated by the prefactor $\frac{1}{2}$. In the sum of the occupied single-particle energies, the band structure energy, the impact of the Coulomb interaction of the electrons, with respect to their contribution to the total energy, is overestimated by a factor 2.

4.3.2 Single-Particle and Neutral Pair Excitations

In a very simplified picture of photoemission experiments a photon with energy $\hbar\omega$ removes an electron from an occupied single-particle state $|\lambda m_s\rangle$ of an N-electron system in its ground state. If the energy $\hbar\omega$ is high enough, this electron leaves the molecule, nanostructure or bulk solid with a certain kinetic energy as illustrated in Fig. 4.1a for a confined $N = 4$ electron system. The smallest possible photon energy measures the ionization energy I (or the work function W in a metal). In the limit of vanishing kinetic energy of the emitted electron, i.e., $\hbar\omega = I$, it results the vacuum level ε_{vac}. The process of photoionization is adiabatic if the escaping electron is slow enough that the remaining $(N-1)$ electrons have enough time to relax, so that the ground state of the $(N-1)$-electron system with a missing electron in $|\lambda m_s\rangle$ is formed but with the constraint that the hole in the single-particle state $|\lambda m_s\rangle$ is conserved, i.e., of a large enough hole lifetime. The single-particle excitation energy for zero kinetic energy of the escaping electrons is given by the Einstein equation of the photoelectric effect [9]

$$\hbar\omega = E_{\Psi_0}(N-1, \lambda m_s) - E_{\Psi_0}(N). \tag{4.36}$$

In the final state the number of remaining electrons is indicated in addition to the level constraint for the hole.

In an inverse photoemission experiment an electron is added to the system. It goes from the vacuum level ε_{vac}, i.e., with zero kinetic energy, into an unoccupied single-particle state $|\lambda' m_s'\rangle$ as illustrated for a small confined $N = 4$ electron system in Fig. 4.1b. The energy gain can be identified with the energy $\hbar\omega$ of an emitted photon. If $|\lambda' m_s'\rangle$ corresponds to the lowest empty state, the energy $\hbar\omega$ corresponds to the electron affinity A of the system (or again the work function W for metals). Allowing relaxation of the other N electrons but keep the state $|\lambda' m_s'\rangle$ occupied with

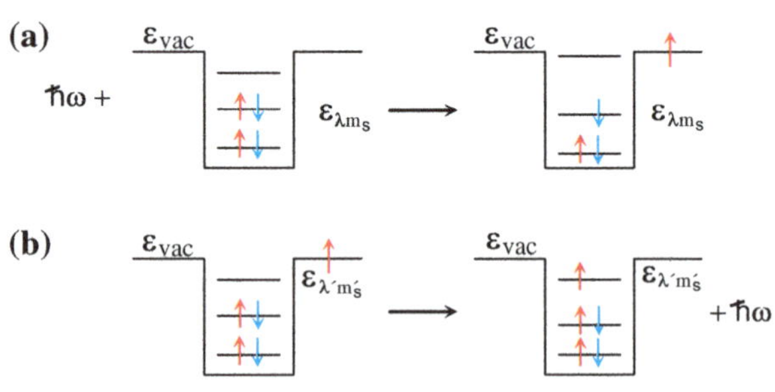

Fig. 4.1 Schematic representations of the Einstein equation in (4.36) and (4.37) for photoemission (**a**) and inverse photoemission (**b**) in a localized system of $N = 4$ electrons

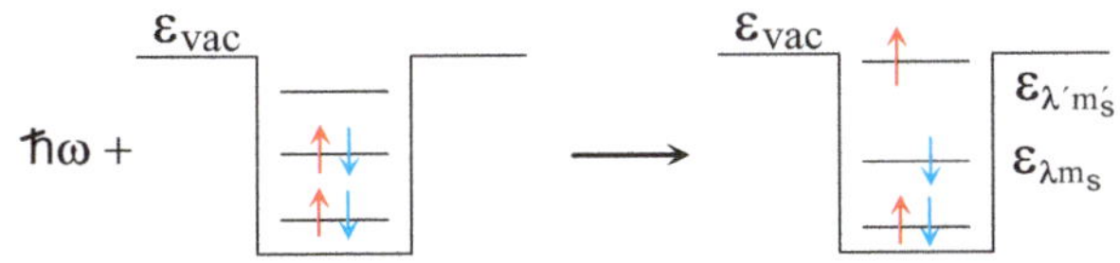

Fig. 4.2 Schematic representation of excitation of a neutral electron-hole pair in a confined system with $N = 4$ electrons in the ground state

the injected electron, it holds the Einstein equation as

$$\hbar\omega = E_{\Psi_0}(N) - E_{\Psi_0}(N+1, \lambda' m_s'). \tag{4.37}$$

If the photon energy $\hbar\omega$ is smaller than the ionization energy I (or the work function W) the system will not be ionized during the excitation process. The absorption of such a photon generates a neutral electron-hole pair in the system as illustrated in Fig. 4.2. If an electron is missing in the single-particle state $|\lambda m_s\rangle$ (what corresponds to a hole), which is occupied in the ground state, the electron will occupy a former empty state $|\lambda' m_s'\rangle$ (with $m_s' = m_s$ omitting spin-flip processes). The N-electron system will change into an excited state $|\Psi, \lambda' m_s', \lambda m_s\rangle$ with the constraints for the hole and the electron. If relaxation of the other electrons is allowed, one may identify the result with the 'ground state' of the system under the constraints of a hole in $|\lambda m_s\rangle$ and an electron in $|\lambda' m_s'\rangle$, which can be simulated by the occupation numbers $n_{\lambda m_s}^{\Psi_0} = 0$ and $n_{\lambda' m_s'}^{\Psi_0} = 1$. The resulting energy conservation reads as

$$\hbar\omega = E_{\Psi_0}(N, \lambda' m_s', \lambda m_s) - E_{\Psi_0}(N). \tag{4.38}$$

4.3.3 Physical Meaning of Lagrange Multipliers $\varepsilon_{\lambda m_s}$

The total energies in (4.36), (4.37), and (4.38) can be described within the Hartree-Fock approximation for the N-, $(N-1)$-, and $(N+1)$-electron systems with and without occupation constraints. Four self-consistent calculations are needed. Here, we apply the frozen-orbital approximation which should be valid in the limit of large numbers of particles $N \to \infty$. We assume that the orbitals $\varphi_{\lambda m_s}(\mathbf{x})$ (and consequently the Lagrange multipliers $\varepsilon_{\lambda m_s}$) are the same in the N-, $(N-1)$-, and $(N+1)$-electron systems with and without occupation constraint. Then, the energy differences can be easily calculated by means of expression (4.32) or (4.35) for the HF total energy of the N-electron system in the ground state. We have only to vary the occupation numbers $n_{\lambda m_s}^{\Psi_0} = 0, 1$ according to the studied many-electron state.

As an example we investigate the difference (4.36) in detail. First, we rewrite (4.32) by means of the expression for the eigenvalues from (4.25)

$$\begin{aligned}\varepsilon_{\lambda m_s} &= \int d^3\mathbf{x}\varphi^*_{\lambda m_s}(\mathbf{x})\left[-\frac{\hbar^2}{2m}\Delta_{\mathbf{x}} + V_{\mathrm{n}}(\mathbf{x})\right]\varphi_{\lambda m_s}(\mathbf{x}) \\ &+ \sum_{\lambda' m'_s} n^{\Psi_0}_{\lambda' m'_s}\int d^3\mathbf{x}\int d^3\mathbf{x}'\mathrm{v}(\mathbf{x}-\mathbf{x}') \\ &\times\left\{\left|\varphi_{\lambda m_s}(\mathbf{x})\right|^2\left|\varphi_{\lambda' m'_s}(\mathbf{x}')\right|^2 - \delta_{m_s m'_s}\varphi^*_{\lambda m_s}(\mathbf{x})\varphi^*_{\lambda' m_s}(\mathbf{x}')\varphi_{\lambda m_s}(\mathbf{x}')\varphi_{\lambda' m_s}(\mathbf{x})\right\}.\end{aligned} \tag{4.39}$$

It yields

$$\begin{aligned}E_{\Psi_0} &= \sum_{\lambda,m_s} n^{\Psi_0}_{\lambda m_s}\int d^3\mathbf{x}\varphi^*_{\lambda m_s}(\mathbf{x})\left[-\frac{\hbar^2}{2m}\Delta_{\mathbf{x}} + V_{\mathrm{n}}(\mathbf{x})\right]\varphi_{\lambda m_s}(\mathbf{x}) \\ &+\frac{1}{2}\sum_{\lambda,m_s}\sum_{\lambda',m'_s} n^{\Psi_0}_{\lambda m_s}n^{\Psi_0}_{\lambda' m'_s}\int d^3\mathbf{x}\int d^3\mathbf{x}'\mathrm{v}(\mathbf{x}-\mathbf{x}') \\ &\times\left\{|\varphi_{\lambda m_s}(\mathbf{x})|^2|\varphi_{\lambda' m'_s}(\mathbf{x}')|^2 - \delta_{m_s m'_s}\varphi^*_{\lambda m_s}(\mathbf{x})\varphi^*_{\lambda' m_s}(\mathbf{x}')\varphi_{\lambda m_s}(\mathbf{x}')\varphi_{\lambda' m_s}(\mathbf{x})\right\}.\end{aligned} \tag{4.40}$$

Taking (4.40) the difference of the total energies (4.36) is simply (with $n^{\Psi_0}_{\lambda m_s} = 1$)

$$\begin{aligned}\hbar\omega &= -\int d^3\mathbf{x}\varphi^*_{\lambda m_s}(\mathbf{x})\left[-\frac{\hbar^2}{2m}\Delta_{\mathbf{x}} + V_{\mathrm{n}}(\mathbf{x})\right]\varphi_{\lambda m_s}(\mathbf{x}) \\ &+\frac{1}{2}\sum_{\lambda',m'_s}\sum_{\lambda'',m''_s} n^{\Psi_0}_{\lambda' m'_s}n^{\Psi_0}_{\lambda'' m''_s}\left[\left(1-\delta_{\lambda\lambda'}\delta_{m_s m'_s}\right)\left(1-\delta_{\lambda\lambda''}\delta_{m_s m''_s}\right)-1\right] \\ &\times\int d^3\mathbf{x}\int d^3\mathbf{x}'\mathrm{v}(\mathbf{x}-\mathbf{x}') \\ &\times\left\{\left|\varphi_{\lambda' m'_s}(\mathbf{x})\right|^2\left|\varphi_{\lambda'' m''_s}(\mathbf{x}')\right|^2 - \delta_{m'_s m''_s}\varphi^*_{\lambda' m'_s}(\mathbf{x})\varphi^*_{\lambda'' m'_s}(\mathbf{x}')\varphi_{\lambda' m'_s}(\mathbf{x}')\varphi_{\lambda'' m'_s}(\mathbf{x})\right\}.\end{aligned}$$

Comparing with expression (4.39) we find

$$\hbar\omega = -\varepsilon_{\lambda m_s}. \tag{4.41}$$

Correspondingly it holds for (4.37)

$$\hbar\omega = -\varepsilon_{\lambda' m'_s}. \tag{4.42}$$

In both cases the HF eigenvalues refer to the vacuum level. The description of an electron-hole pair in the system according to (4.38) is more difficult. With (4.33), (4.34), and (4.35) one finds

$$\hbar\omega = \varepsilon_{\lambda' m'_s} - \varepsilon_{\lambda m_s} - U^{m'_s m_s}_{\lambda'\lambda} + \delta_{m_s m'_s}J^{m_s m_s}_{\lambda'\lambda}. \tag{4.43}$$

Equations (4.41) and (4.42) represent the Koopmans theorem [10], which gives the Hartree-Fock eigenvalues a physical meaning. The negative Lagrange multipliers represent the binding energies of electrons (4.41) in certain occupied HF single-particle states $|\lambda m_s\rangle$ in the N-electron system or the corresponding energies of excess electrons (4.42) in an empty state $|\lambda' m_s'\rangle$. Even the excitation of an electron from an occupied state into an empty state (4.43) can be described within the framework of approximations used, especially the frozen-orbital approximation. The Koopmans theorem means that the Lagrange multipliers of the HF theory $\varepsilon_{\lambda m_s}$ describe single-particle excitation energies of the system. More precisely, the negative eigenvalues $-\varepsilon_{\lambda m_s}$ below the vacuum level ε_{vac} in Fig. 4.1 represent approximations to the energies for addition or removal of electrons, since correlation is omitted. Consequently, the energy $-\varepsilon_{\lambda m_s}$ for the highest-occupied molecular orbital (HOMO) $|\lambda m_s\rangle$ should represent the ionization energy $I = -\varepsilon_{\lambda m_s}^{\text{HOMO}}$ of such a localized system, while the energy $-\varepsilon_{\lambda' m_s'}$ of the lowest unoccupied molecular orbital (LUMO) characterizes the electron affinity $A = -\varepsilon_{\lambda' m_s'}^{\text{LUMO}}$. According to the physical meaning of I and A, their difference

$$E_g = I - A \tag{4.44}$$

defines the fundamental HF gap of the N-electron system without excitonic effects.

Expression (4.43) for the excitation energy of an electron-hole pair is reduced with respect to the expectation $[\varepsilon_{\lambda' m_s'} - \varepsilon_{\lambda m_s}]$ from Koopmans theorem by the Coulomb attraction $-U_{\lambda'\lambda}^{m_s' m_s'}$ of the two particles and the corresponding (repulsive) electron-hole exchange $\delta_{m_s m_s'} J_{\lambda'\lambda}^{m_s m_s}$ for parallel spins. These terms already describe excitonic effects as will be illuminated in Sect. 19.3.2. They are spin-dependent. The electron-hole pair energy (4.43) can be generalized to a matrix $E_{\lambda'\lambda}^{m_s' m_s} \delta_{m_s' m_s'''} \delta_{m_s m_s''} + J_{\lambda'\lambda}^{m_s m_s} \delta_{m_s' m_s} \delta_{m_s''' m_s''}$ in pairs of the spin quantum numbers $m_{s'} m_s = \frac{1}{2}\frac{1}{2}, \frac{1}{2} - \frac{1}{2}, -\frac{1}{2}\frac{1}{2}$, and $-\frac{1}{2} - \frac{1}{2}$ to [11]

$$\begin{pmatrix} E_{\lambda'\lambda}^{\frac{1}{2}\frac{1}{2}} + J_{\lambda'\lambda}^{\frac{1}{2}\frac{1}{2}} & 0 & 0 & J_{\lambda'\lambda}^{\frac{1}{2}-\frac{1}{2}} \\ 0 & E_{\lambda'\lambda}^{\frac{1}{2}-\frac{1}{2}} & 0 & 0 \\ 0 & 0 & E_{\lambda'\lambda}^{-\frac{1}{2}\frac{1}{2}} & 0 \\ J_{\lambda'\lambda}^{-\frac{1}{2}\frac{1}{2}} & 0 & 0 & E_{\lambda'\lambda}^{-\frac{1}{2}-\frac{1}{2}} + J_{\lambda'\lambda}^{-\frac{1}{2}-\frac{1}{2}} \end{pmatrix} \tag{4.45}$$

with the abbreviation $E_{\lambda'\lambda}^{m_s' m_s} = \varepsilon_{\lambda' m_s'} - \varepsilon_{\lambda m_s} - U_{\lambda'\lambda}^{m_s' m_s}$ (see also [12]).

For non-spin-polarized systems with an equal number of spin-down and spin-up electrons, the spin indices $m_{s'} m_s$ can be omitted from the matrix. Its diagonalization leads to two types of eigenvalues, $\hbar\omega_1 = \varepsilon_{\lambda'} - \varepsilon_\lambda - U_{\lambda'\lambda} + 2J_{\lambda'\lambda}$ and $\hbar\omega_{2/3/4} = \varepsilon_{\lambda'} - \varepsilon_\lambda - U_{\lambda'\lambda}$. The threefold-degenerate triplet states $\hbar\omega_{2/3/4}$ with the total electron-hole pair spin quantum numbers $S = 1$ and $M_S = -1, 0, 1$ are lower in energy than the singlet state $\hbar\omega_1$ with $S = 0$ and $M_S = 0$, since the exchange integrals (4.34) are usually positive definite quantities.

At this stage we have to mention two problems. The first one is related to the fact that deriving Koopmans theorem (4.42) as well as (4.43) one uses energies $\varepsilon_{\lambda' m'_s}$ for empty states, which do not occur in the original HF variational problem (4.23) for the N-electron system. For occupied states, the eigenvalues (4.39) are lowered by the exchange term, which also cancels the spurious self-interaction in the Hartree contribution. Generalizing the HF equations (4.25) or (4.28) to empty states their eigenvalues are not affected by the exchange contribution, since the expectation value of the exchange operator vanishes. Thus, the empty-state eigenvalues are not self-interaction corrected [13]. While ionization energies are quite reasonable within the HF approximation, at least for atoms and molecules, electron affinities are not because empty states, taking their derivation seriously, are meaningless and less bound in the HF theory.

As a further consequence, the great asymmetry between occupied and empty orbitals is primarily responsible for the very large overestimate of energy gaps of semiconductors and insulators within the HF approximation (see Table 9.1). Indeed, the HF approximation can be also applied to crystals. According to the Roothaan theorem the single-particle potential of a crystal within the HF approximation obeys the space-group symmetry of the system and not only its translational symmetry. Describing electron and hole excitations in the frozen-orbital approach as the most important effect the reaction of the electron system, its relaxation, is neglected. This approach moves occupied (empty) states toward higher (lower) energies and, thus, shrinks the gap E_g. In finite systems, therefore, significant improvement of the gap E_g is achieved if the energy $E_{\Psi_0}(N, \lambda' m'_s, \lambda m_s)$ in (4.38) is computed from (4.40) but allowing the orbitals to relax, i.e., lifting the approximation of frozen orbitals. Then the method to determine E_g by means of (4.38) contains two self-consistent-field calculations and is called delta self-consistent field (ΔSCF) method [14] or, more precisely, delta Hartree-Fock approximation.

4.4 Homogeneous Electron Gas

4.4.1 Jellium Model

The simplest solids, e.g. simple metals, may be nearly described by a homogeneous electron gas with the uniform density $n = N/\Omega$. Since a metal represents a neutral system the lattice of positively charged ions is replaced by a fixed uniform distribution of positive charge. We investigate the interaction of the electrons in the presence of this background charge. This simplified model of a metal is sometimes known as jellium. Within the jellium model it holds $V_{\text{n}}(\mathbf{x}) + V_{\text{H}}(\mathbf{x}) = 0$ as a consequence of the electrical neutrality. Only the exchange term survives in the HF Hamiltonian (4.26) or (4.29).

In a first step we assume the same number of spin-up and spin-down electrons, so that no spin polarization occurs in the jellium system and the quantum number

Fig. 4.3 Fermi sphere of an electron gas without exchange and correlation in reciprocal space

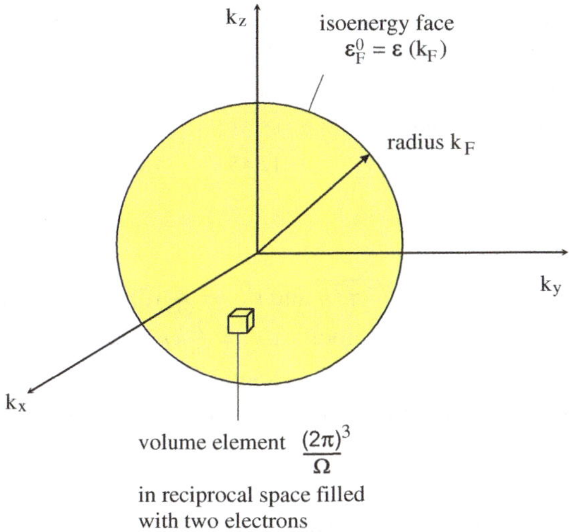

m_s can be removed from the HF equations. The familiar set of free-electron plane waves

$$\varphi_{\mathbf{k}m_s}(\mathbf{x}) = \frac{1}{\sqrt{\Omega}} e^{i\mathbf{kx}} \tag{4.46}$$

with the triple of quantum numbers $\lambda \hat{=} \mathbf{k}$ with $\mathbf{k} \in$ the entire reciprocal space represents self-consistent solutions of (4.25) or (4.28).

We investigate the three-dimensional electron gas at zero temperature $T = 0\,\mathrm{K}$. Then, for a given $\mathbf{k}$ two electrons, one with spin-up and the other one with spin-down, occupy one volume element $\frac{(2\pi)^3}{\Omega}$ of the reciprocal space starting from the origin $\mathbf{k} = 0$ until all electrons are used up. The corresponding largest possible wave vector is $|\mathbf{k}| = k_F$ with the energy $\varepsilon_F^0 = \frac{\hbar^2}{2m} k_F^2$, which corresponds to the Fermi energy in Hartree approximation. The region of the occupied $\mathbf{k}$-space is defined by the Fermi sphere in Fig. 4.3. Consequently the electron density (4.12) is defined as

$$\begin{aligned} n &= \frac{1}{\Omega} \sum_{m_s} \sum_{\substack{\mathbf{k} \\ (|\mathbf{k}| \le k_F)}} \\ &= \frac{2}{\Omega} \frac{\Omega}{(2\pi)^3} \int d^3\mathbf{k}\, \theta(k_F - |\mathbf{k}|) \\ &= \frac{2}{\Omega} \frac{\Omega}{(2\pi)^3} \frac{4\pi}{3} k_F^3 \\ &= \frac{1}{3\pi^2} k_F^3 \end{aligned} \tag{4.47}$$

Table 4.1 Parameters of (valence) electron gases of two metals and a semiconductor

Solid	n (10^{22} cm^{-3})	ε_F^0 (eV)	k_F (10^8 cm^{-1})	v_F (10^6 ms^{-1})	r_s
Na	2.65	3.23	0.92	1.07	3.96
Au	5.90	5.51	1.20	1.39	3.04
Si	19.98	12.45	1.81	2.09	2.00

yielding $k_F = \sqrt[3]{3\pi^2 n}$ and $\varepsilon_F^0 = \frac{\hbar^2}{2m}(3\pi^2 n)^{\frac{2}{3}}$ for spin-paired electrons. Correspondingly, the Fermi velocity $v_F = \frac{\hbar}{m}k_F$ and the dimensionless electron gas parameter $r_s = [3/(4\pi n a_B^3)]^{\frac{1}{3}}$ (see Sect. 3.5.2) are given. Characteristic parameters are listed in Table 4.1 for two metals and one semiconductor.

4.4.2 Exchange Interaction

In the jellium only the exchange potential (4.27)

$$V_{\mathrm{X}}^{m_s}(\mathbf{x},\mathbf{x}') = -\frac{1}{\Omega}\sum_{\substack{\mathbf{k}\\(|\mathbf{k}|\leq k_F)}} \mathrm{v}(\mathbf{x}-\mathbf{x}')e^{i\mathbf{k}(\mathbf{x}-\mathbf{x}')} \tag{4.48}$$

still acts on the electrons. The replacement of the sum by an integral in **k**-space according to (4.47) yields immediately

$$V_{\mathrm{X}}^{m_s}(\mathbf{x},\mathbf{x}') = -\frac{3}{2}n\mathrm{v}(\mathbf{x}-\mathbf{x}')\left.\frac{\sin x - x\cos x}{x^3}\right|_{x=k_F|\mathbf{x}-\mathbf{x}'|}. \tag{4.49}$$

This is indeed a short-range potential which shows an attractive Coulomb behavior $V_{\mathrm{X}}^{m_s}(\mathbf{x},\mathbf{x}') = -\frac{n}{2}\mathrm{v}(\mathbf{x}-\mathbf{x}')$ for small distances $|\mathbf{x}-\mathbf{x}'| \to 0$.

The resulting HF eigenvalues (4.39) of (4.25) are

$$\varepsilon_{m_s}(\mathbf{k}) = \frac{\hbar^2}{2m}\mathbf{k}^2 - \frac{1}{\Omega}\sum_{\substack{\mathbf{k}'\\(|\mathbf{k}'|<k_F)}} \frac{e^2}{\varepsilon_0|\mathbf{k}-\mathbf{k}'|^2}, \tag{4.50}$$

where the Fourier transformation of the bare Coulomb potential (1.2)

$$\mathrm{v}(\mathbf{x}-\mathbf{x}') = \frac{1}{\Omega}\sum_{\mathbf{q}} \frac{e^2}{\varepsilon_0|\mathbf{q}|^2}e^{i\mathbf{q}(\mathbf{x}-\mathbf{x}')},$$

$$\frac{e^2}{\varepsilon_0|\mathbf{q}|^2} = \int d^3\mathbf{x}e^{-i\mathbf{q}\mathbf{x}}\mathrm{v}(\mathbf{x}) \tag{4.51}$$

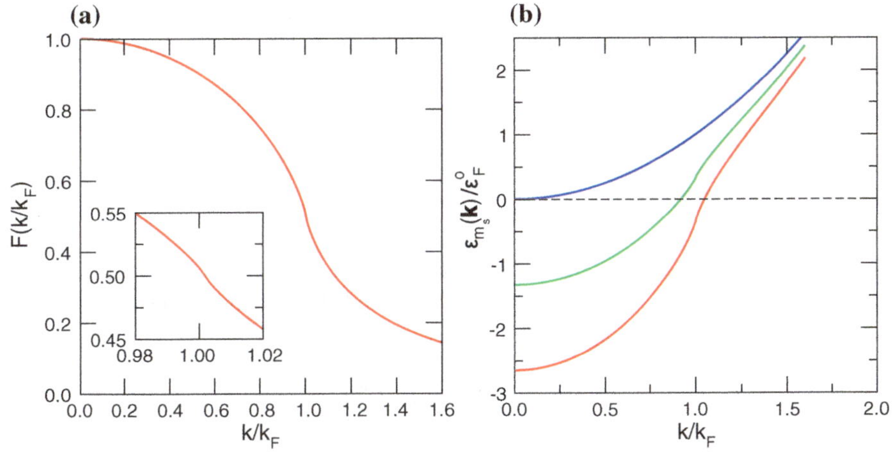

Fig. 4.4 (**a**) Plot of function (4.53) determining exchange and (**b**) dispersion relation (4.52) for electrons in a homogeneous electron gas in Hartree-Fock approximation for various electron densities $r_s = 0$ (*blue line*), $r_s = 2$ (*green line*), and $r_s = 4$ (*red line*)

has been used. With a replacement of the $\mathbf{k}'$-sum in (4.50) one finds after performing the integration [15]

$$\varepsilon_{m_s}(\mathbf{k}) = \frac{\hbar^2 k^2}{2m} - \frac{e^2}{2\pi^2 \varepsilon_0} k_F F\left(\frac{k}{k_F}\right) \tag{4.52}$$

with

$$F(x) = \frac{1}{2} + \frac{1-x^2}{4x} \ln\left|\frac{1+x}{1-x}\right|. \tag{4.53}$$

The exchange contribution to the HF eigenvalue (4.52) is governed by the function $F(x)$ (4.53) (see Fig. 4.4a). This function diverges at $x = 1$, i.e., at $k = k_F$. However, the divergence is logarithmic and cannot be revealed by refining the scale of the plot (see inset). The eigenvalues are displayed in Fig. 4.4b as a function of the wave vector for different electron concentrations, measuring all quantities in units of the Fermi energy ε_F^0 or the Fermi wave vector k_F

$$\varepsilon_{m_s}(\mathbf{k})/\varepsilon_F^0 = x^2 - \frac{4}{\pi}\left(\frac{4}{9\pi}\right)^{\frac{1}{3}} r_s F(x)\bigg|_{x=k/k_F}. \tag{4.54}$$

The curve in Fig. 4.4b for $r_s = 0$ (high-density limit) corresponds to the dispersion relation of free electrons in Hartree approximation. With decreasing electron density (increasing r_s) the electron energy is significantly lowered for small wave vectors indicating the exchange influence. For large wave vectors $x = k/k_F \to \infty$ it holds

$F(x) \to 1/3x^2$ and the Hartree and Hartree-Fock dispersions approach each other. As a consequence, the 'band width' of the HF electron band is increased compared to the Hartree one.

4.4.3 Total Energy

The total HF ground-state energy of the non-spin-polarized electron gas follows from (4.32) with the single-particle energies (4.52) and the exchange electron-electron contribution to avoid its double counting as

$$E_{\Psi_0} = E_{\text{kin}}^{\Psi_0}(n) + E_{\text{X}}^{\Psi_0}(n) \tag{4.55}$$

with the kinetic energy of all electrons

$$\begin{aligned} E_{\text{kin}}^{\Psi_0}(n) &= \sum_{m_s} \sum_{\substack{\mathbf{k} \\ (|\mathbf{k}|\leq k_F)}} \frac{\hbar^2 k^2}{2m} \\ &= N \cdot \frac{3}{5} \frac{\hbar^2 k_F^2}{2m} \\ &= N C_F \frac{\hbar^2}{m} n^{\frac{2}{3}} \\ &= N \cdot \frac{3}{5} (3\pi^2)^{\frac{2}{3}} (n a_B^3)^{\frac{2}{3}} R_\infty \\ &= N \frac{3}{5} \left(\frac{9\pi}{4}\right)^{\frac{2}{3}} \frac{1}{r_s^2} R_\infty \\ &= N \frac{2.21}{r_s^2} R_\infty \end{aligned} \tag{4.56}$$

with the Fermi coefficient $C_F = \frac{3}{10}(3\pi^2)^{\frac{2}{3}}$ and the exchange energy

$$\begin{aligned} E_{\text{X}}^{\Psi_0}(n) &= -\sum_{\substack{\mathbf{k} \\ (|\mathbf{k}|\leq k_F)}} \frac{e^2}{2\pi^2 \varepsilon_0} k_F F\left(\frac{k}{k_F}\right) \\ &= -N \frac{3e^3}{(4\pi)^2 \varepsilon_0} k_F \\ &= -N C_D \frac{e^2}{4\pi\varepsilon_0} n^{\frac{1}{3}} \\ &= -N \frac{3}{2} \left(\frac{3}{\pi}\right)^{\frac{1}{3}} (n a_B^3)^{\frac{1}{3}} R_\infty \end{aligned} \tag{4.57}$$

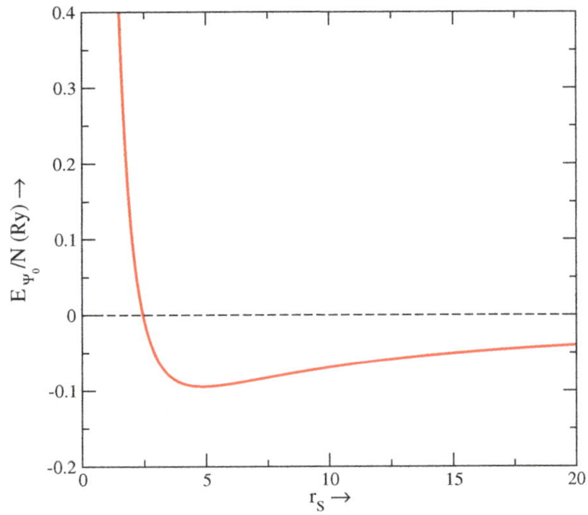

Fig. 4.5 Total energy of a homogeneous electron gas (jellium model) as a function of the density parameter r_s in Hartree-Fock approximation

$$= -N\frac{3}{2}\left(\frac{3}{2\pi}\right)^{\frac{2}{3}}\frac{1}{r_s}R_\infty$$

$$= -N\frac{0.916}{r_s}R_\infty$$

with the Dirac coefficient $C_D = \frac{3}{4}\left(\frac{3}{\pi}\right)^{\frac{1}{3}}$ and the Rydberg energy $R_\infty = 1Ry = 13.605\,\text{eV}$. The sum (4.55) of the two terms illustrates the fact that the total energy depends significantly on the density. The total energy per particle $E_{\Psi_0}(n)/N$ (4.55) is plotted in Fig. 4.5 as a function of the dimensionless electron gas parameter r_s. It demonstrates that only the negative exchange (4.57) stabilizes a HF system. Since $2 \lesssim r_s \lesssim 6$ in ordinary metals (see also Table 4.1), we see that exchange tends to make the total energy negative. High-density systems with $r_s \lesssim 2$ are not stable. Also for large average distances between two electrons, $r_s > 6$, the binding energy is significantly reduced.

Most interesting is the exchange energy contribution (4.57) $\epsilon_X(n) = E_X^{\Psi_0}(n)/N$, per particle, with $\epsilon_X(n) \sim n^{\frac{1}{3}}$. In Sect. 5.3 we will derive its relationship to a density-dependent potential $V_X(n) = \frac{d}{dn}[n\epsilon_X(n)]$. It leads to the Kohn-Sham-Gáspár exchange potential (see e.g. [16–18])

$$V_X^{KSG}(n) = -\frac{e^2}{4\pi\varepsilon_0}\left(\frac{3}{\pi}n\right)^{\frac{1}{3}} \tag{4.58}$$

if the density n is replaced by the space-dependent one of an inhomogeneous electron gas.

Slater [4, 19, 20] derived a similar potential by averaging the non-local/state-dependent exchange potential (4.27) over the free electron states. The result was

$$V_X^S(n) = \alpha \frac{3}{2} V_X^{KSG}(n) \tag{4.59}$$

with $\alpha = 1$. The variation of the parameter α between $\frac{2}{3}$ and 1 suggests the adjustment of such a parameter α to simulate partly correlation effects. This special choice of the X potential with a density-dependent parameter is known as X_α method in the literature [4, 14, 19].

4.4.4 Exchange for Spin-Polarized Systems

While for systems with paired spins the exchange energy depends only on the electron density (4.57), for spin-polarized systems with collinear spins the degree of spin polarization (3.60) has to be taken into account. For a homogeneous electron gas with $N_{\frac{1}{2}}$ ($N_{-\frac{1}{2}} = N - N_{\frac{1}{2}}$) spin-up (spin-down) electrons the two densities $n_{\frac{1}{2}}$ and $n_{-\frac{1}{2}}$ exist which yield

$$n = n_{\frac{1}{2}} + n_{-\frac{1}{2}},$$
$$\xi = \frac{n_{\frac{1}{2}} - n_{-\frac{1}{2}}}{n_{\frac{1}{2}} + n_{-\frac{1}{2}}}. \tag{4.60}$$

The two densities lead to two different Fermi wave vectors $k_{Fm_s} = \sqrt[3]{6\pi^2 n_{m_s}}$ ($m_s = \frac{1}{2}, -\frac{1}{2}$), which generate particle conservation in its own Fermi sphere because of $\frac{4\pi}{3} k_{Fm_s}^3 = \frac{(2\pi)^3}{\Omega} N_{m_s} = (2\pi)^3 n_{m_s}$.

With the wave functions (4.46) the exchange energy (4.15) becomes

$$E_X^{\Psi_0}(n, \xi) = -\frac{1}{2} \sum_{m_s} \int d^3\mathbf{x} \int d^3\mathbf{x}' v(\mathbf{x} - \mathbf{x}') \left| \frac{1}{\Omega} \sum_{\substack{\mathbf{k} \\ (|\mathbf{k}| \le k_{Fm_s})}} e^{i\mathbf{k}(\mathbf{x}-\mathbf{x}')} \right|^2 . \tag{4.61}$$

It contains the spin-polarized pair correlation function (3.61) in spin-orbital representation

$$g_{m_s m_s'}^X(\mathbf{x} - \mathbf{x}') = 1 - \frac{\delta_{m_s m_s'}}{n_{m_s} n_{m_s'}} \left| \frac{1}{\Omega} \sum_{\substack{\mathbf{k} \\ (|\mathbf{k}| \le k_{Fm_s})}} e^{i\mathbf{k}(\mathbf{x}-\mathbf{x}')} \right|^2 . \tag{4.62}$$

Using the result (4.49) of the **k**-summation in (4.48) we obtain

$$g^{\mathrm{X}}_{m_s m'_s}(\mathbf{x}-\mathbf{x}') = 1 - \delta_{m_s m'_s}\left[3\frac{\sin x - x\cos x}{x^3}\right]_{x=k_{Fm_s}|\mathbf{x}-\mathbf{x}'|}. \tag{4.63}$$

Including correlation these functions are displayed in Fig. 3.2. The off-diagonal functions are constant in HF approximation, $g^{\mathrm{X}}_{\frac{1}{2}-\frac{1}{2}}(\mathbf{x}-\mathbf{x}') = g^{\mathrm{X}}_{-\frac{1}{2}\frac{1}{2}}(\mathbf{x}-\mathbf{x}') = 1$, since the classical limit is realized without correlation and antiparallel spins. For parallel spins it holds $g^{\mathrm{X}}_{\frac{1}{2}\frac{1}{2}}(\mathbf{x}-\mathbf{x}') = g^{\mathrm{X}}_{-\frac{1}{2}-\frac{1}{2}}(\mathbf{x}-\mathbf{x}') = \frac{1}{10}k^2_{F\frac{1}{2}}|\mathbf{x}-\mathbf{x}'|^2$ for small distances $|\mathbf{x}-\mathbf{x}'| \to 0$. For large distances $|\mathbf{x}-\mathbf{x}'| \to \infty$ the diagonal pair correlation functions approaches 1 as an inverse power law with the well-known Friedel oscillations [15] due to the sharp Fermi spheres for the two kinds of electrons.

With (4.63) it follows

$$E^{\Psi_0}_{\mathrm{X}}(n,\xi) = -\frac{\Omega}{2}\sum_{m_s}\int d^3x v(\mathbf{x})\left[3n_{m_s}\frac{\sin x - x\cos x}{x^3}\bigg|_{x=k_{Fm_s}|\mathbf{x}|}\right]^2.$$

The space integral can be easily performed so that

$$\begin{aligned} E^{\Psi_0}_{\mathrm{X}}(n,\xi) &= -\frac{3e^2}{(4\pi)^2\varepsilon_0}\Omega\sum_{m_s} n_{m_s}k_{Fm_s} \\ &= -N\frac{3e^2}{(4\pi)^2\varepsilon_0}(6\pi^2)^{\frac{1}{3}}\frac{n^{\frac{4}{3}}_{\frac{1}{2}} + n^{\frac{4}{3}}_{-\frac{1}{2}}}{n_{\frac{1}{2}} + n_{-\frac{1}{2}}}. \end{aligned}$$

With $n_{\pm\frac{1}{2}} = \frac{1}{2}n(1\pm\xi)$ we introduce the spin-polarization function [21]

$$f(\xi) = \frac{(1+\xi)^{\frac{4}{3}} + (1-\xi)^{\frac{4}{3}} - 2}{2(2^{\frac{1}{3}}-1)}, \tag{4.64}$$

and find for the exchange energy per particle, $\epsilon_{\mathrm{X}}(n,\xi) = E^{\Psi_0}_{\mathrm{X}}/N$,

$$\epsilon_{\mathrm{X}}(n,\xi) = -\frac{3e^2}{(4\pi)^2\varepsilon_0}(3\pi^2)^{\frac{1}{3}}n^{\frac{1}{3}}\left[\left(2^{\frac{1}{3}}-1\right)f(\xi)+1\right]. \tag{4.65}$$

The polarization dependence can be better illustrated by the resulting interpolation formula

$$\epsilon_{\mathrm{X}}(n,\xi) = \epsilon_{\mathrm{X}}(n,0) + [\epsilon_{\mathrm{X}}(n,1) - \epsilon_{\mathrm{X}}(n,0)]\,f(\xi) \tag{4.66}$$

between the energy for vanishing spin polarization $\xi = 0$, $f(0) = 0$, and that of a fully spin-polarized system with $\xi = 1$, $f(1) = 1$.

References

1. V. Fock, Näherungsmethode zur Lösung des quantenmechanischen Mehrkörperproblems. Z. Phys. **61**, 126–148 (1930)
2. V. Fock, 'Selfconsistent field' mit Austausch für Natrium. Z. Phys. **61**, 795–805 (1930)
3. J.C. Slater, The theory of complex spectra. Phys. Rev. **34**, 1293–1322 (1929)
4. J.C. Slater, *The Self-Consistent Field for Molecules and Solids*, vol. 2 (Mc-Graw-Hill, New York, 1974)
5. A.S. Davydov, *Quantum Mechanics* (Pergamon Press, University of Michigan, Ann Arbor, 1965)
6. G.C. Wick, The evaluation of the collision matrix. Phys. Rev. **80**, 268–272 (1950)
7. R.G. Parr, W. Yang, *Density Functional Theory of Atoms and Molecules* (Oxford University Press, Oxford, 1989)
8. F. Jensen, *Introduction to Computational Chemistry* (Wiley, Chichester, 1999)
9. A. Einstein, Über einen die Erzeugung und Verwandlung des Lichtes betreffenden heuristischen Gesichtspunkt. Ann. Phys. **322**, 132–148 (1905)
10. T. Koopmans, Über die Zuordnung von Wellenfunktionen und Eigenwerten zu einzelnen Elektronen eines Atoms. Physica **1**, 104–113 (1934)
11. G. Grosso, G.P. Parravicini, *Solid State Physics* (Academic Press, Amsterdam, 2000)
12. C. Rödl, F. Fuchs, J. Furthmüller, F. Bechstedt, Ab initio theory of excitons and optical properties for spin-polarized systems. Phys. Rev. B **77**, 184408 (2008)
13. R.M. Martin, *Electronic Structure. Basic Theory and Practical Methods* (Cambridge University Press, Cambridge, 2004)
14. R.O. Jones, O. Gunnarsson, The density functional formalism, its applications and prospects. Rev. Mod. Phys. **61**, 689–746 (1989)
15. N.W. Ashcroft, N.D. Mermin, *Solid State Physics* (Saunders College, Philadelphia, 1976)
16. L. Hedin, B.I. Lundqvist, S. Lundqvist, Local exchange-correlation potentials. Solid State Commun. **9**, 537–541 (1971)
17. R. Gáspár, Über eine Approximation des Hartree-Fockschen Potentials durch eine universelle Potentialfunktion. Acta Physica Academiae Scientiarum Hungaricae **3**, 263–541 (1954)
18. W. Kohn, L.J. Sham, Self-consistent equations including exchange and correlation effects. Phys. Rev. **140**, A1133–A1138 (1965)
19. J.C. Slater, A simplification of the Hartree-Fock method. Phys. Rev. **81**, 385–390 (1951)
20. J.C. Slater, Magnetic effects and the Hartree-Fock equation. Phys. Rev. **82**, 538–541 (1951)
21. U. von Barth, L. Hedin, A local exchange-correlation potential for the spin-polarized case: I. J. Phys. C **5**, 1629–1642 (1972)

Part II
Electronic Ground State

Chapter 5
Density Functional Theory

Abstract Even with restriction to the longitudinal contribution the treatment of the electron-electron interaction is exceedingly difficult. However, in the case of the ground state any energetic, structural or electronic property of an inhomogeneous electron gas can be viewed as a functional of its local density $n(\mathbf{x})$. This scalar function of the position $\mathbf{x}$, in principle, determines all the information of the many-electron wave function. For a given external potential $V_{\mathrm{ext}}(\mathbf{x})$, for instance that due to the arrangement of the charged nuclei, the proofs of existence of such a functional are given by the Hohenberg-Kohn theorems. The ground state energy is minimized by variation of $n(\mathbf{x})$. Thereby, it decomposes into an external part and a universal Hohenberg-Kohn functional. The latter one fully accounts for the electron-electron interaction, but the theory – the density functional theory – provides no guidance for constructing it. Generalizations of the theory are possible in different directions. The most important one is the spin density functional theory with functionals depending also on the vector of the magnetization density $\mathbf{m}(\mathbf{x})$.

5.1 Ideas

5.1.1 Problem

In Part I we have explored how the mutual interaction of the many electrons in condensed matter can be described, possibly even including relativistic corrections. We have found that the longitudinal electron-electron interaction mediated by the Coulomb potential (1.2) is most important for the properties of a solid or a molecule. However, it has been also found that the treatment of the many-body interactions is exceedingly difficult, even if no external fields (in addition to the potential of the nuclei, which is also an external one from the viewpoint of the inhomogeneous electron gas) are applied to the system. Sum rules to characterize globally the electron-electron interaction can be derived easily. However, how to include systematically and correctly the (longitudinal) electron-electron interaction in calculations of real systems is truly a formidable problem.

F. Bechstedt, *Many-Body Approach to Electronic Excitations*,
Springer Series in Solid-State Sciences 181, DOI 10.1007/978-3-662-44593-8_5

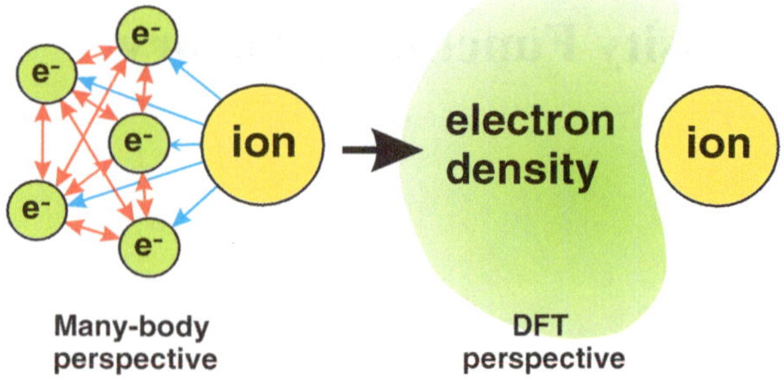

Fig. 5.1 Illustration of the main idea of a density functional theory to describe the individual mutual interactions of the electrons and their interaction with ions by that of an electron ensemble represented only by its density

Why that is the case has been demonstrated investigating the Hartree-Fock approach for an N-electron system in its ground state, although correlation of the electrons is neglected. In the HF approach the expansion coefficient $\Psi(\mathbf{x}_1 s_1 ... \mathbf{x}_N s_N)$ in the many-body state (3.23) is identified as a Slater determinant of N single-particle orbitals which have to be determined self-consistently by solving the problem (4.25) or (4.28). In practice, each of these functions will be expanded in a set of N_k basis functions such as plane waves. The size of the matrices that have to be diagonalized will be then of the order of $\binom{N_k}{N} = \frac{N_k!}{N!(N_k-N)!}$, which grows factorially. In addition, the self-consistency requirement blows up the problem as well as the inclusion of correlation. As a consequence, even though computer power grows exponentially, current state-of-the-art exact numerical diagonalizations have difficulties to handle more than a few tens or hundreds of electrons.

The formulation of the many-body problem in terms of densities and density matrices in Sect. 3.3 paves the way to another treatment of electron systems. The correspondence between these density quantities and the Hilbert-space states $|\Psi\rangle$ suggests to use a philosophy which in a manner of speaking starts from the other end, i.e., a search for the density and hence energy of the electron ensemble. Thereby, a restriction to the ground state $|\Psi_0\rangle$ of the system may be acceptable. The idea of such a density functional theory (DFT) is illustrated in Fig. 5.1. The interacting system of electrons is described via its density and rather than via its many-body wave function. For N electrons which obey the Pauli exclusion principle and repel each other via a Coulomb potential one introduces a basis variable of the system that only depends on three spatial coordinates rather than $3N$ degrees of freedom in the many-body wave function. Individual interactions are substituted by global ones in which the electron ensemble is represented by its density.

5.1.2 Grassroots: Thomas-Fermi-Dirac Theory

Thomas [1] and Fermi [2] independently studied the first three contributions $E_{\mathrm{kin}}^{\Psi_0}$, $E_{\mathrm{pot}}^{\Psi_0}$, and $E_{\mathrm{H}}^{\Psi_0}$ to the total energy (3.45). At that time they were not aware of the exchange energy and neglected the electron correlation. They generalized the result (4.56) for the kinetic energy per electron $C_F n^{\frac{2}{3}} \frac{\hbar^2}{m}$ of a non-interacting, homogeneous, and non-spin-polarized ($\xi = 0$) electron gas to the inhomogeneous case replacing the uniform density n by the inhomogeneous one $n(\mathbf{x})$. A functional of the total energy in the ground state of the inhomogeneous density results as

$$E_{\Psi_0}^{\mathrm{TF}}[n, V_\mathrm{n}] = C_\mathrm{F}\frac{\hbar^2}{m}\int d^3\mathbf{x} n^{\frac{5}{3}}(\mathbf{x}) + \int d^3\mathbf{x} V_\mathrm{n}(\mathbf{x}) n(\mathbf{x}) + \frac{1}{2}\int d^3\mathbf{x}\int d^3\mathbf{x}' \mathrm{v}(\mathbf{x}-\mathbf{x}') n(\mathbf{x}) n(\mathbf{x}'), \quad (5.1)$$

which moreover parametrically depends on the potential $V_\mathrm{n}(\mathbf{x})$ of the nuclei that may be replaced by an arbitrary 'external' potential $V_{\mathrm{ext}}(\mathbf{x})$.

The theory was extended by Dirac [3], who formulated the exchange contribution as a local exchange energy per electron, $\epsilon_\mathrm{X}(n) = -C_\mathrm{D}\frac{e^2}{4\pi\varepsilon_0} n^{\frac{1}{3}}$ (4.57). The improved functional reads as

$$E_{\Psi_0}^{\mathrm{TFD}}[n, V_\mathrm{n}] = E_{\Psi_0}^{\mathrm{TF}}[n, V_\mathrm{n}] - C_\mathrm{D}\frac{e^2}{4\pi\varepsilon_0}\int d^3\mathbf{x} n^{\frac{4}{3}}(\mathbf{x}). \quad (5.2)$$

The ground-state density and the corresponding total energy of the electronic system are found by minimizing the Thomas-Fermi-Dirac functional $E_{\Psi_0}^{\mathrm{TFD}}[n, V_\mathrm{n}]$ for all possible $n(\mathbf{x})$ being subject to the constraint on the total number of electrons

$$N = \int d^3\mathbf{x} n(\mathbf{x}). \quad (5.3)$$

The minimization with the constraint of particle conservation

$$\frac{\delta}{\delta n(\mathbf{x})}\left\{E_{\Psi_0}^{\mathrm{TFD}}[n, V_\mathrm{n}] - \mu\left[\int d^3\mathbf{x} n(\mathbf{x}) - N\right]\right\} = 0, \quad (5.4)$$

where the (zero-temperature) chemical potential μ of the electrons is introduced as Lagrange multiplier, yields an Euler relation between density and potential contributions

$$\frac{5\hbar^2}{3m} C_\mathrm{F} n^{\frac{2}{3}}(\mathbf{x}) + V(\mathbf{x}) - \mu = 0 \quad (5.5)$$

with

$$V(\mathbf{x}) = V_\mathrm{n}(\mathbf{x}) + V_\mathrm{H}(\mathbf{x}) + V_\mathrm{X}(\mathbf{x}) \quad (5.6)$$

and its exchange contribution

$$V_X(\mathbf{x}) = -\frac{4}{3}C_D\frac{e^2}{4\pi\varepsilon_0}n^{\frac{1}{3}}(\mathbf{x}) \tag{5.7}$$

in accordance with the Kohn-Sham-Gáspár potential (4.58). In the case of the jellium with $V_n(\mathbf{x}) + V_H(\mathbf{x}) = 0$, $V(\mathbf{x}) = V_X(\mathbf{x})$, and $n = n(\mathbf{x})$ relation (5.5) yields

$$\mu = \frac{5}{3}\frac{\hbar^2}{m}C_F n^{\frac{2}{3}} - \frac{4}{3}C_D\frac{e^2}{4\pi\varepsilon_0}n^{\frac{1}{3}} = \varepsilon_F^0 - \frac{4}{3}C_D\frac{e^2}{4\pi\varepsilon_0}n^{\frac{1}{3}}. \tag{5.8}$$

This result indicates that for an interacting electron gas the chemical potential is reduced by exchange (and correlation, if included) with respect to its value ε_F^0 of a non-interacting gas. One extension to account for the spatial inhomogeneity has been proposed by von Weizsäcker [4]. It considers the first gradient of the density and improves the kinetic energy.

The condition (5.5) is obviously in accordance with the idea of a density functional theory as sketched in Fig. 5.1. One has to solve only an equation for the density and not a many-body Schrödinger equation (3.30) for N electrons. However, the Thomas-Fermi-Dirac approach is too simple because it fully neglects correlation and describes the kinetic energy of all electrons by an expression that is valid for weak spatial variations of the density. As a consequence essential physics and chemistry are missing, such as the shell structure of atoms and binding of molecules (see no-binding theorem of E. Teller) [5]. An improved description is needed.

5.2 Hohenberg-Kohn Theory

5.2.1 Basics

Instead of improving an approximate theory such as the Thomas-Fermi-Dirac one, Hohenberg and Kohn [6] followed the idea of a rigorous formulation of the DFT as an exact theory of many-body systems, at least for their ground states $|\Psi_0\rangle$.

For the sake of illustrating the DFT concept we investigate the ground-state expectation values (3.35) of the Hamiltonian (3.1) in 'second' quantization (3.17), (3.18), and (3.19) as well as (3.36) of the density operator (3.14). Then, instead of solving the Schrödinger equation (3.29) one has to study

$$\begin{aligned} E_{\Psi_0} &= \langle\Psi_0|\mathscr{H}_0|\Psi_0\rangle, \\ n_{\Psi_0}(\mathbf{x}) &= \langle\Psi_0|\hat{n}(\mathbf{x})|\Psi_0\rangle. \end{aligned} \tag{5.9}$$

The spin structure of the systems is neglected in Sect. 5.2. In a first step we focus the investigations on non-spin-polarized systems with $\xi_{\Psi_0}(\mathbf{x}) \equiv 0$. The potential

$V_n(\mathbf{x})$ due to the nuclei is generalized to an arbitrary external (not depending on the electron distribution) potential $V_{ext}(\mathbf{x})$, in order to follow the standard denotations of the DFT.

It is convenient to decompose the Hamiltonian

$$\mathscr{H}_0 = \mathscr{H}_{int} + \mathscr{H}_{ext} \tag{5.10}$$

into the sum of an internal part, the kinetic energy of the electrons plus the (longitudinal) electron-electron interaction energy,

$$\mathscr{H}_{int} = \hat{T} + \hat{U} \tag{5.11}$$

with the representations (3.17) and (3.19), and an external part (3.18) (in practice with $V_{ext}(\mathbf{x}) \to V_n(\mathbf{x})$ given by the electron-nucleus interaction, where V_{ext} belongs to a set of potentials $\{V_{ext}\}$ that can be also non-Coulombic ones)

$$\begin{aligned} \mathscr{H}_{ext} &= \hat{V} \\ &= \int d^3\mathbf{x} V_{ext}(\mathbf{x})\hat{n}(\mathbf{x}). \end{aligned} \tag{5.12}$$

Despite of the modification of $\mathscr{H}_0$ by $V_n(\mathbf{x}) \to V_{ext}(\mathbf{x})$, we still denote the ground state of the system of N electrons by $|\Psi_0\rangle$. For simplicity, we suppose that the ground state is non-degenerate. In principle, any degeneracy can be removed by an arbitrary small modification of V_{ext} that appropriately lowers the symmetry of the system.

5.2.2 Hohenberg-Kohn Theorem I

The ground state $|\Psi_0\rangle$ of $\mathscr{H}_0$ (5.10) depends on the chosen external potential $V_{ext}(\mathbf{x})$. This fact may be denoted with the short-hand functional denotation $|\Psi_0[V_{ext}]\rangle$. This fact has consequences for the ground-state expectation values, e.g. for that of the density operator

$$n(\mathbf{x}) = \langle\Psi_0[V_{ext}]|\hat{n}(\mathbf{x})|\Psi_0[V_{ext}]\rangle, \tag{5.13}$$

where in contrast to the definition (3.36) the index Ψ_0, to label the ground-state density on the left side of the equation, has been removed. Correspondingly, we do not anymore write this index of the energy eigenvalue in the stationary Schrödinger equation

$$\mathscr{H}_0|\Psi_0\rangle = E|\Psi_0\rangle. \tag{5.14}$$

Summarizing, we have, via solution of the Schrödinger equation (5.14), defined a map

$$C : V_{ext} \to \Psi_0.$$

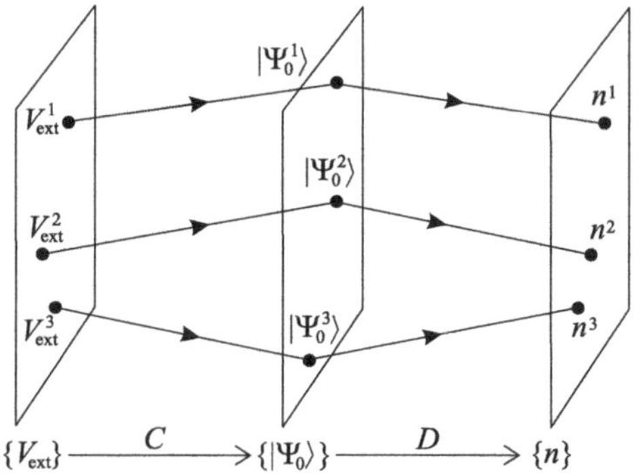

Fig. 5.2 Schematic illustration of the Hohenberg-Kohn maps $\mathcal{C}$ and $\mathcal{D}$

This map is surjective by construction (see Fig. 5.2) since the set $\{\Psi_0\}$ contains no element which is not associated with some element of $\{V_{ext}\}$. For all the ground states $\{|\Psi_0[V_{ext}]\rangle\}$ definition (5.13) establishes a second map,

$$\mathcal{D} : \Psi_0 \to n .$$

This map of the ground-state wave functions onto the ground-state densities is again trivially surjective. The combination of the two maps is illustrated in Fig. 5.2.

The product of the two maps $(\mathcal{C} \cdot \mathcal{D})$ or, more precisely, the existence of its inverse $(\mathcal{C} \cdot \mathcal{D})^{-1} = \mathcal{D}^{-1} \cdot \mathcal{C}^{-1}$ guarantees the *Hohenberg-Kohn theorem I*:

> For a system of interacting electrons in an external potential $V_{ext}(\mathbf{x})$, the potential itself is uniquely determined, except for a constant, by the ground-state density $n(\mathbf{x})$.

In other words, there is a one-to-one correspondence between the ground-state density of a N-electron system and the external potential acting on it. In this sense $n(\mathbf{x})$ becomes the variable of interest. All properties of the system are completely determined by the ground-state density $n(\mathbf{x})$. Therefore, a more specific formulation of the *Hohenberg-Kohn theorem I* could be:

> The total energy $E = E_{V_{ext}}[n]$ in the ground state is a universal functional of the corresponding density $n(\mathbf{x})$.

The proof of the theorem is given in two steps following Dreizler and Gross [7]. Suppose that there are two different external potentials $V_{ext}(\mathbf{x})$ and $V'_{ext}(\mathbf{x})$ with

$$V_{ext}(\mathbf{x}) \neq V'_{ext}(\mathbf{x}) + \text{const.}, \tag{5.15}$$

which differ by more than a constant but lead to the same ground-state density $n(\mathbf{x})$. The two potentials lead to different Hamiltonians $\mathscr{H}_0$ and $\mathscr{H}'_0$ with the ground states $|\Psi_0[V_{ext}]\rangle$ and $|\Psi'_0[V_{ext}]\rangle$ according to (5.14). The assumption of equal ground states $|\Psi_0[V_{ext}]\rangle = |\Psi'_0[V_{ext}]\rangle$ gives

$$
\begin{aligned}
(\mathcal{H}_0 - \mathcal{H}_0')|\Psi_0\rangle &= (\hat{V} - \hat{V}')|\Psi_0\rangle \\
&= \int d^3\mathbf{x}[V_{\text{ext}}(\mathbf{x}) - V_{\text{ext}}'(\mathbf{x})]\hat{n}(\mathbf{x})|\Psi_0\rangle \\
&= (E - E')|\Psi_0\rangle.
\end{aligned}
\tag{5.16}
$$

Since $\hat{V}$ and $\hat{V}'$ are multiplicative operators according to (3.18), (5.16) implies that

$$
\begin{aligned}
\langle\Psi_0|\hat{V} - \hat{V}'|\Psi_0\rangle &= \int d^3\mathbf{x}\left[V_{\text{ext}}(\mathbf{x}) - V_{\text{ext}}'(\mathbf{x})\right]n(\mathbf{x}) \\
&= E - E' \\
&= \text{const.}
\end{aligned}
$$

This relation can be only fulfilled for arbitrary $n(\mathbf{x})$ if

$$
V_{\text{ext}}(\mathbf{x}) - V_{\text{ext}}'(\mathbf{x}) = \text{const.},
$$

in contradiction to the assumption (5.15). This is however only valid for 'well behaved' potentials $V_{\text{ext}}(\mathbf{x})$ and $V_{\text{ext}}'(\mathbf{x})$, if $|\Psi_0\rangle$ does not vanish on a set of positive measure. Consequently, two different potentials $V_{\text{ext}}(\mathbf{x})$ and $V_{\text{ext}}'(\mathbf{x})$ cannot lead to the same ground state $|\Psi_0\rangle$. Provided $|\Psi_0\rangle$ is known, one may conclude on the specific $V_{\text{ext}}(\mathbf{x})$. In other words, the map $\mathcal{C}^{-1}$ exists.

The second part of the proof is related to the map $\mathcal{D}$ (see Fig. 5.2). It is done in the form of *reductio ad absurdum*. One supposes that $|\Psi_0\rangle \neq |\Psi_0'\rangle$ implies $n(\mathbf{x}) = n'(\mathbf{x})$. The Rayleigh-Ritz variational principle leads to

$$
\begin{aligned}
E_{V_{\text{ext}}'} &= \langle\Psi_0'|\mathcal{H}_0'|\Psi_0'\rangle \\
&< \langle\Psi_0|\mathcal{H}_0'|\Psi_0\rangle = \langle\Psi_0|\mathcal{H}_0 + \hat{V}' - \hat{V}|\Psi_0\rangle.
\end{aligned}
\tag{5.17}
$$

With the definition of $\hat{V}$ (3.18) and the stationary Schrödinger equation (5.14) one finds

$$
E_{V_{\text{ext}}'} < E_{V_{\text{ext}}} + \langle\Psi_0|\hat{V}' - \hat{V}|\Psi_0\rangle
$$

or again with equal electron densities

$$
E_{V_{\text{ext}}'} < E_{V_{\text{ext}}} + \int d^3\mathbf{x}\left[V_{\text{ext}}'(\mathbf{x}) - V_{\text{ext}}(\mathbf{x})\right]n(\mathbf{x}). \tag{5.18}
$$

Starting with $E_{V_{\text{ext}}}$ instead of $E_{V_{\text{ext}}'}$ in (5.17), corresponding arguments result in

$$
E_{V_{\text{ext}}} < E_{V_{\text{ext}}'} + \int d^3\mathbf{x}\left[V_{\text{ext}}(\mathbf{x}) - V_{\text{ext}}'(\mathbf{x})\right]n(\mathbf{x}). \tag{5.19}
$$

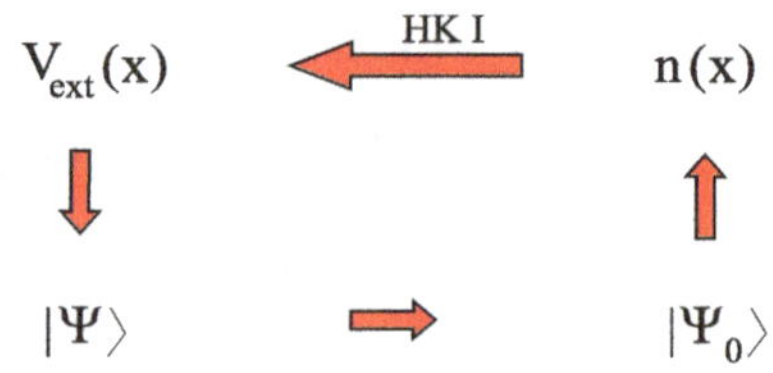

Fig. 5.3 Schematic representation of the role of the Hohenberg-Kohn theorem I. The *short arrows* illustrate the conventional way to solve the many-electron problem starting from the stationary Schrödinger equation. The *long arrow* illustrates the theorem, that closes the *circle*

Addition of the two inequalities (5.18) and (5.19) leads to the contradiction

$$E_{V_{\text{ext}}} + E_{V'_{\text{ext}}} < E_{V'_{\text{ext}}} + E_{V_{\text{ext}}}.$$

From that one concludes that the map $\mathcal{D}$ must be injective. It exists the inverse map $\mathcal{D}^{-1} : n \to \Psi_0$.

The existence of $\mathcal{D}^{-1}$ leads to a generalization of the Hohenberg-Kohn theorem I: The ground-state expectation value of a physical observable $\hat{O}$ is a unique functional of the ground-state density

$$\langle \Psi_0[n] | \hat{O} | \Psi_0[n] \rangle = O[n]. \tag{5.20}$$

Then, the complete inverse map

$$(\mathcal{D}\mathcal{C})^{-1} : n \to V_{\text{ext}}$$

tells us that the knowledge of the ground-state density allows to conclude for the external potential acting on the system (to within a trivial constant) and, consequently, as the kinetic energy and the electron-electron interaction are specified, the lowest eigenvalue of the entire Hamiltonian. The action of the generalized Hohenberg-Kohn theorem I is schematically described in Fig. 5.3.

5.2.3 Hohenberg-Kohn Theorem II

In a second step the variational character of the energy functional

$$E = E_{V_{\text{ext}}}[n] = \langle \Psi_0[n] | \mathscr{H}_0 | \Psi_0[n] \rangle \tag{5.21}$$

with respect to the electron density $n(\mathbf{x})$ (5.13) has to be proven. Its universal character in terms of the density $n(\mathbf{x})$ for a given external potential $V_{\text{ext}}(\mathbf{x})$ has been described above. Thereby, the many-body state $|\Psi_0[n]\rangle$ has been generated via the inverse map $\mathcal{D}^{-1}$. It follows a possible formulation of the *Hohenberg-Kohn theorem II*:

> For a non-degenerate ground state $|\Psi_0\rangle$ and a given external potential $V_{\text{ext}}(\mathbf{x})$ the energy functional $E_{V_{\text{ext}}}[n]$ assumes its global minimum value E_0 varying the density $n(\mathbf{x})$ toward the true ground-state density $n_0(\mathbf{x})$.

Thereby, the density has to fulfill the physical conditions to be positive, $n(\mathbf{x}) \geq 0$, to guarantee particle conservation, $\int d^3\mathbf{x} n(\mathbf{x}) = N$, and to vary continuously.

The proof is trivial. If one assumes that the minimum occurs at a density $n(\mathbf{x}) \neq n_0(\mathbf{x})$ with $n(\mathbf{x}) = \langle\Psi|\hat{n}(\mathbf{x})|\Psi\rangle$, one has

$$E_0 = E_{V_{\text{ext}}}[n_0] > E_{V_{\text{ext}}}[n] = \langle\Psi|\mathscr{H}_0|\Psi\rangle.$$

By virtue of the Rayleigh-Ritz principle, however, it holds

$$E_0 < E_{V_{\text{ext}}}[n].$$

Consequently, the assumption is wrong and, indeed, the exact ground-state density can be determined by minimization of the functional (5.21), in short

$$E_0 = \min_{n \to n_0} E_{V_{\text{ext}}}[n], \tag{5.22}$$

as illustrated in Fig. 5.4.

The ground-state energy can be found by varying the density to minimize the energy, provided we know the form of the functional $E_{V_{\text{ext}}}[n]$, or at least a good approximation for it. Since the contribution of the potential energy $\hat{V}$ of the electrons in the external field $V_{\text{ext}}(\mathbf{x})$ can be exactly described as a linear functional of the density $n(\mathbf{x})$ [see (3.47) or (5.12) with (5.13)], we can write

$$E_{V_{\text{ext}}}[n] = F_{\text{HK}}[n] + \int d^3\mathbf{x} V_{\text{ext}}(\mathbf{x}) n(\mathbf{x}), \tag{5.23}$$

where the Hohenberg-Kohn functional

$$F_{\text{HK}}[n] = \langle\Psi_0[n]|\hat{T} + \hat{U}|\Psi_0[n]\rangle \tag{5.24}$$

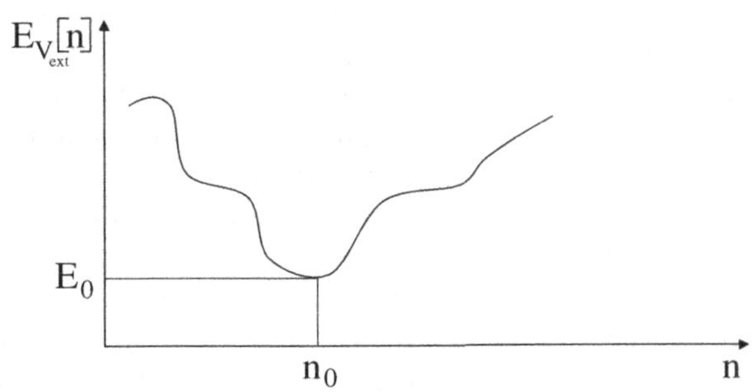

Fig. 5.4 Illustration of the variational principle for the total energy of an ensemble of interacting electrons in its ground state with density $n_0(\mathbf{x})$ and energy E_0 for a given external potential $V_{\text{ext}}(\mathbf{x})$

with the contributions $\hat{T}$ (3.2) and $\hat{U}$ (3.4) [see also (5.11)] to the Hamiltonian (3.1) is indeed a universal functional of the density. By that we mean that (5.24) is the same functional of the density $n(\mathbf{x})$ for all interacting N-electron systems. It has to be determined only once, and can be then applied to all systems. Indeed, $F_{\mathrm{HK}}[n]$ does not depend on a specific physical system characterized by $V_{\mathrm{ext}}(\mathbf{x})$. It is equally valid for atoms, molecules, and solids.

5.2.4 Outlook

The classical formulation of the density functional theory according to Hohenberg and Kohn [6] contains three main messages: (i) the existence of a complete inverse map $(\mathcal{DC})^{-1}$, (ii) the variational character of the ground-state energy functional $E_{V_{\mathrm{ext}}}[n]$ (5.21), and (iii) the universality of the Hohenberg-Kohn functional $F_{\mathrm{HK}}[n]$ (5.24). The two HK theorems formally represent an immense progress. Instead of the many degrees of freedom in the formulation of the many-body problem, e.g. in (3.29) or (3.30), the determination of the ground state of an electronic system can be restricted to the variation of an energy functional that only depends on the electron density which is a function of three space coordinates. However, the theorems proof the existence of a universal functional $F_{\mathrm{HK}}[n]$ but do not say anything about its dependence on $n(\mathbf{x})$ or how this dependence can be constructed.

There are possible generalizations. The original proof of the theorems by Hohenberg and Kohn was restricted to densities $n(\mathbf{x})$ that are ground-state densities of the Hamiltonian $\mathscr{H}_0$ with a certain external potential $V_{\mathrm{ext}}(\mathbf{x})$. Such densities are called "V-representable". However, the Hohenberg-Kohn theorems can be also proven for much more general conditions.

Already Kohn [8] pointed out that the basic formalism can be easily extended to include degenerate ground states. The assumptions of the V-representability and of non-degenerate ground states can be lifted. Especially Levy [9–11] and Lieb [12–14] but also others [15] contributed very much to an alternative definition of the energy functional that lifts several conditions of the Hohenberg-Kohn theory. Avoiding the condition of a direct relationship between density $n(\mathbf{x})$ and external potential $V_{\mathrm{ext}}(\mathbf{x})$ allows generalizations of the density functional theory toward spin-polarized systems [16, 17], finite temperatures and, hence, excited states [18, 19], degenerate ground states [8] as well as non-local potentials $V_{\mathrm{ext}}(\mathbf{x})$ [20] or even time-dependent external potentials [21]. All these generalizations made the definition of the HK functional more tractable, clarified its physical meaning, and provided, at least in principle, a way to determine the exact functional.

The constrained search formulation of the DFT of Levy and Lieb also allows to investigate excited states if those underly a certain symmetry constraint [12, 15, 22]. For example, for crystals with a given point group the electronic states can be classified according to the irreducible representations. The Hohenberg-Kohn theorems can be applied to each irreducible representation of the point group, so that one may search for the lowest-energy state under such a symmetry constraint. If the construction of functionals with defined symmetry would be successful in the presence of spin polarization, even the description of multiplet states could be possible within a DFT [23].

5.3 Spin Density Functional Theory

5.3.1 Electron Spin Density and Magnetization Density

In spin-polarized or magnetic systems the electron density $n(\mathbf{x})$ has to be generalized to a 2×2 Hermitian spin density matrix (3.57) with elements $(s, s' = \uparrow, \downarrow)$

$$n_{ss'}(\mathbf{x}) = \langle \Psi_0 | \psi_s^+(\mathbf{x}) \psi_{s'}(\mathbf{x}) | \Psi_0 \rangle \tag{5.25}$$

for the ground state of the system. Such a spin density matrix is also needed to describe the electronic system in the presence of an external magnetic field $\mathbf{B}_{\text{ext}}(\mathbf{x})$.

In order to generalize the DFT without spin polarization to the spin-polarized case, one has to deal with four densities (5.25) instead of one (5.13). According to their physical meaning the elements of the spin density matrix can be also arranged into the ground-state density (3.58)

$$n(\mathbf{x}) = n_{\uparrow\uparrow}(\mathbf{x}) + n_{\downarrow\downarrow}(\mathbf{x}) \tag{5.26}$$

and a vector of the magnetization density (3.59) in the ground state

$$\mathbf{m}(\mathbf{x}) = -\mu_B \sum_{s,s'} \boldsymbol{\sigma}_{ss'} n_{ss'}(\mathbf{x}). \tag{5.27}$$

Here we focus on the Zeeman interaction of the type (2.28)

$$-\int d^3\mathbf{x} \mathbf{B}_{\text{ext}}(\mathbf{x}) \mathbf{m}(\mathbf{x})$$

of the external magnetic field with the electronic system. Thereby, we restrict ourselves to the interaction mediated by the spin densities or, more precisely, by the spin-related magnetization density (5.27). Effects due to the orbital magnetization are neglected because of their smallness in many cases. The most important consequences for the formulation of the Hohenberg-Kohn theory in Sect. 5.2 are due to a generalization of the potential energy of the electrons in the external fields $\mathscr{H}_{\text{ext}}$ (5.12). With the operator of the magnetization density $\hat{\mathbf{m}}(\mathbf{x})$ (3.33) it holds for the generalized interaction operator

$$\hat{V} = \int d^3\mathbf{x} \sum_{s,s'} \psi_s^+(\mathbf{x}) u_{ss'}(\mathbf{x}) \psi_{s'}(\mathbf{x}) \tag{5.28}$$

with

$$u_{ss'}(\mathbf{x}) = V_{\text{ext}}(\mathbf{x})\delta_{ss'} + \mu_B \mathbf{B}_{\text{ext}}(\mathbf{x})\boldsymbol{\sigma}_{ss'}. \tag{5.29}$$

In the non-collinear limit, e.g. taking the spin-orbit interaction (2.23) into account, the magnetic field in (5.29) has to be further generalized according to

$$\mathbf{B}_{\text{ext}}(\mathbf{x}) \to \mathbf{B}_{\text{ext}}(\mathbf{x}) + \frac{i\hbar^2}{\mu_B(2mc)^2}\{[\nabla_{\mathbf{x}} V_{\text{ext}}(\mathbf{x})] \times \nabla_{\mathbf{x}}\}.$$

5.3.2 Generalized Hohenberg-Kohn Theorems

Besides the ground-state electron density $n(\mathbf{x})$ (5.9) also a vector of the magnetization density $\mathbf{m}(\mathbf{x})$ (5.27) characterizes the ground state and its energy functional [17, 19, 22]

$$E = E_{V_{\text{ext}},\mathbf{B}_{\text{ext}}}[n, \mathbf{m}]. \tag{5.30}$$

This fact together with the still simple form of the potential energy term in the external fields (5.28) suggest an easy generalization of the Hohenberg-Kohn theorems. That will be done below but without giving detailed proofs.

According to the spinless case still the surjective map (see Sect. 5.2.2)

$$\mathcal{C} : V_{\text{ext}}, \mathbf{B}_{\text{ext}} \to \Psi_0$$

is valid. By definition also the surjective map

$$\mathcal{D} : \Psi_0 \to n, \mathbf{m}$$

exists. As a consequence of this map two different non-degenerate ground states Ψ_0 and Ψ_0' always lead to different spin density matrices $n_{ss'}(\mathbf{x}) \neq n'_{ss'}(\mathbf{x})$, or, equivalently to different pairs $n(\mathbf{x}), \mathbf{m}(\mathbf{x}) \neq n'(\mathbf{x}), \mathbf{m}'(\mathbf{x})$. The invertibility of $\mathcal{C}$ cannot be immediately proven for $\mathcal{D}$ in the presence of $\mathbf{B}_{\text{ext}}(\mathbf{x}) \neq 0$. One has to exclude magnetic-field-induced phase transitions. For example, a transition of a system with localized spins from a paramagnetic phase into the ferromagnetic ordering destroys the invertibility of $\mathcal{D}$. On the other hand, the accompanying energy variation is small, so that one may approximately assume that the map is invertible. The proof of the Hohenberg-Kohn theorem I can be done in analogy to the spinless case; only the diagonal matrix $V_{\text{ext}}(\mathbf{x})\delta_{ss'}$ has to be generalized to the off-diagonal matrix $u_{ss'}(\mathbf{x})$ (5.29).

The Hohenberg-Kohn theorem I can be formulated as:

> For a non-degenerate ground state $|\Psi_0\rangle$ the total energy $E = E_{V_{\text{ext}},\mathbf{B}_{\text{ext}}}[n, \mathbf{m}]$ of an inhomogeneous electron gas is a functional of the ground-state density $n(\mathbf{x})$ and magnetization density $\mathbf{m}(\mathbf{x})$.

For a given external perturbation $u_{ss'}(\mathbf{x})$ the map $\mathcal{D}^{-1}$ leads to the ground-state functional

$$E_{V_{\text{ext}},\mathbf{B}_{\text{ext}}}[n,\mathbf{m}] = F[n,\mathbf{m}] + \int d^3\mathbf{x}\,\{V_{\text{ext}}(\mathbf{x})n(\mathbf{x}) - \mathbf{B}_{\text{ext}}(\mathbf{x})\mathbf{m}(\mathbf{x})\} \tag{5.31}$$

with the universal functional

$$F[n,\mathbf{m}] = \langle\Psi_0[n,\mathbf{m}]|\hat{T} + \hat{U}|\Psi_0[n,\mathbf{m}]\rangle \tag{5.32}$$

of $n(\mathbf{x})$ and $\mathbf{m}(\mathbf{x})$.

The variational character of the functional (5.31) is obvious. For fixed external fields, density $n_0(\mathbf{x})$, and magnetization $\mathbf{m}_0(\mathbf{x})$, the ground-state energy

$$E_0 = E_{V_{\text{ext}},\mathbf{B}_{\text{ext}}}[n_0,\mathbf{m}_0] \tag{5.33}$$

obeys the inequality

$$E_0 < E_{V_{\text{ext}},\mathbf{B}_{\text{ext}}}[n,\mathbf{m}]$$

with $n(\mathbf{x}), \mathbf{m}(\mathbf{x}) \neq n_0(\mathbf{x}), \mathbf{m}_0(\mathbf{x})$. A possible formulation of the Hohenberg-Kohn theorem II is:

> For a non-degenerate ground state $|\Psi_0\rangle$ the variation of the energy functional $E_{V_{\text{ext}},\mathbf{B}_{\text{ext}}}[n,\mathbf{m}]$ with respect the density $n(\mathbf{x})$ and the magnetization $\mathbf{m}(\mathbf{x})$ yields the minimum E_0 at the ground-state densities $n_0(\mathbf{x})$ and $\mathbf{m}_0(\mathbf{x})$.

In analogy to (5.22) one can write in short

$$E_0 = \min_{n\to n_0,\mathbf{m}\to\mathbf{m}_0} E_{V_{\text{ext}},\mathbf{B}_{\text{ext}}}[n,\mathbf{m}]. \tag{5.34}$$

For a formulation of the fully relativistic density functional theory the reader is referred to the spin current density functional theory and corresponding review articles [24–28]. This topic is outside the scope of the book.

5.3.3 Collinear Spins

Neglecting spin-orbit interaction and other spin-dependent effects the majority of spin-polarized systems can be studied in the framework of the collinear-spin approximation. Only the z-component of the magnetization density (5.27), $m_z(\mathbf{x})$, plays a role. Spatial rotations of the spin density matrix (5.25) diagonalize it with the two independent components $n_\uparrow(\mathbf{x})$ and $n_\downarrow(\mathbf{x})$. They are directly related to the total density, $n(\mathbf{x}) = n_\uparrow(\mathbf{x}) + n_\downarrow(\mathbf{x})$ (5.26), and the z-component of the spin magnetization density, $m_z(\mathbf{x}) = -\mu_B[n_\uparrow(\mathbf{x}) - n_\downarrow(\mathbf{x})]$ (5.27).

Deriving the HF equations (see Sect. 4.2.1) we have shown that in the collinear case every single-particle spinor can be factorized into a space-dependent orbital $\varphi_{\lambda m_s}(\mathbf{x})$ and a spinor $\chi_{\frac{1}{2}m_s}(s)$. For instance, by means of (4.6) diagonal matrix elements of the spin density operator with the spinors can be introduced. This fact allows us to change from the spin-space representation to densities $n_{m_s}(\mathbf{x})$ (4.19) depending not anymore on the spin variable s but on the spin quantum number $m_s = \pm\frac{1}{2}$ of the two spin channels, majority-spin and minority-spin electrons. The two variables of a collinear spin density functional theory are the densities $n_{\frac{1}{2}}(\mathbf{x})$ and $n_{-\frac{1}{2}}(\mathbf{x})$ in a certain spin channel or two quantities derived according to (3.58), (3.59), and (3.60)

$$\begin{aligned} n(\mathbf{x}) &= n_{\frac{1}{2}}(\mathbf{x}) + n_{-\frac{1}{2}}(\mathbf{x}), \\ m_z(\mathbf{x}) &= -\mu_B\left[n_{\frac{1}{2}}(\mathbf{x}) - n_{-\frac{1}{2}}(\mathbf{x})\right] \end{aligned} \tag{5.35}$$

or

$$\begin{aligned} n(\mathbf{x}) &= n_{\frac{1}{2}}(\mathbf{x}) + n_{-\frac{1}{2}}(\mathbf{x}), \\ \xi(\mathbf{x}) &= \frac{n_{\frac{1}{2}}(\mathbf{x}) - n_{-\frac{1}{2}}(\mathbf{x})}{n_{\frac{1}{2}}(\mathbf{x}) + n_{-\frac{1}{2}}(\mathbf{x})}, \end{aligned} \tag{5.36}$$

the total electron density $n(\mathbf{x})$ and the spin polarization $\xi(\mathbf{x})$ [see also (4.60) for jellium]. In the collinear limit the particle conservation is given by

$$N_{m_s} = \int d^3\mathbf{x} n_{m_s}(\mathbf{x}) \tag{5.37}$$

with the number N_{m_s} of electrons in the spin channel m_s and the total number $N = N_{\frac{1}{2}} + N_{-\frac{1}{2}}$. For paired electrons $N_{\frac{1}{2}} = N_{-\frac{1}{2}}$ the system is globally not spin-polarized, while for $N_{\frac{1}{2}} \neq N_{-\frac{1}{2}}$ magnetic or spin phenomena should occur. In antiferromagnetic systems it holds $N_{\frac{1}{2}} = N_{-\frac{1}{2}}$ globally but $n_{\frac{1}{2}}(\mathbf{x}) \neq n_{-\frac{1}{2}}(\mathbf{x})$ locally. As an illustration the magnetization density of antiferromagnetic NiO near its (001)surface is displayed in Fig. 5.5.

In the limit described above a spin density functional theory can be formulated similar to the Hohenberg-Kohn theory (5.23). An external Zeeman field $\mathbf{B}_{\text{ext}}(\mathbf{x})$ is however not anymore needed. Even in the absence of such a magnetic field magnetic phenomena can be described. One only needs the definition of the z-axis in the system, e.g. in the limit $\mathbf{B}_{\text{ext}}(\mathbf{x}) \to 0$ with $\mathbf{B}_{\text{ext}}(\mathbf{x}) \parallel z$-axis. The lowest energy solution may be spin-polarized, i.e., $n_{\frac{1}{2}}(\mathbf{x}) \neq n_{-\frac{1}{2}}(\mathbf{x})$, which is analogous to the broken symmetry solution of the unrestricted HF (UHF) theory (see Sect. 4.2.1). This may happen in finite systems with an odd number N of electrons and in magnetic solids [16, 22, 30]. However, also in systems with even number N such a solution may occur. An instructive example is a two-level system with bonding interaction and Coulomb repulsion U. For large U the correlated state of the two electrons is

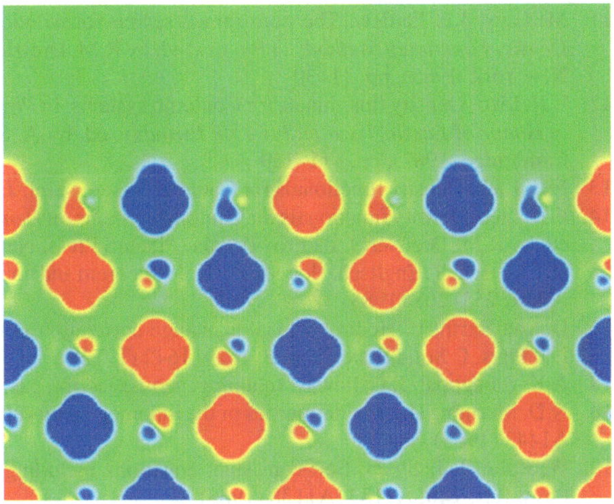

Fig. 5.5 Magnetization density near the NiO(001)2 × 1 surface within collinear spin density functional theory in a (100) plane. The Ni^{2+} (O^{2-}) ions give rise to *large* (*small*) clouds. Their different magnetization is indicated by *red* and *blue colors*. Adapted from [29]. Copyright IOP Publishing. Reproduced with permission of IOP Publishing. All rights reserved

found. Thus the UHF solution rectifies the most serious error of the HF theory [31]. The spin density functional $E_{V_{\text{ext}},0}[n_{\frac{1}{2}}, n_{-\frac{1}{2}}]$ is useful in these cases as well and is also applied in explicit computations. The original Hohenberg-Kohn theorems are valid and the ground state is determined by the total density via a functional $\tilde{E}_{V_{\text{ext}}}[n] = E_{V_{\text{ext}},0}[n, m_z[n]]$, so far $\mathbf{B}_{\text{ext}}(\mathbf{x}) \equiv 0$ [32]. The only modification is that degenerate ground states have to be investigated if the broken symmetry solution is indeed degenerated.

References

1. L.H. Thomas, The calculation of atomic fields. Proc. Cambridge Phil. Roy. Soc. **23**, 542–548 (1927)
2. E. Fermi, Un metodo statistico per la determinazione di alcune priorieta dell'atome. Rend. Accad. Naz. Lincei **6**, 602–607 (1927)
3. P.A.M. Dirac, Note on exchange phenomena in the Thomas-Fermi atom. Proc. Cambridge Phil. Roy. Soc. **26**, 376–385 (1930)
4. C.F. von Weizsäcker, Zur Theorie der Kernmassen. Z. Phys. **96**, 431–458 (1935)
5. E. Teller, On the stability of molecules in the Thomas-Fermi theory. Rev. Mod. Phys. **34**, 627–630 (1962)
6. P. Hohenberg, W. Kohn, Inhomogeneous electron gas. Phys. Rev. **136**, B864–B871 (1964)
7. R.M. Dreizler, E.K.U. Gross, *Density Functional Theory* (Springer, Berlin, 1990)
8. W. Kohn, Density functional theory: fundamentals and applications, in *Highlights of Condensed Matter Theory*, ed. by F. Bassani, F. Fumi, M.P. Tosi (North-Holland, Amsterdam, 1985), pp. 1–15
9. M. Levy, Universal variational functionals of electron densities, first-order matrices, and natural spin-orbitals and solution of the n-representability problem. Proc. Natl. Acad. Sci. USA **76**, 6062–6065 (1979)
10. M. Levy, Electron densities in search of Hamiltonians. Phys. Rev. A **26**, 1200–1208 (1982)

11. M. Levy, J.P. Perdew, The constrained-search formulation of density functional theory, in *Density Functional Methods in Physics*, ed. by R.M. Dreizler, J. da Providencia (Plenum Press, New York, 1985), pp. 11–30
12. E.H. Lieb, Density functionals for Coulomb systems, in *Physics as Natural Philosophy: Essays in Honor of Laszlo Tisza on his 75th Birthday*, ed. by A. Shimony, H. Feshbach (MIT Press, Cambridge, 1982), pp. 111–149
13. E.H. Lieb, Density functionals for Coulomb systems. Int. J. Quant. Chem. **24**, 243–277 (1983)
14. E.H. Lieb, Density functionals for Coulomb systems, in *Density Functional Methods in Physics*, ed. by R.M. Dreizler, J. da Providencia (Plenum Press, New York, 1985), pp. 31–80
15. H. Englisch, R. Englisch, Hohenberg-Kohn theorem and non-V-representable densities. Physica A **121**, 253–268 (1983)
16. U. von Barth, L. Hedin, A local exchange-correlation potential for the spin-polarized case: I. J. Phys. C Solid State Phys. **5**, 1629–1642 (1972)
17. A.K. Rajagopal, J. Calloway, Inhomogeneous electron gas. Phys. Rev. B **7**, 1912–1919 (1973)
18. N.D. Mermin, Thermal properties of the inhomogeneous electron gas. Phys. Rev. **137**, A1441–A1443 (1965)
19. W. Kohn, L.J. Sham, Self-consistent equations including exchange and correlation effects. Phys. Rev. **140**, A1133–A1138 (1965)
20. T.L. Gilbert, Hohenberg-Kohn theorem for nonlocal external potentials. Phys. Rev. B **12**, 2111–2120 (1975)
21. E.K.U. Gross, E. Runge, Density functional theory for time-dependent systems. Phys. Rev. Lett. **52**, 997–1000 (1984)
22. O. Gunnarsson, B.I. Lundqvist, Exchange and correlation in atoms, molecules, and solids by the spin-density-functional formalism. Phys. Rev. B **13**, 4274–4298 (1976)
23. U. von Barth, Local-density theory of multiplet structure. Phys. Rev. A **20**, 1693–1703 (1979)
24. G. Vignale, M. Rasolt, Density functional theory in strong magnetic fields. Phys. Rev. Lett. **59**, 2360–2363 (1987)
25. G. Vignale, M. Rasolt, Current- and spin-density-functional theory for inhomogeneous electronic systems in strong magnetic fields. Phys. Rev. B **37**, 10685–10696 (1988)
26. H. Eschrig, *The Fundamentals of Density Functional Theory* (Teubner-Verlagsgesellschaft, Stuttgart, 1996)
27. E. Engel, Relativistic density functional theory: foundations and basic formalism, in *Relativistic Electronic Structure Theory, Part 1*, ed. by P. Schwerdtfeger (Elsevier, Amsterdam, 2002), pp. 523–621
28. E. Engel, R.M. Dreizler, S. Varga, B. Fricke, Relativistic density functional theory, in *Relativistic Effects in Heavy-Element Chemistry and Physics*, ed. by B.A. Hess (Wiley, New York, 2003), pp. 123–164
29. A. Schrön, M. Granovskij, F. Bechstedt, Influence of on-site Coulomb interaction *U* on properties of MnO(001)2 x 1 and NiO(001)2 x 1 surfaces. J. Phys. Condens. Matter **25**, 094006 (2013)
30. R.O. Jones, O. Gunnarsson, The density functional formalism, its applications and prospects. Rev. Mod. Phys. **61**, 689–746 (1989)
31. W.A. Harrison, *Elementary Electronic Structure* (World Scientific Publishing, Singapore, 1999)
32. R.M. Martin, *Electronic Structure. Basic Theory and Practical Methods* (Cambridge University Press, Cambridge, 2004)

Chapter 6
Kohn-Sham Scheme

Abstract The density functional theory does not provide any useful computational scheme to apply it to real systems such as condensed matter. The way to make it applicable is based on the idea of Kohn and Sham to project the interacting electron gas onto a non-interacting reference or auxiliary system. More in detail, the Kohn-Sham ansatz replaces the problem of interacting electrons by an auxiliary independent-particle problem, where all many-body effects beyond the Hartree term are included in an explicit exchange-correlation functional. The ground state density $n(\mathbf{x})$ of the interacting system is constructed by the solutions of the non-interacting system with an appropriate effective potential. Such a potential can be combined by the true external potential, the Hartree potential, and an exchange-correlation contribution that is defined by the functional derivative of the exchange-correlation functional. It allows the formulation of a single-particle equation, the Kohn-Sham equation, which has to be solved self-consistently. Generalizations for spin-polarized electron gases are straightforward.

6.1 Kohn-Sham Ansatz

6.1.1 Toward New Ideas

While the Hohenberg-Kohn theorems in Sect. 5.2 rigorously establish that we may use the electron density (or the spin densities in Sect. 5.3.3, and not the density alone) as a variable to find the ground-state energy of an N-electron problem, they lead to the density functional theory but do not provide any useful computational scheme to apply the theory to real systems. The difficulties to develop such a computational scheme are mainly related to the (longitudinal) electron-electron interaction term $\hat{U}$ (3.19) in the Hamiltonian $\mathscr{H}_0$ (5.10), especially the exchange-correlation effects (3.48) or (3.62) beyond the classical Hartree repulsion (3.43). Neglecting correlation we have shown how the problem of N interacting electrons can be solved in the framework of the Hartree-Fock approximation (Sect. 4.2). Indeed, the HF equations (4.25) or (4.28) provide an approximate way to treat an inhomogeneous electron gas in the ground state.

F. Bechstedt, *Many-Body Approach to Electronic Excitations*,
Springer Series in Solid-State Sciences 181, DOI 10.1007/978-3-662-44593-8_6

The HF approximation shows that the main difficulties to treat an N-electron system are due to the mutual interaction of the particles. On the other hand, the sole influence of an external potential $V_{\rm ext}(\mathbf{x})$ or the potential $V_{\rm n}(\mathbf{x})$ of an electron in the field of the nuclei (1.4) is easily tractable, since it still leaves the particles interaction-free. These experiences with many-electron systems lead to the idea to project the interacting N-electron system onto a non-interacting 'reference', or auxiliary, system which can be treated more easily than the interacting system described by the Hamiltonian $\mathscr{H}_0$ (3.1). In the non-interacting limit its ground-state wave function $\Psi_0(\mathbf{x}_1 s_1 ... \mathbf{x}_N s_N)$ (3.30) can be described by a single Slater determinant.

The explicit formulation of this idea suggests to look for a certain effective external potential $V_s(\mathbf{x})$ such that the non-interacting system has the same ground-state density $n(\mathbf{x})$ [we omit the label 0] as the real, interacting system. Once one has obtained this density, the energy functional (5.23) or (5.31) can be used or some approximation of it. This concept immediately leads to the ansatz of Kohn and Sham (KS) [1].

6.1.2 Kohn-Sham Assumptions

As stated in the end of Chap. 3 we ignore the spin-orbit and Breit interactions. Scalar-relativistic corrections are explicitly taken into account in the numerical codes based on the Kohn-Sham scheme, however, will not be mentioned during the formulation of the theory. Consequently, the spins of the electrons can be treated in a collinear manner. Similar to the HF approach in Sect. 4.2.3 and the DFT for collinear spins, in Sect. 5.3.3, electrons in two spin channels denoted by the spin quantum number $m_s = \pm\frac{1}{2}$ have to be investigated.

The KS formulation rests upon two assumptions:

(i) In analogy to the more general Hohenberg-Kohn theory now a 'non-interacting-V representability' is assumed in line with the representation of the exact ground-state density by that of an auxiliary system of non-interacting particles.
(ii) The single-particle Hamiltonian for one of these non-interacting particles is chosen to consist of the usual kinetic energy operator and an local potential for each spin channel $V_s^{m_s}(\mathbf{x})$ acting on an electron with the spin quantum number m_s at point $\mathbf{x}$ in the auxiliary system. The local form is not essential but convenient. In Chap. 9 a generalization to non-local potentials will be discussed and applied (see also [1]).

6.1.3 Kinetic Energy of Auxiliary System

Such a single-particle Hamiltonian of the auxiliary system with $N = N_{\frac{1}{2}} + N_{-\frac{1}{2}}$ independent electrons is then

$$\hat{H}_s^{m_s} = -\frac{\hbar^2}{2m}\Delta_{\mathbf{x}} + V_s^{m_s}(\mathbf{x}) \tag{6.1}$$

with the potential energy $V_s^{m_s}(\mathbf{x})$ of an electron in the spin channel m_s. It obeys a Schrödinger-like equation with eigenorbitals $\varphi_{\lambda m_s}(\mathbf{x})$, eigenvalues $\varepsilon_{\lambda m_s}$, quantum numbers λm_s, and occupation numbers $n_{\lambda m_s} = 1, 0$ (see also Sect. 4.2). The latter ones are normalized according to $N_{m_s} = \sum_\lambda n_{\lambda m_s}$. The N_{m_s} orbitals $\varphi_{\lambda m_s}(\mathbf{x})$ with the lowest eigenvalues $\varepsilon_{\lambda m_s}$ are occupied in the ground state. The ground-state density

$$n(\mathbf{x}) = n_{\frac{1}{2}}(\mathbf{x}) + n_{-\frac{1}{2}}(\mathbf{x}) \tag{6.2}$$

is the sum of the two spin densities

$$n_{m_s}(\mathbf{x}) = \sum_\lambda n_{\lambda m_s} \left|\varphi_{\lambda m_s}(\mathbf{x})\right|^2, \tag{6.3}$$

which are given by sums over squares of the orbitals from the corresponding spin channel. The particle conservation

$$N_{m_s} = \int d^3\mathbf{x} n_{m_s}(\mathbf{x}) \tag{6.4}$$

is evident, if the orbitals $\{\varphi_{\lambda m_s}(\mathbf{x})\}$ are orthonormalized. Accordingly the total kinetic energy T_s of the independent electrons is given by

$$T_s = \sum_{m_s}\sum_\lambda n_{\lambda m_s} \int d^3\mathbf{x}\varphi^*_{\lambda m_s}(\mathbf{x}) \left\{-\frac{\hbar^2}{2m}\Delta_{\mathbf{x}}\right\} \varphi_{\lambda m_s}(\mathbf{x}). \tag{6.5}$$

The total energy of the non-interacting particles with the Hamiltonian (6.1) is simply described by

$$E_s[n] = T_s[n] + \sum_{m_s} \int d^3\mathbf{x} V_s^{m_s}(\mathbf{x}) n_{m_s}(\mathbf{x}). \tag{6.6}$$

Without spin polarization in the system, i.e., $V_s^{\frac{1}{2}}(\mathbf{x}) = V_s^{-\frac{1}{2}}(\mathbf{x}) = V_s(\mathbf{x})$, expression (6.6) is a unique functional of the density (HK theorem I). Together with the particle conservation (6.4) and a Lagrange multiplier μ_s the application of the HK theorem II to the auxiliary non-interacting system yields an Euler equation

$$\frac{\delta T_s[n]}{\delta n(\mathbf{x})} + V_s(\mathbf{x}) - \mu_s = 0. \tag{6.7}$$

Since the electrons in the two spin channels are independent, and the energy contributions (6.5) and (6.6) are additive, the generalization of (6.7) to spin-dependent potentials $V_s^{m_s}(\mathbf{x})$ is obvious,

$$\frac{\delta T_s[n]}{\delta n_{m_s}(\mathbf{x})} + V_s^{m_s}(\mathbf{x}) - \mu_s^{m_s} = 0. \tag{6.8}$$

More precisely, the Euler equations (6.7) and (6.8) are only valid for the true ground-state densities which, however, we will not anymore indicate by corresponding labels in detail.

6.1.4 Functional with Interaction

The knowledge of the true electron density (6.2) immediately leads to the Hartree energy in the interacting system in the ground state (3.43) or (4.16)

$$E_{\mathrm{H}} = \frac{1}{2} \int d^3\mathbf{x} \int d^3\mathbf{x}' \mathrm{v}(\mathbf{x} - \mathbf{x}') n(\mathbf{x}) n(\mathbf{x}'). \tag{6.9}$$

Since both quantities $T_s[n]$ (6.5) and $E_{\mathrm{H}}[n]$ (6.9) are functionals of the total electron density $n(\mathbf{x})$ in the ground state which is just the sum of the two densities $n_{m_s}(\mathbf{x})$ in the auxiliary subsystems for a given spin orientation m_s, important contributions to the universal Hohenberg-Kohn functional (5.24) or its generalization (5.32) are known. The rest of the HK functional

$$E_{\mathrm{XC}}[n] = F_{\mathrm{HK}}[n] - T_s[n] - E_{\mathrm{H}}[n] \tag{6.10}$$

contains all many-body effects of exchange and correlation. Comparing $E_{\mathrm{XC}}[n]$ (6.10) with previous definitions (3.48) and (3.62) it obvious that these many-body effects not only arise from the direct Coulomb interaction of the electrons as $\langle \Psi_0[n] | \hat{U} | \Psi_0[n] \rangle - E_{\mathrm{H}}[n]$ but also in an indirect manner via the difference $\langle \Psi_0[n] | \hat{T} | \Psi_0[n] \rangle - T_s[n]$ of the kinetic energy of the interacting system and that of the non-interacting one. The exchange-correlation energy of the Kohn-Sham theory (6.10) can be also written in the more revealing form

$$E_{\mathrm{XC}}[n] = \langle \Psi_0[n] | \hat{T} | \Psi_0[n] \rangle - T_s[n] + \langle \Psi_0[n] | \hat{U} | \Psi_0[n] \rangle - E_{\mathrm{H}}[n]. \tag{6.11}$$

This expression well illustrates the two contributions to exchange and correlation. Thereby, since we allow for systems with a collinear spin polarization, the denotation $[n]$ indicates a functional dependence on both densities $n_{m_s}(\mathbf{x})$.

6.2 Kohn-Sham Equation

6.2.1 Variational Problem

Together with the specified definition of exchange and correlation contributions to the total ground-state energy functional (6.10) or (6.11), the Kohn-Sham approach to an interacting many-electron problem leads to

$$E_{\mathrm{KS}}[n] = T_s[n] + \int d^3\mathbf{x} V_{\mathrm{ext}}(\mathbf{x})n(\mathbf{x}) + E_{\mathrm{H}}[n] + E_{\mathrm{XC}}[n] \tag{6.12}$$

instead of the functional (5.23) of the Hohenberg-Kohn theory. In contrast to (5.28) no spin dependence of the external perturbation is taken into account in (6.12). In the majority of applications of the theory the external potential $V_{\mathrm{ext}}(\mathbf{x})$ is due to the nuclei $V_{\mathrm{n}}(\mathbf{x})$ (1.4) which is independent of the electron spin. The density variation of the Kohn-Sham energy functional (6.12) together with the particle conservation (6.4) yields the Euler equation

$$\frac{\delta T_s[n]}{\delta n(\mathbf{x})} + V_{\mathrm{ext}}(\mathbf{x}) + V_{\mathrm{H}}(\mathbf{x}) + V_{\mathrm{XC}}(\mathbf{x}) - \mu = 0 \tag{6.13}$$

with the Lagrange multiplier μ, the Hartree potential (6.9)

$$V_{\mathrm{H}}(\mathbf{x}) = \int d^3\mathbf{x}' v(\mathbf{x} - \mathbf{x}')n(\mathbf{x}'), \tag{6.14}$$

and the exchange-correlation potential

$$V_{\mathrm{XC}}(\mathbf{x}) = \frac{\delta E_{\mathrm{XC}}[n]}{\delta n(\mathbf{x})}. \tag{6.15}$$

In the presence of spin polarization in the system, instead of (6.13) we find for the two spin channels

$$\frac{\delta T_s[n]}{\delta n_{m_s}(\mathbf{x})} + V_{\mathrm{ext}}(\mathbf{x}) + V_{\mathrm{H}}(\mathbf{x}) + V_{\mathrm{XC}}^{m_s}(\mathbf{x}) - \mu^{m_s} = 0 \tag{6.16}$$

with the XC potential $V_{\mathrm{XC}}^{m_s}(\mathbf{x})$ acting on the electrons with the spin quantum number m_s,

$$V_{\mathrm{XC}}^{m_s}(\mathbf{x}) = \frac{\delta E_{\mathrm{XC}}[n]}{\delta n_{m_s}(\mathbf{x})}. \tag{6.17}$$

By comparing the result (6.13) or (6.16) with (6.7) or (6.8) for the auxiliary system we see that the effective potential $V_s(\mathbf{x})$ or $V_s^{m_s}(\mathbf{x})$ in (6.1) must satisfy the relation

$$V_s(\mathbf{x}) = V_{\text{ext}}(\mathbf{x}) + V_{\text{H}}(\mathbf{x}) + V_{\text{XC}}(\mathbf{x}) + \mu_s - \mu \tag{6.18}$$

or

$$V_s^{m_s}(\mathbf{x}) = V_{\text{ext}}(\mathbf{x}) + V_{\text{H}}(\mathbf{x}) + V_{\text{XC}}^{m_s}(\mathbf{x}) + \mu_s^{m_s} - \mu^{m_s}. \tag{6.19}$$

Apart from a certain constant $\mu_s - \mu$ (or $\mu_s^{m_s} - \mu^{m_s}$) the effective potential is given as the sum of the external potential, the classical repulsion of the electrons in the system, and an XC contribution that, however, is only given as a functional derivative (6.15) or (6.17) of the unknown XC energy (6.10) or (6.11). In the relations (6.18) and (6.19) the electron and spin densities have to be replaced by those in the true ground state of the system.

6.2.2 Eigenvalue Problem

Another open problem is the determination of the spin densities (6.3) and, hence, the total electron density (6.2). We investigate this problem for the case of a spin-polarized system. The external potential $V_{\text{ext}}(\mathbf{x})$ may be replaced by that of the nuclei $V_{\text{n}}(\mathbf{x})$ (1.4). The trivial constant $\mu_s^{m_s} - \mu^{m_s} = 0$ is fixed to be zero by an appropriate choice of the energy zero. Instead of the potentials (6.18) or (6.19) we define the Kohn-Sham potential

$$V_{\text{KS}}^{m_s}(\mathbf{x}) = V_{\text{n}}(\mathbf{x}) + V_{\text{H}}(\mathbf{x}) + V_{\text{XC}}^{m_s}(\mathbf{x}). \tag{6.20}$$

It is uniquely determined by the ground-state density following the Hohenberg-Kohn theorems. Such a potential is illustrated in Fig. 6.1 for a non-spin-polarized system, the diamond(111)2 $\times$ 1 surface. The vacuum region (potential plateaus) and the diamond crystal beneath the surface (potentials oscillating with the atomic bilayer thickness) are clearly visible. The comparison of the bare electrostatic part with the full KS potential (6.20) shows that the exchange-correlation effects significantly increase the bonding of the electrons in the crystal. The figure indicates the general importance of exchange and correlation.

Studying the HK theorems in Sect. 5.2 we found that the HK functional is well defined for V-representable densities. Therefore, the density variations $n(\mathbf{x}) + \delta n(\mathbf{x})$ used e.g. in (6.13) should also belong to the same class of densities. This has to be guaranteed by some potentials $V_s^{m_s}(\mathbf{x}) + \delta V_s^{m_s}(\mathbf{x})$ in (6.1). Since the densities are expressed by normalized single-particle orbitals $\varphi_{\lambda m_s}(\mathbf{x})$ in (6.3), they should be also modified according to $\varphi_{\lambda m_s}(\mathbf{x}) + \delta\varphi_{\lambda m_s}(\mathbf{x})$. Since the kinetic energy of independent particles T_s (6.5) is explicitly expressed by the orbitals and all other contributions to the functional E_{KS} (6.12) are considered to be functionals of the densities and, hence, of the orbitals [see (6.3)], one can also vary the wave functions using the chain rule to derive a variational equation subject to the orthonormalization constraint for each spin channel

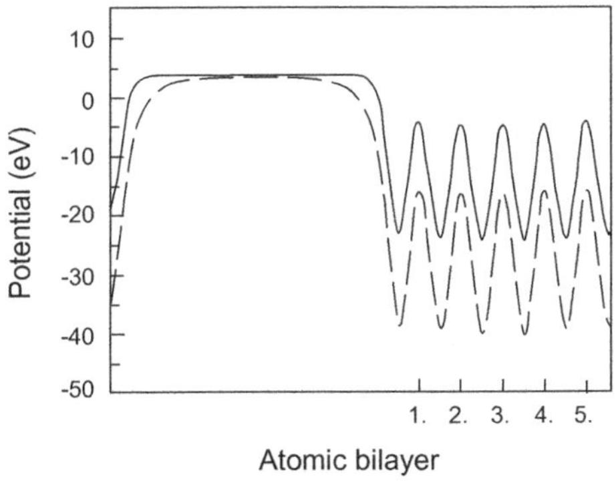

Fig. 6.1 Plane-averaged Kohn-Sham potential $\bar{V}_{\mathrm{KS}}(z)$ for the diamond(111)2 × 1 surface (*dashed line*) and the corresponding averaged electrostatic potential $\bar{V}_{\mathrm{n}}(z)+\bar{V}_{\mathrm{H}}(z)$ without exchange and correlation (*solid line*). The surface normal defines the z-axis. The material slab with its periodic arrangement of atomic layers beneath the surface and the vacuum region are clearly visible. From [2]

$$\int d^3\mathbf{x}\varphi^*_{\lambda m_s}(\mathbf{x})\varphi_{\lambda' m_s}(\mathbf{x}) = \delta_{\lambda\lambda'}. \tag{6.21}$$

Together with the variations

$$\frac{\delta T_s[n]}{\delta\varphi^*_{\lambda m_s}(\mathbf{x})} = -\frac{\hbar^2}{2m}\Delta_{\mathbf{x}}\varphi_{\lambda m_s}(\mathbf{x})$$

and

$$\frac{\delta n_{m_s}(\mathbf{x})}{\delta\varphi^*_{\lambda m_s}(\mathbf{x})} = \varphi_{\lambda m_s}(\mathbf{x}),$$

at least for the occupied orbitals with $n_{\lambda m_s} = 1$, and the Lagrange multiplier method for handling the constraints [see e.g. (4.23) and the derivation of the HF equations], we derive the Kohn-Sham (Schrödinger-like) equations

$$\hat{H}^{m_s}_{\mathrm{KS}}\varphi_{\lambda m_s}(\mathbf{x}) = \varepsilon_{\lambda m_s}\varphi_{\lambda m_s}(\mathbf{x}) \tag{6.22}$$

with the Hamiltonian

$$\hat{H}^{m_s}_{\mathrm{KS}} = -\frac{\hbar^2}{2m}\Delta_{\mathbf{x}} + V^{m_s}_{\mathrm{KS}}(\mathbf{x}) \tag{6.23}$$

similar to that proposed in (6.1). The Kohn-Sham potential $V^{m_s}_{\mathrm{KS}}(\mathbf{x})$ is given in (6.20) for the two spin channels. Its density dependence requires a self-consistent solution of the Kohn-Sham equations (6.22). The self-consistent cycle is illustrated in Fig. 6.2

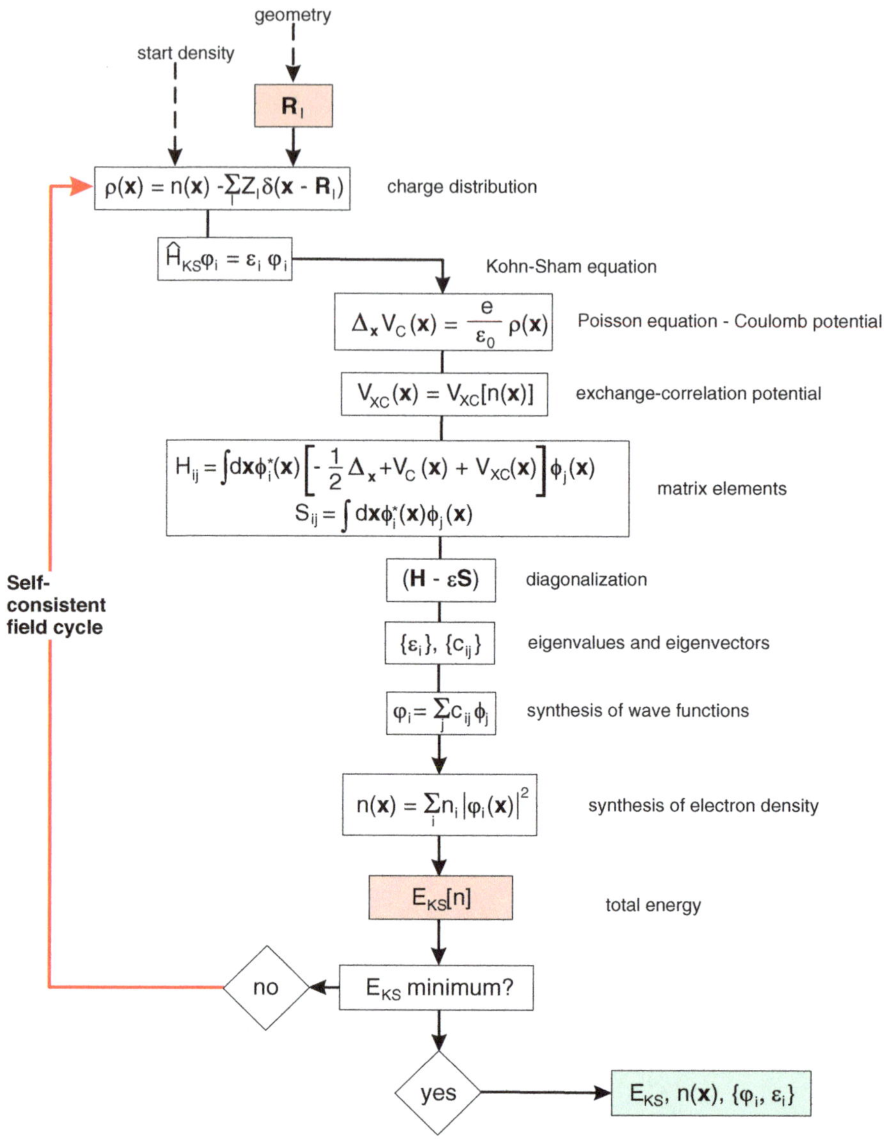

Fig. 6.2 Self-consistent solution of the Kohn-Sham equation without spin polarization and calculation of the total energy of the electronic system. The Kohn-Sham eigenfunctions $\{\varphi_i(\mathbf{x})\}$ are expanded in a complete set of functions $\{\phi_j(\mathbf{x})\}$. The abbreviation $V_C(\mathbf{x}) = V_n(\mathbf{x}) + V_H(\mathbf{x})$ is used

together with the calculation of the density (6.2) and total energy (6.12) for the spinless case.

Finally, we argue how the KS equations have to be generalized for non-collinear spins. The exchange-correlation energy $E_{XC}[n, \mathbf{m}]$ in the functional (5.31) depends on the vector of the magnetization density $\mathbf{m}(\mathbf{x})$. This has two consequences: (i) Instead of their factorization in (4.6) Pauli spinors

$$\phi_\Lambda(\mathbf{x}, s) \neq \varphi_{\lambda m_s}(\mathbf{x})\chi_{\frac{1}{2}m_s}(s)$$

with a generalized set of quantum numbers Λ of the coupled orbital-spin motion have to be investigated. (ii) Besides the exchange-correlation potential $V_{\text{XC}}(\mathbf{x}) = \delta E_{\text{XC}}[n, \mathbf{m}]/\delta n(\mathbf{x})$ (6.15), that depends also on the vector of magnetization, additional variational derivatives

$$B_{\text{XC}_j}(\mathbf{x}) = -\frac{\delta E_{\text{XC}}[n, \mathbf{m}]}{\delta m_j(\mathbf{x})}$$

have to be taken into account. The three components represent an internal magnetic field due to exchange and correlation effects. This internal field has interesting consequences for the effective field acting on the Kohn-Sham particles. The Hartree energy does not directly depend on the magnetization density. However, the XC-induced magnetic field leads to an additional interaction term of the form (5.29). Instead of Kohn-Sham equations for two spin channels (6.22) we have to solve four coupled sets of KS equations

$$\sum_{s'} \left\{ \left[-\frac{\hbar^2}{2m}\Delta_{\mathbf{x}} + V_{\text{ext}}(\mathbf{x}) + V_{\text{H}}(\mathbf{x}) + V_{\text{XC}}(\mathbf{x}) \right] \delta_{ss'} \right.$$
$$\left. -\mu_B \left[\mathbf{B}_{\text{ext}}(\mathbf{x}) + \mathbf{B}_{\text{XC}}(\mathbf{x})\right] \boldsymbol{\sigma}_{ss'} \right\} \phi_\Lambda(\mathbf{x}, s') = \varepsilon_\Lambda \phi_\Lambda(\mathbf{x}, s)$$

for $s, s' = \uparrow, \downarrow$ and the two spinor components.

6.2.3 Summary

We close this section by highlighting a few points about the Kohn-Sham formalism:

(i) The formalism presented is exact to find the ground-state density and the ground-state energy of a system of electrons, supposing that the exact XC potential $V_{\text{XC}}^{m_s}(\mathbf{x})$ or functional $E_{\text{XC}}[n]$ is known.

(ii) One has to cast the solution of the interacting N-electron problem in terms of non-interacting Kohn-Sham electrons in an 'external' potential $V_{\text{KS}}^{m_s}(\mathbf{x})$. The N orbitals $\{\varphi_{\lambda m_s}(\mathbf{x})\}$ of the occupied states follow immediately from the KS equations. Their knowledge allows to construct the ground-state wave function of the non-interacting system which is just a Slater determinant. More important, however, is that they define the true electron density $n(\mathbf{x})$ [or spin densities $n_{m_s}(\mathbf{x})$]. The density and some KS orbitals of few-electron systems are displayed in Fig. 6.3 for illustration.

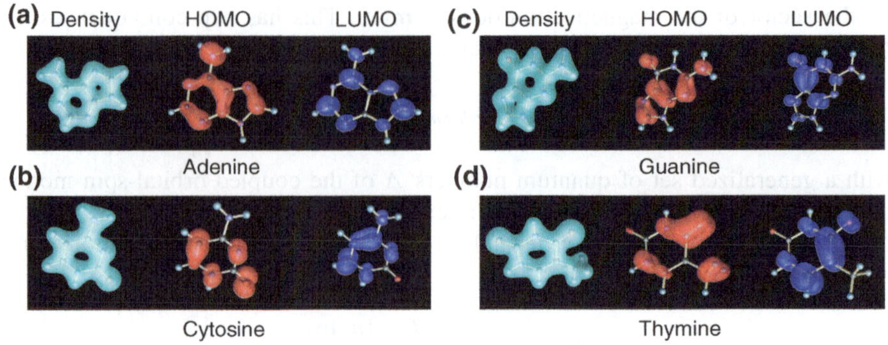

Fig. 6.3 Isosurfaces of the electron density $n(\mathbf{x})$ and orbitals for the HOMO and LUMO states of DNA base molecules adenine, cytosine, guanine, and thymine as derived from a Kohn-Sham scheme implemented in the code VASP [3] with an exchange-correlation functional that includes gradient corrections. The positions of the C, N, O, and H atoms are indicated by *dots*. From [4] and adapted from results in [5]. The IBM Data Explorer [6] has been used for presentation

(iii) The KS wave functions $\{\varphi_{\lambda m_s}(\mathbf{x})\}$ define the kinetic energy T_s of the non-interacting electrons. Together with $E_{\mathrm{XC}}[n]$ the ground state of the system of interacting electrons follows with their density $n(\mathbf{x})$.

(iv) The Kohn-Sham equations look formally very similar to the Hartree-Fock equations (4.28). However, the Hartree-Fock potential in (4.29) is orbital-dependent. The fact, that in the Kohn-Sham equations the effective potential is the same for every orbital labeled by λ, makes their numerical solution much simpler.

6.3 Beyond the Ground-State Energy

6.3.1 Highest-Occupied Kohn-Sham Eigenvalue

Conceptually, the Kohn-Sham equations (6.22) exactly determine the electron density (6.2) [and the corresponding spin densities (6.3)] and the total electronic energy (6.12) of the ground state. However, the orbital energies $\varepsilon_{\lambda m_s}$ in (6.22) are and remain purely Lagrange multipliers. According to the 'pure' KS theory they have no physical meaning. A physical interpretation can be also not rescued by any Koopmans theorem as described for the Hartree-Fock approach in Sect. 4.3.3. Below, in several chapters, we will illustrate that any identification of KS eigenvalues $\varepsilon_{\lambda m_s}$ with occupied or empty one-particle energies is to be justified (and sometimes heuristically corrected). For instance, the comparison with experimental data shows that calculated Kohn-Sham energies (usually independent of the approximation for the used XC energy) tend to underestimate the energy band gap in semiconductors and insulators, while the general trend of the wave-vector dispersion of the valence and conduction

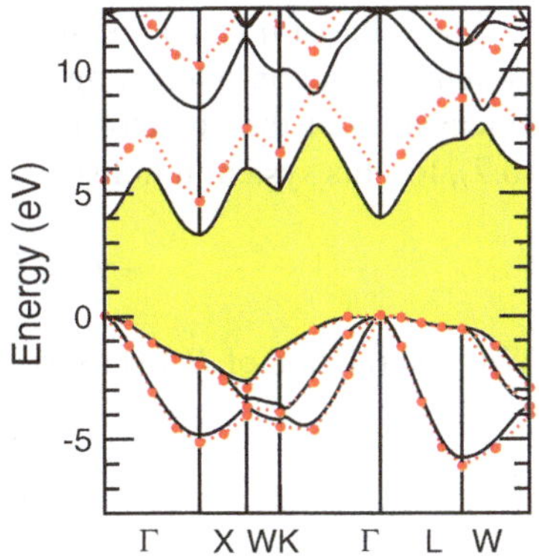

Fig. 6.4 Band structure of zinc-blende AlN as derived from Kohn-Sham eigenvalues (*solid black lines*) and quasiparticle band structure modified by corresponding shifts from the many-body perturbation theory (*red dots* and *dotted lines*). The valence-band maximum is used as energy zero. Adapted from [8]

bands is often represented to reasonable accuracy [7]. An illustration is displayed in Fig. 6.4.

Furthermore, our experience with Schrödinger-like equations such as (6.22) suggests that some eigenvalues possess possible relations to measurable quantities. Indeed, it can be shown that the highest-occupied KS state, the HOMO state, with energy $\varepsilon_{\lambda m_s}^{\mathrm{HOMO}}$ (see e.g. Fig. 6.3) has a physical meaning, at least for systems with large numbers $N \gg 1$ of electrons. In the low-temperature limit this orbital energy (at least for the 'exact' XC functional) can be identified with the negative ionization energy $-I$ (see Sect. 4.3.3) of the interacting electron gas [1, 9, 10].

The proof is here only given for metals with the chemical potential μ of the electrons (or for $T = 0\,\mathrm{K}$ the true Fermi energy). We investigate a system without spin polarization. The situation in both spin channels is identical. Hence, we label the KS states with $i = \lambda m_s$ and $i = 1, ..., N$ for the occupied ones. Despite the spin degeneracy $i = N$ should denote the highest occupied one. For such metallic systems the position of the chemical potential also agrees with the negative electron affinity $-A$. Consequently, the proof can assume either the removal or the addition of an electron. The metal is an extended electronic system with $N \gg 1$ for which the accompanying variation of the electron density $\delta n(\mathbf{x}) = |\varphi_N(\mathbf{x})|^2$ can be related to the KS orbital of the highest-occupied state $i = N$ (here: studying removal of an electron). With these settings the Lagrange parameter μ in the Euler equation (6.13),

$$\frac{\delta E_{\mathrm{KS}}[n]}{\delta n(\mathbf{x})} = \mu,$$

may be identified with the negative ionization energy and, thus, the chemical potential of the electrons. With the normalization condition (6.21) this can be immediately seen from

$$\begin{aligned} -I = E_{\mathrm{KS}}[n] - E_{\mathrm{KS}}[n - \delta n] &\approx \int d^3\mathbf{x} \frac{\delta E_{\mathrm{KS}}[n]}{\delta n(\mathbf{x})} \delta n(\mathbf{x}) \\ &= \int d^3\mathbf{x} \mu \delta n(\mathbf{x}) \\ &= \mu. \end{aligned}$$

The value of μ can be estimated from the total energy $E_s[n]$ (6.6) of the auxiliary system. For a non-interacting system it is simply given by the sum of the occupied eigenvalues of the operator (6.1), $E_s[n] = \sum_{i=1}^{N} \varepsilon_i$. Applying the Euler equation (6.7), for this system it holds

$$-I = E_s[n] - E_s[n - \delta n] \approx \mu_s.$$

With the relation of $E_s[n]$ to the eigenvalues of the occupied states we find

$$-I = \sum_{i=1}^{N} \varepsilon_i - \sum_{i=1}^{N-1} \varepsilon_i = \varepsilon_N \approx \mu_s.$$

For negligible differences $\mu - \mu_s$ it follows $I = -\varepsilon_N$. Similar arguments are valid for the addition of an electron. Because of the high density of electronic states around the Fermi energy in a metal with $N \gg 1$, and the fact that $I = A = W$ (with work function W), the above conclusion should hold for extended systems in general. However, we have to point out that (in particular for non-metals) the approximation used for the XC functional influences the numerical result.

Despite the limited relevance of the Kohn-Sham eigenvalues for the description of excitation energies, they have a definite mathematical meaning within the Kohn-Sham formalism itself. Introducing occupation numbers $n_{\lambda m}^{\Psi_0}$ (4.10) as in the HF description or $n_{\lambda m_s}$ in the definition (6.3) of the spin densities, an eigenvalue $\varepsilon_{\lambda m_s}$ is given as the derivative of the total energy with respect to occupation of state

$$\varepsilon_{\lambda m_s} = \frac{\partial}{\partial n_{\lambda m_s}} E_{\mathrm{KS}}[n].$$

This relation is known as the Janak theorem [11].

6.3.2 ΔSCF Method

The Kohn-Sham scheme has been developed to compute the ground state of an N-electron system. The question arises if this theory can be also applied to study electronic excitations. One trial could be to follow the idea of Koopmans (see Sect. 4.3.2) and to calculate single-particle excitation energies as differences of total energies with varying particle numbers N, e.g. $N \pm 1$. This approach is referred to as ΔSCF scheme, since it is based on the energy difference (i.e.: delta) between two self-consistent-field (SCF) calculations [12]. The application of the KS formalism to compute certain excitation energies can be justified if it is applied to the lowest energy (i.e., ground) state for a given electron number [13]. Examples are again (cf. Sect. 4.3.3) the electron affinity and ionization energy according to

$$I = E(N-1) - E(N),$$
$$A = E(N) - E(N+1) \tag{6.24}$$

with the ground-state total energy $E(M)$ of an $M = N, N+1, N-1$ electron system.

According to our experience the application of (6.24) to extended electron systems fails. This is especially true for using an implementation of the KS scheme in a plane-wave code. For extended systems and many electrons with non-localized wave functions, from (6.24) one more or less obtains the corresponding KS eigenvalues for the applied XC approximation. For localized systems with less electrons and confinement in all directions such as atoms or nanocrystals the results are however promising. Even applications to atomic multiplets give some promising results [14, 15]. Here, as examples ionization energies of the DNA base molecules adenine, cytosine, guanine, and thymine (see Fig. 6.3) are listed in Table 6.1 [5]. Both vertical (keeping the ground-state geometry $\{\mathbf{R}_s\}$) and adiabatic (allowing for atomic relaxation in the $(N-1)$-electron state) ionization energies are compared with measured values [16, 17]. Despite numerical problems with the electrostatics in the used supercell approach and the description of the KS eigenstates within a plane-wave basis set the agreement between theory and experiment is good. The mean-square deviation is less than 0.18 eV. Indeed, this means that the ΔSCF method can be applied to the described class of problems for localized systems. Possible generalizations of the method for solids, e.g. the energy gaps of insulating or semiconducting crystals, ask for the inclusion of electronic relaxation and screening (see e.g. [18]).

Indeed, the ΔSCF method can be also used to discuss the fundamental (quasiparticle [7]) gap of a semiconductor or an insulator. The gap E_g is defined in terms of the ionization potential I and the affinity energy A as

$$E_g = I - A \tag{6.25}$$

(see HF approximation in Sect. 4.3.3).

Interestingly, for fractional numbers of electrons it can be shown [19, 20] that relation (6.25) can be reformulated to

$$E_g = \varepsilon_{\mathrm{LUMO}}(N+\eta') - \varepsilon_{\mathrm{HOMO}}(N-\eta) \tag{6.26}$$

Table 6.1 Vertical and adiabatic ionization energies (in eV) of DNA base molecules as calculated by means of the ΔSCF method using a Kohn-Sham scheme including gradient corrections [1] to the XC functional [5]

Base molecule	Vertical	Adiabatic
Adenine	8.23 (8.44)	8.06 (8.26)
Cytosine	8.75 (8.94)	8.66 (8.68)
Guanine	7.82 (8.24)	7.63 (7.77)
Thymine	9.13 (9.14)	9.08 (8.87)

For comparison experimental values [16, 17] are given in parentheses for $\eta, \eta' \rightarrow +0$ by the eigenvalues of the electron system with a slightly deviating number of electrons. This result can be rewritten to

$$E_g = E_g^{\text{KS}} + \Delta \tag{6.27}$$

with the fundamental gap in the Kohn-Sham system of N electrons

$$E_g^{\text{KS}} = \varepsilon_{\text{LUMO}}(N) - \varepsilon_{\text{HOMO}}(N). \tag{6.28}$$

It underestimates the true gap E_g by a quantity Δ that is due to the discontinuity Δ in the XC potential $V_{\text{XC}}(\mathbf{x})$ (6.15) or (6.17) approaching the number N of electrons from the two sides $N + \eta'$ and $N - \eta$. The discontinuity Δ can be large for atoms [12] but also for solid non-metals. Unfortunately, the usually used XC functionals do not possess a discontinuity so that for computations of fundamental gaps of non-metallic solids the many-body perturbation theory [7] has to be used. This fact is illustrated in Fig. 6.4 for zinc-blende AlN. The band structure obtained by the Kohn-Sham eigenvalues underestimates the fundamental gap but also, in general, the distances between occupied and empty bands in comparison to the quasiparticle band structure. On the other hand, the band dispersion – especially that of the occupied bands – is well described within the KS approach. Only in very special cases XC potentials have been constructed so that their derivative includes a discontinuity at an integer particle number. One example is the XC potential of Kuisma et al. [21] which is based on an earlier functional [22].

In optically excited non-metallic systems, instead of the fundamental gap, the so-called optical gap E_g^{opt} determines the absorption edge. It is renormalized by excitonic effects (see also Sect. 4.4.3). It can be computed according to

$$E_g^{\text{opt}} = E(N, e + h) - E(N), \tag{6.29}$$

where $e + h$ indicates the excitation of the N-electron (here: KS) system by an electron(e)-hole(h) pair. The calculation of the total energy $E(N, e + h)$ of the excited state is done within the KS scheme applying the occupation constraint that the HOMO state of the ground state is unoccupied, i.e., contains a hole. Then, the excited electron resides in the lowest possible single-particle state, e.g. the LUMO of the ground-state system. However, deformations of the orbitals during the self-consistent calculations are possible. The approach (6.29) may be exact if the exact XC functional is known [20].

The ΔSCF method (6.29) works very well for localized electronic systems with a spatial extent below about 3 nm [23, 24]. For illustration of the method the optical gaps E_g^{opt} derived from the ΔSCF method and the Kohn-Sham gaps E_g^{KS} calculated by means of the KS eigenvalues of the ground state are displayed in Fig. 6.5 for silicon nanocrystals embedded in an amorphous SiO_2 matrix or passivated by hydrogen [25]. The figure indicates a strong reduction of the confinement effects with the embedment in an oxide instead of a surface passivation by hydrogen atoms. Surprisingly, the optical gaps are only slightly larger than the KS HOMO-LUMO gaps. This fact

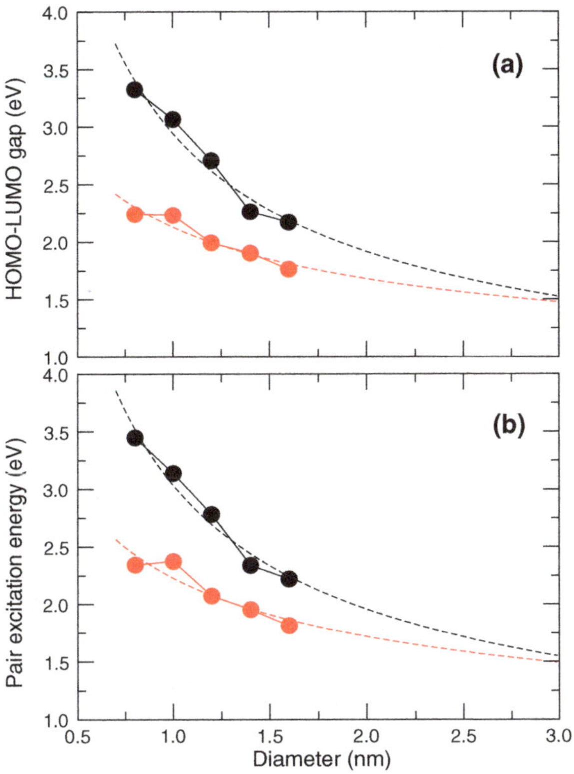

Fig. 6.5 (**a**) Kohn-Sham gap E_g^{KS} (6.28) and (**b**) optical gap E_g^{opt} (6.29) for Si nanocrystals versus diameter. *Red circles* represent results for Si nanocrystals embedded in amorphous SiO_2. For comparison gap energies of hydrogenated Si nanocrystals are displayed as *black dots*. From [25].

indicates an almost cancellation of the many-particle effects, e.g. the blue shift due to the quasiparticle renormalization of the excited electrons and holes and the red shift due to the electron-hole attraction, in strongly localized systems.

References

1. W. Kohn, L.J. Sham, Self-consistent equations including exchange and correlation effects. Phys. Rev. **140**, A1133–A1138 (1965)
2. A. Scholze, Selbstkonsistente *ab-initio* Pseudopotential-Rechnungen für Diamant(111)- und (100)-Oberflächen. Diploma thesis, Friedrich-Schiller-Universität Jena (1996)
3. www.vasp.at/
4. M. Preuss, Ab-initio-Berechnungen von Grund- und Anregungseigenschaften der DNA-Basen Adenin, Cytosin, Guanin und Thymin. Diploma thesis, Friedrich-Schiller-Universität Jena (2003)

5. M. Preuss, W.G. Schmidt, K. Seino, J. Furthmüller, F. Bechstedt, Ground- and excited-state properties of DNA base molecules from plane-wave calculations using ultrasoft pseudopotentials. J. Comput. Chem. **25**, 113–122 (2004)
6. IBM: IBM Data Explorer 3.1.4. http://www.research.ibm.com/dx/ (1997)
7. W.G. Aulbur, L. Jönsson, J.W. Wilkins, Quasiparticle calculations in solids, in *Solid State Physics. Advances in Research and Applications*, vol. 54, ed. by H. Ehrenreich, F. Spaepen (Academic Press, San Diego, 2000), pp. 1–218
8. A. Riefer, F. Fuchs, C. Rödl, A. Schleife, F. Bechstedt, R. Goldhahn, Interplay of excitonic effects and van Hove singularities in optical spectra: CaO and AlN polymorphs. Phys. Rev. B **84**, 075218 (2011)
9. C.O. Ambladh, U. von Barth, Exact results for the charge and spin densities, exchange-correlation potentials, and density-functional eigenvalues. Phys. Rev. B **31**, 3231–3244 (1985)
10. J.P. Perdew, M. Levy, Comment on "Significance of the highest occupied Kohn-Sham eigenvalue". Phys. Rev. B **56**, 16021–16028 (1997)
11. J.F. Janak, Proof that $\partial E/\partial n_i = \varepsilon_i$ in density functional theory. Phys. Rev. B **18**, 7165–7168 (1978)
12. R.O. Jones, O. Gunnarsson, The density functional formalism, its applications and prospects. Rev. Mod. Phys. **61**, 689–746 (1989)
13. O. Gunnarsson, B.I. Lundqvist, Exchange and correlation in atoms, molecules, and solids by the spin-density-functional formalism. Phys. Rev. B **13**, 4274–4298 (1976)
14. T. Ziegler, A. Rauk, E.J. Baerends, On the calculation of multiplet energies by the Hartree-Fock-Slater method. Theor. Chim. Acta **43**, 261–271 (1977)
15. U. von Barth, Local-density theory of multiplet structure. Phys. Rev. A **20**, 1693–1703 (1979)
16. V.M. Orlov, A.N. Smirnov, Y.M. Vasharsky, Ionization potentials and electron-donor ability of nucleic acid bases and their analogues. Tetrahedron Lett. **48**, 4377–4378 (1976)
17. N.S. Hush, A.S. Cheung, Ionization potentials and donor properties of nucleic acid bases and related compounds. Chem. Phys. Lett. **34**, 11–13 (1975)
18. M.K.Y. Chan, G. Ceder, Efficient band gap prediction for solids. Phys. Rev. Lett. **105**, 196403 (2010)
19. J.P. Perdew, M. Levy, Physical content of the exact Kohn-Sham orbital energies: band gaps and derivative discontinuities. Phys. Rev. Lett. **51**, 1884–1887 (1983)
20. L.J. Sham, M. Schlüter, Density-functional theory of the energy gap. Phys. Rev. Lett. **51**, 1888–1891 (1983)
21. M. Kuisma, J. Ojanen, J. Enkovaara, T.T. Rantala, Kohn-Sham potential with discontinuity for band gap materials. Phys. Rev. B **82**, 115106 (2010)
22. O.V. Gritzsenko, R. van Leeuwen, E. van Lenthe, E.J. Baerends, Self-consistent approximation to the Kohn-Sham exchange potential. Phys. Rev. A **51**, 1944–1954 (1995)
23. R. Godby, I.D. White, Density-relaxation part of the self-energy. Phys. Rev. Lett. **80**, 3161–3161 (1998)
24. H.C. Weissker, J. Furthmüller, F. Bechstedt, Excitation energies and radiative lifetimes of $Ge_{1-x}Si_x$ nanocrystals: alloying versus confinement effects. Phys. Rev. Lett. **90**, 085501 (2003)
25. K. Seino, F. Bechstedt, P. Kroll, Influence of SiO_2 matrix on electronic and optical properties of Si nanocrystals. Nanotechology **20**, 135702 (2009)

Chapter 7
Exchange-Correlation Functionals

Abstract Crucial for the application of the density functional theory in the framework of the Kohn-Sham ansatz is the knowledge of the exchange-correlation functional, which usually is formulated in terms of a density- and space-dependent exchange-correlation energy per particle. Such a formulation immediately leads to an explicit expression by replacing the density dependence calculated numerically for a homogeneous electron gas by the dependence on the local density of the inhomogenous electron gas. This is the local density approximation (LDA). Generalizations for spin-polarized systems to a local spin density approximation are obvious. Improvements of the local approximation for exchange and correlation include density gradients. A generalized gradient approximation (GGA) has to fulfill the sum rules. Nevertheless, many different functionals can be formulated, e.g. PW91, PBE, AM05, PBEsol, etc. Explicit formulas for some widely used functionals are given. Their applicability and accuracy are discussed and shown, respectively, for test quantities such as lattice constants, bulk moduli, and binding energies.

7.1 Properties of the Exact XC Functional

7.1.1 General Remarks

The success of the Kohn-Sham scheme, in particular of expression (6.12) presented in Chap. 6, is based on two main supports: (i) It explicitly separates out the independent-particle kinetic energy $T_s[n]$ and the long-range Hartree energy $E_{\mathrm{H}}[n]$ from the universal HK functional [cf. (6.10)]. (ii) The possibility for reasonable approximations of the remaining exchange-correlation energy $E_{\mathrm{XC}}[n]$ (6.10) which, however, contains contributions from the difference $(T - T_s)$ between the kinetic energies in the interacting and non-interacting systems.

The contribution $(T - T_s)$ requires a generalization of the XC energy compared to the expression (3.48) or (3.62). Nevertheless, $E_{\mathrm{XC}}[n]$ (6.10) remains a functional of the density (or spin densities). This means that it can be expressed in the form

F. Bechstedt, *Many-Body Approach to Electronic Excitations*,
Springer Series in Solid-State Sciences 181, DOI 10.1007/978-3-662-44593-8_7

$$E_{\rm XC}[n] = \int d^3\mathbf{x} n(\mathbf{x}) \epsilon_{\rm XC}(\mathbf{x}; [n]), \tag{7.1}$$

where $\epsilon_{\rm XC}(\mathbf{x}; [n])$ is the XC energy per particle at point $\mathbf{x}$ that depends only upon the density $n(\mathbf{x})$ [or densities $n_{m_s}(\mathbf{x})$] in some neighborhood of point $\mathbf{x}$. Fortunately, the total density $n(\mathbf{x})$ appears in (7.1) because of the independence of the Coulomb interaction of spin. In a collinearly spin-polarized system $\epsilon_{\rm XC}(\mathbf{x}; [n])$ incorporates the information on the spin densities. In a non-collinear magnetic system or in insulators with a spontaneous electric polarization field the XC energy per particle is not only a function of the density in a region nearby $\mathbf{x}$. In principle, the magnetization or polarization vectors have to be also taken into consideration.

7.1.2 XC Hole

We analyze the XC contribution to the total KS energy using the method of 'adiabatic connection' [1]. It is based on two ideas: (i) The possible scaling $\lambda\hat{U}$ of the electron-electron interaction (3.19) with a parameter λ $(0 \leq \lambda \leq 1)$ between the interaction-free limit and the full interaction $\hat{U}$. (ii) The assumption that the electron density $n(\mathbf{x}) = n^\lambda(\mathbf{x})$ can be fixed at the same value despite different interaction strengths, i.e., for different values λ. Instead of (3.1) the Hamiltonian [2]

$$\mathscr{H}_0^\lambda = \hat{T} + \hat{V}^\lambda + \lambda\hat{U} \tag{7.2}$$

is considered, where the external potential $V_{\rm n}(\mathbf{x})$ is replaced by another one $V_{\rm n}^\lambda(\mathbf{x})$ which guarantees the same (ground-state) density. The (ground-state) of the Hamiltonian (7.2) is $|\Psi_0^\lambda\rangle$. With the electron density operator (3.14) it should therefore hold

$$n^\lambda(\mathbf{x}) = \langle\Psi_0^\lambda|\hat{n}(\mathbf{x})|\Psi_0^\lambda\rangle = n(\mathbf{x}). \tag{7.3}$$

The general existence of such potentials $V_{\rm n}^\lambda(\mathbf{x})$ is not guaranteed but will be assumed here.

In order to rewrite the total energy we apply the Hellmann-Feynman theorem [3, 4]

$$\frac{\partial}{\partial\lambda}\langle\Psi_0^\lambda|\mathscr{H}_0^\lambda|\Psi_0^\lambda\rangle = \langle\Psi_0^\lambda|\frac{\partial\mathscr{H}_0^\lambda}{\partial\lambda}|\Psi_0^\lambda\rangle. \tag{7.4}$$

Then the original ground-state energy functional of the density functional theory (5.21) can be rewritten step by step

$$
\begin{aligned}
E_0 = E_{V_\mathrm{n}}[n] &= \langle \Psi_0^1 | \mathscr{H}_0^1 | \Psi_0^1 \rangle \\
&= \langle \Psi_0^0 | \mathscr{H}_0^0 | \Psi_0^0 \rangle + \int_0^1 d\lambda \frac{\partial}{\partial \lambda} \langle \Psi_0^\lambda | \mathscr{H}_0^\lambda | \Psi_0^\lambda \rangle \\
&= \langle \Psi_0^0 | \mathscr{H}_0^0 | \Psi_0^0 \rangle + \int d^3\mathbf{x} \left[V_\mathrm{n}^1(\mathbf{x}) - V_\mathrm{n}^0(\mathbf{x}) \right] n(\mathbf{x}) + \int_0^1 d\lambda \langle \Psi_0^\lambda | \hat{U} | \Psi_0^\lambda \rangle .
\end{aligned}
$$

With (7.3), the definition of the non-interacting Hamiltonian, and $V_\mathrm{n}^1(\mathbf{x}) \equiv V_\mathrm{n}(\mathbf{x})$ we find

$$
E_0 = E_{V_\mathrm{n}}[n] = \langle \Psi_0^0 | \hat{T} | \Psi_0^0 \rangle + \int d^3\mathbf{x} V_\mathrm{n}(\mathbf{x}) n(\mathbf{x}) + \int_0^1 d\lambda \langle \Psi_0^\lambda | \hat{U} | \Psi_0^\lambda \rangle .
$$

Since the first term on the right-hand side corresponds to the kinetic energy of non-interacting electrons, $T_s[n]$ (6.5), we find instead of (6.11) the expression

$$
E_\mathrm{XC}[n] = \int_0^1 d\lambda \langle \Psi_0^\lambda | \hat{U} | \Psi_0^\lambda \rangle - E_\mathrm{H}[n] \tag{7.5}
$$

for the XC energy that also contains the difference $(T - T_s)$ of the kinetic energies with and without interaction. The above derivation of the kinetic contribution to the electron correlation energy follows the concept of *adiabatic connection.* It is schematically displayed in Fig. 7.1. The main idea is to switch the electron-electron interaction gradually from the non-interacting reference system toward the fully interacting "real" system, while still conserving the ground-state density (5.9).

With the definition of a pair correlation function (3.42) in the state $|\Psi_0^\lambda\rangle$

$$
g^\lambda(\mathbf{x}, \mathbf{x}'; [n]) = \langle \Psi_0^\lambda | \hat{n}(\mathbf{x}') \hat{n}(\mathbf{x}) - \delta(\mathbf{x} - \mathbf{x}') \hat{n}(\mathbf{x}) | \Psi_0^\lambda \rangle / n(\mathbf{x}) n(\mathbf{x}') \tag{7.6}
$$

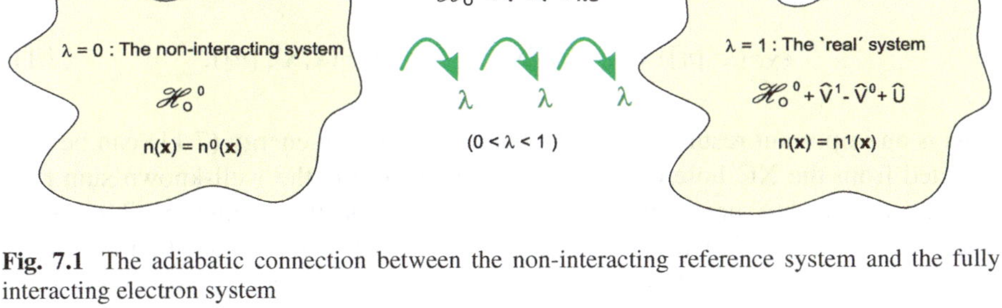

Fig. 7.1 The adiabatic connection between the non-interacting reference system and the fully interacting electron system

the XC energy functional (7.5)

$$E_{XC}[n] = \frac{1}{2}\int d^3\mathbf{x}\int d^3\mathbf{x}' v(\mathbf{x}-\mathbf{x}')n(\mathbf{x})n(\mathbf{x}')\left[\bar{g}(\mathbf{x},\mathbf{x}';[n]) - 1\right] \tag{7.7}$$

can be expressed exactly in terms of an integral over the coupling constant λ

$$\bar{g}(\mathbf{x},\mathbf{x}';[n]) = \int_0^1 d\lambda g^{\lambda}(\mathbf{x},\mathbf{x}';[n]). \tag{7.8}$$

The function $g^{\lambda}(\mathbf{x},\mathbf{x}';[n])$ is the pair correlation function of the system with density $n(\mathbf{x})$ and the Coulomb interaction $\lambda\hat{U}$. The function $\bar{g}(\mathbf{x},\mathbf{x}';[n]))$ is the coupling-constant-averaged one. The occurrence of $\bar{g}$ instead of g (i.e., the pair correlation function for $\lambda = 1$) as in (3.48) is a consequence of the inclusion of $(T - T_s)$. Finally, the XC functional (7.7) takes the form of a Coulomb integral [2, 5, 6]

$$E_{XC}[n] = \frac{1}{2}\int d^3\mathbf{x}\int d^3\mathbf{x}' v(\mathbf{x}-\mathbf{x}')n(\mathbf{x})\bar{n}_{XC}(\mathbf{x},\mathbf{x}';[n]), \tag{7.9}$$

where

$$\bar{n}_{XC}(\mathbf{x},\mathbf{x}';[n])) = n(\mathbf{x}')\left[\bar{g}(\mathbf{x},\mathbf{x}';[n]) - 1\right] \tag{7.10}$$

is the coupling-constant-averaged XC hole density. It again describes the effect of the interelectronic interactions beyond the Hartree repulsion, i.e., the fact that an electron present at point $\mathbf{x}$ reduces the probability of finding one at $\mathbf{x}'$. The XC hole density contains contributions from parallel and antiparallel spins in the same way as discussed in Sect. 3.5.2 for spin-polarized systems and the limit $\lambda = 1$. The density (7.10) fulfills the same sum rule as that (3.50) without $(T - T_s)$ corrections. This means that still one electron is missing in the proximity $\mathbf{x}'$ of an electron at $\mathbf{x}$. Globally an XC hole is present. Thereby, the correlation effects counteract the exchange term locally to some extent, but the net deficit is still precisely one electron. According to the sum rule (4.22) globally it can be traced back to exchange interactions.

Finally, the comparison of the expressions (7.1) and (7.9) shows that the coupling-constant-integrated exchange-correlation energy per particle can be exactly written as

$$\epsilon_{XC}(\mathbf{x};[n]) = \frac{1}{2}\int d^3\mathbf{x}' v(\mathbf{x}-\mathbf{x}')\bar{n}_{XC}(\mathbf{x},\mathbf{x}';[n]). \tag{7.11}$$

This is an important result: (i) The exchange-correlation energy (7.11) can be constructed from the XC hole density (7.10) which fulfills the well-known sum rule. (ii) Together with expression (7.11) the exchange-correlation energy (7.9) can be understood in terms of an interaction energy between the electrons and their XC holes. (iii) The XC potential (6.15) in the Kohn-Sham equations can be easily expressed in terms of the XC energy per particle (7.11) as

$$V_{\mathrm{XC}}(\mathbf{x}) = \epsilon_{\mathrm{XC}}(\mathbf{x}; [n]) + n(\mathbf{x})\frac{\delta\epsilon_{\mathrm{XC}}(\mathbf{x}; [n])}{\delta n(\mathbf{x})}. \tag{7.12}$$

The generalization to the (collinearly) spin-polarized case (6.17) is trivial

$$V_{\mathrm{XC}}^{m_s}(\mathbf{x}) = \epsilon_{\mathrm{XC}}(\mathbf{x}; [n]) + n(\mathbf{x})\frac{\delta\epsilon_{\mathrm{XC}}(\mathbf{x}; [n])}{\delta n_{m_s}(\mathbf{x})}. \tag{7.13}$$

One has to take only into account that $\epsilon_{\mathrm{XC}} \equiv \epsilon_{\mathrm{XC}}\left(\mathbf{x}; \left[n_{\frac{1}{2}}, n_{-\frac{1}{2}}\right]\right)$ is a functional of the two spin densities (6.3).

7.2 Local (Spin) Density Approximation

7.2.1 Relation to Homogeneous Electron Gas

Similar to the Hartree-Fock approximation (4.13) we assume that exchange and correlation effects are additive. Thus, per particle, we decompose the total XC energy (7.11) according to

$$\epsilon_{\mathrm{XC}}(\mathbf{x}; [n]) = \epsilon_{\mathrm{X}}(\mathbf{x}; [n]) + \epsilon_{\mathrm{C}}(\mathbf{x}; [n]) \tag{7.14}$$

into an exchange and a correlation contribution. Thereby, for systems with collinear spins the functional dependence on the density $[n]$ has to be replaced by that $\left[n_{\frac{1}{2}}, n_{-\frac{1}{2}}\right]$ of the two spin densities $n_{m_s}(\mathbf{x})$ (4.19) or (6.3) or, equivalently, by that $[n, \xi]$ of the electron density $n(\mathbf{x})$ and the spin polarization

$$\xi(\mathbf{x}) = \left[n_{\frac{1}{2}}(\mathbf{x}) - n_{-\frac{1}{2}}(\mathbf{x})\right]/n(\mathbf{x})$$

(3.60) or (4.60). The explicit functional dependences are however still not known.

According to the construction leading to (7.11), the XC energy consists of three contributions. The first one is the potential energy of exchange where, however, the Fock term is computed by the Kohn-Sham orbitals. The second term is due to the correlation (4.4), while the third contribution is the $(T - T_s)$ correction. In Sect. (4.4) we have learnt that for a homogeneous non-spin-polarized ($\xi = 0$) electron gas (or more precisely, a jellium model) the HF exchange energy per particle (4.57)

$$\epsilon_{\mathrm{X}}^{\mathrm{hom}}(n) = \epsilon_{\mathrm{X}}^{\mathrm{hom}}(n, \xi = 0) = -\frac{3}{2}\left(\frac{3}{2\pi}\right)^{\frac{2}{3}}\frac{1}{r_s}Ry \tag{7.15}$$

with the dimensionless electron-gas parameter $r_s = \left(4\pi n a_B^3/3\right)^{-1/3}$ is a function of the homogeneous density n. For a better differentiation the subscript 'hom' (from 'homogeneous' electron gas) has been introduced. A corresponding generalization for finite spin polarizations $0 < \xi < 1$ could be found via the interpolation between the para- and ferromagnetic limits as [7]

$$\epsilon_X^{\text{hom}}(n,\xi) = \epsilon_X^{\text{hom}}(n,0) + \left[\epsilon_X^{\text{hom}}(n,1) - \epsilon_X^{\text{hom}}(n,0)\right] f(\xi). \tag{7.16}$$

The interpolation function $f(\xi)$ (4.64) is exact and may be also assumed to be valid for the correlation energy $\epsilon_C^{\text{hom}}(n,\xi)$ [7]. However, meanwhile also generalizations of the interpolation function $f(\xi)$ are used (see e.g. [8]).

For electronic systems with a weakly spatially varying electron density $n(\mathbf{x})$, as e.g. in simple metals, one may think that the exchange-correlation energy

$$\epsilon_{XC}(\mathbf{x};[n,\xi]) \approx \epsilon_{XC}^{\text{hom}}(n,\xi)\Big|_{\substack{n=n(\mathbf{x})\\ \xi=\xi(\mathbf{x})}} \tag{7.17}$$

is described by that of a homogeneous electron gas but replacing the uniform density of the homogeneous electron gas by the spatially varying one of the studied inhomogeneous electron gas. This leads to the local (spin) density approximation (L(S)DA) [9].

Already Kohn and Sham [9] pointed out that solids can be often considered as close to the limit of the homogeneous electron gas. Now, there is a common belief that many effects of exchange and correlation can be described to be local in their character. Then, the spatial dependence of ϵ_{XC} is only due to the strong spatial variation of $n(\mathbf{x})$ [and $\xi(\mathbf{x})$]. This variation is illustrated in Figs. 7.2 and 7.3 for a covalently bonded silicon crystal and a binary compound SiC with partially ionic bonds. Both materials are non-magnetic, i.e., $\xi(\mathbf{x}) = 0$. Their bonds between adjacent atoms are directed to the corners of a tetrahedron. The inhomogeneity of the electron gases in

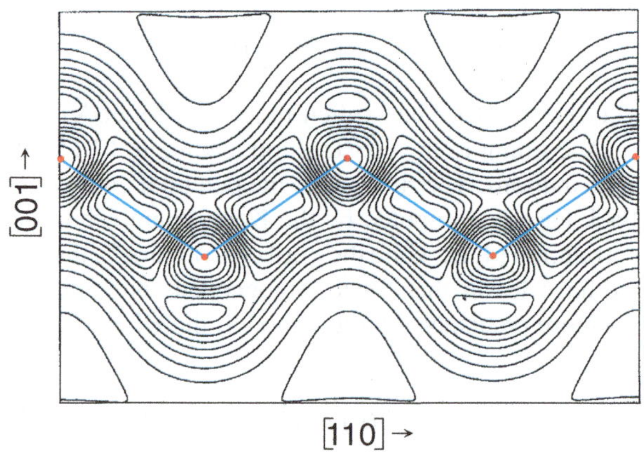

Fig. 7.2 Valence-electron density $n(\mathbf{x})$ of crystalline silicon in a $(1\bar{1}0)$ plane. The *red dots* indicate the positions of the Si atoms, while the *blue lines* characterize the bond directions. The distance of two lines in the contour plot amounts to 0.04 electron/Å^3. Adapted from a DFT-LDA computation [10]

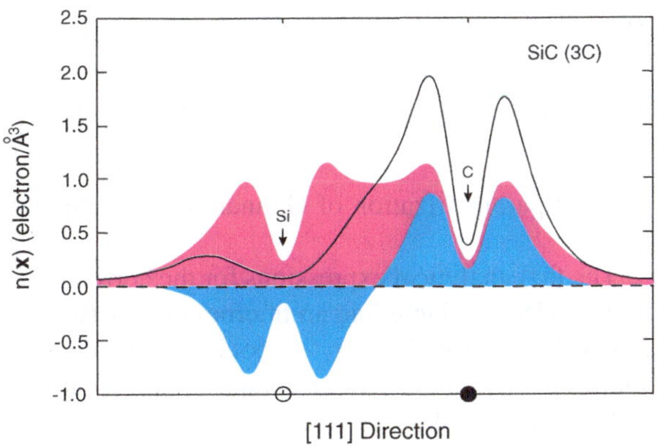

Fig. 7.3 Valence-electron density of cubic SiC along a [111] bond direction (*black solid line*). It is decomposed into a symmetric (*red*) and an antisymmetric (*blue*) part. The positions of cation (Si) and anion (C) are indicated by dots. Drawn using DFT-LDA results from [11]

these semiconductors is obvious. The electron density varies by orders of magnitude from the bond regions to the interstitial regions in Fig. 7.2. Moreover, in the case of partially ionic bonds as in SiC (Fig. 7.3) also the density redistribution from the cation environment to the anion region is obvious. Because of the fact that the majority of the electrons is found along the bonds (see Fig. 7.2) and the electron transfer from cation to anion the electron density $n(\mathbf{x})$ along such a tetrahedron direction can be divided into a symmetric part $n_S(\mathbf{x})$ and an antisymmetric contribution $n_A(\mathbf{x})$ as indicated in Fig. 7.3. They define a charge asymmetry coefficient g by taking the square root of the ratio of the squares of the antisymmetric and symmetric parts of the electron density integrated over the volume [12]. From the contributions plotted in Fig. 7.3 a value $g = 0.476$ is derived for zinc-blende SiC [11]. For covalent materials such as silicon it holds $g = 0$. For strongly ionic compounds such as group-III nitrides large charge asymmetry coefficients $g = 0.794$ (AlN), 0.780 (GaN), and 0.853 (InN) have been computed [13].

7.2.2 Correlation in a Homogeneous Electron Gas

Once one has made the L(S)DA (7.17), parameterfree calculations within the Kohn-Sham scheme and the DFT can be performed. The functional $E_{\mathrm{XC}}[n]$ or $E_{\mathrm{XC}}[n, \xi]$ is universal. It is exactly the same as for the homogeneous gas, only the local density (spin densities) is (are) taken for the inhomogeneous system. However, still there is a missing link, the knowledge of $\epsilon_{\mathrm{C}}^{\mathrm{hom}}(n)$ [or $\epsilon_{\mathrm{C}}^{\mathrm{hom}}(n, \xi)$], the density dependence of the correlation energy per particle. It has been calculated to great accuracy by means of the Quantum Monte Carlo method [14, 15] or some analytical approximations, e.g. the two-parameter model suggested by Hedin et al. [16] and Hedin and Lundqvist [17] on the basis of the local-field results of Singwi et al. [8] for electron-gas parameters $1 \lesssim r_s \lesssim 6$

$$\epsilon_{\mathrm{C}}^{\mathrm{hom}}(n, \xi = 0) = -C\left[(1+x^3)\ln\left(1+\frac{1}{x}\right) + \frac{x}{2} - x^2 - \frac{1}{3}\right]_{x=r_s/A} Ry \quad (7.18)$$

with the parametrization of Gunnarsson and Lundqvist [6] $C = 0.0666$ and $A = 11.4$.

The first analytical expressions for the correlation energy have been suggested by Wigner [18, 19] including an interpolation between the high-density limit, $r_s \to 0$, and the low-density limit, $r_s \to \infty$. In the low-density limit the electrostatic potential dominates, and the electrons condense into what is known as a Wigner crystal, e.g. with electrons at the sites of a bcc crystal. Their correlation energy is proportional to $\frac{1}{r_s}$. Since in the thirties of the last century the exchange energy has been considered to approach a constant in the high-density limit, a resulting interpolation formula [18] was

$$\epsilon_{\mathrm{C}}^{\mathrm{hom}}(n, 0) = -\frac{0.88}{7.8 + r_s} Ry \quad (7.19)$$

with an incorrect high-density limit.

In the high-density limit, $r_s \to 0$, the kinetic energy dominates as visible from (4.56). The random phase approximation (RPA) (see [20]) or the ring approximation becomes exact. In this density limit Gell-Mann and Brückner [21] have calculated the correlation energy by a partial resummation of the perturbation series, incorporating the most diverging energy diagrams, i.e., the ring diagrams, in each order and, hence, introducing a screened Coulomb interaction. Gell-Mann and Brückner improved a result of Macke [22] to

$$\epsilon_{\mathrm{C}}^{\mathrm{hom}}(n, 0) = \left[\frac{2(1-\ln 2)}{\pi^2}\ln r_s - 0.142\right] Ry. \quad (7.20)$$

Further improvements in the high-density expansion gave rise to terms proportional to $r_s \ln r_s$ and to r_s [23, 24].

The manifold analytical results have inspired Perdew and Zunger (PZ) [25] to their frequently used parametrization of the Monte-Carlo results of Ceperley and Alder [14]. In addition, they used the Padé approximant in $\sqrt{r_s}$ which was originally suggested by Ceperley [26] for $r_s \gtrsim 1$. Combining the previous results, in particular the high-density limit of Gell-Mann and Brückner and the scaling relation of the spin-polarized ring approximation, they found

$$\epsilon_{\mathrm{C}}^{\mathrm{hom}}(n, \xi = 0, 1) = \left\{ \begin{array}{l} A(\xi)\ln r_s + B(\xi) + C(\xi) r_s \ln r_s + D(\xi) r_s \quad (\text{for } r_s \le 1) \\ \gamma(\xi)/\left[1 + \beta_1(\xi)\sqrt{r_s} + \beta_2(\xi) r_s\right] \quad (\text{for } r_s \ge 1) \end{array} \right\} Ry \quad (7.21)$$

with

$$\begin{aligned}
&A(0) = 0.0622,\ B(0) = -0.096,\ C(0) = 0.004,\ D(0) = -0.0232,\\
&\gamma(0) = -0.2846,\ \beta_1(0) = 1.0529,\ \beta_2(0) = 0.3334 \text{ and}\\
&A(1) = 0.0311,\ B(1) = -0.0538,\ C(1) = 0.0014,\ D(1) = -0.0096,\\
&\gamma(1) = -0.1686,\ \beta_1(1) = 1.3981,\ \beta_2(1) = 0.2611.
\end{aligned} \tag{7.22}$$

The correlation energy (7.21) is measured in units of Rydberg and not Hartree. For that reason, apart from β_1 and β_2 all constants in (7.22) are twice as large in comparison to other representations in the literature. Together with the parameters in (7.22) expression (7.21) fulfills the requirements that the correlation energy and the resulting correlation potential are continuous at $r_s = 1$. For $\xi = 0$, expression (7.21) for $r_s \leq 1$ matches the RPA result of Gell-Mann and Brückner. Within the RPA it holds the scaling relation $\epsilon_{\mathrm{C}}^{\mathrm{hom}}(n, \xi = 1) = \frac{1}{2}\epsilon_{\mathrm{C}}^{\mathrm{hom}}(16n, \xi = 0)$, just like that for the exchange energy $\epsilon_{\mathrm{X}}^{\mathrm{hom}}(n, \xi)$ (4.65). The interpolation of (7.21) for arbitrary $0 < \xi < 1$ is assumed to have the same functional form as for the exchange energy (7.16).

The density dependences of the correlation energy $\epsilon_{\mathrm{C}}^{\mathrm{hom}}(n, 0)$ and the exchange contribution $\epsilon_{\mathrm{X}}^{\mathrm{hom}}(n, 0)$ are displayed in Fig. 7.4 for a homogeneous electron gas without spin polarization. Both many-body interactions increase the energy gain of the interacting system in contrast to the repulsive Hartree interaction. Thereby, in the high-density limit the effect of exchange is much larger than that due to correlation. In the low-density limit the opposite tendency is seen. For $r_s > 6.7$ the correlation predominates. Already for $r_s = 2$, which corresponds to the average electron density in a silicon crystal, the two energy contributions −3.80 eV (X) and −1.23 eV (C) per particle are of the same order of magnitude. Figure 7.4b also shows that the analytical expression of Gunnarsson and Lundqvist (7.18) [6] overestimates the correlation effects compared to the fit (7.21) [25] to the Monte Carlo results [14].

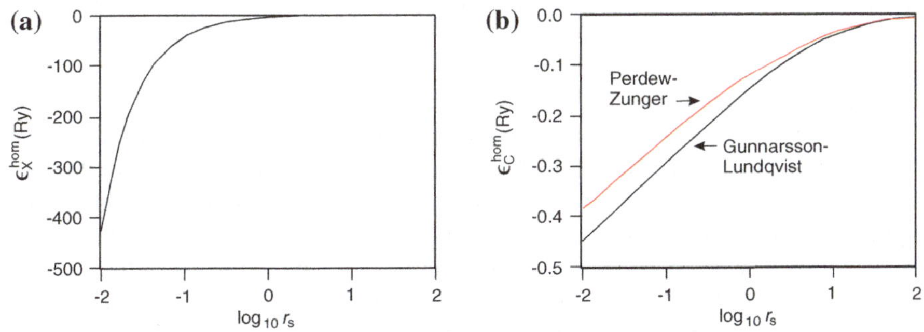

Fig. 7.4 Density dependence of the exchange (**a**) and correlation (**b**) energy per particle for a non-spin-polarized ($\xi = 0$) homogeneous electron gas with parameter r_s. In the correlation case (**b**) the functions (7.18) and (7.21) are plotted, while in (**a**) the well-known $n^{1/3}$ dependence (7.15) is displayed

7.2.3 Interpretation: Advantages and Limits

The DFT in conjunction with the L(S)DA (7.17) has become a very successful tool for the parameterfree investigation of the physical and chemical ground-state properties of electronic systems. The XC functional (7.1) is known within the L(S)DA. In particular, its variation with the density (spin densities) is known. Even analytical formulas or explicit analytical forms fitted to numerical Monte Carlo results are known. Frequently, physicists call the method '*ab initio*' meaning that for a given functional no further input is needed. In previous years, in chemistry it was traditional to refer only the 'standard methods' of the quantum chemistry for directly solving the electronic structure problems such as the coupled-cluster (CC) expansion, the Møller-Plesset perturbation theory, and the quantum Monte Carlo (QMC) method [27] as *ab initio* [28]. These methods are however restricted in general to systems with a few number of atoms. Nevertheless, very recently such quantum-chemical computations have been also performed within a full configuration-interaction quantum Monte Carlo method for solids by theoreticians working in the field of condensed matter physics [29].

The validity of the local approximation for exchange and correlation can be evaluated only in a few cases. In general, more qualitative arguments related to 'slowly varying electron densities' are used. In a mathematical sense, this means a restriction to minor changes of the density $n(\mathbf{x})$ on a length scale of the average distance $r_s a_B$ [with n as the spatially averaged density $n(\mathbf{x})$], where sometimes $r_s a_B$ is also called Wigner-Seitz radius. This might be fulfilled for free-electron-like metals, e.g. Al. In real solids with typical values $r_s = 1 \ldots 4$ the characteristic length $r_s a_B \approx 0.5 \ldots 2$ Å is smaller or of the order of the typical interatomic distances. In a silicon crystal with $r_s = 2$ the distance $r_s a_B = 1.06$ Å is however much smaller than the bond length 2.35 Å. Moreover, the remarkable density variation in Fig. 7.2 indicates significant deviations of the local density from the averaged one. As a clear consequence of the density variation in a real inhomogeneous electron gas a significant spatial modification of the exchange-correlation energy appears as illustrated in Fig. 7.5. Nevertheless, DFT-LDA calculations frequently yield very good results for ground-state properties such as lattice constants and bulk moduli. For systems with not too localized valence electrons the deviations from measured values are of the order of 1 % as we will demonstrate in Chap. 8.

The mathematical assumptions for the applicability of the L(S)DA are less fulfilled for finite systems. The local XC approximation becomes already questionable for surfaces of simple metals. The electron density near metal surfaces varies very rapidly. Nevertheless, the first real calculation using DFT-LDA yielded reasonable values for the surface barriers, the work functions, of simple metals [30].

More questionable should be the application of the DFT-LDA to atomic systems. A good test case to study this problem may be the atoms and ions He, Li^+, and Be^{++} with two spin-paired $1s$ electrons. The spin polarization should not play a role. In these systems the exact wave functions are known. It is possible to determine not only the density, but also the correct forms of E_{XC} and V_{XC}. Almbladh and Pedroza

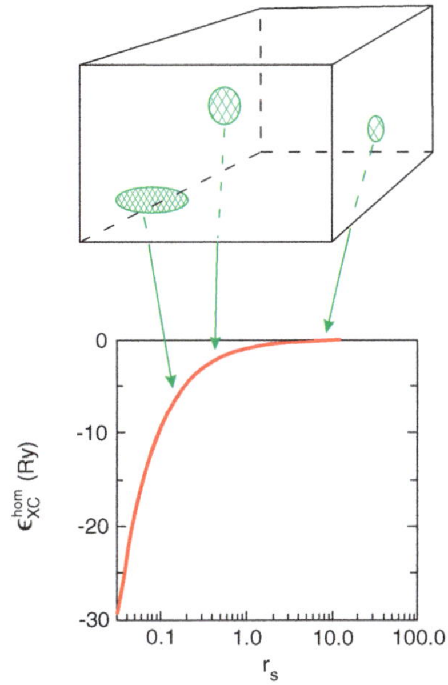

Fig. 7.5 Density variation of the exchange-correlation energy related to the actual density of the inhomogeneous electron gas given in a box

[31] found that $E_{XC}[n]$ is described rather accurately by LDA (the errors are of the order of 10 %). However, errors in V_{XC} are substantially greater. This is illustrated in Fig. 7.6, where the LDA and exact V_{XC} are displayed versus the radius of the spherical objects. The figure shows that in free atoms the XC effects increase the

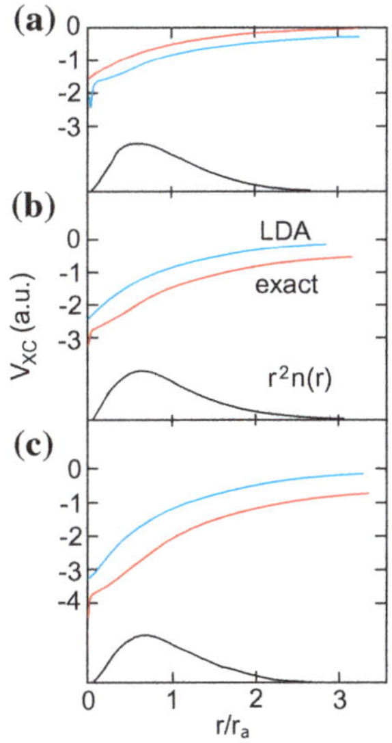

Fig. 7.6 Exchange-correlation potential $V_{XC}(r)$ (*red line*) and corresponding LDA function (*blue line*) for (**a**) He, (**b**) Li^+, and (**c**) Be^{++}. The radius is normalized by the atomic radii $r_a = 0.929, 0.573$, and $0.414\ a_B$, respectively. The *dotted curve* illustrates the radial density of the two electrons. After Almbladh and Pedroza [31]

binding of the electrons to the nucleus. However, the LDA tends to reduce the actual binding and can result in larger errors in the eigenvalues.

Surprisingly the L(S)DA description of exchange and correlation yields reasonable results also for many-electron systems with varying particle density. Some plausibility arguments can be given to explain this observation. Fortunately, the XC energy is a relatively small part of the total energy, although it is by far the largest contribution to chemical bonding, i.e., the largest part of 'nature's glue' that binds atoms together [32]. It is a consequence because electrons avoid each other, thus lowering the total electron-electron interaction. The XC effects are usually described by quantities represented by space integrals. Though the exact spatial distribution of the XC hole density is not correctly described within the L(S)DA, the corresponding integral quantity, e.g. the total energy, is better reproduced due to possible error compensations and the guarantee of the sum rule (see e.g. [33, 34]). The latter arguments can be to a certain extent illustrated rewriting the integral (7.9) to

$$E_{\mathrm{XC}}[n] = \frac{1}{2}\int d^3\mathbf{x}n(\mathbf{x})\int_0^\infty dRR^2\frac{e^2}{\varepsilon_0 R}\bar{\bar{n}}_{\mathrm{XC}}(\mathbf{x},R;[n]) \tag{7.23}$$

with a spherically averaged XC hole density

$$\bar{\bar{n}}(\mathbf{x},R;[n]) = \frac{1}{4\pi}\int d\Omega(\mathbf{R})\bar{n}_{\mathrm{XC}}(\mathbf{x},\mathbf{x}+\mathbf{R};[n]). \tag{7.24}$$

Expression (7.23) shows that the XC energy depends on the spherical average (7.24) of $\bar{n}_{\mathrm{XC}}(\mathbf{x},\mathbf{x}';[n])$, so that approximations for E_{XC} can give an exact value even if the description of the non-spherical parts of $\bar{n}_{\mathrm{XC}}$ is quite inaccurate [6].

The XC energy consists of three contributions. The first is the potential energy of exchange. The second is the potential energy of correlation. Both potential energies are negative (cf. Fig. 7.4) and reduce the Hartree repulsion. The third is a smaller positive kinetic energy due to the extra motion of the electrons as they avoid one another. In L(S)DA the first term is described by $\epsilon_{\mathrm{X}} \sim n^{1/3}$ as proven for a homogeneous electron gas (4.57). For an inhomogeneous electron gas the potential energy of exchange, the first contribution, is described by the Fock term (4.18), here expressed by Kohn-Sham orbitals $\varphi_{\lambda m_s}(\mathbf{x})$ instead of HF ones. This term includes the self-exchange (4.17) which cancels the corresponding contribution to the Hartree energy. As a consequence the Hartree-Fock theory is self-interaction-free. The L(S)DA treatment with the $n^{1/3}$ approximation for exchange is not at all self-interaction-free. This is why sometimes self-interaction corrections (SIC) are taken into account, in particular, describing localized states [25]. A consequence of the self-interaction is illustrated in Fig. 7.7 for the spin-averaged pair correlation function of a non-spin-polarized electron gas with a given density. The LDA does not correctly describe the pair correlation for vanishing distances $|\mathbf{x}-\mathbf{x}'| \to 0$, where one expects a much stronger reduction of the function g.

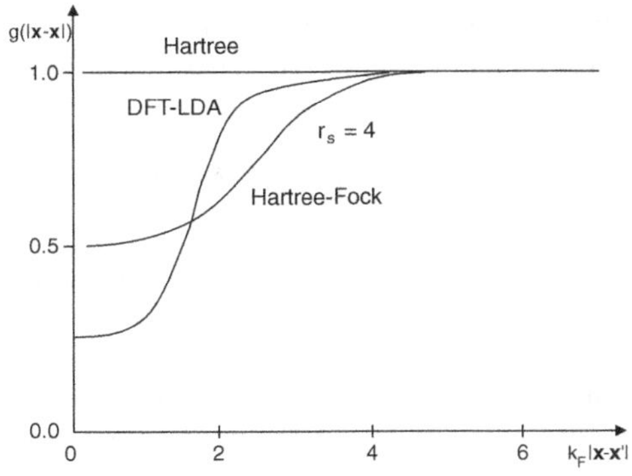

Fig. 7.7 Illustration (*schematic*) of the spin-averaged pair correlation function of a non-spin-polarized electron gas with the parameters k_F and $r_s = 4$ for three different approximations of the exchange-correlation effects

7.3 Gradient Corrections

7.3.1 Density Gradient Expansion

It is tempting to regard the L(S)DA as only the lowest-order term in an expansion of the exchange-correlation energy in powers of the first- and higher-order gradients of the density. The XC energy can be related to an exchange-correlation energy density $g_r[n]$ by

$$E_{\mathrm{XC}}[n] = \int d^3\mathbf{x} g_r[n].$$

This quantity can be systematically expanded according to [35]

$$g_r[n] = g_0(n(\mathbf{x})) + g_1(n(\mathbf{x}))\,[\nabla_{\mathbf{x}} n(\mathbf{x})] + \cdots$$

as suggested in the original paper of Kohn and Sham [9]. The low-order expansion of the exchange energies are known [36]. Unfortunately, the gradient expansion does not lead to systematic improvements over the L(S)DA. This has been demonstrated up to second-order terms [37]. It violates the sum rules and other relevant conditions [35]. The main problem is however that the density gradients can become so large that an expansion up to a certain power breaks down locally.

7.3.2 Generalized Gradient Approximation

Many attempts for improving of the L(S)DA are due to J.P. Perdew and collaborators/coworkers [37–41]. The resulting gradient-approximated XC hole densities are restored to fulfill the sum rule (3.50) and desired properties. These approaches lead to *generalized-gradient approximations* (GGAs). The most popular forms only take first-order gradients $|\nabla_{\mathbf{x}} n_{m_s}(\mathbf{x})|$ into account. In a spin-polarized GGA the XC energy functional has the generalized form [38]

$$E_{XC}\left[n_{\frac{1}{2}}, n_{-\frac{1}{2}}\right] = \int d^3\mathbf{x} n(\mathbf{x}) \epsilon_{XC}\left(n_{\frac{1}{2}}, n_{-\frac{1}{2}}, \left|\nabla n_{\frac{1}{2}}\right|, \left|\nabla n_{-\frac{1}{2}}\right|\right)\Bigg|_{n_{m_s}=n_{m_s}(\mathbf{x})} \tag{7.25}$$
$$= \int d^3\mathbf{x} n(\mathbf{x}) \epsilon_X^{\text{hom}}(n) F_{XC}\left(n_{\frac{1}{2}}, n_{-\frac{1}{2}}, \left|\nabla n_{\frac{1}{2}}\right|, \left|\nabla n_{-\frac{1}{2}}\right|\right)\Bigg|_{n_{m_s}=n_{m_s}(\mathbf{x})}$$

with $\epsilon_X^{\text{hom}}(n)$ (7.15) as the exchange energy per particle of an unpolarized homogeneous electron gas and a dimensionless function F_{XC} of densities and their gradients. Such a functional of type (7.25) is frequently called a semilocal XC functional. For pure exchange instead of (7.25), it is straightforward to show a 'spin-scaling relation', $E_X\left[n_{\frac{1}{2}}, n_{-\frac{1}{2}}\right] = \frac{1}{2}\left\{E_X\left[2n_{\frac{1}{2}}\right] + E_X\left[2n_{-\frac{1}{2}}\right]\right\}$ (see expression (4.65) derived for the homogeneous but spin-polarized electron gas). As a consequence the description of exchange only asks for a factor $F_X(n, |\nabla n|)$ derived for the non-spin-polarized case.

Very popular are the GGAs of Perdew and Wang (PW91) [39, 40] and Perdew, Burke, and Enzerhof (PBE) [41]. For example, the PW91 parametrization is implemented in the VASP code, Version 4.4, while the PBE functional appears in the VASP code beyond the Version 4.5. Here, we give explicit results for the dimensionless factor only for the non-spin-polarized case, $\xi(\mathbf{x}) = 0$. We have to deal only with the local electron density $n(\mathbf{x})$ and a dimensionless parameter

$$s(\mathbf{x}) = \frac{|\nabla_{\mathbf{x}} n(\mathbf{x})|}{2k_F(\mathbf{x})n(\mathbf{x})} = \left(\frac{3}{2\pi}\right)^{\frac{1}{3}} a_B |\nabla_{\mathbf{x}} r_s(\mathbf{x})| \tag{7.26}$$

characterizing the first density gradient normalized to the electron gas parameters k_F and r_s [see (4.47)] for the actual local density. We explicitly describe only the PBE result.

The function $F_{XC} = F_X + F_C$ is again decomposed into an exchange (F_X) and a correlation (F_C) contribution. In the exchange case F_X only depends on the normalized gradient s (7.26) as

$$F_X(n, \nabla n) = 1 + \kappa - \frac{\kappa}{1 + \mu s^2/\kappa} \tag{7.27}$$

with $\kappa = 0.804$ and $\mu = 0.235$. The correlation contribution is

$$F_C(n, \nabla n) = \frac{\epsilon_C^{\text{hom}}(n)}{\epsilon_X^{\text{hom}}(n)}\left[1 + \frac{1}{\epsilon_C^{\text{hom}}(n)} H(t)\right] \tag{7.28}$$

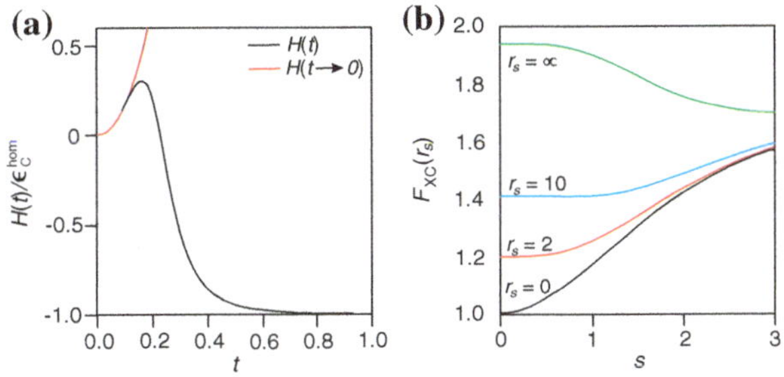

Fig. 7.8 Gradient correction to the correlation energy (7.29) (**a**) and enhancement factor in (7.25) (**b**)

with $t = \sqrt{\frac{4}{\pi} k_F a_B} s$,

$$H(t) = \frac{e^2}{4\pi \varepsilon_0 a_B} \gamma \ln \left\{ 1 + \frac{\beta}{\gamma} t^2 \frac{1 + At^2}{1 + At^2 + A^2 t^4} \right\}, \tag{7.29}$$

and

$$A = \frac{\beta}{\gamma} \left[\exp \left\{ - \frac{\epsilon_C^{\text{hom}}(n)}{\gamma \frac{e^2}{4\pi \varepsilon_0 a_B}} \right\} - 1 \right]^{-1} .$$

The other parameters are $\beta = 0.066726$ and $\gamma = (1 - \ln 2)/\pi^2 = 0.031091$.

This special choice guarantees the correct behavior of the XC energy per particle. For strongly varying densities, i.e., $t \to \infty$, the correlation vanishes as $H(t \to \infty) = -\epsilon_C^{\text{hom}}(n)$. In the limit of high densities the correlation energy nearly follows the result (7.20) of Macke [22]

$$\epsilon_C(n, 0) = \left[2\gamma \ln r_s - 0.093288 \right] Ry.$$

The Lieb-Oxford bound is also fulfilled [41]. The gradient contribution to the correlation and the total enhancement factor are plotted in Fig. 7.8. The gradient corrections $H(t)$ (7.29) to the correlation energy vary significantly with the parameter t. They can be both positive or negative. The total enhancement factor F_{XC} increases with the normalized gradient s (7.26), at least for high densities. For very low densities it may even decrease with rising s.

7.3.3 Influence of Gradient Corrections on Ground-state Properties

First, we focus on cubic solids for simplicity. The relative deviations of their computed lattice constants and cohesive energies from experimental values are displayed in Fig. 7.9. This figure clearly indicates an *overbinding* tendency for the used LDA

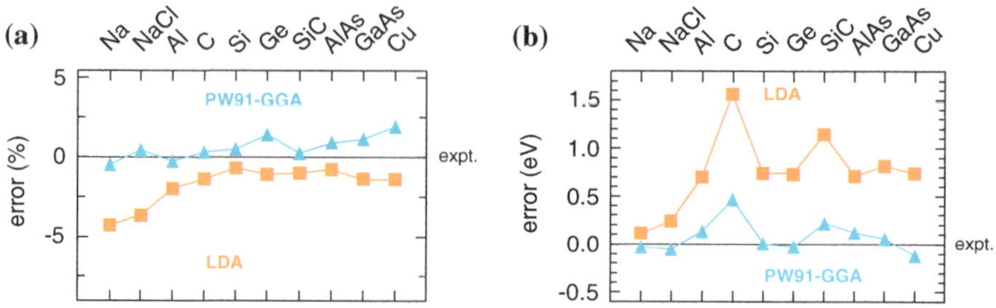

Fig. 7.9 Comparison of cohesive properties computed within LDA and GGA for some cubic metallic and non-metallic crystals. Relative deviations of (**a**) lattice constants and (**b**) cohesive energies from experimental values. Reprinted with permission from [42]. Copyright 1998 by the American Physical Society

irrespective whether covalent, ionic or metallic bonds are modeled. In average, the properties, in particular, the lattice constants, exhibit a minor *underbinding* within GGA (using the PW91 parametrization). Varying the computational details such as the pseudopotentials used and the expansion of the eigenfunctions the same tendencies have been found for further metals and also their bulk moduli [43].

The great strength of the GGA lies in the dramatic improvement it gives over the LDA for properties of molecules such as dissociation energies and bond lengths. These energies may be overestimated within LDA by as much as 100 %, while the GGA gives errors typically of the order of ten percent or less. For a H_2O molecule the PBE-GGA yields a bond length d_{O-H} 1.6 % and a bond angle θ 0.6 % larger than the experimentell values $d_{O-H} = 0.9572$ Å and $\theta = 104.47°$ [44]. Interesting are also water dimers (see Fig. 7.10a). With gradient corrections to the XC functional the energy minimum is displaced toward the measured equilibrium distance of the two oxygen atoms (see Fig. 7.10b). A similar picture is valid for solid water, i.e., ice, as illustrated in Table 7.1, where results are listed for the three XC functionals

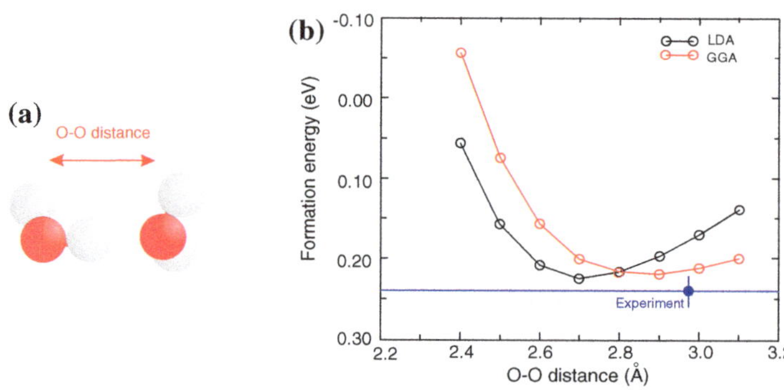

Fig. 7.10 Formation energy of a water dimer (**a**) versus the distance of the two oxygen atoms (**b**) using LDA and GGA. From [45]

Table 7.1 Equilibrium properties of Bernal-Fowler ice computed by means of different (semi)local XC functionals and compared with experimental data [47–49]. From [46]

Functional	Volume (Å^3)	Bulk modulus (GPa)	Sublimation energy (eV)
LDA	26.43	25.3	0.99
PW91	31.35	13.5	0.55
PBE	31.82	12.8	0.53
Exp.	32.05 [47]	10.9 [48]	0.58 [49]

PZ-LDA, PW91-GGA, and PBE-GGA. The strong overbinding effect occurring in LDA disappears in GGA. On the contrary, a weak underbinding is visible.

7.3.4 Improved GGA Functionals

Several attempts have been made to improve the semilocal GGA functionals and to overcome some shortcomings in their application. One obvious idea is to account better for the inhomogeneity of the electron gas. Its realization is based on the study of an 'edge electron gas' by Kohn and Armiento [50]. They considered the distinct variations of the density $n(\mathbf{x})$ in different space regions of an electronic system with edge surface-like regions where the density decays exponentially. To take such effects into account the entire system is decomposed into subsystems with own functionals.

Kohn and Armiento [50] discussed the creation of an XC functional from a surface-oriented model system, e.g. an Airy gas, and its possible combination with another treatment where this model was unsuitable. This approach has been formalized and generalized in the subsystem functional scheme by Armiento and Mattson [51]. Separate functionals from different model systems have been created and merged using a density functional index. The two model systems are bulk-like regions, where the results from the uniform electron gas are used, and surface-like regions, where the functional is derived from an Airy gas combined with a jellium surface.

In the resulting AM05 exchange-correlation functional [51] both regions are combined within the subsystem functional scheme using a density index

$$X = 1/(1 + 2.804\, s^2) \tag{7.30}$$

with the normalized gradient (7.26). The final composed expressions for the AM05 XC functional

$$\epsilon_{\mathrm{XC}}(n, |\nabla n|) = \epsilon_{\mathrm{X}}(n, |\nabla n|) + \epsilon_{\mathrm{C}}(n, |\nabla n|) \tag{7.31}$$

are

$$\epsilon_{\mathrm{X}}(n, |\nabla n|) = \epsilon_{\mathrm{X}}^{\mathrm{hom}}(n)[X + (1 - X)F_{\mathrm{X}}^{\mathrm{LAA}}(s)],$$
$$\epsilon_{\mathrm{C}}(n, |\nabla n|) = \epsilon_{\mathrm{C}}^{\mathrm{hom}}(n)[X + (1 - X)0.8098]. \tag{7.32}$$

The Perdew-Wang parametrization $\epsilon_{\mathrm{C}}^{\mathrm{hom}}(n)$ [39, 40] is used for the correlation part, while the exchange is described by that of the homogeneous electron gas, $\epsilon_{\mathrm{X}}^{\mathrm{hom}}(n)$ (7.15). The factor $F_{\mathrm{X}}^{\mathrm{LAA}}(s)$ has been obtained from a local Airy approximation (LAA) parametrization

$$F_{\mathrm{X}}^{\mathrm{LAA}}(s) = (1 + 0.7168\, s^2)/(1 + 0.7168\, s^2/F_{\mathrm{X}}^{b}). \tag{7.33}$$

The form of $F_{\mathrm{X}}^{\mathrm{LAA}}(s)$ is chosen to impose a correct uniform limit onto F_{X}^{b}, which is constructed on an analytical interpolation between two known limits of the Airy refinement function [51, 52].

The progress using the AM05 XC functional versus the PW91-LDA and PBE-GGA ones is demonstrated in Fig. 7.11 for the structural properties, more precisely the lattice constants, of 20 cubic metallic and non-metallic solids. The mean relative errors $-1.6\,\%$ of LDA and $1.0\,\%$ of GGA surrounds the $0.6\,\%$ obtained for AM05 [52, 53]. Similar improvements are also found for the bulk moduli (which, however, are not plotted in Fig. 7.11). The AM05 XC functional seems to be exceptionally

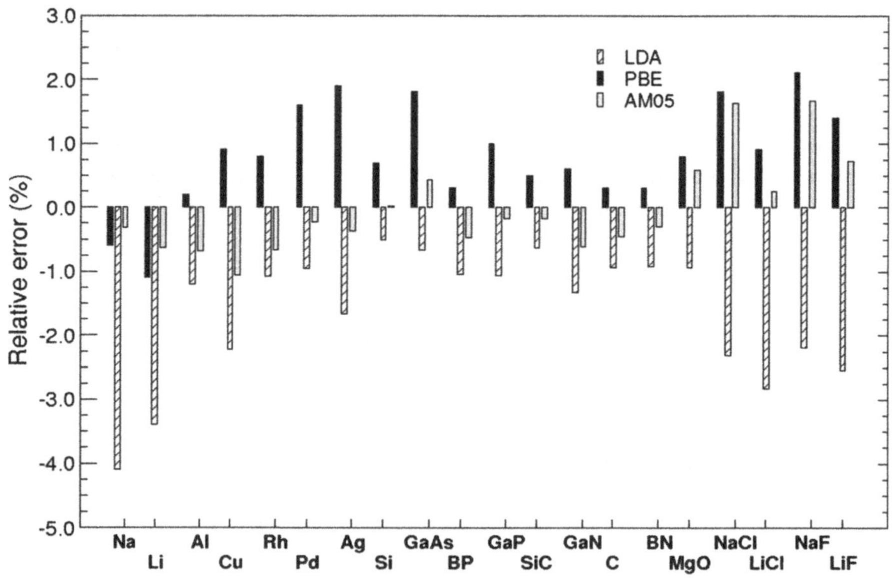

Fig. 7.11 Relative errors in the lattice constants, computed with LDA, PBE-GGA, and AM05 functionals, with respect to experimental values. Adapted from data in [52]. Courtesy of Paier, University of Vienna [53]

Table 7.2 The cubic lattice constant a_0 (in Å) and the hexagonal lattice parameters a, c (in Å) and the cell-internal parameter u for AlN, GaN and InN polytypes, zinc blende and wurtzite

		AM05	LDA	PBE-GGA	Exp.
zb-AlN	a_0	4.374	4.343	4.402	4.37
	B_0	204.7	212.0	193.2	202
	B_0'	4.38	3.22	4.16	
zb-GaN	a_0	4.495	4.465	4.547	4.49
	B_0	182.9	188.8	172.0	190
	B_0'	4.07	4.44	3.36	
zb-InN	a_0	5.005	4.959	5.059	4.98
	B_0	130.8	144.7	120.2	136
	B_0'	4.07	4.95	4.10	
wz-AlN	a	3.112	3.088	3.129	3.11
	c	4.976	4.946	5.018	4.978
	u	0.380	0.379	0.379	0.382
	B_0	202.3	210.8	187.2	185
	B_0'	4.36	3.95	4.02	5.7
wz-GaN	a	3.181	3.158	3.217	3.19
	c	5.180	5.145	5.241	5.166–5.185
	u	0.376	0.376	0.376	0.377
	B_0	183.2	197.4	172.2	188
	B_0'	4.17	4.23	4.63	4.3
wz-InN	a	3.549	3.517	3.587	3.54
	c	5.736	5.685	5.789	5.718
	u	0.378	0.377	0.378	0.375
	B_0	131.3	145.3	120.9	125.5
	B_0'	4.76	4.52	5.37	12.7

In addition, the bulk moduli B_0 (in GPa) and their pressure derivatives B_0' are given. Results are derived from calculations using the LDA, PBE-GGA and AM05 XC functionals. For comparison, experimental values are also listed. From [54]

successful for ionic compounds such as the group-III nitrides [54]. This is clearly demonstrated in Table 7.2 for structural and elastic properties of AlN, GaN, and InN crystallizing in zinc-blende (*zb*) or wurtzite (*wz*) structure. The AM05 values for the lattice parameters a_0 (*zb*), a, c, and u (*wz*) are in between the corresponding values computed by means of the PZ-LDA and PBE-GGA. In the average, there seems to be neither underbinding nor overbinding. Also the agreement with measured values is much better in the AM05 case.

Probably, the XC functional most commonly used in solid-state calculations is the PBE-GGA one [41]. In Sect. 7.3.2 we have described how it employs both the density $n(\mathbf{x})$ and its gradient $\nabla_{\mathbf{x}} n(\mathbf{x})$ at each point $\mathbf{x}$ in space. Such GGA functionals (including the AM05 one [51]) represent a well-tempered balance between computational efficiency, numerical accuracy, and reliability. PBE-GGA also juggles the demands of quantum chemistry and solid-state physics [55]. These demands are somewhat at variance for the construction of a better GGA. Those with an enhanced gradient dependence improve atomization and total energies, but worsen bond lengths. With a reduced gradient dependence the lattice parameters and/or surface energies of solids may be improved. This dilemma cannot be solved; no GGA can do both types of ground-state computations with the same high accuracy. For instance, accurate atomic exchange energies require violating the gradient expansion for shortly-varying densities, which is valid for solids and their surfaces [56].

At the GGA level, one must choose an optimum. A pragmatic approach is needed for the construction of a new XC functional, thereby selecting the range of applications. Recently, Perdew et al. [56] presented a modification of the PBE functional, PBEsol, intended only for solids and surface systems. It is based on a gradient expansion of the exchange energy and a final fit of the XC energy to that of the surface of jellium. Consequently, it improves equilibrium properties of densely packed solids and their surfaces.

The authors showed for 40 solids that the PBEsol lattice parameters are systematically lower than the PBE ones by 1–2 %. The average value of the PBE error on this example of 40 lattice parameters is 0.95 %, and for PBEsol it is −0.31%. We confirm the improvements for wide-gap systems such as BaF_2 and CdF_2. However, this will not be demonstrated here. Rather we refer the reader to the literature [57]. It has been shown for about 30 metals, semiconductors, and insulators that the two GGA XC functionals AM05 and PBEsol lead to equilibrium lattice constants and bulk moduli in excellent agreement [58, 59]. Further improvements of the GGA level for XC are possible. A meta-GGA adds the positive orbital kinetic energy density [60–62] or, almost equivalently [63], the Laplacians of the spin densities. Such a meta-GGA, e.g. [64], may also give better excitation properties as the fundamental gap of semiconductors. Also other modifications of GGA XC functionals such as PBErev [65] are now frequently used in electronic-structure codes.

References

1. J. Harris, Adiabatic-connection approach to Kohn-Sham theory. Phys. Rev. A **29**, 1648–1659 (1984)
2. J. Harris, R.O. Jones, The surface energy of a bounded electron gas. J. Phys. F. Metal Phys. **4**, 1170–1186 (1974)
3. H. Hellmann, *Einführung in die Quantenchemie* (Deuticke, Leipzig, 1937)
4. R.P. Feynman, Forces in molecules. Phys. Rev. **56**, 340–343 (1939)
5. D.C. Langreth, J.P. Perdew, The exchange-correlation energy of a metallic surface. Solid State Commun. **17**, 1425–1429 (1975)
6. O. Gunnarsson, B.I. Lundqvist, Exchange and correlation in atoms, molecules, and solids by the spin-density-functional formalism. Phys. Rev. B **13**, 4274–4298 (1976)
7. U. von Barth, L. Hedin, A local exchange-correlation potential for the spin-polarized case: I. J. Phys. C **5**, 1629–1642 (1972)
8. K.S. Singwi, A. Sjölander, M.P. Tosi, R.H. Land, Electron correlations at metallic densities. IV. Phys. Rev. B **1**, 1044–1053 (1970)
9. W. Kohn, L.J. Sham, Self-consistent equations including exchange and correlation effects. Phys. Rev. **140**, A1133–A1138 (1965)

10. K. Karch, Ab-initio Berechnung von statischen und dynamischen Eigenschaften des Diamanten, Siliziums und Siliziumcarbids. Ph.D. thesis, University of Regensburg (1993)
11. K. Karch, P. Pavone, W. Windl, D. Strauch, F. Bechstedt, Ab initio calculation of structural, lattice dynamical, and thermal properties of cubic silicon carbide. Int. J. Quantum Chem. **56**, 801–817 (1995)
12. A. Garcia, M.L. Cohen, First-principles ionicity scales. I. Charge asymmetry in the solid state. Phys. Rev. B **47**, 4215–4220 (1993)
13. U. Grossner, J. Furthmüller, F. Bechstedt, Bond-rotation versus bond-contraction relaxation of (110) surfaces of group-III nitrides. Phys. Rev. B **58**, R1722–R1725 (1998)
14. D.M. Ceperley, B.J. Alder, Ground state of the electron gas by a stochastic method. Phys. Rev. Lett. **45**, 566–569 (1980)
15. G. Ortiz, P. Ballone, Correlation energy, structure factor, radial distribution function, and momentum distribution of the spin-polarized uniform electron gas. Phys. Rev. B **50**, 1391–1405 (1994)
16. L. Hedin, B.I. Lundqvist, S. Lundqvist, Local exchange-correlation potentials. Solid State Commun. **9**, 537–541 (1971)
17. L. Hedin, B.I. Lundqvist, Explicit local exchange-correlation potentials. J. Phys. C **4**, 2064–2084 (1971)
18. E.P. Wigner, On the interaction of electrons in metals. Phys. Rev. **46**, 1002–1011 (1934)
19. E.P. Wigner, Effects of the electron interaction on the energy levels of electrons in metals. Trans. Faraday Soc. **34**, 678–685 (1938)
20. R.M. Dreizler, E.K.U. Gross, *Density Functional Theory* (Springer, Berlin, 1990)
21. M. Gell-Mann, K.A. Brueckner, Correlation energy of an electron gas at high density. Phys. Rev. **106**, 364–368 (1957)
22. W. Macke, Über die Wechselwirkungen im Fermi-Gas: Polarisationserscheinungen, Korrelationsenergie, Elektronenkondensation. Z. Naturforschung **5a**, 192–208 (1950)
23. D.F. du Bois, Electron interactions: part I. Field theory of a degenerate electron gas. Ann. Phys. **7**, 174–237 (1959)
24. W.J. Carr Jr, A.A. Maradudin, Ground-state energy of a high-density electron gas. Phys. Rev. **133**, A371–A374 (1964)
25. J.P. Perdew, A. Zunger, Self-interaction correction to density-functional approximations for many-electron systems. Phys. Rev. B **23**, 5048–5079 (1981)
26. D.M. Ceperley, Ground state of the fermion one-component plasma: a Monte Carlo study in two and three dimensions. Phys. Rev. B **18**, 3126–3138 (1978)
27. I.N. Levine, *Quantum Chemistry* (Prentice Hall, Upper Saddle River, 2008)
28. K. Burke, Perspective on density functional theory. J. Chem. Phys. **136**, 150901 (2012)
29. G.H. Booth, A. Grüneis, G. Kresse, A. Alavi, Towards an exact description of electronic wave-functions in real solids. Nature **493**, 365–370 (2013)
30. N.D. Lang, W. Kohn, Theory of metal surfaces: work function. Phys. Rev. B **3**, 1215–1223 (1971)
31. C.-O. Almbladh, A.C. Pedroza, Density-functional exchange-correlation potentials and orbital eigenvalues for light atoms. Phys. Rev. A **29**, 2322–2330 (1984)
32. S. Kurth, J.P. Perdew, Role of the exchange-correlation energy: nature's glue. Int. J. Quantum Chem. **77**, 814–818 (2000)
33. O. Gunnarsson, M. Jonson, B.I. Lundqvist, Descriptions of exchange and correlation effects in inhomogeneous electron systems. Phys. Rev. B **20**, 3136–3164 (1979)
34. U. von Barth, A.R. Williams, Applications of density functional theory to atoms, molecules, and solids, in *Theory of the Inhomogeneous Electron Gas*, Chap. 4, Sect. 2.1.3, ed. by S. Lundqvist, N.H. March (Plenum Press, New York 1983), pp. 189–308
35. F. Herman, J.P. Van Dyke, I.P. Ortenburger, Improved statistical exchange approximation for inhomogeneous many-electron systems. Phys. Rev. Lett. **22**, 807–811 (1969)
36. P.S. Svendsen, U. von Barth, Gradient expansion of the exchange energy from second-order density response theory. Phys. Rev. B **54**, 17402–17413 (1996)

37. J.P. Perdew, Generalized gradient approximations for exchange and correlation: a look backward and forward. Physica B. Condens. Matter **172**, 1–6 (1991)
38. J.P. Perdew, K. Burke, Comparison shopping for a gradient-corrected density functional. Int. J. Quant. Chem. **57**, 309–319 (1996)
39. J.P. Perdew, Unified theory of exchange and correlation beyond the local density approximation, in *Electronic Structure of Solids '91*, ed. by P. Ziesche, H. Eschrig (Akademie-Verlag Berlin, 1991), pp. 11–20
40. J.P. Perdew, Y. Wang, Accurate and simple analytic representation of the electron gas correlation energy. Phys. Rev. B **45**, 13244–13249 (1992)
41. J.P. Perdew, K. Burke, M. Ernzerhof, Generalized gradient approximation made simple. Phys. Rev. Lett. **77**, 3865–3868 (1996)
42. M. Fuchs, M. Bockstedte, E. Pehlke, M. Scheffler, Pseudopotential study of binding properties of solids within generalized gradient approximations: the role of core-valence exchange-correlation. Phys. Rev. B **57**, 2134–2145 (1998)
43. A. Khein, D.J. Singh, C.J. Umrigar, All-electron study of gradient corrections to the local-density functional in metallic systems. Phys. Rev. B **51**, 4105–4109 (1995)
44. J.B. Hasted, Liquid water: dielectric properties, in *Water: A Comprehesive Treatise*, vol. 1, ed. by F. Franks (Plenum Press, New York, 1972), pp. 255–309
45. P.H. Hahn, unpublished
46. D.R. Hamann, H_2O hydrogen bonding in density-functional theory. Phys. Rev. B **55**, R10157–R10160 (1997)
47. R. Brill, A. Tippe, Gitterparameter von Eis I bei tiefen temperaturen. Acta Crystallogr. **23**, 343–345 (1967)
48. P.V. Hobbs, *Ice Physics* (Clarendon Press, Oxford, 1974)
49. E. Whalley, The difference in the intermolecular forces of H_2O and D_2O. Trans. Faraday Soc. **53**, 1578–1585 (1957)
50. W. Kohn, R. Armiento, Edge electron gas. Phys. Rev. Lett. **81**, 3487–3490 (1998)
51. R. Armiento, A.E. Mattsson, Functional designed to include surface effects in self-consistent density functional theory. Phys. Rev. B **72**, 085108 (2005)
52. A.E. Mattsson, R. Armiento, J. Paier, G. Kresse, J.M. Wills, T.R. Mattsson, The AM05 density functional applied to solids. J. Chem. Phys. **128**, 084714 (2008)
53. J. Paier, private information
54. L.C. de Carvalho, A. Schleife, F. Bechstedt, Influence of exchange and correlation on structural and electronic properties of AlN, GaN, and InN polytypes. Phys. Rev. B **84**, 195105 (2011)
55. S. Kurth, J.P. Perdew, P. Blaha, Molecular and solid-state tests of density functional approximations: LSD, GGAs, and meta-GGAs. Int. J. Quantum Chem. **75**, 889–909 (1999)
56. J.P. Perdew, A. Ruzsinszky, G.I. Csonka, O.A. Vydrov, G.E. Scuseria, L.A. Constantin, X. Zhou, K. Burke, Restoring the density-gradient expansion for exchange in solids and surfaces. Phys. Rev. Lett. **100**, 136406 (2008)
57. G. Cappellini, J. Furthmüller, E. Cadelano, F. Bechstedt, Electronic and optical properties of cadmium fluoride: the role of many-body effects. Phys. Rev. B **87**, 075203 (2013)
58. A.E. Mattsson, R. Armiento, T.R. Mattsson, Comment on "Restoring the density-gradient expansion for exchange in solids and surfaces". Phys. Rev. Lett. **101**, 239701 (2008)
59. T.R. Mattsson, unpublished
60. J. Tao, J.P. Perdew, V.N. Staroverov, G.E. Scuseria, Climbing the density functional ladder: nonempirical meta-generalized gradient approximation designed for molecules and solids. Phys. Rev. Lett. **91**, 146401 (2003)
61. V.N. Staroverov, G.E. Scusceria, J. Tao, J.P. Perdew, Comparative assessment of a new nonempirical density functional: Molecules and hydrogen-bonded complexes. J. Chem. Phys. **119**, 12129–12137 (2003)
62. F. Furche, J.P. Perdew, The performance of semilocal and hybrid density functionals in 3*d* transition-metal chemistry. J. Chem Phys. **124**, 044103 (2006)
63. J.P. Perdew, L.A. Constantin, Laplacian-level density functionals for the kinetic energy density and exchange-correlation energy. Phys. Rev. B **75**, 155109 (2007)

64. F. Tran, P. Blaha, Accurate band gaps of semiconductors and insulators with a semilocal exchange-correlation potential. Phys. Rev. Lett. **102**, 226401 (2009)
65. Y. Zhang, W. Yang, Comment on "Generalized gradient approximation made simple". Phys. Rev. Lett. **80**, 890–890 (1998)

Chapter 8
Energies and Forces

Abstract The knowledge of the total energy of the electron gas in its ground state as a function of the arrangement of nuclei, the electron density and, hence, the sample volume makes a generalization to an '*ab initio*' thermodynamics possible, which does however not account for the lattice vibrations. Thermodynamic properties can be characterized by the grand canonical potential as a function of the chemical potentials of atoms or molecules forming the matter. Globally equations of states yield average elastic properties and pressure-induced phase transitions. The variations of the total energy with the atomic coordinates give the Hellmann-Feynman forces, whose minimization yields equilibrium coordinates or, at least, atomic positions of a metastable geometry. Stress and strain can be formulated. In addition to an explicit exchange-correlation functional the total energy and electronic structure calculations ask for an expansion of the Kohn-Sham eigenstates in terms of basis functions, e.g. plane waves. Their use is especially convenient for less localized electronic states and suggests to focus the studies on the valence electrons which contribute to the chemical bonding. The division in core and valence electrons asks for the concept of atomic pseudopotentials.

8.1 '*Ab Initio*' Thermodynamics

8.1.1 Thermodynamic Relations

The equilibrium state of a one-component system consisting of N particles, atoms or molecules, at fixed temperature T and pressure p (but without external fields) is the one with the minimum Gibbs free enthalpy $G(T, p, N)$ [1],

$$G = F + p\Omega, \tag{8.1}$$

where $F(T, \Omega, N)$,

$$F = U - ST, \tag{8.2}$$

F. Bechstedt, *Many-Body Approach to Electronic Excitations*,
Springer Series in Solid-State Sciences 181, DOI 10.1007/978-3-662-44593-8_8

is the Helmholtz free energy with volume Ω. It is related to the internal energy $U = U(S, \Omega, N)$ and the entropy S by a Legendre transformation. The energy conservation law and the relationship between heat and work can be written in the form

$$dU = T dS - p\, d\Omega + \mu\, dN \tag{8.3}$$

for an infinitesimal change of the internal energy. A variation of the number of particles N is allowed due to particle exchange with a reservoir characterized by the chemical potential μ of the particles. For an isolated system with no heat exchange ($dS = 0$) and particle exchange ($dN = 0$) the internal energy is a constant at constant volume ($d\Omega = 0$). The corresponding microscopic distribution is the microcanonical ensemble of the statistical mechanics.

The thermodynamic potential G (or F) can be used to derive the thermodynamic quantities of the considered system at constant temperature T, particle number N and pressure p (or volume Ω). Infinitesimal changes of the three variables give rise to infinitesimal changes of the potential, so that

$$dG = -S dT + \Omega dp + \mu\, dN \tag{8.4}$$

or

$$dF = -S dT - p\, d\Omega + \mu\, dN. \tag{8.5}$$

In equilibrium, F is a minimum with respect to the inner variables at constant T, V, and N, whereas G is a minimum at constant T, p, and N. The corresponding microscopic distribution is the canonical ensemble. The chemical potential μ in (8.3), (8.4) or (8.5) is given by

$$\mu = \left(\frac{\partial U}{\partial N}\right)_{S,\Omega} = \left(\frac{\partial F}{\partial N}\right)_{T,\Omega} = \left(\frac{\partial G}{\partial N}\right)_{T,p}. \tag{8.6}$$

Under normal pressure of about 1 atmosphere, the difference between the Helmholtz free energy F and the Gibbs free energy G,

$$\Omega_g = F - G = -p\Omega, \tag{8.7}$$

is insignificant for a bulk solid or liquid. This holds in particular for volume-induced changes $-p\, d\Omega$. Thus, it is sufficient to use F for the description of the most thermodynamic phenomena in solid state physics. The difference (8.7) is the Kramers grand thermodynamic potential $\Omega_g = \Omega_g(T, \Omega, \mu)$ [2]. Despite its smallness, more precisely its vanishing influence on changes in bulk systems including phase transitions, the potential is conveniently usable for system transformations that occur at a constant temperature T, volume V, and chemical potential μ. Together with the Gibbs–Duhem equation, $S dT - \Omega dp + N d\mu = 0$, infinitesimal changes of the variables result in

$$d\Omega_g = -SdT - p\,d\Omega - Nd\mu. \tag{8.8}$$

In this case the thermodynamic properties of a system are governed by the grand canonical statistical operator. The comparison of (8.5) and (8.8) indicates a transformation law

$$\Omega_g = F - \mu N. \tag{8.9}$$

This is a consequence of the fact that the Gibbs free enthalpy varies linearly with the number of particles [cf. (8.7) and (8.9)],

$$G = \mu N \tag{8.10}$$

with the proportionality factor $\mu = \mu(T, p)$ for each homogeneous phase [1].

Solids, which are mainly in our focus, are, in general, multicomponent systems with atomic (or molecular) species $i = A, B, C, \ldots$ with the number of N_i particles of such a species and the corresponding chemical potential μ_i. Therefore, they will be characterized by the free energy $F = F(T, \Omega, N_A, N_B, \ldots)$. In cases where variations of the particle numbers $N_A, N_B, \ldots$ are taken into account, for example, to describe the preparation dependence of solid phases, several reservoirs which define the chemical potentials $\mu_A, \mu_B, \ldots$ are considered. Instead of the free energy the Kramers grand potential

$$\Omega_g(T, \Omega, \mu_A, \mu_B, \ldots) = F(T, \Omega, N_A, N_B, \ldots) - \sum_{i=A,B,\ldots} \mu_i N_i \tag{8.11}$$

determines the ground state of the system.

The free energy (8.2) also gives the equation of state (EOS) of a solid

$$p = -\left(\frac{\partial F}{\partial \Omega}\right)_{T,N_A,N_B,\ldots} \tag{8.12}$$

with the equilibrium condition $p = 0$. It is strongly related to the isothermal bulk modulus

$$\begin{aligned} B &= -\Omega\left(\frac{\partial p}{\partial \Omega}\right)_{T,N_A,N_B,\ldots} \\ &= \Omega\left(\frac{\partial^2 F}{\partial \Omega^2}\right)_{T,N_A,N_B,\ldots}. \end{aligned} \tag{8.13}$$

8.1.2 Equation of State

In order to describe the temperature dependence of structural and elastic properties and even more the thermal properties, the inclusion of the free energy of the lattice vibrations is indispensable. This holds in particular for the entropy contribution $-ST$ in (8.2). The vibrational entropy per atom is of the order of one Boltzmann constant k_B or somewhat larger around or below room temperature [3]. Its changes with the atomic geometry are only fractions of this value. Therefore, in contrast to the pressure influence near the equilibrium the entropy term cannot be neglected, not at room temperature and not at all for higher temperatures. Only for low temperatures, e.g., below $T = 10$ K, it is possible. However, studying only changes of the free energy, e.g. with the atomic geometry, the entropy term can be neglected, and F can be replaced by the internal energy $U(T, \Omega, N_A, N_B, \ldots)$ which is different from the internal energy (8.3) and hence not a thermodynamic potential.

The temperature dependence of U may be negligible but U is still influenced by the contribution of the vibrating lattice. In particular, the zero-point vibrations have to be taken into account. In the Debye approximation this energy contribution per unit cell is given by $\frac{9}{8}k_B\Theta_D$ with Θ_D as the Debye temperature of the solid. The Debye temperature varies with the material, but its value is usually, also for metals, larger than 100 K [4] and, hence, cannot be neglected to determine absolute values of the internal energy. However, for energy differences such as the grand potential (8.9), cohesive energies or formation energies one expects a compensation of the effects of zero-point vibrations on the different energy contributions. Also, in studying the equilibrium positions the contribution of the vibrating lattice may be neglected in a first approximation. Consequently, in explicit parameterfree or '*ab initio*' calculations the internal energy $U(T, \Omega, N_A, N_B, \ldots)$ is frequently replaced by its low-temperature value $U \equiv U(0, \Omega, N_A, N_B, \ldots)$ without taking the vibrating lattice into account.

If the volume dependence of this reduced internal energy U is known as displayed in Fig. 8.1 it can be also fitted to an EOS. The EOS of Murnaghan [8] assumes a linear relation of the isothermal bulk modulus (8.13) to the pressure for zero temperature

$$B_0(p) = B_0 + B_0' p \tag{8.14}$$

with the equilibrium value B_0 and its pressure derivative B_0'. It corresponds to a pressure

$$p = \frac{B_0}{B_0'}\left[\left(\frac{\Omega_0}{\Omega}\right)^{B_0'} - 1\right] \tag{8.15}$$

and to an internal energy

$$U = \frac{\Omega_0 B_0}{B_0'}\left[\frac{1}{B_0' - 1}\left(\frac{\Omega_0}{\Omega}\right)^{B_0'-1} + \frac{\Omega}{\Omega_0}\right] + \text{const.} \tag{8.16}$$

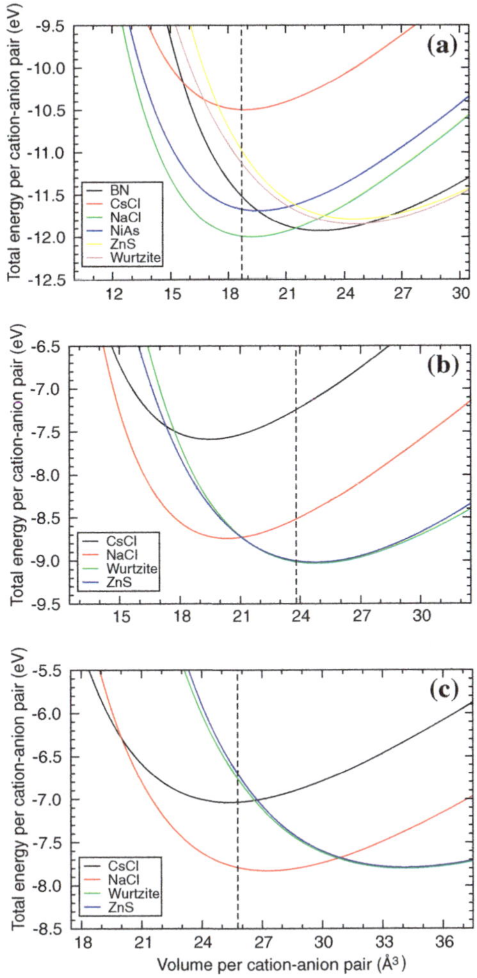

Fig. 8.1 Total energy versus volume for MgO (**a**), ZnO (**b**), and CdO (**c**) in different crystal structures CsCl, NiAs, hexagonal BN, rocksalt (NaCl), wurtzite, and zinc blende (ZnS). The XC functional PW91-GGA has been used within the VASP implementation [5, 6]. Adapted from [7]

with the equilibrium volume Ω_0. For higher pressures the EOS of Birch [9] or the more universal one of Vinet [10, 11] has to be applied. The Vinet equation of state is

$$p = 3B_0 e^{\gamma(1-x)} \left(\frac{1}{x^2} - \frac{1}{x} \right) \tag{8.17}$$

and

$$U = -9\Omega_0 B_0 e^{\gamma(1-x)} \left(\frac{1}{\gamma^2} - \frac{1}{\gamma} + \frac{x}{\gamma} \right) + \text{const.}, \tag{8.18}$$

respectively, with $x = (\Omega/\Omega_0)^{\frac{1}{3}}$ and $\gamma = 3(B_0' - 1)/2$.

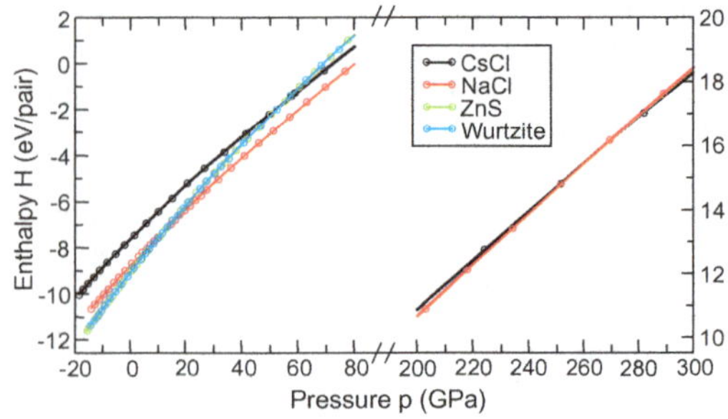

Fig. 8.2 Enthalpy H as a function of pressure for four possible crystal structures of ZnO. Adapted from [7]

Such $U = U(\Omega)$ curves for different crystal structures in Fig. 8.1 do not only give the equilibrium volumes Ω_0 and the parameters B_0 and B_0' characterizing globally the elastic properties of the studied systems but also the internal energies $U(\Omega_0)$ in equilibrium. The $U(\Omega)$ curves for different crystal structures contain further information. By means of the method of common tangent transition pressures p_t and volumes Ω_t can be computed for pressure-induced phase transitions between two crystal structures. More precisely, the Gibbs free enthalpy (8.1) has to be investigated. Within the above described approximations the studies can however be restricted to the enthalpy $H = U(\Omega) + p\Omega$ with an external pressure p. As an example, the enthalpies are displayed for several crystal structures of ZnO in Fig. 8.2 [7].

By means of relations (8.15) or (8.17) the volume dependence can be translated into a pressure dependence $H(p)$. The crossing of $H(p)$ curves for two crystal structures then defines the transition pressure p_t. Under equilibrium conditions the lowest energy configuration of ZnO in Fig. 8.1b is the wurtzite structure. A first phase transition wurtzite → rocksalt appears at $p_t = 11.8$ GPa [7] in Fig. 8.2. A second one between rocksalt and CsCl structure is seen for $p_t = 261$ GPa. The computed first transition pressure agrees nearly with an experimental value of $p_t \approx 10$ GPa [12], while the second phase transition is not confirmed experimentally.

8.1.3 Energy Differences

The equilibrium internal energy $U(\Omega_0, N_A, N_B, \ldots)$ of a system consisting of N_A A atoms, N_B B atoms, ... can be combined with the corresponding energies U_A, U_B, ... of free atoms. One obtains the cohesive energy (or binding energy) per particle as

$$E_{\mathrm{coh}} = -\left\{ U(\Omega_0, N_A, N_B, \ldots) - \sum_{i=A,B,\ldots} N_i U_i \right\} \Big/ \sum_{i=A,B,\ldots} N_i. \tag{8.19}$$

The cohesive energy is required to decompose the solid into neutral atoms at $T = 0$ K at atmospheric pressure. In the computation of the internal energy of many

Table 8.1 Cohesive energies per atom (elements) or cation-anion pair (compounds) (8.19) computed using several implementations of the density functional theory, PW91-GGA [7] and PZ-LDA [13, 14]

Element/ compound	Crystal structure	Cohesive energy (eV)	
		Calc.	Exp.
B	Rhombohedral	6.11 [13]	5.81 [15]
Al	Fcc	4.19 [13]	3.39 [15]
Ga	Orthorhombic	3.58 [13]	2.81 [15]
In	Tetragonal	2.89 [13]	2.52 [15]
C	Diamond	10.12 [14]	7.37 [15]
Si	Diamond	5.94 [14]	4.63 [15]
Ge	Diamond	5.20 [14]	3.85 [15]
MgO	Rocksalt	10.17 [7]	10.26 [16]
ZnO	Wurtzite	7.20 [7]	7.52 [16]
CdO	Rocksalt	6.00 [7]	6.40 [16]

They are compared with experimental values. For the oxides the experimental binding energies are taken from [16] as heat of vaporization or heat of atomization

non-magnetic compounds and elemental solids the spin polarization does hardly play a role. However, the spin polarization significantly reduces the total energy U_i for many atoms of species i. One example is oxygen with a correction of -1.83 eV due to the spin polarization. For group-II atoms, however, this effect is with -0.001 (Mg), -0.003 (Zn), and -0.005 (Cd) eV negligibly small [7]. For illustration, some resulting values are listed in Table 8.1 and compared with measured values. The agreement is reasonable. However, the calculated values for the elemental crystals tend to an overbinding in agreement with the used PZ-LDA XC functional while the oxides have been described by means of the PW91-GGA functional resulting in an underbinding tendency (cf. discussion in Sect. 8.3).

A chemical potential $\mu_i = \mu_i(p, T)$ (8.6) is defined to be the derivative of the Gibbs free enthalpy G for a given phase with respect to the number of particles of type i, $\mu_i = (\partial G/\partial N_i)_{p,T,\{N_j\}}$ and fixed numbers $\{N_j\}$ of the other particles apart from N_i. Since in equilibrium the chemical potential μ_i of a given species is the same in all phases which are in contact, each μ_i can be considered as the free enthalpy per particle in each reservoir for particles of type i [17].

Within the approximation discussed to derive the cohesive energy (8.19) for elemental solids and compounds (see Table 8.1), their negative values can be used to describe approximately the chemical potentials $\mu_A^{\text{solid}}, \mu_B^{\text{solid}}, \ldots, \mu_{AB}^{\text{solid}}, \ldots$ of the reservoirs. Thereby, the chemical potentials of the solid phase of a compound and those of the individual elemental solids define the heat of formation ΔH_f of this compound. For an $A_N B_M$ compound it holds

$$\mu_{A_N B_M}^{\text{solid}} = N\mu_A^{\text{solid}} + M\mu_B^{\text{solid}} - \Delta H_f^{A_N B_M}. \tag{8.20}$$

Table 8.2 Thermochemical data of a few III-V and IV-IV compounds. Chemical potentials $-\mu_{AB}^{\text{solid}}$ (8.19) and heats of formation ΔH_f^{AB} in eV/pair (8.20) of AB compounds crystallizing in zinc-blende or wurtzite structure are listed

Compound	μ_{AB}^{solid}		ΔH_f^{AB}	
	Calc.	Exp.	Calc.	Exp.
AlN	15.98 [13]	11.52 [18]	3.28 [13]	3.30 [19]
GaN	13.61 [13]	8.96 [18]	1.28 [13]	1.11 [19]
InN	12.01 [13]	7.72 [18]	0.38 [13]	–
AlAs	9.14 [20]	7.56 [18]	1.01 [20]	1.20 [16]
GaAs	8.26 [20]	6.52 [18]	0.66 [20]	0.74 [16]
SiC	16.65 [21]	12.68 [18]	0.58 [21]	0.72 [19]

The values calculated in the framework of a density functional theory (calc.) are compared with experimental values (exp.)

This means that computing the heat of formation of an $A_N B_M$ compound also the cohesive energies should be known for the solid elements (see Table 8.1) in addition to the corresponding value for the compound. Relations are illustrated in Table 8.2. The comparison between computed and measured values for total energies of solid compounds again indicates the overbinding effect using an LDA XC functional as discussed for Table 8.1. However, the resulting heats of formation agree much better with measured values indicating a significant error compensation.

The real chemical potentials depend on the actual growth and/or preparation conditions, e.g. the kind of epitaxy used to deposit a certain atomic species or more complex materials. They allow to describe varying stages of the reservoirs. Close to equilibrium with known chemical potentials of the atoms or molecules $A, B, \ldots$ forming a compound $A_N B_M \ldots$, their formation energy can be formulated starting from the grand thermodynamic potential (8.11). Together with the approximations discussed above we write

$$\Omega_g(\Omega_0, \mu_A, \mu_B, \ldots) = U(\Omega_0, N_A, N_B, \ldots) - \sum_{i=A,B,\ldots} N_i \mu_i. \tag{8.21}$$

The formation energy is then given as the value of Ω_g with opposite sign.

In order to illustrate the use of Kramers grand canonical potential to investigate the equilibrium situation by means of '*ab initio*' calculated internal energies U (without lattice vibrations) and the chemical potentials of the reservoirs, we study the adsorption of an atomic species C on the surface of a substrate consisting of an AB compound. The AB system occupies a halfspace or a material slab (as in many surface modeling studies). The chemical potential of one of the species A or B on the surface is assumed to be variable due to preparation, e.g. such in a vacuum chamber combined with effusion cells for A and B atoms. If the surface is in equilibrium with the bulk substrate, pairs of A and B atoms can be exchanged with the substrate, for which the chemical potential is μ_{AB}^{solid} as defined in (8.20). The equilibrium condition

(Gibbs phase rule) then reads as

$$\mu_A + \mu_B = \mu_{AB}^{\text{solid}}. \tag{8.22}$$

In principle, relation (8.22) may be also written in form of a *mass action law*, if the chemical potentials are rewritten as functions of the partial pressures p_A and p_B and the temperature T. The latter dependence is here however not considered but may be important, especially for gas sources of the species A or B.

The variation of the surface preparation conditions may be represented by deviations of the actual chemical potentials from their solid values,

$$\Delta\mu_i = \mu_i - \mu_i^{\text{solid}}, \qquad (i = A, B) \tag{8.23}$$

assuming that in extreme cases $\mu_i = \mu_i^{\text{solid}}$ droplets, clusters, layers, ... of the material i can be formed on the surface. Together with (8.20) it holds for the deviations

$$\Delta\mu_A + \Delta\mu_B = -\Delta H_f^{AB}. \tag{8.24}$$

The A-rich (B-rich) preparation conditions are characterized by $\Delta\mu_A = 0$ ($\Delta\mu_B = 0$). Consequently, the deviations can only vary in the interval

$$-\Delta H_f^{AB} \leq \Delta\mu_i \leq 0 \qquad (i = A, B). \tag{8.25}$$

In a last step many surfaces can be calculated for different stoichiometries, atomic geometries, and two-dimensional translational symmetries. Because of the linear dependence of μ_A and μ_B (8.22) all the resulting grand potentials Ω_g can be plotted versus $\Delta\mu_A$ or $\Delta\mu_B$. For given preparation conditions the lowest-energy surface should be the most stable one. As an example results for the GaAs(001) surface are displayed in Fig. 8.3 [22]. The figure indicates that the stability of different surface phases (labeled by their translational symmetry) varies with the preparation conditions described by variations of the chemical potential of the cations. Electronic structures or spectra are usually only investigated for the stable phases.

Instead by means of a molecular beam epitaxy (MBE) such a surface may be prepared within a metal organic vapor phase epitaxy (MOVPE) process. Then also hydrogen H plays a role. Its chemical potential μ_{H} may be approximated by the chemical potential of a two-atomic ideal gas of H_2 molecules which is temperature- and pressure-dependent. Roughly $\Delta\mu_{\text{H}} = 0$ corresponds to the situation where the surface is exposed to molecular hydrogen at $T = 0$ K. For typical MOVPE growth conditions $\Delta\mu_{\text{H}}$ is estimated to be about -1 eV. In the presence of hydrogen, much more possible surface structures and stoichiometries have to be investigated. The results are summarized in Fig. 8.4 for the InP(001) surface [23]. The surface phase diagram in Fig. 8.4a indicates the stability of a certain surface with the lowest value Ω_g (8.21) for given chemical potentials μ_{In} and μ_{H}. It is clearly evident that the hydrogen-stabilized structure shown in Fig. 8.4b is the most favorable one under typical preparation conditions.

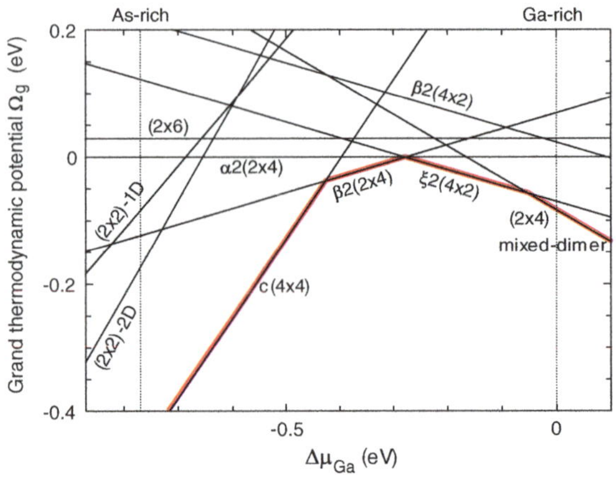

Fig. 8.3 Grand thermodynamic potential Ω_g for nine possible structures of the GaAs(001) surface versus variation of the cation chemical potential μ_{Ga}. The $\alpha 2(2 \times 4)$ structure is used as reference system. The *red lines* indicate the structure favored energetically for a given chemical potential of Ga atoms. From [22]

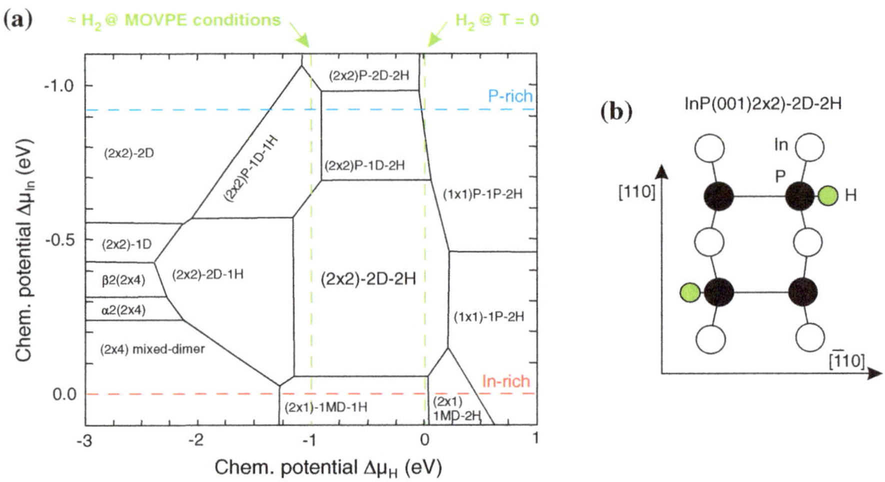

Fig. 8.4 (**a**) Calculated surface phase diagram of the hydrogen-exposed InP(001) surface. The *dashed lines* indicate the approximate range of the thermodynamically allowed values of $\Delta\mu_{\text{In}}$ and $\Delta\mu_{\text{H}}$. (**b**) The geometry and stoichiometry of the most favorable surface phase with H-passivated P dimers. All energies have been computed using a DFT-LDA approach. Adapted from [23]

8.2 Hellmann-Feynman Forces

8.2.1 Total Energy

The details of the computation of the internal energy $U(0, \Omega, N_A, N_B, \ldots)$ in Sect. 8.1 remain somewhat hidden. In Sect. 8.1.2 the internal energy $U = U(0, \Omega, N_A, N_B, \ldots)$ is usually described by the static total energy $E_{\text{tot}}(\{\mathbf{R}_l\})$ of the system of nuclei of species A, B, ... fixed at the positions $\{\mathbf{R}_l\}$ and its electrons moving in the field of the nuclei $V_{\text{n}}(\mathbf{x})$ (1.4). The two most important contributions to the total energy

$$U(0, \Omega, N_A, N_B, \ldots) \Leftrightarrow E_{\text{tot}}(\{\mathbf{R}_l\}) \tag{8.26}$$

are the energy of the electrons in their ground state for a given configuration $\{\mathbf{R}_l\}$ of the nuclei, which is described by the Kohn-Sham energy (6.12) as (here: spin polarization is omitted)

$$\begin{aligned} E_{\text{KS}}[n] &= E_{V_{\text{n}}}[n] = E_{\text{KS}}([n], \{\mathbf{R}_l\}) \\ &= T_s[n] + \int d^3\mathbf{x} V_{\text{n}}(\mathbf{x}) n(\mathbf{x}) + E_{\text{H}}[n] + E_{\text{XC}}[n], \end{aligned} \tag{8.27}$$

and the energy of the Coulomb repulsion of the charged nuclei (see Sect. 1.2)

$$E_{\text{nn}}(\{\mathbf{R}_l\}) = \frac{1}{2} \sum_{\substack{l,l'=1 \\ (l \neq l')}}^{N_{\text{n}}} Z_l Z_{l'} \text{v}(\mathbf{R}_l - \mathbf{R}_{l'}). \tag{8.28}$$

In summary, we have to deal with the energy

$$E_{\text{tot}}(\{\mathbf{R}_l\}) = E_{\text{nn}}(\{\mathbf{R}_l\}) + E_{\text{KS}}([n], \{\mathbf{R}_l\}), \tag{8.29}$$

where in E_{KS} only the potential energy $V_{\text{n}}(\mathbf{x})$ depends explicitly on the coordinates of the nuclei. The effect of lattice vibrations is neglected in expression (8.29).

The total energy E_{tot} (8.29), when studied as a function of the atomic positions $\{\mathbf{R}_l\}$ occupied with $N_A, N_B, \ldots$ atoms [cf. (8.26)], is sometimes called the *potential energy surface* (PES) in the $3\sum_{i=A,B,\ldots} N_i$-dimensional configuration space. It defines the potential energy landscape on which the atoms A, B, ... may 'travel'. Because of the assumptions of low temperature, not very fast motion of the nuclei, and the electrons in the ground state for each configuration $\{\mathbf{R}_l\}$ the adiabatic or Born-Oppenheimer approximation (see Sect. 1.2) is valid, and the terms *adiabatic potential surface* or *Born-Oppenheimer surface* are also used.

Usually a complete PES cannot be represented graphically because of too many coordinates $\{\mathbf{R}_l\}$. However, important information about the atomic geometry and its energetics can be obtained when the energy of a test atom within the nominal atomic

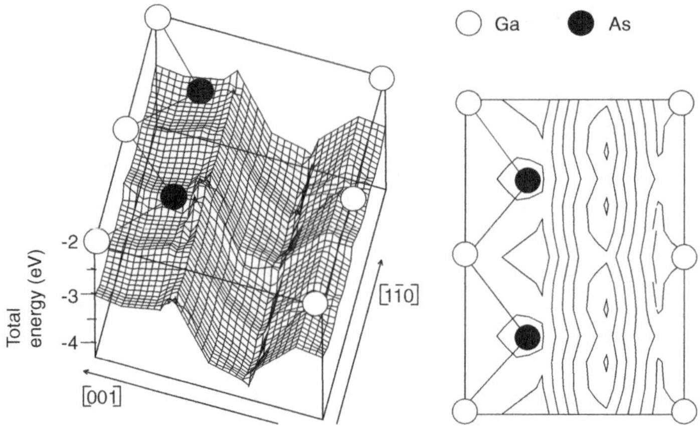

Fig. 8.5 Total energy surface of an Sb test atom on a GaAs(110)1×1 surface plotted over an area of two 1×1 surface unit cells as a three-dimensional perspective view (*left*), and E_{tot} versus xy-adatom coordinates as a contour plot (*right*). The surface Ga and As atoms are indicated by filled and empty *circles*. Drawn in [25] using data from [24]. Republished with permission from Elsevier

geometry is studied. Such a test atom may be a real adatom at a surface or an atom diffusing through a solid. In Fig. 8.5 the case of an adatom on a surface is studied. The coordinates of the test atom are fixed in the surface plane. However, the normal distance of this atom and the coordinates of the surface atoms are allowed to relax. One obtains a special PES, that of a surface with a test atom.

As an example of a resulting total energy surface that of an Sb test atom on a GaAs(110)1×1 cleavage face is plotted in Fig. 8.5. The energy is computed by means of a plane-wave pseudopotential code [24, 25]. The sum of the energy of the clean surface and the energy of the free (i.e., isolated) Sb atom is used as energy zero. The surface atomic structure determines the displayed PES. The most striking feature in the plotted PES is the deep channel which is quite rectilinear and parallel to the [1$\bar{1}$0] direction, i.e., the direction of the buckled Ga-As zig-zag chains. In each 1×1 surface unit cell two equivalent flat minima occur. They represent possible equilibrium positions of the Sb adatoms. There are no isolated minima in front of Ga or As dangling bonds. The visible trench indicates a possible pathway for surface diffusion of atoms.

8.2.2 Forces

Searching for optimal atomic positions $\{\mathbf{R}_l\}$ in the low-energy limit would necessitate the calculation of the electronic ground state for many atomic configurations $\{\mathbf{R}_l\}$ in order to find the global or at least most important minima on the total-energy surface as illustrated in Fig. 8.5. Away from the energy extrema of $E_{\text{tot}}(\{\mathbf{R}_l\})$ (8.29) there are driving forces

$$\mathbf{F}_l = -\nabla_{\mathbf{R}_l} E_{\text{tot}}(\{\mathbf{R}_l\}) \tag{8.30}$$

acting on the nuclei (or ions, or atoms). These forces are usually called Hellmann-Feynman forces [26, 27].

For a given composition $N_A, N_B, \ldots$ and a given configuration $\{\mathbf{R}_l\}$, the magnitude and the direction of an atomic force (8.30) give information about how far a certain atom is from a position in a metastable or stable configuration. The corresponding 'equilibrium' atomic geometry is identified by eliminating all forces (8.30),

$$\mathbf{F}_l\Big|_{\{\mathbf{R}_l\}=\{\mathbf{R}_l^0\}} = 0 \qquad \text{for all } l. \tag{8.31}$$

Thus, the optimal atomic structure $\{\mathbf{R}_l^0\}$ of a molecule or solid corresponds to a minimum of the total energy (8.29). The resulting minimum does, in general, not necessarily need to be a global one. In order to find it, usually several optimal configurations have to be studied and compared with respect to the resulting total-energy value $E_{\text{tot}}(\{\mathbf{R}_l^0\})$. Usually intelligent guesses are needed to generate an appropriate starting geometry. Besides the self-consistent cycle to obtain the electronic energy (8.27) for a given arrangement $\{\mathbf{R}_l\}$ also a second self-consistent cycle is needed to find the equilibrium geometry $\{\mathbf{R}_l^0\}$ of a local energy minimum whose symmetry is often determined by the guess for the starting geometry $\{\mathbf{R}_l\}$. This self-consistent cycle is illustrated in Fig. 8.6. It encapsulates the internal self-consistent cycle with respect to the electronic degrees of freedom as described in Fig. 6.2.

The Hellmann-Feynman forces (8.30) can be easily calculated by means of the explicit expression for the total energy (8.29). There are two contributions [28]

$$\mathbf{F}_l = \mathbf{F}_l^{\text{n}} + \mathbf{F}_l^{\text{el}}. \tag{8.32}$$

The contribution due to the repulsion of the nuclei is

$$\mathbf{F}_l^{\text{n}} = \sum_{\substack{l'=1 \\ (l' \neq l)}}^{N_{\text{n}}} \text{v}(\mathbf{R}_l - \mathbf{R}_{l'}) \frac{\mathbf{R}_l - \mathbf{R}_{l'}}{|\mathbf{R}_l - \mathbf{R}_{l'}|^2} \tag{8.33}$$

because of the corresponding energy contribution (8.28). Studying the pure electronic contribution we first realize that in (8.27) only the potential energy of an electron in the field of the nuclei $V_{\text{n}}(\mathbf{x})$ depends explicitly on the nuclear positions $\{\mathbf{R}_l\}$. The electron density $n(\mathbf{x})$ exhibits an implicit dependence. According to the Hellmann-Feynman theorem [26, 27] the gradient only acts on the Hamiltonian, i.e., precisely on the potential energy contribution $\int d^3\mathbf{x} V_{\text{n}}(\mathbf{x}) n(\mathbf{x})$ in (8.27). However, out of the minimum we find a second term

$$\mathbf{F}_l^{\text{el}} = \mathbf{F}_l^{\text{el}(1)} + \mathbf{F}_l^{\text{el}(2)} \tag{8.34}$$

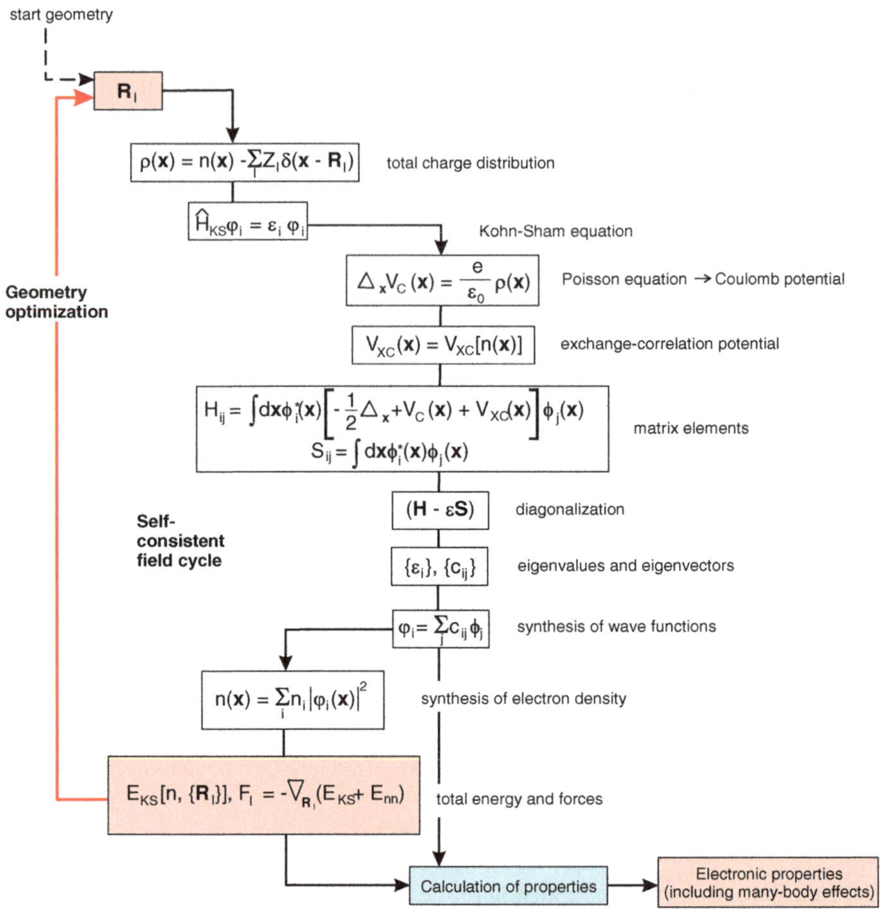

Fig. 8.6 Illustration of the determination of an 'equilibrium' geometry $\{\mathbf{R}_l^0\}$ (*external cycle*) via the calculation of the Kohn-Sham energy E_{KS} of the electronic system for a given configuration $\{\mathbf{R}_l\}$ (*internal cycle*, see Fig. 6.2). Thereby, the Kohn-Sham equation in the non-spin-polarized limit is self-consistently solved by expanding the Kohn-Sham eigenfunctions $\{\varphi_i(\mathbf{x})\}$ in terms of a complete set of functions $\{\phi_j(\mathbf{x})\}$, e.g. plane waves. The determination of the total electrostatic potential $V_{\mathrm{C}}(\mathbf{x}) = V_{\mathrm{n}}(\mathbf{x}) + V_{\mathrm{H}}(\mathbf{x})$ is an important intermediate step. The Hellmann-Feynman forces are calculated as the total energy $E_{\mathrm{KS}} + E_{\mathrm{nn}}$ of the system consisting of electrons and nuclei and, hence, including the repulsion of the nuclei E_{nn}

with

$$\mathbf{F}_l^{\mathrm{el}(1)} = -\int d^3\mathbf{x} n(\mathbf{x}) \nabla_{\mathbf{R}_l} V_{\mathrm{n}}(\mathbf{x}) \tag{8.35}$$

and

$$\mathbf{F}_l^{\mathrm{el}(2)} = -\int d^3\mathbf{x} \frac{\delta E_{\mathrm{KS}}[n]}{\delta n(\mathbf{x})} \nabla_{\mathbf{R}_l} n(\mathbf{x}). \tag{8.36}$$

The two contributions $\mathbf{F}_e^{\mathrm{n}} + \mathbf{F}_l^{\mathrm{el}(1)}$ are the true Hellmann-Feynman forces. According to the Hohenberg-Kohn theorem II and the Euler equation (6.7), the third contribution vanishes for the minimum Kohn-Sham energy when $n(\mathbf{x})$ approaches the corresponding 'ground-state' density and electron conservation is taken into account. In practical numerical treatments, e.g. within the internal self-consistency cycle in Fig. 8.6, variational forces of the type (8.36) may occur. They are due to numerical inaccuracies in the actual electron density. Using the representation of the electron density in terms of the Kohn-Sham orbitals $n(\mathbf{x}) = \sum_i n_i |\varphi_i(\mathbf{x})|^2$ (6.3) and the independent-particle picture (6.1) the variational forces may be rewritten into

$$\mathbf{F}_l^{\mathrm{el}(2)} = -2Re\sum_i n_i \int d^3\mathbf{x}\left[\nabla_{\mathbf{R}_l}\varphi_i^*(\mathbf{x})\right]\left[-\frac{\hbar^2}{2m}\Delta_{\mathbf{x}} + V_{\mathrm{KS}}(\mathbf{x}) - \varepsilon_i\right]\varphi_i(\mathbf{x})$$

or

$$\begin{aligned}\mathbf{F}_l^{\mathrm{el}(2)} = & -2Re\sum_i n_i \int d^3\mathbf{x}\left[\nabla_{\mathbf{R}_l}\varphi_i^*(\mathbf{x})\right]\left[-\frac{\hbar^2}{2m}\Delta_{\mathbf{x}} + \tilde{V}_{\mathrm{KS}}(\mathbf{x}) - \varepsilon_i\right]\varphi_i(\mathbf{x}) \\ & -\int d^3\mathbf{x}\left[V_{\mathrm{KS}}(\mathbf{x}) - \tilde{V}_{\mathrm{KS}}(\mathbf{x})\right]\nabla_{\mathbf{R}_l} n(\mathbf{x}).\end{aligned}$$

with the actual single-particle potential $\tilde{V}_{\mathrm{KS}}(\mathbf{x})$ in a certain stage of the computations leading to the wave functions $\{\varphi_i(\mathbf{x})\}$. There are two different origins of the variational forces. The first term in $\mathbf{F}_l^{\mathrm{el}(2)}$ is zero if the changes of the wave function maintain orthonormality when the atom is displaced. This happens for basis sets independent of atomic positions, e.g. for plane waves, or really complete basis sets. Usually basis sets of localized functions as in the case of atom-centered orbitals are incomplete and so-called Pulay forces appear [29]. The second contribution to $\mathbf{F}_l^{\mathrm{el}(2)}$ measures the non-self-consistency in the solution of the Kohn-Sham equation (see Fig. 8.6).

8.2.3 **k***-space Formalism*

For crystalline solids, however, also for finite objects which are modeled within a supercell or repeated slab method (see Sect. 1.3), plane waves appear to form an appropriate basis set for the expansion of the eigenfunctions of the Kohn-Sham equations (6.22). In the case of free-electron-like metals this is obvious because planes waves (4.46) are even eigenfunctions. For a given translationally invariant system the plane waves (PWs) are

$$\phi_{\mathbf{kG}}(\mathbf{x}) = \frac{1}{\sqrt{\Omega}} e^{i(\mathbf{k}+\mathbf{G})\mathbf{x}} \tag{8.37}$$

with $\mathbf{k} \in$ the Brillouin zone and $\mathbf{G} \in$ the reciprocal lattice (cf. Sect. 1.3).

The set of plane waves $\{\phi_{\mathbf{kG}}(\mathbf{x})\}$ is orthonormal

$$\int d^3\mathbf{x}\phi^*_{\mathbf{kG}}(\mathbf{x})\phi_{\mathbf{k}'\mathbf{G}'}(\mathbf{x}) = \delta_{\mathbf{kk}'}\delta_{\mathbf{GG}'} \tag{8.38}$$

and complete

$$\sum_{\mathbf{k}}\sum_{\mathbf{G}} \phi_{\mathbf{kG}}(\mathbf{x})\phi^*_{\mathbf{kG}}(\mathbf{x}') = \delta(\mathbf{x}-\mathbf{x}'). \tag{8.39}$$

For a translationally invariant system each Kohn-Sham eigenfunction $\varphi_{\nu\mathbf{k}m_s}(\mathbf{x})$ of the Bloch type (1.13) is expanded according to

$$\varphi_{\nu\mathbf{k}m_s}(\mathbf{x}) = \sum_{\mathbf{G}} c_{\nu\mathbf{k}m_s}(\mathbf{G})\phi_{\mathbf{kG}}(\mathbf{x}), \tag{8.40}$$

where ν is the band index in the spin channel m_s. The central quantities of the theory, the periodic spin densities $n_{m_s}(\mathbf{x}) = n_{m_s}(\mathbf{x}+\mathbf{R})$ (6.3) take the simplified form

$$\begin{aligned} n_{m_s}(\mathbf{x}) &= \sum_{\mathbf{G}} e^{-i\mathbf{Gx}}\tilde{n}_{m_s}(\mathbf{G}), \\ \tilde{n}_{m_s}(\mathbf{G}) &= \frac{1}{\Omega}\sum_{\nu,\mathbf{k}} n_{\nu\mathbf{k}m_s}\sum_{\mathbf{G}'} c^*_{\nu\mathbf{k}m_s}(\mathbf{G}'+\mathbf{G})c_{\nu\mathbf{k}m_s}(\mathbf{G}'). \end{aligned} \tag{8.41}$$

The Kohn-Sham equations (6.22) become a system of algebraic equations

$$\sum_{\mathbf{G}'}\left\{\left[\frac{\hbar^2}{2m}(\mathbf{k}+\mathbf{G})^2 - \varepsilon_{\nu m_s}(\mathbf{k})\right]\delta_{\mathbf{GG}'} + V^{m_s}_{\mathrm{KS}}(\mathbf{G}-\mathbf{G}')\right\} c_{\nu\mathbf{k}m_s}(\mathbf{G}') = 0 \tag{8.42}$$

with the Bloch band energies $\varepsilon_{\nu m_s}(\mathbf{k})$. Here the Fourier coefficients $V^{m_s}_{\mathrm{KS}}(\mathbf{G}-\mathbf{G}')$ of a local or semilocal Kohn-Sham potential are considered. In the next section the electron-ion interaction will be generalized to the non-local case. A generalization to non-local XC potentials is studied in Chap. 9. For a non-local potential $V^{m_s}_{\mathrm{KS}}(\mathbf{x},\mathbf{x}')$ a matrix representation $V^{m_s}_{\mathrm{KS}}(\mathbf{k}+\mathbf{G},\mathbf{k}+\mathbf{G}')$ appears in (8.42). The consequences for the total energy E_{KS} (see [28]) will be not investigated here. According to the self-consistent scheme in Fig. 8.6 the eigenvalues and eigenvectors of the system of Kohn-Sham equations (8.42) give the densities (8.41) and, hence, the electronic contribution to the total energy of the inhomogeneous electron gas in the field of the nuclei.

The use of plane waves (8.37) can be interpreted as the use of a grid in the reciprocal space as illustrated in Fig. 8.7a. Only for periodic systems that grid is discrete. For large system volumes Ω a number of plane waves approaching to infinite is used. According to (1.16) the number of $\mathbf{k}$ points within a BZ is given by $\sum_{\mathbf{k}} = \frac{\Omega}{(2\pi)^3}\Omega_{\mathrm{BZ}}$

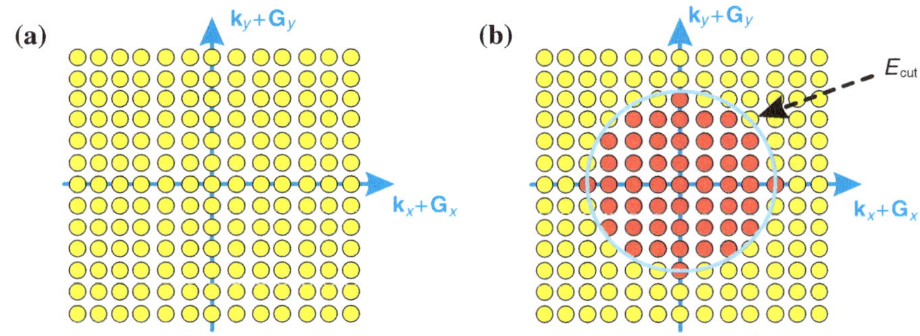

Fig. 8.7 (**a**) Sampling of the reciprocal space when using plane waves. (**b**) Truncation of reciprocal space by means of an isoenergy surface $E = E_{\text{cut}}$ with the cutoff energy (8.43)

with the BZ volume $\Omega_{\text{BZ}} = \frac{(2\pi)^3}{\Omega_0}$. In explicit computations the number of plane waves is limited. Typically the convergence of the results is studied versus a cutoff energy E_{cut} related to the kinetic energy

$$\int d^3\mathbf{x}\phi^*_{\mathbf{kG}}(\mathbf{x}) \left\{-\frac{\hbar^2}{2m}\Delta_{\mathbf{x}}\right\} \phi_{\mathbf{kG}}(\mathbf{x}) = \frac{\hbar^2}{2m}(\mathbf{k}+\mathbf{G})^2 \leq E_{\text{cut}}. \tag{8.43}$$

This energy determines the number of plane waves used in the computations. It corresponds to a truncation of the reciprocal space as illustrated in Fig. 8.7b. Its volume is that of a sphere, $\frac{4\pi}{3}\frac{(2mE_{\text{cut}})^{3/2}}{\hbar^3}$. Then the total number of plane waves per atom N_{pw} is given by [30]

$$N_{\text{pw}} \cdot N_{\text{atom}} \approx \frac{4\pi}{3}\frac{(2mE_{\text{cut}})^{3/2}}{\hbar^3 \Omega_{\text{BZ}}} \tag{8.44}$$

with N_{atom} as the number of atoms in the unit cell. In the case of crystalline silicon with $N_{\text{atom}} = 2$, $\Omega_0 = a_0^3/4$, and $a_0 = 5.43$ Å we find $N_{\text{pw}} \approx 72$ for a typical value of $E_{\text{cut}} = 10$ Ry. This means that the Bloch wave function for each valence state is represented by about 18 plane waves.

The influence of the plane-wave cutoff E_{cut} is illustrated in Fig. 8.8 for three quantities derived from the Murnaghan equation of state (8.16) with the identification of $U \equiv E_{\text{tot}}$, $\Omega = \Omega_0$, and lattice constant $a_0 = \sqrt[3]{4\Omega_0}$ for two different descriptions of the electron-ion interaction using ultrasoft pseudopotentials [30] or those generated within the projector augmented wave (PAW) method [32]. The figure shows that for well-converged ground-state properties of silicon – despite its *sp* bonding – relatively large cutoff energies of about 25 Ry are needed. The tendency for the resulting underbinding is a consequence of the used PW91-GGA XC functional (cf. Sect. 7.3).

For many purposes such a low number of plane waves is sufficient. This holds for *s*- and *p*-like valence states if their behavior in the immediate vicinity of the

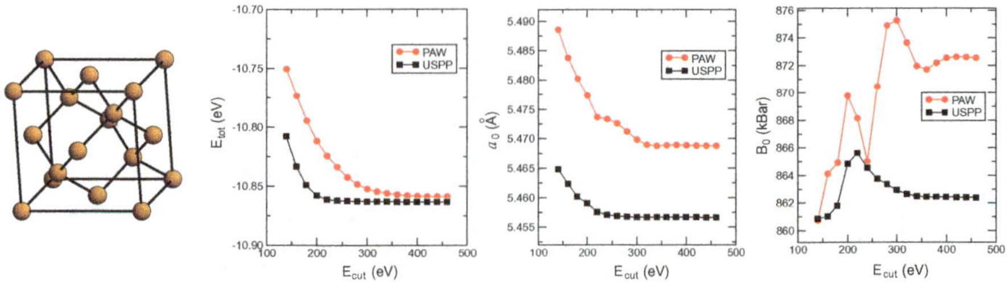

Fig. 8.8 Equilibrium binding energy E_{tot} per unit cell, lattice constant a_0, and bulk modulus B_0 of silicon crystallizing in diamond structure (*left panel*) versus the plane-wave cutoff E_{cut}. Two different pseudopotentials USPP and PAW (see text in Sect. 8.3) are used for the electron-ion interaction. Adapted from [31]

nuclei does not play a role. However, already for semicore d states a plane-wave expansion needs high cutoff energies. The plane-wave description of true strongly localized core states makes less sense. In these cases a better description of the wave functions near the cores and, hence, completely different basis sets are needed. Such refinements or totally different representations of the eigenfunctions are the PAW [32, 33], the linearized augmented plane wave (LAPW) [34], the linearized muffin-tin orbital (LMTO) [35, 36], and the Korringa-Kohn-Rostoker (KKR) [37, 38] methods.

Important quantities of the theory, e.g. the densities [see e.g. (8.41)] and the total energy, involve sums over **k** (or originally integrations over the BZ). In principle, an infinite number of **k** points is needed. In practice, only sums over finite numbers are applied, resulting in the so-called BZ sampling. The number (or the **k**-point density) needed depends on the dispersion of occupied bands. Typically one needs more **k** points for metals, in particular for such with a complex Fermi surface. For insulators with flat bands their number can be reduced. For localized objects such as molecules and nanocrystals described in a supercell arrangement the sampling can be reduced to the BZ center, the Γ point. In explicit computations the convergence of the wanted quantities has to be tested versus the **k**-point sampling.

In the past the influence of the point group symmetry has been exploited. Starting from expansions in symmetrized plane waves Baldereschi [39] could show that one special point $\mathbf{k}^*$ in the irreducible part of the BZ was sufficient. Another method to generate special points has been provided by Chadi and Cohen [40]. Meanwhile, often the sampling scheme according to Monkhorst and Pack (MP) [41] is applied. It usually performs the sampling with an equidistant grid of **k** points with identical weights,

$$\mathbf{k}_{i_1 i_2 i_3} = \sum_{j=1,2,3} u_{i_j} \mathbf{b}_j$$

with the basis vectors $\mathbf{b}_j$ (1.9) of the reciprocal lattice and the coordinates in reciprocal space

$$u_{i_j} = (n_{i_j} - i_j - 1)/n_{i_j} \qquad (i_j = 1, \ldots, n_{i_j})$$

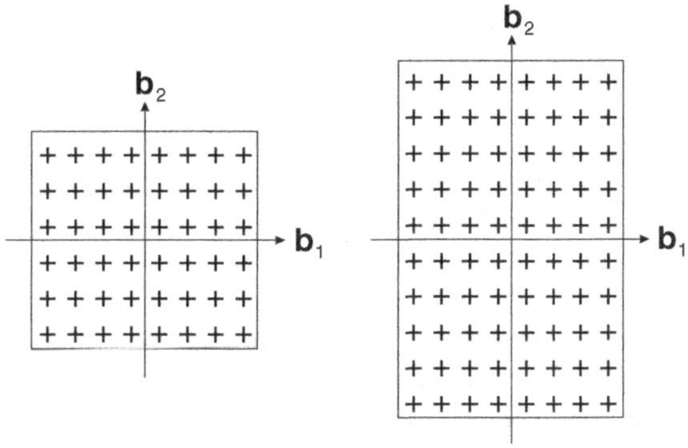

Fig. 8.9 Illustration of the MP **k**-point sampling (*crosses*) of two different Brillouin zones (*solid lines*)

with n_{i_j} as the numbers of **k** points between two reciprocal lattice points connected through $\mathbf{b}_j$. Such a MP mesh is illustrated in Fig. 8.9 for the two-dimensional cases of square and rectangular lattices. One distinguishes between meshes centred on origin, i.e., non-shifted, or not, i.e., shifted.

8.3 Restriction to Valence Electrons

8.3.1 Frozen Core Approximation

If the spatial variation of the total potential due to the nuclei and the other electrons is negligible, then plane waves are exact solutions of the electronic-structure problem (see Sect. 4.4.1). If the potential is reasonably smooth, its spatial variation can be treated as a perturbation. The potential originated in the atomic nuclei, however, is far from smooth. For instance, the potential $V_\mathrm{n}(\mathbf{x})$ (1.4) exhibits Coulomb singularities near the positions $\mathbf{R}_l$ of these nuclei. Therefore, a PW expansion of the wave functions of tightly bound electronic states, especially near the nuclei, is a rather hopeless task, because the number of PW components required to represent strongly localized wave functions is too huge.

This conclusion particularly holds for the wave functions of *core electrons*, the electrons occupying the deep, completely filled shells of atoms. In condensed matter these electrons remain very localized around an atom, whereas the remaining electrons – called *valence electrons* – determine the majority of the properties. A clear distinction between the two classes of electrons is sometimes difficult. Of course, valence electrons participate actively in chemical bonding, while the core electrons are tightly bound to the nuclei and do not participate in bonding. In silicon with an electron configuration $1s^2 2s^2 2p^6 \vdots 3s^2 3p^2$ the core and valence electrons are energetically well separated. The Si $2p$ binding energy approaches a value of about 100 eV. Indeed, the core electrons are less sensitive to the molecular or crystal field.

Their wave functions remain almost frozen. Nevertheless, they show a core level shift. More difficult to classify is a third class of electrons, called *semicore electrons*, which do not participate actively or at least less in chemical bonding. More specific examples are occupied or partly filled d shells. Examples are the In $4d$ electrons which are resonant to the N $2s$ states and, thus, contribute to some extent to the chemical bonding in InN [42]. Therefore, the semicore In $4d$ electrons have to be treated as valence electrons.

In the following we eliminate the core electrons widely from our studies and replace their action within an effective potential or *pseudopotential* (PP). As a consequence we consider a piece of condensed matter as a collection of valence (and semicore) electrons and ion cores. An ion consists of a nucleus and the tightly bound core electrons of the corresponding atom. The positions $\{\mathbf{R}_l\}$ of the nuclei appear now as the coordinates of the ions. Inspired by the orthogonalized plane wave (OPW) method of Herring [43], the origin of the modern pseudopotential approach goes back to Philips and Kleinman [44].

We illustrate the PP idea following Philips and Kleinman [44]. The details of the single-particle equation of the Kohn-Sham type (6.22) are suppressed. We distinguish only between valence (v) and core (c) states. The single-particle Hamiltonian $\hat{H} = \hat{T} + \hat{V}$ contains an effective potential $V(\mathbf{x})$. Using the ket denotation the Schrödinger-like equation reads as ($\lambda = c,\ v$)

$$\hat{H}|\varphi_\lambda\rangle = \varepsilon_\lambda|\varphi_\lambda\rangle.$$

Following the OPW concept we construct pseudo-wave functions $|\tilde{\varphi}_v\rangle$ for the valence electrons

$$|\tilde{\varphi}_v\rangle = |\varphi_v\rangle + \sum_c a_{cv}|\varphi_c\rangle$$

by mixing valence with core states. With $a_{cv} = \langle\varphi_c|\tilde{\varphi}_v\rangle \neq 0$, they are still orthogonal to the core states. The pseudo-wave functions satisfy the Schrödinger-like equation

$$\left[\hat{H} + \sum_c (\varepsilon_v - \varepsilon_c)|\varphi_c\rangle\langle\varphi_c|\right]|\tilde{\varphi}_v\rangle = \varepsilon_v|\tilde{\varphi}_v\rangle$$

for the same eigenvalues $\{\varepsilon_v\}$ but these functions $\{|\tilde{\varphi}_v\rangle\}$ are smooth in the core regions. The above result suggests to construct a pseudo-Hamiltonian $\hat{H}_{ps} = \hat{T} + \hat{V}_{ps}$ with a pseudopotential

$$V_{ps} = V + \sum_c (\varepsilon_v - \varepsilon_c)|\varphi_c\rangle\langle\varphi_c|,$$

that is non-local (more precisely: semilocal) in space and energy-dependent.

For an isolated atom at $\mathbf{R}_l = 0$ with spherical symmetry, i.e., $V(\mathbf{x}) = V(r)$ with $r = |\mathbf{x}|$ and the set of quantum numbers $\lambda = n\ell m$ (n - principal quantum number, ℓm - angular momentum quantum numbers), the electronic states can be separated as

$$\varphi_{n\ell m}(\mathbf{x}) = R_{n\ell}(r)Y_{\ell m}(\theta, \phi), \tag{8.45}$$

where $R_{n\ell}(r)$ represent the radial parts and $\langle \mathbf{x}|\ell m\rangle = Y_{\ell m}(\theta, \phi)$ are spherical harmonics. It is clear that a pseudopotential acts differently on wave functions of different angular momentum, thereby expressing its energy dependence. The most general form of a pseudopotential of this kind is

$$V_{ps} = \sum_{l=0}^{\infty} V_{ps}^{\ell}(r)\hat{P}_{\ell} \tag{8.46}$$

with the partial pseudopotential $V_{ps}^{\ell}(r)$ related to the angular momentum ℓ, and the operator

$$\hat{P}_{\ell} = \sum_{m=-\ell}^{\ell} |\ell m\rangle\langle \ell m|, \tag{8.47}$$

which is a projection operator onto the ℓth angular momentum subspace. It makes obvious that the total pseudopotential (8.46) is a non-local (at least, semilocal) operator in space.

8.3.2 Atomic Pseudopotentials

There are several degrees of freedom in how pseudopotentials are constructed [45–47]. Empirical pseudopotentials are determined by fitting experimental data, e.g. energy bands or transition energies. However, the results lack a very important property, the *transferability*, namely that a pseudopotential constructed for a specific environment can be used for the same atomic species in another environment. In the core region the resulting pseudopotentials should be smooth or ‘soft’ and not ‘hard’, i.e., their spatial variation should be limited. The *hardness* of a pseudopotential should be reduced as much as possible. However, in practice, one has to make a compromise since a ‘softening’ tends to lead to poorer transferability.

The construction of a (‘*ab initio*’) pseudopotential is an inverse problem. However, there are important rules, formulated by Hamann et al. [48] and first applied by Bachelet, Hamann and Schlüter (BHS) [49], which help to solve the problem. Without spin polarization the radial Schrödinger equation (omitting the principal quantum number n and any index referring to the atomic species)

$$\left\{-\frac{\hbar^2}{2m}\left[\frac{d^2}{dr^2}+\frac{\ell(\ell+1)}{2r^2}\right]+V(r)\right\}rR_\ell(\varepsilon,r)=\varepsilon rR_\ell(\varepsilon,r) \tag{8.48}$$

is a second-order linear differential equation. Once ε has been fixed (not necessarily to an eigenvalue ε_ℓ), its solution is uniquely determined by the value of the radial function $R_\ell(\varepsilon,r)$ and its derivative $R'_\ell(\varepsilon,r)$. The potential $V(r)$ is usually taken from the Kohn-Sham theory (6.20) as a sum of the Coulomb potential of the nucleus, the Hartree potential, and the exchange-correlation potential. Typically also scalar relativistic corrections (see Sect. 2.3) are taken into account. The assumption of spherical symmetry of the total potential is not always valid. However, the effect of the deviations from the spherical symmetry are generally small. In the case of silicon atoms with a Ne core it may hold for an excited electron configuration Ne$3s^13p^3$ but not for the ground state Ne$3s^23p^2$. In many crystal phases and compounds the Si atoms are fourfold coordinated and prefer an sp^3 hybridization. For carbon one may start from the ground-state configuration $1s^22s^22p^2$ to construct the pseudopotentials for $\ell=0$ and $\ell=1$. For the construction of the pseudopotential for d states the use of an ionized configuration $1s^22s^{1.00}2p^{1.75}3d^{0.25}$ seems to lead to more reasonable results [50].

The rules to solve the inverse problem and to construct *ab initio* pseudopotentials $V^\ell_{ps}(r)$ are:

(i) The pseudopotentials reproduce energy (pseudo)eigenvalues $\tilde{\varepsilon}_\ell$ in agreement with those ε_ℓ derived from the all-electron calculation for the valence states,

$$\tilde{\varepsilon}_\ell=\varepsilon_\ell.$$

Such atomic eigenvalues are presented in Fig. 8.10.

(ii) Outside some core radius $r_{c\ell}$ also the all-electron wave functions are reproduced,

$$\tilde{R}_\ell(r)=R_\ell(r) \qquad \text{for} \qquad r\geq r_{c\ell}.$$

(iii) Within the core region $r<r_{c\ell}$ the pseudo-radial parts $\tilde{R}_\ell(r)$ are nodeless. Nevertheless, the norm of the all-electron and pseudo-wave functions inside the pseudized core region, $r<r_{c\ell}$, is the same (*norm-conserving condition*)

$$\int_0^{r_{c\ell}} drr^2|\tilde{R}_\ell(r)|^2=\int_0^{r_{c\ell}} drr^2|R_\ell(r)|^2.$$

(iv) Extremely important for the transferability is the correct description of the scattering properties of the partial waves. They are characterized by the scattering phase $\eta_\ell(\varepsilon)$ whose energy derivative is related to the logarithmic derivative [51]

$$D_\ell(\varepsilon,r)=\frac{1}{r}\frac{d\ln R_\ell(\varepsilon,r)}{d\ln r}$$

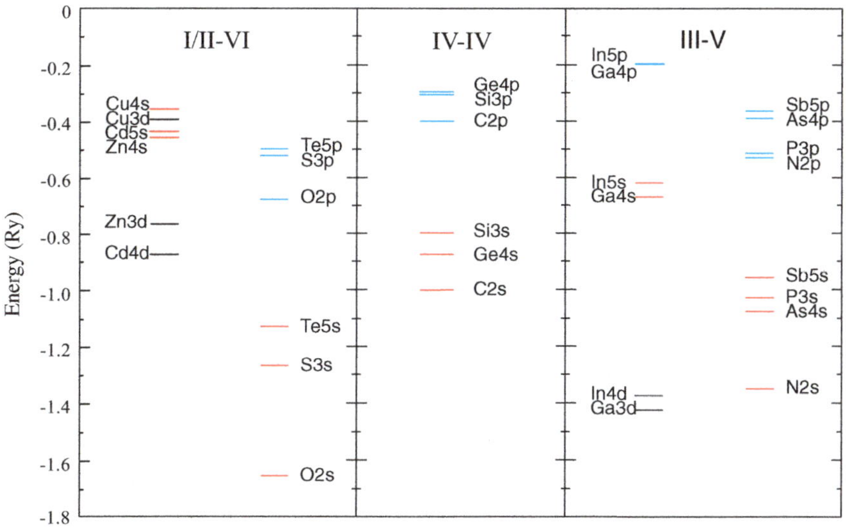

Fig. 8.10 Atomic eigenvalues for valence and semicore states derived within a scalar-relativistic DFT-LDA all-electron calculation. The atomic species form elemental or compound semiconductors indicated by the groups I-VI in the periodic table. Adapted from [50]

of the radial part in the corresponding energy range. The phases should agree for the all-electron and pseudo-wave functions outside the core $r \geq r_{c\ell}$, in particular at the boundary $r = r_{c\ell}$

$$\tilde{D}_\ell(\varepsilon, r_{c\ell}) = D_\ell(\varepsilon, r_{c\ell}).$$

Indeed the conditions (iii) and (iv) are important for the use of $V_{ps}^{\ell}(r)$ in different chemical environments. Requirement (iii) guarantees the same charge density in the core region. Consequently, the electrostatic potentials of the atom and pseudoatom are identical at large distances from the nucleus. The condition (iv) expresses that the scattering properties of the two atoms are rather similar for the energy range ε of the valence electrons. This is related to the identity ($r > r_{c\ell}$)

$$\frac{d}{d\varepsilon} D_\ell(\varepsilon, r) = -\frac{2m}{\hbar^2} \frac{1}{r^2 R_\ell^2(\varepsilon, r)} \int_0^r dr' r'^2 R_\ell^2(\varepsilon, r'), \tag{8.49}$$

which corresponds to the Friedel sum rule [52, 53].

8.3.3 Construction of Pseudopotentials

The general procedure for obtaining a pseudopotential begins with solving the all-electron radial Schrödinger equation (8.48) for a chosen (reference) atomic configuration. Then, a pseudopotential can be constructed according to the conditions (i) – (iv). The resulting pseudopotentials are norm-conserving because of condition (iii). Within the core region $r < r_{c\ell}$ there are still degrees of freedom for the pseudo-wave functions. As a result several construction schemes exist in the literature besides the BHS one [49] or its revisited parametrization by Gonze et al. [54]. Other norm-conserving recipes that improve over the smoothness of the original BHS pseudopotentials have been proposed in the literature, e.g. by Rappe, Rabe, Kaxiras, and Joannopolous (RRKJ) [55] or Troullier and Martins (TM) [56].

The Schrödinger equation of the type (8.48) for the pseudo-wave functions $\tilde{R}_\ell(r)$ is governed by a certain pseudopotential for each angular momentum channel ℓ. Since it is influenced by the interaction with the atomic valence electrons of the density $n_v(\mathbf{x})$ [instead of the total electron density $n(\mathbf{x}) = n_v(\mathbf{x}) + n_c(\mathbf{x})$ with the contribution of core electrons $n_c(\mathbf{x})$], we call it screened (sc) pseudopotential. The inversion of the corresponding Schrödinger equation yields [49]

$$V_{ps}^{(sc)\ell}(r) = \varepsilon_\ell + \frac{\hbar^2}{2m}\left\{-\frac{\ell(\ell+1)}{r^2} + \frac{1}{r\tilde{R}_\ell(r)}\frac{d^2}{dr^2}[r\tilde{R}_\ell(r)]\right\}, \tag{8.50}$$

where the principal quantum number n is still removed to indicate that the pseudization is done for the lowest-lying valence state of each angular momentum ℓ. The inversion (8.50) can be always done because of the nodeless condition. The core states enter the PP generation only through the self-consistent potential $V(r)$ in the all-electron Schrödinger-like (Kohn-Sham) equation. Expression (8.50) clearly indicates three facts: (i) The screened PP is continuous if at least the first and second derivatives of the pseudo-wave function are continuous. (ii) The pseudo-wave function $\tilde{R}_\ell(r)$ should vanish as r^ℓ for $r \to 0$ to avoid a 'hard-core' PP with a singularity at the origin. (iii) The screened PP depends on the angular-momentum state.

The pseudopotential acting on the states of the angular momentum ℓ is finally obtained by substracting the effect of the interaction with the valence electrons distributed according to their pseudo-wave functions

$$V_{ps}^{\ell}(r) = V_{ps}^{sc(\ell)}(r) - \int d^3\mathbf{x}' \mathrm{v}(\mathbf{x}-\mathbf{x}')\tilde{n}_v^{\mathrm{atom}}(\mathbf{x}') - V_{XC}(\mathbf{x}; [\tilde{n}_v^{\mathrm{atom}}]) \tag{8.51}$$

with the spherical atomic valence electron density

$$\tilde{n}_v^{\mathrm{atom}}(\mathbf{x}) = \frac{1}{4\pi}\sum_{\ell=0}^{\ell_{\max}}\sum_{m=-\ell}^{\ell}|\tilde{R}_\ell(r)|^2. \tag{8.52}$$

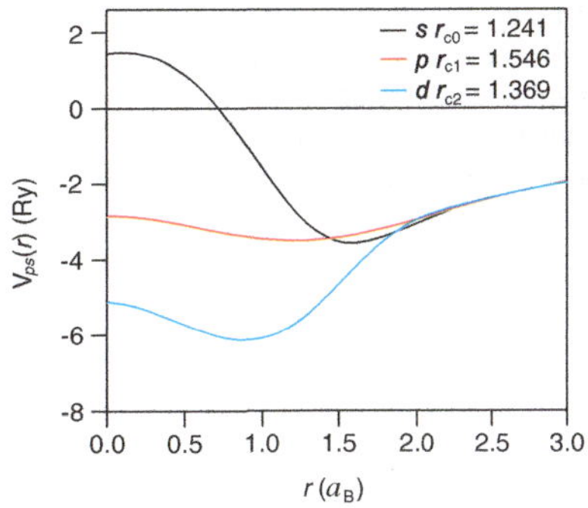

Fig. 8.11 Ionic pseudopotentials $V_{ps}^{\ell}(r)$ of Al [58]. The effective radii $r_{c\ell}$ are given in units of a_B

By substracting the effect of the valence electrons one obtains an (unscreened) 'ionic' pseudopotential (8.51) that does not depend on the chemical environment and, thus is transferable. Ionic pseudopotentials are displayed for Al in Fig. 8.11. The characteristic radii $r_{c\ell}$ are listed in the figure. The potentials $V_{ps}^{\ell}(r)$ are of the Hamann-Schlüter-Chiang type [48]. They have been generated using the PW91-GGA XC functional by means of the fhi99pp code [57]. The equal asymptotic behavior of the pseudopotentials is obvious. The resulting pseudo-wave functions are compared in Fig. 8.12 with all-electron wave functions. The pseudo-wave functions are indeed nodeless and hence labeled by pseudo-principal quantum numbers $\tilde{n}$ starting from $\tilde{n} = 1$. The question how much the XC functional influences the resulting ionic pseudopotential is illustrated in Fig. 8.13 for s states of a Si atom. It is also generated by means of the fhi98pp code [57] using one LDA and four GGA functionals for $\ell = 0$ and Si. The used XC potential influences the resulting pseudopotentials only close to the atomic core.

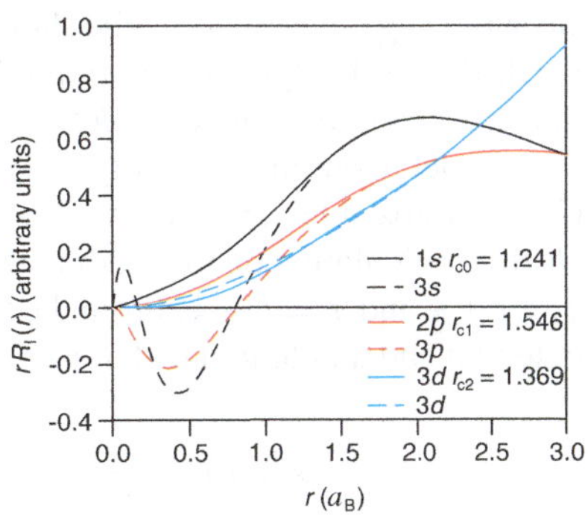

Fig. 8.12 Pseudo-wave functions (*solid lines*) and all-electron ones (*dashed lines*) for the valence states of Al [58]. The effective radii $r_{c\ell}$ are given in units of a_B

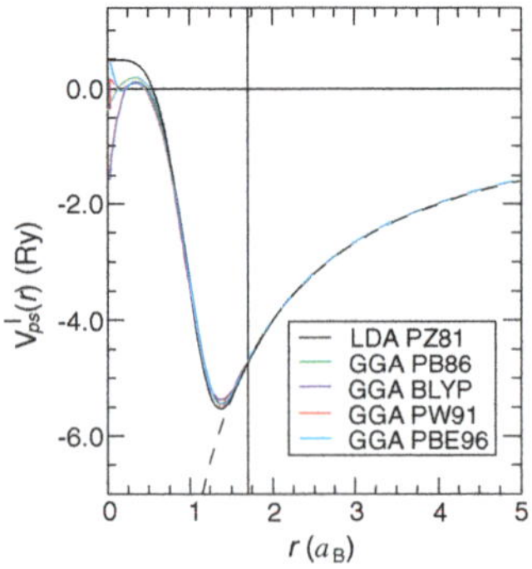

Fig. 8.13 Influence of different LDA and GGA exchange-correlation functionals on the ionic pseudopotential for Si 3s states [31]

8.3.4 Refinements

According to the construction procedure (8.50) each subspace of an angular momentum ℓ has its own pseudopotential. According to expression (8.46) with the projection operator $\hat{P}_\ell$ (8.47) the partial potentials can be combined to a total non-local (in space representation) PP. In the limit $r \to \infty$ it should become the local potential $-Z^{\text{val}}e^2/(4\pi\varepsilon_0 r)$ with Z^{val} as the number of valence electrons of the studied atom. Because of the closure relation of the projection operators $\hat{P}_\ell$, this requires the same behavior for each ℓ, i.e.,

$$V^\ell_{ps}(r) = -\frac{Z^{\text{val}}e^2}{4\pi\varepsilon_0 r} \qquad \text{for} \quad r \to \infty.$$

As a consequence it is useful to decompose $V^\ell_{ps}(r)$ in a long-range, ℓ-independent contribution and an ℓ-dependent short-range potential. The long-range part is local. Because of $\sum_\ell \hat{P}_\ell = 1$ only the short-range contribution to the total potential is non-local. PPs of this type are more precisely called semilocal [48]. They are non-local in the angular coordinates θ, ϕ but still local in the radial coordinate r.

A generalization of BHS-type pseudopotentials to include spin-orbit interaction (2.23) is possible. The first step is the generation of a PP from a relativistic all-electron calculation on the atom for both total angular momentum quantum numbers $j = \ell + \frac{1}{2}$ and $j = \ell - \frac{1}{2}$. From the two potentials we can define [49, 59] an average potential and a potential difference

$$\begin{aligned} V^\ell_{ps}(r) &= \frac{1}{2\ell+1}\left[\ell V^{\ell-1/2}_{ps}(r) + (\ell+1)V^{\ell+1/2}_{ps}\right], \\ \Delta V^\ell_{ps}(r) &= \frac{2}{2\ell+1}\left[V^{\ell+1/2}_{ps}(r) - V^{\ell-1/2}_{ps}(r)\right]. \end{aligned}$$

This arrangement leads to an additional contribution to expression (8.46), which can be represented as [60, 61]

$$\Delta V_{ps}^{so} = \sum_{\ell,m} |\ell m\rangle \Delta V_{ps}^{\ell}(r) \boldsymbol{\ell} \cdot \mathbf{s} \langle \ell m|$$

with the orbital angular momentum operator $\boldsymbol{\ell}$ and the spin operator $\mathbf{s}$ (see Sect. 2.3). An approach in a similar spirit that allows, however, to include spatial anisotropies of the PP was recently suggested [62].

In space representation $V_{ps}(\mathbf{x}, \mathbf{x}')$ the pseudopotential (8.46) is generally non-local, more precisely consists of a local and a semilocal contribution. Kleinman and Bylander [63] recognized that it is possible to construct a separable pseudopotential operator as a sum of products of functions of only $\mathbf{x}$ or $\mathbf{x}'$. They showed that the semilocal PP contribution can be replaced, to a good approximation, by a really non-local but separable operator. It has the numerical advantage to factorize in reciprocal space, e.g. within the plane-wave representation [59].

An undesirable consequence of the introduction of a separable form of norm-conserving pseudopotentials is the possibility that unphysical states, so-called *ghost states*, appear at energies below the eigenvalue ε_ℓ of a specific angular momentum. The reason for such states is that the single-particle Hamiltonian with Kleinman-Bylander pseudopotentials does not always give rise to an ordering of the eigenstates by increasing number of nodes. Solutions with nodes can be lower in energy than nodeless ones. Figure 8.14 shows for Si that ghost states can be observed as divergencies in the energy plot of the logarithmic derivative of the radial part $D_\ell(\varepsilon, r_{c\ell})$ at unphysically low energies. Whereas for ground-state calculations the scattering phase should be especially correct in the energy region of bound states, the computation of electronic excitations also asks for correct phases above the ionization energy.

The question is why the natural ordering of the (pseudo-)eigenstates can be violated. Gonze et al. [64] showed that the existence of ghost states is related basically to the choice of the local component of the pseudopotential [54]. This is indeed illustrated in Fig. 8.14 for the Si s-valence state ($\ell = 0$). The underlying pseudopotentials are generated in the TM and KB schemes with the PBE-GGA XC potential by means of the fhi98pp code [57]. The $\ell = 2$ (d) component has been identified as the local contribution to the PP. By construction (see also Fig. 8.11) it becomes obvious that for a smaller radius $r_{c\ell}$ the potential is more attractive. This fact has enormous consequences as demonstrated for $r_{c0} = 1.25\, a_B$ (left panel) and $r_{c0} = 1.90\, a_B$ (right panel) in Fig. 8.14 for the derivative of the scattering phase $D_0(\varepsilon, r_{c0})$ for energies ε around the eigenvalue $\varepsilon_\ell = \varepsilon_0$. The phases obtained from all-electron, semilocal PP and KB PP calculations agree widely for the energy range of interest below $\varepsilon \approx 2\, Ry$. However, in the left panel a ghost state appears at $\varepsilon = -1.86\, Ry$ in the phase derived from the non-local KB potential. The corresponding pseudopotential computations may lead to strange ground-state properties for materials containing such a Si pseudo-atom. Therefore, test studies are requested when generating fully non-local pseudopotentials.

The generation of norm-conserving pseudopotentials, e.g. of the TM type, for first-row elements such as C and O is difficult because of the strong localization of the $2p$ states near the nucleus. Since the cutoff radii $r_{c\ell}$ should be somewhat smaller

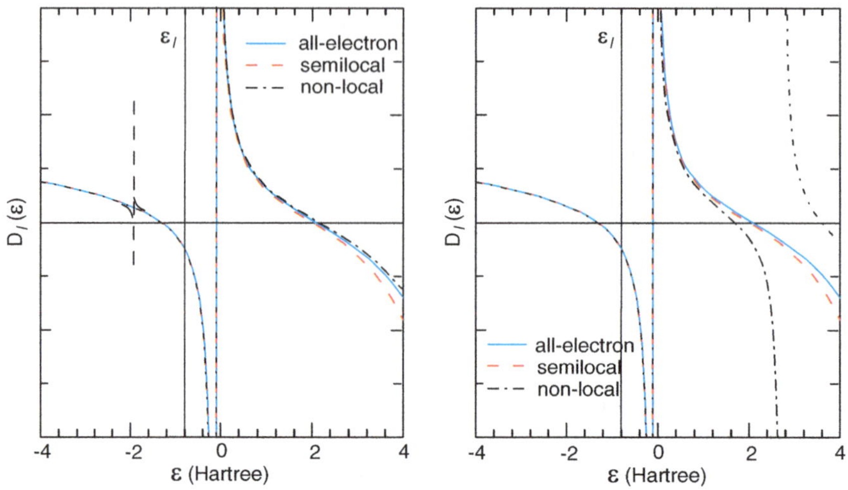

Fig. 8.14 Logarithmic derivative $D_\ell(\varepsilon, r_{c\ell})$ for Si and $\ell = 0$ derived from an all-electron calculation using the PBE-GGA exchange-correlation functional compared with results obtained for the ionic pseudopotential plotted in Fig. 8.13 using the semilocal or non-local Kleinman-Bylander representation for different core radii $r_{c0} = 1.25\ a_B$ (*left panel*) and $r_{c0} = 1.90\ a_B$ (*right panel*). After [31]

than the average radius of these states, it will be small and the generated PPs are 'hard' PPs. The expansion of the corresponding valence states in terms of plane waves therefore needs many basis functions and extremely high cutoff energies E_{cut} (8.43).

The norm-conserving constraint (iii) in Sect. 8.3.2 is the main factor responsible for the hardness of the resulting pseudopotentials. Therefore, efforts to reduce E_{cut} should focus on lifting the norm-conservation condition. This however requires a generalization of the sum rule (8.49). This has been done by Vanderbilt [65], who showed that then much smoother, but still highly transferable, non-norm-conserving pseudopotentials can be obtained. In the literature they are called Vanderbilt or ultrasoft (US) pseudopotentials. For the loss of norm conservation one has to pay a price. Important ingredients of the US pseudopotentials are augmentation charges restoring the correct norm of the pseudocharge.

The concept of Vanderbilt has been later combined with an extreme softening [30]. This procedure allows to treat first-row elements and elements with shallow d shells (considered as valence shells) with plane-wave basis sets with $E_{\text{cut}} = 13-16$ Ry without loss of accuracy of the ground-state properties in comparison to very hard and accurate PPs working at cutoff energies between 60 and 110 Ry. Resulting rather smooth wave functions are displayed for Ga and In in Fig. 8.15. Most interesting are the smooth pseudo-wave functions for the semicore Ga $3d$ and In $4d$ electrons. They even possess larger pseudoization radii than the functions for the valence s and p states.

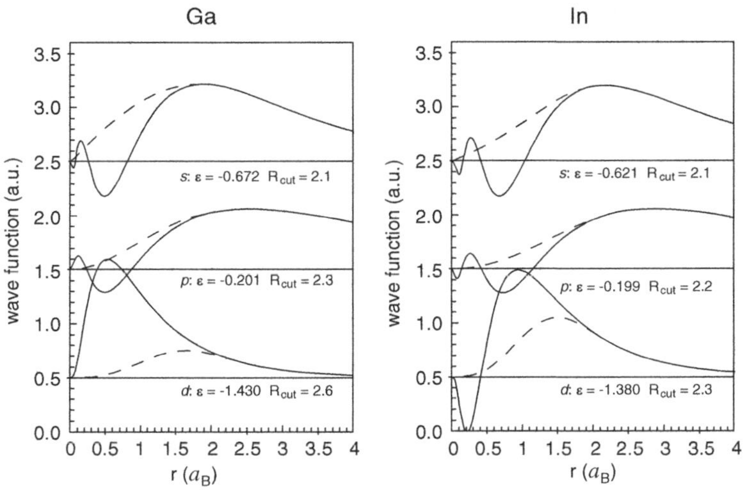

Fig. 8.15 All-electron wave functions (*solid lines*) for ns, np, and $(n-1)d$ states of Ga ($n = 4$) and In ($n = 5$). The pseudo-wave functions (*dashed lines*) of extremely softened US pseudopotentials [30] are shown for comparison. Energy eigenvalues ε (in Ry) and pseudoization radii R_{cut} (in a_B) are also given. Adapted from [66]

A natural step forward away from the original US pseudopotentials was their generation within the PAW method [32, 33], which was possible since projectors and auxiliary localized functions already appear in the ultrasoft pseudopotential method. Instead of an overlap operator a product of transformation operators, that are traced back to projection operators on atomic states within the PAW spheres, is introduced into the theory. As a result, at least in the framework of the frozen-core approximation (see Sect. 8.1), 'exact' all-electron wave functions are constructed. Actually the PAW method is not anymore a pseudopotential theory, at least from the point of view of the valence wave functions.

8.4 Non-linear Core Corrections

The unscreening process (8.51) in the construction of ionic pseudopotentials $V_{ps}^{\ell}(r)$ implicitly assumes that there is no significant overlap between core and valence electron densities. Otherwise the unscreening process leads to an error because the exchange-correlation potential and energy are not linear functions of the total electron density $n = n_v + n_c$. This is particularly valid in the case of systems with few valence electrons such as alkali atoms Na, K, Rb, and Cs [67], transition-metal atoms Fe, Co, ... where d valence states overlap with s valence states [68], and shallow d core states overlapping with s and p valence states as indicated for Ga in Fig. 8.16.

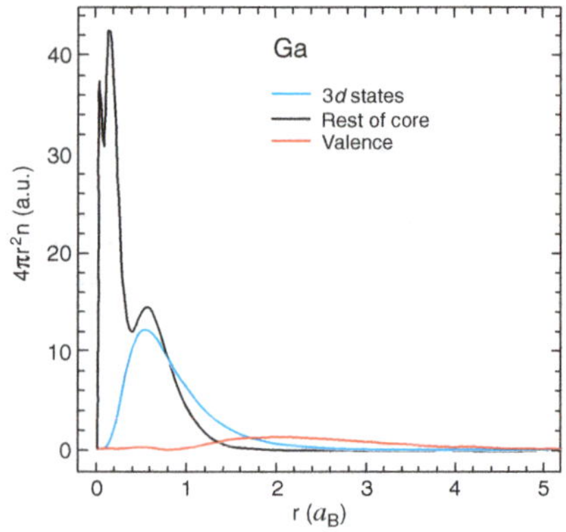

Fig. 8.16 Radial electron distribution of electrons in a Ga atom. The core electron density has been decomposed into the Ga $3d$ contribution (*blue line*) and the rest of the core electrons (*black line*). The valence electron density (*red line*) overlaps with the semicore states for small radii

A similar situation arises when the valence orbitals of one atom overlap with core orbitals of another atom as in II-VI compounds [69].

Since the Hartree potential V_{H} (6.14) is linear in the electron density it can be divided into a core contribution which is added to the screened ionic potential as indicated by (8.51). Restricting to the classical electron-electron interaction the unscreening by substrating the Hartree potential of the valence electrons is correct. However, this is not the case for the exchange and correlation effects. The unscreening by substracting $V_{\mathrm{XC}}(\mathbf{x};[\tilde{n}_v^{\mathrm{atom}}])$ is only valid for non-overlapping core and valence electron densities. Instead a corrected potential [70]

$$\tilde{V}_{\mathrm{XC}}(\mathbf{x}) = V_{\mathrm{XC}}\left(\mathbf{x};\left[\tilde{n}_v\right]\right) + \left[V_{\mathrm{XC}}\left(\mathbf{x};\left[\tilde{n}_v + n_c\right]\right) - V(\mathbf{x};\left[\tilde{n}_v\right])\right] \tag{8.53}$$

has to be applied. The term in square brackets in (8.53) is a 'core' correction. Then the actual electronic structure calculations are performed with the resulting unscreened ionic pseudopotential. The exchange-correlation contribution is computed for the total electron density, $\tilde{n}_v + n_c$, instead of the usual valence density. The frozen core electron density of the isolated atom is not the best description for n_c. Rather, for an adequate plane-wave representability n_c itself is usually replaced by an arbitrary pseudo-charge density for small radii $r \leq r_c$. Thereby, r_c may be chosen to be smaller than the smallest radius where n_v becomes negligible with respect to n_c.

References

1. L. Landau, E.M. Lifshitz, *Statistical Physics*, vol. 5 (Pergamon Press, Oxford, 1959)
2. M.-C. Desjonquires, D. Spanjaard, *Concepts in Surface Physics* (Springer, Berlin, 1996)
3. A. Zywietz, K. Karch, F. Bechstedt, Influence of polytypism on thermal properties of silicon carbide. Phys. Rev. B **54**, 1791–1799 (1996)
4. N.W. Ashcroft, N.D. Mermin, *Solid State Physics* (Saunders College, Philadelphia, 1976)
5. G. Kresse, J. Furthmüller, Efficient iterative schemes for *ab initio* total-energy calculations using a plane-wave basis set. Phys. Rev. B **54**, 11169–11186 (1996)

6. G. Kresse, J. Furthmüller, Efficiency of ab-initio total energy calculations for metals and semiconductors using a plane-wave basis set. Comput. Mater. Sci. **6**, 11–50 (1996)
7. A. Schleife, F. Fuchs, J. Furthmüller, F. Bechstedt, First-principles study of ground- and excited-state properties of MgO, ZnO, and CdO polymorphs. Phys. Rev. B **73**, 245212 (2006)
8. F.D. Murnaghan, The compressibility of media under extreme pressure. Proc. Natl. Acad. Sci. USA **30**, 244–247 (1944)
9. F. Birch, Elasticity and constitution of the Earth's interior. J. Geophys. Res. **57**, 227–286 (1952)
10. P. Vinet, J. Ferrante, J.R. Smith, J.H. Rose, A universal equation of state for solids. J. Phys. C **19**, L467–L473 (1986)
11. P. Vinet, J.H. Rose, J. Ferrante, J.R. Smith, Universal features of the equation of state of solids. J. Phys. Condens. Matter **1**, 1941–1963 (1989)
12. J.M. Recio, M.A. Blanco, V. Luaña, R. Pandey, L. Gerward, J.S. Olsen, Compressibility of the high-pressure rocksalt phase of ZnO. Phys. Rev. B **58**, 8949–8954 (1998)
13. U. Grossner, J. Furthmüller, F. Bechstedt, Initial stages of III-nitride growth. Appl. Phys. Lett. **74**, 3851–3853 (1999)
14. A. Stekolnikov, J. Furthmüller, F. Bechstedt, Absolute surface energies of group-IV semiconductors: dependence on orientation and reconstruction. Phys. Rev. B **65**, 115318 (2002)
15. C. Kittel, *Introduction to Solid State Physics* (Wiley, New York, 2005)
16. D.R. Lide (ed.), *Handbook of Chemistry and Physics*, 79th edn. (CRC, Boca Raton, 1998)
17. R. Reif, *Fundamentals of Statistical and Thermal Physics* (McGraw Hill, New York, 1965)
18. W.A. Harrison, *Electronic Structure and the Properties of Solids* (Dover, New York, 1989)
19. O. Kubaschewski, C.B. Alcock, *Metallurgical Thermochemistry* (Pergamon Press, Oxford, 1979)
20. A. Kley, Theoretische Untersuchungen zur Atomdiffusion auf niederinduzierten Oberflächen von GaAs. Ph.D. thesis, Technical University of Berlin (1997)
21. A. Zywietz, J. Furthmüller, F. Bechstedt, Vacancies in SiC: influence of Jahn-Teller distortions, spin effects, and crystal structure. Phys. Rev. B **59**, 15166–15180 (1999)
22. W.G. Schmidt, III-V compound semiconductor (001) surfaces. Appl. Phys. A. Mater. Sci. Process. **75**, 89–99 (2002)
23. W.G. Schmidt, P.H. Hahn, F. Bechstedt, N. Esser, P. Vogt, A. Wanger, W. Richter, InP(001)-(2×1) surface: a hydrogen stabilized structure. Phys. Rev. Lett. **90**, 126101 (2003)
24. W.G. Schmidt, B. Wenzien, F. Bechstedt, Chemisorption of antimony on GaAs(110). Phys. Rev. B **49**, 4731–4744 (1994)
25. W.G. Schmidt, F. Bechstedt, G.P. Srivastava, Adsorption of group-V elements on III-V(110) surfaces. Surf. Sci. Rep. **25**, 141–225 (1996)
26. R.P. Feynman, Forces in molecules. Phys. Rev. **56**, 340–343 (1939)
27. H. Hellmann, *Einführung in die Quantenchemie* (Deuticke, Leipzig, 1937)
28. G.P. Srivastava, D. Weaire, The theory of the cohesive energy of solids. Adv. Phys. **36**, 463–517 (1987)
29. P. Pulay, Ab initio calculation of force constants and equilibrium geometries in polyatomic molecules. I. Theory. Mol. Phys. **17**, 197–204 (1969)
30. J. Furthmüller, P. Käckell, F. Bechstedt, G. Kresse, Extreme softening of Vanderbilt pseudopotentials: general rules and case studies of first-row and d-electron elements. Phys. Rev. B **61**, 4576–4587 (2000)
31. A. Hermann, Ab intio Untersuchung eines molekularen π-Elektronensystems auf der Si(001)-Oberfläche. Diploma thesis, Friedrich-Schiller-Universität Jena (2004)
32. G. Kresse, D. Joubert, From ultrasoft pseudopotentials to the projector augmented-wave method. Phys. Rev. B **59**, 1758–1775 (1999)
33. P. Blöchl, Projector augmented-wave method. Phys. Rev. B **50**, 17953–17979 (1994)
34. D.J. Singh, *Plane Waves, Pseudopotentials, and the APW Method* (Kluwer Academic Publishers, Boston, 1994)
35. O.K. Andersen, Linear methods in band theory. Phys. Rev. B **12**, 3060–3083 (1975)
36. H. Skriver, *The LMTO Method* (Springer, New York, 1984)

37. J. Korringa, On the calculation of the energy of a Bloch wave in a metal. Physica **13**, 392–400 (1947)
38. W. Kohn, N. Rostocker, Solution of the Schrödinger equation in periodic lattices with an application to metallic lithium. Phys. Rev. **94**, 1111–1120 (1994)
39. A. Baldereschi, Mean-value point in the Brillouin zone. Phys. Rev. B **7**, 5212–5215 (1973)
40. D.J. Chadi, M.L. Cohen, Special points in the Brillouin zone. Phys. Rev. B **8**, 5747–5753 (1973)
41. H.J. Monkhorst, J.D. Pack, Special points for Brillouin-zone integrations. Phys. Rev. B **13**, 5188–5192 (1976)
42. J. Furthmüller, P.H. Hahn, F. Fuchs, F. Bechstedt, Band structures and optical spectra of InN polymorphs: influence of quasiparticle and excitonic effects. Phys. Rev. B **72**, 205106 (2005)
43. C. Herring, A new method for calculating wave functions in crystals. Phys. Rev. **57**, 1169–1177 (1940)
44. J.C. Philips, L. Kleinman, New method for calculating wave functions in crystals and molecules. Phys. Rev. **116**, 287–294 (1959)
45. M.L. Cohen, V. Heine, The fitting of pseudopotentials to experimental data and their subsequent application, in *Solid State Physics*, vol. 27, ed. by H. Ehrenreich, F. Seitz, D. Turnbull (Academic, New York, 1970), pp. 37–248
46. M.L. Cohen, J.R. Chelikowsky, *Electronic Structure and Optical Properties of Semiconductors* (Springer, Berlin, 1988)
47. W.A. Harrison, *Pseudopotentials in the Theory of Metals* (Benjamin, New York, 1966)
48. D.R. Hamann, M. Schlüter, C. Chiang, Norm-conserving pseudopotentials. Phys. Rev. Lett. **43**, 1494–1497 (1979)
49. G.B. Bachelet, D.R. Hamann, M. Schlüter, Pseudopotentials that work: from H to Pu. Phys. Rev. B **26**, 4199–4228 (1982)
50. K. Karch, Ab-initio Berechnung von statischen und dynamischen Eigenschaften des Diamanten, Siliziums und Siliziumcarbids. Ph.D. thesis, University of Regensburg (1993)
51. J.M. Ziman, *Principles of the Theory of Solids* (Cambridge University Press, London, 1972)
52. R.W. Shaw, W.A. Harrison, Reformulation of the screened Heine-Abarenkov model potential. Phys. Rev. **163**, 604–611 (1967)
53. W.C. Topp, J.J. Hopfield, Chemically motivated pseudopotential for sodium. Phys. Rev. B **7**, 1295–1303 (1973)
54. X. Gonze, R. Stumpf, M. Scheffler, Analysis of separable potentials. Phys. Rev. B **44**, 8503–8513 (1991)
55. A.M. Rappe, M. Rabe, E. Kaxiras, J.D. Joannopoulos, Optimized pseudopotentials. Phys. Rev. B **41**, 1227–1230 (1990)
56. N. Troullier, J.L. Martins, Efficient pseudopotentials for plane-wave calculations. Phys. Rev. B **43**, 1993–2006 (1991)
57. M. Fuchs, M. Scheffer, Ab initio pseudopotentials for electronic structure calculations of polyatomic systems using density functional theory. Comput. Phys. Commun. **119**, 67–98 (1999)
58. M. Preuss, Ab-initio-Berechnungen von Grund- und Anregungseigenschaften der DNA-Basen Adenin, Cytosin, Guanin und Thymin. Diploma thesis, Friedrich-Schiller-Universität Jena (2003)
59. W.E. Pickett, Pseudopotential methods in condensed matter applications. Comput. Phys. Rep. **9**, 115–197 (1989)
60. M.S. Hybertsen, S.G. Louie, Spin-orbit splitting in semiconductors and insulators from the ab initio pseudopotential. Phys. Rev. B **34**, 2920–2922 (1986)
61. G. Theurich, N.A. Hill, Self-consistent treatment of spin-orbit coupling in solids using relativistic fully separable ab initio pseudopotentials. Phys. Rev. B **64**, 073106 (2001)
62. U. Gerstmann, N.J. Vollmers, A. Lücke, M. Babilon, W.G. Schmidt, Rashba splitting and relativistic energy shifts in In/Si(111) nanowires. Phys. Rev. B **89**, 165431 (2014)
63. L. Kleinman, D.M. Bylander, Efficacious form for model pseudopotentials. Phys. Rev. Lett. **48**, 1425–1428 (1982)
64. X. Gonze, P. Käckell, M. Scheffler, Ghost states for separable norm-conserving, ab initio pseudopotentials. Phys. Rev. B **41**, 12264–12267 (1990)

65. D. Vanderbilt, Soft self-consistent pseudopotentials in a generalized eigenvalue formation. Phys. Rev. B **41**, 7892–7895 (1990)
66. U. Grossner, Influence of polytypism and surfaces on wide bandgap semiconductors. Ph.D. thesis, Friedrich-Schiller-Universität Jena (2000)
67. J. Hebenstreit, M. Heinemann, M. Scheffler, Atomic and electronic structures of GaAs(110) and their alkali-adsorption-induced changes. Phys. Rev. Lett. **67**, 1031–1034 (1991)
68. J. Zhu, X.W. Wang, S.G. Louie, First-principles pseudopotential calculations for magnetic iron. Phys. Rev. B **45**, 8887–8893 (1992)
69. G.E. Engel, R.J. Needs, Calculations of the structural properties of cubic zinc sulfide. Phys. Rev. B **41**, 7876–7778 (1990)
70. S.G. Louie, S. Froyen, M.L. Cohen, Nonlinear ionic pseudopotentials in spin-density-functional calculations. Phys. Rev. B **26**, 1738–1742 (1982)

Chapter 9
Non-local Exchange and Correlation

Abstract Despite the general success of the local and semilocal approximations for exchange and correlation, limitations or even failures of the density functional theory become obvious for subsystems of localized d or f electrons. In particular, the correlation is not correctly described. One speaks about 'strongly correlated' electrons and materials. One idea to solve the correlation problem is the description of the missing effect by on-site Coulomb interactions. The implementation of this idea is illustrated and explicitly presented for the case, where the interaction beyond the local or semilocal treatment is described by an empirical Hubbard-like parameter U. For the resulting DFT+U approach a special scheme of application and the influence on some properties of atoms with partially filled $3d$ shells are presented. The localization of some electronic states also ask for a better treatment of the non-locality inherent in the exchange interaction and therefore for a generalized density functional theory. The local or semilocal exchange-correlation functional is replaced by a 'hybrid' one, that is a combination of the orbital-dependent Hartree-Fock and a common density functional. Meanwhile, many hybrid functionals exist. They are not only more accurate as far as energetics is concerned but also open the fundamental gaps in the electronic structure of non-metals. The implementation of the adiabatic-connection fluctuation-dissipation theorem is a promising way to describe explicitly the van der Waals interaction.

9.1 Hubbard U Correction to Density Functional Theory

9.1.1 Problem and Idea

Transition metal oxides (TMOs) and rare-earth compounds are characterized by well-localized d or f orbitals. These systems exhibit phenomena associated with electron correlation such as metal-insulator transitions, heavy fermion behavior, and high-temperature superconductivity. Other phenomena related to the localization of electrons are such as the formation of a Luttinger liquid instead of a Fermi liquid in one-dimensional systems. All these phenomena cannot be correctly described within the conventional DFT applying a local (see Sect. 7.2) or a semilocal (see Sect. 7.3)

F. Bechstedt, *Many-Body Approach to Electronic Excitations*,
Springer Series in Solid-State Sciences 181, DOI 10.1007/978-3-662-44593-8_9

approximation to exchange and correlation. Therefore, sometimes these phenomena are roughly explained as effects of "strong correlation of electrons". Some practitioners of this concept probably mean thereby effects due to "strong correlation" or "some things not described correctly by the common local and semilocal density functional approximations" [1].

Many of these "correlation" effects are large and important on the scale of bonding energies, for instance, when they arise from subsystems of localized d or f electrons. Then they are referred to static correlations [2]. The localization of these electrons leads to strong on-site correlations, such that if an electron is occupying a state localized in a particular atomic site $\mathbf{R}_l$ (1.7), placing a second electron in the same site is penalized with an additional repulsive energy U. The central idea of improving the local or semilocal approximations is therefore to push the occupied part of the d or f shell downward in energy, and to make sure that an additional electron added to such a shell has an energy above the Fermi energy of the electron gas.

The idea of an on-site Coulomb repulsion was originally formulated at the level of an empirical Hamiltonian with a site representation of the electron-electron interaction by Hubbard [3, 4]. The study of the Hubbard model is a field in itself [5], being the simplest non-trivial model for correlated electrons. However, this field is beyond the scope of this book and, hence, should not be presented here.

Besides the physical approach toward "strong correlation" one aspect of the interaction of strongly localized electrons such as d and f electrons can be also discussed improving the treatment of exchange and correlation presented in Sects. 7.2 and 7.3 within local or semilocal approximations. In such an approximation an electron moves in a mean field created by all particles in the system including itself. The resulting *self-interaction* is partially taken care of by the exchange-correlation potentials (7.12) or (7.13) with the XC energies per particle (7.17) or (7.25) but not in a complete way like in the Hartree-Fock approximation [see discussion below (4.17)]. There is a common belief [6–8] that the residual self-interaction in L(S)DA or GGA leads to a systematic underestimate of the fundamental gap values in semiconductors and insulators. Although the inclusion of self-energy corrections (SICs) [9, 10] indeed opens the gaps, one has to remember that the DFT is a ground-state theory and not an excited-state theory. Therefore, in the framework of such a theory one cannot request to obtain a correct quantity related to excitation of the electronic system.

9.1.2 Around Mean Field Corrections

The basic idea behind a DFT+U method is to treat the strong correlations of d or f electrons more accurately than in a local or semilocal approximation of exchange and correlation. The intention of DFT+U is to combine conventional DFT calculations with an improved treatment of the electron-electron interaction at one atomic site where the d or f electrons are localized. Such an improvement is usually described by a correction ΔE to the Kohn-Sham energy (6.12) with XC (7.17) or (7.25)

$$E_{\text{DFT}+U} = E_{\text{KS}}\left[n_{\frac{1}{2}}, n_{-\frac{1}{2}}\right] + \Delta E\left[\tilde{n}_{l\frac{1}{2}}, \tilde{n}_{l-\frac{1}{2}}\right] \tag{9.1}$$

with the density matrix $\tilde{n}_{lm_s}$ of the d or f shell in the spin channel $m_s = \pm\frac{1}{2}$ of an atom or ion at the position $\mathbf{R}_l$. The system is assumed to be in the ground state. In principle, locally at the site $\mathbf{R}_l$ a collinear spin polarization is assumed.

The d or f shell of the ion at $\mathbf{R}_l$ can be nearly described within a spherically symmetric potential and hence by quantum numbers $\lambda = n\ell m$ [see (8.45)]. Such a shell is illustrated in Fig. 9.1 for the late transition metal (TM) ions TM^{2+} with $n = 3$ and $\ell = 2$. Since the principal (n) and the angular momentum (ℓ) quan-

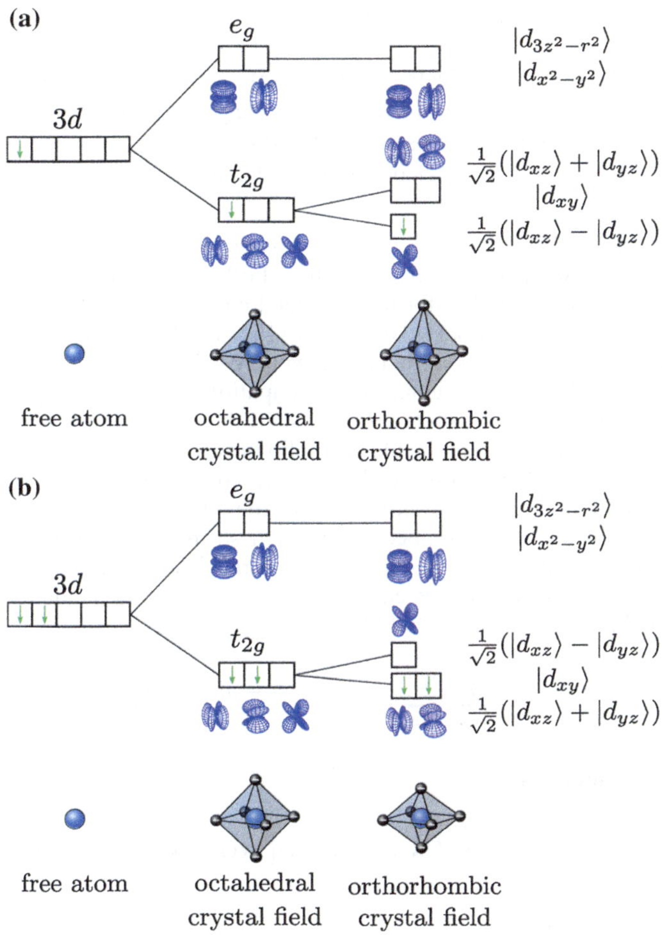

Fig. 9.1 $3d$ level in the minority-spin channel and its splitting in various crystal fields for (**a**) Fe^{2+} and (**b**) Co^{2+} ions. The d orbitals and their irreducible representations e_g and t_{2g} in octahedral environments are indicated. From [12]

tum number are fixed in such a shell we will later replace λ only by the magnetic quantum number m. In general, the density matrices (omitting the quantum numbers n and ℓ) of this shell $\tilde{n}_{lm_s}$ are not diagonal in the quantum numbers m in contrast to the occupation numbers (4.10) in the case of the Hartree-Fock approximation. Rather, in each spin channel we take the full 5×5 (d states) or 7×7 (f states) matrices with elements $n^{m_s}_{mm'}$ into account. Here, the site index l is eliminated. In principle, a generalization to non-collinear spins is possible. Then also the matrix character with respect to the spin quantum numbers m_s and m_s' has to be considered. This is indeed necessary for investigations of the magnetic anisotropy and magnetic-ordering-induced phenomena [11, 12].

The energy correction (9.1) can be formulated as the difference

$$\Delta E\left[\tilde{n}_{l\frac{1}{2}}, \tilde{n}_{l-\frac{1}{2}}\right] = E_{ee}\left[\tilde{n}_{l\frac{1}{2}}, \tilde{n}_{l-\frac{1}{2}}\right] - \bar{E}_{ee}\left[\bar{n}_{l\frac{1}{2}}, \bar{n}_{l-\frac{1}{2}}\right] \tag{9.2}$$

of the energy E_{ee} due to the electron-electron interaction of the localized electrons whereas the double-counting term $\bar{E}_{ee}$ removes the part that was already included in the energy E_{KS} within a local or semilocal approximation of the XC contribution. The averaged occupation number in one spin channel is thereby given by $\bar{n}_{lm_s} = \frac{1}{2\ell+1}\text{Tr}\tilde{n}_{lm_s}$.

According to (3.37) the contribution E_{ee} to the total energy is determined by the spin-averaged two-particle density matrix. Following Anisimov et al. [13, 14] the leading contribution to E_{ee} for the localized electrons in a d or f shell can be approximated in a Hartree-Fock-like fashion [15] based on expressions (4.15) and (4.16) for the Fock and Hartree term, respectively. We further assume that the density matrix of the d or f shell is diagonal with occupation numbers $n_{lmm_s} = 0, 1$. Then, for an isolated atom/ion with such a d or f shell the expressions (4.15) and (4.16) can be directly applied. For a molecule or a solid the arrangement of such atoms/ions belonging to one spin channel on sites $\mathbf{R}_l$ the corresponding wave functions $\varphi_{\lambda m_s}(\mathbf{x})$ have to be rewritten in a linear combination of the wave functions $\phi_{n\ell mm_s}(\mathbf{x}-\mathbf{R}_l)$ of the isolated objects. However, because of their strong localization the overlap of wave functions from different sites and the Coulomb interaction of d ($\ell = 2$) and f ($\ell = 3$) electrons at different sites can be neglected. Omitting the denotation of the shell $n\ell$ one obtains from (4.15) and (4.16) ($-\ell \leq m \leq \ell$)

$$\Delta E_{ee}\left[\tilde{n}_{l\frac{1}{2}}, \tilde{n}_{l-\frac{1}{2}}\right] = \frac{1}{2}\sum_l \sum_{m,m'} \sum_{m_s,m_s'} \left[n_{lmm_s} n_{lm'm_s'} - \bar{n}_{lm_s}\bar{n}_{lm_s'}\right]\left[U^{m_s m_s'}_{mm'} - \delta_{m_s m_s'} J^{m_s m_s'}_{mm'}\right] \tag{9.3}$$

with Coulomb integrals

$$U^{m_s m_{s'}}_{mm'} = \int d^3\mathbf{x} \int d^3\mathbf{x}' \left|\phi_{n\ell mm_s}(\mathbf{x})\right|^2 \text{v}(\mathbf{x}-\mathbf{x}') \left|\phi_{n\ell m'm_s'}(\mathbf{x}')\right|^2 \tag{9.4}$$

and exchange integrals

$$J^{m_s m_s'}_{mm'} = \int d^3\mathbf{x} \int d^3\mathbf{x}' \phi^*_{n\ell mm_s}(\mathbf{x})\phi_{n\ell m'm_s'}(\mathbf{x})\text{v}(\mathbf{x}-\mathbf{x}')\phi^*_{n\ell m'm_s'}(\mathbf{x}')\phi_{n\ell mm_s}(\mathbf{x}'). \tag{9.5}$$

In the Coulomb and exchange integrals (9.4) and (9.5) one may replace the bare Coulomb potential $v(\mathbf{x}-\mathbf{x}')$ by a screened one in order to simulate the influence of an environment in a solid or molecule [16]. The occupation numbers

$$n_{lmm_s} = n_{mm_s}$$

are equal for chemically identical atoms/ions with d or f shells and, therefore, independent of the site $\mathbf{R}_l$. The resulting correction term ΔE_{ee} (9.3) is called around mean field (AFM) corrections [17]. The subtracted contributions proportional to the average occupation numbers $\bar{n}_{lmm_s} = \bar{n}_{lm_s}$ simulate the electron-electron interaction taken into account within the DFT with a local or semilocal approximation of XC on the $n\ell$ shell. The first factor in (9.3) indicates that double counting of interactions will not occur.

9.1.3 Rotationally Invariant Scheme

In a first step of approximations the influence of the spin polarization on the spatial parts $\phi_{n\ell mm_s}(\mathbf{x}) = \phi_{n\ell m}(\mathbf{x})$ of the spinors is neglected, i.e., we assume $U_{mm'}^{m_s m_s'} \approx U_{mm'}$ and $J_{mm'}^{m_s m_{s'}} \approx J_{mm'}$. There are papers in the literature where expressions of the type (9.3) with integrals $U_{mm'}$ and $J_{mm'}$ are interpreted independent of the spin polarization as a consequence of an unrestricted Hartree-Fock approximation for the localized electrons on the d or f shell [16, 18]. In a second step, the difference between the occupation numbers and their average values in (9.3) is reformulated to

$$n_{lmm_s} n_{lm'm_s'} - \bar{n}_{lm_s}\bar{n}_{lm_s'} = n_{lmm_s}\bar{n}_{lm_s'} + n_{lm'm_s'}\bar{n}_{lm_s} - 2\bar{n}_{lm_s}\bar{n}_{lm_s'} + \delta n_{lmm_s}\delta n_{lm'm_s'} \tag{9.6}$$

with the occupation fluctuations for different d or f states

$$\delta n_{lmm_s} = n_{lmm_s} - \bar{n}_{lm_s}. \tag{9.7}$$

Together with the symmetry $U_{mm'} = U_{m'm}$ and $J_{mm'} = J_{m'm}$ it can be demonstrated that summing up over m and m' in (9.3) the contributions of the first three terms on the right-hand side of (9.6) vanish. One obtains

$$\Delta E_{ee}\left[n_{l\frac{1}{2}}, n_{l-\frac{1}{2}}\right] = \frac{1}{2}\sum_{l}\sum_{m,m'}\sum_{m_s,m_{s'}} \delta n_{lmm_s}\delta n_{lm'm_s'}\left[U_{mm'} - \delta_{m_s m_s'} J_{mm'}\right]. \tag{9.8}$$

The correction to the DFT total energy is determined by the occupation fluctuations δn_{lmm_s} on the d or f shell. The result (9.8) motivates the denotation AMF since the fluctuations are not neglected as in a true mean-field theory while the electron-electron interaction is proportional to the occupation numbers themselves as in the HF approximation (4.32).

The summation over the spin quantum numbers $m_s = \pm\frac{1}{2}$ can be rewritten using $1 = (1 - \delta_{m_s m_s'}) + \delta_{m_s m_s'}$ in a contribution due to opposite spins and another one with equal spins

$$\Delta E_{ee}\left[n_{l\frac{1}{2}}, n_{l-\frac{1}{2}}\right] = \frac{1}{2}\sum_{l}\left\{\sum_{m,m'}\sum_{m_s} U_{mm'}\delta n_{lmm_s}\delta n_{lm'-m_s} \right.$$
$$\left. + \sum_{m,m'}{}'\sum_{m_s}(U_{mm'} - J_{mm'})\,\delta n_{lmm_s}\delta n_{lm'm_s}\right\}. \quad (9.9)$$

The contribution (9.9) to the total energy has been derived using the Hartree and Fock expressions (4.15) and (4.16), respectively, of the HF approximation. Nevertheless its interpretation is difficult. Because of the dominant occupation fluctuations it represents some electronic correlation. Discussing neutral electronic excitations, e.g. in (4.38), we have learnt that changes of the occupation with respect to the level occupation in the ground state may be interpreted to give rise to some correlation effects. On the other hand, expression (9.9) contains both direct Coulomb interactions $U_{mm'}$, which operate irrespective of the spin orientation, and exchange integrals $J_{mm'}$, which only act for parallel spins.

The AMF approximation has enormous consequences for the XC potential acting on the electrons on the d or f shells with quantum numbers mm_s at site $\mathbf{R}_l$. The XC contribution $V_{XC}^{m_s}(\mathbf{x})$ (7.13) to the Kohn-Sham potential has to be replaced according to [17]

$$V_{XC}^{m_s}(\mathbf{x}) \rightarrow V_{XC}^{m_s}(\mathbf{x}) + \sum_{m'} U_{mm'}\delta n_{lm'-m_s} + \sum_{\substack{m'\\(m\neq m')}} (U_{mm'} - J_{mm'})\delta n_{lm'm_s}. \quad (9.10)$$

The KS potential is modified by a site- and orbital-dependent non-local potential. It is a consequence of the site- and orbital-dependent AFM energy correction (9.9) to the total XC energy. Expression (9.10) indicates that the conventional density functional theory refers to a charge density of d or f electrons at a site $\mathbf{R}_l$ determined by an average occupation number $\bar{n}_{lm_s}$ of the states in one spin channel. The corrections are however related to the fluctuations δn_{lmm_s} of these numbers.

At each site $\mathbf{R}_l$ a spherical symmetry of the d- or f-electron containing ion/atom is assumed. Then, average values $\frac{1}{2\ell+1}\sum_{m'} U_{mm'} = \bar{U}$ and $\frac{1}{2\ell+1}\sum_{m'}(U_{mm'} - J_{mm'}) = (\bar{U} - \bar{J})$ do not depend on m [17]. The $\sum_{m'}$ summation corresponds to a spherical average. As an additional approximation, in a third step, frequently the Coulomb and exchange matrix elements in (9.9) are replaced by the average values $\bar{U}$ and $(\bar{U} - \bar{J})$ resulting in a rotationally invariant scheme [13, 17, 18]

$$\Delta E_{ee}\left[n_{l\frac{1}{2}}, n_{l-\frac{1}{2}}\right] = \frac{1}{2}\sum_{l}\left\{\bar{U}\sum_{m_s}\sum_{m,m'}\delta n_{lmm_s}\delta n_{lm'-m_s} + (\bar{U}-\bar{J})\sum_{m_s}{\sum_{m,m'}}' \delta n_{lmm_s}\delta n_{lm'm_s}\right\}. \tag{9.11}$$

For a more compact formulation of this result which is obviously not influenced by the double-counting problem, we go back to the original expression (9.3) but now in the rotationally invariant approach. With the number $N_{lm_s} = \sum_m n_{lmm_s} = (2\ell+1)\bar{n}_{lm_s}$ of d or f electrons at site $\mathbf{R}_l$ in the spin channel m_s and guaranteeing that the correction is free of self-interaction, one finds

$$\Delta E_{ee}\left[n_{l\frac{1}{2}}, n_{l-\frac{1}{2}}\right] = \frac{1}{2}\sum_{l}\left\{\bar{U}\sum_{m_s}\left[\sum_{m,m'} n_{lmm_s}n_{lm'-m_s} - N_{lm_s}N_{l-m_s}\right] + (\bar{U}-\bar{J})\sum_{m_s}\left[{\sum_{m,m'}}' n_{lmm_s}n_{lm'm_s} - N_{lm_s}(N_{lm_s}-1)\right]\right\}. \tag{9.12}$$

Applying the definition of the number N_{lm_s} of electrons the disappearance of the first term is obvious. In the rotationally invariant approach the two spin channels are decoupled. In the second term only $m = m'$ diagonal contributions remain. With $N_{lm_s} = \sum_m n_{lmm_s}$ and (9.1) it follows [18]

$$E_{\mathrm{DFT}+U} = E_{\mathrm{KS}}\left[n_{\frac{1}{2}}, n_{-\frac{1}{2}}\right] + \frac{1}{2}\sum_{l} U \sum_{m,m_s}\left(n_{lmm_s} - n^2_{lmm_s}\right). \tag{9.13}$$

The abbreviation

$$U = \bar{U} - \bar{J} \tag{9.14}$$

indicates that only the difference $(\bar{U} - \bar{J})$ between the average on-site repulsion and exchange parameters enters the corrected DFT functional. Therefore, the values for U given in many papers are effective values $(\bar{U} - \bar{J})$. Sometimes the description of the correction $\sim U$ (9.13) is called the Dudarev et al. [18] or, in short, Dudarev scheme.

The effect of the AMF correction term can be now interpreted much easier than in (9.10) by adding the derivative of the functional (9.13) with respect to n_{lmm_s} to the KS potential,

$$V^{m_s}_{\mathrm{XC}}(\mathbf{x}) \to V^{m_s}_{\mathrm{XC}}(\mathbf{x}) + U\left(\frac{1}{2} - n_{lmm_s}\right). \tag{9.15}$$

Its main physical effect is to push filled localized states down and empty ones up in energy, what may effectively open a gap that might be absent in the local or semilocal DFT description. Sometimes, in the literature [17] one finds comments that for the localized levels $U = \bar{U} - \bar{J}$ may be identified with the difference $I - A$, where I and A are the first ionization potential (in solids: ionization energy) and electron affinity, respectively. In this way the correction $\sim U$ in (9.15) mimics the discontinuity of the XC potential (6.27) when going from occupied to unoccupied states.

Finally, we have to mention that the representation of the AMF corrections in (9.13) and (9.15) can be generalized to density matrices $\tilde{n}_{lm_s}$ [18] in such a way that the representations are invariant against unitary transformations in the subspace of the investigated localized orbitals. Such a generalization gives a direct relationship between the orbital-dependent formulation of Anisimov et al. [13] and the rotationally invariant functional suggested by Lichtenstein et al. [15].

9.1.4 Examples

In order to illustrate the action of the AMF energy correction in (9.13) and its consequences for properties of matter as well as interpretations of the underlying physics we consider U (9.14) as an empirical parameter. Then, it can be derived from a fit to obtain correctly a certain property of a material with (partly) filled d- or f-electron shells.

The first example is related to the energetic ordering of different crystal structures of a magnetic system. As a prototypical material the $3d$-transition metal oxide MnO is studied in rocksalt (rs), wurtzite (wz), and zinc-blende (zb) geometries together with the ferromagnetic (FM) and five antiferromagnetic (AF) orderings of the Mn $3d$ spins. These crystal structures and magnetic orderings are illustrated in Fig. 9.2. The total energy calculations using expression (8.29) for E_{tot} and the GGA-PW91

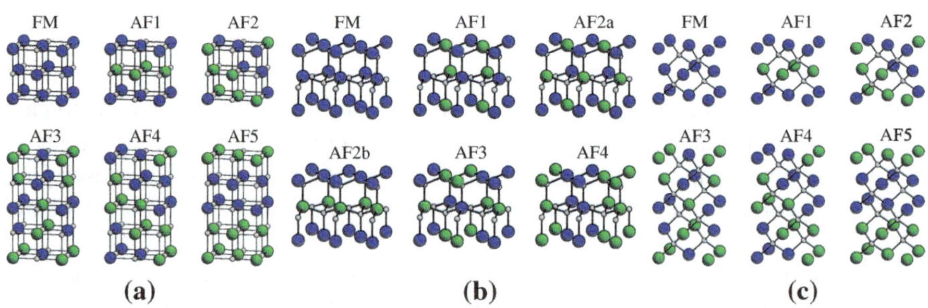

Fig. 9.2 The ferromagnetic and five antiferromagnetic orderings of the Mn $3d$ spins within (**a**) rocksalt, (**b**) wurtzite, and (**c**) zinc-blende crystal structure. The big *green* and *blue dots* represent Mn^{2+} ions with opposite spin directions. The oxygen O^{2-} ions are indicated by small *grey spheres*. From [19]

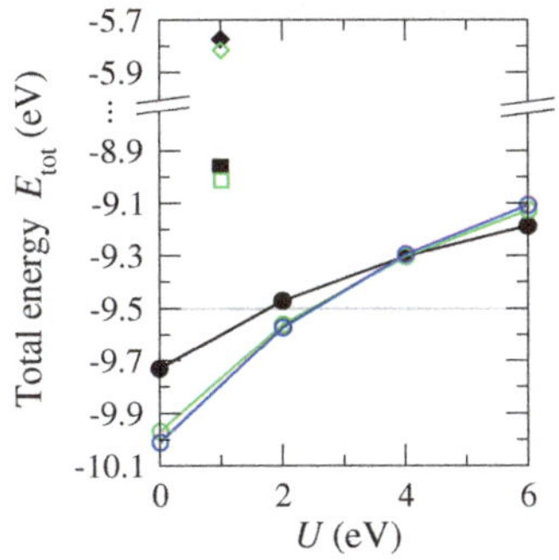

Fig. 9.3 Total energy E_{tot} (8.29) of MnO per cation-anion pair versus the parameter U for the energetically favorable magnetic arrangement of three crystal structures: AF2 in rocksalt (*black filled circles*), AF3 in wurtzite (*green spheres*), and AF3 in zinc blende (*blue spheres*). For comparison total energies in HF approximation (*diamonds*) and a hybrid XC functional HSE03 (*squares*) are also given for rs-AF2 (*black*) and wz-AF3 (*green*). The horizontal *dotted line* indicates the measured energy [20]. From [19]

XC functional (see Sect. 7.3.2) demonstrate that the studied crystal structures lead to local or global minima on the energy surface with the lowest energy for the antiferromagnetic ordering rs-AF2, wz-AF3, and zb-AF3 shown in Fig. 9.2 [19].

The antiferromagnetic ordering is in agreement with experimental findings. However, Fig. 9.3 shows that the spin-polarized total energy calculations with $U = 0$ favor the fourfold coordination of the atoms in wurtzite (or even zinc-blende) structure versus the sixfold one in rocksalt geometry. This wrong energetic ordering is also observed within the HF approximation (4.13), where correlation is omitted, and a hybrid XC functional of the HSE type [21] (see Sect. 9.2). In addition, Fig. 9.3 demonstrates a significant effect of a Hubbard U, i.e., a strong on-site interaction of the Mn $3d$ electrons, added to the GGA XC functional according to (9.13). With increasing U, the total energy increases for the considered three crystal structures, with the two tetrahedrally bonded structures being very close in energy. For values of $U \gtrsim 4$ eV, the rocksalt geometry becomes favorable in comparison to both the zb and wz structures, what is in agreement with experimental findings for ambient conditions, at least below the Néel temperature of 116 K [22]. The inclusion of the on-site repulsion leads to the correct picture of the energetics of MnO. The general overestimation of electron correlation for fractionally occupied subsystems, such as the partially filled d shells in MnO, within a semilocal GGA treatment of XC is reduced by the correction term in (9.13) but in dependence on bonding coordination and, hence, crystal structure.

Among the correlated $3d$-transition metal oxides MnO, FeO, CoO, and NiO that crystallize in ideal rocksalt geometry in their paramagnetic phase above Néel temperature or with small lattice perturbations within the antiferromagnetic AF2 arrangement (see Fig. 9.2) below this critical temperature, MnO has a comparatively easily understandable electronic structure. The Mn $3d$ shell of the Mn^{2+} ions is half-filled.

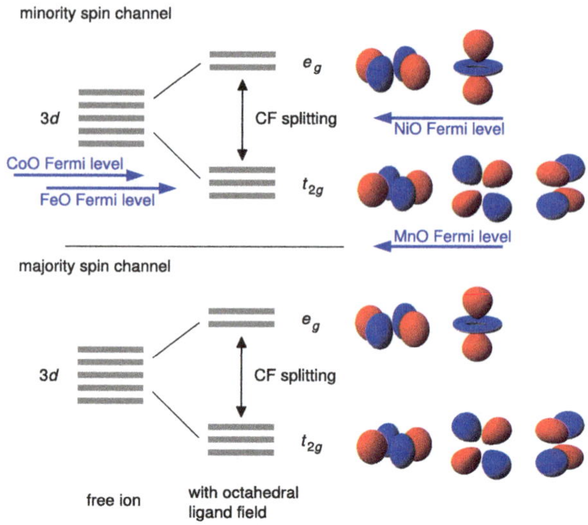

Fig. 9.4 3d energy levels of TM^{2+} ions (schematically) in TM monoxides. The symmetry of the d states, their splitting in the octahedral field, and their occupation in the oxides are indicated

In the octahedral crystal field the $t_{2g}\uparrow$ and $e_g\uparrow$ states of the majority-spin channel are occupied. The three $t_{2g}\downarrow$ and two $e_g\downarrow$ states of the minority-spin channel (cf. Fig. 9.4) are empty. The fundamental gap is related to the exchange splitting between the two spin channels. In the case of NiO the $t_{2g}\downarrow$ subshell is however completely filled with electrons. The fundamental gap is assigned by the crystal-field (CF) splitting in the minority-spin channel. As a consequence the two monoxides possess an insulating band structure with a fundamental gap between occupied e_g and empty t_{2g} states in MnO but between occupied t_{2g} and empty e_g states in NiO even in a local (LDA) or semilocal (GGA) treatment of XC [23] (see also Fig. 9.4).

The situation is much more complicated for FeO and CoO. In their antiferromagnetically ordered rocksalt geometry the minority-spin t_{2g} shell is partially filled with one (Fe^{2+}) or two (Co^{2+}) electrons (cf. Fig. 9.4). Usually in a DFT-LDA or DFT-GGA treatment their Kohn-Sham band structure does not exhibit a fundamental gap. Rather, the corresponding XC treatment leads to a metallic phase where the Fermi energy lies within t_{2g}-derived bands (see Fig. 9.4). Immediately the idea arises to open a fundamental gap by applying the DFT+U method (9.13) and solve the Kohn-Sham equation with U-corrected XC potentials of the type (9.15). Figure 9.5 seems to indicate such a gap opening in the single-particle density of states (DOS) with rising on-site repulsion U. However, this is not completely true. The gaps appearing in this figure for $U \gtrsim 2$ eV have an additional reason.

The reason is illustrated in Fig. 9.6. The total-energy optimization that starts as usual from an electron distribution derived from the atomic ones indeed leads to a local minimum on the total-energy surface versus the electronic degrees of freedom. The results are summarized in the upper graph of Fig. 9.6. This graph shows the expected increase of the total energy with U. However, a gap is still not opened in the corresponding KS band structure (not shown). The FeO material remains a

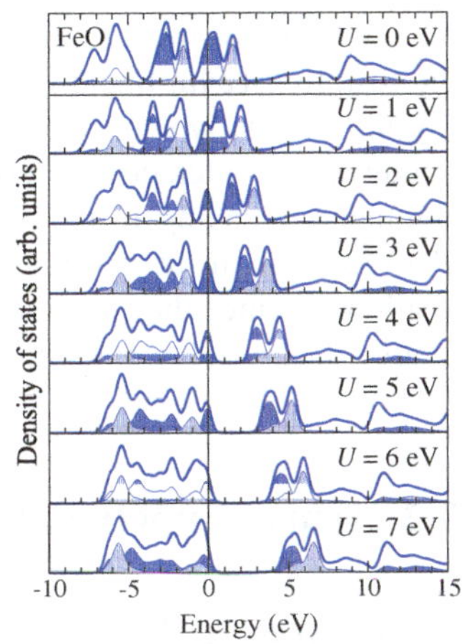

Fig. 9.5 Density of states for *rs*-AF2 FeO calculated with different values of U in the GGA+U scheme. The contributions of the Fe 3d states with t_{2g} (*dark blue*) and e_g (*light blue*) symmetry are indicated. A Gaussian broadening of 0.6 eV full width at half maximum is applied. The top of the valence bands is taken as energy zero. Adapted from [23]

metal for all U investigated. On the contrary, immediately a fundamental gap is opened applying a hybrid (HSE03) functional to describe XC [23]. Typically the d states are more localized using a non-local XC potential. Interestingly the use of the HSE03 electron density as starting point (see iteration scheme in Fig. 6.2) for the iterative determination of the actual electron density and spin densities lowers the total energy within the GGA+U approach. This total energy is plotted in Fig. 9.6 in the lower graph. It indicates that the start of the self-consistent procedure with the HSE electron density leads to an energy gain. The reason is that the lower total-energy minimum found for $U > 2$ eV corresponds to an insulating phase. The band-structure

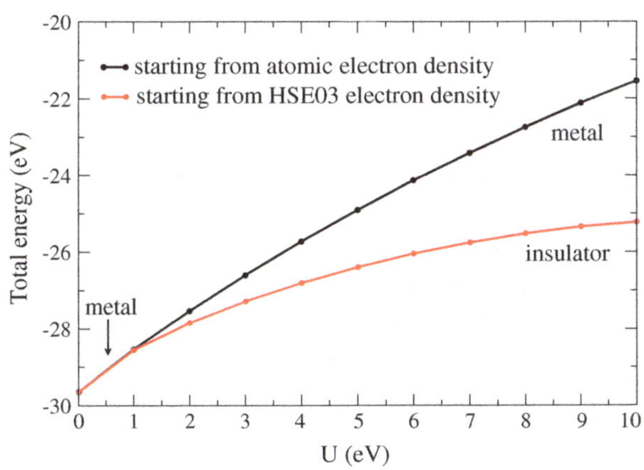

Fig. 9.6 Total energy of *rs*-AF2 FeO per unit cell versus U within the GGA+U approach. Two different starting electron densities have been used: electron density around free atoms as derived from GGA+U, electron density of antiferromagnetic FeO as computed with a non-local XC potential derived from a HSE hybrid XC functional (courtesy of C. Rödl, Friedrich-Schiller Universität Jena)

energy becomes more negative with the down-shift of the occupied levels. Indeed, the resulting Kohn-Sham eigenvalues give rise to the density of states depicted in Fig. 9.5. It clearly shows the opening of a fundamental energy gap for a reasonably strong on-site interaction U. The effect observed may be interpreted as a spontaneous symmetry breaking in the antiferromagnet FeO. Similar effects are observed for CoO. Despite the atoms in ideal rocksalt positions spontaneous symmetry breaking is allowed, since the two oxides are studied in a rhombohedral Bravais lattice with magnetic unit cells containing four atoms.

9.2 Hybrid Functionals

9.2.1 Non-locality

The two-particle density matrix (3.38) and the related pair correlation function $g_\Psi(\mathbf{x}, \mathbf{x}')$ (in a certain state) (3.42) or $\bar{g}(\mathbf{x}, \mathbf{x}'; [n])$ (coupling-constant averaged one) (7.8) are truly non-local objects that depend on electrons at two different space points, $\mathbf{x}$ and $\mathbf{x}'$, and their interaction. These quantities determine exchange and correlation in the studied inhomogeneous electron system and, consequently, many properties of an electron gas. The true dependence on two space coordinates is destroyed in a local (LDA) or semilocal (GGA) XC approximation.

In the LDA the XC energy (7.11) corresponds to the Coulomb interaction between an electron at position $\mathbf{x}$ and the XC hole $\bar{n}_{\mathrm{XC}}(\mathbf{x}, \mathbf{x}', [n])$ at $\mathbf{x}'$ that only depends on the distance $(\mathbf{x} - \mathbf{x}')$. In a GGA the gradient corrections lead to a modification of the XC hole taking inhomogeneities partially into account. In the HFA for the electrons in the ground state (4.18) the true non-locality of the exchange energy (4.15)

$$E_{\mathrm{X}}^{\mathrm{HF}} = -\frac{1}{2}\sum_{m_s}\int d^3\mathbf{x}\int d^3\mathbf{x}' \mathrm{v}(\mathbf{x}-\mathbf{x}')\left|\sum_{\lambda} n_{\lambda m_s}\varphi_{\lambda m_s}^*(\mathbf{x}')\varphi_{\lambda m_s}(\mathbf{x})\right|^2 \tag{9.16}$$

is known, however, not in terms of the density $n(\mathbf{x})$ of the inhomogeneous electron but in terms of orbitals $\{\varphi_{\lambda m_s}(\mathbf{x})\}$ of non-interacting electrons. Here we do not anymore indicate the dependence on the system state Ψ.

The two limiting cases of the treatment of the non-locality of XC or exchange suggest to combine their advantages, i.e., to combine the conventional DFT with the HFA in an appropriate manner, for instance following the idea of an 'adiabatic connection' [24]. An almost linear combination of the XC expressions of type (7.9) or (7.25) with the Fock exchange (9.16) is given by an XC energy ($0 \leq \alpha \leq 1$)

$$E_{\mathrm{XC}}^{\mathrm{hyb}} = \alpha E_{\mathrm{X}}^{\mathrm{HF}} + (1-\alpha)E_{\mathrm{X}}^{\mathrm{DFT}} + E_{\mathrm{C}}^{\mathrm{DFT}}, \tag{9.17}$$

where the correlation energy E_C^{DFT} is still described within the LDA or GGA of the DFT but the exchange contribution combines the Fock operator E_X^{HF} with weight α and the exchange energy E_X^{DFT} of the DFT with weight $(1-\alpha)$. The terms HF and HFA are applied but mean the use of a Fock exchange calculated by means of DFT wave functions. The result is consequently called a *hybrid* XC functional E_{XC}^{hyb}. The coefficient α can be chosen to assume a value derived from physical arguments or is fitted to some properties of a molecular database. The two known limits $\alpha = 1$ (HFA) and $\alpha = 0$ (conventional DFT) are still included.

The first example for such a hybrid functional has been suggested by Becke [24] to take the "half-and-half" form with $\alpha = \frac{1}{2}$. Later he parameterized such hybrid functionals that they become highly accurate for many molecules [24, 25]. Most successful for molecules was the Becke [26], three-parameter, Lee-Yang-Parr [27] (B3LYP) form of (9.17) [25]. This hybrid exchange-correlation functional is described by

$$E_{XC}^{hyb} = E_{XC}^{LSDA} + 0.2(E_X^{HF} - E_X^{LSDA}) + 0.72(E_X^{GGA} - E_X^{LSDA}) + 0.81(E_C^{GGA} - E_C^{LSDA}), \tag{9.18}$$

where the LSDA and GGA exchange energies are taken from widely used functionals. The correlation energy is taken to be the gradient-corrected functional of Lee et al. [26] scaled by a prefactor 0.81. While the functional (9.18) is not convenient within the commonly used plane-wave basis sets (8.37), it can be implemented readily and very efficiently within a Gaussian basis set as used in the GAUSSIAN code [28]. However, meanwhile also plane-wave implementations are successful [29], at least for non-metals [30]. The implementation of (9.18) leads to accurate atomic energies, ionization potentials and proton affinities of a number of molecules. This fact was probably one of the most important reasons for the triumphal procession of the DFT applications in quantum chemistry and for awarding the Nobel Prize in Chemistry 1998 to the solid-state theoretician Walter Kohn "for his development of the density-functional theory" [31]. However, meanwhile the B3LYP functional is also applied to solids, for instance to compute excitation properties, e.g. fundamental gaps of semiconductors and insulators [32, 33].

9.2.2 Inclusion of Screening

The hybrid exchange-correlation energy (9.17) may be constructed by mixing 25 % of Hartree-Fock exchange to 75 % of the well-known PBE-GGA exchange [34] described in Sect. 7.3.2, i.e., by $\alpha = \frac{1}{4}$ (Table 9.1). The electronic correlation is represented by the corresponding proportion of the PBE-GGA density functional. It results the PBE0 functional [35, 36]. Since the reaction of an electron gas in a non-metal may be characterized by a static electronic dielectric constant ϵ_∞, the occurrence of α can be interpreted to take into account static and spatially constant screening by an electronic dielectric constant $\epsilon_\infty = 4$ that represents a mean value for

Table 9.1 Parameters of hybrid functionals

Functional	α	$K_X^{sr}(x)$	μ (Å^{-1})	DFT	References
DFT (LDA/GGA)	0.00	–	–	Arbitrary	–
sX	1.00	$\exp(-x)$	1.55	Arbitrary	[38]
HSE03/06	0.25	$\mathrm{erfc}(x)$	0.3/0.2	PBE	[21, 39]
PBE0	0.25	1	–	PBE	[35, 36]
HF	1.00	1	–	–	–

materials with not too small gaps. From the view point of modeling a more technical argumentation can be given for the use of $\alpha = \frac{1}{4}$ [37].

Further improvements are possible due to the inclusion of spatially varying screening, especially due to the inhomogeneous electron gas in a solid. The obvious generalization is the replacement of the bare Coulomb potential $v(\mathbf{x} - \mathbf{x}')$ (1.2) in the exchange energy (4.15) or (9.16) by a screened one $K_X^{sr}(\mu|\mathbf{x} - \mathbf{x}'|)v(\mathbf{x} - \mathbf{x}')$, e.g. with the Thomas-Fermi screening $K_X^{sr}(\mu|\mathbf{x}|) = e^{-\mu|\mathbf{x}|}$ that results in a short range (sr) Yukawa potential. Indeed, μ represents the Thomas-Fermi screening constant. We generalize expression (9.17) to

$$E_{XC}^{hyb} = \alpha\left[E_X^{HF,sr}(\mu) - E_X^{DFT,sr}(\mu)\right] + E_X^{DFT} + E_C^{DFT}. \tag{9.19}$$

Because of the screening of exchange we call this hybrid functional screened exchange (sX) one [38]. In Table 9.1 we fix the Thomas-Fermi wave vector to $\mu = 1.55$ Å^{-1}.

Depending on the decay of the HF exchange interactions with distance, the evaluation of E_X^{HF} in the PBE0 functional may be computationally very demanding. To avoid the calculation of expensive integrals Heyd et al. [21] proposed to replace the long-range part of the HF exchange in the PBE0 functional by a corresponding DFT counterpart. The resulting HSE03 expression for the XC energy is of type (9.19) but with a function $K_X^{sr}(\mu|\mathbf{x}|) = \mathrm{erfc}(\mu|\mathbf{x}|)$. The DFT XC is described within the PBE-GGA [34]. The decomposition of exchange interactions, HF or PBE exchange, into short-range (sr) and long-range (lr) parts is accomplished through the prefactor 1 of the bare Coulomb potential by $1 = \mathrm{erfc}(\mu|\mathbf{x}|) + \mathrm{erf}(\mu|\mathbf{x}|)$, where μ is the parameter that defines the range separation. μ is related to a characteristic distance $2/\mu$, at which the sr interactions become negligible. Empirically it was shown that the optimum range-separation parameter μ is between 0.2 and 0.3 Å^{-1} [21, 39, 40]. In Table 9.1 the use of $\mu = 0.3$ Å^{-1} (HSE03) or $\mu = 0.2$ Å^{-1} (HSE06) is suggested.

9.2.3 Generalized Kohn-Sham Problems

The question arises if the description of XC by a hybrid functional that depends on the orbitals via the Fock term [see (9.16)] can be related to an eigenvalue problem.

Görling and Levy [41] have shown that hybrid schemes have indeed a rigorous formal justification within the exact DFT scheme, when they are formulated as a generalized Kohn-Sham approach [38].

The Hohenberg-Kohn scheme of the DFT contains a minimization of the energy functional (5.23) with the Hohenberg-Kohn functional (5.24). The basic idea is to replace the calculation of the ground state $|\Psi_0[n]\rangle$ by that of a single Slater determinant with elements obtained from a non-interacting Kohn-Sham model system with the same electron density. In its generalization the reference system is still described by an N-electron Slater determinant Φ that has to be distinguished from the true many-body wave function Ψ. However, at variance with the usual Kohn-Sham approach, it does not correspond to the minimization of the expectation value of the kinetic energy $F_{\mathrm{KS}} = \langle\Phi|\hat{T}|\Phi\rangle$. Instead, $F_{g\mathrm{KS}} = \langle\Phi|\hat{T} + \hat{U}|\Phi\rangle$, containing in part the electron-electron interaction under the constraint that Φ can be written as a Slater determinant, is used. It may be called generalized Kohn-Sham (gKS) functional. If this functional is chosen to be

$$F_{\mathrm{HF-KS}} = \langle\Phi|\hat{T} + \hat{U}|\Phi\rangle = F_{\mathrm{KS}} + E_{\mathrm{H}}[\Phi] + E_{\mathrm{X}}[\Phi], \tag{9.20}$$

the scheme resembles the Hartree-Fock method, at least if the elements of the Slater determinant $\varphi_{\lambda m_s}(\mathbf{x})\chi_{\frac{1}{2}m_s}(s)$ in (4.6) are the single-particle orbitals in the HF ground state Ψ_0. In principle, the functional $F_{\mathrm{HF-KS}}$ contains an unknown, formally exact correlation term that is absent in the standard HFA. Such an approach is known as the Hartree-Fock-Kohn-Sham (HF-KS) scheme [42].

A scheme based on a hybrid functional, for instance, that of the type (9.17), also corresponds to a realization of a generalized Kohn-Sham scheme, where the functional

$$F^{\alpha}_{\mathrm{HF-KS}} = \langle\Phi|\hat{T} + \alpha\hat{U}|\Phi\rangle = F_{\mathrm{KS}} + \alpha E_{\mathrm{H}}[\Phi] + \alpha E_{\mathrm{X}}[\Phi] \tag{9.21}$$

is minimized with respect to the N-electron wave function Φ of the determinantal form, while the rest of the energy is treated within the usual LDA and/or GGA approximations to the Kohn-Sham problem. In the case of a range parameter μ that distinguishes short-range and long-range contributions a further extension of the functional to $F^{\alpha,\mu}_{\mathrm{HF-KS}}$ is needed.

The Kohn-Sham idea in Sect. 6.2.1 can be also applied to the orbital-dependent hybrid functionals, as long as a functional $F^{\alpha,\mu}_{\mathrm{HF-KS}}$ can be minimized in such a way that the elements of the Slater determinant Φ describe the true electron density of the system. Then, the gKS equations are derived in a similar way as the KS equations (6.22). However, the exact form of the difference between the Hohenberg-Kohn functional (5.24) and the functional $F^{\alpha,\mu}_{\mathrm{HF-KS}}$ and, thus, of its functional derivative, resulting in an additional potential, is not known. Only suitable approximations can be found [38]. They are based on the fact that the parameters α and μ are fixed and do not depend on the density. Then, the generalized Kohn-Sham equations can be written in the form of the usual KS equations (6.22)

$$\hat{H}_{\mathrm{KS}}^{m_s}\varphi_{\lambda m_s}(\mathbf{x}) + \int d^3\mathbf{x}'\Delta V_{\mathrm{X}}^{m_s}(\mathbf{x},\mathbf{x}')\varphi_{\lambda m_s}(\mathbf{x}') = \varepsilon_{\lambda m_s}\varphi_{\lambda m_s}(\mathbf{x}) \tag{9.22}$$

supplemented with a non-local correction term. In the case of the sX hybrid functional (9.19) (see also Table 9.1) it may be written as [41]

$$\Delta V_{\mathrm{X}}^{m_s}(\mathbf{x},\mathbf{x}') = \alpha\left\{-\sum_{\lambda} n_{\lambda m_s}\mathrm{v}(\mathbf{x}-\mathbf{x}')e^{-\mu|\mathbf{x}-\mathbf{x}'|}\varphi_{\lambda m_s}^{*}(\mathbf{x}')\varphi_{\lambda m_s}(\mathbf{x}) - \delta E_{\mathrm{X}}^{\mathrm{DFT,sr}}(\mu)/\delta n_{m_s}(\mathbf{x})\delta(\mathbf{x}-\mathbf{x}')\right\}.$$

Hybrid functionals can be used in different ways. Generally their use requires to treat spatially non-local potentials in the KS equations. However, such potentials can be brought into the conventional Kohn-Sham family by constructing additional local density-dependent XC potentials. Such approaches have come to be known as optimized effective potential (OEP) methods [43, 44]. One special case is the exact-exchange (EXX) approach [45, 46]. The construction of the local exchange potential starts from the Fock operator (4.27) or the exchange energy (9.16) to have an explicit dependence on the orbitals. Details of the OEP and EXX constructions can be found elsewhere [47].

9.2.4 Examples/Applications

In Chaps. 7 and 8 the conventional DFT has been proven to be a very powerful tool for the quantitative prediction of materials properties, both in computational solid state physics and quantum chemistry. Thereby, exchange and correlation have been described within LDA or GGA that results in a highly efficient but still surprisingly accurate description of ground-state properties. This is particularly true for refinements of the PBE-GGA, e.g. the AM05 and PBEsol functionals. Still, though, present local and semilocal XC functionals show significant errors, for instance, in the energetics of small molecules (see Fig. 7.10) but also in the description of band gaps and interband energies of extended systems (see Figs. 6.4 and 9.5), the majority of ground-state calculations are performed in the framework of LDA or GGA functionals.

During the years hybrid functionals, i.e., XC functionals that admix a certain amount of unscreened or screened HF exchange to (a part of) a local or semilocal density functional $E_{\mathrm{XC}}^{\mathrm{hyb}}$ (9.19), have been shown to remedy several deficiences of the local or semilocal approaches. This will be illustrated here for three properties. Figure 9.7 shows the relative errors in the calculated lattice constants with respect to experimental values. The same 20 crystals as in Fig. 7.11 are studied. Besides the known PBE-GGA values also results computed in the framework of the three hybrid

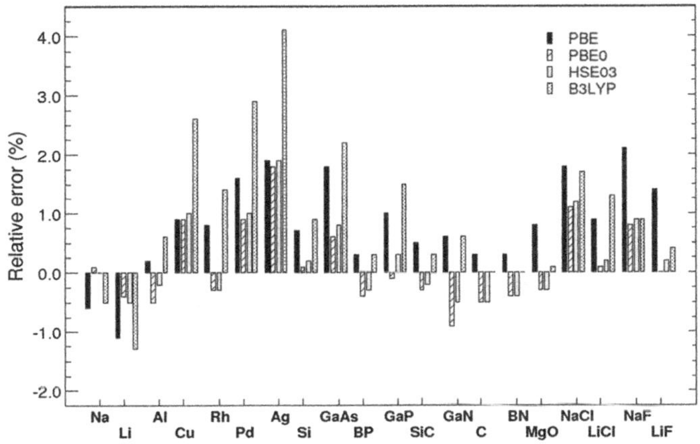

Fig. 9.7 Relative errors in calculated lattice constants with respect to experimental data. Besides the semilocal PBE-GGA functional (Sect. 7.3.2) the three hybrid XC functionals PBE0, HSE03 (9.19) (with parameters in Table 9.1), and B3LYP (9.18) have been applied. From [29]. Copyright IOP Publishing. Reproduced by permission of IOP Publishing. All rights reserved

functionals PBE0, HSE03, and B3LYP are depicted. Apart from Na and Li there is a general trend for overestimation of the lattice constants from the B3LYP functional, probably, because it has been optimized for molecules [48]. For lattice parameters obtained by means of the PBE0 and HSE03 hybrid functionals the situation is much better. The mean absolute relative errors (MARE) only amounts to 0.5 % compared to the 1.0 % for PBE-GGA or 1.2 % for B3LYP [29]. This is really a promising result. The accuracy of these two hybrid functionals is even better than that of the AM05 functional discussed in Sect. 7.3.4.

Another example concerns the thermochemistry, more precisely, the energetics of various conformers of complex molecules. We study the most important conformations of the amino acid cysteine (cys). They are displayed in Fig. 9.8. The different conformers are a consequence of the six rotational degrees of freedom of the molecule. The six most stable conformations, at least on a level of accurate quantum-chemistry studies within the Møller-Plesset perturbation theory (MP2) [50] with a special basis set, MP2/6-31+G* [51], are those labeled with 1,. . .,6. The energy of the cysteine conformation 1 in Fig. 9.8 in this approximation is used as energy zero in Fig. 9.9. The figure indicates a small increase of the total energy of cysteine with the label 1, . . ., 6 of the conformation. The MP2 values are compared with energies obtained by means of (semi)local and hybrid XC functionals. The comparison shows that the local or semilocal XC functionals fail, especially for the conformations 1 and 6. The values obtained by means of the hybrid functionals indicate a stronger bonding as within the MP2 approach. For the other conformations 2, 3, 4 and 5 the energy differences due to the different approximations are negligibly small. They even vanish for the conformation 2. The main reason for these findings is related to the existence or non-existence of an intramolecular O-H...N bond. It is missing (see Fig. 9.8) in the conformations 2, 3, 4 and 5. However, for the two geometries

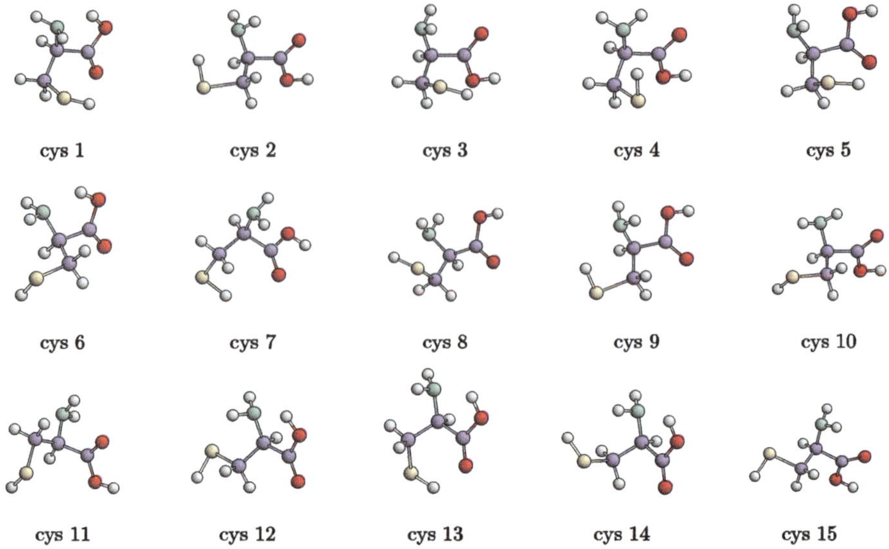

Fig. 9.8 Most important conformations of a cysteine molecule NH_2-C_2H_3SH-COOH. Different atoms are indicated by different colors: carbon (*blue*), oxygen (*red*), nitrogen (*green*), sulphur (*yellow*), and hydrogen (*white*). From [49]. Copyright Wiley-VCH Verlag GmbH & Co. KGaA. Reproduced with permission

1 and 6 such a bond appears. Their total energies are significantly reduced with reduced non-locality of XC. Thus, in comparison to the MP2 level of calculation but also to the computations with hybrid functionals, the DFT optimizations with local or semilocal XC functionals stabilize the O-H...N bond-containing conformations over the other ones.

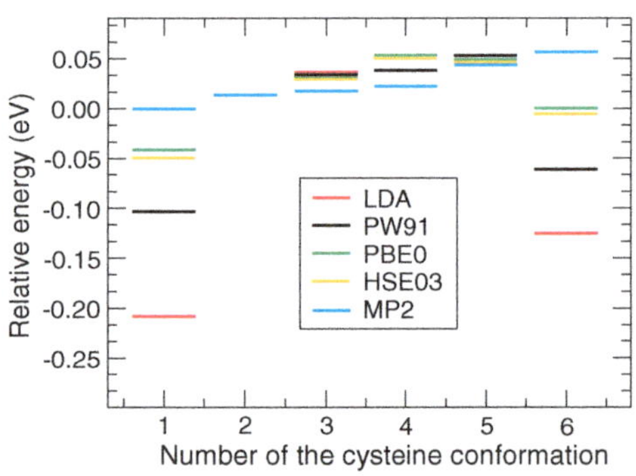

Fig. 9.9 Conformational energies of the cysteine conformations 1-6 (see Fig. 9.8) as calculated in the framework of different (semi)local or hybrid XC functionals: LDA-PZ (*red*), GGA-PW91 (*black*), HSE03 (*yellow*), and PBE0 (*green*). The energies are compared with MP2 values (*blue*) [51]. From [49]. Copyright Wiley-VCH Verlag GmbH & Co. KGaA. Reproduced with permission

In the last years it became very popular to apply hybrid XC functionals also for the calculation of excited electronic states [29, 52–54]. Spin-orbit interaction can be easily taken into account. The α parameter of the mixing of HF-like exchange in the HSE expression (9.19) is sometimes adjusted to obtain the correct fundamental quasiparticle gap of a semiconductor in order to start studies of defects or surfaces within a supercell or slab approximation with a gap close to its experimental value [55–57]. In any case, there is a clear tendency for a further opening of fundamental gaps of semiconductors and insulators toward the experimental values by applying hybrid XC functionals in comparison with the Kohn-Sham values computed in a local or semilocal XC approximation (for reasons see Sect. 6.3.2). This is demonstrated in Fig. 9.10. However, not only the fundamental gaps are opened. All interband energies between occupied and empty bands are widened as shown for the example of zinc-blende AlN in Fig. 9.11 but also for nanostructures, as a Si nanocrystal with 1.2 nm diameter embedded in an amorphous SiO_2 matrix in Fig. 9.12. The down (up) shift of occupied (empty) states is not only visible in the total density of states but also in the ones projected onto Si atoms in the core, interface and matrix regions.

The reasons for the success of all these approximations for fundamental gaps will be derived in Part III of the book as a consequence of the screened exchange contribution to occupied and empty band positions which opens the fundamental gaps. Here we remember that we are still doing density functional theories for ground states. Only the XC functional has been generalized. For that reason we discuss only fundamental gaps for the cases where the DFT in LDA or GGA totally fails in the sense to yield a metal instead of an insulator or semiconductor. Examples are InN in wurtzite or zinc-blende structure [60], CdO in rocksalt geometry [61], and the antiferromagnets FeO (see also Fig. 9.5) or CoO in rocksalt structure [23, 29]. The reasons vary. In the InN and CdO cases the low-lying $5s$ levels together with

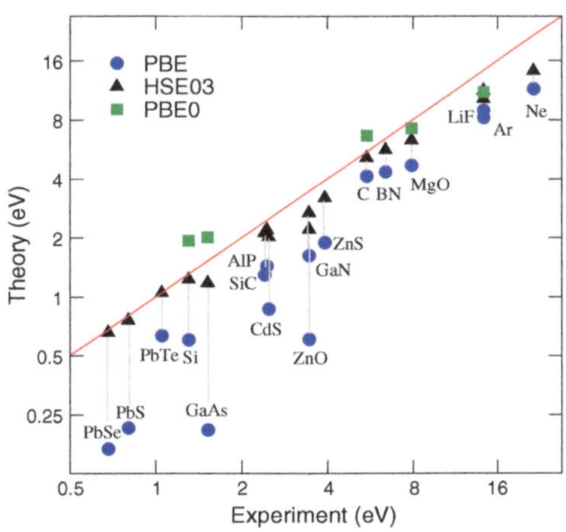

Fig. 9.10 Fundamental band gaps of non-metallic solids. Values derived from the KS theory with the PBE-GGA functional or the gKS theory with the HSE03 and PBE0 hybrid functionals are plotted against experimental values. From [29]. Copyright IOP Publishing. Reproduced by permission of IOP Publishing. All rights reserved

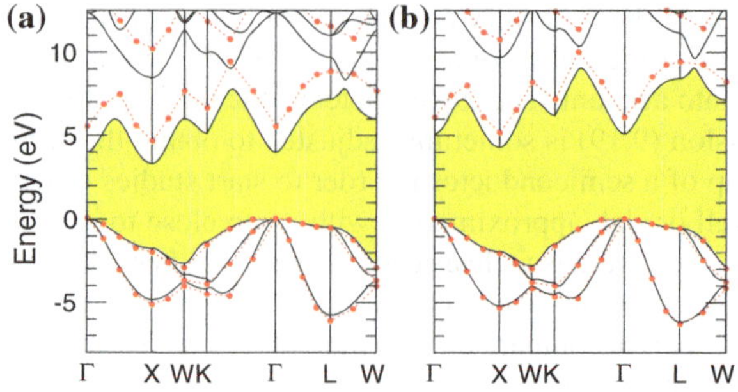

Fig. 9.11 Band structure of zinc-blende AlN as derived from Kohn-Sham eigenvalues (*solid black lines*) and measurable quasiparticle band structures (*red triangles* and *dotted lines*). Two different XC functionals are applied: (**a**) PBE-GGA (*left panel*), see also Fig. 6.4, and (**b**) HSE03/06 (*right panel*). In (**b**) the missing upward shifts of the conduction bands to reach the quasiparticle bands are much smaller than in (**a**). Based on band structures in [58]

the *pd* repulsion of the *p* valence electrons and the shallow 4*d* semicore electrons close the gap. For FeO and CoO the partially occupied t_{2g} shell in the minority-spin channel makes the metallic character (see also discussion in Sect. 9.1.4). Apart from InN the other crystals possess indirect gaps. Results are summarized in Table 9.2. The inclusion of non-local exchange in the XC functional of the HSE03/06 type [21, 39] indeed opens a gap in agreement with the non-metallic character of the

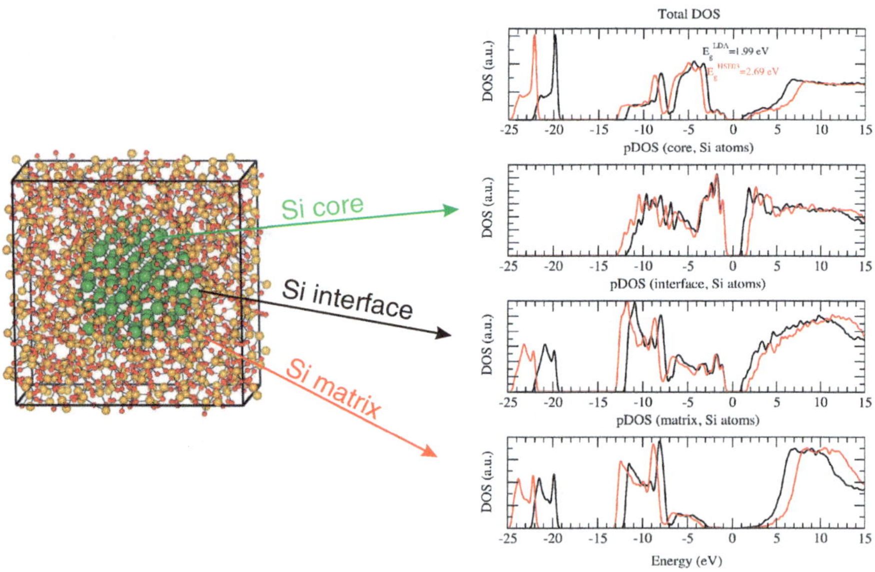

Fig. 9.12 Density of states (DOS) of a Si nanocrystal, its interface with the SiO_2 matrix, and the matrix is displayed in the right panels. Two XC functionals, PZ-LDA (*black line*) and HSE03 (*red line*), are applied. The *left panel* illustrates the arrangement of Si atoms (*green* in core, *red* in matrix) and O atoms (*yellow*). Adapted from [59]

Table 9.2 Fundamental gaps of compounds with first-row anions as calculated by means of the HSE03 hybrid XC functional

Compound	HSE03 gap (eV)	Exp. gap (eV)	References
wz/zb-InN	0.49/0.21	0.7/0.61	[60]
rs-CdO	0.57	0.84	[61]
rs-FeO	2.1	2.4	[23]
rs-CoO	3.2	2.5 ± 0.3, 3.6 ± 0.5, 2.8, 5.43	[23]

For comparison experimental values are listed. While the direct semiconductor InN crystallizes in wurtzite (wz) or zinc-blende (zb) structure, the atomic geometry of the oxides is described by rocksalt (rs). All geometries have been optimized using a GGA XC functional, i.e., possess slightly overestimated lattice constants

four compounds. In this respect the local or semilocal description of the electronic structure in the ground state truly fails. We have to mention that hybrid XC functionals do not give the true conducting or insulating phases for all oxides. The rutile structure of VO_2 is found to be not (but should be) metallic [62].

9.3 Van der Waals Interaction

9.3.1 The Missing Link

The success of the DFT has been facilitated by the computational efficiency of the LDA (Sect. 7.2) or GGA (Sect. 7.3) of the exchange-correlation functional. These approximations make DFT applicable to polyatomic systems containing up to several thousand atoms, at least if s and p electrons mainly contribute to the covalent, metallic or ionic bonding. However, these approximations are also subject to several deficiencies. Two of them and their elimination have been discussed in the last two sections. Other improvements and refinements have been proposed over the years. Some of them have been classified in the "Jacob's ladder" hierarchy [1, 63] (see Fig. 9.13). Still challenges are around. Despite the inclusion of gradient corrections (rung 2), the second derivative of the density (rung 3) or even spatial non-localities described by occupied orbitals $\{\varphi_{\lambda m_s}(\mathbf{x})\}$ with $n_{\lambda m_s} = 1$ to the exchange (rung 4), in all the energy functionals, representing the rungs 1, 2, 3 and 4 of the "Jacob's ladder" in Fig. 9.11, the long-range tails in the bonding forces of the van der Waals (vdW) type are lacking. Indeed, there is a need to treat van-der-Waals-bonded systems with sufficient accuracy but, if possible, numerical efficiency in rung 5 in Fig. 9.13.

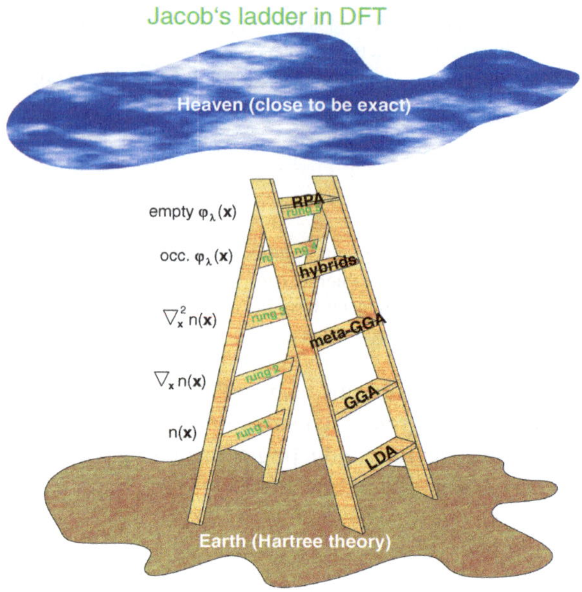

Fig. 9.13 Practical approximations to E_{XC} in DFT and generalized DFT. View of John Perdew on users of DFT who climb the ladder to gain greater accuracy at greater cost depending upon their needs. The resulting Jacob's ladder is in accordance with some vision in the Bible (Genesis 28.10–17): Jacob had a dream in which he saw a ladder descending from Heaven to Earth and angels climbing the ladder. Drawn using ideas from [1, 63]

9.3.2 Adiabatic-Connection Fluctuation-Dissipation Theorem

One way that one can follow for inclusion of vdW forces is to start from exchange and correlation in the coupling-constant-integrated description (see Sect. 7.1.2). Following the concept of adiabatic connection (AC) we found in (7.9) the result (not indicating the functional dependence on density) [64]

$$E_{\mathrm{XC}} = \frac{1}{2}\int_0^1 d\lambda \int d^3\mathbf{x} \int d^3\mathbf{x}' \mathrm{v}(\mathbf{x}-\mathbf{x}')n(\mathbf{x})n^{\lambda}_{\mathrm{XC}}(\mathbf{x},\mathbf{x}') \tag{9.23}$$

for the XC energy with the XC hole (7.10) [see also (3.49)]

$$n^{\lambda}_{\mathrm{XC}}(\mathbf{x},\mathbf{x}') = \langle \Psi^{\lambda}_0 \left| \Delta\hat{n}(\mathbf{x}')\Delta\hat{n}(\mathbf{x}) \right| \Psi^{\lambda}_0 \rangle / n(\mathbf{x}) - \delta(\mathbf{x}-\mathbf{x}') \tag{9.24}$$

with the operator $\Delta\hat{n}(\mathbf{x})$ of density fluctuations (3.53).

The density-density correlations in (9.24) are linked to the response properties, the dissipation, of the system through the (zero-temperature) fluctuation-dissipation theorem (FDT) (see Sect. 13.1.1). It states that the linear response of a system at thermodynamic equilibrium to an external perturbation is the same as its response to the spontaneous internal fluctuations in the absence of the perturbation [65]. The FDT is manifested in many physical properties. A key example is the dielectric formulation of the many-body problem by Nozieres and Pines [66]. In the studied spin-averaged description and $T \to 0$ K, it yields (see also Sect. 13.1.1)

$$\langle \Psi_0^\lambda \left| \Delta\hat{n}(\mathbf{x}')\Delta\hat{n}(\mathbf{x}) \right| \Psi_0^\lambda \rangle = \frac{\hbar}{\pi} \int_0^\infty d\omega \hat{L}^\lambda(\mathbf{x}\mathbf{x}', \omega) \tag{9.25}$$

with the spectral function $\hat{L}^\lambda(\mathbf{x}\mathbf{x}', \omega)$ of the density response function $L^\lambda(\mathbf{x}\mathbf{x}', \omega)$ of the λ-scaled system [for comparison see (3.56)]. With the relations (9.24), (9.25), and spectral representation (13.16) of $L^\lambda(\mathbf{x}\mathbf{x}', \omega)$ the ACFDT expression of the XC energy is [67]

$$\begin{aligned} E_{\mathrm{XC}} &= \frac{1}{2} \int_0^1 d\lambda \int d^3\mathbf{x} \int d^3\mathbf{x}' \mathrm{v}(\mathbf{x} - \mathbf{x}') \\ &\quad \times \left[\frac{\hbar}{\pi} \int_0^\infty d\omega \hat{L}^\lambda(\mathbf{x}\mathbf{x}', \omega) - \delta(\mathbf{x} - \mathbf{x}')n(\mathbf{x}) \right] \\ &= \frac{1}{2} \int_0^1 d\lambda \int d^3\mathbf{x} \int d^3\mathbf{x}' \mathrm{v}(\mathbf{x} - \mathbf{x}') \\ &\quad \times \left[-\frac{2\hbar}{\pi} \int_0^\infty d\omega L^\lambda(\mathbf{x}\mathbf{x}', i\omega) - \delta(\mathbf{x} - \mathbf{x}')n(\mathbf{x}) \right]. \end{aligned} \tag{9.26}$$

The above frequency integration can be performed along the imaginary axis because of the analytical structure of $L^\lambda(\mathbf{x}\mathbf{x}', \omega)$ (see Sect. 13.1) and the fact that it is real on the imaginary frequency axis. A more physical interpretation of the convience to take the integral along the imaginary frequency axis can be found in the textbook of Landau and Lifshitz [68].

The above formulation is exact. A convenient approximation is the random phase approximation (RPA) that leads to a Bethe-Salpeter equation (here: at imaginary frequencies)

$$\begin{aligned} L_{\mathrm{RPA}}^\lambda(\mathbf{x}\mathbf{x}', i\omega) &= L^0(\mathbf{x}\mathbf{x}', i\omega) \\ &\quad + \int d^3\mathbf{x}'' \int d^3\mathbf{x}''' L^0(\mathbf{x}\mathbf{x}'', i\omega)\lambda \mathrm{v}(\mathbf{x}'' - \mathbf{x}''') L_{\mathrm{RPA}}^\lambda(\mathbf{x}'''\mathbf{x}', i\omega). \end{aligned} \tag{9.27}$$

The spin-averaged independent-particle response function $L^0(\mathbf{x}\mathbf{x}', i\omega)$ at $\lambda = 0$ is known explicitly in terms of the single-particle wave functions $\{\varphi_{\lambda m_s}(\mathbf{x})\}$, energies $\{\varepsilon_{\lambda m_s}\}$, and occupation numbers $\{n_{\lambda m_s}\}$ of the non-interacting inhomogeneous electron gas [see (12.70)]

$$L^0(\mathbf{xx}', i\omega) = \frac{1}{2} \sum_{m_s} \sum_{\lambda,\lambda'} (n_{\lambda m_s} - n_{\lambda' m_s}) \frac{\varphi_{\lambda m_s}(\mathbf{x}) \varphi^*_{\lambda' m_s}(\mathbf{x}) \varphi_{\lambda' m_s}(\mathbf{x}') \varphi^*_{\lambda m_s}(\mathbf{x}')}{\varepsilon_{\lambda m_s} - \varepsilon_{\lambda' m_s} - i\hbar\omega}. \quad (9.28)$$

The original RPA was not only related to a factorization of the density-density correlation function (9.24) in a product of two single-particle Green functions. Rather, in addition these functions should be calculated within the Hartree approach. However, in all modern explicit computations the KS reference system with solutions of the KS equations with a certain XC potential are applied to compute the $L^0(\mathbf{xx}', i\omega)$ function and not solutions for $\lambda = 0$. Still the denotation RPA is kept.

By means of the relation (9.26) and the closure relation for the orbitals

$$\sum_{\lambda} \varphi_{\lambda m_s}(\mathbf{x}) \varphi^*_{\lambda m_s}(\mathbf{x}') = \delta(\mathbf{x} - \mathbf{x}') \quad (9.29)$$

the exact exchange (EX or EXX) term, i.e., the Fock operator (9.16) expressed by the given occupied orbitals $\{\varphi_{\lambda m_s}(\mathbf{x})\}$, can be related to the response function $L^0(\mathbf{xx}', i\omega)$ as

$$E_{\mathrm{X}} = \frac{1}{2} \int d^3\mathbf{x} \int d^3\mathbf{x}' \mathrm{v}(\mathbf{x} - \mathbf{x}') \left[-\frac{2\hbar}{\pi} \int_0^\infty d\omega L^0(\mathbf{xx}', i\omega) - \delta(\mathbf{x} - \mathbf{x}') n(\mathbf{x}) \right]. \quad (9.30)$$

The last expression allows an analytical derivation for the ACFDT correlation energy in RPA

$$E_{\mathrm{C}}^{\mathrm{RPA}} = -\frac{\hbar}{\pi} \int d^3\mathbf{x} \int d^3\mathbf{x}' \mathrm{v}(\mathbf{x}-\mathbf{x}') \int_0^\infty d\omega \left[\int_0^1 d\lambda L^\lambda_{\mathrm{RPA}}(\mathbf{xx}', i\omega) - L^0(\mathbf{xx}', i\omega) \right]. \quad (9.31)$$

Within the RPA the integral equation (9.27) can be formally solved applying the spatial Fourier transformations of the Coulomb potential (13.36) and the density correlation function (13.34). The integral over λ can be performed explicitly to give the compact form [69]

$$E_{\mathrm{C}}^{\mathrm{RPA}} = \frac{\hbar}{2\pi} \int_0^\infty d\omega \mathrm{Tr} \left[\ln(1 - 2\mathrm{v}L^0) + 2\mathrm{v}L^0 \right]. \quad (9.32)$$

For brevity the convention

$$\mathrm{Tr}[AB] = \int d^3\mathbf{x} \int d^3\mathbf{x}' A(\mathbf{x}, \mathbf{x}') B(\mathbf{x}', \mathbf{x})$$

is used. The factor 2 in $2\mathrm{v}L^0$ in (9.32) appears because of the spin average (12.59) of the L^0 function. Expression (9.32) can be interpreted as the fifth rung of the

"Jacob's ladder" in Fig. 9.13 to the highest accuracy in ground-state calculations. Because of the independent-particle representation (9.28) of the $L^0(\mathbf{xx}', i\omega)$ function also wave functions of empty states are needed. A fully non-local XC treatment arises. In explicit numerical treatments only the diagonal elements of spatial Fourier representations (13.34) of the L^0 functions are used.

9.3.3 Exact-Exchange Plus Correlation in RPA

In the context of the "Jacob's ladder" hierarchy in Fig. 9.13 the treatment of exchange and correlation in the random-phase approximation offers a promising avenue to highly precise computations of materials properties. This is largely due to three features [67]: (i) The exact-exchange energy (9.30) cancels the spurious self-interaction error present in the Hartree energy (6.9) exactly, although the RPA correlation (9.32) itself may contain some "self-correlation" (sometimes also called "self-screening" [70]). (ii) The RPA correlation energy (9.32) is fully non-local and includes long-range vdW interactions automatically and seamlessly. (iii) Dynamic electronic screening is taken into expression (9.32) as indicated by the frequency integral.

The ACFDT expression of the exact-exchange plus correlation in the random-phase approximation (EX-cRPA) is meanwhile implemented in several codes, e.g. in FHI-aims [67, 71] as well as VASP [72–75]. The progress of such implementations can be easily demonstrated for vdW-bonded systems. As an example Fig. 9.14 displays the interaction energy of two benzene molecules in a stacked dimer that is known as a vdW complex versus the vertical separation of the molecules. The two different GGA approximations studied do not give bonding of the complex. The two

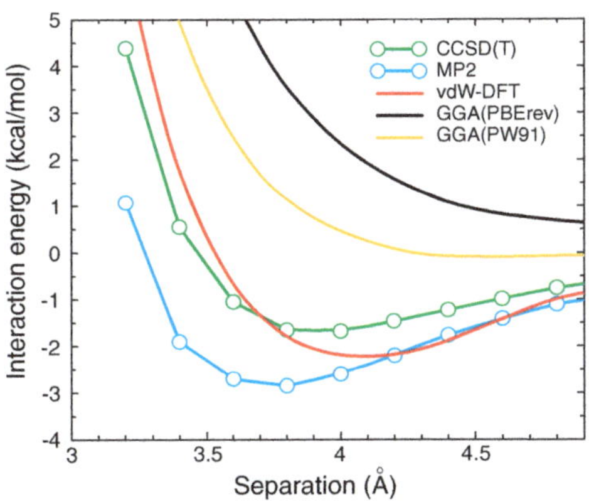

Fig. 9.14 Full interaction energy between two benzene molecules in the atop-parallel configuration using an ACFDT implementation (called vdW-DFT). For comparison coupled-cluster (CCSD(T)) and perturbation-theory (MP2) results as well as the prediction of two flavors of GGA (PW91, PBErev) are shown. Reprinted with permission from [76]. Copyright 2004 by the American Physical Society

quantum-chemical descriptions using coupled-cluster (CCSD(T)) and perturbation-theory (MP2) methods give rise to pronounced minima of their energy curves near molecule separations of about 3.8 Å. The ACFDT correlation energy description, called vdW-DFT in the figure, yields a reasonable description of the complex energetics. The energy curve approaches the MP2 results for large distances, while it is close to the CCSD(T) values for small separations. The attractive interaction for larger distances clearly indicate that the ACFDT approach to the correlation energy includes dispersion forces.

However, improvements of the predictions due to the RPA correlation energy also appear for solids, e.g. for their lattice constants. In Fig. 9.15a the aforementioned overbinding (underbinding) effects are clearly visible for the local (semilocal) XC treatment. The exact-exchange approach performs well for covalently bonded systems, however, exhibits huge errors for metals. The inclusion of correlation within the RPA remedies these problems entirely, in particular, for those materials where EXX yields too small lattice constants. The atomization energies depicted in Fig. 9.15b confirm the previous observations. The importance of correlation increases with increasing polarizability in full agreement with the mathematical expression (9.32). For strongly polarizable metals correlation accounts for more than 80 % of the exchange-correlation energy, whereas weakly polarizable insulators are reasonably described by EXX. Similar results are observed for the small diatomic molecules H_2, O_2, F_2, and Cl_2. Apart from H_2 the RPA correlation is most important. Very recently the success of the RPA-ACFDT [72] has been demonstrated for the polymorphic energy ordering of oxides and nitrides [77]. This approach correctly recovers the rocksalt structure of MnO as the ground-state phase, as observed experimentally, whereas previous density and hybrid functional methods obtained the wrong energy ordering (see Fig. 9.3). Also a better description of the electron

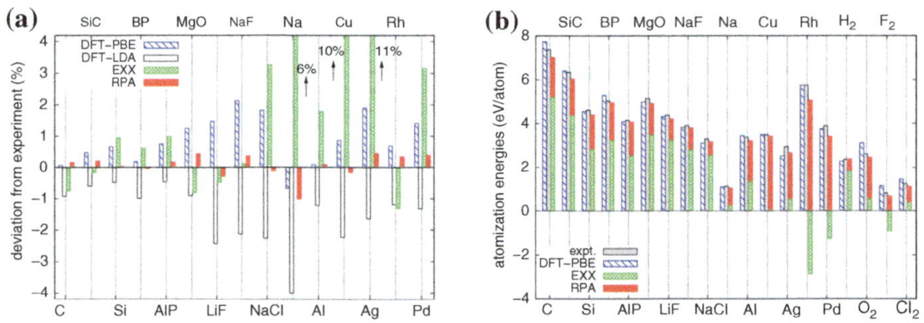

Fig. 9.15 Relative error of the theoretical lattice constants (**a**) and absolute atomization energies (**b**) of insulators, semiconductors, and metals as obtained from the RPA approach to the correlation energy. In (**b**) additional values for the diatomic molecules H_2, F_2, O_2 and Cl_2 are given. The results are compared with those obtained within local or semilocal XC approaches. One set of values has been generated within the EXX approach. In (**b**) instead of DFT-LDA atomization energies experimental values are displayed. Reprinted with permission from [73]. Copyright 2009 by the American Physical Society

correlation on the Mn $3d$ shell by a Hubbard U parameter (see Sect. 9.1.3) is not anymore needed to obtain the correct ground state.

9.3.4 Further Developments

Based on the ACFDT/RPA starting points two opposite developments are obvious in recent years. One direction aims at further improvements of the EX-cRPA treatment beyond RPA. "Single excitation" (SE) contributions [78] or even "double excitations" are included [67]. Another type of corrections is related to a second-order screened exchange (SOSEX) treatment [79].

In principle, the exact ACFDT for the correlation energy based on (9.26) and (9.31) may seamlessly pave the way toward a vdW density functional of an inhomogeneous electron gas. However, the numerical difficulties restrict its use to the RPA framework with some corrections. Even in these cases the benchmark calculations, see e.g. results in Fig. 9.15, are restricted to solids with small unit cells or diatomic molecules. Systems with 12 C and 12 H atoms as studied in Fig. 9.14 are already close to the limits. Larger (partly) vdW-bonded systems as the DNA base molecule adenine on a graphite surface (see Fig. 9.16) cannot be treated with the full RPA functional. The adenine molecule $C_5H_5N_5$ whose electron density and most important orbitals are displayed in Fig. 6.3 possesses 15 atoms. It occupies a $\begin{pmatrix} 9 & 0 \\ -4 & 8 \end{pmatrix}$ surface unit cell (for denotation see [81]) of graphite or graphene. Consequently, the substrate is modeled by a periodic supercell containing 144 C atoms in a single graphene layer. A minimum number of 159 atoms has to be taken into account to simulate the interaction between DNA base molecule and substrate. Other examples for vdW-bonded molecular systems are described in several books, e.g. in [82].

Unfortunately, due to the sheer adsorbate-dictated system size the use of correlation functionals applying RPA in the context of the ACFDT is hopeless. Semiempirical corrections are needed to describe the dispersion forces [83–85]. The basic idea is related to break a complex system in subunits, even in isolated

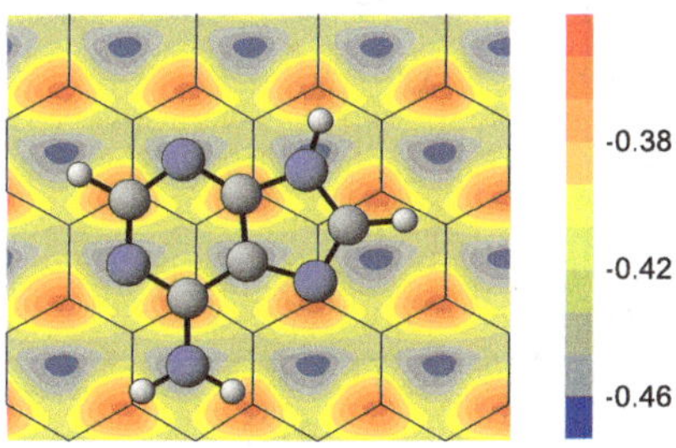

Fig. 9.16 Potential energy surface of an adenine molecule on a graphene layer. Hereby the lateral positions of one C atom of graphene and of the amino-group nitrogen atom were fixed. The vertical spacing as well as the coordinates of all other atoms were free to relax. The figure indicates the adsorption position with the lowest energy. Adapted from [80]

atoms, and study the long-range interaction between such subunits starting from ACDFT/RPA-like treatments [86–88], if possible. Then, the well-known distance dependence of the vdW interaction between two atoms, an atom and a surface, two half spaces, etc. can be derived. In some cases the dispersion coefficients are computed for the interaction between two atoms by an *ab initio* method [85] or taken from known interaction strengths [83, 84]. The transferability of such coefficients is assumed. The resulting vdW energy is added to the DFT energy functional with a local or semilocal description of XC. The dispersion-corrected energy functional may therefore be generalized as

$$E_{\mathrm{KS}}[n] \rightarrow E_{\mathrm{KS}}[n] + E_{\mathrm{vdW}} \tag{9.33}$$

compared to the DFT expression (6.12). In the spirit of the London dispersion formula [89, 90] the density-independent correction term can be formulated as a sum of attractive pair interactions

$$E_{\mathrm{vdW}} = -\sum_{l,l'}{}' f_{ll'}(|\mathbf{R}_l - \mathbf{R}_{l'}|)\frac{C_6^{ll'}}{|\mathbf{R}_l - \mathbf{R}_{l'}|^6}, \tag{9.34}$$

of atoms at $\mathbf{R}_l$ and $\mathbf{R}_{l'}$, where $f_{ll'}(R)$ is a damping function which equals one for large distances R and zero for small values of R. The asymptotic form of (9.34) for $R \rightarrow \infty$ is chosen in agreement with the original description of the dispersion energy as derived by London [89, 90]. One possibility to choose the virial coefficients $C_6^{ll'}$ is [80, 83]

$$C_6^{ll'} = \frac{3}{2}\alpha_l\alpha_{l'}\frac{I_l I_{l'}}{I_l + I_{l'}}, \tag{9.35}$$

where I_l is the ionization potential and α_l the polarizability of the atom at $\mathbf{R}_l$. For a critical discussion of the choice of the damping function and the virial coefficients

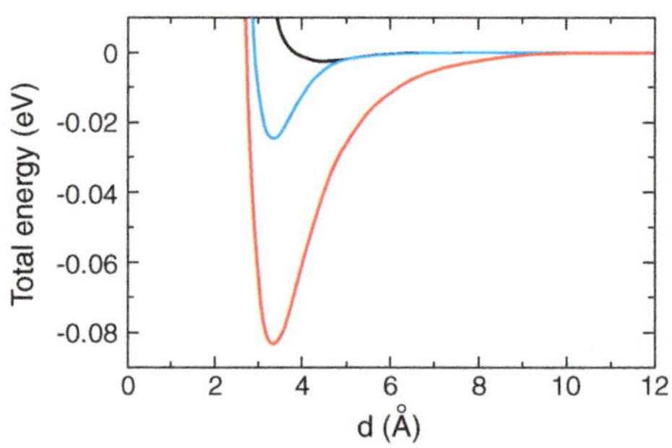

Fig. 9.17 Variation of total energy of graphite as a function of the distance d between two graphene layers. Three approximations have been used to describe exchange and correlation: PW91 – GGA (*black line*), PZ – LDA (*blue line*), and GGA + vdW (*red line*) with the correction (9.34). Adapted from a figure in [83]

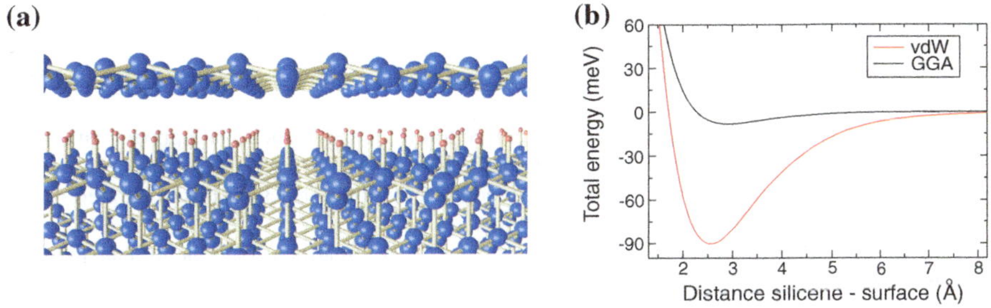

Fig. 9.18 (**a**) Resulting minimum lateral and vertical positions of silicene atoms (*blue dots*) with respect to the H (*red*) and Si (*blue circles*) atoms in the substrate surface. (**b**) Total energy per silicene atom of the silicene-Si(111):H-1×1 adsorbate system versus distance using a GGA functional [75] and a functional including vdW [75, 76]. Adapted from [92]. Copyright Wiley-VCH Verlag GmbH & Co.KGaA. Reproduced with permission

the reader is referred to a recent publication [91]. The effect of the vdW correction on the total energy of graphite is illustrated in Fig. 9.17. It is most important for the PW91-GGA treatment that only gives rise to an almost vanishing bonding between the graphene layers in the graphite crystal in large distances of 9 Å in contrast to the PZ-LDA approximation of the XC functional. The GGA+vdW functional yields lattice constants $a = 2.455$ Å, $c = 6.69$ Å, the elastic constant $C_{33} = 41.7$ GPa, and the exfoliation energy (minimum in Fig. 9.17) of the layers $E_{ex} = 83.5$ meV in good agreement with experimental values $a = 2.459$ Å, $c = 6.672$ Å, $C_{33} = 41$ GPa, and $E_{ex} = 35 - 52$ meV [83]. Improvements of the vdW functional according to Dion et al. [76] together with an appropriate combination of LDA correlation and GGA exchange [75] leads to similar energy versus distance curves for vdW-bonded layers as displayed in Fig. 9.17. As an example results for a graphene-like 2D honeycomb crystal silicene, i.e., a silicon allotrope, adsorbed on a H-passivated Si(111)1×1 surface are shown in Fig. 9.18.

References

1. J.P. Perdew, A. Ruzsinzky, L.A. Constantin, J. Sun, G.I. Csouka, Some fundamental issues in ground-state density functional theory: a guide for the perplexed. J. Chem. Theory Comput. **5**, 902–908 (2009)
2. N.C. Handy, A.J. Cohen, Left-right correlation energy. Molec. Phys. **99**, 403–412 (2001)
3. J. Hubbard, Electron correlations in narrow energy bands. Proc. Roy. Soc. London A **276**, 238 (1963)
4. J. Hubbard, Electron correlations in narrow energy bands. IV. The atomic representation. Proc. Roy. Soc. London **285**, 542–560 (1965)
5. D. Baeriswyl, D.K. Campell, J.M.P. Carmelo, F. Guinea, *The Hubbard Model* (Plenum Press, New York, 1995)
6. D. Sarma, N. Shanthi, S. Barman, N. Hamada, H. Sawada, K. Terakura, Band theory for ground-state properties and excitation spectra of perovskite $LaMO_3$ (M = Mn, Fe Co, Ni). Phys. Rev. Lett. **75**, 1126–1129 (1995)

7. A. Svane, O. Gunnarsson, Transition-metal oxides in the self-energy-corrected density functional formalism. Phys. Rev. Lett. **65**, 1148–1151 (1990)
8. M.M. Rieger, P. Vogl, Self-interaction corrections in semiconductors. Phys. Rev. B **52**, 16567–16574 (1995)
9. J.P. Perdew, A. Zunger, Self-interaction correction to density-functional approximations for many-electron systems. Phys. Rev. B **23**, 5048–5079 (1981)
10. A. Svane, O. Gunnarsson, Localization in self-interaction-corrected density-functional formalism. Phys. Rev. B **37**, 9919–9922 (1988)
11. E. Bonsquet, N. Spaldin, Dependence in the LSDA+U treatment of noncollinear magnets. Phys. Rev. B **82**, 220402(R) (2010)
12. A. Schrön, F. Bechstedt, Crystalline and magnetic anisotropy of the $3d$-transition metal monoxides MnO, FeO, CoO, and NiO. Phys. Rev. B **86**, 115134 (2012)
13. V.I. Anisimov, J. Zaanen, O.K. Andersen, Band theory and Mott insulators: Hubbard U instead of stoner I. Phys. Rev. B **44**, 943–954 (1991)
14. V.I. Anisimov, I.V. Solovyev, M.A. Korotin, M.T. Czyzyk, G.A. Sawatzky, Density-functional theory and NiO photoemission spectra. Phys. Rev. B **48**, 16929–16934 (1993)
15. A.I. Lichtenstein, V.I. Anisimov, J. Zaanen, Density-functional theory and strong interactions: orbital ordering in Mott-Hubbard insulators. Phys. Rev. B **52**, R5467–R5470 (1995)
16. W.A. Harrison, *Electronic Structure and the Properties of Solids* (Dover, Mineola, 1989)
17. M.T. Czyźyk, G.A. Sawatzky, Local-density functional and on-site correlations: the electronic structure of La_2CuO_4 and $LaCuO_3$. Phys. Rev. B **49**, 14211–14228 (1994)
18. S.L. Dudarev, G.A. Botton, S.Y. Savrasov, C.J. Humphreys, A.P. Sutton, Electron-energy-loss spectra and the structural stability of nickel oxide: an LSDA+U study. Phys. Rev. B **57**, 1505–1509 (1998)
19. A. Schrön, C. Rödl, F. Bechstedt, Energetic stability and magnetic properties of MnO in the rocksalt, wurtzite, and zinc-blende structures: influence of exchange and correlation. Phys. Rev. B **82**, 165109 (2010)
20. I. Bahrin, *Thermo-chemical Data of Pure Substances*, 3rd edn. (VCH, Weinheim, 1995)
21. J. Heyd, J. Scuseria, M. Ernzerhof, Hybrid functionals based on a screened Coulomb potential. J. Chem. Phys. **118**, 8207–8215 (2003)
22. C. Kittel, *Introduction to Solid State Physics*, 8th edn. (Wiley, Hoboken, 2005)
23. C. Rödl, F. Fuchs, J. Furthmüller, F. Bechstedt, Quasiparticle band structures of the antiferromagnetic transition-metal oxides MnO, FeO, CoO, and NiO. Phys. Rev. B **79**, 235114 (2009)
24. A.D. Becke, A new mixing of Hartree-Fock and local density-functional theories. J. Chem. Phys. **98**, 1372–1377 (1993)
25. A.D. Becke, Density-functional thermochemistry: the role of exact exchange. J. Chem. Phys. **98**, 5648–5652 (1993)
26. A.D. Becke, Density-functional exchange-energy approximation with correct asymptotic behavior. Phys. Rev. A **38**, 3098–3110 (1988)
27. C. Lee, W. Yang, R.G. Parr, Development of the Colle-Salvetti correlation energy formula into a functional of the density. Phys. Rev. B **37**, 785–789 (1988)
28. http://www.gaussian.com
29. M. Marsman, J. Paier, A. Stroppa, G. Kresse, Hybrid functionals applied to extended systems. J. Phys. Condens. Matter **20**, 064201 (2008)
30. J. Paier, M. Marsman, G. Kresse, Why does the B3LYP hybrid functional fail for metals? J. Chem. Phys. **127**, 024103 (2007)
31. W. Kohn, Nobel lecture: electronic structure of matter—wave functions and density functionals. http://www.nobelprize.org/nobel_prizes/chemistry/laureates/1998/kohn-lecture.html
32. J. Muscat, A. Wander, N.M. Harrison, On the prediction of band gaps from hybrid functional theory. Chem. Phys. Lett. **342**, 397–401 (2001)
33. S. Tomić, B. Montanari, N.M. Harrison, The group III-V's semiconductor energy gaps predicted using the B3LYP hybrid functional. Physica E **40**, 2125–2127 (2008)

34. J.P. Perdew, K. Burke, M. Ernzerhof, Generalized gradient approximation made simple. Phys. Rev. Lett. **77**, 3865–3868 (1996)
35. M. Ernzerhof, G.E. Scuseria, Assessment of the Perdew-Burke-Ernzerhof exchange-correlation functional. J. Chem. Phys. **110**, 5029–5036 (1999)
36. C. Adamo, V. Barone, Toward reliable density functional methods without adjustable parameters: the PBE0 model. J. Chem. Phys. **110**, 6158–6170 (1999)
37. J.P. Perdew, M. Ernzerhof, K. Burke, Rationale for mixing exact exchange with density functional approximations. J. Chem. Phys. **105**, 9982–9985 (1996)
38. A. Seidl, A. Görling, P. Vogl, J.A. Majewski, M. Levy, Generalized Kohn-Sham schemes and the band-gap problem. Phys. Rev. B **53**, 3764–3774 (1996)
39. J. Heyd, G.E. Scuseria, M. Ernzerhof, Erratum: "hybrid functionals based on a screened Coulomb potential" [J. Chem. Phys. 118, 8207 (2003)]. J. Chem. Phys. **124**, 219906 (2006)
40. A.V. Krukau, O.A. Vydrov, A.F. Ismaylov, G.E. Scuseria, Influence of the exchange screening parameter on the performance of screened hybrid functionals. J. Chem. Phys. **125**, 224106 (2006)
41. A. Görling, M. Levy, Hybrid schemes combining the Hartree-Fock method and density-functional theory: underlying formalism and properties of correlation functionals. J. Chem. Phys. **106**, 2675–2680 (1997)
42. S. Baroni, E. Tuncel, Exact-exchange extension of the local-spin-density approximation in atoms: calculation of total energies and electron affinities. J. Chem. Phys. **79**, 6140–6144 (1983)
43. M.E. Casida, Generalization of the optimized-effective-potential model to include electron correlation: a variational derivation of the Sham-Schlüter equation for the exact exchange-correlation potential. Phys. Rev. A **51**, 2005–2013 (1995)
44. T. Grabo, T. Kreibich, S. Kurth, E.K.U. Gross, Orbital functionals in density functional theory: the optimized effective potential method, in *Strong Coulomb Correlations in Electronic Structure Calculations: Beyond the Local Density Approximation*, ed. by V.I. Anisimov (Gordon and Breach, New York, 2000), pp. 203–311
45. M. Städele, J.A. Majewski, P. Vogl, A. Görling, Exact Kohn-Sham exchange potential for semiconductors. Phys. Rev. Lett. **79**, 2089 (1997)
46. P. Rinke, A. Qteish, J. Neugebauer, M. Scheffler, Exciting prospects for solids: exact-exchange based functionals meet quasiparticle energy calculations. Phys. Stat. Solidi B **245**, 929–945 (2008)
47. J. Kohanoff, *Electronic Structure Calculations for Solids and Molecules: Theory and Computational Methods* (Cambridge University Press, Cambridge, 2006)
48. S. Kurth, J.P. Perdew, P. Blaha, Molecular and solid-state tests of density functional approximations: LSD, GGAs and meta-GGAs. Int. J. Quantum Chem. **75**, 889–909 (1999)
49. R. Maul, F. Ortmann, M. Preuss, K. Hannewald, F. Bechstedt, DFT studies using supercells and projector-augmented waves for structure, energetics, and dynamics of glycine, alanine, and cysteine. J. Comput. Chem. **28**, 1817–1833 (2007)
50. C. Møller, M.S. Plesset, Note on an approximation treatment for many-electron systems. Phys. Rev. **46**, 618–622 (1934)
51. S. Gronert, R.A.J. O'Hair, Ab initio studies of amino acid conformations. 1. The conformers of alanine, serine, and cysteine. J. Am. Chem. Soc. **117**, 2071–2081 (1995)
52. S.H. Rhim, M. Kim, A.J. Freeman, R. Asahi, Fully first-principles screened-exchange LDA calculations of excited states and optical properties of III–V semiconductors. Phys. Rev. B **71**, 045202 (2005)
53. J. Wróbel, K.J. Kurzydłowski, K. Hummer, G. Kresse, J. Piechota, Calculations of ZnO properties using the Heyd-Scuseria-Ernzerhof screened hybrid density functional. Phys. Rev. B **80**, 155124 (2009)
54. S.J. Clark, J. Robertson, Screened exchange density functional applied to solids. Phys. Rev. B **82**, 085208 (2010)
55. A. Janotti, C.G. Van de Walle, Fundamentals of zinc oxide as a semiconductor. Rep. Prog. Phys. **72**, 126501 (2009)

56. P.G. Moses, M. Miao, Q. Yan, C.G. Van de Walle, Hybrid functional investigations of band gaps and band alignements for AlN, GaN, InN, and InGaN. J. Chem. Phys. **134**, 084703 (2011)
57. J.L. Lyons, A. Janotti, C.G. Van de Walle, Shallow versus deep nature of Mg acceptors in nitride semiconductors. Phys. Rev. Lett. **108**, 156403 (2012)
58. A. Riefer, F. Fuchs, C. Rödl, A. Schleife, F. Bechstedt, R. Goldhahn, Interplay of excitonic effects and van Hove singularities in optical spectra: CaO and AlN polymorphs. Phys. Rev. B **84**, 075218 (2011)
59. K. Seino, F. Bechstedt, P. Kroll, Tunneling of electrons between Si nanocrystals embedded in a SiO_2 matrix. Phys. Rev. B **86**, 075312 (2012)
60. L.C. de Carvalho, A. Schleife, F. Bechstedt, Influence of exchange and correlation on structural and electronic properties of AlN, GaN, and InN polytypes. Phys. Rev. B **84**, 195105 (2011)
61. A. Schleife, C. Rödl, F. Fuchs, J. Furthmüller, F. Bechstedt, Optical and energy-loss spectra of MgO, ZnO, and CdO from ab initio many-body theory. Phys. Rev. B **80**, 035112 (2009)
62. R. Grau-Crespo, H. Wang, U. Schwingenschlögl, Why the Heyd-Scuseria-Ernzerhof hybrid functional description of VO_2 phases is not correct. Phys. Rev. B **86**, 081101(R) (2012)
63. J.P. Perdew, K. Schmidt, Jacob's ladder of density functional approximations for exchange-correlation energy, in *Density Functional Theory and Its Application to Materials*, ed. by V. Van Doren, C. Van Alsenoy, P. Geerlings (American Institute of Physics, Melville, 2001), pp. 1–20
64. D.C. Langreth, J.P. Perdew, Exchange-correlation energy of a metallic surface: wave-vector analysis. Phys. Rev. B **15**, 2884–2901 (1977)
65. R. Kubo, The fluctuation-dissipation theorem. Rep. Prog. Phys. **29**, 255–284 (1966)
66. D. Pines, P. Nozières, *The Theory of Quantum Liquids* (W.A. Benjamin Inc., New York, 1966)
67. X. Ren, P. Rinke, C. Joas, M. Scheffler, Random-phase approximation and its applications in computational chemistry and materials science. J. Mater. Sci. **47**, 7447–7471 (2012)
68. L.D. Landau, E.M. Lifshitz, *Electrodynamics of Continua* (Pergamon Press, Oxford, 1989)
69. S. Doniach, E.H. Sondheimer, *Green's Functions for Solid State Physicists* (Imperial College Press, London, 1998)
70. P. Romaniello, F. Bechstedt, L. Reining, Insights in the T-matrix approximation beyond the GW approximation: combining correlation channels. Phys. Rev. B **85**, 155131 (2012)
71. X. Ren, P. Rinke, V. Blum, J. Wieferink, A. Tkatchenko, A. Sanfilippo, K. Reuter, M. Scheffler, Resolution-of-identity approach to Hartree-Fock, hybrid density functionals, RPA, MP2 and GW with numeric atom-centered orbital basis functions. New J. Phys. **14**, 053020 (2012)
72. J. Harl, G. Kresse, Cohesive energy curves for noble gas solids calculated by adiabatic connection fluctuation-dissipation theory. Phys. Rev. B **77**, 045136 (2008)
73. J. Harl, G. Kresse, Accurate bulk properties from approximate many-body techniques. Phys. Rev. Lett. **103**, 056401 (2009)
74. J. Harl, L. Schimka, G. Kresse, Assessing the quality of the random phase approximation for lattice constants and atomization energies of solids. Phys. Rev. B **81**, 115126 (2010)
75. J. Klimĕs, D.R. Bowler, A. Michaelides, Van der Waals density functionals applied to solids. Phys. Rev. B **83**, 195131 (2011)
76. M. Dion, H. Rydberg, E. Schröder, D.C. Langreth, B.I. Lundqvist, Van der Waals density functional for general geometries. Phys. Rev. Lett. **92**, 246401 (2004)
77. H. Peng, S. Lany, Polymorphic energy ordering of MgO, ZnO, GaN, and MnO within the random phase approximation. Phys. Rev. B **87**, 174113 (2013)
78. X. Ren, A. Tkatchenko, P. Rinke, M. Scheffler, Beyond the random-phase approximation for the electron correlation energy: the importance of single excitations. Phys. Rev. Lett. **106**, 153003 (2011)
79. J. Paier, B.G. Janesko, T.M. Henderson, G.E. Scuseria, A. Grüneis, G. Kresse, Hybrid functionals including random phase approximation correlation and second-order screened exchange. J. Chem. Phys. **132**, 094103 (2010) (erratum: ibid. 133, 179902 (2010))
80. F. Ortmann, W.G. Schmidt, F. Bechstedt, Attracted by long-range electron correlation: adenine on graphite. Phys. Rev. Lett. **95**, 186101 (2005)
81. F. Bechstedt, *Principles of Surface Physics* (Springer, Berlin, 2003)

82. A.Y. Kipnis, B.E. Yavelov, J.S. Rowlinson, *Van der Waals and Molecular Sciences* (Oxford University Press, New York, 1996)
83. F. Ortmann, F. Bechstedt, W.G. Schmidt, Semiempirical van der Waals correction to the density functional description of solids and molecular structures. Phys. Rev. B **73**, 205101 (2006)
84. S. Grimme, Semiempirical GGA-type density functional constructed with a long-range dispersion correction. J. Comput. Chem. **27**, 1787–1799 (2006)
85. A. Tkatchenko, M. Scheffler, Accurate molecular van der Waals interactions from ground-state electron density and free-atom reference data. Phys. Rev. Lett. **102**, 073005 (2009)
86. B.I. Lundqvist, Y. Andersson, H. Shao, S. Chan, D.C. Langreth, Density functional theory including van der Waals forces. Int. J. Quantum Chem. **56**, 247–255 (1955)
87. Y. Andersson, E. Hult, H. Rydberg, P. Apell, B.I. Lundqvist, D.C. Langreth, Van der Waals interactions in density functional theory, in *Electronic Density Functional Theory: Recent Progress and New Directions*, ed. by J.F. Dobson, G. Vignale, M.P. Das (Plenum Press, New York, 1998), pp. 243–260
88. M. Lein, J.F. Dobson, E.K.U. Gross, Toward the description of van der Waals interactions within density functional theory. J. Comput. Chemistry **20**, 12–22 (1999)
89. F. London, Zur Theorie und Systematik der Molekularkräfte. Z. Phys. **63**, 245–279 (1930)
90. F. London, Über einige Eigenschaften und Anwendungen der Molekularkräfte. Z. Phys. Chem. Abt. B **11**, 222–251 (1931)
91. E.R. McNellis, J. Meyer, K. Reuter, Azobenzene at coinage metal surfaces: role of dispersive van der Waals interactions. Phys. Rev. B **80**, 205414 (2009)
92. S. Kokott, L. Matthes, F. Bechstedt, Silicene on hydrogen-passivated Si(111) and Ge(111) substrates. Phys. Stat. Solidi RRL **7**, 538–541 (2013)

Part III
Single-Particle Excitations: Quasielectrons and Quasiholes

Chapter 10
Description of Electron Ensemble

Abstract The description of excited-state properties of inhomogeneous electron gases in solids, nanostructures and molecules asks for concepts beyond the density functional theory. The treatment of the motion and interaction of individual electrons by means of creation and annihilation operators suggests to introduce appropriate expectation values. The time evolution of the fermion field operators is described by Heisenberg equations of motion. The mutual Coulomb interaction of the electrons turns out to cause serious problems because of the resulting non-linearity with respect to the field operators. Nevertheless, particle conservation can be easily formulated by means of an equation of continuity for the operators of electron density and paramagnetic current density, at least without spin-orbit coupling. The ensemble properties are reasonably described by the grand canonical ensemble, since it allows for particle and energy exchange. Arbitrary numbers of electrons and temperatures of the electron gas can be studied.

10.1 Dynamical Characterization

10.1.1 Time Evolution

We study a system of N electrons in the volume Ω. The field operators $\psi_s^+(\mathbf{x})$ and $\psi_s(\mathbf{x})$ represent creation and annihilation, respectively, of an electron with the spin variable s at the space point $\mathbf{x}$. These fermion operators obey the (anti)commutation relations (3.12). The unperturbed system is described by the Hamiltonian $\mathscr{H}_0$ (3.1). It does not explicitly depend on time t. The time evolution of the system is therefore characterized by the unitary canonical transformation

$$U_0(t) = \exp\left(-\frac{i}{\hbar}\mathscr{H}_0 t\right). \tag{10.1}$$

F. Bechstedt, *Many-Body Approach to Electronic Excitations*,
Springer Series in Solid-State Sciences 181, DOI 10.1007/978-3-662-44593-8_10

The time evolution of the field operators is then given by

$$\psi_s^+(\mathbf{x}, t) = U_0^+(t)\psi_s^+(\mathbf{x})U_0(t),$$
$$\psi_s(\mathbf{x}, t) = U_0^+(t)\psi_s(\mathbf{x})U_0(t) \tag{10.2}$$

in the Heisenberg picture. Each of the operators obeys a Heisenberg equation of motion

$$i\hbar\frac{\partial}{\partial t}\psi_s^+(\mathbf{x}, t) = \left[\psi_s^+(\mathbf{x}, t), \mathscr{H}_0\right]_-,$$
$$i\hbar\frac{\partial}{\partial t}\psi_s(\mathbf{x}, t) = \left[\psi_s(\mathbf{x}, t), \mathscr{H}_0\right]_-. \tag{10.3}$$

Instead of (3.12) the time-dependent operators fulfill generalized anticommutation relations [1]

$$\left[\psi_s(\mathbf{x}, t), \psi_{s'}^+(\mathbf{x}', t)\right]_+ = \delta_{ss'}\delta(\mathbf{x} - \mathbf{x}'),$$
$$\left[\psi_s(\mathbf{x}, t), \psi_{s'}(\mathbf{x}', t)\right]_+ = \left[\psi_s^+(\mathbf{x}, t), \psi_{s'}^+(\mathbf{x}', t)\right]_+ = 0. \tag{10.4}$$

10.1.2 Interaction with Nuclei and Between Electrons

Without relativistic effects the interactions in the electron system are described by the Hamiltonian (3.1)

$$\mathscr{H}_0 = \hat{T} + \hat{V} + \hat{U}. \tag{10.5}$$

Despite the fact that the Hamiltonian $\mathscr{H}_0$ does not explicitly depend on time, we add an implicit time dependence mediated by the field operators (10.2). The operator of the kinetic energy (3.17) reads as

$$\hat{T}(t) = \sum_s \int d^3\mathbf{x}\psi_s^+(\mathbf{x}, t)\left\{-\frac{\hbar^2}{2m}\Delta_\mathbf{x}\right\}\psi_s(\mathbf{x}, t). \tag{10.6}$$

The scalar-relativistic effects (2.22) that are routinely taken into account in the electronic-structure codes will be not explicitly mentioned here. They remain hidden in (10.6) and the total Hamiltonian (10.5). We will not come back to this point throughout the book. However, all numerical examples presented contain these effects.

The ensemble of the nuclei is assumed to be fixed at certain coordinates $\{\mathbf{R}_l\}$, e.g. the equilibrium ones. Thereby the nuclei generate a potential energy $V_\mathrm{n}(\mathbf{x})$ (1.4) felt by an electron at space point $\mathbf{x}$. We do not distinguish between valence electrons and core electrons as in Chap. 8. That is why we do also not discuss the spatial variation of the electrostatic electron-nuclei interaction (1.4), for instance in the limit of valence electrons and pseudopotentials (see Sect. 8.3), in detail. The total operator of the potential energy (3.18) is

$$\hat{V}(t) = \sum_s \int d^3\mathbf{x} \psi_s^+(\mathbf{x}, t) V_\mathrm{n}(\mathbf{x}) \psi_s(\mathbf{x}, t). \tag{10.7}$$

For the investigation of electronic excitations an accurate treatment of the electron-electron interaction is crucial. We have argued in Sect. 2.4 that for not too heavy elements spin-orbit interaction (2.23) can be omitted from the studies. In addition, we have shown that the transverse electron-electron interaction, the Breit interaction (2.37), can be neglected. It remains the longitudinal electron-electron interaction (3.19)

$$\hat{U}(t) = \frac{1}{2} \sum_{s,s'} \int d^3\mathbf{x} \int d^3\mathbf{x}' \psi_{s'}^+(\mathbf{x}', t) \psi_s^+(\mathbf{x}, t) \mathrm{v}(\mathbf{x} - \mathbf{x}') \psi_s(\mathbf{x}, t) \psi_{s'}(\mathbf{x}', t) \tag{10.8}$$

mediated by the bare Coulomb potential (1.2).

In a few cases, for instance, where the lineshape of the spectra is significantly influenced by the spin-orbit interaction or in magnetic systems in which internal magnetic fields influence spectra via their coupling to the electron spins, we have to discuss the impact of the relativistic effects. In the case of heavy elements, e.g. gold, the spin-orbit coupling can alter the band structure of the corresponding metal significantly, by about 1 eV [2]. In anisotropic systems the Rashba effect [3] may induce wave-vector-induced band splittings but also spin currents in the absence of external magnetic fields [4, 5] and modify the spin lifetimes [6, 7]. Even in cubic crystals such band-splitting effects due to the spin-orbit interaction may occur [8]. In Chap. 6 we have learnt that the potential energy of an electron at $\mathbf{x}$ is not only affected by the potential $V_\mathrm{n}(\mathbf{x})$ due to the nuclei. Rather, as a consequence of the electron-electron interaction an effective potential [see e.g. (6.20)] including the Hartree potential and contributions from exchange and correlation acts on the electrons. The negative gradient of the effective potential represents an internal electric field that may be denoted by $\mathbf{E}(\mathbf{x})$. It is responsible for the coupling between orbital and spin motion (2.33)

$$\frac{\mu_B}{2mc^2} (\mathbf{E}(\mathbf{x}) \times \mathbf{p}) \boldsymbol{\sigma},$$

where the electron spin is expressed by the vector of Pauli spin matrices $\boldsymbol{\sigma}$ (1.1) to avoid the introduction of the spin density operator $\hat{\mathbf{s}}(\mathbf{x})$ (3.32). A magnetic field $\mathbf{B}(\mathbf{x})$ may occur in systems with magnetic ordering. It might be mainly caused by the spin of the electrons and their orbital motion. Surely it is modified by the (longitudinal) electron-electron interaction, for instance by an XC contribution, as indicated in the Kohn-Sham equations for non-collinear spins in Sect. 6.2.2. As in these equations we assume that the main effect of this magnetic field stems from the Zeeman coupling to the electron spin

$$\mu_B \mathbf{B}(\mathbf{x})\boldsymbol{\sigma}.$$

The effect of internal electromagnetic fields including their modifications due to the electron-electron interaction in a mean-field approximation can be taken into account by a spin-dependent correction to the potential energy of the electrons. Instead of the major spin-less contribution (10.7) one may generalize the potential energy operator to

$$\hat{V}(t) = \sum_{s,s'} \int d^3\mathbf{x} \psi_s^+(\mathbf{x}, t) \tilde{V}_{ss'}(\mathbf{x}) \psi_{s'}(\mathbf{x}, t) \tag{10.9}$$

with the generalized spin-dependent potential

$$\tilde{V}_{ss'}(\mathbf{x}) = V_{\mathrm{n}}(\mathbf{x})\delta_{ss'} + \mu_B \mathbf{B}(\mathbf{x})\boldsymbol{\sigma}_{ss'} + \frac{\mu_B}{2mc^2}(\mathbf{E}(\mathbf{x}) \times \mathbf{p})\boldsymbol{\sigma}_{ss'}. \tag{10.10}$$

The explicit spin dependence of the Hamiltonain $\mathscr{H}_0$ (10.5) via the generalized potential energy (10.10) can be interpreted in terms of a mean-field approximation for the internal magnetic field in the direct coupling to the spin and the internal electric field appearing in the spin-orbit interaction. There are also attempts in the literature to start from a generalized formulation of the spin-spin and spin-orbit interactions as two-particle interactions [9] as discussed in Sect. 2.4. Here, however, we proceed without these interactions by including their action, at least partially, in the one-particle potential term.

10.1.3 Equations of Motion

With the Hamiltonian (10.5) and the anticommutation rules (10.4) we find

$$\begin{aligned} i\hbar\frac{\partial}{\partial t}\psi_s(\mathbf{x}, t) =& -\frac{\hbar^2}{2m}\Delta_{\mathbf{x}}\psi_s(\mathbf{x}, t) + \sum_{s'} \tilde{V}_{ss'}(\mathbf{x})\psi_{s'}(\mathbf{x}, t) \\ &+ \sum_{s'} \int d^3\mathbf{x}' \mathrm{v}(\mathbf{x} - \mathbf{x}')\psi_{s'}^+(\mathbf{x}', t)\psi_{s'}(\mathbf{x}', t)\psi_s(\mathbf{x}, t), \\ -i\hbar\frac{\partial}{\partial t}\psi_s^+(\mathbf{x}, t) =& -\frac{\hbar^2}{2m}\Delta_{\mathbf{x}}\psi_s^+(\mathbf{x}, t) + \sum_{s'} \tilde{V}_{s's}(\mathbf{x})\psi_{s'}^+(\mathbf{x}, t) \\ &+ \sum_{s'} \int d^3\mathbf{x}' \mathrm{v}(\mathbf{x} - \mathbf{x}')\psi_s^+(\mathbf{x}, t)\psi_{s'}^+(\mathbf{x}', t)\psi_{s'}(\mathbf{x}', t) \end{aligned} \tag{10.11}$$

for the time evolution of the electron field operators (10.3). These are the fundamental dynamical equations of the theory to treat the influence of the longitudinal electron-electron interaction, including spin effects. Formally, they are similar to the time-dependent Hartree equations. This becomes obvious if the time-dependent density or particle number operators

$$\begin{aligned} \hat{n}_{ss'}(\mathbf{x},t) &= \psi_s^+(\mathbf{x},t)\psi_{s'}(\mathbf{x},t), \\ \hat{n}(\mathbf{x},t) &= \sum_s \hat{n}_{ss}(\mathbf{x},t), \\ \hat{N}(t) &= \int d^3\mathbf{x}\hat{n}(\mathbf{x},t) \end{aligned} \tag{10.12}$$

on the right-hand side are replaced by their expectation values according to (3.36).

Such a rough approximation, however, nurtures illusions about the simplicity of the problem. In reality, the equations of motion (10.11) are non-linear in the field operators $\psi_s^+(\mathbf{x},t)$ and $\psi_s(\mathbf{x},t)$ due to the electron-electron interaction in the operator $\hat{U}(t)$ (10.8). The terms in (10.11) proportional to three field operators illustrate the dilemma of the full treatment of the longitudinal electron-electron interaction. The equations of motion (10.11) are not exactly solvable. Approximations will be needed.

The left- or right-hand-side multiplication of the equations (10.11) with $\frac{1}{i\hbar}\psi_s^+(\mathbf{x},t)$ and $-\frac{1}{i\hbar}\psi_s(\mathbf{x},t)$, respectively, together with the addition of the two resulting equations yields

$$\begin{aligned} &\frac{\partial}{\partial t}\left\{\psi_s^+(\mathbf{x},t)\psi_s(\mathbf{x},t)\right\} \\ &\quad+\frac{1}{i\hbar}\sum_{s'}\left\{\psi_{s'}^+(\mathbf{x},t)\tilde{V}_{s's}(\mathbf{x})\psi_s(\mathbf{x},t)-\psi_s^+(\mathbf{x},t)\tilde{V}_{ss'}(\mathbf{x})\psi_{s'}(\mathbf{x},t)\right\} \\ &\quad+\mathrm{div}\frac{\hbar}{2mi}\left\{\psi_s^+(\mathbf{x},t)\left[\nabla_{\mathbf{x}}\psi_s(\mathbf{x},t)\right]-\left[\nabla_{\mathbf{x}}\psi_s^+(\mathbf{x},t)\right]\psi_s(\mathbf{x},t)\right\}=0, \end{aligned}$$

or after summation over the spin variable s, the equation of continuity

$$\frac{\partial}{\partial t}\hat{n}(\mathbf{x},t)+\mathrm{div}\,\hat{\mathbf{j}}(\mathbf{x},t)=0 \tag{10.13}$$

with the electron density operator $\hat{n}(\mathbf{x},t)$ (10.12) and the (paramagnetic) current density operator (3.32)

$$\hat{\mathbf{j}}(\mathbf{x},t)=\frac{\hbar}{2mi}\sum_s\left\{\psi_s^+(\mathbf{x},t)\left[\nabla_{\mathbf{x}}\psi_s(\mathbf{x},t)\right]-\left[\nabla_{\mathbf{x}}\psi_s^+(\mathbf{x},t)\right]\psi_s(\mathbf{x},t)\right\}. \tag{10.14}$$

Because of the neglect of the vector-potential-mediated electron-electron interaction essentially the paramagnetic contribution (2.15) determines the current density operator. The resulting equation of continuity (10.13) represents particle conservation in the electron system.

10.2 Statistical Characterization

10.2.1 Grand Canonical Ensemble

For an electron system in thermodynamic equilibrium the calculation of the expectation value $\langle \hat{\mathcal{O}} \rangle$, the statistical average of a quantum-mechanical operator $\hat{\mathcal{O}}$, requires a statistical operator $\hat{W}$. The cases of solids and other larger electron systems suggest to represent the physical system by a grand canonical ensemble of the statistical mechanics. This is an extension of the canonical ensemble at a given temperature T due to the possibility to exchange energy with its environment, the thermostat in Fig. 10.1. The particle number in the Fock space (see Sect. 3.2.1) is replaced by the average number of particles $N = \langle \hat{N} \rangle$ with $\langle \hat{N} \rangle$ as the statistical expectation value of the electron number operator (10.12). A grand canonical ensemble is in equilibrium with external reservoirs with respect not only to energy exchange but also particle exchange as illustrated in Fig. 10.1 [10]. The chemical potential (or fugacity) μ is introduced to specify the fluctuations of the number of electrons while the particle number N is the thermodynamic conjugate (see Sect. 8.1.1). The thermodynamic state of the system is characterized by the parameters μ and T. The temperature is usually replaced by the inverse thermal energy

$$\beta = \frac{1}{k_B T} \tag{10.15}$$

with k_B as the Boltzmann constant. Zero temperature, or $\beta \to \infty$, describes the ground state of the system. In this limit the results of the presented theoretical apparatus are also applicable to molecular and nanosystems for which the temperature is not defined.

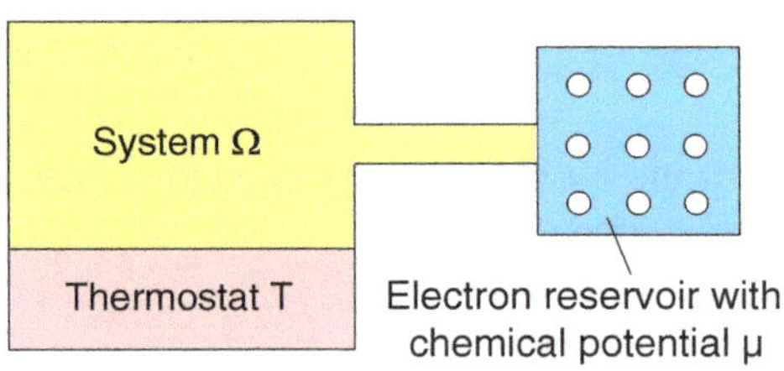

Fig. 10.1 Schematic description of the grand canonical ensemble of electrons in volume Ω with possible exchange of energy and particles

10.2.2 Expectation Values

A compact way of writing the (quantum-)statistical average is

$$\langle\hat{\mathscr{O}}\rangle = \mathrm{Tr}\left(\hat{W}_0\hat{\mathscr{O}}\right) \tag{10.16}$$

with the statistical operator in Gibbs form

$$\begin{aligned}\hat{W}_0 &= \frac{1}{Z_g}e^{-\beta(\mathscr{H}_0-\mu\hat{N})},\\ Z_g &= \mathrm{Tr}\left\{e^{-\beta(\mathscr{H}_0-\mu\hat{N})}\right\},\end{aligned} \tag{10.17}$$

where Z_g is the grand partition function and Tr denotes the trace.

The ensemble in question is stationary, i.e., it does not change in time, as long as no external perturbation is applied. Therefore, by the Liouville theorem $[\hat{W}_0, \mathscr{H}_0]_- = 0$ the statistical operator is diagonal in the stationary eigenstates $|\Psi\rangle$ of the Hamiltonian $\mathscr{H}_0$ (10.5) described by the Schrödinger equation (3.29). It holds

$$\hat{W}_0|\Psi\rangle = W_\Psi|\Psi\rangle \tag{10.18}$$

with the eigenvalues of the statistical operator

$$W_\Psi = \frac{1}{Z_g}e^{-\beta(E_\Psi-\mu N_\Psi)}, \tag{10.19}$$

defined by the eigenenergy E_Ψ and the number N_Ψ of particles in the Fock-space state $|\Psi\rangle$. It holds $\sum_\Psi W_\Psi = 1$. The statistical weights express that, in general, the grand canonical ensemble is in a mixed state. One has to remember that for a grand partition the states $|\Psi\rangle$ are states with multiple particles in Fock space, and the trace $\mathrm{Tr}(\hat{W}_0\cdots)$ (10.16) sums over all of them. As a result, for a system in thermodynamic equilibrium, the expectation value (10.16) of any operator $\hat{\mathscr{O}}$ may be computed using (10.18) as [11]

$$\begin{aligned}\langle\hat{\mathscr{O}}\rangle &= \sum_\Psi \langle\Psi|\hat{W}_0\hat{\mathscr{O}}|\Psi\rangle\\ &= \sum_{\Psi,\Psi'} \langle\Psi|\hat{W}_0|\Psi'\rangle\langle\Psi'|\hat{\mathscr{O}}|\Psi\rangle\\ &= \sum_\Psi W_\Psi\langle\Psi|\hat{\mathscr{O}}|\Psi\rangle.\end{aligned} \tag{10.20}$$

10.2.3 Relation to Thermodynamics

The grand partition function Z_g in (10.17) can be reformulated introducing the grand (canonical) thermodynamic potential $\Omega_g = \Omega_g(T, \Omega, \mu)$ (8.7) as

$$Z_g = e^{-\beta\Omega_g}. \tag{10.21}$$

Infinitesimal changes of the variables result in its total differential in analogy to (8.8)

$$d\Omega_g = -SdT - pd\Omega - Nd\mu, \tag{10.22}$$

where N is given by $\langle\hat{N}\rangle$. Different names are found in the literature for Ω_g, e.g. Kramers grand potential [12] or Landau potential [13]. By means of the definition of the grand partition function (10.17) one can easily show the validity of the relation [14]

$$\frac{\partial}{\partial\mu}\Omega_g(T, \Omega, \mu) = -\langle\hat{N}\rangle = -N.$$

A direct relationship between the grand canonical thermodynamic potential Ω_g and well-defined thermodynamic quantities is difficult to find. In order to do so, one may follow a similar idea as used to derive the coupling-constant-integrated pair correlation function (7.8). However, instead of a potential energy operator $\hat{V}^\lambda$ in (7.2) we use $\lambda\hat{V}$ with $\hat{V}$ as defined in (10.9). The resulting Hamiltonian $\mathscr{H}_0^\lambda$ is only linear in λ. For an arbitrary interaction strength $0 \leq \lambda \leq 1$ the thermodynamic potential obeys the differential equation

$$\frac{\partial}{\partial\lambda}\Omega_g^\lambda(T, \Omega, \mu) = \left\langle\frac{\partial}{\partial\lambda}\mathscr{H}_0^\lambda\right\rangle_\lambda, \tag{10.23}$$

where the statistical operator in $\langle\cdots\rangle_\lambda$ is determined by $\mathscr{H}_0^\lambda$. Using the equations of motion (10.11) for λ-scaled electron-electron and electron-ion interactions and multiplying them with $\psi_{s'}^+(\mathbf{x}', t)$ or $\psi_{s'}(\mathbf{x}', t)$, one finds [11, 15]

$$\begin{aligned}\lambda\hat{V}(t) + \lambda\hat{U}(t) = \frac{1}{4}\sum_{s,s'}\int d^3\mathbf{x}\Bigg\{\Bigg[\left(i\hbar\frac{\partial}{\partial t} - i\hbar\frac{\partial}{\partial t'} + \frac{\hbar^2}{2m}\Delta_\mathbf{x} + \frac{\hbar^2}{2m}\Delta_{\mathbf{x}'}\right)\delta_{ss'}\\ + 2\lambda\tilde{V}_{s's}(\mathbf{x})\Bigg]\psi_{s'}^+(\mathbf{x}', t')\psi_s(\mathbf{x}, t)\Bigg\}_{\substack{\mathbf{x}'=\mathbf{x}\\ t'=t}}\end{aligned} \tag{10.24}$$

taking advantage of the symmetry of the potentials with respect to space coordinates and spin variables. The last relation can be used to find

$$\begin{aligned}\Omega_g(T,\Omega,\mu) = \ & \Omega_g^1(T,\Omega,\mu) = \Omega_g^0(T,\Omega,\mu) \\ & +\frac{1}{4}\int_0^1 d\lambda \frac{1}{\lambda}\sum_{s,s'}\int d^3\mathbf{x}\left\{\left[\left(i\hbar\frac{\partial}{\partial t} - i\hbar\frac{\partial}{\partial t'} + \frac{\hbar^2}{2m}\Delta_\mathbf{x} + \frac{\hbar^2}{2m}\Delta_{\mathbf{x}'}\right)\delta_{ss'}\right.\right. \\ & \left.\left. +2\lambda\tilde{V}_{s's}(\mathbf{x})\right]\langle\psi_{s'}^+(\mathbf{x}',t')\psi_s(\mathbf{x},t)\rangle_\lambda\right\}_{\substack{\mathbf{x}'=\mathbf{x}\\ t'=t}},\end{aligned}$$

i.e., to trace back the grand potential Ω_g to that of the non-interacting electrons Ω_g^0 and the spin densities in a system with a λ-scaled interaction in the statistical operator.

The total energy of the electronic system $\langle\mathscr{H}_0\rangle$ is generally more important than the grand canonical thermodynamic potential. By setting $\lambda = 1$ in (10.24) and adding the kinetic energy operator $\hat{T}(t)$ (10.6) [11] one obtains

$$\langle\mathscr{H}_0\rangle = -\frac{1}{2}\sum_{s,s'}\int d^3\mathbf{x}\left\{\left[\left(\frac{\hbar^2}{2m}\Delta_\mathbf{x} - i\hbar\frac{\partial}{\partial t}\right)\delta_{ss'} - \tilde{V}_{ss'}(\mathbf{x})\right]\langle\psi_{s'}^+(\mathbf{x}',t')\psi_s(\mathbf{x},t)\rangle\right\}_{\substack{\mathbf{x}'=\mathbf{x}\\ t'=t}}. \tag{10.25}$$

In the low-temperature limit $T \to 0\,\mathrm{K}$ this quantity equals the ground-state energy defined in (5.9). Expression (10.25) leads to the Galitskii-Migdal formula (11.35) [16].

References

1. G.D. Mahan, *Many-Particle Physics* (Plenum Press, New York, 1990)
2. P. Romaniello, P.L. de Boeij, The role of relativity in the optical response of gold within the time-dependent current-density-functional theory. J. Chem. Phys. **122**, 164303 (2005)
3. Y.A. Bychkov, E.I. Rashba, Oscillatory effects and the magnetic susceptibility of carriers in inversion layers. J. Phys. C **17**, 6039–6045 (1984)
4. P. Sharma, How to create a spin current. Science **307**, 531–533 (2005)
5. E. Rashba, Electric fields drive spins. Nat. Phys. **2**, 149–150 (2006)
6. D. Awschalom, N. Samarth, Spintronics without magnetism. Physics **2**, 50–54 (2009)
7. F. Pezzalo, F. Bottegoni, D. Trivedi, F. Ciccacci, A. Giorgioni, P. Li, S. Cecchi, E. Grilli, Y. Song, M. Guzzi, H. Dery, G. Isella, Optical spin injection and spin lifetime in Ge heterostructures. Phys. Rev. Lett. **108**, 156603 (2012)
8. G. Dresselhaus, Spin-orbit coupling effects in zinc blende structures. Phys. Rev. **100**, 580–586 (1955)
9. F. Aryasetiawan, S. Biermann, Generalized Hedin's equation for quantum many-body systems with spin-dependent interactions. Phys. Rev. Lett. **100**, 116402 (1998)
10. L. Landau, E.M. Lifshitz, *Statistical Physics* (Pergamon Press, Oxford, 1959)
11. L.P. Kadanoff, G. Baym, *Quantum Statistical Mechanics: Green's Function Methods in Equilibrium and Nonequilibrium Problems* (W.A. Benjamin Inc, New York, 1962)
12. M.-C. Desjonquères, D. Spanjaard, *Concepts in Surface Physics* (Springer, Berlin, 1996)

13. D. Goodstein, *States of Matter* (Dover Publications Inc, New York; Pennsylvania State University, State College, 1985)
14. A.A. Abrikosov, L.P. Gorkov, I.E. Dzyaloshinski, *Methods of Quantum Field Theory in Statistical Physics* (Dover Publ. Inc, New York, 1975)
15. E.N. Economou, *Green's Functions in Quantum Physics* (Springer, Berlin, 1990)
16. V.M. Galitskii, A.B. Migdal, Application of quantum field theory methods to the many-body problem. Zh. Eksp. Teor. Fiz. **34**, 139–150 (1958) [Sov. Phys. JETP (English Transl.) **7**, 96–104 (1958)]

Chapter 11
Thermodynamic Green Functions

Abstract The quantum-field theory is designed to deal with an infinite number of degrees of freedom. This is exactly what the description of electronic excitations in condensed matter needs, at least, together with the quantum-statistical approach. Nevertheless, we start with a single quantum particle embedded in the electron gas. Propagators of electrons and holes are studied as expectation values of pairs of field operators. They allow the introduction of Green functions. The poles in frequency domain of their Fourier transforms contain information about the electronic excitations. Because of the dependence of the grand canonical statistical operator on the Hamiltonian and the inverse temperature a generalization for complex times is possible. Then one speaks about thermodynamic or Matsubara Green functions. On the single-particle level they contain the complete information about the spectral properties mediated solely by the spectral-weight function. The successive application of the equation of motion leads to a hierarchy of equations for N-particle Green functions. In the single-particle case it is closed introducing a self-energy of an electron that accounts for the entire electron-electron interaction. It allows the formulation of an integral equation, a Dyson equation, instead of the differential equation of motion.

11.1 Definition

11.1.1 Propagators

Studying the energetic and thermodynamic properties of the electron gas in (10.23) and (10.25) we meet the correlation function

$$\begin{aligned} G^{<}_{ss'}(\mathbf{x}t, \mathbf{x}'t') &= -\frac{1}{i\hbar}\langle \psi^{+}_{s'}(\mathbf{x}', t')\psi_s(\mathbf{x}, t)\rangle \\ &= -\frac{1}{i\hbar}\mathrm{Tr}\left\{\hat{W}_0\psi^{+}_{s'}(\mathbf{x}', t')\psi_s(\mathbf{x}, t)\right\}. \end{aligned} \tag{11.1}$$

F. Bechstedt, *Many-Body Approach to Electronic Excitations*,
Springer Series in Solid-State Sciences 181, DOI 10.1007/978-3-662-44593-8_11

A similar quantity is

$$
\begin{aligned}
G^{>}_{ss'}(\mathbf{x}t, \mathbf{x}'t') &= \frac{1}{i\hbar}\langle \psi_s(\mathbf{x}, t)\psi^{+}_{s'}(\mathbf{x}', t')\rangle \\
&= \frac{1}{i\hbar}\mathrm{Tr}\left\{\hat{W}_0\psi_s(\mathbf{x}, t)\psi^{+}_{s'}(\mathbf{x}', t')\right\}. \qquad (11.2)
\end{aligned}
$$

The two correlation functions represent a generalization of the one-particle density matrix of type (3.39). This is obvious for the first function (11.1).

The sequence of the operators $\hat{W}_0$, $\psi_s(\mathbf{x}, t)$, and $\psi^{+}_{s'}(\mathbf{x}', t')$ under the trace can be interchanged due to the cyclic invariance. Together with the definition of the time evolution of the field operators (10.2) this cyclic invariance guarantees in thermal equilibrium

$$
G^{\gtrless}_{ss'}(\mathbf{x}t, \mathbf{x}'t') = G^{\gtrless}_{ss'}(\mathbf{x}\mathbf{x}', t - t'), \qquad (11.3)
$$

i.e., the two correlation functions only depend on the difference of the time arguments. This is in contrast to the case of non-equilibrium Green functions where all propagators depend on two time variables along a Keldysh contour [1, 2].

The physical meaning of the correlation functions as particle propagators is illustrated in Fig. 11.1. The first quantity represents a *hole* propagator $G^{<}_{ss'}$ (see Fig. 11.1a). An electron with spin s is annihilated at $\mathbf{x}$ and t. This is equivalent to the creation of a hole. At later time t', the hole is filled by an electron with spin s' at $\mathbf{x}'$. In principle, a hole propagates from $\mathbf{x}, s, t$ to $\mathbf{x}', s', t'$. In the second correlation function, the *electron* propagator $G^{>}_{ss'}$, first an electron with $\mathbf{x}'$, s', t' is injected (see Fig. 11.1b). At a later time t, an electron with spin s is annihilated at $\mathbf{x}$. In a more classical picture, this could be interpreted as the propagation of an electron from $\mathbf{x}', s', t'$ to $\mathbf{x}, s, t$. However, in the case of identical quantum particles, one cannot say that is the "same" electron. Rather, at time t, one finds the system in a quantum state where an electron with spin s can be annihilated at $\mathbf{x}$. The two propagators can be also interpreted as probability amplitudes to restore a hole or an electron.

The spin structure of the propagators significantly simplifies in systems which are not magnetically ordered or influenced by any magnetic field, or in general, in non-spin-polarized systems. If also the spin-orbit interaction is supposed to be negligible,

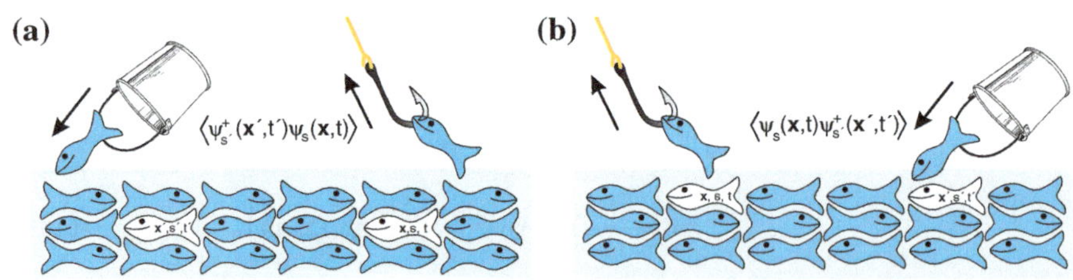

Fig. 11.1 Illustration of the physical meaning of a hole propagator (**a**) and an electron propagator (**b**). The electrons in a Fermi sea are represented by fish in a lake. Following an idea of Zagoskin [3]

then, it holds $\tilde{V}_{ss'}(\mathbf{x}) = V_{\mathrm{n}}(\mathbf{x})\delta_{ss'}$ in (10.10). No spin-dependent interaction occurs in the Hamiltonian $\mathcal{H}_0$. With (3.32) the total spin operator of the system of electrons

$$\hat{\mathbf{S}}(t) = \frac{\hbar}{2}\sum_{s,s'}\int d^3\mathbf{x}\psi^+_{s'}(\mathbf{x},t)\boldsymbol{\sigma}_{ss'}\psi_s(\mathbf{x},t) \tag{11.4}$$

obeys the commutation relations

$$\left[\mathcal{H}_0, \hat{\mathbf{S}}\right]_- = 0,$$
$$\left[\hat{W}_0, \hat{\mathbf{S}}\right]_- = 0.$$

By means of infinitesimal rotations in spin subspace one can show the validity of [4]

$$\left\langle\left[\hat{\mathbf{S}}, \psi_s(\mathbf{x},t)\psi^+_{s'}(\mathbf{x}',t')\right]_-\right\rangle = 0.$$

The evaluation of the commutator needs the identity

$$[\hat{A}\hat{B}, \hat{C}\hat{D}]_- = \hat{A}[\hat{C},\hat{B}]_+\hat{D} - \hat{A}\hat{C}[\hat{D},\hat{B}]_+ + [\hat{C},\hat{A}]_+\hat{D}\hat{B} - \hat{C}[\hat{D},\hat{A}]_+\hat{B}$$

and the anticommutation relations (10.4). They yield

$$\sum_{s''}\left\{\boldsymbol{\sigma}_{s''s'}G^>_{ss''}(\mathbf{xx}', t-t') - \boldsymbol{\sigma}_{ss''}G^>_{s''s'}(\mathbf{xx}', t-t')\right\} = 0.$$

With the vector of the Pauli spin matrices (1.1) one obtains

$$G^>_{\uparrow\uparrow}(\mathbf{xx}', t-t') - G^>_{\downarrow\downarrow}(\mathbf{xx}', t-t') = 0,$$
$$G^>_{\uparrow\downarrow}(\mathbf{xx}', t-t') - G^>_{\downarrow\uparrow}(\mathbf{xx}', t-t') = 0,$$
$$iG^>_{\uparrow\downarrow}(\mathbf{xx}', t-t') + iG^>_{\downarrow\uparrow}(\mathbf{xx}', t-t') = 0$$

from the x-component ($s=\uparrow, s'=\downarrow$), the x-component ($s=\uparrow, s'=\uparrow$), and the y-component ($s=\uparrow, s'=\uparrow$) of the spin matrix vector. In the non-spin-polarized case described above, the electron propagator

$$G^>_{ss'}(\mathbf{xx}', t-t') = \delta_{ss'}G^>(\mathbf{xx}', t-t') \tag{11.5}$$

is independent of the spin orientation. A similar procedure leads to the same result

$$G^<_{ss'}(\mathbf{xx}', t-t') = \delta_{ss'}G^<(\mathbf{xx}', t-t') \tag{11.6}$$

for the hole propagator.

11.1.2 Time Structure

The discussion of the physics of the correlation functions (11.1) and (11.2) as hole and electron propagators indicates that they are highly suitable for the description of single-particle excitations in electronic systems. Therefore, it makes sense to unify hole and electron propagation in one function. This is a (single-particle) Green function. Green functions are used in all fields of physics (and also mathematics) but they are defined in different ways [3, 5, 6]. In the case of electrons the time ordering of creation and annihilation operators suggests the introduction of a Green function as

$$G_{ss'}(\mathbf{x}\mathbf{x}', t-t') = \begin{cases} \theta(t-t')G^{>}_{ss'}(\mathbf{x}\mathbf{x}', t-t') + \theta(t'-t)G^{<}_{ss'}(\mathbf{x}\mathbf{x}', t-t') \\ \quad \text{for } \mathrm{Im}(t-t') = 0 \\ \theta(\mathrm{Im}(t'-t))G^{>}_{ss'}(\mathbf{x}\mathbf{x}', t-t') + \theta(\mathrm{Im}(t-t'))G^{<}_{ss'}(\mathbf{x}\mathbf{x}', t-t') \\ \quad \text{for } \mathrm{Im}(t-t') \neq 0. \end{cases} \tag{11.7}$$

The notation $>$ and $<$ is intended as a reminder that for $t > t'$, $G = G^{>}$, while for $t < t'$, $G = G^{<}$, at least on the real time axis. In the definition (11.7) the difference $(t-t')$ can take complex values. Correspondingly, a time ordering on the imaginary $(t-t')$-axis has to be introduced. The time-development operator (10.1) bears a strong formal similarity to the canonical part of the statistical operator (10.17) that occurs in the statistical average. Apart from a normalization factor the two operators are the same for $t = -i\hbar\beta$.

The generalization to complex times is easily realized. We investigate the two propagators appearing in (11.7) separately. With the orthonormal and complete set of eigenvectors $|\Psi\rangle$ and energies E_Ψ of the stationary Schrödinger equation (3.29) of the many-body problem the absolute convergence of the Ψ-sums appearing in the propagators can be also demonstrated for complex times [4]. One can show that the functions $G^{<}_{ss'}(\mathbf{x}\mathbf{x}', t-t')$ and $G^{>}_{ss'}(\mathbf{x}\mathbf{x}', t-t')$ are analytic functions of their complex argument $(t-t')$ in the interval $0 \leq \mathrm{Im}(t-t') < \hbar\beta$ or $-\hbar\beta < \mathrm{Im}(t-t') \leq 0$ in the complex $(t-t')$-plane.

The two propagators can be related to each other in the complex time plane. Because $\psi_s(\mathbf{x}, t)$ removes a particle it holds

$$\psi_s(\mathbf{x}, t)f(\hat{N}) = f(\hat{N}+1)\psi_s(\mathbf{x}, t),$$

where $f(\hat{N})$ is any function of the particle number operator $\hat{N}$. This can be easily proven using a power series expansion of $f(\hat{N})$ and the definition (10.12). In particular, it holds [7]

$$e^{-\beta\mu\hat{N}}\psi_s(\mathbf{x}, t)e^{\beta\mu\hat{N}} = e^{\beta\mu}\psi_s(\mathbf{x}, t).$$

Complex times lead to a generalization of the field operators to Matsubara (field) operators [3, 8]

$$e^{\beta\mathscr{H}_0}\psi_s(\mathbf{x},t)e^{-\beta\mathscr{H}_0} = \psi_s(\mathbf{x},t-i\hbar\beta).$$

Together with the cyclic property of the trace $\mathrm{Tr}(\hat{A}\hat{B}\hat{C}) = \mathrm{Tr}(\hat{B}\hat{C}\hat{A}) = \mathrm{Tr}(\hat{C}\hat{A}\hat{B})$, we find for the hole propagator (11.1) [7]

$$\begin{aligned}
G^{<}_{ss'}(\mathbf{xx}',t-t') &= -\frac{1}{i\hbar}e^{\beta\Omega_g}\mathrm{Tr}\left\{\psi_s(\mathbf{x},t)e^{-\beta(\mathscr{H}_0-\mu\hat{N})}\psi^{+}_{s'}(\mathbf{x}',t')\right\} \\
&= -\frac{1}{i\hbar}e^{\beta\mu}\langle e^{\beta\mathscr{H}_0}\psi_s(\mathbf{x},t)e^{-\beta\mathscr{H}_0}\psi^{+}_{s'}(\mathbf{x}',t')\rangle \\
&= -\frac{1}{i\hbar}e^{\beta\mu}\langle\psi_s(\mathbf{x},t-i\hbar\beta)\psi^{+}_{s'}(\mathbf{x}',t')\rangle \\
&= -e^{\beta\mu}G^{>}_{ss'}(\mathbf{xx}',t-t'-i\hbar\beta).
\end{aligned}$$

The relation [9]

$$G^{<}_{ss'}(\mathbf{xx}',t-t') = -e^{\beta\mu}G^{>}_{ss'}(\mathbf{xx}',t-t'-i\hbar\beta) \tag{11.8}$$

is called Martin-Schwinger relation. It links the two propagators defined on different stripes in the complex $(t-t')$-plane. $G^{>}_{ss'}(xx',t-t')$ is an analytic function for complex values of the time difference in the region $0 > \mathrm{Im}(t-t') > -\hbar\beta$. Similarly $G^{<}_{ss'}(xx',t-t')$ is an analytic function in the region $0 < \mathrm{Im}(t-t') < \hbar\beta$. Therefore, relation (11.8) can be also interpreted as a boundary condition at the boundaries of the imaginary time domain [7]. The representation (11.8) also sheds light on the frequently used term "thermodynamic" Green functions. Because of the formal equivalence of time evolution operator and statistical operator, imaginary time differences $(t-t')$ correspond to inverse temperatures and vice versa.

The definition (11.7) of the thermodynamic Green function and the Martin-Schwinger relation (11.8) show that $G_{ss'}(\mathbf{xx}',t-t')$ is an analytic function in the complex $(t-t')$-plane excepting parts of the horizontal lines with $\mathrm{Im}(t-t') = k\hbar\beta$ (k - integer number) as indicated in Fig. 11.2. The definition (11.7) can be rewritten in a more compact form

$$G_{ss'}(\mathbf{xx}',t-t') = \frac{1}{i\hbar}\langle\mathcal{T}\psi_s(\mathbf{x},t)\psi^{+}_{s'}(\mathbf{x}',t')\rangle \tag{11.9}$$

with a generalization of the Wick time-ordering operator $\mathcal{T}$ to complex times. It is called Dyson time-ordering operator in the case of pure imaginary times. It orders the field operators in such a way that real time arguments decrease from left to right, while for complex time arguments Re(it) decreases. The further down the imaginary axis a time is, the 'later' it is. With the analytic properties of the Green

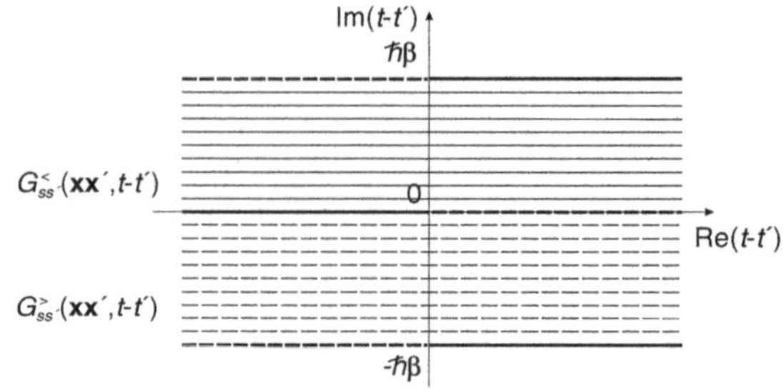

Fig. 11.2 Analytic regions of the thermodynamic Green function $G_{ss'}(\mathbf{xx}', t-t')$ and the relation to electron and hole propagators. The analytic regions are displayed by *thin solid lines* (hole propagator) or *thin dashed lines* (electron propagator). Because of the time ordering in (11.7) the Green function is not analytic on lines with fixed imaginary part of $\mathrm{Im}(t-t') = k\hbar\beta$ (k - integer) and $(t-t')$ has to be shifted infinitesimally along the imaginary axis

function in the complex $(t-t')$-plane the Martin-Schwinger relation (11.8) can be generalized to

$$G_{ss'}(\mathbf{xx}', t-t') = -e^{\beta\mu} G_{ss'}(\mathbf{xx}', t-t'-i\hbar\beta). \tag{11.10}$$

11.1.3 Spectral-(Weight) Function

In agreement with the analytic properties of the propagators and the Green function Fourier transforms with respect to the variable $(t-t')$ are possible with

$$G^{\gtrless}_{ss'}(\mathbf{xx}', t-t') = \pm\frac{1}{2\pi i}\int\limits_{-\infty}^{+\infty} d\omega e^{-i\omega(t-t')} G^{\gtrless}_{ss'}(\mathbf{xx}', \omega),$$

$$G^{\gtrless}_{ss'}(\mathbf{xx}', \omega) = \pm i \int\limits_{-\infty}^{+\infty} dt e^{i\omega(t-t')} G^{\gtrless}_{ss'}(\mathbf{xx}', t-t'). \tag{11.11}$$

Because of the analytic properties of the function $G^{>}_{ss'}(\mathbf{xx}', t-t')$ in the interval $-\hbar\beta \leqq \mathrm{Im}\,(t-t') \leqq 0$ the integration path can be displaced by $-i\hbar\beta$. The Martin-Schwinger relation (11.8) implies

$$\begin{aligned} G^{>}_{ss'}(\mathbf{xx}', \omega) &= -ie^{-\beta\mu}\int\limits_{-\infty}^{+\infty} dt e^{i\omega(t-t')} G^{<}_{ss'}(\mathbf{xx}', t-t'+i\hbar\beta) \\ &= -e^{\beta(\hbar\omega-\mu)} i \int\limits_{-\infty}^{+\infty} dt e^{i\omega(t-t')} G^{<}_{ss'}(\mathbf{xx}', t-t'), \end{aligned}$$

or more directly

$$G^{>}_{ss'}(\mathbf{xx}', \omega) = e^{\beta(\hbar\omega-\mu)} G^{<}_{ss'}(\mathbf{xx}', \omega). \tag{11.12}$$

The sum of the Fourier-transformed particle propagators

$$A_{ss'}(\mathbf{xx}', \omega) = G^{>}_{ss'}(\mathbf{xx}', \omega) + G^{<}_{ss'}(\mathbf{xx}', \omega) \tag{11.13}$$

yields the spectral-(weight) function $A_{ss'}(\mathbf{xx}', \omega)$ of the time-ordered thermodynamic Green function $G_{ss'}(\mathbf{xx}', t-t')$ (11.9). It has three important properties:

(i) It is directly related to the particle propagators. The solutions of the two algebraic equations (11.12) and (11.13) are

$$\begin{aligned} G^{<}_{ss'}(\mathbf{xx}', \omega) &= f(\hbar\omega) A_{ss'}(\mathbf{xx}', \omega), \\ G^{>}_{ss'}(\mathbf{xx}', \omega) &= [1 - f(\hbar\omega)] A_{ss'}(\mathbf{xx}', \omega) \end{aligned} \tag{11.14}$$

with the Fermi distribution function

$$f(\hbar\omega) = \frac{1}{e^{\beta(\hbar\omega-\mu)} + 1}. \tag{11.15}$$

The expressions (11.14) indicate that in thermal equilibrium the spectral properties can be separated from the statistical properties. However, generalizations for non-equilibrium and transport phenomena are possible [1–3, 7].

(ii) The spectral function is a Hermitian matrix

$$A_{ss'}(\mathbf{xx}', \omega) = A^{*}_{s's}(\mathbf{x}'\mathbf{x}, \omega). \tag{11.16}$$

This can be easily proven using the properties of the Fourier transforms (11.11) and the definitions of $G^{\lessgtr}_{ss'}(\mathbf{xx}', t-t')$ (11.1) and (11.2).

(iii) The spectral-weight function fulfills the sum rule

$$\begin{aligned} \frac{1}{2\pi} \int_{-\infty}^{+\infty} d\hbar\omega A_{ss'}(\mathbf{xx}', \omega) &= \langle \psi_s(\mathbf{x}, t) \psi^{+}_{s'}(\mathbf{x}', t) \rangle + \langle \psi^{+}_{s'}(\mathbf{x}', t) \psi_s(\mathbf{x}, t) \rangle \\ &= \delta_{ss'} \delta(\mathbf{x} - \mathbf{x}') \end{aligned} \tag{11.17}$$

with the anticommutation relation (10.4).

11.1.4 Spectral Representations

Unfortunately, it is difficult to derive a spectral representation of the time-ordered Green function (11.9) at finite temperature for real frequencies. For that reason, we investigate in a first step only imaginary time differences, i.e., $\mathrm{Re}(t-t')=0$. Then, the Martin-Schwinger relation describes a quasi-periodicity on the imaginary $(t-t')$-axis. It suggests a Fourier series

$$G_{ss'}(\mathbf{x},\mathbf{x}',t-t')=\frac{1}{-i\hbar\beta}\sum_n G_{ss'}(\mathbf{xx}',z_n)e^{-iz_n(t-t')} \tag{11.18}$$

with the Fourier coefficients

$$G_{ss'}(\mathbf{xx}',z_n)=\int_0^{-i\hbar\beta} dt G_{ss'}(\mathbf{xx}',t-t')e^{iz_n(t-t')} \tag{11.19}$$

at the fermionic Matsubara frequencies

$$\hbar z_n=\mu-\pi n/i\beta \qquad (n=\pm1,\pm3,\pm5,\ldots), \tag{11.20}$$

that are the poles of the Fermi distribution (11.15) in the complex ω-plane. They are equidistant to the imaginary axis. The distance μ is a consequence of the Martin-Schwinger relation and the finite chemical potential of the electrons. Because of (11.20) the thermodynamic Green functions are sometimes also called Matsubara Green functions [6].

Formally we introduce a spectral representation

$$G_{ss'}(\mathbf{xx}',z)=\int_{-\infty}^{+\infty}\frac{d\omega}{2\pi}\frac{A_{ss'}(\mathbf{xx}',\omega)}{z-\omega} \tag{11.21}$$

in the complex z-plane with the exception of the real axis, which agrees at $z=z_n$ with the Fourier coefficients (11.19). The function $G_{ss'}(\mathbf{xx}',z)$ is analytic in the complex z-plane and has singularities (branch cuts, in general) along those portions of the real axis where $A_{ss'}(\mathbf{xx}',\omega)\neq0$. In principle, the representation (11.21) defines two different Green functions, the retarded (advanced) Green function for $\mathrm{Im}\,z>0$ ($\mathrm{Im}\,z<0$) in the upper (lower) z-half-plane. This can be immediately seen using a Fourier transformation at real time differences $(t-t')\gtrless0$.

From (11.21) and (11.14) it follows that the discontinuity of the Fourier-transformed Green function along the real axis yields the spectral function, i.e., it holds

$$\begin{aligned}&i\lim_{\eta\to+0}\left\{G_{ss'}(\mathbf{xx}',\omega+i\eta)-G_{ss'}(\mathbf{xx}',\omega-i\eta)\right\}\\&=i\lim_{\eta\to+0}\int_{-\infty}^{+\infty}\frac{d\omega'}{2\pi}A_{ss'}(\mathbf{xx}',\omega')\left[\frac{1}{\omega+i\eta-\omega'}-\frac{1}{\omega-i\eta-\omega'}\right]\end{aligned}$$

$$= \int_{-\infty}^{\infty} d\omega' A_{ss'}(\mathbf{x}\mathbf{x}', \omega') \lim_{\eta \to +0} \frac{1}{\pi} \frac{\eta}{(\omega - \omega')^2 + \eta^2} = A_{ss'}(\mathbf{x}\mathbf{x}', \omega), \quad (11.22)$$

using the Lorentzian representation of the Dirac δ-function.

Despite the mentioned difficulties with real frequencies ω, one may follow the definition (11.7) for real time differences. Then, the Fourier transform of the time-ordered Green function reads as

$$G_{ss'}(\mathbf{x}\mathbf{x}', \omega) = \int_{-\infty}^{+\infty} d\tau \left\{\theta(\tau) G^{>}_{ss'}(\mathbf{x}\mathbf{x}', \tau) + \theta(-\tau) G^{<}_{ss'}(\mathbf{x}\mathbf{x}', \tau)\right\} e^{i\omega\tau}.$$

With the Heaviside theta function

$$\theta(\tau) = \lim_{\eta \to +0} \frac{i}{2\pi} \int_{-\infty}^{+\infty} d\omega \frac{e^{-i\omega\tau}}{\omega + i\eta}$$

and the transformations (11.11) one finds the Lehmann representation [10]

$$G_{ss'}(\mathbf{x}\mathbf{x}', \omega) = \lim_{\eta \to +0} \int_{-\infty}^{+\infty} \frac{d\omega'}{2\pi} \left\{ \frac{G^{>}_{ss'}(\mathbf{x}\mathbf{x}', \omega)}{\omega - \omega' + i\eta} + \frac{G^{<}_{ss'}(\mathbf{x}\mathbf{x}', \omega)}{\omega - \omega' - i\eta} \right\} \quad (11.23)$$

or with (11.14)

$$G_{ss'}(\mathbf{x}\mathbf{x}', \omega) = \lim_{\eta \to +0} \frac{1}{2\pi} \int_{-\infty}^{+\infty} d\omega' \left\{ \frac{1 - f(\hbar\omega')}{\omega - \omega' + i\eta} + \frac{f(\hbar\omega')}{\omega - \omega' - i\eta} \right\} A_{ss'}(\mathbf{x}\mathbf{x}', \omega').$$

The latter relation is convenient to study the low-temperature limit. With the Weierstrass formula

$$\lim_{\eta \to +0} \frac{1}{\omega \pm i\eta} = \frac{\mathcal{P}}{\omega} \mp i\pi\delta(\omega), \quad (11.24)$$

where $\mathcal{P}$ denotes the Cauchy principal value, the Fourier transform becomes

$$G_{ss'}(\mathbf{x}\mathbf{x}', \omega)$$
$$= \frac{1}{2\pi} \int_{-\infty}^{+\infty} d\omega' \left\{ \frac{\mathcal{P}}{\omega - \omega'} - i\pi\delta(\omega - \omega') \tanh\left[\frac{\beta}{2}(\hbar\omega' - \mu)\right]\right\} A_{ss'}(\mathbf{x}\mathbf{x}', \omega').$$

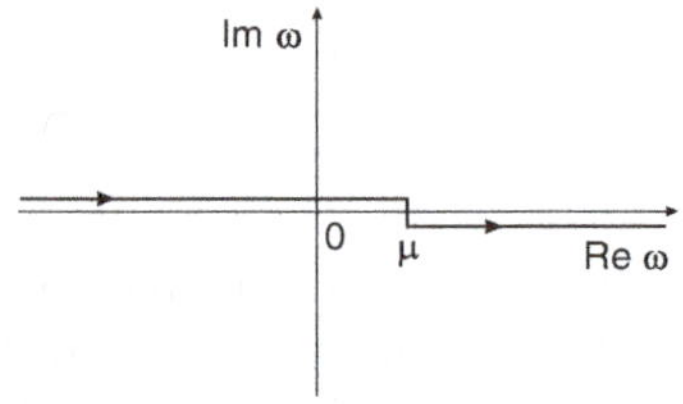

Fig. 11.3 Integration path for the contour integral (11.21) to obtain the spectral representation (11.25) in the complex ω-plane

With the sign function

$$\lim_{\beta\to\infty} \tanh\left[\frac{\beta}{2}(\hbar\omega - \mu)\right] = \operatorname{sgn}(\hbar\omega - \mu)$$

the last frequency integral can be rewritten. Applying again the Weierstrass formula (11.24) we find a compact expression

$$\lim_{T\to 0\,\mathrm{K}} G_{ss'}(\mathbf{xx}', \omega) = \lim_{\eta\to +0} \frac{1}{2\pi} \int_{-\infty}^{+\infty} d\omega' \frac{A_{ss'}(\mathbf{xx}', \omega')}{\omega - \omega' + i\eta\operatorname{sgn}(\hbar\omega' - \mu)}, \tag{11.25}$$

which is also called causal Green function $G^c_{ss'}(\mathbf{xx}', \omega)$. This spectral representation can be used to derive an expression similar to (11.22) for the discontinuity $A_{ss'}(\mathbf{xx}', \omega)$ of the zero-temperature Green function (11.25) on the real axis.

The analytic continuation of expression (11.25) into the complex ω-plane is possible using the integration path displayed in Fig. 11.3. In principle, expression (11.25) defines two analytic functions $G^r_{ss'}(\mathbf{xx}', \omega)$ and $G^a_{ss'}(\mathbf{xx}', \omega)$, the retarded and advanced Green functions, depending on $\operatorname{Im}\omega > 0$ (upper half-plane) or $\operatorname{Im}\omega < 0$

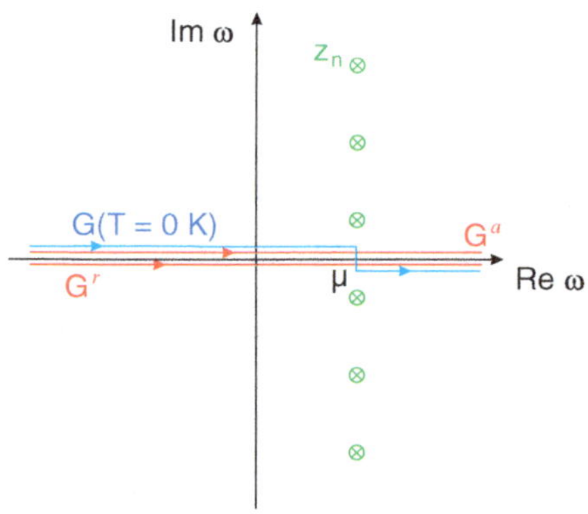

Fig. 11.4 Integration paths for the contour integral (11.21) in the complex frequency plane to define different Green functions

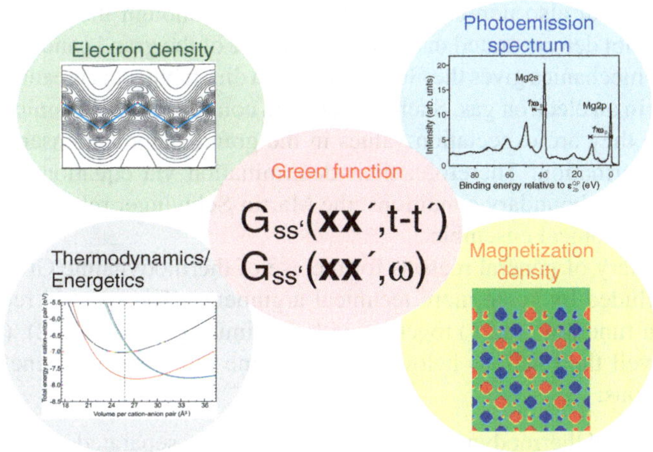

Fig. 11.5 Physical information that can be extracted from the thermodynamic *Green function* (an illustration)

(lower half-plane), more precisely above or below the contour in Fig. 11.3. Formally it holds

$$G^r_{ss'}(\mathbf{xx}', \omega) = \lim_{\eta \to +0} G_{ss'}(\mathbf{xx}', \omega + i\eta)$$

and

$$G^a_{ss'}(\mathbf{xx}', \omega) = \lim_{\eta \to +0} G_{ss'}(\mathbf{xx}', \omega - i\eta).$$

The relations between the three Green functions $G(T = 0\,\mathrm{K})$, G^r, and G^a are indicated in Fig. 11.4 by their integration paths.

According to the derivation of the spectral representation expression (11.21) is only valid for Matsubara frequencies $z = z_n$ (11.20). In the following, however, we analytically continue this expression into the entire complex z-plane. The expression (11.21) indeed defines the true thermodynamic Green function that characterizes many important properties of the inhomogeneous system of interacting electrons under consideration. A few of them will be discussed in the next section. Some of these are illustrated in Fig. 11.5.

11.1.5 Advantages of Thermodynamic Green Functions

The thermodynamic Green functions provide a method for discussing electronic excitations at finite temperatures with no more conceptual difficulties than ground-state problems at zero temperature. In principle, their concept can be applied not only to equilibrium but also non-equilibrium problems, although the capability for the latter ones is not demonstrated in this book. The use of the grand canonical ensemble of statistical mechanics gives the Green function a direct physical meaning as particle propagators in an electron gas. Such propagators contain much dynamic information but, because they are expectation values in the grand canonical ensemble, also all statistical information. Therefore, their determination via equations of motion is supplemented by boundary conditions, the Martin-Schwinger relations, appropriate to the grand canonical ensemble.

The summary of general reasons for the use of thermodynamic Green functions can be concluded by some more technical arguments. The spectral representation of the Green function (11.21) together with the limiting cases (11.22), (11.23), and (11.25) as well the relations below illustrate some more specific benefits of these Green functions:

(i) Spectral and thermodynamic properties are clearly separated in the equilibrium case.
(ii) The central quantities of the theory are spectral functions. Besides the spectral representation of Matsubara Green functions their knowledge also allows the determination of the corresponding causal, retarded and advanced Green functions irrespective of the temperature.
(iii) Along the imaginary axis the time-dependent variations of a single-particle Green function can be traced back to Fourier sums (11.18) over discrete frequencies z_n. As we will see later, e.g. in Sect. 13.1.3, this also holds for two-particle Green functions, although the discrete frequencies $\tilde{z}_m$ are different.
(iv) In all characteristic integral equations such as Dyson and Bethe-Salpeter equations to describe the dynamic reaction and interaction the Fourier representations take simple forms. Instead of integrating over dummy frequencies in the vertices from minus to plus infinity, one has to sum over the discrete sets of Matsubara frequencies z_n or $\tilde{z}_m$. This is generally less troublesome than integration in a complex plane as all discrete mathematics goes and immediately leads to Fermi and Bose functions.
(v) Metals and insulators can be treated on the same footing independent of the position of the chemical potential μ characterizing the electron reservoir. In heavily doped or even stationarily pumped semiconductors the effect of electron-electron interaction on excitations can be easily treated as illustrated in Sects. 19.2 and 22.3. Only the parameter μ has to be changed.
(vi) Additional time dependencies, e.g. due to dynamical screening in vertex corrections in (22.3), can be treated by algebraic sets of equations.
(vii) The interplay of electron-electron interaction and temperature effects is inherent in the theoretical treatment. A modification of all equations to determine electronic excitations and spectra by an additional temperature-dependent gas of bosonic lattice vibrations seems to be easily possible.

11.2 Relation to Observables

11.2.1 Density of States

The spectral(-weight) function $A_{ss'}(\mathbf{xx}', \omega)$ (11.13) contains the full information about the spectral properties of the interacting electron gas, at least on the single-particle level. For equal space coordinates $\mathbf{x} = \mathbf{x}'$ the spin-summed quantity yields the local (single-particle) density of states (LDOS) of the electronic system

$$D(\mathbf{x}, \hbar\omega) = \frac{1}{2\pi} \sum_{s} A_{ss}(\mathbf{xx}, \omega), \tag{11.26}$$

which is measurable by means of scanning tunneling microscopy (STM) or, more precisely, scanning tunneling spectroscopy (STS) [11].

The space integral over the sample volume Ω yields the total density of states (DOS)

$$D(\hbar\omega) = \frac{1}{2\pi} \sum_{s} \int d^3\mathbf{x} A_{ss}(\mathbf{xx}, \omega). \tag{11.27}$$

This quantity determines many features in single-particle excitation spectra as indicated in Fig. 11.5.

11.2.2 Magnetization, Electron, and Current Densities

The 2×2 matrix of the spin densities of the electron gas are generalized according to (5.25) as

$$\begin{aligned} n_{ss'}(\mathbf{x}) &= \int_{-\infty}^{+\infty} \frac{d\hbar\omega}{2\pi} f(\hbar\omega) A_{ss'}(\mathbf{xx}, \omega) \\ &= \int_{-\infty}^{+\infty} \frac{d\hbar\omega}{2\pi} G^{<}_{ss'}(\mathbf{xx}, \omega) \\ &= -i\hbar G^{<}_{ss'}(\mathbf{xx}, t = t') \\ &= \langle \psi^{+}_{s'}(\mathbf{x}, t) \psi_s(\mathbf{x}, t) \rangle \end{aligned} \tag{11.28}$$

with (11.1), (11.3), (11.11), and (11.14) for the system in thermal equilibrium at finite temperature T. It allows us to describe other densities, e.g. the vector of the magnetization density (5.27)

$$\mathbf{m}(\mathbf{x}) = -\mu_B \sum_{s,s'} \boldsymbol{\sigma}_{ss'} n_{ss'}(\mathbf{x}). \tag{11.29}$$

This quantity makes the description of important magnetic properties of the system, e.g. the magnetization density of a NiO(001) surface (Fig. 11.5), possible.

The sum of the diagonal elements of (11.28) represents the electron density (5.26) (see also Fig. 11.5)

$$\begin{aligned} n(\mathbf{x}) &= \sum_s n_{ss}(\mathbf{x}) \\ &= \int_{-\infty}^{+\infty} \frac{d\hbar\omega}{2\pi} f(\hbar\omega) \sum_s A_{ss}(\mathbf{xx}, \omega) \\ &= -i\hbar \sum_s G^<_{ss}(\mathbf{xx}, t = t') \\ &= \sum_s \langle \psi_s^+(\mathbf{x}, t)\psi_s(\mathbf{x}, t)\rangle. \end{aligned} \tag{11.30}$$

Despite its temperature dependence it is normalized. With the sum rule (11.17) it holds

$$N = \int d^3\mathbf{x} n(\mathbf{x}),$$

which defines the average density of the inhomogeneous electron gas

$$n = N/\Omega. \tag{11.31}$$

The relation (11.30) between density and Green function allows us, in turn, to express the thermodynamic properties of a system through its Green function. We illustrate this conclusion for a non-spin-polarized homogeneous electron gas with uniform density n, more precisely a jellium system (Sect. 4.4.1), and vanishing temperature $T \to 0\,\mathrm{K}$. The grand potential of the system (10.22) satisfies the equation [12]

$$d\Omega_g = -N d\mu$$

for fixed volume Ω at $T = 0\,\mathrm{K}$, since the entropy $S(T = 0\,\mathrm{K}) = 0$. This equation can thus be integrated, remembering that $\Omega_g(\mu = 0) = 0$,

$$\Omega_g = -\int_0^{\mu} d\mu' N(\mu'),$$

where one can substitute the expression $N(\mu)$ using (11.30) and (11.31), since the thermodynamic Green function depends on μ. For a non-interacting gas and $T = 0\,\mathrm{K}$,

$\mu = \varepsilon_F = \varepsilon_F^0$ and $N = \frac{1}{3\pi^2}\left(\frac{2m\varepsilon_F^0}{\hbar^2}\right)^{\frac{3}{2}} \Omega$ hold. It implies

$$\Omega_g = -\frac{2}{5} N \varepsilon_F^0. \tag{11.32}$$

This result is in agreement with $\Omega_g = E_0 - N\varepsilon_F^0$ (8.9) and the ground-state energy $E_0 = \frac{3}{5}N\varepsilon_F^0$ of the homogeneous electron gas [13]. With $\Omega_g = -p\Omega$ the last expression leads to the pressure $p = \frac{2}{5}n\varepsilon_F^0$ exerted by the electron gas.

Finally, the (quantum-)statistical expectation value of the current density operator (10.14)

$$\mathbf{j}(\mathbf{x}) = \langle \hat{\mathbf{j}}(\mathbf{x}, t) \rangle$$

can be easily reformulated by means of a "hair splitting" trick (introducing two different space coordinates) to

$$\begin{aligned}
\mathbf{j}(\mathbf{x}) &= \frac{\hbar}{2mi} \lim_{\mathbf{x}' \to \mathbf{x}} (\nabla_{\mathbf{x}} - \nabla_{\mathbf{x}'}) \sum_s \langle \psi_s^+(\mathbf{x}', t) \psi_s(\mathbf{x}, t) \rangle_{\mathbf{x}'=\mathbf{x}} \\
&= -\frac{\hbar^2}{2m} \sum_s \lim_{\mathbf{x}' \to \mathbf{x}} (\nabla_{\mathbf{x}} - \nabla_{\mathbf{x}'}) G_{ss}^<(\mathbf{x}\mathbf{x}', 0) \\
&= \frac{\hbar}{2mi} \sum_s \lim_{\mathbf{x}' \to \mathbf{x}} (\nabla_{\mathbf{x}} - \nabla_{\mathbf{x}'}) \int_{-\infty}^{+\infty} \frac{d\hbar\omega}{2\pi} f(\hbar\omega) A_{ss}(\mathbf{x}\mathbf{x}', \omega) \\
&= \frac{\hbar}{2mi} \lim_{\mathbf{x}' \to \mathbf{x}} (\nabla_{\mathbf{x}} - \nabla_{\mathbf{x}'}) n(\mathbf{x}, \mathbf{x}')
\end{aligned} \tag{11.33}$$

with the one-particle density matrix $n(\mathbf{x}, \mathbf{x}')$ following (3.39). The (paramagnetic) current density can be expressed by the gradients of the off-diagonal elements of the single-particle density matrix.

11.2.3 Galitskii-Migdal Formula

According to the findings in (10.25) the total energy of the system of electrons or its internal energy $U(\Omega, T, \mu)$, which is not a thermodynamic potential in the natural variables of the grand potential, is directly related to the hole propagator $G_{ss'}^<(\mathbf{x}\mathbf{x}', t - t')$ (11.1). It holds [14]

$$U(\Omega, T, \mu) = \tag{11.34}$$

$$i\hbar \frac{1}{2} \sum_{s,s'} \int d^3\mathbf{x} \lim_{\mathbf{x}' \to \mathbf{x}} \lim_{t' \to t} \left\{ \left[\frac{\hbar^2}{2m} \Delta_{\mathbf{x}} - i\hbar \frac{\partial}{\partial t} \right] \delta_{ss'} - \tilde{V}_{ss'}(\mathbf{x}) \right\} G_{ss'}^<(\mathbf{x}\mathbf{x}', t - t').$$

With the Fourier transformation (11.11) and relation (11.14) we find

$$\begin{aligned}
U(\Omega, T, \mu) \\
&= \frac{1}{2} \int_{-\infty}^{+\infty} \frac{d\hbar\omega}{2\pi} \sum_{s,s'} \int d^3\mathbf{x} \lim_{\mathbf{x}' \to \mathbf{x}} \left\{ \left(\hbar\omega - \frac{\hbar^2}{2m} \Delta_{\mathbf{x}} \right) \delta_{ss'} + \tilde{V}_{ss'}(\mathbf{x}) \right\} G^<_{ss'}(\mathbf{x}\mathbf{x}', \omega) \\
&= \frac{1}{2} \int_{-\infty}^{+\infty} \frac{d\hbar\omega}{2\pi} f(\hbar\omega) \sum_{s,s'} \int d^3\mathbf{x} \lim_{\mathbf{x}' \to \mathbf{x}} \left\{ \left(\hbar\omega - \frac{\hbar^2}{2m} \Delta_{\mathbf{x}} \right) + \tilde{V}_{ss'}(\mathbf{x}) \right\} A_{ss'}(\mathbf{x}\mathbf{x}', \omega).
\end{aligned} \tag{11.35}$$

This expression is called the Galitskii-Migdal formula [15]. It represents a surprising relationship (see also Fig. 11.5). The single-particle excitation properties, i.e., the single-particle energy spectrum represented by the spectral function $A_{ss'}(\mathbf{x}\mathbf{x}', \omega)$, or the Green function, are directly connected to some ground-state properties of the system, in particular, its energetics, at least for $T \to 0\,\mathrm{K}$.

The idea to use the exact single-particle Green function $G_{ss'}(\mathbf{x}\mathbf{x}', \omega)$ and to compute the total energy of an electronic system in Sect. 8.2.1 via the Galitskii-Migdal formula (11.35) has been implemented only in a few cases. At least for vanishing temperature and suppressing the particle exchange, this is possible. In the past, the relation of spectral properties and energy via the Galitskii-Migdal formula has been mainly applied to model systems [16]. However, currently there are indeed attempts to study the energetics of a system starting from its single-particle Green function [17, 18].

11.3 Dyson Equation

11.3.1 Equation of Motion

An equation of motion can be formulated for the time-ordered Green function (11.9) by means of the equations of motion of the field operators (10.11). In addition, the time evolution is influenced by that of the time-ordering operator $\mathcal{T}$ as described in (11.7). We use the definition

$$\begin{aligned}
\mathcal{T}\psi_s(\mathbf{x}, t)\psi^+_{s'}(\mathbf{x}', t') =& \theta(\alpha(t - t'))\psi_s(\mathbf{x}, t)\psi^+_{s'}(\mathbf{x}', t) \\
&- \theta(\alpha(t' - t))\psi^+_{s'}(\mathbf{x}', t)\psi_s(\mathbf{x}, t)
\end{aligned} \tag{11.36}$$

with $\alpha = 1$ for real time differences and $\alpha = i$ for imaginary time differences. With the relation of the Dirac δ-function and the Heaviside θ-function, $\delta(\tau) = \frac{d}{d\tau}\theta(\tau)$, one finds

$$\left(\frac{\partial}{\partial t}\mathcal{T}\right)\psi_s(\mathbf{x},t)\psi^+_{s'}(\mathbf{x}',t') = \alpha\delta(\alpha(t-t'))\left[\psi_s(\mathbf{x},t),\psi^+_{s'}(\mathbf{x}',t')\right]_+$$
$$= \delta(t-t')\delta(\mathbf{x}-\mathbf{x}')\delta_{ss'}, \qquad (11.37)$$

because of the anticommutation relation (10.4).

The combination of (10.11) and (11.37) yields the equation of motion for the single-particle Green function

$$\sum_{s''}\left\{\left(i\hbar\frac{\partial}{\partial t}+\frac{\hbar^2}{2m}\Delta_{\mathbf{x}}\right)\delta_{ss''}-\tilde{V}_{ss''}(\mathbf{x})\right\}G_{s''s'}(\mathbf{x}\mathbf{x}',t-t')$$
$$-\frac{1}{i\hbar}\sum_{s''}\int d^3\mathbf{x}''\mathrm{v}(\mathbf{x}-\mathbf{x}'')\langle\mathcal{T}\left\{\psi^+_{s''}(\mathbf{x}'',t)\psi_{s''}(\mathbf{x}'',t)\psi_s(\mathbf{x},t)\right\}\psi^+_{s'}(\mathbf{x}',t')\rangle$$
$$= \delta(\mathbf{x}-\mathbf{x}')\delta(t-t')\delta_{ss'}. \qquad (11.38)$$

The interaction term can be reformulated with the help of the time-ordered two-particle Green function

$$G_{s_1s_2,s'_1s'_2}(\mathbf{x}t_1\mathbf{x}_2t_2,\mathbf{x}'_1t'_1\mathbf{x}'_2t'_2) = \frac{1}{(i\hbar)^2}\langle\mathcal{T}\psi_{s_1}(\mathbf{x}_1,t_1)\psi_{s_2}(\mathbf{x}_2,t_2)\psi^+_{s'_2}(\mathbf{x}'_2,t'_2)\psi^+_{s'_1}(\mathbf{x}'_1,t'_1)\rangle. \qquad (11.39)$$

The product of the field operators is ordered in such a way that earlier real times appear to the right. For imaginary times the reversed ordering is applied as indicated in (11.36).

A more compact description is possible with the abbreviations

$$1 = \mathbf{x}_1t_1,$$
$$\mathrm{v}(1-2) = \mathrm{v}(\mathbf{x}_1-\mathbf{x}_2)\delta(t_1-t_2),$$
$$\int d1 = \int d^3\mathbf{x}_1\int_0^{-i\hbar\beta} dt_1,$$
$$\delta(1-2) = \delta(\mathbf{x}_1-\mathbf{x}_2)\delta(t_1-t_2). \qquad (11.40)$$

The time integral is restricted to a finite interval on the imaginary time axis because of the quasi-periodicity (11.10). The Coulomb potential $\mathrm{v}(1-2)$ is a spinless quantity. Since the generalized spin-dependent potential $\tilde{V}_{ss'}(\mathbf{x})$ occurs in (11.38), the spin variable is not included in the abbreviations (11.40). In order to guarantee the correct ordering of the field operators at equal imaginary times $t = t'$ a small shift $i\eta$ ($\eta \to +0$) has been taken into account. In this case, we write the abbreviation of space and time variables as

$$1^+ = \mathbf{x}_1t_1 - i\eta. \qquad (11.41)$$

Considering the time ordering and the sign changes due to commuting field operators, the interaction contribution in (11.38) can be rewritten by means of the two-particle Green function $G_{s_1s_2,s_1's_2'}(12,1'2')$ (11.39). We obtain

$$\sum_{s'}\left\{\left(i\hbar\frac{\partial}{\partial t_1}+\frac{\hbar^2}{2m}\Delta_{\mathbf{x}_1}\right)\delta_{s_1s'}-\tilde{V}_{s_1s'}(\mathbf{x}_1)\right\}G_{s's_1'}(11')$$
$$+i\hbar\sum_{s_2}\int d2\mathrm{v}(1-2)G_{s_1s_2,s_1's_2}(12,1'2^+)=\delta(1-1')\delta_{s_1s_1'}. \tag{11.42}$$

Even if the two-particle Green function is known, the equation of motion (11.42) is not sufficient to determine G unambiguously. It is a first-order differential equation in time, and thus a supplementary boundary condition is required to fix its solution. This condition is, of course, the Martin-Schwinger relation (11.10).

For vanishing Coulomb interaction between the particles, (11.42) really represents an equation of motion for the single-particle Green function. With interaction an additional term ruled by the two-particle Green function (11.39) appears. As a characteristic consequence of the Coulomb interaction (10.8) a whole hierarchy of coupled equations of motion can be derived for the higher-order Green functions starting with the two-particle, three-particle, etc. Green function. Such a procedure is however numerically not tractable. More efficient procedures are needed to solve the interaction problem, at least approximately.

11.3.2 Self-energy

In a first step we want to trace back the many-body problem (11.42) to an effective single-particle problem by introducing a self-energy $\tilde{\Sigma}$ (or self-energy operator or a so-called mass operator in the early days of the many-body theory [3, 6, 19–21]). We introduce the self-energy $\tilde{\Sigma}$ of a particle characterized by the single-particle Green function according to

$$\sum_{s_2}\int d2\tilde{\Sigma}_{s_1s_2}(12)G_{s_2s_1'}(21')=-i\hbar\sum_{s_2}\int d2\mathrm{v}(1-2)G_{s_1s_2,s_1's_2}(12,1'2^+). \tag{11.43}$$

To find a direct expression we define the inverse G^{-1} of a single-particle Green function (11.9) by

$$\sum_{s_2}\int d2G_{s_1s_2}(12)G^{-1}_{s_2s_1'}(21')=\sum_{s_2}\int d2G^{-1}_{s_1s_2}(12)G_{s_2s_1'}(21')=\delta(1-1')\delta_{s_1s_1'}. \tag{11.44}$$

The formal solution of the integral equation (11.43) reads as

$$\tilde{\Sigma}_{s_1 s_1'}(11') = -i\hbar \sum_{s_2, s_3} \int d2 \int d3 \mathrm{v}(1-2) G_{s_1 s_2, s_3 s_2}(12, 32^+) G^{-1}_{s_3 s_1'}(31'). \quad (11.45)$$

It indicates important properties of the self-energy operator: (i) $\tilde{\Sigma}$ represents a potential that is non-local in space, time and spin. It significantly modifies the potential $\tilde{V}_{ss'}(\mathbf{x})$ (10.10). (ii) It contains the complete (longitudinal) electron-electron interaction on the level of an excited single particle. (iii) $\tilde{\Sigma}$ determines the Green function G via (11.43). It, however, also depends on G [see (11.45)]. A self-consistent procedure is requested for its determination. (iv) The self-energy operator $\tilde{\Sigma}$ is not Hermitian. The homogeneous differential equation corresponding to (11.42) does in general not have real eigenvalues. The non-Hermitian part of $\tilde{\Sigma}$ leads to a finite lifetime of the single-particle excitation and finally, together with its time dependence, to the formation of a *quasiparticle*, instead of an electron or hole with infinite lifetime.

In Sect. 3.3 we have seen that for many purposes it is useful to divide the effect of the electron-electron interaction into Hartree and exchange-correlation contributions, i.e., for the self-energy

$$\tilde{\Sigma}_{s_1 s_2}(12) = \Sigma^{\mathrm{H}}_{s_1 s_2}(12) + \Sigma_{s_1 s_2}(12). \quad (11.46)$$

With the relation (11.30) between the single-particle Green function and the electron density, it holds [see also (3.44)] for the Hartree self-energy

$$\Sigma^{\mathrm{H}}_{s_1 s_2}(12) = -i\hbar\delta(1-2)\delta_{s_1 s_2} \sum_{s_3} \int d3 \mathrm{v}(1-3) G_{s_3 s_3}(33^+), \quad (11.47)$$

or more specifically,

$$\Sigma^{\mathrm{H}}_{s_1 s_2}(12) = \delta(1-2)\delta_{s_1 s_2} V_{\mathrm{H}}(\mathbf{x}_1) \quad (11.48)$$

with

$$\begin{aligned} V_{\mathrm{H}}(\mathbf{x}_1) &= -i\hbar \int d3 \mathrm{v}(1-3) \left\{ G_{\uparrow\uparrow}(33^+) + G_{\downarrow\downarrow}(33^+) \right\} \\ &= \int d^3\mathbf{x}_3 \mathrm{v}(\mathbf{x}_1 - \mathbf{x}_3) \left\{ n_{\uparrow\uparrow}(\mathbf{x}_3) + n_{\downarrow\downarrow}(\mathbf{x}_3) \right\} \\ &= \int d^3\mathbf{x}_3 \mathrm{v}(\mathbf{x}_1 - \mathbf{x}_3) n(\mathbf{x}_3). \end{aligned} \quad (11.49)$$

Here the relations (11.28) and (11.30) have been applied to describe the spin and electron densities.

11.3.3 Integral Equation Versus Differential Equation

The division of the total self-energy in (11.45) into the exchange-correlation self-energy $\Sigma_{s_1 s_2}(12)$ in (11.46) and the Hartree contribution (11.48) allows to rewrite the equation of motion (11.42) as

$$\sum_{s_2} \left\{ \left[i\hbar \frac{\partial}{\partial t_1} + \frac{\hbar^2}{2m} \Delta_{\mathbf{x}_1} - V_{\mathrm{H}}(\mathbf{x}_1) \right] \delta_{s_1 s_2} - \tilde{V}_{s_1 s_2}(\mathbf{x}_1) \right\} G_{s_2 s_1'}(11') - \sum_{s_2} \int d2 \Sigma_{s_1 s_2}(12) G_{s_2 s_1'}(21') = \delta(1-1')\delta_{s_1 s_1'}. \tag{11.50}$$

The structure of this equation suggests the introduction of a single-particle Green function $G^{\mathrm{H}}_{s_1 s_2}(12)$ in Hartree approximation that obeys the equation

$$\sum_{s_3} \left\{ \left[i\hbar \frac{\partial}{\partial t_1} + \frac{\hbar^2}{2m} \Delta_{\mathbf{x}_1} - V_{\mathrm{H}}(\mathbf{x}_1) \right] \delta_{s_1 s_3} - \tilde{V}_{s_1 s_3}(\mathbf{x}_1) \right\} G^{\mathrm{H}}_{s_3 s_2}(12) = \delta(1-2)\delta_{s_1 s_2}. \tag{11.51}$$

The integro-differential equation (11.50) can be now rewritten into an integral equation

$$G_{s_1 s_1'}(11') = G^{\mathrm{H}}_{s_1 s_1'}(11') + \sum_{s_2, s_3} \int d2 \int d3 G^{\mathrm{H}}_{s_1 s_2}(12) \Sigma_{s_2 s_3}(23) G_{s_3 s_1'}(31'), \tag{11.52}$$

i.e., into a Dyson equation [22] for the single-particle Green function. The Dyson equation is illustrated in Fig. 11.6a by diagrams. The right panel b of the figure represents the barely classical interaction in the electron gas, i.e., the Hartree approach.

By means of the inverse Green function defined according to (11.44) the Dyson equation (11.52) can be rewritten in a simple way as

$$G^{-1}_{s_1 s_1'}(11') = \left[G^{\mathrm{H}}_{s_1 s_1'}(11') \right]^{-1} - \Sigma_{s_1 s_1'}(11'), \tag{11.53}$$

where the inverse Green function in Hartree approximation (11.51) is formally given by

$$\left[G^{\mathrm{H}}_{s_1 s_1'}(11') \right]^{-1} = \left\{ \left[i\hbar \frac{\partial}{\partial t_1} + \frac{\hbar^2}{2m} \Delta_{\mathbf{x}_1} - V_{\mathrm{H}}(\mathbf{x}_1) \right] \delta_{s_1 s_1'} - \tilde{V}_{s_1 s_1'}(\mathbf{x}_1) \right\} \delta(1-1'). \tag{11.54}$$

The formulation of a Dyson equation (11.52) or (11.53) depends on how much of the electron-electron interaction is taken into account in the reference Green function. Instead of the Hartree Green function given in (11.51) also a function $G^0_{s_1 s_1'}(11')$

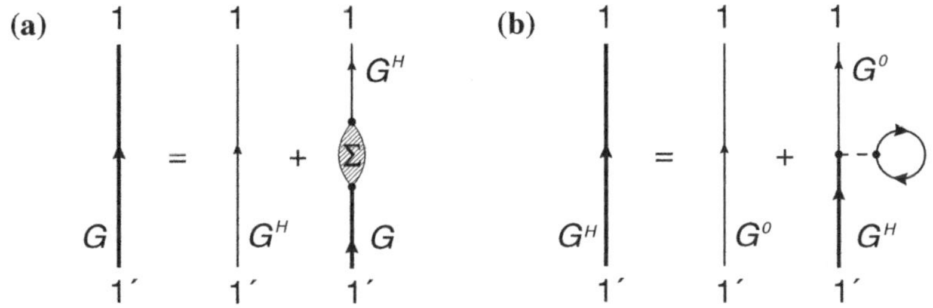

Fig. 11.6 (**a**) Graphical relation between Green function G (*solid line*) and the Green function G^{H} in Hartree approximation (*thin line*) and the proper exchange-correlation self-energy Σ. (**b**) A corresponding representation of the Hartree Green function. The *thin line* represents the Green function G^0 without electron-electron interaction. The *dashed line* illustrates the bare Coulomb repulsion, while the *bubble* represents the density. According to the rules for Feynman diagrams integrations/summations over the variables are indicated by *dots*. For more clarity the spin variable is not displayed

without any interaction, i.e., the Green function of non-interacting electrons, can be used. It obeys an equation of motion (11.51) with $V_{\mathrm{H}}(\mathbf{x}_1) \equiv 0$. The same holds for its inverse (11.54). The resulting Dyson equation is

$$G^{-1}_{s_1 s_1'}(11') = \left[G^0_{s_1 s_1'}(11')\right]^{-1} - \tilde{\Sigma}_{s_1 s_1'}(11') \tag{11.55}$$

with the complete self-energy $\tilde{\Sigma}$ (11.46) that contains Hartree, exchange and correlation contributions.

The relations (11.53) and (11.55) indicate that several Dyson equations of the same type can be formulated for G^{-1}. They differ by the amount of electron-electron interaction that is taken into account in the reference Green function. Besides the Hartree potential $V_{\mathrm{H}}(\mathbf{x}_1)$ in (11.54) to define $[G^{\mathrm{H}}]^{-1}$ or a vanishing potential $V_{\mathrm{H}}(\mathbf{x}_1) = 0$ in (11.54) to define $[G^0]^{-1}$ also potentials where the Hartree potential is supplemented by an exchange-correlation potential, i.e., $V_{\mathrm{H}}(\mathbf{x}_1) + V_{\mathrm{XC}}(\mathbf{x}_1)$, may be applied. We illustrate this procedure here for a spinless XC potential (6.15) occurring in the Kohn-Sham equation. It leads to an inverse Green function $[G^{\mathrm{KS}}]^{-1}$ in (11.54). Also non-local XC potentials $V_{\mathrm{XC}}(\mathbf{x}_1, \mathbf{x}_1')$, as appear in the gKS equation due to the use of hybrid functionals as described in Sect. 9.2.3, will be shown below to be appropriate. The use of the potential $V_{\mathrm{H}}(\mathbf{x}_1)\delta(\mathbf{x}_1 - \mathbf{x}_1') + V_{\mathrm{XC}}(\mathbf{x}_1, \mathbf{x}_1')$ in (11.54) defines the inverse reference Green function $[G^{\mathrm{gKS}}]^{-1}$. A special case could be the Fock operator $V_{\mathrm{X}}(\mathbf{x}_1, \mathbf{x}_1')$ (4.27) which may also depend on the spin coordinates. However, in agreement with (11.53) and (11.55) the consideration of a certain exchange-correlation potential requires a modification of the self-energy in the Dyson equation. For instance, in the case of $V_{\mathrm{XC}}(\mathbf{x})$ (6.15) the net self-energy in the Dyson equation becomes $\Sigma_{s_1 s_1'}(11') - V_{\mathrm{XC}}(\mathbf{x}_1)\delta_{s_1 s_1'}\delta(\mathbf{x}_1 - \mathbf{x}_1')\,\delta(t_1 - t_1')$. The relationship between the potential chosen to represent the electron-electron interaction in a first approach and the reference Green function appearing in the Dyson equation is obvious.

References

1. L.V. Keldysh, Diagram technique for nonequilibrium processes. Zh. Eksp. Teor. Fiz. **47**, 1515–1527 (1965) [Sov. Phys. JETP (English Transl.) **20**, 1018–1026 (1965)]
2. J. Rammer, H. Smith, Quantum field-theoretical methods in transport theory of metals. Rev. Mod. Phys. **58**, 323–359 (1986)
3. A.M. Zagoskin, *Quantum Theory of Many-Body Systems: Techniques and Applications* (Springer, New York, 1998)
4. H. Stolz, *Einführung in die Vielelektronentheorie der Kristalle* (Akademie-Verlag, Berlin, 1974)
5. A.L. Fetter, J.D. Walecka, *Quantum Theory of Many-Particle Systems* (McGraw-Hill, New York, 1971)
6. G.D. Mahan, *Many-Particle Physics* (Plenum Press, New York, 1990)
7. L.P. Kadanoff, G. Baym, *Quantum Statistical Mechanics: Green's Function Methods in Equilibrium and Nonequilibrium Problems* (W.A. Benjamin Inc, New York, 1962)
8. T. Matsubara, A new approach to quantum-statistical mechanics. Prog. Theor. Phys. **14**, 351–378 (1955)
9. P.C. Martin, J. Schwinger, Theory of many-particle systems. I. Phys. Rev. **115**, 1342–1373 (1959)
10. H. Lehmann, Über Eigenschaften von Ausbreitungsfunktionen und Renormierungskonstanten quantisierter Felder. Nuovo Cim. **11**, 342–357 (1954)
11. F. Bechstedt, *Principles of Surface Physics* (Springer, Berlin, 2003)
12. E.M. Lifshitz, L.P. Pitaevskii, *Statistical Physics. Part II (Landau and Lifshitz: Course of Theoretical Physics)* (Pergamon Press, Oxford, 1980)
13. N.W. Ashcroft, N.D. Mermin, *Solid State Physics* (Saunders College, Philadelphia, 1976)
14. L. Hedin, S. Lundqvist, Effects of electron-electron and electron-phonon interactions on the one-electron states of solids, in *Solid State Physics*, vol. 23, ed. by F. Seitz, D. Turnbull, H. Ehrenreich (Academic Press, New York, 1969), pp. 1–181
15. V.M. Galitskii, A.B. Migdal, Application of quantum field theory methods to the many-body problem. Zh. Eksp. Teor. Fiz. **34**, 139–150 (1958) [Sov. Phys. JETP (English Transl.) **7**, 96–104 (1958)]
16. A. Schindlmayr, T.J. Pollehn, R.W. Godby, Spectra and total energies from self-consistent many-body perturbation theory. Phys. Rev. B **58**, 12684–12690 (1998)
17. F. Caruso, P. Rinke, X. Ren, M. Scheffler, A. Rubio, Unified description of ground and excited states of finite systems: the self-consistent GW approach. Phys. Rev. B **86**, 081102(R) (2012)
18. X. Ren, P. Rinke, V. Blum, J. Wiefernik, A. Tkatchenko, A. Sanfilippo, K. Reuter, M. Scheffler, Resolution-of-identity approach to Hartree-Fock, hybrid density functionals, RPA, MP2 and GW with numeric atom-centered orbital basis functions. New J. Phys. **14**, 053020 (2012)
19. G. Strinati, Application of the Green's function method to the study of the optical properties of semiconductors. Riv. Nuovo Cim. **11**, 1–80 (1988)
20. D.N. Zubarev, Double-time Green functions in statistical physics. Uspekhi Fiz. Nauk **71**, 71–116 (1960) [Sov. Phys. Usp. (English Transl.) **3**, 320–345 (1960)]
21. D.N. Zubarev, Zweizeitige Greensche Funktionen in der statistischen Physik. Fortschr. d. Physik (German Transl.) **9**, 275–328 (1961)
22. F. Dyson, The S matrix in quantum electrodynamics. Phys. Rev. **75**, 1736–1755 (1949)

Chapter 12
Set of Fundamental Equations

Abstract The Schwinger functional derivative technique allows the formulation of the electron self-energy in terms of the single-particle Green function G. All quantities depend on a vanishing external perturbation, the response to it results in their functional derivatives. After linearization important physics enters the mathematical representation by the occurrence of response functions, the density correlation function as a variational derivative of the single-particle Green function for the perturbation and a polarization function as derivative to the screened perturbation. The latter one can be directly related to the vertex function. As the central quantity of the theory the dynamically screened Coulomb potential W appears. The equations for the Green function, the self-energy, the screened potential, the polarization function, and the vertex function form the fundamental set of Hedin equations accounting fully for the longitudinal electron-electron interaction. Physical intuition suggests the GW approximation to be appropriate to calculate the exchange-correlation contribution to the self-energy. Vertex corrections are neglected and the polarization function and, consequently, the dielectric function are described within the approximation of independent quasiparticles, the random-phase approximation.

12.1 Schwinger Functional Derivative Technique

12.1.1 External Perturbations

Instead of solving a hierarchy of equations of motion for N-electron Green functions, we apply the functional derivative technique of Schwinger [1–3] to calculate the exchange-correlation self-energy in (11.46). In principle, an expression is needed that relates the two-particle Green function in (11.43) to the one-particle Green function (11.9). Usually, a mathematical trick, but with a clear physical background, is used to do so. The response to a small external perturbation is investigated. It will be set to zero at the end. In most cases an external scalar potential is studied [4, 5].

F. Bechstedt, *Many-Body Approach to Electronic Excitations*,
Springer Series in Solid-State Sciences 181, DOI 10.1007/978-3-662-44593-8_12

Here, however, since we are sometimes also interested in some magnetic response, we introduce external fields, which include a small magnetic field [6] but do not explicitly depend on time.

Under the action of the external perturbation field $\phi_{ss'}(\mathbf{x})$, which depends on both space and spin coordinates but is non-diagonal in the spin coordinates, the Hamiltonian $\mathscr{H}$ of the perturbed inhomogeneous electron gas is divided into two contributions

$$\mathscr{H}(t) = \mathscr{H}_0(t) + \mathscr{H}_{\text{ext}}(t) \tag{12.1}$$

with $\mathscr{H}_0(t)$ (10.5), but applying the generalized potential energy operator (10.9), and the perturbation

$$\mathscr{H}_{\text{ext}}(t) = \sum_{s,s'} \int d^3\mathbf{x} \psi_s^+(\mathbf{x}, t)\phi_{ss'}(\mathbf{x})\psi_{s'}(\mathbf{x}, t). \tag{12.2}$$

The spin-dependent perturbation potential is assumed to be

$$\phi_{ss'}(\mathbf{x}) = \phi_{\text{ext}}(\mathbf{x})\delta_{ss'} + \mu_B \mathbf{B}_{\text{ext}}(\mathbf{x})\boldsymbol{\sigma}_{ss'}, \tag{12.3}$$

i.e., it has a similar form as the generalized potential $\tilde{V}_{ss'}(\mathbf{x})$ (10.10). The influence of an external electric field, acting on the charge of an electron, is described by the scalar potential $\phi_{\text{ext}}(\mathbf{x})$. The external magnetic field $\mathbf{B}_{\text{ext}}(\mathbf{x})$ determines a paramagnetic interaction via the electron spin represented by the vector of the Pauli matrices (1.1). A coupling of the spin to the orbital motion is not considered in (12.3).

The generalized Hamiltonian with perturbation (12.1) determines the time evolution of the field operators according to the Heisenberg equations of motion (10.3). We label the modified solutions (10.2)

$$\psi_{\phi s}^+(\mathbf{x}, t),\ \psi_{\phi s}(\mathbf{x}, t)$$

with an additional index ϕ, representing the external perturbation. In the same spirit the grand canonical statistical operator (10.17) is generalized to $\hat{W}_\phi$ and the statistical expectation value (10.16) to $\langle \ldots \rangle_\phi$. It results a generalized single-particle Green function

$$G_{ss'}(11'|\phi) = \frac{1}{i\hbar}\langle \mathcal{T} \psi_{\phi s}(1)\psi_{\phi s'}^+(1')\rangle_\phi. \tag{12.4}$$

We change over to the interaction picture, first suggested by Dirac, by generalizing the time evolution operator U_ϕ (10.1). The operator $\mathcal{T}$ remains the time-ordering operator that rearranges the field operators in chronological order with later (real) times to the left. Then it holds

$$U_\phi(t) = \exp\left[-\frac{i}{\hbar}(\mathscr{H}_0 + \mathscr{H}_{\rm ext})\,t\right]$$
$$= U_0(t)\mathcal{S}(t, t_0). \tag{12.5}$$

The scattering matrix $S(t, t_0)$ [7] obeys the differential equation

$$i\hbar\frac{\partial}{\partial t}\mathcal{S}(t, t_0) = \mathscr{H}_{\rm ext}(t)\mathcal{S}(t, t_0).$$

with the formal solution

$$\mathcal{S}(t, t_0) = \mathcal{T}\exp\left[\frac{1}{i\hbar}\int_{t_0}^{t} dt'\,\mathscr{H}_{\rm ext}(t')\right], \tag{12.6}$$

if the perturbation is switched on at t_0. The Dirac picture is applied. The $\mathcal{S}$ operator is unitary for real times t. However, one has to keep in mind that for imaginary times t it holds $\mathcal{S}^+(t, 0) = \mathcal{T}\exp\left[\frac{i}{\hbar}\int_0^t dt'\,\mathscr{H}_{\rm ext}(t')\right]$, where $t_0 = 0$ is chosen for the switch-on time of the perturbation.

With (12.5) the Green function (12.4) takes the form

$$G_{ss'}(11'|\phi) = \frac{\mathrm{Tr}\left\{\hat{W}_0\mathcal{T}\mathcal{S}(-i\hbar\beta, t_1)\psi_{Ds}(1)\mathcal{S}(t_1 - t_1', 0)\psi^+_{Ds'}(1')\mathcal{S}(t_1', 0)\right\}}{i\hbar\mathrm{Tr}\{\hat{W}_0 S(-i\hbar\beta, 0)\}}$$
$$= \frac{1}{i\hbar}\frac{\langle\mathcal{T}\mathcal{S}(-i\hbar\beta, 0)\psi_{Ds}(1)\psi^+_{Ds'}(1')\rangle}{\langle\mathcal{S}(-i\hbar\beta, 0)\rangle}. \tag{12.7}$$

The Dirac operators $\psi_{Ds}(1)$ evolve with $\mathscr{H}_0$. A generalization of the scattering matrix to imaginary time intervals is considered resulting in $\mathcal{S}(-i\hbar\beta, 0)$. Thereby, by analogy with (12.5) we have used

$$\mathrm{Tr}\left\{\exp\left[-\beta(\mathscr{H}_0 + \mathscr{H}_{\rm ext} - \mu\hat{N})\right]\right\} = \mathrm{Tr}\left\{\exp\left[-\beta(\mathscr{H}_0 - \mu\hat{N})\right]\mathcal{S}(-i\hbar\beta, 0)\right\}$$
$$= \langle\mathcal{S}(-i\hbar\beta, 0)\rangle.$$

This equation is only valid, if also the perturbed Hamiltonian $\mathscr{H}$ and the particle number operator $\hat{N}$ commute, i.e., the perturbation should be particle-conserving.

12.1.2 Method of Variational Derivative

The interesting quantity in (12.7) is the $\mathcal{S}$ operator (12.6) for imaginary times in the Dirac picture

$$\mathcal{S}(-i\hbar\beta, 0) = \mathcal{T} \exp\left[\frac{1}{i\hbar}\int_{0}^{-i\hbar\beta} dt \mathcal{H}_{\text{ext}}(t)\right]. \tag{12.8}$$

It contains the imaginary-time ordering $\mathcal{T}$ which guarantees that values (it) decrease from left to right. All operators are defined along imaginary time intervals $(0, -i\hbar\beta)$ [3]. The external perturbations appear in (12.7) only via the operator (12.8). Small external perturbations mean small deviations of $\mathcal{S}$ from the unity operator. The variation of ϕ corresponds to a variation of $\mathcal{S}$ (12.8) [3],

$$\delta\mathcal{S} = \mathcal{T}\sum_{s_2,s_2'}\int d2\mathcal{S}\frac{1}{i\hbar}\psi^+_{Ds_2}(2^+)\psi_{Ds_2'}(2)\delta\phi_{Ds_2s_2'}(2)$$

and finally to

$$\frac{\delta\mathcal{S}}{\delta\phi_{ss'}(1)} = \mathcal{T}\mathcal{S}\frac{1}{i\hbar}\psi^+_{Ds}(1^+)\psi_{Ds'}(1).$$

With the chain rule the linear variation of the Green function (12.7) becomes

$$\begin{aligned}\delta G_{ss'}(11'|\phi) = \sum_{s_2,s_2'}\int d2 \Big\{ &\langle\mathcal{T}\mathcal{S}\psi^+_{Ds_2}(2^+)\psi_{Ds_2'}(2)\psi_{Ds}(1)\psi^+_{Ds'}(1')\rangle / \left[(i\hbar)^2\langle\mathcal{S}\rangle\right] \\ &-\langle\mathcal{T}\mathcal{S}\psi_{Ds}(1)\psi^+_{Ds'}(1')\rangle\langle\mathcal{T}\mathcal{S}\psi^+_{Ds_2}(2^+)\psi_{Ds_2'}(2)\rangle / \left[(i\hbar)^2\langle\mathcal{S}\rangle^2\right]\Big\} \\ &\times \delta\phi_{Ds_ss_2'}(2),\end{aligned}$$

where 2^+ is defined in (11.41).

The result can be summarized to

$$\frac{\delta G_{ss'}(11'|\phi)}{\delta\phi_{s_2s_2'}(2^+)} = G_{ss'}(11'|\phi)G_{s_2's_2}(22^+|\phi) - G_{ss_2',s's_2}(12, 1'2^+|\phi). \tag{12.9}$$

Here the two-particle Green function (11.39) has been generalized for the presence of a spin-dependent external perturbation $\phi_{ss'}(\mathbf{x})$ (12.3) in analogy to the single-particle case (12.4). The consequence of the fact, that the external perturbation is not only spin-dependent but also non-diagonal with respect to the spin variables, is a two-particle Green function in (12.9) that depends on four different spin variables. We follow Rödl [8] and take all these spin variables explicitly into consideration. Furthermore, expression (12.9) suggests an interpretation of the two-particle motion represented by the two-particle Green function as the motion of two independent particles expressed by the product of the two one-particle Green functions and a response of the system to the presence of the two particles related to the variational derivative of the single-particle Green function. We will see later how this term relates to exchange and correlation in excited electronic systems.

We have to mention that the replacement of the two-particle Green function in (11.42) using (12.9) leads to an equation of motion for the single-particle Green function G in which the influence of the electron-electron interaction is characterized by the functional derivative of G. Equations of this type are sometimes called Kadanoff-Baym equations [4].

12.1.3 Exchange and Correlation Contributions to Self-energy

The representation (12.9) of the two-particle Green function allows to rewrite the self-energy (11.45) as a functional derivative of the single-particle Green function G, at least in the presence of the external perturbation ϕ (12.3). With the division (11.46) and the definition (11.45) the exchange-correlation self-energy reads

$$\Sigma_{s_1 s_1'}(11'|\phi) = i\hbar \sum_{s_2, s_2', s_3} \int d2 \int d3 \mathrm{v}(1-2)\delta_{s_2 s_2'} \frac{\delta G_{s_1 s_3}(13|\phi)}{\delta \phi_{s_2' s_2}(2^+)} G^{-1}_{s_3 s_1'}(31'|\phi). \tag{12.10}$$

The variation of the generalized definition (11.44) of an inverse Green function leads to

$$\sum_{s_3} \int d3 \frac{\delta G_{s_1 s_3}(13|\phi)}{\delta \phi_{s_2' s_2}(2^+)} G^{-1}_{s_3 s_1'}(31|\phi) = -\sum_{s_3} \int d3 G_{s_1 s_3}(13|\phi) \frac{\delta G^{-1}_{s_3 s_1'}(31'|\phi)}{\delta \phi_{s_2' s_2}(2^+)}. \tag{12.11}$$

By means of the Dyson equation (11.55) with the reference Green function G^0 of non-interacting electrons defined by (11.54) with $V_{\mathrm{H}}(\mathbf{x}_1) \equiv 0$, it holds

$$\frac{\delta \left[G^0_{s_3 s_1'}(31'|\phi) \right]^{-1}}{\delta \phi_{s_2' s_2}(2^+)} = -\delta(3-1')\delta(3-2^+)\delta_{s_3 s_2'}\delta_{s_{1'} s_2}$$

or more generally, using the Dyson equation (11.55) with the self-energy $\tilde{\Sigma}$ (11.46),

$$\frac{\delta G^{-1}_{s_3 s_1'}(31'|\phi)}{\delta \phi_{s_2' s_2}(2^+)} = -\left\{ \delta(3-1')\delta(3-2^+)\delta_{s_3 s_2'}\delta_{s_{1'} s_2} + \frac{\delta \tilde{\Sigma}_{s_3 s_1'}(31'|\phi)}{\delta \phi_{s_2' s_2}(2^+)} \right\}. \tag{12.12}$$

Using the two variational relations (12.11) and (12.12) in expression (12.10) the XC self-energy decomposes in

$$\Sigma_{s_1s_1'}(11'|\phi) = \Sigma^{\mathrm{X}}_{s_1s_1'}(11'|\phi) + \Sigma^{\mathrm{C}}_{s_1s_1'}(11'|\phi) \tag{12.13}$$

with the exchange self-energy

$$\Sigma^{\mathrm{X}}_{s_1s_1'}(11'|\phi) = i\hbar \mathrm{v}(1-1')G_{s_1s_1'}(11'^{+}|\phi) \tag{12.14}$$

and the correlation self-energy

$$\Sigma^{\mathrm{C}}_{s_1s_1'}(11'|\phi) = i\hbar \sum_{s_2,s_2',s_3} \int d2 \int d3 \mathrm{v}(1-2)\delta_{s_2s_2'}G_{s_1s_3}(13|\phi)\frac{\delta\tilde{\Sigma}_{s_3s_1'}(31'|\phi)}{\delta\phi_{s_2's_2}(2^{+})}. \tag{12.15}$$

The latter one is determined by the variation of the complete self-energy $\tilde{\Sigma}$ (11.46) with respect to the external perturbation ϕ. Therefore, to lowest order in the bare Coulomb potential $\mathrm{v}(1-2)$ the correlation contribution Σ^{C} vanishes. In this lowest order, assuming that the relations are also valid in the limit of vanishing external perturbations $\phi \to 0$, it results the Hartree-Fock approximation [see Sect. 4.2] with a self-energy

$$\tilde{\Sigma}_{s_1s_1'}(11') = \Sigma^{\mathrm{H}}_{s_1s_1'}(11') + \Sigma^{\mathrm{X}}_{s_1s_1'}(11'), \tag{12.16}$$

whose two contributions are given in (11.47) and (12.14) with a Green function defined by (11.52) but replacing Σ by Σ^{X}, i.e., the non-local exchange self-energy (12.14).

12.1.4 Modified Equation of Motion

Equations of motion of the type (11.42) or (11.50) can be also derived for the Green function $G_{ss'}(11'|\phi)$ in the presence of the external perturbation (12.2). However, one has to bear in mind that the Green function and the self-energy are functionals of the perturbation $\phi_{ss'}(\mathbf{x})$ (12.3). In addition, the potential energy acting on an electron has to be modified by this quantity. Instead of (11.50) it holds

$$\sum_{s_2}\left\{\left[i\hbar\frac{\partial}{\partial t_1} + \frac{\hbar^2}{2m}\Delta_{\mathbf{x}_1} - V_{\mathrm{H}}(\mathbf{x}_1|\phi)\right]\delta_{s_1s_2} - \tilde{V}_{s_1s_2}(\mathbf{x}_1) - \phi_{s_1s_2}(\mathbf{x}_1)\right\} G_{s_2s_1'}(11'|\phi)$$
$$-\sum_{s_2}\int d2\,\Sigma_{s_1s_2}(12|\phi)G_{s_2s_1'}(21'|\phi) = \delta(1-1')\delta_{s_1s_1'}. \tag{12.17}$$

Here the occurring Hartree potential (11.49) has to be calculated with the electron density $n(\mathbf{x}|\phi)$ in the presence of the perturbation.

In the spirit of the variational procedure described in Sect. 12.1.2 the response of the system should be studied in the limit $\phi_{ss'}(\mathbf{x}) \to 0$. Thereby, we have to guarantee that in this limit the system goes back into the unperturbed state, i.e.,

$$\lim_{\phi\to 0} G_{ss'}(11'|\phi) = G_{ss'}(11'),$$
$$\lim_{\phi\to 0} \Sigma_{ss'}(11'|\phi) = \Sigma_{ss'}(11'). \tag{12.18}$$

The first relation also implies

$$\lim_{\phi\to 0} n(\mathbf{x}|\phi) = n(\mathbf{x}).$$

Similar relations should be valid for the two-particle Green function and the spin densities.

Relations (12.18) sometimes represent non-trivial requirements. To fulfill them one has to assume that the symmetry of the Hamiltonian $\mathscr{H}$ is not lower than the symmetry of the unperturbed one $\mathscr{H}_0$. Symmetry can be broken spontaneously. Such an effect may also be desirable. One example could be a spin-polarized system without spin-orbit interaction in the absence of an external magnetic field. Then, the Hamiltonian $\mathscr{H}_0$ contains the spin-independent potential $\tilde{V}_{ss'}(\mathbf{x}) = V_{\text{n}}(\mathbf{x})\delta_{ss'}$, whereas the perturbation contains an external magnetic field. We consider the model potential

$$\phi_{ss'}(\mathbf{x}) = \delta_{ss'}\mu_B B_{\text{ext}}(\delta_{s\uparrow} - \delta_{s\downarrow})$$

with $\mathbf{B}_{\text{ext}} \parallel z$-axis and $B_{\text{ext}} \to 0$. This small perturbation destroys the invariance of the system under rotations in spin subspace. For $B_{\text{ext}} \neq 0$ it results a magnetization density (11.29)

$$m_z(\mathbf{x}|\phi) = -\mu_B \sum_{s,s'} \sigma^z_{ss'} n_{ss'}(\mathbf{x}|\phi).$$

In the limit $\phi \to 0$ not in all cases it holds

$$\lim_{\phi\to 0} m_z(\mathbf{x}|\phi) = 0.$$

Exceptions are (anti)ferromagnetic systems, subject to a transition from the paramagnetic to the (anti)ferromagnetic phase due to the perturbation. The perturbation gives rise to an alignment of the electron spins that is conserved for $\phi = 0$. Nevertheless, we assume the validity of the relations (12.18). One may argue that an unperturbed Hamiltonian $\mathscr{H}_0'$ can be constructed with an infinitesimally small magnetic field that leads to a symmetry lowering in agreement with the properties of the system that should be studied.

With these formal considerations the Hartree term in (12.17) can be reformulated. With the original potential $V_{\mathrm{H}}(\mathbf{x})$ we find

$$\sum_{s_2}\left\{\left[i\hbar\frac{\partial}{\partial t_1}+\frac{\hbar^2}{2m}\Delta_{\mathbf{x}_1}-V_{\mathrm{H}}(\mathbf{x}_1)\right]\delta_{s_1s_2}-\tilde{V}_{s_1s_2}(\mathbf{x}_1)-\phi^{\mathrm{eff}}_{s_1s_2}(\mathbf{x}_1)\right\}G_{s_2s_1'}(11'|\phi)$$
$$-\sum_{s_2}\int d2\,\Sigma_{s_1s_2}(12|\phi)G_{s_2s_1'}(21'|\phi)=\delta(1-1')\delta_{s_1s_1'} \tag{12.19}$$

with the effective perturbation potential

$$\begin{aligned}\phi^{\mathrm{eff}}_{ss'}(1)&=\phi_{ss'}(1)-i\hbar\delta_{ss'}\sum_{s_2,s_2'}\int d2\mathrm{v}(1-2)\delta_{s_2s_2'}\left[G_{s_2s_2'}(33^+|\phi)-G_{s_2s_2'}(33^+)\right]\\&=\phi_{ss'}(1)+\delta_{ss'}\int d^3\mathbf{x}_2\mathrm{v}(\mathbf{x}_1-\mathbf{x}_2)\left[n(\mathbf{x}_2|\phi)-n(\mathbf{x}_2)\right].\end{aligned} \tag{12.20}$$

It is modified by the perturbation-induced change of the spin-independent Hartree potential.

12.2 Response Functions

12.2.1 Density Correlation Function

In the limit $\phi\to 0$ the response of the system to the perturbation ϕ in (12.20) can be linearized [9] as described in Sect. 12.1.2. As a result, instead of (12.20), one derives the relation

$$\phi^{\mathrm{eff}}_{s_1s_1'}(1)=\sum_{s_2,s_2'}\int d2\epsilon^{-1}_{s_1s_1',s_2s_2'}(12)\phi_{s_2s_2'}(2) \tag{12.21}$$

with a (linear) response function

$$\epsilon^{-1}_{s_1s_1',s_2s_2'}(12)=\left.\frac{\delta\phi^{\mathrm{eff}}_{s_1s_1'}(1)}{\delta\phi_{s_2s_2'}(2)}\right|_{\phi=0}, \tag{12.22}$$

that will be later linked to the inverse dielectric function after spin summations. The quantity (12.22) includes spin flip excitations due to the x- and y-components of the external field $\mathbf{B}_{\mathrm{ext}}$. Using relation (12.9) expression (12.20) is related to the variation of the single-particle Green function by

$$\epsilon^{-1}_{s_1 s_1', s_2 s_2'}(12) = \delta(1-2)\delta_{s_1 s_2}\delta_{s_1' s_2'} + \delta_{s_1 s_1'} \sum_{s_3, s_3'} \int d3 \mathrm{v}(1-3)\delta_{s_3 s_3'} L_{s_3 s_3', s_2' s_2}(32) \tag{12.23}$$

with the spin-dependent *density correlation function* [4] or *two-particle correlation function* [10]

$$L_{s_1 s_1', s_2 s_2'}(12) = -i\hbar \frac{\delta G_{s_1 s_1'}(11^+|\phi)}{\delta \phi_{s_2' s_2}(2)}\bigg|_{\phi=0} \tag{12.24}$$

taken in the limit $\phi \to 0$. It is essentially given by the variational derivative of the electron density with respect to the external perturbation. Expression (12.24) is somewhat generalized by starting from the spin density matrix (11.28). The denotation of the quantity (12.23) and the arrangement of the spin indices become more clear below.

A more explicit form of the density correlation function (12.24) follows with (12.9) as

$$L_{s_1 s_1', s_2 s_2'}(12) = -i\hbar \left\{ G_{s_1 s_1'}(11^+) G_{s_2 s_2'}(22^+) - G_{s_1 s_2, s_1' s_2'}(12, 1^+ 2^+) \right\}. \tag{12.25}$$

With the definitions (11.9) and (11.39) and a reordering of the fermion field operators in compliance with the resulting sign changes we may formulate an explicit expression as [11]

$$\begin{aligned} L_{s_1 s_1', s_2 s_2'}(12) =& \frac{1}{i\hbar} \langle \mathcal{T} \psi^+_{s_1'}(\mathbf{x}_1, t_1^+) \psi_{s_1}(\mathbf{x}_1, t_1) \psi^+_{s_2'}(\mathbf{x}_2, t_2^+) \psi_{s_2}(\mathbf{x}_2, t_2) \rangle \\ &- \frac{1}{i\hbar} \langle \mathcal{T} \psi^+_{s_1'}(\mathbf{x}_1, t_1^+) \psi_{s_1}(\mathbf{x}_1, t_1) \rangle \langle \mathcal{T} \psi^+_{s_2'}(\mathbf{x}_2, t_2^+) \psi_{s_2}(\mathbf{x}_2, t_2) \rangle \\ =& \frac{1}{i\hbar} \langle \mathcal{T} \Delta \hat{n}_{s_1 s_1'}(\mathbf{x}_1, t_1) \Delta \hat{n}_{s_2 s_2'}(\mathbf{x}_2, t_2) \rangle \end{aligned} \tag{12.26}$$

with the operator of spin density fluctuations

$$\Delta \hat{n}_{ss'}(\mathbf{x}, t) = \hat{n}_{ss'}(\mathbf{x}, t) - \langle \hat{n}_{ss'}(\mathbf{x}, t) \rangle \tag{12.27}$$

according to (10.12) and (11.28). The generalized (spin) density correlation function (12.26) is the statistical expectation value of the time-ordered fluctuations of spin density operators.

The diagonal elements with respect to the spin variables represent the charge density response. It is the leading effect on the electron system with longitudinal electron-electron interaction (10.8) due to the external potential (12.20). The restriction to the spin-diagonal part of the perturbation, $\phi_{ss'}(\mathbf{x}) = \delta_{ss'}\phi_{ss}(\mathbf{x})$, makes obvious that due to (12.21) the spin-averaged (two-point) density correlation function

$$L(\mathbf{x}\mathbf{x}', t-t') = \frac{1}{2} \sum_{s_1,s_1'} \sum_{s_2,s_2'} \delta_{s_1 s_1'} \delta_{s_2 s_2'} L_{s_1 s_1', s_2 s_2'}(11')$$

describes the response of the system. The off-diagonal elements are related to a magnetic (i.e., magnetization density) response. With the definition of the spin density operator (3.32) the fluctuations of the spin density are defined as

$$\Delta\hat{\mathbf{s}}(\mathbf{x}, t) = \frac{\hbar}{2} \sum_{s,s'} \boldsymbol{\sigma}_{ss'} \Delta\hat{n}_{ss'}(\mathbf{x}, t). \tag{12.28}$$

In terms of the generalized density response function they define a magnetic susceptibility with tensor character [6, 12]

$$\chi^{\text{mag}}_{\alpha\beta}(\mathbf{x}\mathbf{x}', t-t') = -\sum_{s_1,s_1'} \sum_{s_2,s_2'} \sigma^{\alpha}_{s_1 s_1'} \sigma^{\beta}_{s_2 s_2'} L_{s_1 s_1', s_2 s_2'}(11'),$$

in which however contributions of the orbital magnetism are not considered. The poles in the frequency plane of the Fourier-transformed susceptibility correspond to spin-wave excitations of the electronic system.

The last two relations express that two types of response functions are needed, spin-averaged or spin-summed two-point functions and four-point functions similarly to the two-particle Green function. In agreement with the definition (12.25) for the density response we relate these functions according to

$$L_{s_1 s_1', s_2 s_2'}(11^+, 22^+) \equiv L_{s_1 s_1', s_2 s_2'}(12). \tag{12.29}$$

We will see below that four-point response functions are needed to compute two-point functions.

12.2.2 Polarization and Vertex Functions

The effective perturbation potential ϕ^{eff} (12.20) suggests to introduce another response function following formally the definition (12.24). The variation of the single-particle Green function with respect to the effective potential yields the *polarization function*

$$P_{s_1 s_1', s_2 s_2'}(12) = -i\hbar \left. \frac{\delta G_{s_1 s_1'}(11^+|\phi)}{\delta \phi^{\text{eff}}_{s_2' s_2}(2)} \right|_{\phi=0} \tag{12.30}$$

with its four-point generalization

$$P_{s_1 s_1', s_2 s_2'}(11^+, 22^+) \equiv P_{s_1 s_1', s_2 s_2'}(12). \tag{12.31}$$

Sometimes P is also called *irreducible polarization function* since it is the sum of all irreducible polarization insertions in the diagrammatic language [13, 14]. The polarization insertion is the part of a diagram that is connected to the rest of it only by two interaction lines.

As a consequence of the relation (12.21) and the response function (12.22) the definition (12.24) can be rewritten by means of the chain rule

$$\begin{aligned} L_{s_1 s_1', s_2 s_2'}(12) &= -i\hbar \sum_{s_3, s_3'} \int d3 \frac{\delta G_{s_1 s_1'}(11^+|\phi)}{\delta \phi^{\text{eff}}_{s_3' s_3}(3)} \frac{\delta \phi^{\text{eff}}_{s_3' s_3}(3)}{\delta \phi_{s_2' s_2}(2)} \bigg|_{\phi=0} \\ &= \sum_{s_3, s_3'} \int d3 P_{s_1 s_1', s_3 s_3'}(13) \epsilon^{-1}_{s_3' s_3, s_2' s_2}(32). \end{aligned} \tag{12.32}$$

This equation relates the two response functions L and P to each other via a quantity that will later lead to the inverse dielectric function.

The use of (12.32) in the definition (12.23) leads to an integral equation

$$\begin{aligned} &\epsilon^{-1}_{s_1 s_1', s_2 s_2'}(12) = \\ &\delta(1-2)\delta_{s_1 s_2}\delta_{s_1' s_2'} + \delta_{s_1 s_1'} \sum_{s_3, s_3'} \sum_{s_4, s_4'} \int d3 \int d4 \mathrm{v}(1-3) \delta_{s_3 s_3'} P_{s_3 s_3', s_4 s_4'}(34) \epsilon^{-1}_{s_4' s_4, s_2 s_2'}(42) \end{aligned}$$

for the response function ϵ^{-1}. The inversion of the response function according to

$$\sum_{s_3, s_3'} \int d3 \epsilon^{-1}_{s_1 s_1', s_3 s_3'}(13) \epsilon_{s_3 s_3', s_2 s_2'}(32) = \delta(1-2)\delta_{s_1 s_2}\delta_{s_1' s_2'}$$

yields an expression for another response quantity, ϵ,

$$\epsilon_{s_1 s_1', s_2 s_2'}(12) = \delta(1-2)\delta_{s_1 s_2}\delta_{s_1' s_2'} - \delta_{s_1 s_1'} \sum_{s_3, s_3'} \int d3 \mathrm{v}(1-3) \delta_{s_3 s_3'} P_{s_3 s_3', s_2' s_2}(32). \tag{12.33}$$

This function will be later related to the dielectric function of the system.

With the definition of the inverse Green function (11.44) and the chain rule for the variational derivatives the polarization function (12.30) becomes

$$P_{s_1 s_1', s_2 s_2'}(12) = i\hbar \sum_{s_3, s_4} \int d3 \int d4 G_{s_1 s_3}(13) \frac{\delta G^{-1}_{s_3 s_4}(34|\phi)}{\delta \phi^{\text{eff}}_{s_2' s_2}(2)} \bigg|_{\phi=0} G_{s_4 s_1'}(41^+).$$

At this point we introduce the *vertex function* (sometimes called scalar irreducible vertex function [10])

$$\Gamma_{s_1s_2,s_3s_3'}(12,3) = \left.\frac{\delta G^{-1}_{s_1s_2}(12|\phi)}{\delta\phi^{\text{eff}}_{s_3's_3}(3)}\right|_{\phi=0} \tag{12.34}$$

and the polarization function of independent (quasi)particles

$$L^0_{s_1s_1',s_2s_2'}(11',22') = -i\hbar G_{s_1s_2'}(12')G_{s_2s_1'}(21'). \tag{12.35}$$

The resulting relation

$$P_{s_1s_1',s_2s_2'}(12) = -\sum_{s_3,s_4}\int d3\int d4 L^0_{s_1s_1',s_4s_3}(11^+,43)\Gamma_{s_3s_4,s_2s_2'}(34,2) \tag{12.36}$$

indicates that the vertex function mediates between the non-interacting (quasi) particles and the system with interaction on the two-particle level. So it contains the interaction between the two particles represented by single-particle Green functions in (12.35). Omitting its spin dependence the vertex function shows a three-point character. However, with respect to the spin dependence it may be even a four-point function if spin-dependent perturbations (12.3) are taken in consideration.

The definition (12.34) of the vertex function asks for a generalization of the Dyson equation (11.53) in the presence of the perturbation ϕ. The response of the system in (12.20), i.e., the difference $\phi^{\text{eff}}_{ss'}(\mathbf{x}) - \phi_{ss'}(\mathbf{x})$, is essentially given by the difference of the Hartree potentials of the perturbed and unperturbed systems linearized in $\phi_{ss'}(\mathbf{x})$. As a consequence, in the presence of ϕ, in (11.54) for $[G^{\text{H}}_{s_1s_1'}(11'|\phi)]^{-1}$ an additional potential $\phi^{\text{eff}}_{s_1s_1'}(\mathbf{x}_1)$ appears. This linear term makes the variation of the Dyson equation (11.53) simple with (12.34). We find

$$\begin{aligned}\Gamma_{s_1s_2,s_3s_3'}(12,3) =& \left.\frac{\delta[G^{\text{H}}_{s_1s_2}(12|\phi)]^{-1}}{\delta\phi^{\text{eff}}_{s_3's_3}(3)}\right|_{\phi=0} - \left.\frac{\delta\Sigma_{s_1s_2}(12|\phi)}{\delta\phi^{\text{eff}}_{s_3's_3}(3)}\right|_{\phi=0}\\ =& -\delta(1-2)\delta(1-3)\delta_{s_1s_3'}\delta_{s_2s_3}\\ &-\sum_{s_4,s_5}\int d4\int d5 \left.\frac{\delta\Sigma_{s_1s_2}(12|\phi)}{\delta G_{s_4s_5}(45|\phi)}\right|_{\phi=0}\left.\frac{\delta G_{s_4s_5}(45|\phi)}{\delta\phi^{\text{eff}}_{s_3's_3}(3)}\right|_{\phi=0}.\end{aligned}$$

With the definition (12.30) of the polarization function it results

$$\begin{aligned}\Gamma_{s_1s_2,s_3s_3'}(12,3) =& -\delta(1-2)\delta(1-3)\delta_{s_1s_3'}\delta_{s_2s_3}\\ &+\frac{1}{i\hbar}\sum_{s_4,s_5}\int d4\int d5 \left.\frac{\delta\Sigma_{s_1s_2}(12|\phi)}{\delta G_{s_4s_5}(45|\phi)}\right|_{\phi=0} P_{s_4s_5,s_3s_3'}(45,33^+).\end{aligned} \tag{12.37}$$

In addition to the first summand that characterizes non-interacting (quasi)particles, there are complex contributions to the vertex function which are determined by the variation of the XC self-energy with respect to the Green function and by the polarization function. By means of (12.36), (12.37) can be transformed into an integral equation, a Bethe-Salpeter equation, for the vertex function Γ or the polarization function P.

12.2.3 XC Self-energy and Screened Potential

With the Hartree self-energy (11.47), the inversion of the single-particle Green function (11.44), the total self-energy (11.45), and the decomposition (11.46), one finds the XC self-energy to be

$$\begin{aligned}\Sigma_{s_1 s_2}(12) = &- i\hbar \sum_{s_3, s_4} \int d3 \int d4 \mathrm{v}(1-3) \left\{ G_{s_1 s_3, s_4 s_3}(13, 43^+) \right. \\ &\left. - G_{s_1 s_4}(14) G_{s_3 s_3}(33^+) \right\} G^{-1}_{s_4 s_2}(42) \\ = &i\hbar \sum_{s_3, s_3', s_4} \int d3 \int d4 \mathrm{v}(1-3) \delta_{s_3 s_3'} \left. \frac{\delta G_{s_1 s_4}(14|\phi)}{\delta \phi_{s_3' s_3}(3^+)} \right|_{\phi=0} G^{-1}_{s_4 s_2}(42) \\ = &- \sum_{s_3, s_4} \int d3 \int d4 \mathrm{v}(1-3) L_{s_1 s_4, s_3 s_3}(14, 33^+) G^{-1}_{s_4 s_2}(42).\end{aligned} \tag{12.38}$$

Here, also the variational derivative of the single-particle Green function (12.9) and its relation to the density correlation function (12.24) have been applied. It is interesting to note the appearance of the Kronecker symbol $\delta_{s_3 s_3'}$ for the two inner spin variables. That means that only a change in the spin-diagonal elements $\phi_{ss}(\mathbf{x})$ of the external perturbation has to be considered when calculating the electron self-energy. The observation that in the variational derivative in (12.38) only the electric-field contribution affects the self-energy and not the variation with respect to the magnetic-field components in (12.3) has been first made by Aryasetiawan and Karlsson [6].

The relation between the bare Coulomb potential v and the density correlation function L suggests to follow Hubbard [15] and to introduce a screened potential W. By means of the chain rule and (12.24) the variational derivative in (12.38) can be rewritten as

$$
\begin{aligned}
\Sigma_{s_1s_2}(12) =& -i\hbar \sum_{s_3,s_3',s_4} \int d3 \int d4 \mathrm{v}(1-3) G_{s_1s_4}(14)\delta_{s_3s_3'} \left.\frac{\delta G^{-1}_{s_4s_2}(42|\phi)}{\delta\phi_{s_3's_3}(3^+)}\right|_{\phi=0} \\
=& -i\hbar \sum_{s_3,s_3',s_4} \sum_{s_5,s_5'} \int d3 \int d4 \int d5 \mathrm{v}(1-3) G_{s_1s_4}(14)\delta_{s_3s_3'} \\
& \times \left.\frac{\delta G^{-1}_{s_4s_2}(42|\phi)}{\delta\phi^{\mathrm{eff}}_{s_5's_5}(5)}\right|_{\phi=0} \left.\frac{\delta\phi^{\mathrm{eff}}_{s_5's_5}(5)}{\delta\phi_{s_3's_3}(3^+)}\right|_{\phi=0} \\
=& -i\hbar \sum_{s_4,s_5,s_5'} \int d4 \int d5 G_{s_1s_4}(14)\Gamma_{s_4s_2,s_5s_5'}(42,5) \\
& \times \sum_{s_3,s_3'} \int d3 \mathrm{v}(1-3)\epsilon^{-1}_{s_5's_5,s_3',s_3}(53^+)\delta_{s_3s_3'}
\end{aligned}
\tag{12.39}
$$

using the definitions (12.22) and (12.34) in addition.

We investigate the last factor under the sums and integrals in more detail by means of the relation (12.23)

$$
\begin{aligned}
&\sum_{s_3,s_3'} \int d3 \mathrm{v}(1-3)\epsilon^{-1}_{s_5's_5,s_3's_3}(53^+)\delta_{s_3s_3'} \\
&= \sum_{s_3,s_3'} \int d3 \mathrm{v}(1-3) \Big[\delta(5-3^+)\delta_{s_5's_3'}\delta_{s_5s_3} \\
&\quad + \delta_{s_5's_5} \sum_{s_6,s_6'} \int d6 \mathrm{v}(5-6)\delta_{s_6s_6'} L_{s_6s_6',s_3s_3'}(63^+) \Big] \delta_{s_3s_3'} \\
&= \delta_{s_5s_5'} \int d3 \mathrm{v}(1-3) \left[\delta(5-3^+) + \sum_{s_3,s_6} \int d6 \mathrm{v}(5-6) L_{s_6s_6,s_3s_3}(63^+) \right] \\
&= \delta_{s_5s_5'} \left[\mathrm{v}(1^+-5) + \sum_{s_3,s_6} \int d3 \int d6 \mathrm{v}(5-6) L_{s_6s_6,s_3s_3}(63) \mathrm{v}(1^+-3) \right].
\end{aligned}
\tag{12.40}
$$

Apart from the Kronecker symbol in the spin variables s_5 and $s_{5'}$ the right-hand side represents a quantity that does not depend on spin. Also the density correlation function contains only electron density fluctuations in (12.26) because of the sums over parallel spin pairs.

Therefore and because of the relation (12.23) between L and ϵ^{-1} we introduce the dynamically screened Coulomb potential (see Fig. 12.1a)

$$
W(11') = \mathrm{v}(1-1') + \sum_{s_2,s_3} \int d2 \int d3 \mathrm{v}(1-2) L_{s_2s_2,s_3s_3}(23) \mathrm{v}(3-1') \tag{12.41}
$$

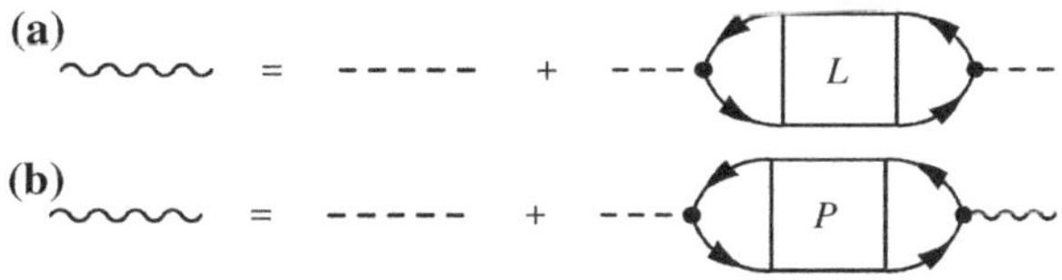

Fig. 12.1 Representations of the dynamically screened Coulomb potential $W(11')$ (*wavy* or *wiggling line*), (**a**) by the bare Coulomb potential $\mathrm{v}(1-1')$ (*dashed line*) and the spin-summed density correlation function $L(23)$ (12.41) and (**b**) by the integral equation (12.49) where the kernel is determined by the spin-summed polarization function P in (12.46)

in agreement with Hedin [16]. This function is symmetric

$$W(11') = W(1'1) \tag{12.42}$$

in the space and time variables. It does not depend on spin. Only a spin-summed density correlation function appears in (12.41). The symmetry of W is also obvious from the corresponding Feynman diagrams in Fig. 12.1a. Formally, using the symmetry of the two-particle Green function (11.39), the symmetry of the bare Coulomb interaction $\mathrm{v}(1-1')$ (11.40), and the definition of the density correlation function (12.25), the screened interaction can be also written as

$$W(11') = \mathrm{v}(1-1') + \sum_{s_2,s_3} \int d2 \int d3 \mathrm{v}(1'-3) L_{s_3 s_3, s_2 s_2}(32) \mathrm{v}(1-2).$$

With the dynamically screened interaction (12.41) the self-energy (12.39) takes a simple form [13, 16]

$$\Sigma_{s_1 s_2}(12) = -i\hbar \sum_{s_3,s_4,s_4'} \int d3 \int d4 G_{s_1 s_3}(13) \Gamma_{s_3 s_2, s_4 s_4'}(32,4) \delta_{s_4 s_4'} W(1^+4). \tag{12.43}$$

Because of the spin Kronecker symbol only a true three-point vertex function appears in reality. This self-energy is illustrated in Fig. 12.2 by a Feynman diagram.

In (12.43) we have replaced the last factor in (12.39) by

$$W(11') = \sum_{s_2} \int d2 \mathrm{v}(1-2) \epsilon^{-1}_{s_1' s_1', s_2 s_2}(1'2).$$

At first glance this relation suggests that the screened potential depends on the spin variable s_1'. However, it can be proven that it is not the case. The function ϵ^{-1} (12.23), which only depends on two instead of four spin variables due to the aforementioned pure density response, is transformed into a spin-less quantity ϵ^{-1} with the definition (12.24) of L

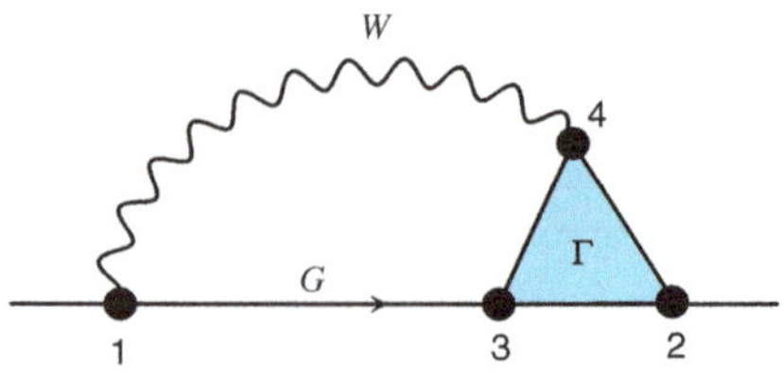

Fig. 12.2 Schematic representation of self-energy (12.43). The electron line G (*solid line*) and the dynamically screened potential W (*wiggly line*) are displayed together with the vertex function Γ (with the indicated *three-point* character)

$$\epsilon^{-1}(1'2) = \sum_{s_2} \epsilon^{-1}_{s'_1 s'_1, s_2 s_2}(1'2) = \delta(1'-2) - i\hbar \sum_{s_2, s_3} \int d3 \mathrm{v}(1'-3) \frac{\delta G_{s_3 s_3}(33^+|\phi)}{\delta \phi_{s_2 s_2}(2)}\bigg|_{\phi=0}. \tag{12.44}$$

Because of the sum over the spins s_2 the right-hand side is independent of s'_1. This can be shown by taking the spin independence of the bare Coulomb potential into account. This spin-summed quantity is indeed the inverse dielectric function of an inhomogeneous electron gas. It is solely determined by density fluctuations and not by spin density fluctuations. The inverse dielectric function is related to the dielectric function $\epsilon(12)$ itself via the relations

$$\int d2 \epsilon^{-1}(12)\epsilon(21') = \int d2 \epsilon(12)\epsilon^{-1}(21') = \delta(1-1'). \tag{12.45}$$

For the dielectric function $\epsilon(12)$ it holds an expression similar to (12.44) as for the inverse quantity,

$$\epsilon(12) = \sum_{s_2} \epsilon_{s_1 s_1, s_2 s_2}(12) = \delta(1-2) - \sum_{s_2, s_3} \int d3 \mathrm{v}(1-3) P_{s_3 s_3, s_2 s_2}(33^+, 22^+), \tag{12.46}$$

if (12.33) is applied. This quantity does also not depend on the spin variable s_1 for the same reasons as discussed for the inverse dielectric function (12.44).

The inverse dielectric function leads to the representations

$$W(11') = \int d2 \mathrm{v}(1-2)\epsilon^{-1}(1'2) = \int d2 \epsilon^{-1}(12)\mathrm{v}(2-1'), \tag{12.47}$$

which are in obvious agreement with the denotation of a dynamically screened Coulomb potential. The electron density fluctuations give rise to a reduction of the strength of the interaction between the electrons. Its symmetric formulation (12.47) follows from the symmetry of the bare Coulomb potential $\mathrm{v}(1-1')$ (11.40).

For the pure electron density response the relation (12.32) between L and P takes the form

$$L_{s_1s_1,s_2s_2}(12) = \sum_{s_3,s_3'} \int d3 P_{s_1s_1,s_3s_3'}(13)\epsilon^{-1}_{s_3's_3,s_2s_2}(32). \qquad (12.48)$$

Inserting this relation into (12.41) and using again (12.47) a Dyson equation (see Fig. 12.1b)

$$W(11') = \mathrm{v}(1-1') + \sum_{s_3,s_4} \int d3 \int d4 \mathrm{v}(1-3) P_{s_3s_3,s_4s_4}(34) W(41') \qquad (12.49)$$

is derived for the screened potential. The kernel of this inhomogeneous integral equation is ruled by the polarization function (12.30). It can be represented by the diagrams depicted in Fig. 12.1b in analogy to the graphical representation of the Dyson equation (11.52) for the Green function with the XC self-energy.

12.3 Hedin Equations

12.3.1 Summary of Important Relations

The central quantity of the many-body theory for charged electronic excitations, electrons or holes, is the single-particle Green function G. It obeys the equation of motion (11.50)

$$\sum_s \left\{ \left[i\hbar\frac{\partial}{\partial t_1} + \frac{\hbar^2}{2m}\Delta_{\mathbf{x}_1} - V_{\mathrm{H}}(\mathbf{x}_1) \right] \delta_{s_1s} - \tilde{V}_{s_1s}(\mathbf{x}_1) \right\} G_{ss_1'}(11')$$
$$- \sum_{s_2} \int d2 \Sigma_{s_1s_2}(12) G_{s_2s_1'}(21') = \delta(1-1')\delta_{s_1s_1'}. \qquad (12.50)$$

It can be reformulated in a Dyson equation (11.52)

$$G_{s_1s_1'}(11') = G^{\mathrm{H}}_{s_1s_1'}(11') + \sum_{s_2,s_3} \int d2 \int d3 G^{\mathrm{H}}_{s_1s_2}(12) \Sigma_{s_2s_3}(23) G_{s_3s_1'}(31') \qquad (12.51)$$

with the Green function G^{H} within the Hartree approximation (11.51).

Therein the exchange-correlation self-energy (12.43)

$$\Sigma_{s_1s_2}(12) = -i\hbar \sum_{s_4,s_5,s_5'} \int d4 \int d5 G_{s_1s_4}(14)\Gamma_{s_4s_2,s_5s_5'}(42,5)\delta_{s_5s_5'} W(1^+5) \tag{12.52}$$

is determined by the unknown Green function G, the vertex function Γ, and the dynamically screened potential W. The screened Coulomb potential fulfills the Dyson equation (12.49)

$$W(11') = \mathrm{v}(1-1') + \sum_{s_3,s_4} \int d3 \int d4 \mathrm{v}(1-3) P_{s_3s_3,s_4s_4}(34)W(41') \tag{12.53}$$

with the polarization function (12.36)

$$P_{s_1s_1',s_2s_2'}(12) = -\sum_{s_3,s_4} \int d3 \int d4 L^0_{s_1s_1',s_4s_3}(11^+,43)\Gamma_{s_3s_4,s_2s_2'}(34,2). \tag{12.54}$$

The corresponding quantity L^0 for non-interacting (quasi)particles is given by a product of two single-particle Green functions as $-i\hbar GG$ (12.35). The two-particle interaction in the vertex function (12.37)

$$\begin{aligned}\Gamma_{s_1s_2,s_3s_3'}(12,3) = &-\delta(1-2)\delta(1-3)\delta_{s_1s_3'}\delta_{s_2s_3} \\ &+\frac{1}{i\hbar}\sum_{s_4,s_5}\int d4 \int d5 \frac{\delta\Sigma_{s_1s_2}(12)}{\delta G_{s_4s_5}(45)} P_{s_4s_5,s_3s_3'}(45,33^+)\end{aligned} \tag{12.55}$$

goes back to the variational derivative of the XC self-energy Σ with respect to the Green function G and the polarization function P.

The set of these five coupled integral equations (12.51)–(12.55) is well-known as the set of Hedin equations [13, 16]. Like G in (12.51), the quantities Γ and W satisfy Dyson or Bethe-Salpeter equations (12.53) and (12.55), respectively, at least after replacement of the irreducible polarization function P in (12.53) by relation (12.54). The equations can be solved in a self-consistent manner as displayed in the Hedin magic pentagon in Fig. 12.3. Starting from a given approximation for Σ the set of Hedin equations can be used to generate higher-order approximations for the quantities G, Γ, P, and W as well as the self-energy itself. Although the equations are exact, a straightforward expansion for the self-energy in powers of the screened potential may yield unphysical results such as negative spectral functions [17, 18]. Such an expansion is only conditionally convergent due to the long-range nature of the Coulomb potential. So far, there is no systematic way for a complete solution of the set of equations. Approximations are needed. Their choice is usually dictated by physical situation as we will see below.

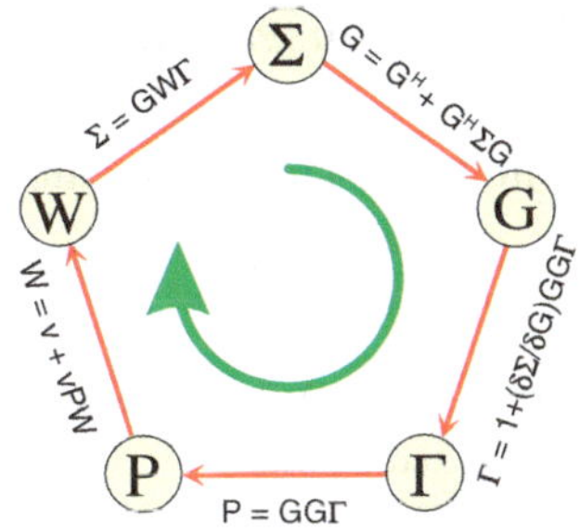

Fig. 12.3 Magic pentagon: schematic representation of an iterative solution of the set of Hedin equations for G, Σ, W, P, and Γ

12.3.2 GW Approximation

From the Hedin equations the exchange-correlation self-energy Σ and with it the Green function G can be determined iteratively (see Fig. 12.3). The simplest but consistent version of the solution starts from $G = G^{\mathrm{H}}$ or $\Sigma \equiv 0$. This starting point sets the vertex function (12.55) to unity, more precisely to the interaction-free form

$$\Gamma_{s_1 s_2, s_3 s_3'}(12, 3) = -\delta(1-2)\delta(1-3)\delta_{s_1 s_3'}\delta_{s_2 s_3}, \tag{12.56}$$

which, with (12.52), yields the *Hedin GW approximation*

$$\Sigma_{s_1 s_2}(12) = i\hbar G_{s_1 s_2}(12) W(1^+2) \tag{12.57}$$

for the XC self-energy, in which the spin dependence is only mediated by the unknown single-particle Green function. In the old days of many-body theory the approximation (12.57) has been denoted as *shielded potential approximation* [4, 5] with W as the shielded potential. The self-energy is essentially given by the product of the single-particle propagator G and the dynamically screened interaction W. Thereby, W has to be computed with $P = L^0 = -i\hbar GG$ in the kernel of the Dyson equation (12.53). All these quantities have to be determined in a self-consistent manner. In comparison to the Hartree-Fock approximation, where Σ is replaced by the Fock operator Σ^{X} (12.14), in (12.57) the effect of correlation is described by the dynamically screened, spatially non-local Coulomb potential. The GW approximation is consistent in the spirit that it is a particle- and energy-conserving approximation. Indeed the approximation (12.57) is a conserving approximation as shown by Baym and Kadanoff [19, 20]. Thereby, conservation of particle number means that the continuity equation (10.13) is fulfilled for the statistically averaged quantities (11.30) and (11.33). One possible manifestation of the energy conservation is given by the Galitskii-Migdal formula (11.35).

12.3.3 Consequences for Dielectric Properties and Screening

With (12.35) the polarization function (12.54) becomes

$$P_{s_1s_1',s_2s_2'}(12) = L^0_{s_1s_1',s_2s_2'}(11^+,22) = -i\hbar G_{s_1s_2'}(12)G_{s_2s_1'}(21^+), \quad (12.58)$$

within the GW approximation. The resulting dielectric function (12.46) can be expressed by

$$\begin{aligned}\epsilon(12) &= \delta(1-2) - 2\int d3\mathrm{v}(1-3)L^0(32),\\ L^0(12) &= -i\hbar\frac{1}{2}\sum_{s_1,s_2} G_{s_1s_2}(12)G_{s_2s_1}(21^+)\end{aligned} \quad (12.59)$$

with L^0 as the spin-averaged polarization function of independent (quasi)particles.

According to the setting $G = G^{\mathrm{H}}$ expression (12.59) indeed yields the RPA in its original meaning [21, 22]. The physical meaning of the RPA is that the electrons respond to the total effective (external + induced) field as if they were non-interacting. Sometimes the resulting screened potential is therefore denoted by $W = W^{\mathrm{RPA}}$ [5, 13, 23]. It is a good approximation in the limit of large electron densities, i.e., $r_s \ll 1$. Today, the single-particle Green functions in (12.59) are replaced by Green functions within a certain approximation for the XC self-energy, e.g. G^{GW} in GW approximation. In practice, in the majority of numerical calculations the Green function G is replaced by a Green function of the KS, gKS or HF problems solved in (6.22), (9.22), and (4.25). Therefore, since XC is partially taken into account, the approximation of P by a product of Green functions, $-i\hbar GG$, is frequently called independent-particle or independent-quasiparticle approach [24–26]. It goes beyond the original RPA. All the mentioned starting points for the electronic structure include exchange and correlation on a certain level in the Green function and, therefore, are however somewhat inconsistent with the approximation (12.56) of the vertex function. Such an approximation of the XC influence on the single-particle spectrum results in eigenvalues ε_Λ and Pauli eigenspinors $\phi_\Lambda(\mathbf{x}, s)$ and, consequently, in a spectral function (11.13)

$$A_{ss'}(\mathbf{xx}', \omega) = 2\pi\sum_\Lambda \delta(\hbar\omega - \varepsilon_\Lambda)\phi_\Lambda(\mathbf{x}, s)\phi^+_\Lambda(\mathbf{x}', s') \quad (12.60)$$

of the approximated single-particle Green function G. We will see below that in practice, in particular within self-consistent GW treatments (14.41), the eigenvalues ε_Λ are replaced by quasiparticle eigenvalues $\varepsilon^{\mathrm{QP}}_\Lambda$.

On the imaginary time axis the Green function is given as

$$G_{ss'}(\mathbf{xx}', t-t') = \frac{1}{-i\hbar\beta}\sum_n \int_{-\infty}^{+\infty} \frac{d\omega}{2\pi}\frac{A_{ss'}(\mathbf{xx}', \omega)}{z_n - \omega} e^{-iz_n(t-t')} \quad (12.61)$$

because of the Fourier transformation (11.18), the spectral representation (11.21), and the fermionic Matsubara frequencies (11.20). The resulting dielectric function (12.59) reads as

$$\epsilon(\mathbf{x}\mathbf{x}', t-t') = \delta(\mathbf{x}-\mathbf{x}')\delta(t-t') + \frac{i\hbar}{(-i\hbar\beta)^2}\int d^3\mathbf{x}'' \mathrm{v}(\mathbf{x}-\mathbf{x}'')$$
$$\times \sum_{s_1,s_2}\sum_{n,n'}\int_{-\infty}^{+\infty}\frac{d\omega}{2\pi}\int_{-\infty}^{+\infty}\frac{d\omega'}{2\pi}\frac{A_{s_2s_1}(\mathbf{x}''\mathbf{x}',\omega)}{z_n-\omega}\frac{A_{s_1s_2}(\mathbf{x}'\mathbf{x}'',\omega')}{z_{n'}-\omega'}e^{-i(z_n-z_{n'})(t-t')}.$$

This expression makes obvious that the dielectric function on the imaginary time axis possesses a Fourier representation

$$\epsilon(\mathbf{x}\mathbf{x}', t-t') = \frac{1}{-i\hbar\beta}\sum_m \epsilon(\mathbf{x}\mathbf{x}', \tilde{z}_m)e^{-i\tilde{z}_m(t-t')} \tag{12.62}$$

with Fourier coefficients at the bosonic Matsubara frequencies [4, 14]

$$\tilde{z}_m = \frac{\pi m}{-i\hbar\beta} \qquad (m = 0, \pm 2, \pm 4, \ldots), \tag{12.63}$$

which are poles of the Bose distribution

$$g(\hbar\tilde{\omega}) = \frac{1}{e^{\beta\hbar\tilde{\omega}} - 1} \tag{12.64}$$

in the complex $\tilde{\omega}$-plane. They are situated on the imaginary axis. Since the frequencies $\tilde{z}_m = z_n - z_{n'}$ correspond to differences of fermionic Matsubara frequencies, the Fourier coefficients of the dielectric function are

$$\epsilon(\mathbf{x}\mathbf{x}', \tilde{z}_m) = \delta(\mathbf{x}-\mathbf{x}') - 2\int d^3\mathbf{x}'' \mathrm{v}(\mathbf{x}-\mathbf{x}'')L^0(\mathbf{x}''\mathbf{x}', \tilde{z}_m), \tag{12.65}$$

$$L^0(\mathbf{x}''\mathbf{x}', \tilde{z}_m) = \frac{1}{2}\sum_{s',s''}\frac{1}{\beta}\sum_n \int_{-\infty}^{+\infty}\frac{d\omega}{2\pi}\int_{-\infty}^{+\infty}\frac{d\omega'}{2\pi}\frac{A_{s''s'}(\mathbf{x}''\mathbf{x}',\omega)A_{s's''}(\mathbf{x}'\mathbf{x}'',\omega')}{(z_n-\omega)(z_n-\omega'-\tilde{z}_m)}.$$

Under the frequency integrals it holds (14.61)

$$\frac{1}{\hbar\beta}\sum_n \frac{1}{z_n-\omega} = f(\hbar\omega) - \frac{1}{2},$$
$$f(\hbar\omega - \hbar\tilde{z}_m) = f(\hbar\omega). \tag{12.66}$$

By means of a partial fraction decomposition expression (12.65) for the spin-averaged polarization function of independent (quasi)particles becomes

$$L^0(\mathbf{x}\mathbf{x}', \tilde{z}_m) = \hbar \int_{-\infty}^{+\infty} \frac{d\omega}{2\pi} \int_{-\infty}^{+\infty} \frac{d\omega'}{2\pi} \frac{f(\hbar\omega) - f(\hbar\omega')}{\omega - \omega' - \tilde{z}_m} \frac{1}{2} \sum_{s,s'} A_{ss'}(\mathbf{x}\mathbf{x}', \omega) A_{s's}(\mathbf{x}'\mathbf{x}, \omega'). \tag{12.67}$$

With the explicit representation (12.60) of the spectral function one obtains

$$L^0(\mathbf{x}\mathbf{x}', \tilde{z}_m) = \sum_{\Lambda,\Lambda'} \frac{f(\varepsilon_\Lambda) - f(\varepsilon_{\Lambda'})}{\varepsilon_\Lambda - \varepsilon_{\Lambda'} - \hbar\tilde{z}_m} \frac{1}{2} \sum_s \phi_\Lambda(\mathbf{x}, s)\phi^+_{\Lambda'}(\mathbf{x}, s) \sum_{s'} \phi_{\Lambda'}(\mathbf{x}', s')\phi^+_\Lambda(\mathbf{x}', s'). \tag{12.68}$$

In the limit of a non-collinear spin system (4.6), the Pauli spinors factorize in orbitals and spin functions ($\Lambda = \lambda m_s$)

$$\phi_\Lambda(\mathbf{x}, s) = \varphi_{\lambda m_s}(\mathbf{x}) \chi_{\frac{1}{2} m_s}(s). \tag{12.69}$$

The orbitals $\varphi_{\lambda m_s}(\mathbf{x})$ satisfy a Schrödinger-like equation with a potential energy which depends on the XC treatment in the reference system. Using the closure relation for the spin functions (4.7) the Fourier coefficients of the polarization function are

$$L^0(\mathbf{x}\mathbf{x}', \tilde{z}_m) = \frac{1}{2} \sum_{m_s} \sum_{\lambda,\lambda'} \varphi_{\lambda m_s}(\mathbf{x})\varphi^*_{\lambda' m_s}(\mathbf{x})\varphi_{\lambda' m_s}(\mathbf{x}')\varphi^*_{\lambda m_s}(\mathbf{x}') \frac{f(\varepsilon_{\lambda m_s}) - f(\varepsilon_{\lambda' m_s})}{\varepsilon_{\lambda m_s} - \varepsilon_{\lambda' m_s} - \hbar\tilde{z}_m}, \tag{12.70}$$

in agreement with the low-temperature findings (9.28). In the collinear framework the frequency-dependent electronic polarizations arising from the two different spin channels simply have to be added. Representations of this kind have been derived first by Ehrenreich and Cohen [27, 28]. Therefore, the expression (12.65) together with (12.70) is frequently called Ehrenreich-Cohen formula for the dielectric function. It describes the polarization in a spin-polarized system of independent particles with collinear spins. It can be used to characterize important effects of the dynamical screening in the electron gas or even to calculate in a first approximation the frequency dependence of the dielectric function, i.e., the optical properties.

References

1. J. Schwinger, On Green's functions of quantized fields I + II. Proc. Natl. Acad. Sci. USA **37**, 452–459 (1951)
2. J. Schwinger, *Particles, Sources and Fields*, vols. I and II (Addison-Wesley, Reading, 1973)
3. P.C. Martin, J. Schwinger, Theory of many-particle systems. I. Phys. Rev. **115**, 1342–1373 (1959)
4. L.P. Kadanoff, G. Baym, *Quantum Statistical Mechanics: Green's Function Methods in Equilibrium and Nonequilibrium Problems* (W.A. Benjamin Inc, New York, 1962)

5. H. Stolz, *Einführung in die Vielelektronentheorie der Kristalle* (Akademie-Verlag, Berlin, 1974)
6. F. Aryasetiawan, K. Karlsson, Green's function formalism for calculating spin-wave spectra. Phys. Rev. B **60**, 7419–7428 (1999)
7. F. Dyson, The S matrix in quantum electrodynamics. Phys. Rev. **75**, 1736–1755 (1949)
8. C. Rödl, *Spinabhängige GW-Approximation*, Diplomarbeit, Friedrich-Schiller-Universität Jena, 2005
9. R. Kubo, Statistical mechanical theory of irreversible process. I. J. Phys. Soc. Jpn. **12**, 570–586 (1957)
10. G. Strinati, Application of the Green's functions method to the study of the optical properties of semiconductors. Riv. Nuovo Cimento **11**, 1–86 (1988)
11. C. Rödl, *Elektronische und exzitonische Anregungen in magnetischen Isolatoren*, Ph.D. thesis, Friedrich-Schiller-Universität, Jena, 2009
12. A.L. Fetter, J.D. Walecka, *Quantum Theory of Many-Particle Systems* (McGraw-Hill, San Francisco, 1971)
13. L. Hedin, S. Lundqvist, Effects of electron-electron and electron-phonon interactions on the one-electron states of solids, in *Solid State Physics*, vol. 23, ed. by F. Seitz, D. Turnbull, H. Ehrenreich (Academic Press, New York, 1969), pp. 1–181
14. A.M. Zagoskin, *Quantum Theory of Many-Body Systems: Techniques and Applications* (Springer, New York, 1998)
15. J. Hubbard, The description of collective motions in terms of many-body perturbation theory. Proc. Royal Soc. London A **240**, 539–560 (1957)
16. L. Hedin, New method for calculating the one-particle Green's function with application to the electron-gas problem. Phys. Rev. **139**, A796–A823 (1965)
17. P. Minnhagen, Vertex correction calculations for an electron gas. J. Phys. C **7**, 3013–3019 (1974)
18. A. Schindlmayr, R.W. Godby, Spectra and total energies from self-consistent many-body perturbation theory. Phys. Rev. B **58**, 12684–12690 (1998)
19. G. Baym, L.P. Kadanoff, Conservation laws and correlation functions. Phys. Rev. **124**, 287–299 (1961)
20. G. Baym, Self-consistent approximations in many-body systems. Phys. Rev. **127**, 1391–1401 (1962)
21. D. Pines, D. Bohm, A collective description of electron interactions: II. Collective vs individual particle aspects of the interactions. Phys. Rev. **85**, 338–353 (1952)
22. D. Bohm, D. Pines, A collective description of electron interactions: III. Coulomb interactions in a degenerate electron gas. Phys. Rev. **92**, 609–625 (1953)
23. W.G. Aulbur, L. Jonsson, J.W. Wilkins, Quasiparticle calculations in solids, in *Solid State Physics*. Advances in Research and Applications, vol. 54, ed. by H. Ehrenreich, F. Spaepen (Academic Press, San Diego, 2000), pp. 1–218
24. R. Del Sole, R. Girlanda, Optical properties of semiconductors within the independent-quasiparticle approximation. Phys. Rev. B **48**, 11789–11795 (1993)
25. B. Adolph, V.I. Gavrilenko, K. Tenelsen, F. Bechstedt, R. Del Sole, Nonlocality and many-body effects in the optical properties of semiconductors. Phys. Rev. B **53**, 9797–9808 (1996)
26. G. Onida, L. Reining, A. Rubio, Electronic excitations: density-functional versus many-body Green's-function approaches. Rev. Mod. Phys. **74**, 601–659 (2002)
27. H. Ehrenreich, M.H. Cohen, Self-consistent field approach to the many-electron problem. Phys. Rev. **115**, 786–790 (1959)
28. H. Ehrenreich, Electromagnetic Transport in Solids. Optical Properties and Plasma Effects, in *The Optical Properties of Solids*. Proceedings of International School of Physics 'E. Fermi', ed. by J. Tauc (Academic Press, New York, 1966), pp. 106–154

Chapter 13
Density Correlation and Electronic Polarization

Abstract The spin-summed density correlation (polarization) function is directly related to the inverse dielectric function (dielectric function) of the electron gas. It obeys a spectral representation with a spectral function that represents the spectral behavior of collective and/or two-particle excitations. It fulfills the oscillator-strength and screening sum rules. Special care has to be taken for the spatial inhomogeneity of the system that gives rise to local-field effects. By means of the spectral functions of the density correlation and polarization functions the validity of the Kramers-Kronig relations can be easily proven for the inverse dielectric function as well as the dielectric function. For the model system of a homogeneous gas of non-interacting electrons the density correlation function is analytically calculated as a wave-vector- and frequency-dependent function. It allows the discussion of electron-hole pair excitations and excitation of plasmons. The latter ones suggest a plasmon pole approximation as a reasonable description of the screening properties. Improvements to include spatial inhomogeneities and image potential effects at surfaces as well as for layered systems are also discussed.

13.1 Inverse Dielectric Function

13.1.1 Spectral Function of Density Correlation Function

In Sect. 12.2.3 we have demonstrated that the dynamical screening of the Coulomb potential [see e.g. (12.47)] and the dielectric function [see e.g. (12.46)] are determined by spin-averaged response functions

$$L(12) = \frac{1}{2} \sum_{s_1, s_2} L_{s_1 s_1, s_2 s_2}(12),$$
$$P(12) = \frac{1}{2} \sum_{s_1, s_2} P_{s_1 s_1, s_2 s_2}(12), \tag{13.1}$$

F. Bechstedt, *Many-Body Approach to Electronic Excitations*,
Springer Series in Solid-State Sciences 181, DOI 10.1007/978-3-662-44593-8_13

i.e., by true two-point functions. They depend on two space points $\mathbf{x}_1$ and $\mathbf{x}_2$ and the time difference $t_1 - t_2$, though, at least L, with (12.25) they go back to a two-particle Green function (12.26), that is actually a four-point quantity.

Because of the dependence on a reduced number of variables we follow some descriptions developed for the single-particle Green function in Sect. 11.1.2. We restrict the time differences to the imaginary axis. The time ordering of the density fluctuations in (12.26) is then defined by [1, 2]

$$L(12) = \theta(\mathrm{Im}(t_2 - t_1))L^>(12) + \theta(\mathrm{Im}(t_1 - t_2))L^<(12) \tag{13.2}$$

with the propagators

$$\begin{aligned} L^>(12) &= \frac{1}{2i\hbar}\langle \Delta\hat{n}(1)\Delta\hat{n}(2)\rangle, \\ L^<(12) &= \frac{1}{2i\hbar}\langle \Delta\hat{n}(2)\Delta\hat{n}(1)\rangle \end{aligned} \tag{13.3}$$

and the operator $\Delta\hat{n}(1) = \sum_{s_1} \Delta\hat{n}_{s_1 s_1}(1)$ of the density fluctuations (12.27). The propagator $L^>$ is an analytic function on the lower strip in the complex time plane defined by the interval $(-\hbar\beta, 0)$ on the imaginary axis. By contrast the propagator $L^<$ is defined on the strip with imaginary time arguments $\mathrm{Im}(t - t')$ in $(0, \hbar\beta)$.

Again Martin-Schwinger-like relations

$$\begin{aligned} L^>(\mathbf{x}_1\mathbf{x}_2, t_1 - t_2 - i\hbar\beta) &= L^<(\mathbf{x}_1\mathbf{x}_2, t_1 - t_2), \\ L(\mathbf{x}_1\mathbf{x}_2, t_1 - t_2 - i\hbar\beta) &= L(\mathbf{x}_1\mathbf{x}_2, t_1 - t_2) \end{aligned} \tag{13.4}$$

can be derived [2]. Because of the four field operators appearing in the definition of the two-particle Green function, they indicate that the two-point density correlation function L is a periodic function on the imaginary time axis where $L(t_1 - t_2)$ obeys the periodic boundary condition $L(t_1 - t_2)|_{t_1=0} = L(t_1 - t_2)|_{t_1=-i\hbar\beta}$ [1]. Fourier transformations

$$L^{\gtrless}(\mathbf{x}_1\mathbf{x}_2, \omega) = i \int_{-\infty}^{+\infty} dt_1 L^{\gtrless}(\mathbf{x}_1\mathbf{x}_2, t_1 - t_2)e^{i\omega(t_1-t_2)} \tag{13.5}$$

are introduced in accordance with those for the electron and hole propagators in (11.11). As a consequence of the periodicity (13.4) it holds in the frequency domain

$$L^>(\mathbf{x}_1\mathbf{x}_2, \omega) = e^{\beta\hbar\omega}L^<(\mathbf{x}_1\mathbf{x}_2, \omega). \tag{13.6}$$

The two fluctuation propagators define a spectral function by

$$\hat{L}(\mathbf{x}_1\mathbf{x}_2, \omega) = L^>(\mathbf{x}_1\mathbf{x}_2, \omega) - L^<(\mathbf{x}_1\mathbf{x}_2, \omega). \tag{13.7}$$

The two algebraic equations (13.6) and (13.7) have the solutions

$$L^<(\mathbf{x}_1\mathbf{x}_2,\omega) = g(\hbar\omega)\hat{L}(\mathbf{x}_1\mathbf{x}_2,\omega),$$
$$L^>(\mathbf{x}_1\mathbf{x}_2,\omega) = [1+g(\hbar\omega)]\hat{L}(\mathbf{x}_1\mathbf{x}_2,\omega) \tag{13.8}$$

with the Bose distribution (12.64). Because of the singularity of this distribution function for $\omega \to 0$, we have to assume that $\lim_{\omega\to 0}\hat{L}(\mathbf{x}_1\mathbf{x}_2,\omega) \sim \omega$. This is in agreement with the condition $\lim_{|t_1-t_2|\to\infty}\langle[\hat{n}(1),\hat{n}(2)]_-\rangle = 0$, i.e., that for large time differences density fluctuations are not anymore correlated. With (13.8) it also holds

$$L^>(\mathbf{x}\mathbf{x}',\omega) + L^<(\mathbf{x}\mathbf{x}',\omega) = \coth\left(\frac{1}{2}\beta\hbar\omega\right)\hat{L}(\mathbf{x}\mathbf{x}',\omega)$$
$$= \frac{1}{2}\coth\left(\frac{1}{2}\beta\hbar\omega\right)S(\mathbf{x}\mathbf{x}',\omega),$$

which shows that the sum of the two fluctuation propagators is also directly related to the dynamic structure factor (3.54). The above relation is therefore one form of the fluctuation-dissipation theorem [3, 4]. It correlates fluctuations with dissipation. Here it relates density fluctuations in the system described by a correlation function to the dissipation described by the dynamic structure factor or, as we will see below, by the imaginary part of the inverse dielectric function, i.e., energy losses.

Because of (13.5) and (13.7) the spectral function can be written as

$$\hat{L}(\mathbf{x}_1\mathbf{x}_2,\omega) = \frac{1}{2\hbar}\int_{-\infty}^{+\infty} dt_1 \big\langle\big[\Delta\hat{n}(1),\Delta\hat{n}(2)\big]_-\big\rangle e^{i\omega(t_1-t_2)}$$
$$= \frac{1}{2\hbar}\int_{-\infty}^{+\infty} dt_1 \big\langle\big[\hat{n}(1),\hat{n}(2)\big]_-\big\rangle e^{i\omega(t_1-t_2)}, \tag{13.9}$$

i.e., as a Fourier-transformed correlation function of the density fluctuations or densities. These representations suggest the validity of the symmetry relations

$$\hat{L}(\mathbf{x}_1\mathbf{x}_2,\omega) = \hat{L}^*(\mathbf{x}_2\mathbf{x}_1,\omega) = -\hat{L}(\mathbf{x}_2\mathbf{x}_1,-\omega) = -\hat{L}(\mathbf{x}_1\mathbf{x}_2,-\omega). \tag{13.10}$$

The first relation is a two-particle generalization of the symmetry relation (11.16) for the spectral function of the single-particle Green function. The second and third ones guarantee that $\hat{L}(\mathbf{x}_1\mathbf{x}_2,0) = 0$ as discussed above. The validity of the last relation is a consequence of the spin summations in (13.1). The same relations are valid for the spectral function $\hat{P}$ of the spin-averaged polarization function P.

13.1.2 *f*-Sum Rule

In the low-temperature limit we have already studied the dynamic structure factor (3.54) for the electronic system in a certain many-electron state. The spatial Fourier transformations of the spectral function (13.9)

$$\hat{L}(\mathbf{Q},\mathbf{Q}',\omega) = \frac{1}{\Omega}\int d^3\mathbf{x}_1 e^{-i\mathbf{Q}\mathbf{x}_1}\int d^3\mathbf{x}_2 e^{i\mathbf{Q}'\mathbf{x}_2}\hat{L}(\mathbf{x}_1\mathbf{x}_2,\omega)$$
$$= \frac{1}{2\Omega}\frac{1}{\hbar}\int_{-\infty}^{+\infty} dt_1 \left\langle\left[\hat{n}(\mathbf{Q},t_1),\hat{n}^+(\mathbf{Q}',t_2)\right]_-\right\rangle e^{i\omega(t_1-t_2)} \tag{13.11}$$

are related to the Fourier-transformed density operator

$$\hat{n}(\mathbf{Q},t) = \int d^3\mathbf{x}\hat{n}(\mathbf{x},t)e^{-i\mathbf{Q}\mathbf{x}}.$$

We investigate the frequency integral

$$\int_{-\infty}^{+\infty}\frac{d\omega}{2\pi}\omega\hat{L}(\mathbf{Q},\mathbf{Q}',\omega) = \frac{i}{\hbar}\frac{1}{2\Omega}\lim_{t_1\to t_2}\left\langle\left[\frac{\partial}{\partial t_1}\hat{n}(\mathbf{Q},t_1),\hat{n}^+(\mathbf{Q}',t_2)\right]_-\right\rangle \tag{13.12}$$

by means of (13.11) following the proof in [5]. With the equation of continuity (10.13) it holds

$$\frac{\partial}{\partial t_1}\left[\hat{n}(\mathbf{x}_1,t_1),\hat{n}(\mathbf{x}_2,t_2)\right]_- = -\nabla_{\mathbf{x}_1}\left[\hat{\mathbf{j}}(\mathbf{x}_1,t_2),\hat{n}(\mathbf{x}_2,t_1)\right]_-.$$

Applying the anticommutation rules (10.4) for the operators $\psi_s(\mathbf{x},t)$ and $\psi_s^+(\mathbf{x},t)$ with identical time arguments and the relation of $[\hat{A}\hat{B},\hat{C}\hat{B}]_-$ to anticommutators given in the proof below (11.4), we obtain

$$\lim_{t_1\to t_2}\frac{\partial}{\partial t_1}\left[\hat{n}(\mathbf{x}_1,t_1),\hat{n}(\mathbf{x}_2,t_2)\right]_-$$
$$= \frac{i}{m}\nabla_{\mathbf{x}_1}\left\{\hat{n}(\mathbf{x}_1,t_1)\nabla_{\mathbf{x}_1}\delta(\mathbf{x}_1-\mathbf{x}_2) - \frac{1}{2}\delta(\mathbf{x}_1-\mathbf{x}_2)\nabla_{\mathbf{x}_1}\hat{n}(\mathbf{x}_1,t_1)\right\}$$

or after spatial Fourier transformations

$$i\hbar\lim_{t_1\to t_2}\frac{\partial}{\partial t_1}\left[\hat{n}(\mathbf{Q},t_1),\hat{n}^+(\mathbf{Q}',t_2)\right]_- = \frac{\hbar^2}{m}\mathbf{Q}\mathbf{Q}'\hat{n}(\mathbf{Q}-\mathbf{Q}',t_1).$$

Then the integral (13.12) takes the form

$$\int_{-\infty}^{+\infty} \frac{d\omega}{2\pi} \omega \hat{L}(\mathbf{Q}, \mathbf{Q}', \omega) = \frac{1}{2m\Omega} \mathbf{Q}\mathbf{Q}' \langle \hat{n}(\mathbf{Q} - \mathbf{Q}', 0) \rangle. \tag{13.13}$$

The dynamic structure factor $S(\mathbf{Q}, \omega) = 2\hat{L}(\mathbf{Q}, \mathbf{Q}, \omega)$ (3.54) is directly related to the diagonal elements of the spatially Fourier-transformed spectral function of the density correlation function. It fulfills the *f-sum rule* [6, 7]

$$\int_{-\infty}^{+\infty} \frac{d\omega}{2\pi} \omega S(\mathbf{Q}, \omega) = \frac{n}{m} \mathbf{Q}^2, \tag{13.14}$$

since $\hat{n}(\mathbf{Q} - \mathbf{Q}', t)|_{\mathbf{Q}=\mathbf{Q}'}$ corresponds to the operator $\hat{N}(t)$ of the total number of electrons (3.15). The average electron density of the inhomogeneous electron gas is $n = N/\Omega$. The quantity $S(\mathbf{Q}, \omega)$ we have already used in Sect. 3.4 to discuss the relationship between correlation and screening effects in arbitrary many-electron states.

13.1.3 Screening Sum Rule

Similar to the single-particle Green function G (11.18), from the Martin-Schwinger-like relations (13.4) we conclude a Fourier representation

$$L(\mathbf{x}_1\mathbf{x}_2, t_1 - t_2) = \frac{1}{-i\hbar\beta} \sum_m L(\mathbf{x}_1\mathbf{x}_2, \tilde{z}_m) e^{-i\tilde{z}_m(t_1 - t_2)} \tag{13.15}$$

with the bosonic Matsubara frequencies $\tilde{z}_m$ (12.63), also in agreement with the finding (12.62). By analogy with (11.21) the Fourier coefficients obey a spectral representation

$$L(\mathbf{x}_1\mathbf{x}_2, \tilde{z}_m) = \int_{-\infty}^{+\infty} \frac{d\omega}{2\pi} \frac{\hat{L}(\mathbf{x}_1\mathbf{x}_2, \omega)}{\tilde{z}_m - \omega} \tag{13.16}$$

with the spectral function (13.7). It may be analytically continued into the whole z-plane to derive the function $L(\mathbf{x}_1\mathbf{x}_2, z)$ that agrees at the Matsubara frequencies with (13.16) (see Fig. 13.1).

The properties of the density correlation function convert into the properties of the inverse dielectric function [cf. (12.23)]

$$\epsilon^{-1}(\mathbf{x}_1\mathbf{x}_2, t_1 - t_2) = \delta(\mathbf{x}_1 - \mathbf{x}_2)\delta(t_1 - t_2) + 2 \int d^3\mathbf{x}' \mathrm{v}(\mathbf{x}_1 - \mathbf{x}') L(\mathbf{x}'\mathbf{x}_2, t_1 - t_2) \tag{13.17}$$

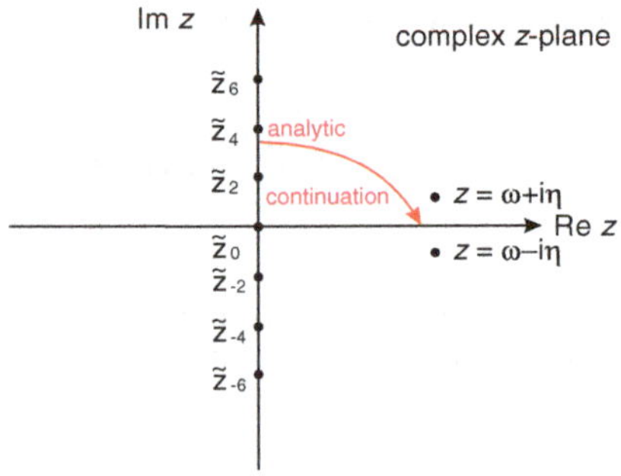

Fig. 13.1 Analytic continuation of $L(\mathbf{x}_1\mathbf{x}_2, \tilde{z}_m)$ into the complex z-plane to $L(\mathbf{x}_1\mathbf{x}_2, z)$

with the spin-averaged density correlation function (13.1). With (13.15) it also holds

$$\epsilon^{-1}(\mathbf{x}_1\mathbf{x}_2, \tilde{z}_m) = \delta(\mathbf{x}_1 - \mathbf{x}_2) + 2\int d^3\mathbf{x}' \mathrm{v}(\mathbf{x}_1 - \mathbf{x}')L(\mathbf{x}'\mathbf{x}_2, \tilde{z}_m) \tag{13.18}$$

and, consequently,

$$\epsilon^{-1}(\mathbf{x}_1\mathbf{x}_2, z) = \delta(\mathbf{x}_1 - \mathbf{x}_2) + \int\limits_{-\infty}^{+\infty} \frac{d\omega}{2\pi} \frac{\hat{\epsilon}^{-1}(\mathbf{x}_1\mathbf{x}_2, \omega)}{z - \omega},$$

$$\hat{\epsilon}^{-1}(\mathbf{x}_1\mathbf{x}_2, \omega) = 2\int d^3\mathbf{x}' \mathrm{v}(\mathbf{x}_1 - \mathbf{x}')\hat{L}(\mathbf{x}'\mathbf{x}_2, \omega), \tag{13.19}$$

where the spectral function $\hat{\epsilon}^{-1}$ of the inverse dielectric function has been introduced.

As in (13.11) we study the spatial Fourier transformations

$$\hat{\epsilon}^{-1}(\mathbf{Q}, \mathbf{Q}', \omega) = \frac{1}{\Omega}\int d^3\mathbf{x}_1 e^{-i\mathbf{Q}\mathbf{x}_1}\int d^3\mathbf{x}_2 e^{i\mathbf{Q}'\mathbf{x}_2}\hat{\epsilon}^{-1}(\mathbf{x}_1\mathbf{x}_2, \omega). \tag{13.20}$$

Substituting the bare Coulomb potential (4.51)

$$\mathrm{v}(\mathbf{x} - \mathbf{x}') = \int \frac{d^3\mathbf{Q}}{(2\pi)^3} e^{i\mathbf{Q}(\mathbf{x}-\mathbf{x}')}\tilde{\mathrm{v}}(Q),$$

$$\tilde{\mathrm{v}}(Q) = \frac{e^2}{\varepsilon_0 Q^2}, \tag{13.21}$$

the transformation (13.11), and the definition (13.19) into (13.20) we obtain

$$\begin{aligned}\hat{\epsilon}^{-1}(\mathbf{Q}, \mathbf{Q}, \omega) &= 2\tilde{\mathrm{v}}(Q)\hat{L}(\mathbf{Q}, \mathbf{Q}, \omega) \\ &= \tilde{\mathrm{v}}(Q)S(\mathbf{Q}, \omega).\end{aligned} \tag{13.22}$$

The spectral function of the inverse dielectric function is directly related to the dynamic structure factor (3.54), apart from a prefactor, the Fourier-transformed bare Coulomb potential.

The spectral representation (13.19) takes the form

$$\epsilon^{-1}(\mathbf{QQ}, z) = 1 + \tilde{v}(Q) \int\limits_{-\infty}^{+\infty} \frac{d\omega}{2\pi} \frac{S(\mathbf{Q}, \omega)}{z - \omega}. \tag{13.23}$$

The quantity ϵ_∞

$$\frac{1}{\epsilon_\infty} = \lim_{\mathbf{Q}\to 0} \epsilon^{-1}(\mathbf{Q}, \mathbf{Q}, 0) = 1 - \kappa \tag{13.24}$$

defines the (macroscopic) static electronic dielectric constant ϵ_∞ of a solid. It characterizes the compensation of excess charges, that is complete in metals, i.e., it holds $\kappa = 1$ and $1/\epsilon_\infty = 0$. In non-metals one finds $\kappa < 1$ and $\epsilon_\infty > 1$. One can formulate the screening sum rule [8, 9] as

$$\lim_{\mathbf{Q}\to 0} \tilde{v}(Q) \int\limits_{-\infty}^{+\infty} \frac{d\omega}{2\pi} \frac{S(\mathbf{Q}, \omega)}{\omega} = \kappa = 1 - \frac{1}{\epsilon_\infty} \tag{13.25}$$

or, using the relation (13.22), also in terms of the spectral function of the inverse dielectric function

$$\lim_{\mathbf{Q}\to 0} \int\limits_{-\infty}^{+\infty} \frac{d\omega}{2\pi} \frac{\hat{\epsilon}^{-1}(\mathbf{Q}, \mathbf{Q}, \omega)}{\omega} = \kappa = 1 - \frac{1}{\epsilon_\infty}. \tag{13.26}$$

The relation between dynamic structure factor and spectral function of the inverse dielectric function (13.22) allows a reformulation of the f-sum rule (13.14) as

$$\int\limits_{-\infty}^{+\infty} \frac{d\omega}{2\pi} \omega \hat{\epsilon}^{-1}(\mathbf{Q}, \mathbf{Q}, \omega) = \omega_p^2, \tag{13.27}$$

where the plasma frequency of the electrons

$$\omega_p = \left(\frac{e^2 n}{\varepsilon_0 m}\right)^{\frac{1}{2}} \tag{13.28}$$

is introduced by means of the spatially averaged electron density n.

Other rewritings are possible applying the spectral representations of type (13.16) or (13.19) and their analytic continuations. For instance, with the symmetry relations

(13.10) for the spectral function the density correlation function obeys the integral representation

$$L(\mathbf{x}_1\mathbf{x}_2, z) = \int_{-\infty}^{+\infty} \frac{d\omega}{2\pi} \frac{\omega \hat{L}(\mathbf{x}_1\mathbf{x}_2, \omega)}{z^2 - \omega^2}. \tag{13.29}$$

With the Fourier transformation (13.11) and the relation (13.13) the f-sum rule (13.14) becomes

$$\lim_{z\to\infty} z^2 L(\mathbf{Q}, \mathbf{Q}', z) = \frac{1}{2m\Omega} \mathbf{Q}\mathbf{Q}' \langle \hat{n}(\mathbf{Q} - \mathbf{Q}', 0) \rangle. \tag{13.30}$$

13.2 Kramers-Kronig Relations

13.2.1 Inversion

The dielectric function can be derived from the inverse one via the definition (12.45). In the case of the pure charge density response also relation (12.33) can be employed. After spin average, Fourier transformation, and analytic continuation the polarization function P is related to the discussed quantities L and ϵ^{-1} via the relation

$$L(\mathbf{x}_1\mathbf{x}_2, z) = \int d^3\mathbf{x}' P(\mathbf{x}_1\mathbf{x}', z) \epsilon^{-1}(\mathbf{x}'\mathbf{x}_2, z). \tag{13.31}$$

With the existence of the dielectric function ϵ and the spectral representations (13.16) and (13.19) for L and ϵ^{-1}, respectively, the same spectral representations and f-sum or screening sum rules (apart from a sign) can be derived for the polarization function and the dielectric function. This should not be shown here in detail.

We only remember that ϵ and P fulfill the same relations (13.17) or (13.18) between ϵ^{-1} and L apart from a negative sign,

$$\epsilon(\mathbf{x}_1\mathbf{x}_2, t_1 - t_2) = \delta(\mathbf{x}_1 - \mathbf{x}_2)\delta(t_1 - t_2) - 2\int d^3\mathbf{x}' \mathrm{v}(\mathbf{x}_1 - \mathbf{x}') P(\mathbf{x}'\mathbf{x}_2, t_1 - t_2),$$

$$\epsilon(\mathbf{x}_1\mathbf{x}_2, z) = \delta(\mathbf{x}_1 - \mathbf{x}_2) - 2\int d^3\mathbf{x}' \mathrm{v}(\mathbf{x}_1 - \mathbf{x}') P(\mathbf{x}'\mathbf{x}_2, z). \tag{13.32}$$

13.2.2 Fourier Representations

In Sect. 1.3 we argued that crystals are very important model systems and that artificial crystals, i.e., periodic arrangements of slabs and superlattices, are frequently used to model the properties of nanostructures. For all these systems it holds

$$\begin{aligned} L(\mathbf{x}_1 + \mathbf{R}\,\mathbf{x}_2 + \mathbf{R}, t_1 - t_2) &= L(\mathbf{x}_1\mathbf{x}_2, t_1 - t_2), \\ P(\mathbf{x}_1 + \mathbf{R}\,\mathbf{x}_2 + \mathbf{R}, t_1 - t_2) &= P(\mathbf{x}_1\mathbf{x}_2, t_1 - t_2) \end{aligned} \tag{13.33}$$

as a consequence of the translational symmetry represented by a Bravais lattice $\{\mathbf{R}\}$.

As a consequence of the periodicity the wave vectors $\mathbf{Q} = \mathbf{q}+\mathbf{G}$ and $\mathbf{Q}' = \mathbf{q}+\mathbf{G}'$ appear, and the inverse Fourier representations of (13.11) and (13.20) can be written as sums

$$\begin{aligned} L(\mathbf{x}_1\mathbf{x}_2, t - t') &= \frac{1}{\Omega}\sum_{\mathbf{q}}\sum_{\mathbf{G},\mathbf{G}'} e^{i[(\mathbf{q}+\mathbf{G})\mathbf{x}_1 - (\mathbf{q}+\mathbf{G}')\mathbf{x}_2]} L(\mathbf{q}+\mathbf{G}, \mathbf{q}+\mathbf{G}', t - t'), \\ P(\mathbf{x}_1\mathbf{x}_2, t - t') &= \frac{1}{\Omega}\sum_{\mathbf{q}}\sum_{\mathbf{G},\mathbf{G}'} e^{i[(\mathbf{q}+\mathbf{G})\mathbf{x}_1 - (\mathbf{q}+\mathbf{G}')\mathbf{x}_2]} P(\mathbf{q}+\mathbf{G}, \mathbf{q}+\mathbf{G}', t - t') \end{aligned} \tag{13.34}$$

with the wave vector $\mathbf{q} \in$ BZ and $\mathbf{G}, \mathbf{G}' \in$ the reciprocal lattice $\{\mathbf{G}\}$.

With the modification of the Fourier representation (13.21) of the bare Coulomb potential

$$\mathrm{v}(\mathbf{x} - \mathbf{x}') = \frac{1}{\Omega}\sum_{\mathbf{q}}\sum_{\mathbf{G}} e^{i(\mathbf{q}+\mathbf{G})(\mathbf{x}-\mathbf{x}')}\tilde{\mathrm{v}}(|\mathbf{q}+\mathbf{G}|), \tag{13.35}$$

and the Fourier representation of the Dirac δ-function in three dimensions

$$\delta(\mathbf{x} - \mathbf{x}') = \frac{1}{\Omega}\sum_{\mathbf{q}}\sum_{\mathbf{G}} e^{i(\mathbf{q}+\mathbf{G})(\mathbf{x}-\mathbf{x}')}, \tag{13.36}$$

the spectral representations of the dielectric function and its inverse read as

$$\begin{aligned} \epsilon^{-1}(\mathbf{q}+\mathbf{G}, \mathbf{q}+\mathbf{G}', z) &= \delta_{\mathbf{G}\mathbf{G}'} + 2\tilde{\mathrm{v}}(|\mathbf{q}+\mathbf{G}|)\int_{-\infty}^{+\infty}\frac{d\omega}{2\pi}\frac{\hat{L}(\mathbf{q}+\mathbf{G}, \mathbf{q}+\mathbf{G}', \omega)}{z - \omega}, \\ \epsilon(\mathbf{q}+\mathbf{G}, \mathbf{q}+\mathbf{G}', z) &= \delta_{\mathbf{G}\mathbf{G}'} - 2\tilde{\mathrm{v}}(|\mathbf{q}+\mathbf{G}|)\int_{-\infty}^{+\infty}\frac{d\omega}{2\pi}\frac{\hat{P}(\mathbf{q}+\mathbf{G}, \mathbf{q}+\mathbf{G}', \omega)}{z - \omega}. \end{aligned} \tag{13.37}$$

Taking the inhomogeneity of the electron gas into account, the f-sum rule of the type (13.30) can be rewritten into forms such as

$$\lim_{z\to\infty}\epsilon^{-1}(\mathbf{q}+\mathbf{G}, \mathbf{q}+\mathbf{G}', z) = \delta_{\mathbf{G}\mathbf{G}'} + \lim_{z\to\infty}\frac{\omega_p^2}{z^2}\frac{(\mathbf{q}+\mathbf{G})(\mathbf{q}+\mathbf{G}')}{|\mathbf{q}+\mathbf{G}|^2}\frac{\tilde{n}(\mathbf{G}-\mathbf{G}')}{\tilde{n}(0)},$$

$$\lim_{z\to\infty} \epsilon(\mathbf{q}+\mathbf{G},\mathbf{q}+\mathbf{G}',z) = \delta_{\mathbf{GG}'} - \lim_{z\to\infty} \frac{\omega_p^2}{z^2} \frac{(\mathbf{q}+\mathbf{G})(\mathbf{q}+\mathbf{G}')}{|\mathbf{q}+\mathbf{G}|^2} \frac{\tilde{n}(\mathbf{G}-\mathbf{G}')}{\tilde{n}(0)}, \tag{13.38}$$

where the Fourier representation of the periodic electron density

$$n(\mathbf{x}) = \frac{1}{\Omega} \sum_{\mathbf{G}} e^{i\mathbf{Gx}} \tilde{n}(\mathbf{G}) \tag{13.39}$$

has been used to reformulate the generalized f-sum rule (13.13). Thereby the zeroth Fourier component $\tilde{n}(0)/\Omega = N/\Omega = n$ represents the spatially averaged electron density of the system. Sometimes relations of type (13.38) are called Johnson sum rule [10, 11].

13.2.3 Consequences of Analytic Properties

The analytic continuation of $\epsilon^{-1}(\mathbf{q}+\mathbf{G},\mathbf{q}+\mathbf{G}',z)$ and $\epsilon(\mathbf{q}+\mathbf{G},\mathbf{q}+\mathbf{G}',z)$ into the entire complex frequency plane, as illustrated in Fig. 13.1 in principle, defines two functions $\epsilon_r^{-1}(\mathbf{q}+\mathbf{G},\mathbf{q}+\mathbf{G}',z)$ and $\epsilon_a^{-1}(\mathbf{q}+\mathbf{G},\mathbf{q}+\mathbf{G}',z)$ for ϵ^{-1} and correspondingly two for ϵ. ϵ_r^{-1} (ϵ_a^{-1}) is an analytic function in the upper (lower) frequency half plane. The continuation onto the real axis $z \to \omega + i\eta$ ($z \to \omega - i\eta$) (see Fig. 13.1) gives the corresponding retarded (advanced) function. Thereby, $\epsilon(\mathbf{xx}',\omega) = \lim_{\eta\to+0} \epsilon_r(\mathbf{xx}',\omega+i\eta)$ describes the response of the electron system to a real perturbing potential. We note that in the limit $T \to 0\,\mathrm{K}$ the physical (causal) response function ϵ_r is related to the time-ordered one by $\mathrm{Re}\epsilon(\omega) = \mathrm{Re}\epsilon_r(\omega)$ and $\mathrm{Im}\epsilon(\omega)\mathrm{sgn}\omega = \mathrm{Im}\epsilon_r(\omega)$ (see also discussion in Sect. 11.1.4 for G) [12, 13].

These conclusions are valid for the functions depending on the space coordinates, e.g. $\epsilon^{-1}(\mathbf{xx}',z)$. We investigate this function that is analytic in the upper half plane. In addition, we remember that $[\epsilon^{-1}(\mathbf{xx}',z) - \delta(\mathbf{x}-\mathbf{x}')]$ vanishes for large frequency arguments $z \to \infty$ as $\frac{1}{z^2}$ [see behavior of density correlation function (13.29)]. Then, the Cauchy theorem holds for any contour within this region, e.g. that in Fig. 13.2 with the semicircle in the upper half plane at infinity and a contour segment that

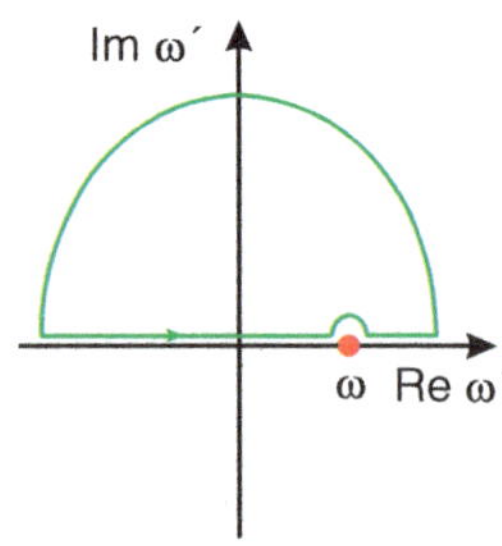

Fig. 13.2 Integral contour in complex ω'-plane for deriving Kramers-Kronig relations

traces the real axis with a hump at $\omega = \omega'$. It remains

$$\int_{-\infty}^{+\infty} d\omega' \frac{[\epsilon^{-1}(\mathbf{xx}', \omega') - \delta(\mathbf{x} - \mathbf{x}')]}{\omega' - \omega + i\eta} = 0$$

or using the Weierstrass formula (11.24)

$$\mathcal{P} \int_{-\infty}^{+\infty} d\omega' \frac{[\epsilon^{-1}(\mathbf{xx}', \omega') - \delta(\mathbf{x} - \mathbf{x}')]}{\omega' - \omega} - i\pi \left[\epsilon^{-1}(\mathbf{xx}', \omega) - \delta(\mathbf{x} - \mathbf{x}')\right] = 0.$$

Because of the asymmetry of $\epsilon^{-1}(\mathbf{xx}', \omega)$ with respect to the arguments $\mathbf{x}$ and $\mathbf{x}'$ we introduce the Hermitian and anti-Hermitian components

$$\begin{aligned} \epsilon_{\mathrm{H}}^{-1}(\mathbf{xx}', \omega) &= \frac{1}{2} \left\{\epsilon^{-1}(\mathbf{xx}', \omega) + \epsilon^{-1*}(\mathbf{x}'\mathbf{x}, \omega)\right\}, \\ \epsilon_{\mathrm{A}}^{-1}(\mathbf{xx}', \omega) &= \frac{1}{2i} \left\{\epsilon^{-1}(\mathbf{xx}', \omega) - \epsilon^{-1*}(\mathbf{x}'\mathbf{x}, \omega)\right\}, \end{aligned} \tag{13.40}$$

where $\epsilon^{-1}(\mathbf{xx}', \omega)$ is related to the retarded and advanced functions as introduced above.

The rearrangement of the Cauchy integral leads to the *Kramers-Kronig relations* [14, 15]

$$\begin{aligned} \epsilon_{\mathrm{H}}^{-1}(\mathbf{xx}', \omega) &= \delta(\mathbf{x} - \mathbf{x}') + \mathcal{P} \int_{-\infty}^{+\infty} \frac{d\omega'}{\pi} \frac{\epsilon_{\mathrm{A}}^{-1}(\mathbf{xx}', \omega')}{\omega' - \omega}, \\ \epsilon_{\mathrm{A}}^{-1}(\mathbf{xx}', \omega) &= -\mathcal{P} \int_{-\infty}^{+\infty} \frac{d\omega'}{\pi} \frac{\epsilon_{\mathrm{H}}^{-1}(\mathbf{xx}', \omega') - \delta(\mathbf{x} - \mathbf{x}')}{\omega' - \omega}. \end{aligned} \tag{13.41}$$

The same relations are valid for the dielectric function $\epsilon(\mathbf{xx}', \omega)$.

The anti-Hermitian parts of the two response function can be used to reformulate the f-sum rule. We do so for the spatially Fourier-transformed functions, e.g. the spectral function of the density correlation function. We use the Fourier transformations (13.34) with (13.11) and derive

$$\epsilon_{\mathrm{A}}^{-1}(\mathbf{xx}', \omega) = -\int d^3\mathbf{x}'' \mathrm{v}(\mathbf{x} - \mathbf{x}'') \hat{L}(\mathbf{x}''\mathbf{x}', \omega),$$

$$\epsilon_{\mathrm{A}}^{-1}(\mathbf{Q}, \mathbf{Q}', \omega) = -\tilde{\mathrm{v}}(Q) \hat{L}(\mathbf{Q}, \mathbf{Q}', \omega) \equiv -\frac{1}{2} \hat{\epsilon}^{-1}(\mathbf{Q}, \mathbf{Q}', \omega).$$

For the diagonal elements the f-sum rule reads as

$$\int_{-\infty}^{+\infty} \frac{d\omega}{2\pi} \omega \epsilon_{\mathrm{A}}^{-1}(\mathbf{Q}, \mathbf{Q}, \omega) = -\frac{\omega_p^2}{2},$$

$$\int_{-\infty}^{+\infty} \frac{d\omega}{2\pi} \omega \epsilon_{\mathrm{A}}(\mathbf{Q}, \mathbf{Q}, \omega) = \frac{\omega_p^2}{2} \tag{13.42}$$

with the plasma frequency (13.28). The off-diagonal elements fulfill the Johnson sum rule [10, 11] according to

$$\int_{-\infty}^{+\infty} \frac{d\omega}{2\pi} \omega \epsilon_{\mathrm{A}}^{-1}(\mathbf{q}+\mathbf{G}, \mathbf{q}+\mathbf{G}', \omega) = -\frac{\omega_p^2}{2} \frac{(\mathbf{q}+\mathbf{G})(\mathbf{q}+\mathbf{G}')}{|\mathbf{q}+\mathbf{G}||\mathbf{q}+\mathbf{G}'|} \frac{\tilde{n}(\mathbf{G}-\mathbf{G}')}{\tilde{n}(0)}$$

in this symmetric form for translationally invariant systems.

13.3 Approximate Screening Functions

13.3.1 Inhomogeneous and Homogeneous Electron Gases

The starting point for screening studies may be a system of electrons with collinear spins in single-particle states characterized by orbitals $\varphi_{\lambda m_s}(\mathbf{x})$ and energies $\varepsilon_{\lambda m_s}$ in which the electrons do not further interact. Its dielectric properties are described by expression (12.70). We assume a true or artificial translational symmetry expressed by the reciprocal lattice $\{\mathbf{G}\}$ and the Bloch character of the single particle states with $\lambda = \nu\mathbf{k}$ and $\varepsilon_{\lambda m_s} \equiv \varepsilon_{\nu m_s}(\mathbf{k})$ (ν - band index, $\mathbf{k}$ - Bloch wave vector $\in$ BZ). As a consequence of the assumption that the Bloch electrons do not interact, the polarization function of independent particles (12.70) has to be investigated. However, we will see below that in self-consistent approaches, for instance within the independent-QP approximation, the eigenvalues $\varepsilon_{\nu m_s}(\mathbf{k})$ have to be replaced by QP ones $\varepsilon^{\mathrm{QP}}_{\nu m_s}(\mathbf{k})$. Accordingly, the dielectric function is described in the RPA framework as discussed in Sect. 12.3.3. After Fourier transformation according to (13.11) or (13.20) and (13.35) the dielectric function reads as

$$\epsilon(\mathbf{q}+\mathbf{G}, \mathbf{q}+\mathbf{G}', z) = \delta_{\mathbf{G}\mathbf{G}'} - 2\tilde{v}(|\mathbf{q}+\mathbf{G}|) L^0(\mathbf{q}+\mathbf{G}, \mathbf{q}+\mathbf{G}', z),$$

$$L^0(\mathbf{q}+\mathbf{G}, \mathbf{q}+\mathbf{G}', z) = \frac{1}{2\Omega} \sum_{m_s} \sum_{\nu,\nu'} \sum_{\mathbf{k},\mathbf{k}'} B^{\mathbf{k}\mathbf{k}'*}_{\substack{m_s m_s \\ \nu\nu'}}(\mathbf{q}+\mathbf{G}) B^{\mathbf{k}\mathbf{k}'}_{\substack{m_s m_s \\ \nu\nu'}}(\mathbf{q}+\mathbf{G}')$$

$$\times \frac{f(\varepsilon_{\nu m_s}(\mathbf{k})) - f(\varepsilon_{\nu' m_s}(\mathbf{k}'))}{\varepsilon_{\nu m_s}(\mathbf{k}) - \varepsilon_{\nu' m_s}(\mathbf{k}') - \hbar z}. \tag{13.43}$$

When we use the fact, that, because of the time-reversal symmetry, for every $\varphi_{\nu\mathbf{k}m_s}(\mathbf{x})$ there is a function $\varphi^*_{\nu-\mathbf{k}m_s}(\mathbf{x})$ with the same eigenvalue $\varepsilon_{\nu m_s}(\mathbf{k}) = \varepsilon_{\nu m_s}(-\mathbf{k})$, several similar expressions can be generated with the Bloch integrals

$$B\,{}^{\mathbf{kk}'}_{\substack{m_s m_s \\ \nu\nu'}}(\mathbf{q}+\mathbf{G}) = \int d^3\mathbf{x}\varphi^*_{\nu\mathbf{k}m_s}(\mathbf{x})e^{i(\mathbf{q}+\mathbf{G})\mathbf{x}}\varphi_{\nu'\mathbf{k}'m_s}(\mathbf{x}). \tag{13.44}$$

Expression (13.43) represents the original Ehrenreich-Cohen formula [16] generalized for a spin-polarized system with the two spin channels $m_s = \pm\frac{1}{2}$. Because of the different components $\mathbf{G}$ and $\mathbf{G}'$ in (13.43), $\epsilon(\mathbf{q}+\mathbf{G}, \mathbf{q}+\mathbf{G}', z)$ is sometimes also called (frequency-dependent) dielectric matrix. The inversion of this matrix yields the corresponding (RPA) inverse dielectric matrix $\epsilon^{-1}(\mathbf{q}+\mathbf{G}, \mathbf{q}+\mathbf{G}', z)$ that fulfills the relations

$$\begin{aligned}
&\sum_{\mathbf{G}''}\epsilon(\mathbf{q}+\mathbf{G}, \mathbf{q}+\mathbf{G}'', z)\epsilon^{-1}(\mathbf{q}+\mathbf{G}'', \mathbf{q}+\mathbf{G}', z)\\
&\quad = \sum_{\mathbf{G}''}\epsilon^{-1}(\mathbf{q}+\mathbf{G}, \mathbf{q}+\mathbf{G}'', z)\epsilon(\mathbf{q}+\mathbf{G}'', \mathbf{q}+\mathbf{G}', z) = \delta_{\mathbf{GG}'}.
\end{aligned}$$

The dielectric matrix (13.43) fulfills the oscillator-strength and screening sum rules. In the retarded limit $z = \omega + i\eta$ the diagonal components of the anti-Hermitian part are

$$\begin{aligned}
\epsilon_{\mathrm{A}}(\mathbf{q}+\mathbf{G}, \mathbf{q}+\mathbf{G}, \omega) &= -\pi\tilde{\mathrm{v}}\left(|\mathbf{q}+\mathbf{G}|\right)\sum_{m_s}\sum_{\nu,\nu'}\frac{1}{\Omega}\sum_{\mathbf{k},\mathbf{k}'}\left|B\,{}^{\mathbf{kk}'}_{\substack{m_s m_s \\ \nu\nu'}}(\mathbf{q}+\mathbf{G})\right|^2\\
&\times\left[f\left(\varepsilon_{\nu m_s}(\mathbf{k})\right) - f\left(\varepsilon_{\nu' m_s}(\mathbf{k}')\right)\right]\delta\left(\varepsilon_{\nu m_s}(\mathbf{k}) - \varepsilon_{\nu' m_s}(\mathbf{k}') - \hbar\omega\right).
\end{aligned} \tag{13.45}$$

The frequency integral for the diagonal components, for which the anti-Hermitian part is identical with the imaginary part, yields

$$\begin{aligned}
&\int_{-\infty}^{+\infty}\frac{d\omega}{2\pi}\omega\epsilon_{\mathrm{A}}(\mathbf{q}+\mathbf{G}, \mathbf{q}+\mathbf{G}, \omega) = \int_{-\infty}^{+\infty}\frac{d\omega}{2\pi}\omega\mathrm{Im}\epsilon(\mathbf{q}+\mathbf{G}, \mathbf{q}+\mathbf{G}, \omega)\\
&= -\frac{1}{2\hbar^2\Omega}\tilde{\mathrm{v}}(|\mathbf{q}+\mathbf{G}|)\sum_{m_s}\sum_{\nu,\nu'}\sum_{\mathbf{k},\mathbf{k}'}\left|B\,{}^{\mathbf{kk}'}_{\substack{m_s m_s \\ \nu\nu'}}(\mathbf{q}+\mathbf{G})\right|^2\\
&\times\left[\varepsilon_{\nu m_s}(\mathbf{k}) - \varepsilon_{\nu' m_s}(\mathbf{k}')\right]\left[f(\varepsilon_{\nu m_s}(\mathbf{k})) - f(\varepsilon_{\nu' m_s}(\mathbf{k}'))\right]\\
&= -\frac{1}{2\hbar^2\Omega}\tilde{\mathrm{v}}(\mathbf{q}+\mathbf{G})\sum_{m_s}\sum_{\nu,\nu'}\sum_{\mathbf{k},\mathbf{k}'}\left[\left|B\,{}^{\mathbf{kk}'}_{\substack{m_s m_s \\ \nu\nu'}}(\mathbf{q}+\mathbf{G})\right|^2 + \left|B\,{}^{\mathbf{kk}'}_{\substack{m_s m_s \\ \nu\nu'}}(-\mathbf{q}-\mathbf{G})\right|^2\right]\\
&\times\left[\varepsilon_{\nu m_s}(\mathbf{k}) - \varepsilon_{\nu' m_s}(\mathbf{k}')\right]f\left(\varepsilon_{\nu m_s}(\mathbf{k})\right).
\end{aligned}$$

The proof of the f-sum rule is easier for a local potential in the single-particle Hamiltonian $\hat{H} = -\frac{\hbar^2}{2m}\nabla_{\mathbf{x}}^2 + V(\mathbf{x})$ used to generate eigenfunctions and eigenvalues. With the commutator

$$[e^{-i\mathbf{Q}\mathbf{x}}, \hat{H}]_- = -\frac{\hbar^2}{2m} e^{-i\mathbf{Q}\mathbf{x}}(\mathbf{Q}^2 + 2i\mathbf{Q}\nabla_{\mathbf{x}})$$

it holds

$$\left[\left|B\,{}^{\mathbf{k}\mathbf{k}'}_{m_s m_s}{}_{\nu\nu'}(-\mathbf{q}-\mathbf{G})\right|^2 + \left|B\,{}^{\mathbf{k}\mathbf{k}'}_{m_s m_s}{}_{\nu\nu'}(\mathbf{q}+\mathbf{G})\right|^2\right]\left[\varepsilon_{\nu m_s}(\mathbf{k}) - \varepsilon_{\nu' m_s}(\mathbf{k})\right]$$
$$= -\frac{\hbar^2}{2m}(\mathbf{q}+\mathbf{G})^2\left[\left|B\,{}^{\mathbf{k}\mathbf{k}'}_{m_s m_s}{}_{\nu\nu'}(\mathbf{q}+\mathbf{G})\right|^2 + \left|B\,{}^{\mathbf{k}\mathbf{k}'}_{m_s m_s}{}_{\nu\nu'}(-\mathbf{q}-\mathbf{G})\right|^2\right].$$

The application of this result yields

$$\int_{-\infty}^{+\infty} \frac{d\omega}{2\pi}\omega\epsilon_{\mathrm{A}}(\mathbf{q}+\mathbf{G}, \mathbf{q}+\mathbf{G}, \omega)$$
$$= \frac{e^2}{4m\varepsilon_0\Omega}\sum_{m_s}\sum_{\nu,\nu'}\sum_{\mathbf{k},\mathbf{k}'}\left|B\,{}^{\mathbf{k}\mathbf{k}'}_{m_s m_s}{}_{\nu\nu'}(\mathbf{q}+\mathbf{G})\right|^2\left[f(\varepsilon_{\nu m_s}(\mathbf{k})) + f(\varepsilon_{\nu' m_s}(\mathbf{k}'))\right]$$
$$= \frac{e^2}{2m\varepsilon_0\Omega}\sum_{m_s}\sum_{\nu}\sum_{\mathbf{k}} f(\varepsilon_{\nu m_s}(\mathbf{k}))$$
$$= \frac{e^2 n}{2m\varepsilon_0} = \frac{\omega_p^2}{2}$$

with the orthonormalization and closure relation of the wave functions (6.21) as well as the definition of the average electron density (11.31).

In the case of a non-spin-polarized homogeneous electron gas of density n, more precisely for the jellium model (see Sect. 4.4.1), with plane-wave eigenfunctions (4.46) and eigenenergies $\varepsilon(\mathbf{k}) = \frac{\hbar^2}{2m}\mathbf{k}^2$ with $\mathbf{k} \in$ the entire reciprocal space, the Ehrenreich-Cohen formula (13.43) turns into the Lindhard formula [17] which is diagonal in the wave vectors $\mathbf{Q} = \mathbf{q}+\mathbf{G}$ and $\mathbf{Q}' = \mathbf{q}+\mathbf{G}'$. For the retarded function $\epsilon(\mathbf{Q}, \omega) \equiv \epsilon(\mathbf{Q}, \mathbf{Q}, \omega + i\eta)$ it holds

$$\epsilon(\mathbf{Q}, \omega) = 1 - 2\tilde{v}(Q)L^0(\mathbf{Q}, \omega),$$
$$L^0(\mathbf{Q}, \omega) = \frac{1}{\Omega}\sum_{\mathbf{k}}\frac{f(\varepsilon(\mathbf{k}+\mathbf{Q})) - f(\varepsilon(\mathbf{k}))}{\varepsilon(\mathbf{k}+\mathbf{Q}) - \varepsilon(\mathbf{k}) - \hbar(\omega + i\eta)}. \tag{13.46}$$

For $T = 0\,\mathrm{K}$ the evaluation yields

$$\mathrm{Re}L^0(Q,\omega) = -\frac{3n}{4\varepsilon_F^0}\left\{\frac{1}{2} + \frac{k_F}{4Q}\left[\left(1 - \left(\frac{\omega}{Qv_F} - \frac{Q}{2k_F}\right)^2\right)\ln\left|\frac{\frac{\omega}{Qv_F} - \frac{Q}{2k_F} - 1}{\frac{\omega}{Qv_F} - \frac{Q}{2k_F} + 1}\right| \right.\right.$$
$$\left.\left. -\left(1 - \left(\frac{\omega}{Qv_F} + \frac{Q}{2k_F}\right)^2\right)\ln\left|\frac{\frac{\omega}{Qv_F} + \frac{Q}{2k_F} - 1}{\frac{\omega}{Qv_F} + \frac{Q}{2k_F} + 1}\right|\right]\right\}, \tag{13.47}$$

$$\mathrm{Im}L^0(Q,\omega) = -\frac{3n}{4\varepsilon_F^0}\begin{cases}\frac{\pi}{2}\frac{k_F}{2Q}\left[1 - \left(\frac{\omega}{Qv_F} - \frac{Q}{2k_F}\right)^2\right] & \text{in I,}\\ \frac{\pi}{2}\frac{\omega}{Qv_F} & \text{in II,}\\ 0 & \text{otherwise}\end{cases}$$

with the electron-gas parameters defined in (4.47) and the Thomas-Fermi wave vector $q_{TF} = \sqrt{\frac{e^2}{\varepsilon_0}\frac{dn}{d\varepsilon_F^0}} = \left(\frac{12}{\pi}\right)^{\frac{1}{3}}/(r_s^{1/2}a_B) = \sqrt{3}\omega_p/v_F$. For two wave vectors corresponding spectra are plotted in Fig. 13.3. They exhibit significant variations with wave vector and frequency. Expression (13.47) also gives the dynamic structure factor of the homogeneous gas of non-interacting electrons by $S^0(Q,\omega) = 2\hat{L}^0(Q,\omega)$. In the frequency-wave-vector plane (see Fig. 13.4) regions I and II introduced in (13.47) are indicated. The two regions characterize the electron-hole pair excitation spectrum of a non-interacting homogeneous system defined by $\mathrm{Im}\epsilon(Q,\omega) \neq 0$. For $Q \leq 2k_F$, the particle-hole continuum extends from $\omega = 0$ up to $\omega = \omega_+(Q)$ (see Fig. 13.4). For $Q > 2k_F$, there is no pair or collective excitation at low energy and the particle-hole continuum extends from $\omega_-(Q)$ to $\omega_+(Q)$. The collective excitation is a plasmon mode defined by the zeros $\omega = \Omega(Q)$ of

$$\mathrm{Re}\epsilon(Q,\omega) = 0.$$

The plasmons are collective excitations of the electron gas which are directly related to the density fluctuations and hence to the screening of charged particles in the

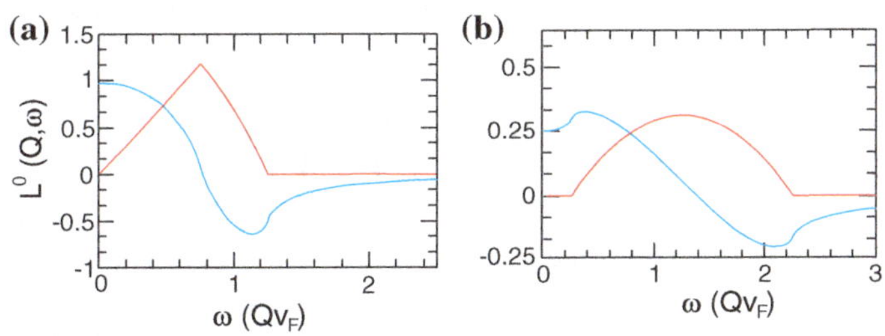

Fig. 13.3 Real (*blue line*) and imaginary (*red line*) parts of the Lindhard polarization function (13.47) divided by $(-3n/4\varepsilon_F^0)$ for wave vectors (**a**) $Q = 0.5\,k_F$ and (**b**) $Q = 2.5\,k_F$

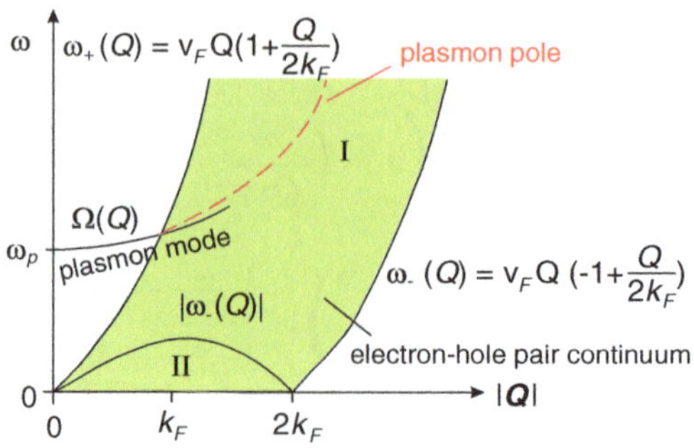

Fig. 13.4 Excitation energies $\hbar\omega$ and transferred momenta $\hbar Q$ in a three-dimensional homogeneous electron gas. The electron-hole pair continuum is indicated as *green regions I* and *II*. In the other regions $\text{Im}\epsilon(Q,\omega) = 0$ holds. The plasmon curve $\hbar\Omega(Q)$ corresponds to the zeros of $\text{Re}\epsilon(Q,\omega) = 0$

electron gas. Such a plasmon has an infinite lifetime as long $\text{Im}\epsilon(Q,\omega) = 0$, i.e., as long as it does not overlap with the particle-hole pair continuum. Within this continuum the plasmon becomes strongly damped (Landau damping [18]) and is not a well-defined collective excitation of the system anymore.

Indeed, for $Q \to 0$ with fixed ω the Drude behavior

$$\lim_{Q\to 0} \epsilon(Q,\omega) = 1 - \frac{\omega_p^2}{\omega^2},$$

holds and, hence, the zero recovers $\omega = \omega_p = \Omega(0)$ as the excitation energy of the plasmon mode in the long-wavelength limit. This expression also gives the correct asymptotics for $\omega \to \infty$ in agreement with the f-sum rule (13.38). The expansion of (13.47) to higher orders in Q^2 yields the long-wavelength plasmon dispersion [19]

$$\Omega^2(Q) = \omega_p^2 + \frac{\alpha}{3} v_F^2 Q^2 + \mathcal{O}(Q^4)$$

with the plasmon stiffness $\alpha = 1$ in the case of Thomas-Fermi screening and $\alpha = \frac{9}{5}$ in a somewhat more sophisticated approach [19]. With the function (4.53) introduced to characterize the exchange contribution to the HF eigenvalues, the static limit of (13.46) is

$$\epsilon(Q,0) = 1 + \left(\frac{q_{TF}}{Q}\right)^2 F\left(\frac{Q}{2k_F}\right).$$

Thereby $F(x)$ (4.53) describes the Lindhard correction to the Thomas-Fermi result. The static dielectric function $\epsilon(Q,0)$ is not analytic at $Q = 2k_F$. As a result it can be shown that at large distances $r = |\mathbf{x} - \mathbf{x}'|$ the screened potential W (12.47) has a contribution that varies as $\sim\cos(2k_F r)/r^3$. The corresponding oscillations are known under the name Friedel oscillations or Ruderman-Kittel oscillations [22]. The large-wave vector limit is [21, 22]

$$\lim_{Q\to\infty} \epsilon(Q,0) = 1 + \frac{\omega_p^2}{\left(\frac{\hbar Q^2}{2m}\right)^2} = \frac{4}{3}\left(\frac{q_{TF}k_F}{Q^2}\right)^2 .$$

The Q^4-term is important for the description of the short-wavelength density fluctuations in the response. The existence of collective excitations with a plasmon dispersion relation suggests to approximate the (retarded) dielectric function by a frequency-dependent function with one pole for $\omega > 0$ [22],

$$\epsilon(Q,\omega) = 1 + \frac{\omega_p^2}{\Omega^2(Q) - \omega_p^2 - (\omega + i\eta)^2}, \tag{13.48}$$

that fulfills the important limits and sum rules setting the plasmon dispersion relation to be

$$\Omega^2(Q) = \omega_p^2\left[1 + \alpha\left(\frac{Q}{q_{TF}}\right)^2\right] + \left(\frac{\hbar Q^2}{2m}\right)^2, \tag{13.49}$$

where α is not fixed to the value $\alpha = 9/5 = 1.8$ but may vary down to $\alpha = 1$ to recover the Thomas-Fermi screening in the short-wavelength limit.

The inversion of the dielectric function yields

$$\frac{1}{\epsilon(Q,\omega)} = 1 - \frac{\omega_p^2}{\Omega^2(Q) - (\omega + i\eta)^2}. \tag{13.50}$$

For positive (or negative) frequencies it contains only one pole at the plasmon frequency for a given momentum Q. Expressions (13.48) and (13.50) can be therefore called single-plasmon-pole (SPP) approximation for the (inverse) dielectric function. It recovers important limits. For instance, the inverse dielectric function (13.50) gives rise to the Thomas-Fermi screening ($\alpha = 1$) [20] for small wave vectors and the static limit $\omega = 0$.

13.3.2 Electron Gas in Non-metals

We investigate a non-metal at $T = 0\,\text{K}$ with occupied valence bands $\nu = v$ and empty conduction bands $\nu = c$. All electron spins should be paired. There is no spin polarization in the system. Expression (13.43) of the dielectric tensor is rewritten for the diagonal elements to

$$\epsilon(\mathbf{q}+\mathbf{G},\mathbf{q}+\mathbf{G},z) = 1 + \tilde{v}(|\mathbf{q}+\mathbf{G}|)\frac{4}{\Omega}\sum_{c,v}\sum_{\mathbf{k},\mathbf{k}'}\left|B_{cv}^{\mathbf{k}\mathbf{k}'}(\mathbf{q}+\mathbf{G})\right|^2 \times \frac{\varepsilon_c(\mathbf{k})-\varepsilon_v(\mathbf{k}')}{[\varepsilon_c(\mathbf{k})-\varepsilon_v(\mathbf{k}')]^2-\hbar^2 z^2}. \tag{13.51}$$

A rather crude approximation of the frequency and wave-vector dependence follows replacing the interband transition energies $[\varepsilon_c(\mathbf{k})-\varepsilon_v(\mathbf{k}')]$ by an average energy distance, an average gap $\hbar\bar{\omega}_g(\mathbf{q}+\mathbf{G})$, that depends on the wave vector $\mathbf{q}+\mathbf{G}$ appearing originally in the Bloch integrals. The result is the Penn model [23]

$$\epsilon(\mathbf{q}+\mathbf{G},\mathbf{q}+\mathbf{G},z) = 1 + \frac{\omega_p^2}{\bar{\omega}_g^2(\mathbf{q}+\mathbf{G})-z^2}, \tag{13.52}$$

which fulfills the f-sum rule as expression (13.45). With the definition of the static electronic dielectric constant ϵ_∞ (13.25) the Penn gap [23] reads as

$$\bar{\omega}_g(0) = \omega_p/\sqrt{\epsilon_\infty - 1} \tag{13.53}$$

for vanishing wave vector.

The inversion of (13.52) yields a single-plasmon-pole approximation for the inverse dielectric function of a non-metal, which however can be also applied to metals,

$$\epsilon^{-1}(\mathbf{q}+\mathbf{G},\mathbf{q}+\mathbf{G},z) = 1 - \frac{\omega_p^2}{\Omega^2(\mathbf{q}+\mathbf{G})-z^2} \tag{13.54}$$

with a dispersion relation

$$\Omega^2(Q) = \left[\frac{1}{1-\frac{1}{\epsilon_\infty}} + \alpha\frac{Q^2}{q_{TF}^2}\right]\omega_p^2 + \left(\frac{\hbar Q^2}{2m}\right)^2. \tag{13.55}$$

In the electron-gas limit $\epsilon_\infty \to \infty$ it equals the dispersion relation of the plasmon (13.49) derived for jellium. The renormalization constant α accounts for a possible treatment of the electron-electron interaction beyond RPA [22] and modifications due to the finite static electronic polarization at vanishing wave vector. Dispersion relations similar to (13.55) have been derived in several publications (see e.g. [24–26]). The parameters of the model can be easily generalized for a valence electron gas with the homogeneous density n or a metal, where the Fermi surface is given by bands ν with $\varepsilon_\nu(\mathbf{k}) = \varepsilon_F$. In this case the parameters have to be modified to $q_{TF}^2 = \frac{2e^2}{\varepsilon_0\Omega}\sum_\nu\sum_{\mathbf{k}}\frac{\partial f(\varepsilon_\nu(\mathbf{k}))}{\partial\varepsilon_\nu(\mathbf{k})}$ and $\omega_p^2 = \frac{2e^2}{\varepsilon_0\Omega}\sum_\nu\sum_{\mathbf{k}} f(\varepsilon_\nu(\mathbf{k}))\hat{\mathbf{q}}m_\nu^{-1}(\mathbf{k})\hat{\mathbf{q}}$ with the

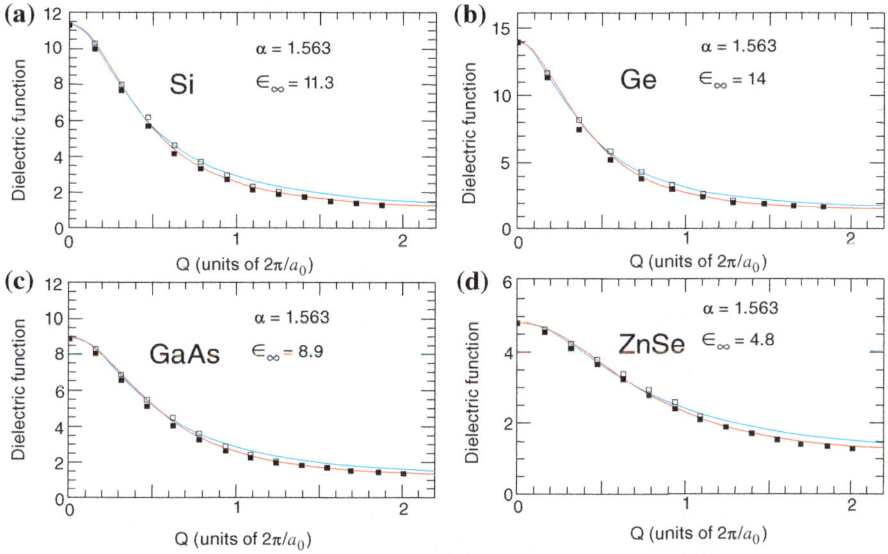

Fig. 13.5 Diagonal elements of the static dielectric function $\epsilon(\mathbf{Q}, \mathbf{Q}, 0)$ as a function of the wave vector $|\mathbf{Q}|$ for four semiconductors. The plasmon-pole approximation (13.54) with the dispersion relation (13.55) for $\alpha = 1.563$ (*red lines*) [27] is compared with results of RPA calculations (13.43) applying eigenvalues and eigenfunctions of an empirical-pseudopotential approach (*empty boxes* $\mathbf{Q}\|[100]$, *filled boxes* $\mathbf{Q}\|[111]$) [28]. In addition, results obtained by means of the Levine-Louie model function (*blue lines*) [29] are displayed. The same static electronic dielectric constants have been used. Adapted from [27]

tensor of the inverse effective mass $\left\{m_\nu^{-1}(\mathbf{k})\right\}_{ij} = \frac{1}{\hbar^2}\frac{\partial^2}{\partial k_i \partial k_j}\varepsilon_\nu(\mathbf{k})$ and the direction vector $\hat{\mathbf{q}} = \mathbf{q}/|\mathbf{q}|$.

In Fig. 13.5 the reliability of the plasmon-pole approximation (13.54) using the dispersion relation with $\alpha = 1.563$ in (13.55) is displayed. The wave-vector dependence of the static diagonal dielectric matrix $\epsilon(\mathbf{QQ}, 0)|_{\mathbf{Q}=\mathbf{q}+\mathbf{G}}$ [27] agrees well with results of RPA calculations (13.43) using eigenvalues and eigenfunctions of an empirical pseudopotential approach [28] for four tetrahedrally coordinated semiconductors, in particular for wave vectors along the [111] direction. The RPA values of the static electronic dielectric constant ϵ_∞ have been used in the plasmon-pole approximation. The Levine-Louie dielectric function [29] leads to a similar wave-vector dependence with a tendency for slightly larger values of the dielectric function at large wave vectors. RPA calculations with an electronic structure derived from a DFT approach [30] also seem to indicate these slightly larger values of the dielectric function for $Q \to \infty$ [31]. In the same spirit as the Penn model (13.52), the Levine-Louie dielectric function modifies the Lindhard function (13.46) by an *ad-hoc* introduction of an energy gap in the spectrum of the homogeneous electron gas. Interestingly the static model dielectric function (13.52) with $\alpha = 1$ agrees well with RPA results obtained in the framework of a more sophisticated hybrid-functional HSE03/06

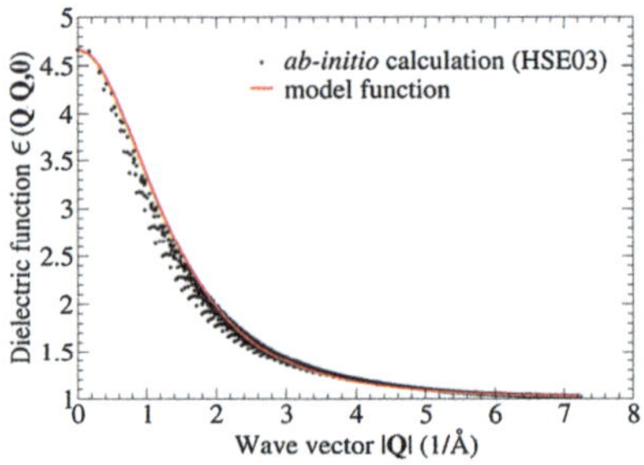

Fig. 13.6 Similar representation as in Fig. 13.5 for antiferromagnetic NiO. The model function (13.52) (*red line*) with the dispersion (13.55) for $\alpha = 1$ is compared with RPA calculations (13.43) using eigenvalues and eigenfunctions of a hybrid-functional HSE03/06 approach. The corresponding *black dots* represent wave-vector orientations in various directions (courtesy of C. Rödl, Universität Jena)

approach (9.22). This is clearly demonstrated in Fig. 13.6 for rocksalt NiO. The most important influence of the approach to the electronic structure is given by the actual value of ϵ_∞. The wave vector dependence is mainly described by the average electron density, which depends less on the electronic-structure calculation applied.

Usually the inverse dielectric function, more precisely, its negative imaginary part and, hence, the energy loss function, must describe (i) collective plasmon-like excitations, which dominate the screening for small wave vectors, and (ii) electron-hole pair excitations at lower frequencies, which guarantee that the f-sum rule is also fulfilled at large wave vectors where the screening is less effective. Therefore, at first glance, one expects significant shortcomings of the plasmon-pole approximation (13.54) for the dielectric function. This is certainly true if one would apply expressions of the type (13.52) to describe for instance optical phenomena. A single oscillator leads to absorption of light at one frequency and cannot describe the rich spectra with complex lineshapes measured for solids and molecules. The situation is however much better in the case of the inverse dielectric function, in particular, studying its negative imaginary part, the loss function. This is demonstrated by the electron energy loss function measured for bulk Si in Fig. 13.7a [32]. The electron-hole pair loss peak visible for $Q = 0$ vanishes immediately for finite wave vectors. There is a pronounced plasmon peak at $\hbar\Omega(0) = 16.7\,\text{eV}$ that shifts to higher energies with increasing wave vector in agreement with the approximation (13.55). As expected, the position $\hbar\Omega(Q)$ of this peak shows a linear Q^2 behavior in Fig. 13.7b up to $Q \sim 1.2\,\text{Å}^{-1}$. The saturation of the peak position at larger wave vectors cannot be described by the model dispersion (13.55). However, the experimental zero-wave vector result $\hbar\Omega(0) = 16.7\,\text{eV}$ can be nearly explained by treating the valence electrons of silicon as free electrons via formula (13.55).

Nevertheless, there is a common belief that physical effects related to the XC self-energy, e.g. that in (12.57), where only integrals over the entire frequency spectrum in the dynamically screened Coulomb potential play a role, is less influenced by the details of the frequency dependence of the inverse dielectric function.

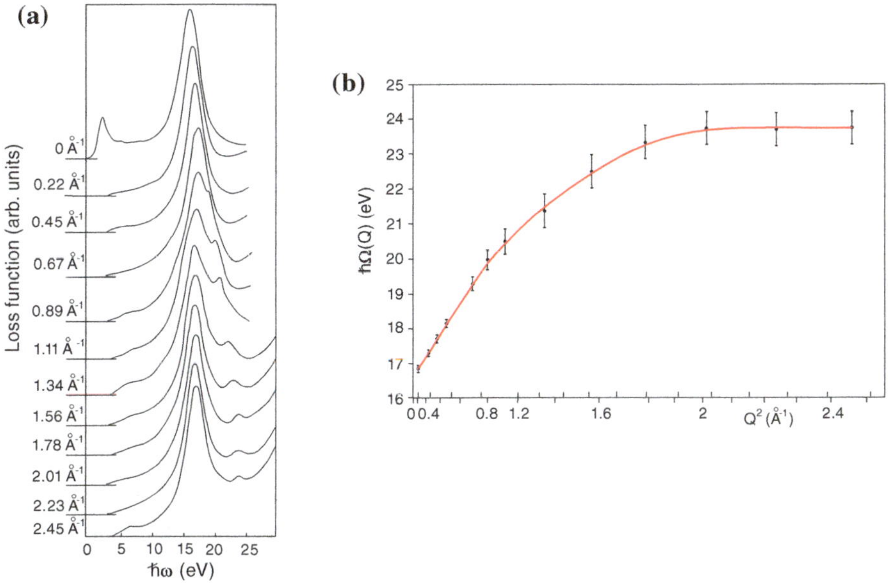

Fig. 13.7 (**a**) Electron energy loss function for bulk Si with increasing wave vector Q up to $Q \sim 2.5\,\text{Å}^{-1}$. (**b**) Position $\hbar\Omega(Q)$ of the main plasmon peak in (**a**) versus squared wave vector Q. Reprinted with permission from [32]. Copyright 1978 by the American Physical Society

Already Lundqvist [22, 33] showed that the electron-hole pair excitations are of minor importance for the determination of the self-energy for a homogeneous electron gas. In inhomogeneous electron gases, e.g. sp-bonded semiconductors such as Si, quasiparticle gap energies determined using plasmon-pole models differ by no more than 50 meV compared to those computed with the full frequency dependence [34, 35]. The deviations of absolute quasiparticle energy values may be however much bigger.

Another problem appears for systems without pronounced bulk single-plasmon peak in the loss function. As an example the loss spectrum of pristine graphene is displayed in Fig. 13.8 for a finite wave vector $\mathbf{Q} \| \Gamma M$ line. Theoretical [36] and experimental [37] data are compared. One observes the π and $\pi + \sigma$ plasmon modes in a free-standing single sheet near the zero wave-vector values $\hbar\Omega(0) = 4.7$ and 14.6 eV, respectively. At least, a two-plasmon-pole model is necessary to describe the loss function. Indeed, the use of N-parameter plasmon-pole models have been suggested in the literature to improve numerical self-energy studies [38, 39].

13.3.3 Spatial Inhomogeneity

For inhomogeneous electron gases improvements of the single-plasmon-pole approximation are needed. Their results should obey the generalized f-sum rules (13.38).

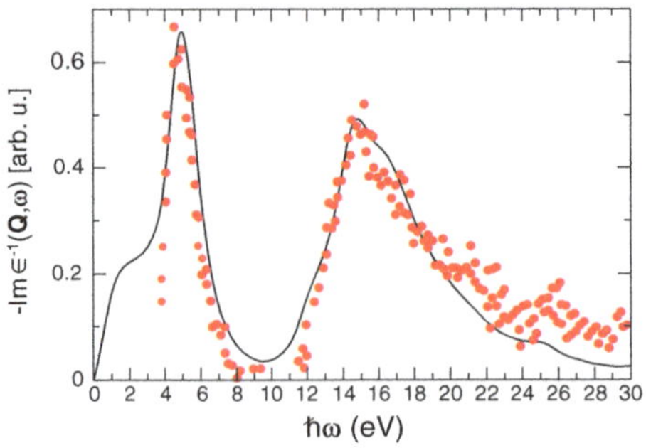

Fig. 13.8 Energy loss spectrum in pristine graphene for a wave vector along the ΓM direction. The theoretical spectrum (*black solid line*) [36] has been calculated for $Q = 0.087\,\text{Å}^{-1}$ whereas the experimental results (*red dots*) [37] have been taken for $Q = 0.100\,\text{Å}^{-1}$. Reprinted with permission from [36]. Copyright 2013 by the American Physical Society

Many explicit plasmon-pole treatments have been suggested in the literature. Model dielectric matrices for real crystals $\epsilon(\mathbf{q}+\mathbf{G},\mathbf{q}+\mathbf{G}',\omega)$ must however capture important features that determine the screening, such as density inhomogeneities and many-body effects, in addition to the inclusion of an energy gap in the excitation spectrum (13.52). Non-zero off-diagonal elements with $\mathbf{G}\neq\mathbf{G}'$ are needed to account for local-field effects [40, 41].

Combining the plasmon-pole approximation (13.52) with the sum rule (13.38) a straightforward generalization is [26, 42]

$$\begin{aligned}&\epsilon(\mathbf{q}+\mathbf{G},\mathbf{q}+\mathbf{G}',z)\\&=\delta_{\mathbf{G}\mathbf{G}'}+\frac{\omega_p^2}{\bar{\omega}_g^2\left(\left|\mathbf{q}+\frac{1}{2}(\mathbf{G}+\mathbf{G}')\right|\right)-z^2}\,\frac{(\mathbf{q}+\mathbf{G})(\mathbf{q}+\mathbf{G}')}{|\mathbf{q}+\mathbf{G}|^2}\,\frac{\tilde{n}(\mathbf{G}-\mathbf{G}')}{\tilde{n}(0)},\end{aligned}\tag{13.56}$$

where the wave vector $\mathbf{q}+\mathbf{G}$ in the Penn gap in (13.52) is replaced by the arithmetic average of the two wave vectors $\mathbf{q}+\mathbf{G}$ and $\mathbf{q}+\mathbf{G}'$. In some cases it is more convenient to generalize the inverse dielectric function (13.54) instead of the dielectric one in a similar way as in (13.56) but with an effective plasmon frequency which depends on both wave vectors $\mathbf{q}+\mathbf{G}$ and $\mathbf{q}+\mathbf{G}'$. Rohlfing [43] suggests another generalization where the square of the Penn gap is replaced by $\bar{\omega}_g(|\mathbf{q}+\mathbf{G}|)\bar{\omega}_g(|\mathbf{q}+\mathbf{G}'|)$. The advantage of this approach is the possibility for a generalization of (13.56) to mesoscopically inhomogeneous systems. In its original form expression (13.56) is formulated for systems which are characterized by one common dielectric constant without spatial variation. This makes it difficult to employ it for systems with spatially varying screening, like interfaces, heterostructures, molecules in gas phase or adsorbates, etc., on a length scale larger than a typical bond length.

We have to mention that in the majority of the numerical quasiparticle studies in the 80s and 90s of the last century the static dielectric matrices $\epsilon(\mathbf{q}+\mathbf{G}, \mathbf{q}+\mathbf{G}', 0)$ have been calculated within an *ab initio* density functional theory (see e.g. [44, 45]), while the frequency dependence was described in the framework of the plasmon-pole approximation. Still this approach is implemented in some codes such as BerkeleyGW and YAMBO to compute self-energies and quasiparticle electronic structures [46, 47]. For bulk Si a comparison of the full frequency dependence within the SPP approximation with results of an empirical-pseudopotential calculation [28] is shown in Fig. 13.9. The plasmon-pole model of Hybertsen and Louie [48] is used.

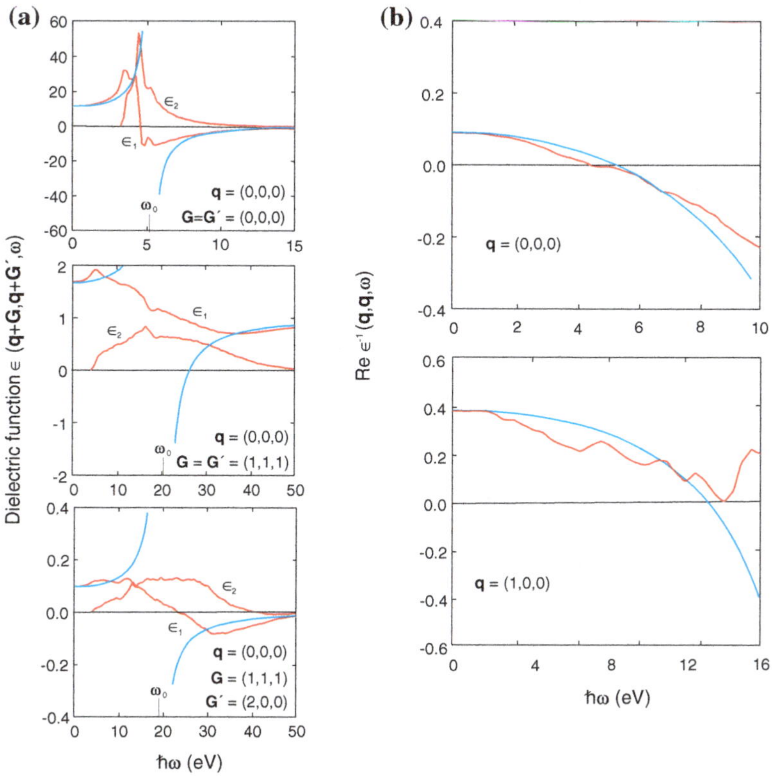

Fig. 13.9 (**a**) Comparison of the frequency dependence of components of the dielectric tensor $\epsilon(\mathbf{q}+\mathbf{G}, \mathbf{q}+\mathbf{G}', \omega)$ as calculated in the framework of an empirical-pseudopotential method (*red line*) [48] within the SPP approximation (*blue line*, real part) for bulk Si. The denotation ϵ_1 (ϵ_2) replaces $\epsilon_{\mathrm{H}}(\mathbf{q}+\mathbf{G}, \mathbf{q}+\mathbf{G}', \omega)$ [$\epsilon_{\mathrm{A}}(\mathbf{q}+\mathbf{G}, \mathbf{q}+\mathbf{G}', \omega)$]. The abbreviation ω_0 means $\bar{\omega}_g(|\mathbf{q}+\frac{1}{2}(\mathbf{G}+\mathbf{G}')|)$. In the *left panel*, for different reciprocal lattice vectors $\mathbf{G} = (1, 1, 1)$ and $\mathbf{G}' = (2, 0, 0)$ symmetrized results are displayed. (**b**) In the reciprocal case only *curves* for the head element $\epsilon^{-1}(\mathbf{q}, \mathbf{q}, \omega)$ are shown. All wave vectors are in units of $2\pi/a_0$. Adapted from [48]

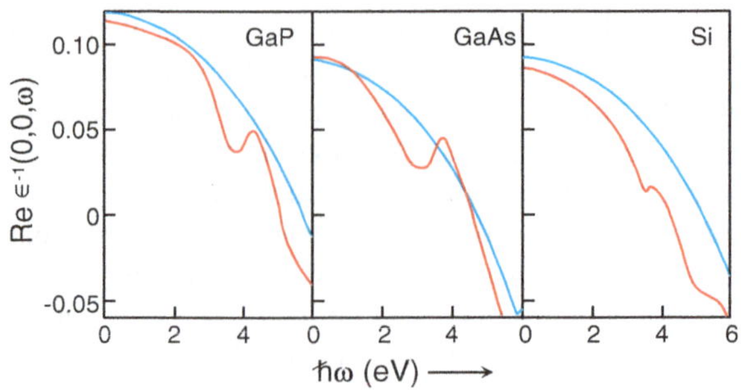

Fig. 13.10 Real part of the diagonal inverse dielectric function $\text{Re}\epsilon^{-1}(0, 0, \omega)$ as constructed from reflectance data (*red lines*) [49, 50] for the three semiconductors GaP, GaAs, and Si. For comparison the results of the plasmon-pole model (13.54) (*blue lines*) are plotted. Adapted from [25]

In Fig. 13.9a the model, that replaces the peak-like structure in $\epsilon_A(\mathbf{q}+\mathbf{G}, \mathbf{q}+\mathbf{G}', \omega)$ by a δ-function, is constructed to describe the limits $\omega \to 0$ and $\omega \to \infty$ of $\epsilon_H(\mathbf{q}+\mathbf{G}, \mathbf{q}+\mathbf{G}', \omega)$ correctly, but may break down for intermediate frequencies. In Fig. 13.9b the average behavior of the inverse dielectric function, the real part of the head element of the dielectric matrix is captured rather well by the plasmon-pole model for energies below the plasmon energy of Si, $\hbar\Omega(0) = 16.7\,\text{eV}$. The latter conclusion is confirmed by Fig. 13.10 [25] in which results of the single-plasmon-pole model (13.54) for the inverse dielectric function are compared with spectra derived from reflectance data [49, 50].

13.3.4 Image Potential Effects

An inhomogeneous electron gas in a more macroscopic sense occurs when two polarizable media are combined to a heterosystem. As a prototypical system we investigate a heterostructure where its both materials 1 and 2 fill one of the two half spaces separated by an interface defined by the plane $z = 0$ as illustrated in Fig. 13.11. We do not consider the spatial inhomogeneities of the materials due to their

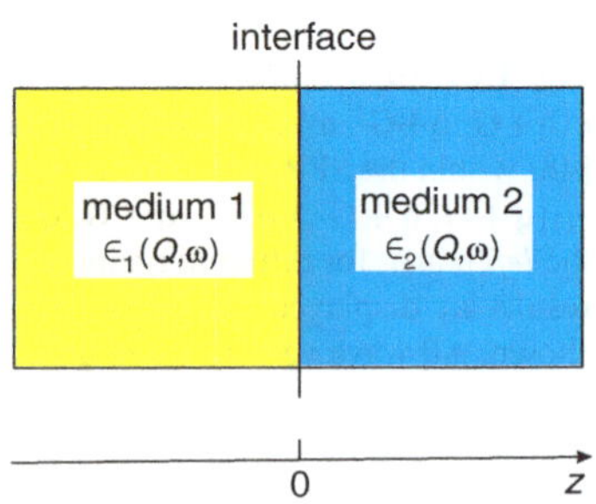

Fig. 13.11 Interface between two polarizable media 1 and 2 (schematic)

atomic geometry. Consequently, their electronic polarizabilities are characterized by diagonal dielectric functions $\epsilon_j(Q,\omega)$ $(j=1,2)$ as indicated in (13.46) or (13.48) with a wave vector $\mathbf{Q}=\mathbf{q}+\mathbf{G}$ running through the entire reciprocal space.

The cylindrical symmetry of the composite system suggests the use of the Fourier transformation in two dimensions instead of a three-dimensional one of the type (13.35). Then, it holds

$$W(\mathbf{x}\mathbf{x}',\omega)=\int\frac{d^2\mathbf{Q}_{||}}{(2\pi)^2}e^{i\mathbf{Q}_{||}(\mathbf{x}_{||}-\mathbf{x}'_{||})}W(Q_{||},z,z';\omega) \tag{13.57}$$

for the screened potential of the composite system that is homogeneous in the planes parallel to the interface as indicated by the $Q_{||}$ dependence. With the definition of the screened potential (12.47) the inverse dielectric function of the inhomogeneous total system is given by

$$\epsilon^{-1}(Q_{||},z,z';\omega)=-\frac{\varepsilon_0}{e^2}\left[\frac{d^2}{dz^2}-Q_{||}^2\right]W(Q_{||},z,z';\omega). \tag{13.58}$$

The screened potential can be easily computed in terms of the macroscopic electrodynamics [51] taking two additional approximations into account: (i) Neglect of the interface polarization (in agreement with the macroscopic electrodynamics), and (ii) application of the condition of specular electron reflection at the interface (see e.g. [52, 53] and references therein). The screened potential $W(Q_{||},z,z';\omega)$ fulfills an integro-differential equation. Its solution is given as

$$\begin{aligned}
W(Q_{||},z,z';\omega)&=\frac{e^2}{2\varepsilon_0 Q_{||}}\\
&\times\Bigg\{\theta(-z)\theta(-z')\bigg[a_1(Q_{||},z-z';\omega)+a_1(Q_{||},z+z';\omega)\\
&-\frac{2a_1(Q_{||},z;\omega)a_1(Q_{||},z';\omega)}{a_1(Q_{||},0;\omega)+a_2(Q_{||},0;\omega)}\bigg]\\
&+\theta(-z)\theta(z')\frac{2a_1(Q_{||},z;\omega)a_2(Q_{||},z';\omega)}{a_1(Q_{||},0;\omega)+a_2(Q_{||},0;\omega)}\\
&+\theta(z)\theta(-z')\frac{2a_2(Q_{||},z;\omega)a_1(Q_{||},z';\omega)}{a_1(Q_{||},0;\omega)+a_2(Q_{||},0;\omega)}\\
&+\theta(z)\theta(z')\bigg[a_2(Q_{||},z-z';\omega)+a_2(Q_{||},z+z';\omega)\\
&-\frac{2a_2(Q_{||},z;\omega)a_2(Q_{||},z';\omega)}{a_1(Q_{||},0;\omega)+a_2(Q_{||},0;\omega)}\bigg]\Bigg\},
\end{aligned} \tag{13.59}$$

where, apart from a factor $e^2/2\varepsilon_0 Q_{||}$,

$$a_j(Q_{||}, z - z'; \omega) = \frac{2Q_{||}}{\pi} \int_0^\infty dQ_z \frac{\cos Q_z(z - z')}{Q^2 \epsilon_j(Q, \omega)} \tag{13.60}$$

represents the screened Coulomb potential of the infinite medium j ($j = 1, 2$).

One important special case is a semi-infinite polarizable solid with a (macroscopic) surface at $z = 0$, i.e., the interface between the vacuum with $\epsilon_2(Q, \omega) = 1$ for $z > 0$ and this medium with $\epsilon_1(Q, \omega) = \epsilon(Q, \omega)$. We present the result only for a wave-vector-independent screening $\epsilon(Q, \omega) \equiv \epsilon(0, \omega)$. It holds

$$a_j(Q_{||}, z - z'; \omega) = \left[\frac{1}{\epsilon(0, \omega)} \delta_{j1} + \delta_{j2}\right] e^{-Q_{||}|z - z'|}. \tag{13.61}$$

The screened potential becomes

$$\begin{aligned} W(Q_{||}, z, z', \omega) = \frac{e^2}{2\varepsilon_0 Q_{||}} \Bigg\{ & \left[\frac{1}{\epsilon(0, \omega)} \theta(-z)\theta(-z') + \theta(z)\theta(z')\right] e^{-Q_{||}|z - z'|} \\ & + \left[\frac{\epsilon(0, \omega) - 1}{\epsilon(0, \omega) + 1} \frac{1}{\epsilon(0, \omega)} \theta(-z)\theta(-z') + \frac{2}{\epsilon(0, \omega) + 1} \left[\theta(-z)\theta(z') + \theta(z)\theta(-z')\right] \right. \\ & \left. - \frac{\epsilon(0, \omega) - 1}{\epsilon(0, \omega) + 1} \theta(z)\theta(z')\right] e^{-Q_{||}(|z| + |z'|)} \Bigg\}. \end{aligned} \tag{13.62}$$

Interestingly, the screened Coulomb interaction of two particles outside the polarizable medium, in the vacuum $z, z' > 0$, is still influenced by the polarization in the other half space $z, z' < 0$. According to (13.57) one finds

$$W(\mathbf{x}\mathbf{x}', \omega) = \mathrm{v}(\mathbf{x} - \mathbf{x}) - \frac{e^2}{4\pi\varepsilon_0} \frac{\epsilon(0, \omega) - 1}{\epsilon(0, \omega) + 1} \frac{1}{\left[\left|\mathbf{x}_{||} - \mathbf{x}'_{||}\right|^2 + (z + z')^2\right]^{\frac{1}{2}}} \tag{13.63}$$

in the vacuum. The bare Coulomb interaction is modified by an image potential term $\sim[\epsilon(0, \omega) - 1]/[\epsilon(0, \omega) + 1]$ for $z, z' > 0$ [51]. As a consequence the correlation of electrons near the surface in the vacuum is significantly changed. Thereby, the image potential effect vanishes for large distances from the surface according to

$$-\frac{e^2}{4\pi\varepsilon_0} \frac{\epsilon(0, \omega) - 1}{\epsilon(0, \omega) + 1} \frac{1}{|z + z'|}. \tag{13.64}$$

As a consequence, surface QP shifts may be enlarged in comparison to corresponding bulk values [54].

The theory developed for heterostructures and the presence of interfaces (13.59) can be easily generalized to double heterostructures [55] and superlattices [56]. We focus on an infinite, periodic arrangement of two polarizable media 1 and 2 with layer thicknesses d_1 and d_2 and bulk dielectric functions $\epsilon_1(Q,\omega)$ and $\epsilon_2(Q,\omega)$, respectively. The lattice constant of the resulting 1D Bravais lattice (of the superlattice arrangement) is $L = d_1 + d_2$. Within the same approximations as used above for an isolated heterointerface, one finds the general results given in [56].

Here we only give results in the limit, in which the wave-vector dependence of the dielectric function can be neglected, i.e., we assume $\epsilon_j(Q,\omega) = \epsilon_j(0,\omega)$ $(j = 1, 2)$. It holds [56, 57]

$$W(Q_{||}, z, z'; \omega) = \frac{e^2}{2\epsilon_2(0,\omega)Q_{||}} \frac{1}{c_- - c_+} \left[\theta(z-z')\phi_-(z)\phi_+(z') + \theta(z'-z)\phi_+(z)\phi_-(z')\right] \tag{13.65}$$

with

$$\begin{aligned}
\phi_\pm(z) &= \sum_{i=1}^{2} \sum_{m=-\infty}^{+\infty} X_i(z - mL - d_1\delta_{i2}) \\
&\quad \times \left[A_i^\pm e^{Q_{||}(z-mL-d_1)} + B_i^\pm e^{-Q_{||}(z-mL-d_1)}\right] e^{\pm m\kappa} \\
X_i(z) &= \theta(z)\theta(d_i - z), \\
A_1^\pm &= a + bc_\pm, \quad B_1^\pm = b + ac_\pm, \quad A_2^\pm = 1, \quad B_2^\pm = c_\pm, \\
a &= \frac{1}{2}\left[1 + \epsilon_2(0,\omega)/\epsilon_1(0,\omega)\right], \quad b = \frac{1}{2}\left[1 - \epsilon_2(0,\omega)/\epsilon_1(0,\omega)\right], \\
c_\pm &= -\left[e^{Q_{||}d_2} - e^{\pm\kappa}(ae^{-Q_{||}d_1} + be^{Q_{||}d_1})\right] \\
&\quad / \left[e^{-Q_{||}d_2} - e^{\pm\kappa}(ae^{Q_{||}d_1} + be^{-Q_{||}d_1})\right].
\end{aligned} \tag{13.66}$$

The actual frequency- and wave-vector-dependent value κ follows from the transcendental equation

$$\begin{aligned}
\cosh(\kappa) &= \frac{[\epsilon_1(0,\omega) + \epsilon_2(0,\omega)]^2}{4\epsilon_1(0,\omega)\epsilon_2(0,\omega)} \cosh\left[Q_{||}(d_1 + d_2)\right] \\
&\quad - \frac{[\epsilon_1(0,\omega) - \epsilon_2(\omega)]^2}{4\epsilon_1(0,\omega)\epsilon_2(0,\omega)} \cosh\left[Q_{||}(d_1 - d_2)\right].
\end{aligned} \tag{13.67}$$

The screened potential (13.65) is directly related to a frequency-dependent inverse dielectric matrix $\epsilon^{-1}(Q_{||}, q+G, q+G'; \omega)$ with G, G' as reciprocal lattice vectors in one dimension. Its diagonal elements are

$$
\begin{aligned}
&\epsilon^{-1}(Q_{||}, q+G, q+G; \omega) \\
&= \frac{1}{L}\sum_{i=1,2}\frac{d_i}{\epsilon_i(0,\omega)} - \left[\frac{1}{\epsilon_1(0,\omega)} - \frac{1}{\epsilon_2(0,\omega)}\right]^2 \frac{1}{Q_{||}L}\frac{Q_{||}^2}{Q_{||}^2+(q+G)^2} \\
&\times \left\{\epsilon_1(0,\omega)\sinh(Q_{||}d_1)\left[\cosh(Q_{||}d_2) - \cosh((q+G)d_2)\right]\right. \\
&\left. + \epsilon_2(0,\omega)\sinh(Q_{||}d_2)\left[\cosh(Q_{||}d_1) - \cosh((q+G)d_1)\right]\right\} \\
&\Big/ \left\{\frac{1}{2}\left[\frac{\epsilon_1(0,\omega)}{\epsilon_2(0,\omega)} + \frac{\epsilon_2(0,\omega)}{\epsilon_1(0,\omega)}\right]^2 \sinh(Q_{||}d_1)\sinh(Q_{||}d_2)\right. \\
&\left. + \cosh(Q_{||}d_1)\cosh(Q_{||}d_2) - \cos(qL)\right\}.
\end{aligned}
\tag{13.68}
$$

The result (13.65) can be applied to describe the screening in two-dimensional systems, e.g. in novel 2D honeycomb sheet crystals graphane, silicane, and germanane, which represent alternately hydrogenated graphene-like group-IV structures [58]. As an example the top and side views of such a sheet crystal are illustrated in Fig. 13.12. Their modeling in *ab initio* studies usually uses superlattice arrangements of such sheets as indicated in Fig. 13.13. In a subsequent step the dielectric properties of 2D isolated crystals are computed via the in-plane component $\epsilon_\perp(\omega)$ of the dielectric tensor of the superlattice arrangement, which is estimated using the RPA-like expression (12.70). The result can be expressed by a 2D electronic polarizability of an isolated sheet as $\alpha_{2D}(\omega) = L[\epsilon_\perp(\omega) - 1]/4\pi$. It can be related to the dielectric function $\epsilon_1(0,\omega) = 1 + 4\pi\alpha_{2D}(\omega)/d$ of the sheet crystal that is identified as medium 1 with a thickness $d_1 = d$. The second medium 2 is the vacuum with $\epsilon_2(0,\omega) = 1$. Numerical values are $\alpha_{2D} = 2.08$, 5.86, and 6.61 a_B for graphane, silicane, and germanane, respectively, sheet crystals (see Fig. 13.12) [58].

Apart from the modeling by means of superlattices, in the majority of studies we are only interested in the screening and the screened potential in an isolated sheet, e.g. that in the superlattice unit cell $m = 0$. Two interacting particles are situated in

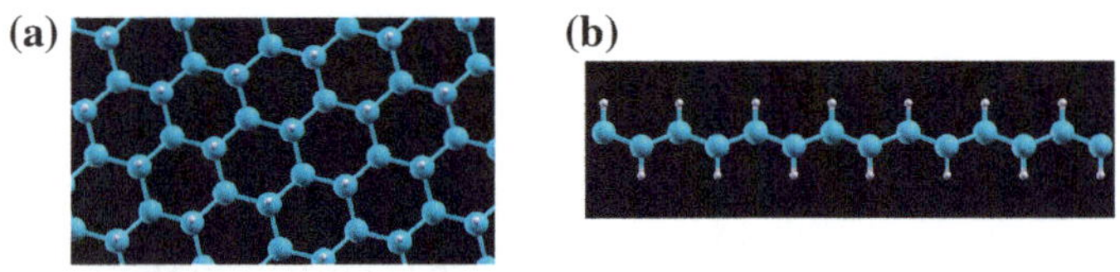

Fig. 13.12 Top (**a**) and side (**b**) view of a hydrogenated honeycomb group-IV crystal, e.g. silicane. Group-IV atoms: *blue circles*, hydrogen atoms: *white dots*

Fig. 13.13 1D superlattice arrangement of 2D sheet crystals (schematic)

the sheet with $0 < z, z' < d$. Because of its two-dimensionality it holds $d \to 0$. Together with the assumptions $\epsilon_1(0,\omega)Q_{\|}d \ll 1$ and $Q_{\|}L \gg 1$, one finds

$$\begin{aligned} W(Q_{\|}, z, z'; \omega) &\approx \frac{e^2}{2\varepsilon_0 Q_{\|}} \frac{1}{\sqrt{1 + [\epsilon_1(0,\omega)Q_{\|}d/2]^2 + [\epsilon_1(0,\omega)Q_{\|}d]\coth(Q_{\|}L)}} \\ &\approx \frac{e^2}{2\varepsilon_0 Q_{\|}} \frac{1}{1 + \epsilon_1(0,\omega)dQ_{\|}/2} \\ &\approx \frac{e^2}{2\varepsilon_0 Q_{\|}} \frac{1}{1 + 2\pi\alpha_{2D}(\omega)Q_{\|}}, \end{aligned} \tag{13.69}$$

expressing the dielectric function of the sheet by the 2D electronic polarizability $\alpha_{2D}(\omega)$. The final expression is derived in the literature [57–60] using different ways. The interaction of two charged particles in such a sheet is almost unscreened. The image potential effects caused by the adjacent vacuum vanish. Only the electronic polarizability of the sheet itself slightly modifies the Coulomb potential in two dimensions.

In real space the screened potential (13.57) of a sheet depends only on the relative in-plane coordinate $|\mathbf{x}_{\|} - \mathbf{x}'_{\|}|$ by

$$W(\mathbf{x}\mathbf{x}', \omega) = \frac{e^2}{4\pi\varepsilon_0|\mathbf{x}_{\|} - \mathbf{x}'_{\|}|} \int_0^{\infty} ds \frac{J_0(s)}{1 + 2\pi\alpha_{2D}(\omega)s/|\mathbf{x}_{\|} - \mathbf{x}'_{\|}|} \tag{13.70}$$

with the Bessel function $J_0(s)$. The integral can be analytically performed [61]. It results [59]

$$W(\mathbf{x}\mathbf{x}', \omega) = \frac{e^2}{16\pi\varepsilon_0\alpha_{2D}(\omega)} \left[H_0\left(\frac{|\mathbf{x}_{\|} - \mathbf{x}'_{\|}|}{2\pi\alpha_{2D}(\omega)}\right) - N_0\left(\frac{|\mathbf{x}_{\|} - \mathbf{x}'_{\|}|}{2\pi\alpha_{2D}(\omega)}\right)\right],$$

where $N_0(s)$ and $H_0(s)$ are the Struve and Neumann functions. For large in-plane distances $|\mathbf{x}_{\|} - \mathbf{x}'_{\|}|$ it becomes the bare 2D Coulomb potential, whereas for small distances it instead describes a potential with a logarithmic singularity.

References

1. L.P. Kadanoff, G. Baym, *Quantum Statistical Mechanics: Green's Function Methods in Equilibrium and Nonequilibrium Problems* (W.A. Benjamin Inc, New York, 1962)
2. H. Stolz, *Einführung in die Vielelektronentheorie der Kristalle* (Akademie-Verlag, Berlin, 1974)
3. H.B. Callen, R.T. Welton, Irreversibility and generalized noise. Phys. Rev. B **83**, 34–40 (1951)
4. R. Kubo, The fluctuation-dissipation theorem. Rep. Prog. Phys. **29**, 255–284 (1966)
5. A.A. Abrikosov, L.P. Gorkov, I.E. Dzyaloshinski, *Methods of Quantum Field Theory in Statistical Physics* (Prentice-Hall Inc, Englewood Cliffs, 1963)
6. M. Cohen, R.P. Feynman, Theory of inelastic scattering of cold neutrons from liquid helium. Phys. Rev. **107**, 13–24 (1957)
7. P. Noziéres, D. Pines, Electron interaction in solids: general formulation. Phys. Rev. **109**, 741–761 (1958)
8. D. Pines, P. Noziéres, *The Theory of Quantum Liquids* (W.A. Benjamin Inc, New York, 1966)
9. A. Sjölander, Spatial correlations and sum rules for a Fermi liquid. Nuovo Cim. B **23**, 124–134 (1974)
10. D.L. Johnson, Local field effects and the dielectric response matrix of insulators: a model. Phys. Rev. B **9**, 4475–4484 (1974)
11. M. Taut, Frequency moments of the dielectric function for an inhomogeneous electron gas. J. Phys. C Sol. State Phys. **18**, 2677–2690 (1985)
12. A.L. Fetter, J.D. Walecka, *Quantum Theory of Many-Particle Systems* (Dover Publ. Inc, Mineola, 2003)
13. L. Hedin, S. Lundqvist, Effects of electron-electron and electron-phonon interactions on the one-electron states of solids, in *Solid State Physics*, vol. 23, ed. by F. Seitz, D. Turnbull, H. Ehrenreich (Academic Press, New York, 1969), pp. 1–181
14. R.L. de Kronig, On the theory of the dispersion of X-rays. J. Opt. Soc. Am. **12**, 547–557 (1926)
15. H.A. Kramers, La diffiusion de la lumiere par les atomes. Atti Cong. Intern. Fisica (Transactions of Volta Centenary Congress) Como **2**, 545–557 (1927)
16. H. Ehrenreich, M.H. Cohen, Self-consistent field approach to the many-electron problem. Phys. Rev. **115**, 786–790 (1959)
17. J. Lindhard, On the properties of a gas of charged particles. Det Kongelige Danske Videnskabernes Selskab, Matematisk-Fysiske Meddelelser **28**, 8–57 (1954)
18. L.D. Landau, On the vibrations of the electronic plasma. Zh. Eksp. Teor. Fiz.**16**, 574–586 (1946) [English Translation: J. Phys. (USSR) **10**, 25–34 (1946)]
19. W. Jones, N.H. March, *Theoretical Solid State Physics*, vol. 1 (Dover Publ. Inc, New York, 1985)
20. N.W. Ashcroft, N.D. Mermin, *Solid State Physics* (Saunders College, Philadelphia, 1976)
21. B.I. Lundqvist, Single-particle spectrum of the degenerate electron gas. I. The structure of the spectral weight function. Phys. kondens. Materie **6**, 193–205 (1967)
22. B.I. Lundqvist, Single-particle spectrum of the degenerate electron gas. II. Numerical results for electrons coupled to plasmons. Phys. kondens. Materie **6**, 206–217 (1967)
23. D.R. Penn, Wave-number-dependent dielectric function of semiconductors. Phys. Rev. **128**, 2093–2097 (1962)
24. J.C. Inkson, Many-body effects at metal-semiconductor junctions. I. Surface plasmons and the electron-electron screened interaction. J. Phys. C **5**, 2599–2610 (1972)
25. F. Bechstedt, R. Enderlein, Electronic polarization (relaxation) effects in the core level spectra of semiconductors. I. General theory of electronic polarization (relaxation) in semiconductors. Phys. Status Solidi B **94**, 239–248 (1979)
26. F. Bechstedt, R. Enderlein, R. Wischnewski, Binding energies and chemical shifts of least bound core electron excitations in cubic $A_N B_{B-N}$ semiconductors. Phys. Status Solidi B **107**, 637–651 (1981)
27. G. Cappellini, R. Del Sole, L. Reining, F. Bechstedt, Model dielectric function for semiconductors. Phys. Rev. B **47**, 9892–9895 (1993)

28. J.P. Walter, M.L. Cohen, Wave-vector-dependent dielectric function for Si, Ge, GaAs, and ZnSe. Phys. Rev. B **2**, 1821–1826 (1979)
29. Z.H. Levine, S.G. Louie, New model dielectric function and exchange-correlation potential for semiconductors and insulators. Phys. Rev. B **25**, 6310–6316 (1982)
30. A. Baldereschi, E. Tosatti, Mean-value point and dielectric properties of semiconductors and insulators. Phys. Rev. B **17**, 4710–4717 (1978)
31. W.G. Aulbur, L. Jönsson, J.W. Wilkins, Quasiparticle calculations in solids, in *Solid State Physics. Advances in Research and Applications*, vol. 54, ed. by H. Ehrenreich, F. Spaepen (Academic Press, San Diego, 2000), pp. 1–218
32. J. Stiebling, H. Raether, Dispersion of the volume plasmon of silicon (16.7 eV) at large wave vectors. Phys. Rev. Lett. **40**, 1293–1295 (1978)
33. B.I. Lundqvist, Single-particle spectrum of the degenerate electron gas. III. Numerical results in the random phase approximation. Phys. kondens. Materie **7**, 117–123 (1968)
34. A. Fleszar, W. Hanke, Spectral properties of quasiparticles in a semiconductor. Phys. Rev. B **67**, 10228–10232 (1997)
35. M.M. Rieger, L. Steinbeck, I.D. White, H.N. Rojas, R.W. Godby, The GW space-time method for the self-energy of large systems. Comput. Phys. Commun. **117**, 211–228 (1999)
36. V. Despoja, D. Novko, K. Dekanič, M. Šunjić, L. Marušić, Two-dimensional and π plasmon spectra in pristine and doped graphene. Phys. Rev. B **87**, 075447 (2013)
37. T. Eberlein, U. Bangert, R.R. Nair, R. Jones, M. Gass, A.L. Bleloch, K.S. Novoselov, A. Geim, P.R. Briddon, Plasmon spectroscopy of free-standing graphene films. Phys. Rev. B **77**, 233406 (2008)
38. W. von der Linden, P. Horsch, Precise quasiparticle energies and Hartree-Fock bands of semiconductors and insulators. Phys. Rev. B **37**, 8351–8362 (1988)
39. G.E. Engel, B. Farid, Generalized plasmon-pole model and plasmon band structures of crystals. Phys. Rev. B **47**, 15931–15934 (1993)
40. S.L. Adler, Quantum theory of the dielectric constant in real solids. Phys. Rev. **126**, 413–420 (1962)
41. N. Wiser, Dielectric constant with local-field effects included. Phys. Rev. **129**, 62–69 (1963)
42. F. Bechstedt, M. Fiedler, G. Kress, R. Del Sole, Model for inverse dielectric matrices of semiconductors. Solid State Commun. **89**, 669–672 (1994)
43. M. Rohlfing, Electronic excitations from a perturbative LDA+GdW approach. Phys. Rev. B **82**, 205127 (2010)
44. M.S. Hybertsen, S.G. Louie, *Ab initio* static dielectric matrices from the density-functional approach. I. Formulation and application to semicondcutors and insulators. Phys. Rev. B **35**, 5585–5601 (1987)
45. M.S. Hybertsen, S.G. Louie, *Ab initio* static dielectric matrices from the density-functional approach. II. Calculation of the screening response in diamond, Si, Ge, and LiCl. Phys. Rev. B **35**, 5602–5610 (1987)
46. http://www.berkeleygw.org
47. http://www.yambo-code.org
48. M.S. Hybertsen, S.G. Louie, Electron correlation in semiconductors and insulators: band gaps and quasiparticle energies. Phys. Rev. B **34**, 5390–5413 (1986)
49. H.R. Philipp, H. Ehrenreich, Optical properties of semiconductors. Phys. Rev. **129**, 1550–1560 (1963)
50. H.R. Philipp, Optical properties of non-crystalline Si, SiO, SiO_x and SiO_2. J. Phys. Chem. Solids **32**, 1935–1945 (1971)
51. L.D. Landau, E.M. Lifshitz, *Electrodynamics of Continua*, vol. 8 (Pergamon Press, Oxford, 1959)
52. F. Bechstedt, R. Enderlein, D. Reichardt, Inverse dielectric function for a semi-infinite solid. Phys. Status Solidi B **117**, 261–270 (1983)
53. F. Bechstedt, R. Enderlein, D. Reichardt, Electronic relaxation effects in core level spectra of surfaces and interfaces. Phys. Status Solidi B **118**, 327–336 (1983)

54. F. Bechstedt, R. Del Sole, Giant quasiparticle shifts of semiconductor surface states. Solid State Commun. **74**, 41–44 (1990)
55. F. Bechstedt, R. Enderlein, Dielectric screening, polar phonons, and longitudinal electronic excitations of quantum well double heterostructures. Phys. Status Solidi B **131**, 53–66 (1985)
56. F. Bechstedt, R. Enderlein, Inverse dielectric function of a superlattice including local field effects and spatial dispersion. Superlattices Microstruct. **2**, 543–549 (1986)
57. R.R. Guseinov, Coulomb interaction and excitons in superlattices. Phys. Status Solidi B **125**, 237–243 (1984)
58. O. Pulci, P. Gori, M. Marsili, V. Garbuio, R. Del Sole, F. Bechstedt, Strong excitons in novel two-dimensional crystals: silicane and germanane. Europhys. Lett. **98**, 37004 (2012)
59. L.V. Keldysh, Coulomb interaction in thin semiconductor and semimetal films. Pis'ma Zh. Eksp. Teor. Fiz. **29**, 716–719 (1979) [English translation: JETP Lett. **29**, 658–661 (1980)]
60. P. Cudazzo, C. Attaccalite, I.V. Tokatly, A. Rubio, Strong charge-transfer excitonic effects and the Bose-Einstein exciton condensate in graphane. Phys. Rev. Lett. **104**, 226804 (2010)
61. I.S. Gradstein, I.M. Ryshik, *Sum, Product and Integral Tables*, vol. 1 (Verlag Harri Deutsch, Thun, 1981)

Chapter 14
Self-energy

Abstract Single-particle excitations cannot be described by non-interacting particles with infinite lifetime. Rather, due to interactions with other particles they are 'dressed' as expressed by their self-energy. In contrast to the solutions of the Hartree-Fock and Kohn-Sham equations, the Dyson equation leads to quasiparticles. The dynamics of the screening reaction, i.e., the frequency-dependent correlation contribution to the self-energy, is responsible for spectral functions which differ from a Dirac δ-function at a certain energy. Rather, a Lorentzian-broadened peak at a shifted energy with reduced spectral weight may occur. It represents a quasiparticle with finite lifetime. The rest of the spectral weight appears in incoherent spectral contributions at other energies. The description of quasiparticles requires a self-consistent procedure since the self-energy and the screening/vertex functions appearing therein depend on the unknown Green function. If one is mainly interested in the energy position and spectral weight of the main quasiparticle peak, the standard approach is based on the GW approximation and a starting electronic structure derived from a (generalized) Kohn-Sham equation. We demonstrate that this procedure leads to single-particle excitation energies in good agreement with measured values. This holds especially for the opening of the fundamental gap of non-metals.

14.1 Quasiparticle Picture

14.1.1 Reference System

We use the solutions of the Kohn-Sham equations or generalized Kohn-Sham equations. In the most general form they describe electron systems with non-collinear spins as discussed at the end of Sect. 6.2.2 with a certain spin-dependent potential as derived in (12.3). We also allow for a spatially non-local XC potential (4.27) as derived in the Hartree-Fock approach (Sect. 4.2.3) or in the generalized Kohn-Sham description (9.22). In this most general case the XC potential could be also spin-dependent. We arrive at gKS equations

F. Bechstedt, *Many-Body Approach to Electronic Excitations*,
Springer Series in Solid-State Sciences 181, DOI 10.1007/978-3-662-44593-8_14

$$\sum_{s'}\int d^3x' \left\{\left[-\frac{\hbar^2}{2m}\Delta_{\mathbf{x}}\delta_{ss'} + \tilde{V}_{ss'}(\mathbf{x}) + V_{\mathrm{H}}(\mathbf{x})\delta_{ss'}\right]\delta(\mathbf{x}-\mathbf{x}') + V_{\mathrm{XC}}^{ss'}(\mathbf{x},\mathbf{x}')\right\}\phi_\Lambda(\mathbf{x}',s')$$
$$= \varepsilon_\Lambda \phi_\Lambda(\mathbf{x},s') \tag{14.1}$$

with the orthonormalized and complete set of single-particle eigenstates

$$\sum_s \int d^3\mathbf{x}\phi_\Lambda^+(\mathbf{x},s)\phi_{\Lambda'}(\mathbf{x},s) = \delta_{\Lambda\Lambda'},$$
$$\sum_\Lambda \phi_\Lambda(\mathbf{x},s)\phi_\Lambda^+(\mathbf{x}',s') = \delta(\mathbf{x}-\mathbf{x}')\delta_{ss'} \tag{14.2}$$

and their eigenvalues ε_Λ.

Nevertheless, in order to reduce the formalism and to focus more on the physics of excitations in interacting systems, we make some simplifying assumptions: (i) The non-collinearity is omitted. Then the Pauli spinors factorize according to (4.6) or (12.69) as

$$\phi_\Lambda(\mathbf{x},s) = \varphi_{\lambda m_s}(\mathbf{x})\chi_{\frac{1}{2}m_s}(s) \tag{14.3}$$

with the set of quantum numbers $\Lambda = \lambda m_s$, the orbitals $\varphi_{\lambda m_s}(\mathbf{x})$, and the spin functions $\chi_{\frac{1}{2}m_s}(s)$. (ii) In accordance, the off-diagonal elements of the spin-dependent potential $\tilde{V}_{ss'}(\mathbf{x})$ (10.10) are omitted in (14.1), i.e., we set $V_{ss}(\mathbf{x})\delta_{ss'}$. (iii) The non-locality of the XC potential with respect to space and spin coordinates is omitted formally but still may be included in explicit computations. Then Kohn-Sham/generalized KS/KS-like equations of the type

$$\left[-\frac{\hbar^2}{2m}\Delta_{\mathbf{x}} + V^{m_s}(\mathbf{x}) + V_{\mathrm{H}}(\mathbf{x}) + V_{\mathrm{XC}}^{m_s}(\mathbf{x})\right]\varphi_{\lambda m_s}(\mathbf{x}) = \varepsilon_{\lambda m_s}\varphi_{\lambda m_s}(\mathbf{x}), \tag{14.4}$$

with e.g. $V^{m_s}(\mathbf{x}) = V_{\mathrm{n}}(\mathbf{x})$ in (6.20), define the spatial orbitals of the simplified system.

Another specification may be again the consideration of translationally invariant systems with $\lambda = \nu\mathbf{k}$ (ν - band index, $\mathbf{k}$ - Bloch wave vector $\in$ BZ) and Bloch energies $\varepsilon_{\lambda m_s} \to \varepsilon_{\nu m_s}(\mathbf{k})$. Each field operator can be represented in analogy to (4.6) in terms of these Bloch functions as

$$\psi_s(\mathbf{x},t) = \sum_{\nu,\mathbf{k},m_s} \varphi_{\nu\mathbf{k}m_s}(\mathbf{x})\chi_{\frac{1}{2}m_s}(s)\hat{a}_{\nu\mathbf{k}m_s}(t) \tag{14.5}$$

with an annihilation operator $\hat{a}_{\nu\mathbf{k}m_s}$ of an electron in a reference single-particle state that is characterized by the quantum numbers $\nu\mathbf{k}m_s$. Similar expansions hold for creation operators.

Because of the known time dependence $\hat{a}_{\nu\mathbf{k}m_s}(t) = \hat{a}_{\nu\mathbf{k}m_s}\exp\left[-\frac{i}{\hbar}\varepsilon_{\nu m_s}(\mathbf{k})t\right]$ in the Heisenberg representation, the corresponding single-particle Green function can be easily determined. The Fourier coefficients of the corresponding thermodynamic Green function, referred to as $\tilde{G}_{ss'}(\mathbf{xx}', z)$ (11.21), obey the representation

$$\tilde{G}_{ss'}(\mathbf{xx}', z) = \sum_{\nu,\nu'} \sum_{\mathbf{k}} \sum_{m_s,m_s'} \tilde{G}_{\nu\nu'}^{m_s m_s'}(\mathbf{k}, z) \varphi_{\nu\mathbf{k}m_s}(\mathbf{x}) \varphi^*_{\nu'\mathbf{k}m_s'}(\mathbf{x}') \chi_{\frac{1}{2}m_s}(s) \chi^+_{\frac{1}{2}m_s'}(s') \tag{14.6}$$

with the expansion coefficients

$$\tilde{G}_{\nu\nu'}^{m_s m_s'}(\mathbf{k}, z) = \frac{\delta_{\nu\nu'} \delta_{m_s m_s'}}{\hbar z - \varepsilon_{\nu m_s}(\mathbf{k})}, \tag{14.7}$$

which are diagonal in the set of the single-particle quantum numbers $\nu\mathbf{k}m_s$ for different reasons. The diagonality in the Bloch wave vectors (which is taken into account in (14.6) from the very beginning) is a consequence of the translational symmetry $\tilde{G}_{ss'}(\mathbf{x} + \mathbf{R}\mathbf{x}' + \mathbf{R}, z) = \tilde{G}_{ss'}(\mathbf{xx}', z)$, that in the spin-quantum numbers is due to the assumed collinearity, and finally that in the band indices is related to the fact that the single-particle Hamiltonian in the KS/gKS/KS-like (14.4) rules the equation of motion for $\tilde{G}_{ss'}$.

The XC self-energy (12.52) of a fermion can be easily studied for time differences along the imaginary time axis. As a consequence its times dependence can be represented by a Fourier expansion

$$\Sigma_{ss'}(\mathbf{xx}', t - t') = \frac{1}{-i\hbar\beta} \sum_n \Sigma_{ss'}(\mathbf{xx}', z_n) e^{-iz_n(t-t')}, \tag{14.8}$$

similar to that for the Green function (11.18) with the Matsubara frequencies (11.20). In a further step the Fourier frequencies in (14.8) can be analytically continued into the complex z-plane to define a function $\Sigma_{ss'}(\mathbf{xx}', z)$ following the same procedure as in Sect. 11.1.4 for $G_{ss'}(\mathbf{xx}', z)$. The spatial behavior of the spin-averaged self-energy is illustrated in Fig. 14.1 for silicon with $\mathbf{x}$ fixed at a bond center and a tetrahedral interstitial site (see Fig. 3.1). The frequency is chosen in a midgap position [1].

With the Green function $\tilde{G}$ (14.6) that includes exchange and correlation in a certain approximation, e.g. by $V_{\mathrm{XC}}^{ss'}(\mathbf{x}, \mathbf{x}')$ as in (14.1), the Dyson equation of type (12.51) turns into a Dyson equation

$$\begin{aligned} G_{ss'}(\mathbf{xx}', z) &= \tilde{G}_{ss'}(\mathbf{xx}', z) \\ &+ \sum_{s_1,s_2} \int d^3\mathbf{x}_1 \int d^3\mathbf{x}_2 \tilde{G}_{ss_1}(\mathbf{xx}_1, z) \delta\Sigma_{s_1 s_2}(\mathbf{x}_1\mathbf{x}_2, z) G_{s_2 s'}(\mathbf{x}_2\mathbf{x}', z) \end{aligned} \tag{14.9}$$

with the reduced XC kernel

$$\delta\Sigma_{ss'}(\mathbf{xx}', z) = \Sigma_{ss'}(\mathbf{xx}', z) - V_{\mathrm{XC}}^{ss'}(\mathbf{x}, \mathbf{x}') \tag{14.10}$$

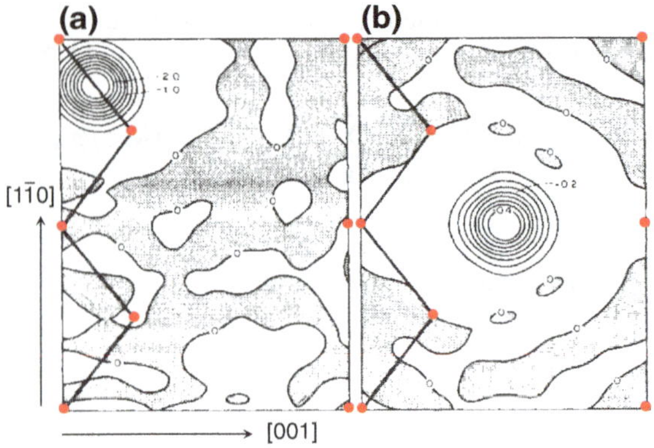

Fig. 14.1 Contour plots of the spin-averaged self-energy $\Sigma(\mathbf{r}, \mathbf{r}', \omega = \text{midgap})$ for silicon with $\mathbf{x}$ fixed at (**a**) a bond center and (**b**) a tetrahedral interstitial site, and varying $\mathbf{x}'$ shown in the (110) plane containing a bond chain of Si atoms indicated by *red dots* and *solid lines*. Adapted from [1]

for the time-Fourier-transformed single-particle Green function G. Expression (14.10) indicates that XC is already included in the function $\tilde{G}$ on a certain level defined by $V_{\text{XC}}^{ss'}(\mathbf{x}, \mathbf{x}')$ as starting point to treat XC in the excited states.

Under the simplifying assumptions leading to the representation (14.3) with the orbitals $\varphi_{\nu\mathbf{k}m_s}(\mathbf{x})$ (14.4) one finds a Bloch-spin representation similar to (14.6) for the XC self-energy (12.52)

$$\Sigma_{\nu\nu'}^{m_s m_s'}(\mathbf{k}, z) = \sum_{s,s'} \int d^3\mathbf{x} \int d^3\mathbf{x}' \varphi^*_{\nu\mathbf{k}m_s}(\mathbf{x}) \chi^+_{\frac{1}{2}m_s}(s) \Sigma_{ss'}(\mathbf{x}\mathbf{x}', z) \varphi_{\nu'\mathbf{k}m_s'}(\mathbf{x}') \chi_{\frac{1}{2}m_s'}(s') \tag{14.11}$$

and the same representation $\delta\Sigma_{\nu\nu'}^{m_s m_s'}(\mathbf{k}, z)$ for the self-energy difference (14.10).

Within the representation (14.6) for the Green function $\tilde{G}$ and (14.11) for the self-energy difference $\delta\Sigma$ it can be shown that the full Green function G can be also represented by an expression similar to (14.6) but with unknown and off-diagonal coefficients $G_{\nu\nu'}^{m_s m_s'}(\mathbf{k}, z)$. Then, the Dyson equation (14.9) transforms into a matrix equation

$$G_{\nu\nu'}^{m_s m_s'}(\mathbf{k}, z) = \tilde{G}_{\nu\nu}^{m_s m_s}(\mathbf{k}, z) \left\{ \delta_{\nu\nu'} \delta_{m_s m_s'} + \sum_{\nu'', m_s''} \delta\Sigma_{\nu\nu''}^{m_s m_s''}(\mathbf{k}, z) G_{\nu''\nu'}^{m_s'' m_s'}(\mathbf{k}, z) \right\}. \tag{14.12}$$

Because of (11.21) the matrix elements of the Green function fulfill a spectral representation

$$G_{\nu\nu'}^{m_s m_s'}(\mathbf{k}, z) = \int_{-\infty}^{+\infty} \frac{d\omega}{2\pi} \frac{A_{\nu\nu'}^{m_s m_s'}(\mathbf{k}, \omega)}{z - \omega} \tag{14.13}$$

with a Bloch-spin representation of the spectral-weight function

$$A_{\nu\nu'}^{m_s m_s'}(\mathbf{k},\omega) = \sum_{s,s'} \int d^3\mathbf{x} \int d^3\mathbf{x}' \varphi^*_{\nu\mathbf{k}m_s}(\mathbf{x}) \chi^+_{\frac{1}{2}m_s}(s) A_{ss'}(\mathbf{x}\mathbf{x}',\omega) \varphi_{\nu'\mathbf{k}m_s'}(\mathbf{x}') \chi_{\frac{1}{2}m_s'}(s'). \tag{14.14}$$

The definition of the spectral function as the discontinuity of the Green function along the real axis in the complex z-plane (11.22) leads to a similar relation for its matrix elements

$$A_{\nu\nu'}^{m_s m_s'}(\mathbf{k},\omega) = i \lim_{\eta\to+0} \left\{ G_{\nu\nu'}^{m_s m_s'}(\mathbf{k},\omega+i\eta) - G_{\nu\nu'}^{m_s m_s'}(\mathbf{k},\omega-i\eta) \right\}. \tag{14.15}$$

14.1.2 Approximate Spectral Function in Insulators

The XC self-energy Σ (12.13) as well as the self-energy difference $\delta\Sigma$ (14.11) can be divided into a static background contribution, e.g. due to exchange, and a dynamical one, mostly due to correlation. In the case of their matrix elements a possible description could be

$$\delta\Sigma_{\nu\nu'}^{m_s m_s'}(\mathbf{k},z) = \delta_{\nu\nu'}^{m_s m_s'}(\mathbf{k}) + \int_{-\infty}^{+\infty} \frac{d\omega}{2\pi} \frac{\Gamma_{\nu\nu'}^{m_s m_s'}(\mathbf{k},\omega)}{z-\omega}, \tag{14.16}$$

with a static contribution $\delta_{\nu\nu'}^{m_s m_s'}(\mathbf{k})$ and a spectral function $\Gamma_{\nu\nu'}^{m_s m_s'}(\mathbf{k},\omega)$ of the dynamical contribution. Thereby, for physical reasons it should hold $\Gamma_{\nu\nu}^{m_s m_s}(\mathbf{k},\omega) > 0$ for its diagonal elements. Together with the expansion coefficients (14.7) the Dyson equation (14.12) can be transformed into

$$\begin{aligned} &\left\{\hbar\omega \pm i\eta - \varepsilon_{\nu m_s}(\mathbf{k})\right\} G_{\nu\nu'}^{m_s m_s'}(\mathbf{k},\omega\pm i\eta) \\ &\quad - \sum_{\nu'',m_s''} \left\{ V_{\nu\nu''}^{m_s m_s''}(\mathbf{k},\omega) \mp \frac{i}{2}\Gamma_{\nu\nu''}^{m_s m_s''}(\mathbf{k},\omega) \right\} G_{\nu''\nu'}^{m_s'' m_s'}(\mathbf{k},\omega\pm i\eta) \\ &= \delta_{\nu\nu'}\delta_{m_s m_s'} \end{aligned} \tag{14.17}$$

using the Weierstrass formula (11.24) and the abbreviation

$$V_{\nu\nu'}^{m_s m_s'}(\mathbf{k},\omega) = \delta_{\nu\nu'}^{m_s m_s'}(\mathbf{k}) + \mathcal{P}\int_{-\infty}^{+\infty} \frac{d\omega'}{2\pi} \frac{\Gamma_{\nu\nu'}^{m_s m_s'}(\mathbf{k},\omega')}{\omega-\omega'}. \tag{14.18}$$

For a discussion of the spectral function we choose such a XC potential so that the interband matrix elements of the self-energy difference are negligibly small.

We assume that the eigenfunctions (14.3) and eigenvalues in (14.4) nearly describe the single-particle states in the fully interacting electron system. Then, the interband contributions $\nu \neq \nu'$ and the coupling terms $m_s \neq m_s'$ of the two spin channels can be neglected, at least on the single-particle level. It holds

$$\left.\begin{aligned} \left|\Gamma_{\nu\nu'}^{m_s m_s'}(\mathbf{k},\omega)\right| &\ll \left|\varepsilon_{\nu m_s}(\mathbf{k}) - \varepsilon_{\nu' m_s'}(\mathbf{k})\right|, \\ \left|V_{\nu\nu'}^{m_s m_s'}(\mathbf{k},\omega)\right| &\ll \left|\varepsilon_{\nu m_s}(\mathbf{k}) - \varepsilon_{\nu' m_s'}(\mathbf{k})\right| \end{aligned}\right\} \quad \text{for } \nu \neq \nu' \text{ and/or } m_s \neq m_s'.$$

In the case of degenerate states one has to perform a proper transformation in the subspace of degenerate states, in order to still decouple the equations in (14.17). Neglecting these coupling terms the Dyson equation (14.17) can be formally solved for $G_{\nu\nu}^{m_s m_s}(\mathbf{k}, \omega \pm i\eta)$. With (14.15) the diagonal elements of the spectral function obey the expression

$$A_{\nu\nu}^{m_s m_s}(\mathbf{k},\omega) = \frac{\Gamma_{\nu\nu}^{m_s m_s}(\mathbf{k},\omega)}{\left[\hbar\omega - \varepsilon_{\nu m_s}(\mathbf{k}) - V_{\nu\nu}^{m_s m_s}(\mathbf{k},\omega)\right]^2 + \left[\frac{1}{2}\Gamma_{\nu\nu}^{m_s m_s}(\mathbf{k},\omega)\right]^2}. \tag{14.19}$$

This result is consistent with the assumption that the Bloch spinors (14.3) diagonalize the spectral function $A_{ss'}(\mathbf{x}\mathbf{x}',\omega)$ (11.22). For a non-spin-polarized system, for which the spectral-weight function does not depend on the spin quantum number m_s and the Hedin GW approximation (12.57) for the XC self-energy, the spectral function, resulting for a bulk silicon crystal, is displayed in Fig. 14.2 for varying Bloch wave vector $\mathbf{k}$ versus the particle energy $\hbar\omega$. The variations of the positions of the main peaks with $\mathbf{k}$ represent quasiparticle energies for a given band ν. The broad features at lower energies in the case of the occupied valence states indicate the appearance of satellite structures in the spectral function. The four valence bands at negative energies are clearly recognizable.

In general, the spectral variation of a spectral function $A_{ss'}(\mathbf{x}\mathbf{x}',\omega)$ of an interacting electron gas deviates from that of a "non-interacting" system described by a representation of the type (12.60). Instead of a weighted sum of Dirac δ-functions at a defined single-particle energy ε_A or $\varepsilon_{\nu m_s}(\mathbf{k})$ sums of distribution functions (14.19) appear. The corresponding single-particle excitations are not anymore described by a defined energy and an infinite lifetime. Rather, they represent the excitation of so-called *quasiparticles* (QPs) [3–5], whose physical character may be intuitively discussed [5]. A better approach is a description based on the Green function G and the spectral-weight function A. On the other hand, the spectral distribution described by A (14.19) can be also better interpreted in terms of isolated poles of the Green function.

We begin such an interpretation with a fictitious time-dependent QP wave function $\varphi_{\nu\mathbf{k}m_s}^{\mathrm{QP}}(\mathbf{x},t)$. For this purpose we add a bare particle in the single-particle state $|\nu\mathbf{k}m_s\rangle$ to the system. It will gather a screening cloud around it, and become a quasiparticle, more precisely a quasielectron, because of the particle addition. The addition process is best described by a propagator of a particle in the state $|\nu\mathbf{k}m_s\rangle$. With the creation operator $\hat{a}_{\nu\mathbf{k}m_s}^{+}$ (14.5), for a particle that is added at time t to the system in a state

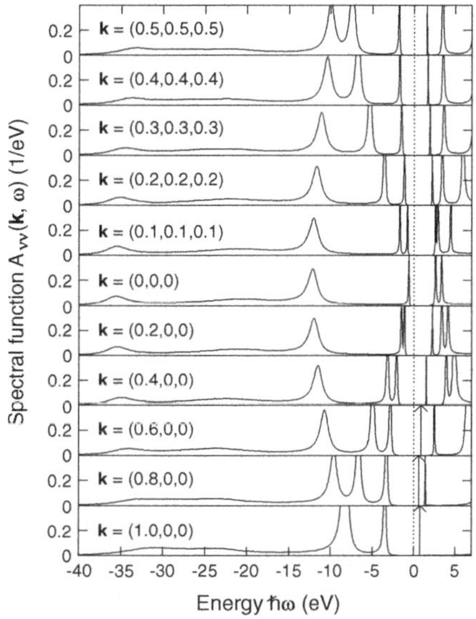

Fig. 14.2 The spectral-weight functions for quasielectrons and quasiholes in bulk Si crystallizing in diamond structure. The Bloch wave vector **k** varies along the $L\Gamma$ and ΓX directions. It is given in units of $2\pi/a_0$. The energy zero (*dotted vertical line*) is fixed at the energy of the topmost occupied DFT-LDA KS state at Γ. *Arrows* indicate undamped quasiparticle peaks. They are δ-functions with a spectral weight smaller than 1. The main peaks occur at quasiparticle energies while the broad features at lower energies represent satellite structures. Reprinted with permission from [2].

$|\nu\mathbf{k}m_s\rangle$, one finds the propagator

$$\begin{aligned}\langle\psi_s(\mathbf{x},t)\hat{a}^+_{\nu\mathbf{k}m_s}(t')\rangle &= i\hbar\sum_{s'}\int d^3\mathbf{x}' G^>_{ss'}(\mathbf{x}\mathbf{x}', t-t')\varphi_{\nu\mathbf{k}m_s}(\mathbf{x}')\chi_{\frac{1}{2}m_s}(s') \\ &= \varphi_{\nu\mathbf{k}m_s}(\mathbf{x})\chi_{\frac{1}{2}m_s}(s)\int\limits_{-\infty}^{+\infty}\frac{d\hbar\omega}{2\pi}\left[1-f(\hbar\omega)\right]A^{m_sm_s}_{\nu\nu}(\mathbf{k},\omega)e^{-i\omega(t-t')}.\end{aligned}$$

Here the expressions and definitions (11.1), (11.11), (11.14), and (14.5) have been applied.

We simplify the discussion for $T = 0\,\mathrm{K}$, the spin-averaged case, $t' = 0$, and $t > 0$. The right-hand side represents a time-dependent wave function of a quasiparticle, more precisely, of a quasielectron with energy $\hbar\omega > \mu$. Its orbital part is given by

$$\varphi^{\mathrm{QP}}_{\nu\mathbf{k}m_s}(\mathbf{x},t) = \sum_s\chi^+_{\frac{1}{2}m_s}(s)\langle\psi_s(\mathbf{x},t)a^+_{\nu\mathbf{k}m_s}\rangle = \varphi_{\nu\mathbf{k}m_s}(\mathbf{x})\int\limits_{\mu}^{\infty}\frac{d\hbar\omega}{2\pi}A^{m_sm_s}_{\nu\nu}(\mathbf{k},\omega)e^{-i\omega t}. \tag{14.20}$$

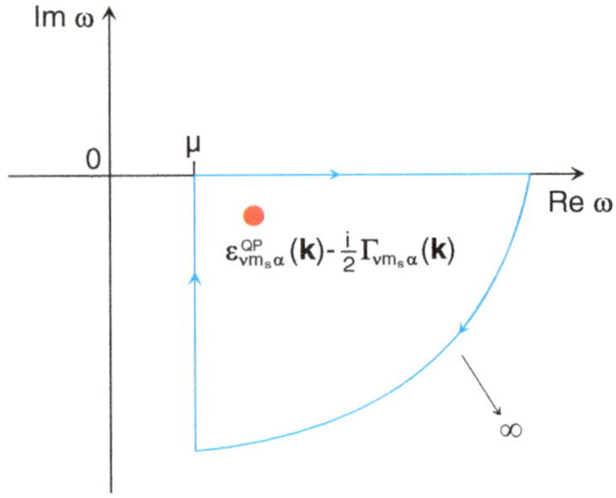

Fig. 14.3 Complex ω-plane with integration contour to determine the propagator in (14.20). One of the poles of the spectral function (14.19) is indicated

The frequency integral in (14.20) can be replaced by a contour integral in the complex ω-plane with the contour as given in Fig. 14.3, since the segment of the circle with infinite radius in the fourth quadrant does not give a finite contribution. For positive t the closed contour contains poles labeled by the index α

$$\varepsilon^{QP}_{\nu m_s\alpha}(\mathbf{k}) - \frac{i}{2}\Gamma_{\nu m_s\alpha}(\mathbf{k}) \tag{14.21}$$

of the spectral-weight function (14.19) with the residues $iz_{\nu m_s\alpha}(\mathbf{k})$. Applying the Cauchy theorem (14.20) becomes

$$\varphi^{QP}_{\nu\mathbf{k}m_s}(\mathbf{x},t) = \varphi_{\nu\mathbf{k}m_s}(\mathbf{x})\left\{\sum_\alpha \theta\left(\varepsilon^{QP}_{\nu m_s\alpha}(\mathbf{k})-\mu\right) z_{\nu m_s\alpha}(\mathbf{k}) e^{-\frac{i}{\hbar}\left[\varepsilon^{QP}_{\nu m_s\alpha}(\mathbf{k})-\frac{i}{2}\Gamma_{\nu m_s\alpha}(\mathbf{k})\right]t} + \int_\mu^{\mu-i\infty} \frac{d\hbar\omega}{2\pi} A^{m_s m_s}_{\nu\nu}(\mathbf{k},\omega) e^{-i\omega t}\right\}. \tag{14.22}$$

The result (14.22) shows that, in general, and in contrast to the reference electronic system with eigenfunctions $\varphi_{\nu\mathbf{k}m_s}(\mathbf{x})$ and eigenvalues $\varepsilon_{\nu m_s}(\mathbf{k})$, the QP wave function does not have the character of an unperturbed single-particle function. Its time dependence is much more complex.

We discuss the time dependence in (14.22) under some simplifying assumptions. The pole with the smallest value $\Gamma_{\nu m_s}(\mathbf{k})$ of all $\Gamma_{\nu m_s\alpha}(\mathbf{k})$, characterized by the energy $\varepsilon^{QP}_{\nu m_s}(\mathbf{k})$ and the residue $iz_{\nu m_s}(\mathbf{k})$, says that for times

$$t \approx \frac{\hbar}{\Gamma_{\nu m_s}(\mathbf{k})}$$

the contributions of the other poles can be only approximately taken into account. Because of the fact that $A^{m_s m_s}_{\nu\nu}(\mathbf{k},\omega)$ is real, the pole $\varepsilon^{QP}_{\nu m_s}(\mathbf{k}) - \frac{i}{2}\Gamma_{\nu m_s}(\mathbf{k})$ is accompanied by a complex conjugated pole $\varepsilon^{\mathrm{QP}}_{\nu m_s}(\mathbf{k}) + \frac{i}{2}\Gamma_{\nu m_s}(\mathbf{k})$. Together with the assumption that the other poles $\varepsilon^{\mathrm{QP}}_{\nu m_s \alpha}(\mathbf{k})$ are more far away from μ than $\varepsilon^{\mathrm{QP}}_{\nu m_s}(\mathbf{k})$, the spectral-weight function (14.19) can be approximately described by

$$A^{m_s m_s}_{\nu\nu}(\mathbf{k},\omega) = z_{\nu m_s}(\mathbf{k}) \frac{\Gamma_{\nu m_s}(\mathbf{k})}{\left[\hbar\omega - \varepsilon^{\mathrm{QP}}_{\nu m_s}(\mathbf{k})\right]^2 + \left[\frac{1}{2}\Gamma_{\nu m_s}(\mathbf{k})\right]^2} + a^{m_s m_s}_{\nu\nu}(\mathbf{k},\omega) \quad (14.23)$$

with some (incoherent) background $a^{m_s m_s}_{\nu\nu}(\mathbf{k},\omega)$ that guarantees the sum rule (11.17). The representation (14.23) allows to evaluate the second contribution in (14.22) as [6]

$$\begin{aligned}
\int\limits_{\mu}^{\mu-i\infty} \frac{d\hbar\omega}{2\pi} A^{m_s m_s}_{\nu\nu}(\mathbf{k},\omega) e^{-i\omega t} &= \frac{1}{2\pi i} e^{-\frac{i}{\hbar}\mu t} \int\limits_0^{\infty} ds \frac{z_{\nu m_s}(\mathbf{k}) \Gamma_{\nu m_s}(\mathbf{k}) e^{-\frac{st}{\hbar}}}{\left[\mu - \varepsilon^{\mathrm{QP}}_{\nu m_s}(\mathbf{k}) - is\right]^2 + \left[\frac{1}{2}\Gamma_{\nu m_s}(\mathbf{k})\right]^2} \\
&\quad + \int\limits_{\mu}^{\mu-i\infty} \frac{d\hbar\omega}{2\pi} a^{m_s m_s}_{\nu\nu}(\mathbf{k},\omega) e^{-i\omega t} \\
&\approx \frac{\hbar z_{\nu m_s}(\mathbf{k})}{2\pi i t} \frac{\Gamma_{\nu m_s}(\mathbf{k})}{\left[\mu - \varepsilon^{\mathrm{QP}}_{\nu m_s}(\mathbf{k})\right]^2 + \left[\frac{1}{2}\Gamma_{\nu m_s}(\mathbf{k})\right]^2} e^{-\frac{i}{\hbar}\mu t}
\end{aligned}$$

for $t \gg \frac{\hbar}{\left|\varepsilon^{\mathrm{QP}}_{\nu m_s}(\mathbf{k}) - \mu\right|}$. This relation should be fulfilled in insulators with the chemical potential in the forbidden energy region, the fundamental gap. The second contribution is smaller than the first term by the factor $\Gamma_{\nu m_s}(\mathbf{k}) / \left|\varepsilon^{\mathrm{QP}}_{\nu m_s}(\mathbf{k}) - \mu\right|$. Together with an aforementioned condition of large times, in the time interval

$$\frac{\hbar}{\left|\varepsilon^{\mathrm{QP}}_{\nu m_s}(\mathbf{k}) - \mu\right|} \ll t \lesssim \frac{\hbar}{\Gamma_{\nu m_s}(\mathbf{k})}$$

we derive the time dependence of the QP wave function (14.22) of a quasielectron. It takes the approximate form

$$\varphi^{\mathrm{QP}}_{\nu\mathbf{k}m_s}(\mathbf{x},t) \approx \theta\left(\varepsilon^{\mathrm{QP}}_{\nu m_s}(\mathbf{k}) - \mu\right) z_{\nu m_s}(\mathbf{k}) \varphi_{\nu\mathbf{k}m_s}(\mathbf{x}) e^{-\frac{i}{\hbar}\left[\varepsilon^{\mathrm{QP}}_{\nu m_s}(\mathbf{k}) - \frac{i}{2}\Gamma_{\nu m_s}(\mathbf{k})\right]t}. \quad (14.24)$$

Its time dependence obviously differs from that of a Bloch state of the reference system,

$$\varphi_{\nu\mathbf{k}m_s}(\mathbf{x},t) = \varphi_{\nu\mathbf{k}m_s}(\mathbf{x}) e^{-\frac{i}{\hbar}\varepsilon_{\nu m_s}(\mathbf{k})t},$$

in three characteristic features:

(i) The energy $\varepsilon_{\nu m_s}(\mathbf{k})$ is replaced by the excitation energy $\varepsilon^{\mathrm{QP}}_{\nu m_s}(\mathbf{k})$ that contains exchange and correlation in a better approach (14.18).

(ii) The Bloch wave is damped by the factor $e^{-\Gamma_{\nu m_s}(\mathbf{k})t/2\hbar}$ which indicates that the electron at $t = 0$ in the state $|\nu\mathbf{k}m_s\rangle$ is scattered by the additional XC effects into other states at time $t > 0$. The quasielectron has a finite lifetime $\hbar/\Gamma_{\nu m_s}(\mathbf{k})$.

(iii) For $t \ll \hbar/\Gamma_{\nu m_s}(\mathbf{k})$ the wave function (14.24) is not normalized. The total probability

$$\int d^3\mathbf{x} \left|\varphi^{\mathrm{QP}}_{\nu\mathbf{k}m_s}(\mathbf{x}, t)\right|^2 \approx \left|z_{\nu m_s}(\mathbf{k})\right|^2 \theta\left(\varepsilon^{\mathrm{QP}}_{\nu m_s}(\mathbf{k}) - \mu\right)$$

to find an electron in the state $|\nu\mathbf{k}m_s\rangle$ is reduced by the factor $|z_{\nu m_s}(\mathbf{k})|^2 < 1$, since the incoherent background in (14.23) has not been considered.

According to the three pecularities (i), (ii), and (iii) one interprets $\varphi^{\mathrm{QP}}_{\nu\mathbf{k}m_s}(\mathbf{x}, t)$ (14.24) as the wave function of a quasielectron in which an electron is transformed in a real, interacting many-particle system. We note that an expression similar to (14.24) can be derived for hole excitations and quasiholes with energies $\varepsilon^{\mathrm{QP}}_{\nu m_s}(\mathbf{k}) < \mu$.

Quasielectrons and quasiholes are summarized into quasiparticles. Their spectral functions are pictured in Fig. 14.4. The spectral functions in Fig. 14.4 with a pronounced quasiparticle peak and the three pecularities illustrate the definition of quasiparticles: The single-particle electronic excitations of a system of strongly interacting particles are described in terms of weakly interacting quasiparticles. In a solid or molecule a bare electron repels the other electrons via the Coulomb potential v and, in effect, surrounds itself with a positively charged polarization cloud due to the positively charged background due to the nuclei (see Sect. 4.4.1). The positive charge and the bare electron mainly form a quasiparticle that weakly interacts with other quasiparticles via a screened potential W as introduced in Sect. 12.2.3. According to (12.52) the renormalization is determined by exchange and correlation effects. This situation is graphically illustrated in Fig. 14.5. For hole excitations a polarization cloud with opposite sign appears.

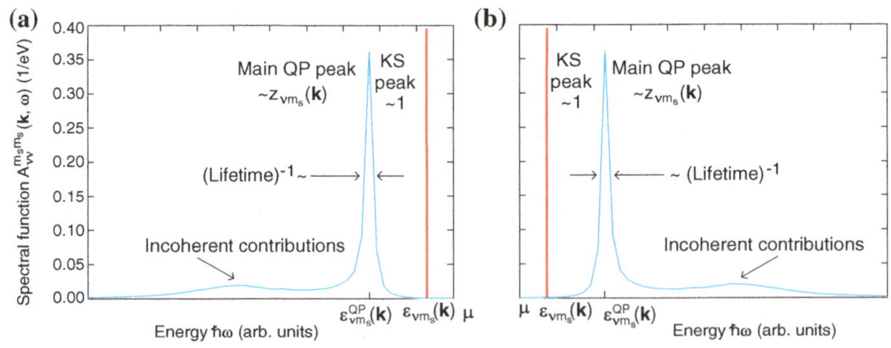

Fig. 14.4 Schematic spectral-weight function of (**a**) hole and (**b**) electron excitations in an interacting electron gas (*blue*) compared to the δ-like one (*red*) in a KS reference system. The incoherent contributions, mentioned in (14.23), are also indicated

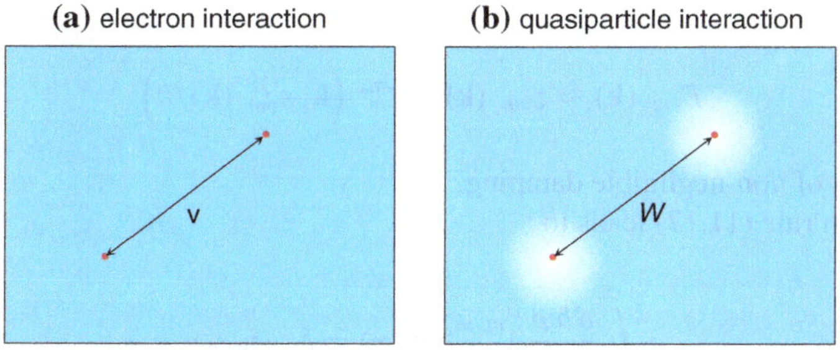

Fig. 14.5 Interaction of (a) electrons and (b) corresponding quasiparticles in a many-electron system (schematically)

In the limit of small damping $\Gamma_{\nu m_s}(\mathbf{k})$ of the quasiparticles, near $\hbar\omega \approx \varepsilon^{\mathrm{QP}}_{\nu m_s}(\mathbf{k})$ the spectral function (14.23) is approximately represented by a Dirac δ-function with reduced spectral weight,

$$A^{m_s m_s}_{\nu\nu}(\mathbf{k}, \omega) \approx z_{\nu m_s}(\mathbf{k}) 2\pi \delta\left(\varepsilon^{\mathrm{QP}}_{\nu m_s}(\mathbf{k}) - \hbar\omega\right). \tag{14.25}$$

The neglect of the damping $\Gamma^{m_s m_s}_{\nu\nu}(\mathbf{k}, \omega)$ in (14.19) is justified if in the vicinity of the solution $\hbar\omega = \varepsilon^{\mathrm{QP}}_{\nu m_s}(\mathbf{k})$ of

$$\hbar\omega - \varepsilon_{\nu m_s}(\mathbf{k}) - V^{m_s m_s}_{\nu\nu}(\mathbf{k}, \omega) = 0 \tag{14.26}$$

the spectral function $\Gamma^{m_s m_s}_{\nu\nu}(\mathbf{k}, \omega)$ only weakly depends on ω, so that it can be replaced by $\Gamma^{m_s m_s}_{\nu\nu}(\mathbf{k}, \varepsilon^{\mathrm{QP}}_{\nu m_s}(\mathbf{k})/\hbar)$, and, moreover, the inequality

$$\Gamma^{m_s m_s}_{\nu\nu}\left(\mathbf{k}, \varepsilon^{\mathrm{QP}}_{\nu m_s}(\mathbf{k})/\hbar\right) \ll \left|\varepsilon^{\mathrm{QP}}_{\nu m_s}(\mathbf{k}) - \mu\right|$$

is fulfilled. This is likely if $|\varepsilon^{\mathrm{QP}}_{\nu m_s}(\mathbf{k}) - \mu|$ is of the order the fundamental gap. Together with the spectral weight

$$z_{\nu m_s}(\mathbf{k}) = \frac{1}{\left|1 - \left[\partial V^{m_s m_s}_{\nu\nu}(\mathbf{k}, \omega)/\partial(\hbar\omega)\right]_{\hbar\omega = \varepsilon^{\mathrm{QP}}_{\nu m_s}(\mathbf{k})}\right|} \tag{14.27}$$

the spectral function (14.19) or (14.23) takes the elegant form [5, 7]

$$A^{m_s m_s}_{\nu\nu}(\mathbf{k}, \omega) = z_{\nu m_s}(\mathbf{k}) \frac{\Gamma_{\nu m_s}(\mathbf{k})}{\left[\varepsilon^{\mathrm{QP}}_{\nu m_s}(\mathbf{k}) - \hbar\omega\right]^2 + \left[\frac{1}{2}\Gamma_{\nu m_s}(\mathbf{k})\right]^2} + a^{m_s m_s}_{\nu\nu}(\mathbf{k}, \omega) \tag{14.28}$$

with

$$\Gamma_{\nu m_s}(\mathbf{k}) \approx z_{\nu m_s}(\mathbf{k})\Gamma_{\nu\nu}^{m_s m_s}\left(\mathbf{k}, \varepsilon_{\nu m_s}^{\mathrm{QP}}(\mathbf{k})/\hbar\right) \tag{14.29}$$

in the limit of non-negligible damping.

The sum rule (11.17) leads to

$$\int_{-\infty}^{+\infty} \frac{d\hbar\omega}{2\pi} A_{\nu\nu'}^{m_s m_s'}(\mathbf{k}, \omega) = \delta_{\nu\nu'}\delta_{m_s m_s'} \tag{14.30}$$

for the Bloch-Fourier representation (14.14) of the spectral function. In contrast to the representation (14.28) the approximate spectral function (14.25) yields

$$\int_{-\infty}^{+\infty} \frac{d\hbar\omega}{2\pi} A_{\nu\nu}^{m_s m_s}(\mathbf{k}, \omega) = z_{\nu m_s}(\mathbf{k}) < 1,$$

since the incoherent contributions (see (14.28)) with the spectral weight $[1-z_{\nu m_s}(\mathbf{k})]$ are not taken into consideration. In order to keep one pole but do not lose spectral weight, sometimes renormalized QP wave functions and spectral functions according to

$$\begin{aligned}
\tilde{\varphi}_{\nu\mathbf{k}m_s}^{\mathrm{QP}}(\mathbf{x}, t) &= \frac{1}{z_{\nu m_s}(\mathbf{k})}\varphi_{\nu m_s}^{\mathrm{QP}}(\mathbf{x}, t), \\
\tilde{A}_{\nu\nu}^{m_s m_s}(\mathbf{k}, \omega) &= \frac{1}{z_{\nu m_s}(\mathbf{k})}A_{\nu\nu}^{m_s m_s}(\mathbf{k}, \omega) = 2\pi\delta\left(\varepsilon_{\nu m_s}^{\mathrm{QP}}(\mathbf{k}) - \hbar\omega\right)
\end{aligned} \tag{14.31}$$

are introduced for the discussion of some excitations and the screening in the XC self-energy.

Expression (14.28) clearly indicates that the quasiparticle description requires that at least more than 50 % of the spectral weight are included in the main QP peak (see Fig. 14.4). Indeed for *sp* valence semiconductors or insulators the QP residues $z_{\nu m_s}(\mathbf{k})$ are in the range 0.6–0.9 [3, 8, 9]. In this case one may speak about weakly correlated systems [10], instead of strongly correlated ones with, for instance, $z_{\nu m_s}(\mathbf{k}) \to 0.5$ as in the case of a simple two-band Hubbard model [5, 11].

For weakly correlated systems the resulting Green function $\tilde{\tilde{G}}_{ss'}(\mathbf{x}\mathbf{x}', z)$ approximately obeys a similar Bloch-Fourier representation as the Green function (14.6) of the reference system,

$$\tilde{\tilde{G}}_{ss'}(\mathbf{x}\mathbf{x}', z) = \sum_{\nu,\nu'}\sum_{\mathbf{k}}\sum_{m_s,m_s'} \tilde{\tilde{G}}_{\nu\nu'}^{m_s m_s'}(\mathbf{k}, z)\tilde{\varphi}_{\nu\mathbf{k}m_s}^{\mathrm{QP}}(\mathbf{x})\tilde{\varphi}_{\nu'\mathbf{k}m_s'}^{\mathrm{QP}*}(\mathbf{x}')\chi_{\frac{1}{2}m_s}(s)\chi_{\frac{1}{2}m_s'}^{+}(s') \tag{14.32}$$

with

$$\tilde{\tilde{G}}_{\nu\nu'}^{m_s m_s'}(\mathbf{k}, z) = \frac{\delta_{\nu\nu'}\delta_{m_s m_s'}}{\hbar z - \varepsilon_{\nu m_s}^{\text{QP}}(\mathbf{k})} \tag{14.33}$$

and

$$\tilde{\varphi}_{\nu\mathbf{k}m_s}^{\text{QP}}(\mathbf{x}) \approx \varphi_{\nu\mathbf{k}m_s}(\mathbf{x}).$$

In some applications of the theory the approximation (14.32) may serve as an acceptable description of the interacting electron system, at least in the calculations of the self-energy itself (see Sect. 14.2) or the RPA screening (see Sect. 12.3.3).

14.1.3 Bloch-Landau Quasiparticles in Metals

In Sect. 14.1.2 we have demonstrated that the solution $\varepsilon_{\nu m_s}^{\text{QP}}(\mathbf{k})$ of (14.26) can be interpreted as a QP energy with a normalized wave function in a certain time interval, if

$$\Gamma_{\nu m_s}(\mathbf{k}) \ll \left|\varepsilon_{\nu m_s}^{\text{QP}}(\mathbf{k}) - \mu\right| \tag{14.34}$$

holds. While this inequality should be fulfilled for systems with not too small fundamental energy gaps, it has to be carefully investigated for energies $\varepsilon_{\nu m_s}^{\text{QP}}(\mathbf{k}) = \mu$ or close to μ in metals. We do this for $T = 0\,\text{K}$ with $\mu = \varepsilon_F$ as the Fermi energy, i.e., neglecting the temperature dependence of the chemical potential, and

$$\varepsilon_{\nu m_s}^{\text{QP}}(\mathbf{k}) = \varepsilon_F \tag{14.35}$$

as the definition of the Fermi surface of the metal [12]. The corresponding solutions in $\mathbf{k}$ space are denoted by $\mathbf{k} = \mathbf{k}_F$. In a rotationally invariant system the Fermi surface is a sphere of radius k_F (4.47), called the Fermi momentum or Fermi wave vector.

The condition (14.34) can be only fulfilled near the Fermi surface if

$$\lim_{\mathbf{k}\to\mathbf{k}_F} \Gamma_{\nu m_s}(\mathbf{k}) = 0. \tag{14.36}$$

Close to the Fermi surface the QP quantities $\varepsilon_{\nu m_s}^{\text{QP}}(\mathbf{k})$ and $\Gamma_{\nu m_s}(\mathbf{k})$ can be expanded up to first non-vanishing order according to

$$\left|\varepsilon^{\rm QP}_{\nu m_s}(\mathbf{k}) - \varepsilon_F\right| \approx \left|\nabla_{\mathbf{k}}\left[\varepsilon^{\rm QP}_{\nu m_s}(\mathbf{k})\right]_{\mathbf{k}=\mathbf{k}_F}(\mathbf{k}-\mathbf{k}_F)\right|, \tag{14.37}$$

$$\Gamma_{\nu m_s}(\mathbf{k}) \approx \left[\nabla_{\mathbf{k}}\Gamma_{\nu m_s}(\mathbf{k})\right]_{\mathbf{k}=\mathbf{k}_F}(\mathbf{k}-\mathbf{k}_F) + \frac{1}{2}(\mathbf{k}-\mathbf{k}_F)\left\{\left[\nabla^2_{\mathbf{k}}\Gamma_{\nu m_s}(\mathbf{k})\right]_{\mathbf{k}=\mathbf{k}_F}(\mathbf{k}-\mathbf{k}_F)\right\}.$$

One immediately sees that the condition (14.34) is only fulfilled if also

$$\lim_{\mathbf{k}\to\mathbf{k}_F} \nabla_{\mathbf{k}}\Gamma_{\nu m_s}(\mathbf{k}) = 0 \tag{14.38}$$

holds. A system of interacting fermions for which the QP quantities $\varepsilon^{\rm QP}_{\nu m_s}(\mathbf{k})$ and $\Gamma_{\nu m_s}(\mathbf{k})$ obey the conditions (14.37) and (14.38) is called *normal*. The inverse lifetime of the QPs

$$\Gamma_{\nu m_s}(\mathbf{k}) \sim \left[\varepsilon^{\rm QP}_{\nu m_s}(\mathbf{k}) - \varepsilon_F\right]^2 \tag{14.39}$$

depends quadratically on its energy deviation from the Fermi surface. Thus, in normal metals quasiparticles close to the Fermi surface are stable elementary excitations [3] with long lifetimes.

The spectral function contains the full information about the average occupation of the single-particle states. We investigate the occupation number of a state $|\nu\mathbf{k}m_s\rangle$ at $T = 0$ K according to the derivation of (14.20) as

$$n_{\nu m_s}(\mathbf{k}) = \langle \hat{a}^+_{\nu\mathbf{k}m_s}\hat{a}_{\nu\mathbf{k}m_s}\rangle = \int_{-\infty}^{\mu}\frac{d\hbar\omega}{2\pi}A^{m_s m_s}_{\nu\nu}(\mathbf{k},\omega). \tag{14.40}$$

In a similar manner as used to prove (14.22) one finds

$$n_{\nu m_s}(\mathbf{k}) = z_{\nu m_s}(\mathbf{k})\theta\left(\mu - \varepsilon^{\rm QP}_{\nu m_s}(\mathbf{k})\right) - \int_{\mu}^{\mu+i\infty}\frac{d\hbar\omega}{2\pi}A^{m_s m_s}_{\nu\nu}(\mathbf{k},\omega).$$

The occupation number has a discontinuity at the Fermi surface $\mathbf{k} = \mathbf{k}_F$, because the quasiparticle peak comes outside the region of integration for $\varepsilon^{\rm QP}_{\nu m_s}(\mathbf{k}) > \mu$. The magnitude of the discontinuity equals the strength $z_{\nu m_s}(\mathbf{k})$ of the quasiparticle pole (14.21) (or here the complex conjugated energy). Thus, the occupation number versus energy has the form as indicated in Fig. 14.6, which is different from that of a non-interacting system.

The spectral function of an interacting electron gas in a normal metal is schematically illustrated for the excitation of quasiparticles close to the Fermi surface in Fig. 14.7 [3]. It is zero for $\hbar\omega = \mu$ in agreement with the vanishing damping. It possesses a sharp peak at the position $\hbar\omega = \varepsilon^{\rm QP}_{\nu m_s}(\mathbf{k})$. Its sharpness is more pronounced for smaller values of the spectral broadening $\Gamma^{m_s m_s}_{\nu\nu}(\mathbf{k},\omega)$ (14.16) and its weaker

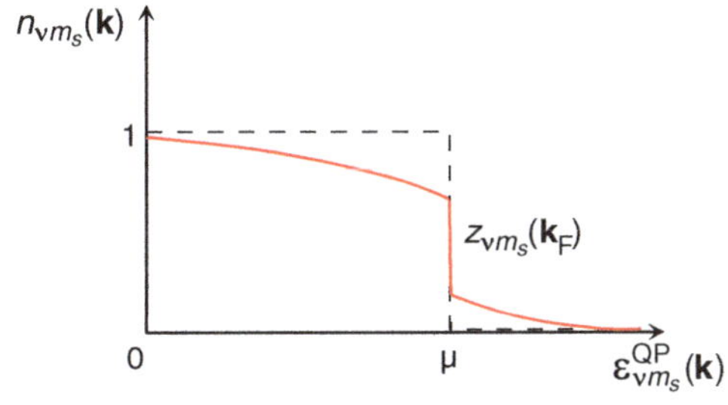

Fig. 14.6 Occupation number $n_{\nu m_s}(\mathbf{k})$ of a single-particle state $|\nu \mathbf{k} m_s\rangle$ versus the quasiparticle energy $\varepsilon^{QP}_{\nu m_s}(\mathbf{k})$ in an interacting system (*red line*). For comparison, the behavior in a non-interacting system is indicated by the *dashed line*

dependence on ω in the vicinity of $\hbar\omega \approx \varepsilon^{QP}_{\nu m_s}(\mathbf{k})$. Thereby, the conditions (14.36) and (14.38) mean that for energies $\varepsilon^{QP}_{\nu m_s}(\mathbf{k})$ near μ the QP peak becomes a Dirac δ-function with the spectral weight $z_{\nu m_s}(\mathbf{k})$. Close to the Fermi surface the spectral function can be approximately described by the incoherent contribution in (14.28). In order to guarantee the sum rule (14.30), however, a renormalization according to (14.31) is required.

The excitations near the Fermi surface of a normal metal represent Landau quasiparticles [13, 14] or, here, because of the translational symmetry, Bloch-Landau quasiparticles. They have to fulfill the conditions (14.36) and (14.38). Because of $T = 0\,\mathrm{K}$ and the Coulomb interaction the collision of an additional electron leads to the excitation of another electron out of the Fermi sea and the lowering of the energy of the electron above μ. For an energy of the primary electron near the Fermi surface the available phase space for damping processes vanishes according to $(\mathbf{k} - \mathbf{k}_F)^2$ as $\mathbf{k} \to \mathbf{k}_F$ [12]. The lifetime becomes infinite. The existence of long-lived quasiparticles and the available phase space $\sim (\mathbf{k} - \mathbf{k}_F)^2$ in strongly interacting electron systems are based upon the conservation of energy and momentum.

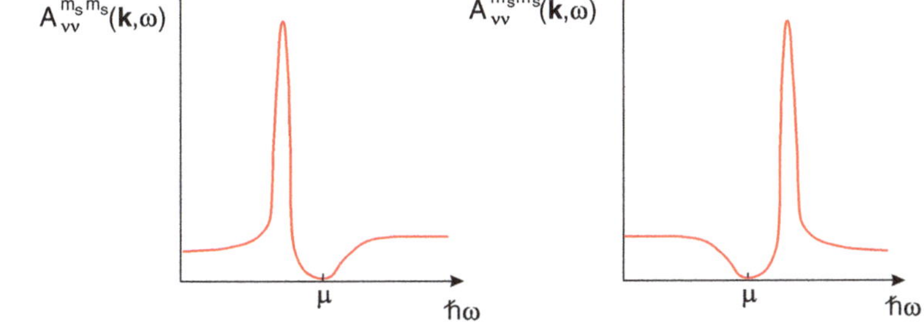

Fig. 14.7 Spectral function in a normal metal for quasiparticle excitation energies near the Fermi surface (schematically)

14.2 Self-consistency

14.2.1 Quasiparticle Shifts and Strengths

The XC self-energy Σ (12.52) depends directly and indirectly via the screened potential W (12.53) on the unknown single-particle Green function G (12.51). An additional dependence on W enters the problem via the vertex function Γ (12.55), mainly via the variational derivative with respect to G of Σ itself. These complex dependencies are illustrated by the magic pentagon in Fig. 12.3. The complete determination of Σ and G requires a complicated self-consistent treatment and suggests an iterative treatment of the Hedin equations (12.51–12.55). In the following chapters we will see that at least five aspects of the theory may be distinguished in relation to the self-consistency problem [15, 16]:

(i) Shift of quasiparticle energies
(ii) Modification of quasiparticle wave functions
(iii) Modification of quasiparticle residua
(iv) Modification of quasiparticle lifetimes
(v) Modification of screening.

We will focus on the first two aspects (i) and (ii) but also discuss their consequences for the screening properties of the system (v). To do so, we start from the Green function $\tilde{\tilde{G}}_{ss'}(\mathbf{xx}', z)$ (14.32) of non-interacting quasiparticles with unknown QP energies $\varepsilon^{\text{QP}}_{\nu m_s}(\mathbf{k})$ and wave functions $\varphi^{\text{QP}}_{\nu\mathbf{k}m_s}(\mathbf{x})$ but with their full spectral weight $z_{\nu m_s}(\mathbf{k}) = 1$ in the spectral function (14.31) and eventually some renormalization of the wave functions. In a first approach we follow an idea of Blomberg, Bergersen, and Kus [17, 18]. We assume that the quasiparticle wave functions $\tilde{\varphi}^{\text{QP}}_{\nu\mathbf{k}m_s}(\mathbf{x})$ are identical with those in (14.3), obtained for a certain KS/gKS/HF reference system. Despite this assumption, then only diagonal elements in the band index appear in the Dyson equation (14.12). The Bloch-Fourier coefficients of the Green function (14.33) are

$$\tilde{\tilde{G}}^{m_s m_s'}_{\nu\nu'}(\mathbf{k}, z) = \frac{\delta_{\nu\nu'}\delta_{m_s m_s'}}{\hbar z - \varepsilon^{\text{QP}}_{\nu m_s}(\mathbf{k})} \tag{14.41}$$

with unknown QP energies

$$\varepsilon^{\text{QP}}_{\nu m_s}(\mathbf{k}) = \varepsilon_{\nu m_s}(\mathbf{k}) + \Delta_{\nu m_s}(\mathbf{k}) \tag{14.42}$$

with a QP shift $\Delta_{\nu m_s}(\mathbf{k})$ of the eigenvalues $\varepsilon_{\nu m_s}(\mathbf{k})$ of the reference system. As a consequence of the new starting point $\tilde{\tilde{G}}$ with unknown single-quasiparticle energies we obtain a new Dyson equation

$$G_{\nu\nu}^{m_s m_s}(\mathbf{k}, z) = \tilde{\tilde{G}}_{\nu\nu}^{m_s m_s}(\mathbf{k}, z)\left\{1 + \left[\delta\Sigma_{\nu\nu}^{m_s m_s}(\mathbf{k}, z) - \Delta_{\nu m_s}(\mathbf{k})\right] G_{\nu\nu}^{m_s m_s}(\mathbf{k}, z)\right\}, \tag{14.43}$$

in the diagonal approximation, instead of (14.12) with the reference Green function $\tilde{G}$ (14.7). The Dyson equation (14.43) can be rewritten in the form of a geometric series by subsequent replacing G on the right-hand side by $\tilde{\tilde{G}} + \tilde{\tilde{G}}[\delta\Sigma - \Delta]G$, which leads to, symbolically written,

$$\begin{aligned} G = \tilde{\tilde{G}} &+ \tilde{\tilde{G}}[\delta\Sigma - \Delta]\tilde{\tilde{G}} + \tilde{\tilde{G}}[\delta\Sigma - \Delta]\tilde{\tilde{G}}[\delta\Sigma - \Delta]\tilde{\tilde{G}} \\ &+ \tilde{\tilde{G}}[\delta\Sigma - \Delta]\tilde{\tilde{G}}[\delta\Sigma - \Delta]\tilde{\tilde{G}}[\delta\Sigma - \Delta]\tilde{\tilde{G}} + \ldots . \end{aligned}$$

This is a typical equation of the scattering theory, where the different terms of the geometric series describe single, double, triple, etc., scattering processes, and $[\delta\Sigma - \Delta]$ is the scattering potential. Such a succession of scattering processes can be illustrated by Feynman diagrams [19].

In (14.43) the difference in the matrix elements $[\delta\Sigma_{\nu\nu}^{m_s m_s}(\mathbf{k}, z) - \Delta_{\nu m_s}(\mathbf{k})]$ characterizes the dynamical XC effects which are taken into account beyond their inclusion in the (unknown) QP energies $\varepsilon_{\nu m_s}^{\mathrm{QP}}(\mathbf{k})$. We assume that this perturbation, i.e., the difference $\left[\mathrm{Re}\delta\Sigma_{\nu\nu}^{m_s m_s}(\mathbf{k}, z) - \Delta_{\nu m_s}(\mathbf{k})\right]$ is small against the QP shift $\Delta_{\nu m_s}(\mathbf{k})$ itself. The perturbation depends also on Σ and G in a self-consistent manner. We restrict the investigation to the first non-vanishing order in the perturbation

$$G_{\nu\nu}^{m_s m_s}(\mathbf{k}, z) = \tilde{\tilde{G}}_{\nu\nu}^{m_s m_s}(\mathbf{k}, z) + \left[\delta\tilde{\tilde{\Sigma}}_{\nu\nu}^{m_s m_s}(\mathbf{k}, z) - \Delta_{\nu m_s}(\mathbf{k})\right]\left[\tilde{\tilde{G}}_{\nu\nu}^{m_s m_s}(\mathbf{k}, z)\right]^2, \tag{14.44}$$

where $\delta\tilde{\tilde{\Sigma}}$ means that its determining Green function G is replaced by the QP reference one $\tilde{\tilde{G}}$. The second term on the right-hand side is quadratic in $\tilde{\tilde{G}}$. According to its definition (14.41) the resulting Green function would contain an unphysical double pole at $\hbar z = \varepsilon_{\nu m_s}^{\mathrm{QP}}(\mathbf{k})$ near the real axis. In order to avoid such an unphysical pole one has to choose

$$\Delta_{\nu m_s}(\mathbf{k}) = \mathrm{Re}\delta\tilde{\tilde{\Sigma}}_{\nu\nu}^{m_s m_s}\left(\mathbf{k}, \varepsilon_{\nu m_s}^{\mathrm{QP}}(\mathbf{k})/\hbar\right). \tag{14.45}$$

A graphical solution of the two equations (14.42) and (14.45) is pictured in Fig. 14.8 for the energy levels of the water molecule H_2O. The crossings of the linear function of the single-particle energy and the real parts of the self-energy difference shifted by the reference energy levels yield the quasiparticle energies. The quasiparticle shifts $\Delta_{\nu m_s}(\mathbf{k}) = -4.71, -5.15, -5.68$ eV resulting for the three highest occupied levels $1b_2$, $3a_1$, and $1b_1$ [20] significantly modify the reference levels.

The unknown QP shift in (14.42) is indeed defined by (14.45) which, however, has to be treated also self-consistently with respect to $\varepsilon_{\nu m_s}^{\mathrm{QP}}(\mathbf{k})$. The restriction to

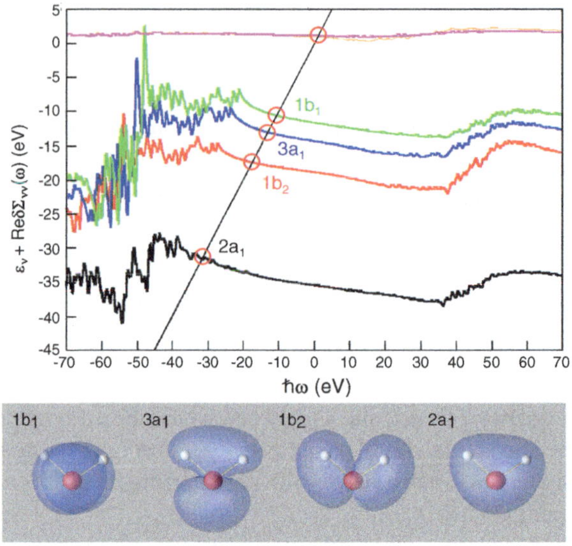

Fig. 14.8 Graphical solution $\varepsilon_\nu^{\mathrm{QP}}$ of a H_2O molecule in vacuum without spin polarization and wave-vector dispersion. For that reason the quantum numbers m_s and $\mathbf{k}$ do not appear. The energy variation of the self-energy is clearly visible for the four valence levels $1b_1$, $3a_1$, $1b_2$, and $2a_1$, whose molecule orbitals are also displayed. It almost vanishes for the two lowest empty levels (not labeled). The *straight line* represents the linear function $\hbar\omega$. Its crossings (*red circles*) with the other *curves* define the QP energies $\varepsilon_\nu^{\mathrm{QP}}$. The reference electronic structure is derived as solutions ε_ν of the Kohn-Sham equation with an LDA XC functional (7.15) and (7.21). Adapted from [20]

the Green function $\tilde{\tilde{G}}$ (14.41) can be identified within a first-order perturbation-theory treatment of the perturbation $\delta\Sigma$ (14.10) of the reference system $\tilde{G}$ (14.7). In any case, the relation (14.45) gives an explicit definition of the QP shift, which corrects the eigenvalues of the reference system toward some QP eigenvalues in a self-consistent manner. The (14.44) and (14.15) also clearly indicate the request of doing approximations for quantities on the same footing.

In addition, in many practical calculations the energy dependence of the self-energy difference (14.16) is assumed to be linear around $\hbar z = \varepsilon_{\nu m_s}^{\mathrm{QP}}(\mathbf{k})$ (see confirmation in Fig. 14.8) and, hence, $\varepsilon_{\nu m_s}(\mathbf{k})$ if the QP corrections $\Delta_{\nu m_s}(\mathbf{k})$ are small. Then, instead of (14.45), it approximately holds

$$\Delta_{\nu m_s}(\mathbf{k}) = z_{\nu m_s}(\mathbf{k})\mathrm{Re}\delta\tilde{\tilde{\Sigma}}_{\nu\nu}^{m_s m_s}(\mathbf{k}, \varepsilon_{\nu m_s}(\mathbf{k})/\hbar), \tag{14.46}$$

where

$$z_{\nu m_s}(\mathbf{k}) = \left[1 - \frac{\partial}{\partial\hbar\omega}\mathrm{Re}\delta\tilde{\tilde{\Sigma}}_{\nu\nu}^{m_s m_s}(\mathbf{k}, \omega)|_{\hbar\omega=\varepsilon_{\nu m_s}(\mathbf{k})}\right]^{-1}, \tag{14.47}$$

i.e., with a spectral weight nearly equal to that $z_{\nu m_s}(\mathbf{k})$ (14.27) of the main QP peak in the spectral function. These findings clearly indicate that the QP shifts and strengths are influenced by dynamical effects in the screening and vertex functions.

In contrast to the QP shift (14.45), the spectral function $A_{\nu\nu}^{m_s m_s}(\mathbf{k}, \omega)$ of the Green function G is much more influenced by the dynamics of screening and, hence, dynamical correlation. A careful handling of the limits $z = \omega \pm i\eta$ $(\eta \to +0)$ is needed in the definition (14.15). The first iteration (14.44) of the Dyson equation (14.43) leads to [21]

$$A_{\nu\nu}^{m_s m_s}(\mathbf{k}, \omega) = 2\pi \left\{ \left[1 + \frac{\partial}{\partial \hbar\omega} \mathrm{Re}\delta \tilde{\tilde{\Sigma}}_{\nu\nu}^{m_s m_s}(\mathbf{k}, \omega) \right] \delta\left(\hbar\omega - \varepsilon_{\nu m_s}^{\mathrm{QP}}(\mathbf{k}) \right) + \left[\frac{\partial}{\partial \hbar\omega} \frac{\mathcal{P}}{\hbar\omega - \varepsilon_{\nu m_s}^{\mathrm{QP}}(\mathbf{k})} \right] \frac{1}{\pi} \mathrm{Im}\delta \tilde{\tilde{\Sigma}}_{\nu\nu}^{m_s m_s}(\mathbf{k}, \omega) \right\} \tag{14.48}$$

using the representation (14.41), the definition (11.22), and the decomposition

$$\lim_{\eta \to +0} \delta \tilde{\tilde{\Sigma}}_{\nu\nu'}^{m_s m_s'}(\mathbf{k}, \omega \pm i\eta) = \mathrm{Re}\delta \tilde{\tilde{\Sigma}}_{\nu\nu'}^{m_s m_s'}(\mathbf{k}, \omega) \pm i \mathrm{Im}\delta \tilde{\tilde{\Sigma}}_{\nu\nu'}^{m_s m_s'}(\mathbf{k}, \omega).$$

Instead of the not well defined double pole in (14.44) the derivative

$$\frac{\partial}{\partial \hbar\omega} \frac{\mathcal{P}}{\hbar\omega - \varepsilon_{\nu m_s}^{\mathrm{QP}}(\mathbf{k})} = \lim_{\eta \to +0} \frac{\left(\hbar\omega - \varepsilon_{\nu m_s}^{\mathrm{QP}}(\mathbf{k}) \right)^2 - (\hbar\eta)^2}{\left[\left(\hbar\omega - \varepsilon_{\nu m_s}^{\mathrm{QP}}(\mathbf{k}) \right)^2 + (\hbar\eta)^2 \right]^2}$$

of the principal value enters the spectral function. Such as spectral function (14.48) represents a sharp peak at the QP energy $\varepsilon_{\nu m_s}^{\mathrm{QP}}(\mathbf{k})$ with a reduced spectral weight

$$z_{\nu m_s}(\mathbf{k}) = [1 + \frac{\partial}{\partial \hbar\omega} \mathrm{Re}\delta \tilde{\tilde{\Sigma}}_{\nu\nu}^{m_s m_s}(\mathbf{k}, \omega)|_{\hbar\omega = \varepsilon_{\nu m_s}^{\mathrm{QP}}(\mathbf{k})}],$$

and additional broad structures $\sim \mathrm{Im}\delta \tilde{\tilde{\Sigma}}_{\nu\nu}^{m_s m_s}(\mathbf{k}, \omega)$, which describe incoherent contributions to the spectral function (14.23). The modified spectral weight $z_{\nu m_s}(\mathbf{k})$ compared to (14.27) is a consequence of using the first iteration (14.44). It may be considered as the first term in the series expansion of the denominator in (14.47).

14.2.2 Quasiparticle Wave Functions

The discussion of the QP energies in Sect. 14.2.1 are based upon the approximation of the QP wave functions $\tilde{\varphi}_{\nu\mathbf{k}m_s}^{\mathrm{QP}}(\mathbf{x})$ by those $\varphi_{\nu\mathbf{k}m_s}^{\mathrm{QP}}(\mathbf{x}) = \varphi_{\nu\mathbf{k}m_s}(\mathbf{x})$ of the reference system which diagonalize the Green function $\tilde{G}$ (14.6). The improvement of the wave functions to $\tilde{\varphi}^{\rm QP}_{\nu\mathbf{k}m_s}(\mathbf{x})$ asks for generalizations. One way was illustrated by Hedin and Lundqvist [3, 10, 22] using a generalized Lehmann representation [see (11.23)] of the true Green function G applying energy-dependent Lehmann amplitudes and energy-dependent complex eigenvalues. Despite the elegance of the formulation of the problem, it is difficult to use such quantities in explicit numerical calculations.

Here, we follow the idea to investigate these wave functions for energies close to the quasiparticle ones studied above. With an approximate Green function $\tilde{\tilde{G}}$ (14.32) and an effective XC contribution, even a non-local one as in (14.10), to the total single-particle potential $\tilde{V}_{ss'}(\mathbf{x})$ introduced in (10.10) or (14.1), which may be also extended toward non-local XC contributions as in (14.10), the equation of motion for the Green function of the type (11.50) can be rewritten to a so-called quasiparticle equation. We restrict its investigation to the case of collinear spins. Then, at least for the spin-less potential $\tilde{V}_{ss'}(\mathbf{x}) = V_{\rm n}(\mathbf{x})\delta_{ss'}$ in (14.4), it results a QP equation for each spin channel m_s

$$\left\{-\frac{\hbar^2}{2m}\Delta_{\mathbf{x}} + V_{\rm n}(\mathbf{x}) + V_{\rm H}(\mathbf{x})\right\}\tilde{\varphi}^{\rm QP}_{\nu\mathbf{k}m_s}(\mathbf{x}) + \int d^3\mathbf{x}'\,\Sigma^{m_s m_s}\left(\mathbf{x}\mathbf{x}'; \varepsilon^{\rm QP}_{\nu m_s}(\mathbf{k})/\hbar\right)\tilde{\varphi}^{\rm QP}_{\nu\mathbf{k}m_s}(\mathbf{x}')$$
$$= \varepsilon^{\rm QP}_{\nu m_s}(\mathbf{k})\tilde{\varphi}^{\rm QP}_{\nu\mathbf{k}m_s}(\mathbf{x}) \qquad (14.49)$$

with $\Sigma^{m_s m_s}$ as the matrix element of the XC self-energy taken with single-particle spinors (14.3). The XC influence is included by the non-local, non-Hermitian, and energy-dependent self-energy $\Sigma^{m_s m_s}$ in the spin channel m_s.

In explicit calculations it is convenient to neglect the lifetime of the quasiparticles and to study solely real eigenvalues $\varepsilon^{\rm QP}_{\nu m_s}(\mathbf{k})$ by taking only the Hermitian part of $\Sigma^{m_s m_s}$ into account in (14.49) [23]. In this framework numerical procedures are even implemented in modern versions of some codes as e.g. VASP [24, 25], which allow for two types of self-consistent computations, only eigenvalues or alternatively eigenvalues and orbitals, within of the GW approximation (12.57). The quasiparticle wave functions can be expanded in a series using the complete set of orthonormalized functions $\varphi_{\nu\mathbf{k}m_s}(\mathbf{x})$ of the reference system (14.4) according to

$$\tilde{\varphi}^{\rm QP}_{\nu\mathbf{k}m_s}(\mathbf{x}) = \sum_{\nu'} c^{m_s}_{\nu\nu'}(\mathbf{k})\varphi_{\nu'\mathbf{k}m_s}(\mathbf{x}). \qquad (14.50)$$

The QP equation (14.49) is transformed into a set of algebraic equations

$$\sum_{\nu'}\left\{\left[\varepsilon_{\nu m_s}(\mathbf{k}) - \varepsilon^{\rm QP}_{\nu m_s}(\mathbf{k})\right]\delta_{\nu\nu'} + \delta\tilde{\tilde{\Sigma}}^{m_s m_s}_{\nu\nu'}\left(\mathbf{k}, \varepsilon^{\rm QP}_{\nu m_s}(\mathbf{k})/\hbar\right)\right\}c^{m_s}_{\nu\nu'}(\mathbf{k}) = 0.$$

The self-energy term couples different Bloch states of the reference system. For a translationally invariant and collinear electron system the coupling is restricted to the band indices. The replacement of $\delta\Sigma$ by $\delta\tilde{\tilde{\Sigma}}$ indicates that only Green functions with defined QP peaks are used to compute the self-energy difference [see (14.44)].

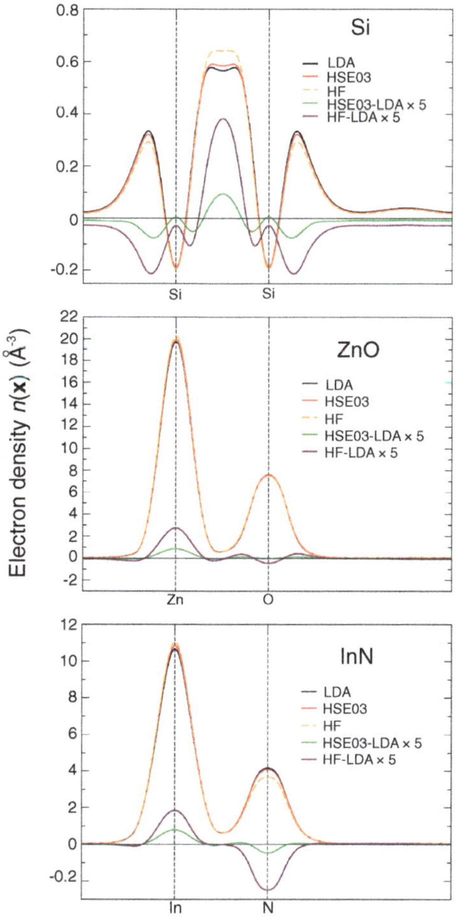

Fig. 14.9 Electron density $n(\mathbf{x})$ along a [111] bond direction in the tetrahedrally coordinated semiconductors Si, ZnO, and InN crystallized in diamond or zinc-blende structure. Three different XC approximations are used: local approximation PZ-LDA (7.21), hybrid functional HSE (9.19), and Hartree-Fock HF (4.27). Their differences are also displayed. Adapted from [26]

The localization of certain states $|\nu\mathbf{k}m_s\rangle$ and the applied treatment of XC in the calculation of the reference system determine the deviations between the reference wave functions $\varphi_{\nu\mathbf{k}m_s}(\mathbf{x})$ and the QP ones $\tilde{\varphi}^{\mathrm{QP}}_{\nu\mathbf{k}m_s}(\mathbf{x})$. Thereby, the description of exchange and correlation in the reference system plays an important role. In Fig. 14.9 this fact is globally illustrated for the valence states by the electron density along a bond direction in the semiconductors Si, ZnO, and InN with varying degree of ionic bonding. While in the compounds the density variations with the XC treatment are relatively small, they become more important in the case of covalently bonded systems, especially in the bond region. The figure shows that Hartree-Fock-derived wave functions seem to be hardly applicable for systems with strong band dispersion such as Si but also InN, because of the increased bond charge in Si and the modified localization of the N-derived wave functions in InN, respectively.

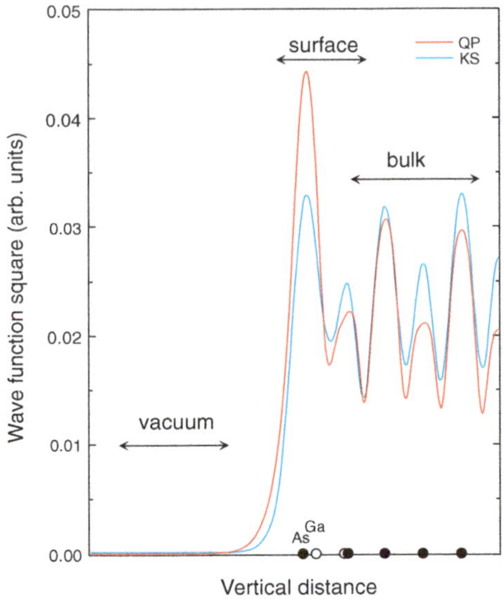

Fig. 14.10 Quasiparticle and KS wave functions (square modulus integrated over the surface plane) along the surface normal for the lowest empty state at Γ of a GaAs(110)1 × 1 surface. The reference KS wave function is calculated in the framework of PZ-LDA for XC (7.21). From [23]

For sp-bonded materials the KS wave functions $\varphi_{\nu\mathbf{k}m_s}(\mathbf{x})$ in (6.22) within a local or semilocal description of the XC potential (7.17) or (7.31) give some reasonable approximations of the quasiparticle wave functions $\tilde{\varphi}^{\mathrm{QP}}_{\nu\mathbf{k}m_s}(\mathbf{x})$. This holds also for the results of hybrid-functional calculations. Only the HF functions deviate significantly in the regions of large electron density. However, if in such systems localized states appear, e.g. near surfaces or around point defects, larger deviations may occur. As an example Fig. 14.10 shows the wave function square for the lowest unoccupied state at Γ of the cleavage (110)1 × 1 surface of GaAs [23]. A redistribution of the probability to find an electron in this surface state due to off-diagonal self-energy effects is obvious.

A strong influence of the self-consistent procedure (14.52) to compute QP wave functions starting from a KS reference system also occurs for unoccupied states in systems with flat bands, e.g. solid argon. As an example the squared modulus of the wave function belonging to the second conduction band at $\mathbf{k} = \left(-\frac{1}{8}, -\frac{3}{8}, \frac{1}{4}\right)\frac{2\pi}{a_0}$ is plotted in Fig. 14.11. Significant differences between self-consistently calculated (called QPscGW) and KS (using the LDA functional) wave functions are demonstrated in Fig. 14.11. The figure also shows that other reference systems that account better for the non-locality of XC are closer to the final self-consistent result. For valence states such differences almost vanish (not shown here), i.e., the KS reference wave functions can be approximately used for the quasiholes [27].

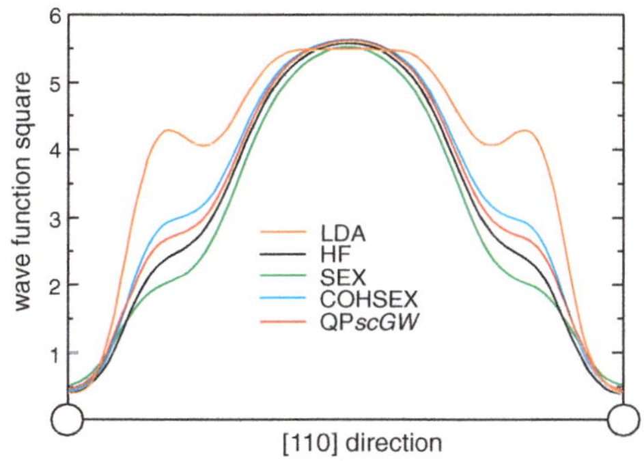

Fig. 14.11 Squared modulus of the second-conduction band wave function at $\mathbf{k} = \left(-\frac{1}{8}, -\frac{3}{8}, \frac{1}{4}\right)\frac{2\pi}{a_0}$ of solid argon along the [110] direction. Besides the KS reference system (LDA) and the self-consistently computed wave function (QPscGW) also some reference functions (HF, SEX, COHSEX) are plotted for other functionals which account better for the non-locality of XC. Reprinted with permission from [27]. Copyright 2006 by the American Physical Society

14.3 Standard Treatment

14.3.1 Bloch-Fourier Representation

The standard treatment of the XC self-energy [1, 8, 28] is based on the Hedin GW approximation, $\Sigma = i\hbar GW$ (12.57), with a screened potential W (12.53) that is ruled by the polarization function $P = L_0 = -i\hbar GG$ (12.54) of independent quasiparticles. According to the discussion in Sects. 14.1.2 and 14.2.1 Σ has to be computed in a self-consistent manner applying the self-consistent QP Green function $\tilde{\tilde{G}}$ with defined poles at $\hbar z = \varepsilon^{\mathrm{QP}}_{\nu m_s}(\mathbf{k})$. In the standard approximation for QP shifts we restrict ourselves to the reference electronic structure described by the Green function $\tilde{G}$ (14.6) computed by means of the eigenvalues $\varepsilon_{\nu m_s}(\mathbf{k})$ and eigenfunctions $\varphi_{\nu\mathbf{k}m_s}(\mathbf{x})$ of a starting electronic structure, e.g. given by (14.4). This approach is frequently denoted one-shot approach and called G_0W_0 [10].

While in the time domain the XC self-energy Σ (12.57) factorizes

$$\Sigma_{ss'}(\mathbf{xx}', t-t') = i\hbar G_{ss'}(\mathbf{xx}', t-t')W(\mathbf{xx}', t^+ - t'), \tag{14.51}$$

its Fourier coefficients (14.8) at fermionic Matsubara frequencies z_n (11.20), which follow the same definition as those of the Green function (11.19),

$$\Sigma_{ss'}(\mathbf{xx}', z_n) = -\frac{1}{\beta}\lim_{\eta\to+0}\sum_{n'} e^{-(z_n - z_{n'})\eta} W(\mathbf{xx}', z_n - z_{n'})G_{ss'}(\mathbf{xx}', z_{n'}) \quad (14.52)$$

with

$$W(\mathbf{xx}', \tilde{z}_m) = \int_0^{-i\hbar\beta} dt e^{i\tilde{z}_m(t-t')} W(\mathbf{xx}', t - t') \tag{14.53}$$

at bosonic Matsubara frequencies $\tilde{z}_m$ (12.63), represent a convolution in the frequency domain. The exponential factor in (14.52) containing the infinitesimal η is a consequence of the slightly shifted time $t \to t^+$ (see 11.41) in the argument of the screened potential in (14.51).

To account for the dependence on space ($\mathbf{x}$) and spin (s) variables, we use the Bloch-spin representations (14.6) and (14.11) of an electron system with collinear spins. Instead of (14.52) it holds

$$\Sigma_{\nu\mu}^{m_s m_s'}(\mathbf{k}, z_n) = -\frac{1}{\beta} \lim_{\eta \to +0} \sum_{n'} e^{-(z_n - z_{n'})\eta} \sum_{\nu', \mu', \mathbf{k}'} W^{m_s m_s'}_{\substack{\mathbf{kk}' \\ \nu\nu' \\ \mu\mu'}}(z_n - z_{n'}) G_{\nu'\mu'}^{m_s m_s'}(\mathbf{k}', z_{n'}). \tag{14.54}$$

The matrix elements of the screened potential are given as generalized Coulomb integrals

$$W^{m_s m_s'}_{\substack{\mathbf{kk}' \\ \nu\nu' \\ \mu\mu'}}(\tilde{z}_m) = \int d^3\mathbf{x} \int d^3\mathbf{x}' \varphi^*_{\nu\mathbf{k}m_s}(\mathbf{x}) \varphi_{\nu'\mathbf{k}'m_s}(\mathbf{x}) W(\mathbf{xx}', \tilde{z}_m) \varphi_{\mu\mathbf{k}m_s'}(\mathbf{x}') \varphi^*_{\mu'\mathbf{k}'m_s'}(\mathbf{x}'). \tag{14.55}$$

In the collinear approximation we have demonstrated that the Green functions $\tilde{G}$ (14.6) and $\tilde{\tilde{G}}$ (14.32) are diagonal in the spin quantum numbers m_s and m_s'. Therefore, the same diagonality holds for the self-energy (14.54) in GW approximation. As a consequence the self-energy as well as the Green function are separately defined for each spin channel. However, the screening in the screened potential W (12.53) is due to electrons in both spin channels because of the spin summations in the dielectric function (12.65). The dependence of the matrix elements (14.55) on the spin quantum number m_s is solely due to the used wave functions belonging to a certain spin channel.

With the spatial Fourier representations of the density correlation function (13.34) and the bare Coulomb potential (13.35), as well as the spectral representation (13.37), the screened potential can be written as

$$W(\mathbf{xx}', \tilde{z}_m) = \frac{1}{\Omega} \sum_{\mathbf{q}} \sum_{\mathbf{G}, \mathbf{G}'} e^{i[(\mathbf{q}+\mathbf{G})\mathbf{x} - (\mathbf{q}+\mathbf{G}')\mathbf{x}']} \epsilon^{-1}(\mathbf{q}+\mathbf{G}, \mathbf{q}+\mathbf{G}', \tilde{z}_m) \tilde{v}(|\mathbf{q}+\mathbf{G}'|).$$

Together with the definition of the Bloch integrals (13.44) the matrix elements (14.55) become

$$W\begin{smallmatrix} m_s m_s' \\ \mathbf{kk}' \\ \nu\nu' \\ \mu\mu' \end{smallmatrix}(\tilde{z}_m)$$
$$= \frac{1}{\Omega}\sum_{\mathbf{q}}\sum_{\mathbf{G},\mathbf{G}'} B\begin{smallmatrix}\mathbf{kk}' \\ m_s m_s \\ \nu\nu'\end{smallmatrix}(\mathbf{q}+\mathbf{G})B\begin{smallmatrix}\mathbf{kk}' \\ m_s' m_s'{}^* \\ \mu\mu'\end{smallmatrix}(\mathbf{q}+\mathbf{G}')\epsilon^{-1}(\mathbf{q}+\mathbf{G},\mathbf{q}+\mathbf{G}',\tilde{z}_m)\tilde{\mathrm{v}}(|\mathbf{q}+\mathbf{G}'|). \tag{14.56}$$

Introducing the imaginary part of the symmetrized inverse dielectric matrix $\tilde{\epsilon}^{-1}(\mathbf{q}+\mathbf{G},\mathbf{q}+\mathbf{G}',z) = \frac{|\mathbf{q}+\mathbf{G}|}{|\mathbf{q}+\mathbf{G}'|}\epsilon^{-1}(\mathbf{q}+\mathbf{G},\mathbf{q}+\mathbf{G}',z)$ the spectral representation (13.37) changes into

$$\tilde{\epsilon}^{-1}(\mathbf{q}+\mathbf{G},\mathbf{q}+\mathbf{G}',z) = \delta_{\mathbf{GG}'} + \int_{-\infty}^{+\infty}\frac{d\omega}{\pi}\frac{\mathrm{Im}\tilde{\epsilon}^{-1}(\mathbf{q}+\mathbf{G},\mathbf{q}+\mathbf{G}',\omega)}{\omega - z}, \tag{14.57}$$

where the abbreviation

$$\mathrm{Im}\,\tilde{\epsilon}^{-1}(\mathbf{q}+\mathbf{G},\mathbf{q}+\mathbf{G}',\omega) = -\tilde{\mathrm{v}}(\sqrt{|\mathbf{q}+\mathbf{G}||\mathbf{q}+\mathbf{G}'|})\hat{L}(\mathbf{q}+\mathbf{G},\mathbf{q}+\mathbf{G}',\omega)$$

is introduced. As a consequence the matrix elements of the screened potential (14.55) decompose into contributions of the bare Coulomb potential $\sim\tilde{\mathrm{v}}(|\mathbf{q}+\mathbf{G}|)$ and contributions which are modified by the screening in the electronic system. Therefore, the matrix elements of the self-energy (14.54) can be divided according to $\Sigma = \Sigma^{\mathrm{X}}+\Sigma^{\mathrm{C}}$ (12.13) into matrix elements of a pure static exchange part

$$\Sigma^{\mathrm{X}}\,{}^{m_s m_s'}_{\nu\mu}(\mathbf{k}) = -\lim_{\eta\to+0}\frac{1}{\Omega}\sum_{\mathbf{q},\mathbf{G}}\sum_{\mathbf{k}}\sum_{\nu',\mu'} B\begin{smallmatrix}\mathbf{kk}' \\ m_s m_s \\ \nu\nu'\end{smallmatrix}(\mathbf{q}+\mathbf{G})B\begin{smallmatrix}\mathbf{kk}' \\ m_s' m_s'{}^* \\ \mu\mu'\end{smallmatrix}(\mathbf{q}+\mathbf{G})$$
$$\times\,\tilde{\mathrm{v}}(|\mathbf{q}+\mathbf{G}|)\frac{1}{\beta}\sum_{n'} e^{z_{n'}\eta}G^{m_s m_s'}_{\nu'\mu'}(\mathbf{k}',z_{n'}), \tag{14.58}$$

which however are slightly different from the matrix elements of the Fock operator, due to the used wave and Green functions, and matrix elements of the correlation contribution

$$\Sigma^{\mathrm{C}}\,{}^{m_s m_s'}_{\nu\mu}(\mathbf{k},z_n) = \frac{1}{\beta}\sum_{n'}\frac{1}{\Omega}\sum_{\mathbf{q},\mathbf{G},\mathbf{G}'}\sum_{\nu',\mu'} B\begin{smallmatrix}\mathbf{kk}' \\ m_s m_s \\ \nu\nu'\end{smallmatrix}(\mathbf{q}+\mathbf{G})B\begin{smallmatrix}\mathbf{kk}' \\ m_s' m_s'{}^* \\ \mu\mu'\end{smallmatrix}(\mathbf{q}+\mathbf{G}')$$
$$\times\,\tilde{\mathrm{v}}\left(\sqrt{|\mathbf{q}+\mathbf{G}||\mathbf{q}+\mathbf{G}'|}\right)\int_{-\infty}^{+\infty}\frac{d\omega}{\pi}\frac{\mathrm{Im}\tilde{\epsilon}^{-1}(\mathbf{q}+\mathbf{G},\mathbf{q}+\mathbf{G}',\omega)}{z_n - z_{n'} - \omega}G^{m_s m_s'}_{\nu'\mu'}(\mathbf{k}',z_{n'}), \tag{14.59}$$

that contains the complete, static and dynamical, screening response to the excitation of a (quasi)particle in the system. The exponential $e^{z_{n'}\eta}$ in (14.58) is a remainder of the slight shift of the time argument t^+ in the screened potential in (14.51). It guarantees that only the hole propagator (11.1) in the definition of the Green function contributes to the exchange. In contrast to the X contribution (14.58) the correlation one C (14.59) contains sums over unoccupied bands ν' and μ'. They give rise to some numerical difficulties to perform converged numerical calculations. While for infinite solids the convergence can be widely reached, this difficulty mainly holds for the investigation of excitations in molecules, nanocrystals and other finite objects, where continuum states occur above an ionization edge.

14.3.2 First Iteration

Following the above discussion the G function in the self-energy is not only replaced by the QP one $\tilde{\tilde{G}}$ with one pole. Rather, we apply the Green function $\tilde{G}$ of the reference system (14.6) in the spirit of a first iteration. To reformulate the exchange and correlation contributions we have to apply some theorems [4, 7, 29]. A useful method to compute the Fourier sums in (14.58) and (14.59) is to represent them as contour integrals in the complex frequency plane [29]. We have to study functions $F(z)$ that are regular at the fermionic Matsubara frequencies z_n (n - odd integer). We take the integration contour $\mathcal{C}$ that encircles all poles $\hbar z_n$ of the Fermi function $f(\hbar z)$ in the negative sense, but none of the poles of $F(z)$. Since the residue of $f(\hbar z)$ at $z = z_n$ is $-\frac{1}{\hbar\beta}$, it holds

$$\frac{1}{-i\hbar\beta}\sum_n F(z_n) = \oint_{\mathcal{C}} \frac{dz}{2\pi} f(\hbar z) F(z).$$

If the function $F(z)$ obeys the condition $\lim_{|z|\to\infty} z f(\hbar z) F(z) = 0$ the contour $\mathcal{C}$ can be deformed into a contour $\mathcal{C}'$ that encircles all poles of $F(z)$ in the positive sense. In the special case of $\tilde{G}^{m_s m_s'}_{\nu\mu}(\mathbf{k}, z) = \delta_{m_s m_s'}\delta_{\nu\mu}/[\hbar z - \varepsilon_{\nu m_s}(\mathbf{k})]$ (14.7) it holds in (14.58) [7, 29]

$$\frac{1}{\beta}\sum_n e^{z_n\eta}\frac{1}{\hbar z_n - \varepsilon} = \hbar\oint_{\mathcal{C}} \frac{dz}{2\pi i}\frac{f(\hbar z)e^{z\eta}}{\hbar z - \varepsilon} = \hbar\oint_{\mathcal{C}'} \frac{dz}{2\pi i}\frac{f(\hbar z)}{\hbar z - \varepsilon} = f(\varepsilon)$$

applying the Cauchy theorem. The exchange contribution (14.58) becomes

$$\Sigma^{\mathrm{X}\, m_s m_s'}_{\;\nu\mu}(\mathbf{k}) = -\delta_{m_s m_s'}\frac{1}{\Omega}\sum_{\mathbf{q},\mathbf{G}}\tilde{\mathrm{v}}(|\mathbf{q}+\mathbf{G}|)\sum_{\nu',\mathbf{k}'} f(\varepsilon_{\nu' m_s}(\mathbf{k}'))B^{\mathbf{kk}'}_{\substack{m_s m_s \\ \nu\nu'}}(\mathbf{q}+\mathbf{G})B^{\mathbf{kk}'\,*}_{\substack{m_s m_s \\ \mu\nu'}}(\mathbf{q}+\mathbf{G}). \tag{14.60}$$

The spectral sum $\sum_{n'}$ in (14.59) can be performed in a similar way as in the case of the exchange. Nevertheless, we use another procedure. It holds

$$\begin{aligned}\frac{1}{\hbar\beta}\sum_n \frac{1}{z_n-\omega} &= \sum_n \frac{1}{\beta(\mu-\hbar\omega)+i\pi n} = \sum_{n>0} \frac{2\beta(\mu-\hbar\omega)}{\beta^2(\mu-\hbar\omega)^2+\pi^2 n^2} \\ &= -\frac{1}{2}\frac{4}{\pi}\sum_{k=1}^{\infty} \frac{\frac{x}{\pi}}{(2k-1)^2+\left(\frac{x}{\pi}\right)^2}\Bigg|_{x=\beta(\hbar\omega-\mu)}\end{aligned}$$

with $n = 2k-1$ ($k = 1, 2, 3, \ldots$) for positive n. The series on the right-hand side represents a hyperbolic tangent function [30], more precisely

$$\frac{1}{\hbar\beta}\sum_n \frac{1}{z_n-\omega} = -\frac{1}{2}\tanh\left(\frac{x}{2}\right)\Bigg|_{x=\beta(\hbar\omega-\mu)} = f(\hbar\omega)-\frac{1}{2}. \tag{14.61}$$

The argument of the Fermi function in (14.61) has sometimes to be shifted by bosonic or fermionic Matsubara frequencies defined in (11.20) and (12.63), respectively. Thereby, it holds

$$\begin{aligned} f(\hbar\omega+\hbar\tilde{z}_m) &= f(\hbar\omega), \\ f(\hbar\omega+\hbar z_n) &= -g(\hbar\omega). \end{aligned} \tag{14.62}$$

These formulas can be applied to perform the $\sum_{n'}$ sum in the correlation expression (14.59) with $G_{\nu\mu}^{m_s m_s'}(\mathbf{k}, z_n) = \tilde{G}_{\nu\mu}^{m_s m_s'}(\mathbf{k}, z_n) = \delta_{m_s m_s'}\delta_{\nu\mu}/[\hbar z_n - \varepsilon_{\nu m_s}(\mathbf{k})]$. With a partial fraction decomposition we find

$$\begin{aligned} &\Sigma^{\mathrm{C}\, m_s m_s'}_{\ \nu\mu}(\mathbf{k}, z) \\ &= -\delta_{m_s m_s'}\frac{1}{\Omega}\sum_{\mathbf{q},\mathbf{G},\mathbf{G}'} \tilde{\mathrm{v}}\left(\sqrt{|\mathbf{q}+\mathbf{G}||\mathbf{q}+\mathbf{G}'|}\right)\sum_{\nu',\mathbf{k}'} B^{\mathbf{k}\mathbf{k}'}_{m_s m_s\,\nu\nu'}(\mathbf{q}+\mathbf{G})B^{\mathbf{k}\mathbf{k}'\,*}_{m_s m_s\,\mu\nu'}(\mathbf{q}+\mathbf{G}') \\ &\quad\times\int_{-\infty}^{+\infty}\frac{d\hbar\omega}{\pi}\mathrm{Im}\tilde{\epsilon}^{-1}(\mathbf{q}+\mathbf{G},\mathbf{q}+\mathbf{G}',\omega)\frac{f(\varepsilon_{\nu' m_s}(\mathbf{k}'))+g(-\hbar\omega)}{\varepsilon_{\nu' m_s}(\mathbf{k}')+\hbar\omega-\hbar z}. \end{aligned} \tag{14.63}$$

Thereby, the results obtained for $z = z_n$ have been analytically continued into the entire complex z-plane. This self-energy contribution contains all static and dynamical screening actions and, hence, the correlation of the electrons.

In the low-temperature limit $T = 0\,\mathrm{K}$ it holds

$$\begin{aligned} \lim_{T\to 0K} f(\varepsilon) &= \theta(\mu-\varepsilon), \\ \lim_{T\to 0K} g(\varepsilon) &= -\theta(-\varepsilon). \end{aligned} \tag{14.64}$$

The antisymmetry of the spectral function $\hat{L}$ (13.10) of the density correlation function leads to the same property of the anti-Hermitian component of the inverse dielectric function and, consequently, for the imaginary part of the corresponding symmetrized quantity

$$\mathrm{Im}\tilde{\epsilon}^{-1}(\mathbf{q}+\mathbf{G},\mathbf{q}+\mathbf{G}',\omega) = -\mathrm{Im}\tilde{\epsilon}^{-1}(\mathbf{q}+\mathbf{G},\mathbf{q}+\mathbf{G}',-\omega).$$

Therefore, together with (14.64), the low-temperature expression of the XC self-energy in Bloch-Fourier representation reads as [8, 9, 20, 28]

$$\begin{aligned}
&\Sigma_{\nu\mu}^{m_s m_s'}(\mathbf{k},z)\\
&= -\delta_{m_s m_s'}\frac{1}{\Omega}\sum_{\mathbf{q},\mathbf{G},\mathbf{G}'}\tilde{v}\left(\sqrt{|\mathbf{q}+\mathbf{G}||\mathbf{q}+\mathbf{G}'|}\right)\sum_{\nu',\mathbf{k}'}B\,{}^{\mathbf{kk}'}_{\substack{m_s m_s\\ \nu\nu'}}(\mathbf{q}+\mathbf{G})B\,{}^{\mathbf{kk}'\,*}_{\substack{m_s m_s\\ \mu\nu'}}(\mathbf{q}+\mathbf{G}')\\
&\quad\times\left\{\theta\left(\mu-\varepsilon_{\nu' m_s}(\mathbf{k}')\right)\delta_{\mathbf{GG}'}-\int_0^\infty\frac{d\hbar\omega}{\pi}\frac{\mathrm{Im}\tilde{\epsilon}^{-1}(\mathbf{q}+\mathbf{G},\mathbf{q}+\mathbf{G}',\omega)}{\varepsilon_{\nu' m_s}(\mathbf{k}')-\hbar z+\mathrm{sgn}\left(\varepsilon_{\nu' m_s}(\mathbf{k}')-\mu\right)\hbar\omega}\right\}.
\end{aligned} \tag{14.65}$$

The XC self-energy in a collinear electron system is different for the two spin channels, if the system is spin-polarized. The channels are however coupled via the screening reaction and, hence, the density fluctuations in the system. Expression (14.65) can be divided into a bare exchange contribution and a correlation part that depends on the screening dynamics. In the standard approach the screening is described by the inverse of the dielectric function in independent-particle approximation (13.43). Moreover, the off-diagonal elements with respect to the band indices ν and μ are usually omitted in the first-shot approximation used to solve equations (14.42) and (14.45).

The variation of the diagonal self-energy operator $\Sigma_{\nu\nu}^{m_s m_s}(\mathbf{k},\varepsilon/\hbar)$ (14.65), more precisely of its real part, versus the single-particle energy ε is drawn in Fig. 14.12 for four face-centered cubic semiconductors or insulators, diamond C, Si, Ge, and LiCl, and selected Bloch states near their fundamental band gap. A single-plasmon-pole approximation (13.50) is used in the computations [8]. The figure illustrates important properties of the XC self-energy for Bloch states X_{1c} (L_{1c}), Γ_{15c} (Γ_{1c}), $\Gamma_{25'v}$ (Γ_{15v}), and X_{4v} $(X_{4'v})$ near the band gap. In this energy region the energy variation is nearly linear in contrast to the regions extended by energies of the order of the plasmon energy, where more rich spectra, even resonances, appear (see e.g. Fig. 14.8). In these distant regions also the imaginary part of the self-energy significantly influences the QP properties. The energy curves in Fig. 14.12 have a negative slope near the QP energies. They are slightly concave upwards for occupied (hole) states and concave downwards for empty (electron) states. The non-zero slope at the QP energy is related to the renormalization constant $z_{\nu m_s}(\mathbf{k})$ (14.27). For the band edges its values vary between 0.78 and 0.87 [8]. The plots in Fig. 14.12 summarize concisely further important results: (i) The magnitude of the self-energies depends significantly on the

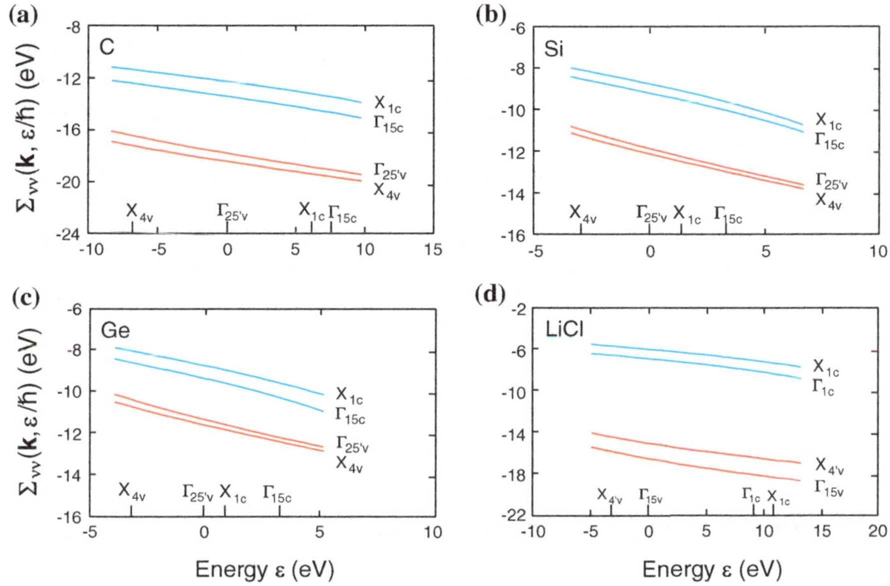

Fig. 14.12 Matrix elements (real part) of the XC self-energy as a function of the single-particle energy for selected Bloch states near the band gap. Results for (**a**) diamond, (**b**) silicon, (**c**) germanium, and (**d**) LiCl are displayed. Empty (occupied) states are indicated by *blue* (*red*) *solid lines*. The resulting QP energy levels are indicated by ticks on the *horizontal line*. The valence band maximum is used as energy zero. Adapted from [8]

Bloch state and the material. (ii) Their energy variations confirm the above discussion. (iii) The self-energies of the electron states are much higher in energy, i.e., are smaller on an absolute scale, than those of the hole states. As a consequence, there is a substantial XC-induced contribution to the QP gap between conduction and valence band states.

The absolute energy value of the XC self-energy (14.65) in Fig. 14.12 is mainly determined by its bare exchange part (14.60) while its energy variation is only represented by the correlation contribution (14.63). This is illustrated in Fig. 14.13

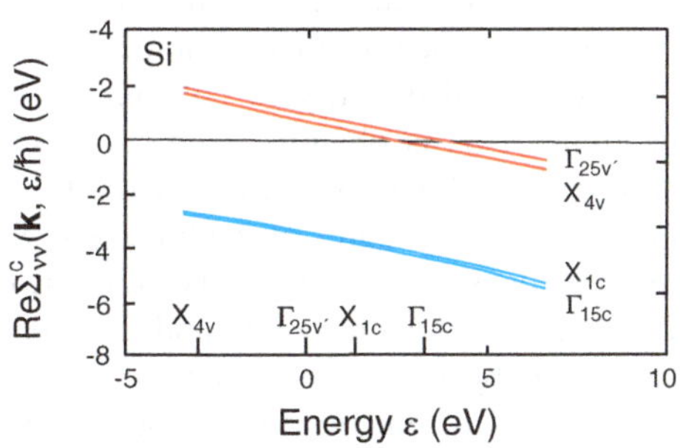

Fig. 14.13 Real part of matrix elements of the correlation self-energy (14.63) as a function of single-particle energy for some Bloch states near the fundamental gap in silicon. *Blue* (*red*) *lines* indicate empty (occupied) states. Adapted from [8]

for near-gap states in silicon [8]. The trends for other materials are similar: (i) The correlation energy varies around zero for hole states but is negative for electron states. Correlation is small for holes but more important for electrons. Indeed, the electron-hole asymmetry is evident as the magnitude of the correlation energy is substantially larger for electron states. We conclude that the electron-hole differences of the XC self-energies in Fig. 14.12 of about 4 eV are mainly due to correlation effects. The differences in the bare exchange contributions only slightly shrink this gap by about 1 eV. The correlation of electrons (or holes) does hardly vary with the symmetry of the Bloch state. The behavior of the correlation self-energy suggests its possible modeling near the band-gap region by a constant shift plus an energy-dependent term for electron states and hole states, respectively [8].

14.4 Quasiparticle Shifts

14.4.1 Physical and Numerical Approaches

The large negative values of the XC self-energies in Fig. 14.12 indicate that the details of the physical and numerical treatment of the self-energy expression (14.65) may substantially influence the explicit results. Among the physical approximations are (i) the neglect of vertex corrections, i.e., the GW approximation (12.57), (ii) the RPA, or more precisely, the independent-particle or -quasiparticle approach for the polarization function (12.58), (iii) the treatment of the frequency dependence in (14.65) via a plasmon-pole approximation or a full frequency integration, and (iv), of course, the used reference electronic structure.

Numerically, the reciprocal space-real frequency method illustrated above and the real space-imaginary time approach [31] seem to suggest to give totally different numerical results. Convergence with respect to the sampling of the reciprocal or real space as well as of the frequency or time interval play an important role. The number of bands and the representation of the wave functions also influence the self-energy results (for details see [28]). Also the used basis functions and pseudopotentials have a certain influence on the exact numerical values [28, 32]. However, fortunately converged calculations approach to similar values for the matrix elements, at least describe the correct trends, but give very promising results for relative quantities such as differences of matrix elements, especially QP shifts.

One critical quantity in both real space and reciprocal space is the Coulomb potential. This is obvious from reciprocal-space expression (14.65) studying the contributions for small wave vectors $\mathbf{G} = \mathbf{G}' = 0$ and $\mathbf{q} \to 0$. In this limit one needs to compute sums of the type $\frac{1}{\Omega}\Sigma_{\mathbf{q}}\tilde{v}(q)$ with $\mathbf{q} \in \mathrm{BZ}$. Converged calculations require a dense mesh of $\mathbf{q}$ points including the corresponding screening functions and Bloch matrix elements. Gygi and Baldereschi [33] suggested to remove the $\sim 1/q^2$ singularity by an auxiliary function $F(\mathbf{q})$. A special choice for $F(\mathbf{q})$ has been tested for the fcc case. Functions $F(\mathbf{q})$ appropriate for other crystal symmetries than fcc have been suggested by other authors [34]. Such an auxiliary function should (i) reflect the translational symmetry of the considered Bravais lattice and (ii) diverge like $1/q^2$ as q vanishes. The main idea is to compute $\frac{1}{\Omega}\Sigma_{\mathbf{q}} F(\mathbf{q})$ before the self-energy calculation and study later only non-divergent contributions $\frac{1}{\Omega}\Sigma_{\mathbf{q}}\left[\tilde{\mathrm{v}}(q) - F(\mathbf{q})\right]$. In practice, also more simplified functions $F(\mathbf{q})$, e.g. the Coulomb potential multiplied with a Gaussian function, can be used, for instance in the description of self-energy effects in some nanostructures [35]. Also more sophisticated treatments of the Coulomb singularity based on the Ewald method [36] can be applied [32].

The quasiparticle shifts described within the GW approximation (12.57) and a one-shot approach (14.45) are usually computed by means of the representation (14.46) with prefactor $z_{\nu m_s}(\mathbf{k})$ (14.47). However in cases, where the starting single-particle energies $\varepsilon_{\nu m_s}(\mathbf{k})$ are almost identical with the desired QP ones $\varepsilon^{\mathrm{QP}}_{\nu m_s}(\mathbf{k})$, i.e., for vanishing QP shifts, the prefactor may be omitted and a representation

$$\Delta_{\nu m_s}(\mathbf{k}) = \mathrm{Re}\delta\tilde{\Sigma}^{m_s m_s}_{\nu\nu}(\mathbf{k}, \varepsilon_{\nu m_s}(\mathbf{k})/\hbar) \tag{14.66}$$

is useful. In other words, one uses the reference Green function $\tilde{\tilde{G}} = \tilde{G}$ and eigenvalues $\varepsilon^{\mathrm{QP}}_{\nu m}(\mathbf{k}) = \varepsilon_{\nu m_s}(\mathbf{k})$ in the explicit calculations. Indeed, in many recent computations, despite the replacement of $\tilde{\tilde{G}}$ by $\tilde{G}$, the energy ε at which the self-energy has to be taken, is still chosen as $\varepsilon = \varepsilon^{\mathrm{QP}}_{\nu m_s}(\mathbf{k})$ (see e.g. [8]) but simultaneously expanded in terms of small QP shifts using (14.42). In the latter case, the resulting QP shift has to be reduced by a factor $z_{\nu m_s}(\mathbf{k})$. Usually a certain local or semilocal approximation is used for the XC potential in the Kohn-Sham equation. However, there are also cases where reference electronic structures are computed using appropriate non-local XC potentials (see Sect. 9.2), which yield eigenvalues close to the final QP result. We examine *a posteriori* the form of such a correction

$$\Delta_{\nu m_s}(\mathbf{k}) = \Sigma^{m_s m_s}_{\nu\nu}(\mathbf{k}, \varepsilon_{\nu m_s}(\mathbf{k})/\hbar) - \int d^3\mathbf{x} \int d^3\mathbf{x}' \varphi^*_{\nu\mathbf{k}m_s}(\mathbf{x}) V^{m_s}_{\mathrm{XC}}(\mathbf{x}, \mathbf{x}')\varphi_{\nu\mathbf{k}m_s}(\mathbf{x}'), \tag{14.67}$$

where an appropriate spatially non-local XC potential of a collinear-spin system is considered to determine the reference electronic structure. This formula suggests a partial error compensation while subtracting the self-energy and potential matrix elements computed with eigenvalues and eigenfunctions of a gKS problem.

14.4.2 Influence of State Symmetry and Occupation

To discuss the dependence on the single-particle state and its occupation, we investigate the indirect semiconductor silicon crystallizing in diamond structure with an fcc Bravais lattice as prototypical material. Because of the pure covalent sp^3 bonds, one finds a pronounced plasmon-pole peak in the loss function (see Fig. 13.7a), and a finite gap between occupied and empty Kohn-Sham eigenvalues, independent of the actually applied local or semilocal XC approximation but noticeable below the experimental fundamental gap of about 1.1 eV. The resulting quasiparticle corrections and quasiparticle band structures should however be computable in the framework of (14.65) and (14.66). To illustrate this fact some results of early QP calculations [8, 9, 35] are listed in Table 14.1 for Bloch band states at high-symmetry points Γ, X, and L of silicon. All these early studies are based on the spin-less LDA XC functional (7.15) and (7.21). They mainly differ with respect to the description of the frequency dependence in the inverse dielectric function, the plasmon-pole approach [8, 35] or imaginary-time axis integration [9].

The QP energies in Table 14.1 support the general conclusion that absolute values for QP energies of semiconductors obtained in different calculations vary between 0.1 and 0.6 eV because of different techniques, different degrees of convergence, different dynamical screening, state localization etc. [28]. The slightly larger deviations of the localized Si 3s states Γ_{1v} are a consequence of the used model screening [35]. The Si example in Table 14.1 shows that the QP effects significantly open the energy distances between valence and conduction band states. This is especially obvious for the fundamental gap. The underestimation of the indirect gap with the conduction band minimum near $0.85\Gamma X$ is about 60 % using the KS eigenvalues of the ground state treated in DFT-LDA. Nevertheless, the resulting QP gaps approach the measured values with an uncertainty of about 0.1 eV. That means, already for such a simple sp-bonded material the quasiparticle gap opening is larger than the reference gap.

Table 14.1 Single-particle energies $\varepsilon_\nu(\mathbf{k})$ of reference system and quasiparticle energies $\varepsilon_\nu^{\mathrm{QP}}(\mathbf{k})$ for important band states at high-symmetry points Γ, X, and L in the BZ for (non-spin-polarized) Si

Band state	$\varepsilon_\nu(\mathbf{k})$	$\Delta_\nu(\mathbf{k})$	$\varepsilon_\nu^{\mathrm{QP}}(\mathbf{k})$	$\varepsilon_\nu^{\mathrm{exp}}(\mathbf{k})$
Γ_{1v}	−11.93, –, −12.08	−0.11, –, −0.81	−12.04, –, −12.89	−12.5 ± 0.6
$\Gamma_{25'v}$	0	0	0	0
Γ_{15c}	2.57, 2.57, 2.53	0.78, 0.73, 0.68	3.35, 3.30, 3.21	3.4
$\Gamma_{2'c}$	–, 3.56, 3.35	–, 0.71, 0.73	4.08, 4.27, 4.08	4.19
X_{1v}	–, –, −7.89	–, –, −0.52	–, –, −8.41	–
X_{4v}	–, –, −2.91	–, –, −0.10	−2.99, –, −3.01	−2.9, −3.3 ± 0.2
X_{1c}	–, –, 0.58	–, –, 0.64	1.44, –, 1.22	1.3
Δ_{1c}	0.52, 0.52, 0.44	0.77, 0.72, 0.62	1.29, 1.24, 1.06	1.17
$L_{3'v}$	–, −1.22, –	–, 0.03, –	–, −1.19, –	−1.2 ± 0.2, −1.5
L_{1c}	–, 1.53, –	–, 0.77, –	–, 2.30, –	2.1, 2.4
L_{3c}	–, 3.37, –	–, 0.74, –	–, 4.11, –	4.3 ± 0.2, 4.0

Δ_{1c} denotes a conduction band minimum on a Δ line near an X point. The quasiparticle shifts $\Delta_\nu(\mathbf{k})$ are also listed. Results of three pseudopotential-plane wave calculations with an LDA XC potential [8, 9, 35] are listed. The use of the imaginary-time axis integration [9] avoids the need of a model ansatz for the frequency dependence of the inverse dielectric function within the single-plasmon-pole model. The experimental band energy values are taken from the collections in [8, 9]. The $\Gamma_{25'v}$ VBM is used as energy zero. All energy values are in eV

However, for other conduction band states such as Γ_{15c}, $\Gamma_{2'c}$, X_{1c}, L_{1c}, and L_{3c} but also valence band states Γ_{1v}, X_{4v}, and $L_{3'v}$ the agreement between QP values and measured band energies is good, taking also the experimental uncertainties into account. Except for the lowest valence states the deviations are typically smaller than 0.2 eV. In any case this reasonable agreement indicates that the QP treatment based on the GW approximation and a reference KS electronic structure is a promising method to predict the energy positions of excited electronic states, at least of semiconductors with not too large gaps. Consequently, the presented QP theory should have predictive power.

As a summary the KS and QP band structures of silicon [28] are compared in Fig. 14.14 along the ΓX and ΓL directions with results of photoemission and inverse photoemission experiments [37–41]. The typical experimental resolution is 0.27 eV in energy and 0.1 Å^{-1} in momentum (see [37]). In numerical calculations the momentum is well defined, while the error in the theoretical band energies is estimated to be about ±0.1 to 0.2 eV [28]. The agreement between QP theory and PES/IPES experiments is good along the ΓL direction. The lowest conduction band shows the largest discrepancies outside L toward Γ. Along the ΓX direction the agreement between experiment and theory is satisfactory. Nevertheless, due to the large experimental momentum uncertainty, theory and experiment still agree to within the above discussed uncertainties. On the other hand, the lowest valence band and the

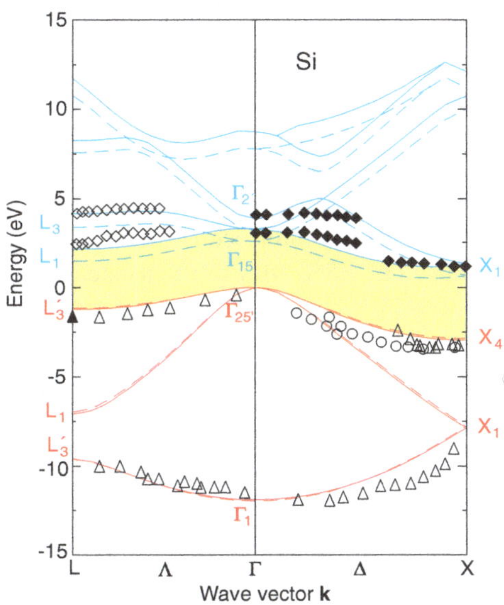

Fig. 14.14 Comparison of KS (*dashed lines*) and QP (*solid lines*) bands of silicon along the ΓX and ΓL directions [28] with results of photoemission and inverse photoemission experiments. Experimental data are taken from [37] (*full diamonds*), [38] (*open circles*), [39] (*full triangles*), [40] (*open triangles*), and [41] (*open diamonds*). The VBM is used as energy zero. The fundamental gap region is indicated by the *yellow* color. Conduction (valence) bands are plotted as *blue* (*red*) *lines*. Adapted from [28]. Copyright (2000), with permission from Elsevier

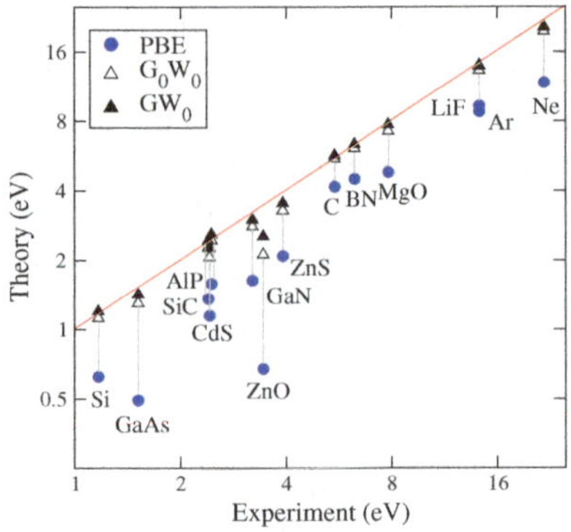

Fig. 14.15 QP fundamental gaps of semiconductors and insulators versus measured values. A logarithmic scale is used for both axes. The reference electronic structure is computed in the framework of a semilocal PBE-GGA XC functional. The one-shot GW (G_0W_0) corrections generate QP gaps close to the experimental ones. A self-consistent treatment of the eigenvalues in the Green function (GW_0) slightly improves the agreement. Reprinted with permission from [42]. Copyright 2006 by the American Physical Society

second lowest conduction band show a weaker dispersion as expected from the QP calculations.

As another summary the fundamental gaps of semiconductors and insulators are displayed in Fig. 14.15 as results of the standard QP approach versus measured values [42]. In this case the standard one-shot GW approach starts from a reference electronic structure obtained using the PBE-GGA XC functional. Values modified by a self-consistent treatment of the energies in the Green function are also shown. The quasiparticle corrections significantly open the gaps toward the experimental values. A further improvement seems to be possible going beyond the standard approach and taking partially self-consistency into account. Indeed the MARE of 8.5 % in the standard approach is reduced to 4.5 % with self-consistency.

14.4.3 Influence of Reference Electronic Structure

In Sect. 14.2 we have discussed that the perturbation operator (14.10) between the XC self-energy and the local or non-local XC potential used to compute the reference electronic structure to determine $\tilde{G}$ (14.6) should be small. In this case a kind of first iteration or a one-shot approximation should be sufficient to describe the unknown QP electronic structure. With other words, the reference electronic structure should be not too far from the "true" quasiparticle one. In order to study the influence of the starting reference electronic structure on the final QP energy bands or energy levels and, hence, on the QP shifts, we consider five different local and non-local XC potentials entering a KS equation (6.22) or gKS equation (9.22) [or even HF equation (4.25)].

More precisely, we apply the five XC functionals described in Table 9.1 to derive corresponding XC potentials. Thereby, the LDA limit is modeled by (7.15) and (7.21).

An implementation of the XC self-energy and the QP corrections [42] in the VASP code [43] is used in all studies. It is based on a projector-augmented wave (PAW) method [44] for the representation of eigenfunctions and the generation of pseudopotentials. Details of the GW implementation can be found in the literature [25, 42]. One problem in the presence of shallow core, e.g. semicore d levels, is the strong core-valence XC interaction [45, 46]. It can be estimated within a local (e.g. LDA) or a non-local (e.g. HF) approximation, where the latter is expected more reliable since the GW self-energy (14.65) approaches the bare Fock exchange operator (14.60) in the short-wave length regime, i.e., at large electron binding energies. Therefore, in many cases the HF approximation to the core-valence XC self-energy is applied in GW calculations. This technical point, mainly due to the use of pseudopotentials, should not be discussed in the following.

One-shot G_0W_0 results [26, 47] are presented in Fig. 14.16 for the quasiparticle shifts $\Delta_\nu(\mathbf{k})$ in diamond-Si, zb-ZnO, and zb-InN of band energies $\varepsilon_\nu(\mathbf{k})$ of a reference electronic structure computed by means of the XC potentials in PZ-LDA, sX, HSE03/06, PBE0, and HF (see Table 9.1). The screening parameter $\mu = 0.3\,\text{Å}^{-1}$ in the HSE functional has been slightly increased with respect to the original value (see Sect. 9.2.2). The PZ-LDA starting point confirms the results for Si in Table 14.1 and Fig. 14.14. The sign of the QP shifts of the three semiconductors depends on the state occupation, i.e., valence bands acquire negative shifts while conduction bands are shifted upward by positive QP shifts. In the case of the two hybrid functionals HSE03/06 and PBE0, which essentially mix 25 % HF and 75 % DFT exchange, this picture (apart from the lower valence and higher conduction bands of Si) is confirmed but with small QP shifts. This fact highlights that the $\alpha = 0.25$ recipe is indeed a remarkable good and robust choice for one-electron QP energies in semiconductors. Starting energy values $\varepsilon_{\nu m_s}(\mathbf{k})$ and first-shot QP ones $\varepsilon^{\mathrm{QP}}_{\nu m_s}(\mathbf{k})$ are rather close. On the other hand, a reference HF electronic structure is a less suitable starting point. Large QP shifts with opposite signs appear in order to compensate the significant overestimation (underestimation) of binding energies of occupied (empty) states with respect to vacuum level. Close to the band gap the performance of the sX functional as starting XC description is similar to that of HSE03/06 or PBE0, except for the conduction bands of ZnO and the semicore d-state positions.

The general good performance of the HSE03/06 and PBE0 starting points on the fundamental energy gaps of many semiconductors and insulators is illustrated in Fig. 14.17. In the average the gKS gap values are already close to the experimental ones in contrast to KS values based on the PBE-GGA functional. This fact underlines the quality of the hybrid-functional starting points versus the KS gaps within PBE-GGA, at least for wide-band gap materials. More in detail the influence of the starting point on quasiparticle gaps and In 4d or Zn 3d QP binding energies obtained within the standard GW approach is illustrated in Table 14.2 for three tetrahedrally coordinated semiconductors. In addition, the static electronic dielectric constants ϵ_∞ (13.24) computed within the independent-particle approximation are listed. For

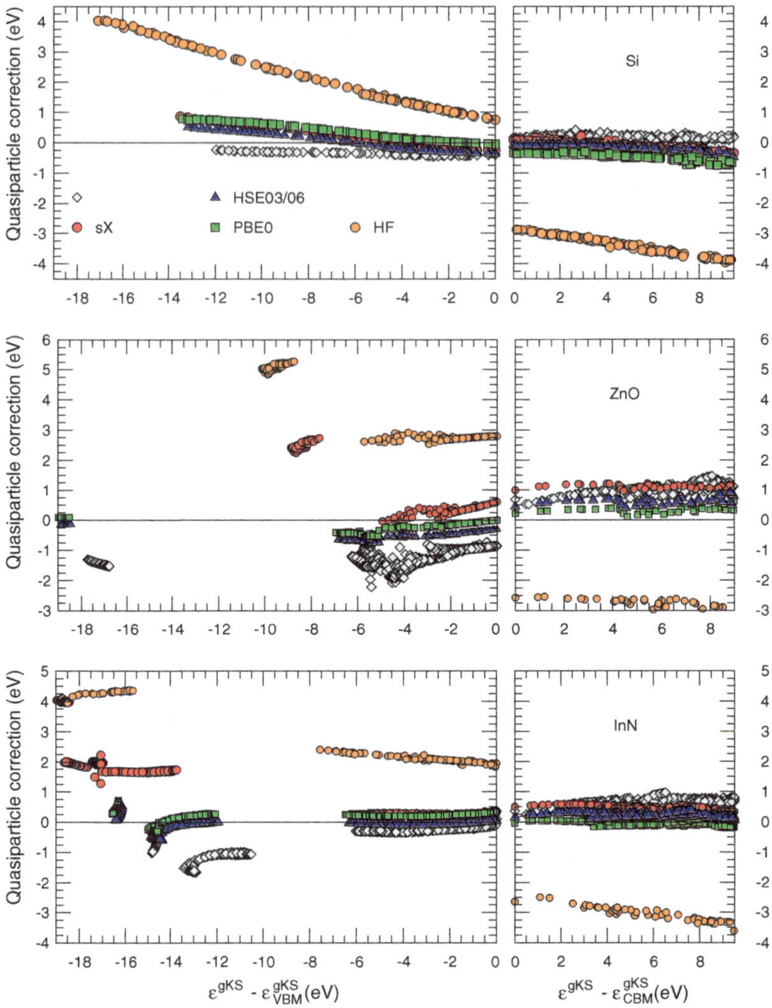

Fig. 14.16 Quasiparticle corrections $\Delta_\nu(\mathbf{k})$ for valence (*left panels*) and conduction (*right panels*) band states of non-spin-polarized diamond-Si, *zb*-ZnO, and *zb*-InN versus the eigenvalues of the reference electronic structures. The QP calculations are performed in the one-shot framework of the Hedin GW approach. Four different KS or gKS approximations with XC functionals LDA, sX, HSE03/06, PBE0, and HF [see (7.21), Table 9.1, and (4.25)] are applied. The VBM and CBM have been used as energy zeros. Adapted from [26, 47]

the covalently bonded silicon, in a one-shot approach the KS starting point with the PZ-LDA functional yields excellent QP gap values. With a maximum deviation of 0.15 eV also the sX and HSE03/06 starting points lead to reasonable agreement with experiment. The other non-local functionals PBE0 and HF tend to significant gap overestimations. For InN and ZnO, the trend discussed for Si still holds, but

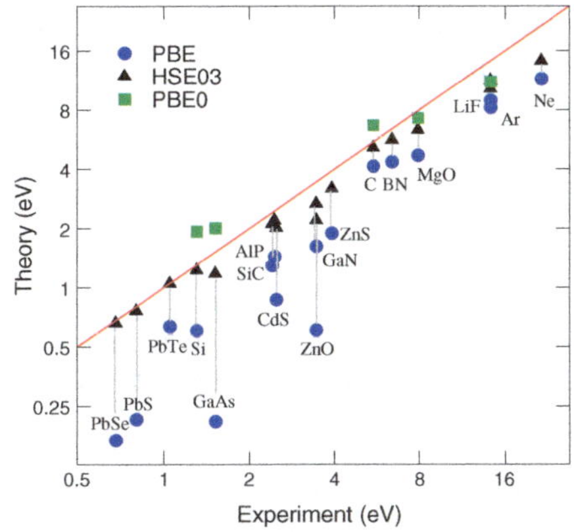

Fig. 14.17 Fundamental gap values from KS/gKS eigenvalues that are computed using the semi-local PBE-GGA functional (7.25) and the hybrid functionals HSE03 and PBE0 (see Sect. 9.2.2). From [48]. Copyright IOP Publishing. Reproduced by permission of IOP Publishing. All rights reserved

the actual benefit of a gKS starting point becomes more apparent. In contrast to the PZ-LDA functional with a negative gap, the gKS starting points generate the correct ordering of the band-edge states for InN [49] and yield more meaningful dielectric constants. This is also valid for sX and HSE03/06. The other gKS functionals give rise to a gap overestimation. In the case of ZnO the sX and PBE0 hybrid functionals give reasonable gaps while the HSE03/06 value is too small. Thereby, we have to take into consideration that the calculations have been performed for the zinc-blende polytype, whereas the experimental values are measured for the wurtzite structure (which tends to a gap increase by 0.1–0.2 eV). The QP results in Table 14.2 for the

Table 14.2 Direct (d) and indirect (i) GW QP band gaps E_g^{QP} in one-shot approximation, average d-band binding energies E_d^{QP}, and static electronic dielectric constants ϵ_∞

	Energy	PZ-LDA	sX	HSE03/06	PBE0	HF	Expt.
Si	$E_{g,i}^{QP}$	1.08	1.31	1.32	1.65	2.93	1.17
	$E_{g,d}^{QP}$	3.18	3.49	3.48	3.72	5.21	3.40
	ϵ_∞	13.9	10.8	9.8	7.8	3.4	11.90
InN	E_g^{QP}	0.00	0.55	0.47	0.78	2.56	0.61
	E_d^{QP}	15.1	15.6	15.2	15.3	16.6	16.0 – 16.9
	ϵ_∞	12.2	6.6	6.8	4.9	2.4	7.96
ZnO	E_g^{QP}	2.14	3.36	2.87	3.24	5.71	3.44
	E_d^{QP}	5.6	6.2	6.1	6.2	7.0	7.5 – 8.8
	ϵ_∞	5.3	3.0	3.4	3.0	1.8	3.74

All energy values are given in eV. Experimental results are listed for comparison. From [47]

semicore binding energies in InN and ZnO underestimate the measured values aside those using the HF starting point.

References

1. R.W. Godby, M. Schlüter, L.J. Sham, Trends in self-energy operators and their corresponding exchange-correlation potentials. Phys. Rev. B **36**, 6497–6500 (1987)
2. A. Fleszar, W. Hanke, Spectral properties of quasiparticles in a semiconductor. Phys. Rev. B **56**, 10228–10232 (1997)
3. L. Hedin, S. Lundqvist, Effects of electron-electron and electron-phonon interactions on the one-electron states of solids, in *Solid State Physics*, vol 23, ed. by F. Seitz, D. Turnbull, H. Ehrenreich (Academic Press, New York 1969), pp. 1–181
4. A.M. Zagoskin, *Quantum Theory of Many-Body Systems. Techniques and Applications* (Springer, New York, 1998)
5. W. Jones, N.H. March, *Theoretical Solid State Physics. Perfect Lattices in Equilibrium*, vol. 1 (Dover Publications Inc, New York, 1973)
6. A.A. Abrikosov, L.P. Gorkov, I.E. Dzyaloshinski, *Methods of Quantum Field Theory in Statistical Physics* (Prentice-Hall Inc, Englewood Cliffs, 1963)
7. H. Stolz, *Einführung in die Vielelektronentheorie der Kristalle* (Akademie, Berlin, 1974)
8. M.S. Hybertsen, S.G. Louie, Electron correlation in semiconductors and insulators: band gaps and quasiparticle energies. Phys. Rev. B **34**, 5390–5413 (1986)
9. R.W. Godby, M. Schlüter, L.J. Sham, Self-energy operators and exchange-correlation potentials in semiconductors. Phys. Rev. B **37**, 10159–10175 (1988)
10. G. Onida, L. Reining, A. Rubio, Electronic excitations: density-functional versus many-body Green's-function approaches. Rev. Mod Phys. **74**, 601–659 (2002)
11. J. Hubbard, Electron correlations in narrow energy bands. Proc. Roy. Soc. London A **276**, 238–257 (1963)
12. N.W. Ashcroft, N.D. Mermin, *Solid State Physics*, (Saunders College, Philadelphia, 1976)
13. L.D. Landau, The theory of a Fermi liquid. Zh. Eksp. Teor. Fiz. **30**, 1058–1064 (1956), [Soviet Phys. JETP (English Transl.) **3**, 920–925 (1956)]
14. L.D. Landau, Oscillations in a Fermi liquid. Zh. Eksp. Teor. Fiz. **32**, 59–66 (1957), [Soviet Phys. JETP (English Transl.) **5**, 101–108 (1957)]
15. E.L. Shirley, Self-consistent GW and higher-order calculations of electron states in metals. Phys. Rev. B **54**, 7758–7764 (1996)
16. U. von Barth, B. Holm, Self-consistent GW_0 results for the electron gas: fixed screened potential W_0 within the random-phase. Phys. Rev. B **54**, 8411–8419 (1996)
17. C. Blomberg, B. Bergersen, Spurious structure from approximations to the Dyson equation. Canadian J. Phys. **50**, 2286–2293 (1972)
18. B. Bergersen, F.W. Kus, C. Blomberg, Single-particle Green's function in the electron-plasmon approximation. Canadian J. Phys. **51**, 102–110 (1973)
19. R.D. Mattuck, *A Guide to Feynman Diagrams in the Many-Body Problem* (Dover Publ. Inc, New York, 1992)
20. P.H. Hahn, W.G. Schmidt, F. Bechstedt, Molecular electronic excitations calculated from a solid-state approach. Phys. Rev. B **72**, 245425 (2005)
21. R. Zimmermann, H. Stolz, The mass action law in two-component fermi systems revisited excitons and electron-hole pairs. Phys. Status Solidi B **131**, 151–164 (1985)
22. L. Hedin, New method for calculating the one-particle Green's function with application to the electron-gas problem. Phys. Rev. **139**, A796–A823 (1965)
23. O. Pulci, F. Bechstedt, G. Onida, R. Del Sole, L. Reining, State mixing for quasiparticles at surfaces: nonperturbative GW approximation. Phys. Rev. B **60**, 16758–16761 (1999)
24. http://cms.mpi.univie.ac.at/vasp/

25. M. Shishkin, G. Kresse, Self-consistent GW calculations for semiconductors and insulators. Phys. Rev. B **75**, 235102 (2007)
26. F. Fuchs, Ab-initio-Methoden zur Berechnung der elektronischen Anregungseigenschaften von Halbleitern und Isolatoren unter Berücksichtigung von Vielteilcheneffekten. Ph.D. thesis, Friedrich-Schiller-Universität Jena (2008)
27. F. Bruneval, N. Vast, L. Reining, Effect of self-consistency on quasiparticles in solids. Phys. Rev. B **74**, 045102 (2006)
28. W.G. Aulbur, L. Jönsson, J.W. Wilkins, Quasiparticle calculations in solids, in *Solid State Physics*. Advances in Research and Applications, vol. 54, ed. by H. Ehrenreich, F. Spaepen (Academic Press, San Diego, 2000), pp. 1–218
29. L.P. Kadanoff, G. Baym, *Quantum Statistical Mechanics. Green's Function Methods in Equilibrium and Nonequilibrium Problems* (W.A. Benjamin Inc, New York, 1962)
30. S. Gradstein, I.M. Ryshik, *Tables of Series, Products, and Integrals* (Harri Deutsch, Frankfurt, 1981)
31. M.M. Rieger, L. Steinbeck, I.D. White, H.N. Rojas, R.W. Godby, The GW space-time method for the self-energy of large systems. Comput. Phys. Commun. **117**, 211–228 (1999)
32. F. Aryasetiawan, The GW approximation and vertex corrections, in *Strong Coulomb Correlations in Electronic Structure Calculations. Beyond the Local Density Approximation*, ed. by V.I. Anisimov (Gordon and Breach Science Publishers, Amsterdam. 2000), pp. 1–95
33. F. Gygi, A. Baldereschi, Self-consistent Hartree-Fock and screened-exchange calculations in solids: application to silicon. Phys. Rev. B **34**, 4405–4408 (1986)
34. B. Wenzien, G. Cappellini, F. Bechstedt, Efficient quasiparticle band-structure calculations for cubic and noncubic crystals. Phys. Rev. B **51**, 14701–14704 (1995)
35. J. Furthmüller, G. Cappellini, H.-Ch. Weissker, F. Bechstedt, GW self-energy calculations for systems with huge supercells. Phys. Rev. B **66**, 045110 (2002)
36. P. Ewald, Die Berechnung optischer und elektrostatischer Gitterpotentiale. Ann. Phys. **369**, 253–287 (1921)
37. J.E. Ortega, F.J. Himpsel, Inverse-photoemission study of Ge(100), Si(100), and GaAs(100): bulk bands and surface states. Phys. Rev. B **47**, 2130–2137 (1993)
38. A.L. Wachs, T. Miller, T.C. Hsieh, A.P. Shapiro, T.-C. Chiang, Angle-resolved photoemission studies of Ge(111)-c(2×8), Ge(111)-(1×1)H, Si(111)-(7×7), and Si(100)-(2×1). Phys. Rev. **B 32**, 2326–2333 (1985) (as presented in [8])
39. F.J. Himpsel, P. Heimann, D.E. Eastmann, Surface states on Si(111)-(2×1). Phys. Rev. B **24**, 2003–2008 (1981)
40. D.H. Rich, T. Miller, G.E. Franklin, T.C. Chiang, Sb-induced bulk band transitions in Si(111) and Si(001) observed in synchrotron photoemission studies. Phys. Rev. B **39**, 1438–1441 (1989)
41. D. Straub, L. Ley, F.J. Himpsel, Inverse-photoemission study of unoccupied electronic states in Ge and Si: bulk energy bands. Phys. Rev. B **33**, 2607–2614 (1986)
42. M. Shishkin, G. Kresse, Implementation and performance of the frequency-dependent GW method within the PAW framework. Phys. Rev. B **74**, 035101 (2006)
43. http://cms.mpi.univie.ac.at/wiki/index.php/GW-recipes
44. G. Kresse, D. Joubert, From ultrasoft pseudopotentials to the projector augmented-wave method. Phys. Rev. B **59**, 1758–1775 (1999)
45. W. Ku, A.G. Eguiluz, Band-gap problem in semiconductors revisited: Effects of core states and many-body self-consistency. Phys. Rev. Lett. **89**, 126401 (2002)
46. S. Sharma, J.K. Dewhurst, C. Ambrosch-Draxl, All-electron exact exchange treatment of semiconductors: effect of core-valence interaction on band-gap and d-band position. Phys. Rev. Lett. **95**, 136402 (2005)
47. F. Fuchs, J. Furthmüller, F. Bechstedt, M. Shishkin, G. Kresse, Quasiparticle band structure based on generalized Kohn-Sham scheme. Phys. Rev. B **76**, 115109 (2007)
48. M. Marsman, J. Paier, A. Stroppa, G. Kresse, Hybrid functionals applied to extended systems. J. Phys. Condens. Matter **20**, 064201 (2008)
49. F. Bechstedt, J. Furthmüller, Do we know the fundamental energy gap of InN?. J. Crystal Growth **246**, 315–319 (2002)

Chapter 15
Model GW Studies

Abstract The decomposition of the GW self-energy into a screened exchange (SEX) and a Coulomb hole (COH) term allows for deeper insight in the formation of quasiparticles. The SEX term has a structure similar to that known from the exchange in the Hartree-Fock theory. Only the bare Coulomb potential is reduced by screening. The COH term is dominated by a state-independent, local potential that expresses the effect of electronic polarization. Consequently, the opening of gaps and interband transitions is mainly due to the SEX term, whereas the COH one significantly influences the absolute position of eigenvalues. This is demonstrated for the gap shrinkage due to the presence of an additional degenerate electron gas. The differences of matrix elements of the GW self-energy and the exchange-correlation potential may be further approximated. As most simplified approach the description of the quasiparticle effects by a scissors operator is derived.

15.1 Coulomb Hole and Screened Exchange

15.1.1 Decomposition in Real Space

A physically appealing way of expressing the GW self-energy (14.52) is to divide it into some screened exchange (SEX) contribution and a rest that will be later interpreted as Coulomb hole (COH) term [1, 2]. To do so, we restrict the Green function G in the self-energy to that $\tilde{G}$ (14.6) of the reference system. It results

$$\begin{aligned}&\Sigma_{ss'}(\mathbf{xx}', z_n) \qquad (15.1)\\ &= -\lim_{\eta\to+0}\frac{1}{\beta}\sum_{n'} e^{z_{n'}\eta}W(\mathbf{xx}', z_n - z_{n'})\sum_{\nu,m_s}\sum_{\mathbf{k}}\frac{\varphi_{\nu\mathbf{k}m_s}(\mathbf{x})\chi_{\frac{1}{2}m_s}(s)\varphi^*_{\nu\mathbf{k}m_s}(\mathbf{x}')\chi^+_{\frac{1}{2}m_s}(s')}{\hbar z_{n'} - \varepsilon_{\nu m_s}(\mathbf{k})}.\end{aligned}$$

The idea of a decomposition of the self-energy is suggested in order to distinguish between contributions from the poles in the screened potential W and the Green function $\tilde{G}$. The screened potential obeys a spectral representation according to that of the inverse dielectric function (13.19)

F. Bechstedt, *Many-Body Approach to Electronic Excitations*,
Springer Series in Solid-State Sciences 181, DOI 10.1007/978-3-662-44593-8_15

$$W(\mathbf{xx}', \tilde{z}_m) = \mathrm{v}(\mathbf{x} - \mathbf{x}') + \int_{-\infty}^{+\infty} \frac{d\omega}{2\pi} \frac{\hat{W}(\mathbf{xx}', \omega)}{\tilde{z}_m - \omega} \tag{15.2}$$

with the spectral function

$$\begin{aligned} \hat{W}(\mathbf{xx}', \omega) &= \int d^3\mathbf{x}'' \hat{\epsilon}^{-1}(\mathbf{xx}'', \omega)\mathrm{v}(\mathbf{x}'' - \mathbf{x}') \\ &= 2\int d^3\mathbf{x}'' \int d^3\mathbf{x}''' \mathrm{v}(\mathbf{x} - \mathbf{x}'')\hat{L}(\mathbf{x}''\mathbf{x}''', \omega)\mathrm{v}(\mathbf{x}''' - \mathbf{x}'). \end{aligned} \tag{15.3}$$

It fulfills the same symmetry relations (13.10) as the spectral function $\hat{L}(\mathbf{xx}', \omega)$ of the density correlation function.

Expression (15.1) can be easily decomposed using the rules derived for the bare exchange (14.58) and the $\sum_{n'}$ sum (14.61). Together with a partial fraction decomposition of the product of two energy denominators one finds

$$\begin{aligned} &\Sigma_{ss'}(\mathbf{xx}', z_n) = \Sigma^{\mathrm{COH}}_{ss'}(\mathbf{xx}', z_n) + \Sigma^{\mathrm{SEX}}_{ss'}(\mathbf{xx}', z_n), \\ &\Sigma^{\mathrm{COH}}_{ss'}(\mathbf{xx}', z_n) \\ &= \sum_{\nu,m_s}\sum_{\mathbf{k}} \varphi_{\nu\mathbf{k}m_s}(\mathbf{x})\chi_{\frac{1}{2}m_s}(s)\varphi^*_{\nu\mathbf{k}m_s}(\mathbf{x}')\chi^+_{\frac{1}{2}m_s}(s') \int_{-\infty}^{+\infty} \frac{d\hbar\omega}{2\pi} \frac{g(-\hbar\omega)\hat{W}(\mathbf{xx}', \omega)}{\varepsilon_{\nu m_s}(\mathbf{k}) + \hbar\omega - \hbar z_n}, \\ &\Sigma^{\mathrm{SEX}}_{ss'}(\mathbf{xx}', z_n) \\ &= -\sum_{\nu,m_s}\sum_{\mathbf{k}} f(\varepsilon_{\nu m_s}(\mathbf{k}))\varphi_{\nu\mathbf{k}m_s}(\mathbf{x})\chi_{\frac{1}{2}m_s}(s)\varphi^*_{\nu\mathbf{k}m_s}(\mathbf{x}')\chi^+_{\frac{1}{2}m_s}(s')W(\mathbf{xx}', z_n - \frac{1}{\hbar}\varepsilon_{\nu m_s}(\mathbf{k})). \end{aligned} \tag{15.4}$$

The second term Σ^{SEX} is simply the exchange term (14.58) but with a frequency-dependent screening of the bare Coulomb potential. The physical interpretation of the first term Σ^{COH} becomes clear in the static approximation due to Hedin [1] neglecting the dynamics of the screening reaction. One may argue that one is interested in quasiparticle states near energies $\hbar z_n \approx \varepsilon_{\nu m_s}(\mathbf{k})$. Then most of the spectral weight comes from states close to this energy. The energy difference $\hbar z_n - \varepsilon_{\nu m_s}(\mathbf{k})$ in (15.4) is small compared to the energy of the main contributions to the energy loss function in $\hat{W}$, which is centered around the plasmon energy $\hbar\Omega(0)$ (13.55) that, e.g., approaches a value of 16.7 eV for the valence electrons of silicon.

If we set $\varepsilon_{\nu m_s}(\mathbf{k}) - \hbar z_n \approx 0$ in the denominator of the spectral representation (15.4), one obtains a space dependence that is dominated by the wave functions of the reference electronic structure in the start approximation. In the zero temperature limit $g(-\hbar\omega) = -\theta(\omega)$ (14.64) using the symmetry $\hat{W}(\mathbf{xx}', \omega) = -\hat{W}(\mathbf{xx}', -\omega)$ (13.10), the spectral representation (15.2), and the closure relations for the wave functions and spinors (14.2), one finds

$$\Sigma^{\mathrm{COH}}_{ss'}(\mathbf{xx}', z_n) = \frac{1}{2}\delta_{ss'}\delta(\mathbf{x} - \mathbf{x}')\left\{W(\mathbf{xx}', 0) - \mathrm{v}(\mathbf{x} - \mathbf{x}')\right\}. \tag{15.5}$$

This is simply the interaction energy of the quasiparticle with the induced potential due to the static screening in the electron gas around the quasiparticle. The factor $\frac{1}{2}$ in (15.5) arises from the adiabatic growth of the interaction. In this static approximation Σ^{COH} becomes local in space and independent of spin. It equally acts on empty and occupied states, electrons and holes, independent of their spin orientation.

The simplification of the SEX contribution in (15.4) is also obvious in the static screening limit. It holds

$$
\begin{aligned}
\Sigma_{ss'}^{\mathrm{SEX}}(\mathbf{xx}', z_n) &= -n_{\mathrm{X}}(\mathbf{x}s, \mathbf{x}'s')W(\mathbf{xx}', 0), \qquad (15.6)\\
n_{\mathrm{X}}(\mathbf{x}s, \mathbf{x}'s') &= \sum_{\nu, m_s}\sum_{\mathbf{k}} f\left(\varepsilon_{\nu m_s}(\mathbf{k})\right)\varphi_{\nu\mathbf{k}m_s}(\mathbf{x})\chi_{\frac{1}{2}m_s}(s)\varphi^*_{\nu\mathbf{k}m_s}(\mathbf{x}')\chi^+_{\frac{1}{2}m_s}(s'),
\end{aligned}
$$

where a spin-dependent exchange density $n_{\mathrm{X}}(\mathbf{x}s, \mathbf{x}'s')$ according to (4.20) is introduced. In non-spin-polarized systems the m_s-sum can be carried out using the closure relation of the spin functions (4.6) with the result

$$
\Sigma_{ss'}^{\mathrm{SEX}}(\mathbf{xx}', z_n) = -\delta_{ss'}\sum_{\nu,\mathbf{k}} f(\varepsilon_\nu(\mathbf{k}))\varphi_{\nu\mathbf{k}}(\mathbf{x})\varphi^*_{\nu\mathbf{k}}(\mathbf{x}')W(\mathbf{xx}', 0). \qquad (15.7)
$$

The static COHSEX approximation (15.5) and (15.7) derived in the framework of the Hedin GW approximation has a distinct computational advantage. It yields a Hermitian self-energy and, hence, is a reasonable starting self-energy for solving the quasiparticle equation (14.49). Indeed, this approximation is known to be a reliable approach for self-consistent calculations, the purpose of which is to compute updated wave functions close to the QP ones [3]. Of course, dynamical correlation is missing, i.e., its effect on the QP shifts and the redistribution of the spectral strength, e.g. due to the appearance of satellites.

All expressions of the contributions to the XC self-energy contain the spatial inhomogeneity of the electron gas indicated by $W(\mathbf{xx}', 0)$ and $n_{\mathrm{X}}(\mathbf{x}s, \mathbf{x}'s')$. In the early literature there were suggestions to describe the inhomogeneity in the screening by applying expressions for the homogeneous electron gas, e.g. by [4]

$$
W(\mathbf{xx}', 0) = \frac{1}{2}\left[W^h(\mathbf{x}-\mathbf{x}', n(\mathbf{x})) + W^h(\mathbf{x}-\mathbf{x}', n(\mathbf{x}'))\right], \qquad (15.8)
$$

i.e., by the statically screened potential in a homogeneous electron gas $W^h(\mathbf{x}-\mathbf{x}', n)$ but replacing the uniform electron density n by the local one, $n(\mathbf{x})$ or $n(\mathbf{x}')$, in the spirit of the LDA (cf. Sect. 7.2).

15.1.2 Matrix Elements

The COHSEX result can be also transformed into the Bloch-Fourier representation (see Sect. 14.3.1). The matrix elements of the low-temperature self-energy (14.65) decompose into

$$
\begin{aligned}
\Sigma_{\nu\mu}^{m_s m_s'}(\mathbf{k},z) &= \Sigma^{\mathrm{COH}}{}_{\nu\mu}^{m_s m_s'}(\mathbf{k},z) + \Sigma^{\mathrm{SEX}}{}_{\nu\mu}^{m_s m_s'}(\mathbf{k},z),\\
\Sigma^{\mathrm{COH}}{}_{\nu\mu}^{m_s m_s'}(\mathbf{k},z) &= \delta_{m_s m_s'}\frac{1}{\Omega}\sum_{\mathbf{q},\mathbf{G},\mathbf{G}'}\tilde{\mathrm{v}}\left(\sqrt{|\mathbf{q}+\mathbf{G}||\mathbf{q}+\mathbf{G}'|}\right)\\
&\times\sum_{\nu',\mathbf{k}'} B^{\mathbf{kk}'}_{\substack{m_s m_s\\ \nu\nu'}}(\mathbf{q}+\mathbf{G}) B^{\mathbf{kk}'*}_{\substack{m_s m_s\\ \mu\nu'}}(\mathbf{q}+\mathbf{G}')\\
&\times\int_0^\infty \frac{d\hbar\omega}{\pi}\frac{\mathrm{Im}\tilde{\epsilon}^{-1}(\mathbf{q}+\mathbf{G},\mathbf{q}+\mathbf{G}',\omega)}{\varepsilon_{\nu' m_s}(\mathbf{k}')-\hbar z+\hbar\omega}, \qquad (15.9)\\
\Sigma^{\mathrm{SEX}}{}_{\nu\mu}^{m_s m_s'}(\mathbf{k},z) &= -\,\delta_{m_s m_s'}\frac{1}{\Omega}\sum_{\mathbf{q},\mathbf{G},\mathbf{G}'}\tilde{\mathrm{v}}\left(\sqrt{|\mathbf{q}+\mathbf{G}||\mathbf{q}+\mathbf{G}'|}\right)\\
&\times\sum_{\nu',\mathbf{k}'} B^{\mathbf{kk}'}_{\substack{m_s m_s\\ \nu\nu'}}(\mathbf{q}+\mathbf{G}) B^{\mathbf{kk}'*}_{\substack{m_s m_s\\ \mu\nu'}}(\mathbf{q}+\mathbf{G}')\theta\left(\mu-\varepsilon_{\nu' m_s}(\mathbf{k}')\right)\\
&\times\left\{\delta_{\mathbf{GG}'}-\int_0^\infty\frac{d\hbar\omega}{\pi}\frac{2\hbar\omega\mathrm{Im}\tilde{\epsilon}^{-1}(\mathbf{q}+\mathbf{G},\mathbf{q}+\mathbf{G}',\omega)}{\left[\varepsilon_{\nu' m_s}(\mathbf{k}')-\hbar z\right]^2-\hbar^2\omega^2}\right\},
\end{aligned}
$$

where the same decomposition into contributions from the different poles as mentioned in the previous Sect. 15.1.1 has been used.

The interpretation of these expressions is much easier if the dynamics of the screening is neglected and the static COHSEX approximation [1, 2] is applied. For that reason we separate the dynamical (dyn) contribution from the static (st) COHSEX self-energy according to

$$
\Sigma_{\nu\mu}^{m_s m_s'}(\mathbf{k},z) = \Sigma_{\mathrm{st}}^{\mathrm{COH}}{}_{\nu\mu}^{m_s m_s'}(\mathbf{k},z) + \Sigma_{\mathrm{st}}^{\mathrm{SEX}}{}_{\nu\mu}^{m_s m_s'}(\mathbf{k},z) + \Sigma_{\mathrm{dyn}}{}_{\nu\mu}^{m_s m_s'}(\mathbf{k},z). \qquad (15.10)
$$

The corresponding matrix elements read as

$$
\begin{aligned}
\Sigma_{\mathrm{st}}^{\mathrm{COH}}{}_{\nu\mu}^{m_s m_s'}(\mathbf{k},z) &= \delta_{m_s m_s'}\frac{1}{2}\frac{1}{\Omega}\sum_{\mathbf{q},\mathbf{G},\mathbf{G}'}\tilde{\mathrm{v}}\left(\sqrt{|\mathbf{q}+\mathbf{G}||\mathbf{q}+\mathbf{G}'|}\right) B^{\mathbf{kk}}_{\substack{m_s m_s\\ \nu\mu}}(\mathbf{G}-\mathbf{G}')\\
&\times\left[\tilde{\epsilon}^{-1}(\mathbf{q}+\mathbf{G},\mathbf{q}+\mathbf{G}',0)-\delta_{\mathbf{GG}'}\right],\\
\Sigma_{\mathrm{st}}^{\mathrm{SEX}}{}_{\nu\mu}^{m_s m_s'}(\mathbf{k},z) &= -\,\delta_{m_s m_s'}\frac{1}{\Omega}\sum_{\mathbf{q},\mathbf{G},\mathbf{G}'}\tilde{\mathrm{v}}\left(\sqrt{|\mathbf{q}+\mathbf{G}||\mathbf{q}+\mathbf{G}'|}\right)\sum_{\nu',\mathbf{k}'} B^{\mathbf{kk}'}_{\substack{m_s m_s\\ \nu\nu'}}(\mathbf{q}+\mathbf{G})\\
&\times B^{\mathbf{kk}'*}_{\substack{m_s m_s\\ \mu\nu'}}(\mathbf{q}+\mathbf{G}')\theta(\mu-\varepsilon_{\nu' m_s}(\mathbf{k}'))\tilde{\epsilon}^{-1}(\mathbf{q}+\mathbf{G},\mathbf{q}+\mathbf{G}',0),\\
\Sigma_{\mathrm{dyn}}{}_{\nu\mu}^{m_s m_s'}(\mathbf{k},z) &= -\,\delta_{m_s m_s'}\frac{1}{\Omega}\sum_{\mathbf{q},\mathbf{G},\mathbf{G}'}\tilde{\mathrm{v}}\left(\sqrt{|\mathbf{q}+\mathbf{G}||\mathbf{q}+\mathbf{G}'|}\right)\sum_{\nu',\mathbf{k}'} B^{\mathbf{kk}'}_{\substack{m_s m_s\\ \nu\nu'}}(\mathbf{q}+\mathbf{G})
\end{aligned}
$$

$$\times B^{\mathbf{kk}'*}_{\substack{m_s m_s \\ \mu\nu'}}(\mathbf{q}+\mathbf{G}') \int_0^\infty \frac{d\omega}{\pi} \frac{\varepsilon_{\nu' m_s}(\mathbf{k}') - \hbar z}{\omega}$$

$$\times \frac{\mathrm{Im}\tilde{\epsilon}^{-1}(\mathbf{q}+\mathbf{G}, \mathbf{q}+\mathbf{G}', \omega)}{\mathrm{sgn}(\varepsilon_{\nu' m_s}(\mathbf{k}') - \mu)[\varepsilon_{\nu' m_s}(\mathbf{k}') - \hbar z] + \hbar\omega}, \tag{15.11}$$

where the closure relation of the single-particle wave functions has been used to rewrite the static COH contribution.

Expressions (15.11) show that, aside the treatment of the Coulomb singularity for small wave vectors [5], the frequency integration to account for the dynamics of the screening reaction is a challenge for the numerical calculation of the matrix elements of the self-energy. Several approaches are used in the literature. We have already mentioned an imaginary-time integration [6]. Also deformations of the integration path in the complex frequency plane have been suggested [7]. The general experience is that one of the major computational efforts in the self-energy calculations is the calculation of the frequency dependence of the screened interaction W, in particular of its spectral function $\hat{W}$ (15.3).

On the other hand, in Sect. 13.3 we have seen that the general features of $\hat{W}$ are well known, in particular in electron gases with extended states. Because of the frequency dependence of the energy loss function, the spectral function $\hat{W}$ is characterized by a strong peak corresponding to a plasmon excitation at the plasmon frequency. We have demonstrated this fact in Fig. 13.7 for the semiconductor Si. This behavior is particularly evident in the case of the homogenous electron gas or simple metals such as Na with $\hbar\Omega(0) = 5.7\,\mathrm{eV}$ and Al with $\hbar\Omega(0) = 15.3\,\mathrm{eV}$. The single-plasmon-pole (SPP) approximation (13.54) assumes that all the spectral weight in $\hat{\epsilon}^{-1}(\mathbf{q}+\mathbf{G}, \mathbf{q}+\mathbf{G}', \omega)$ (13.22) and hence in $\hat{W}$ resides in the plasmon excitation [8]. This is of course strictly only true in the limit of long wavelengths $|\mathbf{q}+\mathbf{G}| \to 0$ as demonstrated in Fig. 13.7 for silicon. Neglecting the deviations for large wave vectors and the modifications due to electron-hole excitations at lower energies, from (13.54) and (13.56) (applied to the inverse dielectric function) we find in the symmetrized case

$$\begin{aligned}\mathrm{Im}\tilde{\epsilon}^{-1}(\mathbf{q}+\mathbf{G}, \mathbf{q}+\mathbf{G}', \omega) = &-\frac{\pi}{2} \frac{\omega_p^2}{\Omega(\mathbf{q}+\mathbf{G}, \mathbf{q}+\mathbf{G}')} \frac{(\mathbf{q}+\mathbf{G})(\mathbf{q}+\mathbf{G}')}{|\mathbf{q}+\mathbf{G}||\mathbf{q}+\mathbf{G}'|} \frac{\tilde{n}(\mathbf{G}-\mathbf{G}')}{\tilde{n}(0)} \\ &\times \left\{\delta(\omega - \Omega(\mathbf{q}+\mathbf{G}, \mathbf{q}+\mathbf{G}')) \right. \\ &\left. -\delta(\omega + \Omega(\mathbf{q}+\mathbf{G}, \mathbf{q}+\mathbf{G}'))\right\}. \end{aligned} \tag{15.12}$$

The corresponding real part is given by

$$\begin{aligned}&\mathrm{Re}\tilde{\epsilon}^{-1}(\mathbf{q}+\mathbf{G}, \mathbf{q}+\mathbf{G}', \omega) \\ &= \delta_{\mathbf{GG}'} - \frac{\omega_p^2}{\Omega^2(\mathbf{q}+\mathbf{G}, \mathbf{q}+\mathbf{G}') - \omega^2} \frac{(\mathbf{q}+\mathbf{G})(\mathbf{q}+\mathbf{G}')}{|\mathbf{q}+\mathbf{G}||\mathbf{q}+\mathbf{G}'|} \frac{\tilde{n}(\mathbf{G}-\mathbf{G}')}{\tilde{n}(0)}. \end{aligned} \tag{15.13}$$

The generalized plasmon frequency is

$$\Omega^2(\mathbf{q}+\mathbf{G},\mathbf{q}+\mathbf{G}') = \frac{\omega_p^2}{\delta_{\mathbf{GG}'} - \mathrm{Re}\tilde{\epsilon}^{-1}(\mathbf{q}+\mathbf{G},\mathbf{q}+\mathbf{G}',0)} \frac{(\mathbf{q}+\mathbf{G})(\mathbf{q}+\mathbf{G}')}{|\mathbf{q}+\mathbf{G}||\mathbf{q}+\mathbf{G}'|} \frac{\tilde{n}(\mathbf{G}-\mathbf{G}')}{\tilde{n}(0)}. \tag{15.14}$$

Its dispersion can be therefore related to the wave-vector dependence of the Penn gap, e.g. $\bar{\omega}_g(|\mathbf{q}+\frac{1}{2}(\mathbf{G}+\mathbf{G}')|)$ in (13.56). A straightforward generalization of the dispersion relation (13.55) to $\Omega(|\mathbf{q}+\frac{1}{2}(\mathbf{G}+\mathbf{G}')|)$ may be also applied to simulate important features of the spatial inhomogeneity. However, independent of the details of this dispersion relation expression (15.12) fulfills the Johnson f-sum rule [9]

$$\int_0^\infty \frac{d\omega}{\pi}\omega \mathrm{Im}\tilde{\epsilon}^{-1}(\mathbf{q}+\mathbf{G},\mathbf{q}+\mathbf{G}',\omega) = -\frac{\omega_p^2}{2}\frac{(\mathbf{q}+\mathbf{G})(\mathbf{q}+\mathbf{G}')}{|\mathbf{q}+\mathbf{G}||\mathbf{q}+\mathbf{G}'|}\frac{\tilde{n}(\mathbf{G}-\mathbf{G}')}{\tilde{n}(0)}. \tag{15.15}$$

By means of the SPP expression (15.12) the corresponding frequency behavior of the spectral function $\hat{W}$ (15.3) of the dynamically screened potential can be easily described. The frequency integrals in (15.2) and (15.4) can be carried out in the frequency-dependent COH and SEX contributions to the XC self-energy. Here we only focus on the total dynamical contribution in (15.11)

$$\begin{aligned} \Sigma_{\mathrm{dyn}}{}^{m_s m_s'}_{\nu\mu}(\mathbf{k},z) &= \delta_{m_s m_s'}\frac{1}{2}\sum_{\mathbf{q},\mathbf{G},\mathbf{G}'}\tilde{v}\left(\sqrt{|\mathbf{q}+\mathbf{G}||\mathbf{q}+\mathbf{G}'|}\right) \\ &\times \sum_{\nu',\mathbf{k}'} B^{\mathbf{kk}'}_{\substack{m_s m_s \\ \nu\nu'}}(\mathbf{q}+\mathbf{G}) B^{\mathbf{kk}'*}_{\substack{m_s m_s \\ \mu\nu'}}(\mathbf{q}+\mathbf{G}')\left[\delta_{\mathbf{GG}'} - \mathrm{Re}\tilde{\epsilon}^{-1}(\mathbf{q}+\mathbf{G},\mathbf{q}+\mathbf{G}',0)\right] \\ &\times \frac{\varepsilon_{\nu' m_s}(\mathbf{k}') - \hbar z}{\mathrm{sgn}\left(\varepsilon_{\nu' m_s}(\mathbf{k}')-\mu\right)\left[\varepsilon_{\nu' m_s}(\mathbf{k}')-\hbar z\right] + \hbar\Omega(\mathbf{q}+\mathbf{G},\mathbf{q}+\mathbf{G}')}. \end{aligned} \tag{15.16}$$

This expression confirms the discussion that the magnitude of the plasmon energy determines the dynamical contribution to the XC self-energy. If the inequality

$$\left|\varepsilon_{\nu m_s}(\mathbf{k}) - \varepsilon_{\nu' m_s}(\mathbf{k}')\right| < \hbar\Omega(\mathbf{q}+\mathbf{G},\mathbf{q}+\mathbf{G}')$$

is fulfilled, the dynamical screening only plays a minor role. In any case (15.16) shows that the plasmon-pole approximation may be sufficient to compute quasiparticle energy shifts. One drawback of this approximation is however that the imaginary part of the self-energy is zero except at the plasmon poles. As a consequence, it is difficult to calculate precise spectral functions and lifetimes of the quasiparticles. Another drawback is its limited applicability to systems with a less pronounced plasmon pole in the energy loss function. For more complex systems, where the plasmon excitations merge with the electron-hole pair excitations, and systems with only localized states, it is not anymore clear if the plasmon-pole approximation is appropriate. A generalization to a multiple plasmon-pole model may partially overcome the drawbacks [10–12].

15.1.3 Validity of COHSEX Approximation

The validity of the static COHSEX approximation relies on whether the energy $\hbar z - \varepsilon_{\nu' m_s}(\mathbf{k}')$ in (15.9) is small compared to the energy of the main excitations in the screened interaction, which is essentially the plasmon energy in solids. For its illustration we study the conduction band minimum near the X point and the valence band maximum at Γ, i.e., the two band-edge states X_{1c} and $\Gamma_{25'v}$, in the diagonal approximation $\nu = \mu$ of the XC self-energy (15.11) in GW approximation for the prototypical solid, crystalline Si (see Fig. 14.14) [13]. In order to shine more light on the details of the screening reaction, we distinguish between the reaction in a homogeneous electron gas $\mathbf{G} = \mathbf{G}'$ and the local-field (LF) corrections $\mathbf{G} \neq \mathbf{G}'$ in the static limit. The results computed for the contributions to the self-energy are listed in Table 15.1. The inclusion of dynamical screening means that the (real part of the) self-energy matrix elements is taken at $\hbar z = \varepsilon_{\nu m_s}(\mathbf{k})$. To get some feeling why the resulting QP shifts $\Delta_{\nu m_s}(\mathbf{k})$ are relatively small, also the matrix elements $\langle \nu \mathbf{k} m_s | V_{XC} | \nu \mathbf{k} m_s \rangle$ of the XC potential described within the LDA and used to compute the starting (reference) electronic structure are listed. A part of the results is graphically described in Fig. 15.1. Here, in addition, the bare exchange self-energy (14.58) calculated with wave functions of the starting electronic structure, i.e., an approximate Fock operator $\Sigma^X = -5.28\,\text{eV}$ (X_{1c}) and $= -12.54\,\text{eV}$ ($\Gamma_{25'v}$), is also displayed. Interestingly all self-energies in Table 15.1 and Fig. 15.1 are negative in agreement with the attractive action of exchange and correlation.

Table 15.1 and Fig. 15.1 indicate that, in principle, all contributions have to be taken into account in order to generate QP shifts which correct the band edges around the fundamental gap almost correctly from the starting DFT-LDA KS eigenvalues toward the QP ones. Even for the gap opening the COH contribution cannot be

Table 15.1 The static COH and SEX contributions to the matrix elements of the XC self-energy for states at the conduction and valence band edges in Si

Band state	Σ^{COH}_{stat} (without LFs)	Σ^{SEX}_{stat} (without LFs)	Σ^{COH}_{stat} (with LFs)	Σ^{SEX}_{stat} (with LFs)	Σ^{COH}	Σ^{SEX}	Σ	V_{XC}	Δ
X_{1c}	−8.72	−2.37	−8.70	−2.08	−7.40	−1.65	−9.05	−9.60	0.55
$\Gamma_{25'v}$	−8.72	−4.44	−10.30	−3.85	−8.41	−3.56	−11.97	−11.76	−0.21

Results for the static COHSEX approximation are given without and with LF effects. In addition, the contributions Σ^{COH} and Σ^{SEX} including dynamical screening as well as the resulting total XC self-energy are listed. For comparison, the diagonal matrix elements of the local XC potential V_{XC}, used to compute the starting electronic structure, and the resulting QP shifts Δ are given. All values are in eV. The single-plasmon-pole approximation is used. From [13]

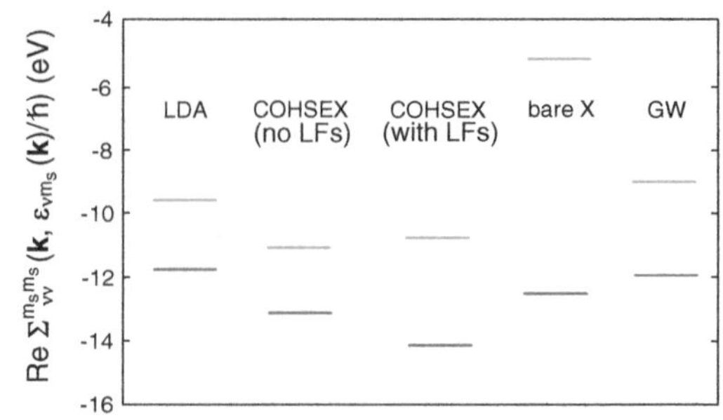

Fig. 15.1 Matrix elements of the XC self-energy in five different approximations for the CBM states X_{1c} (*blue lines*) and VBM states $\Gamma_{25'v}$ (*red lines*) of Si: LDA, COHSEX (without LFs), COHSEX (with LFs), bare exchange, and full self-energy in GW approximation taken in single-plasmon-pole approximation. Adapted from [13]

neglected. Including screening by local-field effects the occupied state is drastically shifted down in energy by about 1.6 eV. Therefore, these effects contribute much to the QP gap opening between X_{1c} and $\Gamma_{25'v}$ in Fig. 14.14. Local-field effects cannot be neglected.

The same holds essentially also for the dynamics in the screening. The absolute values of the self-energy contributions are reduced as a consequence of the reduction of the electron-gas reaction due to its dynamics. Its influence on the QP gap widening is however much weaker than the LF effects. Nevertheless, it tends to reduce the gap opening and, hence, has to be taken into account. On an absolute scale the static COHSEX approximation leads to a consistent overestimation of the magnitude by about 20 %, more precisely -10.78 eV versus -9.05 eV for X_{1c} or -14.15 eV versus -11.97 eV for $\Gamma_{25'v}$. For the resulting QP energy of an occupied state, most of the error resides in the Coulomb-hole term Σ^{COH}. The static approximation is more severe for the COH term than for the SEX term, because the Coulomb-hole term involves a sum over unoccupied states as well as over occupied states whereas the screened-exchange contribution only involves a sum over occupied states. Interestingly, for the occupied states $\Gamma_{25'v}$ the pure bare exchange self-energy of -12.54 eV is encased by these values. However, the bare exchange energy of -5.28 eV is much smaller than the XC value of the empty X_{1c} states. One may conclude that static and dynamic correlations make Σ much more negative for unoccupied states compared to filled states. In other words, electron excitations are more sensitive to correlation than hole excitations. This fact has been already pointed out discussing Fig. 14.13.

Another source of errors from the neglect of the dynamics comes from the quasiparticle renormalization factor $z_{vm_s}(\mathbf{k})$ (14.27). Some characteristic values are listed in Table 15.2 for the CBM and VBM of bulk semiconductors and insulators. The average value $z_{vm_s}(\mathbf{k}) \approx 0.8$ indicates a significant influence of the dynamics on the quasiparticle renormalization. Only about 80 % of the spectral weight are included in the main QP peak in the spectral functions for electrons and holes. In other words, 20 % of the spectral strength of an electron or hole excitation appear in form of satellites.

Table 15.2 Spectral weights of the quasiparticle peak (14.27) or renormalization constants (14.47) $z_{\nu m_s}(\mathbf{k})$ for electron (hole) states at the CBM (VBM) for diamond, Si, Ge, and LiCl [13]

State	Diamond	Si	Ge	LiCl
CBM	0.86	0.80	0.70	0.82
VBM	0.86	0.78	0.79	0.83

The quasiparticle shifts $\Delta_{\nu m_s}(\mathbf{k})$ (14.46) of the initial eigenvalues $\varepsilon_{\nu m_s}(\mathbf{k})$ are ruled by the difference $\delta\Sigma$ (14.10) of the total dynamical self-energy Σ and the XC potential V_{XC} used in the starting electronic-structure calculations. Corresponding matrix elements are given in Table 15.1 and Fig. 15.1. The local XC potential has been derived from a RPA description of correlation [14]. Hybertsen and Louie [13] pointed out a private communication of M. Schlüter that the use of a consistent XC potential in LDA is essential for the correct QP shifts. Indeed, the XC parametrization of von Barth and Hedin [14] is based on a GW approximation for the self-energy operator. The matrix elements of V_{XC} in Table 15.1 and Fig. 15.1 really approach the corresponding values of the self-energy. As a result only a small positive (negative) QP shift of the CBM (VBM) appears in Table 15.1. In the case of Si the effective gap opening of 0.76 eV by QP effects with respect to the DFT-LDA KS gap value 0.53 eV is mainly due to a QP shift of the CBM toward higher energies, i.e., smaller binding energies.

As a summary fundamental QP gaps of four semiconductors and insulators are listed in Table 15.3. The XC potential in LDA gives band gaps which are too small by 0.5 to 2.0 eV, i.e., by about 30 % in the average, as compared to experiment. The COHSEX approximation with homogeneous screening, i.e., without local-field effects, gives band gaps that are in better agreement with experiment, although significant deviations from measured values remain. The inclusion of LF effects dramatically opens up the band gaps. In any case the inclusion of dynamical screening as in the GW description is required to achieve quantitative agreement with the measured gap values.

Table 15.3 Fundamental gap of diamond, Si, Ge, and LiCl in four approximations to the XC self-energy

Material	DFT-LDA	Statistics COHSEX (no LFs)	Statistics COHSEX (with LFs)	GW	Expt.
Diamond	3.9	5.1	6.6	5.6	5.48
Si	0.52	0.50	1.70	1.29	1.17
Ge	0.07	0.33	1.09	0.75	0.744
LiCl	6.0	8.2	10.4	9.1	9.4

A single-plasmon-pole approximation is used. The results are compared with experimental values. All energies are in eV. From [13]

15.1.4 Gap Shrinkage Due to Free Carriers

Strong optical excitation or heavy doping of semiconductors may change the QP band structure of the intrinsic material. Apart from band-filling effects one important consequence is the band-gap shrinkage or narrowing with increasing densities of free carriers. Such effects have been indeed observed experimentally, for example for highly n-doped Si and ZnO (see e.g. [15, 16]) and optically pumped ZnO (see e.g. [17]). The effects can also be described in the framework of the GW or, more specifically, COHSEX approximation for the XC self-energy [15, 18].

We study the free-carrier influence on the QP band structure for a model system, a two-band model semiconductor as illustrated in Fig. 15.2a. The two bands are isotropic and parabolic with extrema at $\mathbf{k} = 0$, i.e., it holds $\varepsilon_c^{\text{QP}}(\mathbf{k}) = E_g(0) + \hbar^2 k^2/2m_c^*$ and $\varepsilon_v^{\text{QP}}(\mathbf{k}) = -\hbar^2 k^2/2m_v^*$. They are characterized by the QP gap $E_g(0)$ and an effective electron (hole) mass m_c^* (m_v^*) [19] in the absence of free carriers. The model semiconductor has the static electronic dielectric constant ϵ_∞. All spins are paired. In the case of moderate or high n-doping a degenerate electron gas with density n_e occurs in the conduction band that is characterized by a Fermi energy $\varepsilon_F = \mu - \varepsilon_c(0) = \frac{\hbar^2}{2m_c^*} k_F^2$ and a Fermi wave vector $k_F = \sqrt[3]{3\pi^2 n_e}$ (see Fig. 15.2b). We are mainly interested in the variation $\Delta E_g(n_e)$ of the fundamental QP gap $E_g(n_e)$ with the density n_e of the free electrons, i.e., in the free-carrier-induced shifts of the band edges.

Since the density n_e is by orders of magnitude smaller than the density of the valence electrons, the carrier-induced modifications of the band-edge energies $\varepsilon_{c/v}^{\text{QP}}(\mathbf{k})|_{n_e}$ are given by the corresponding changes of the XC self-energy (14.65).

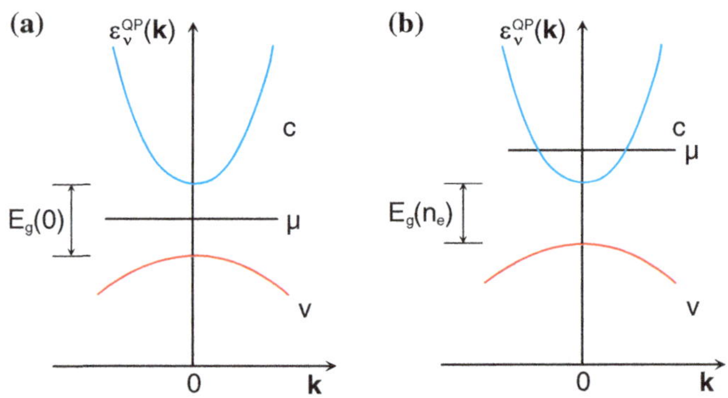

Fig. 15.2 Quasiparticle band structure of a two-band model semiconductor with one conduction (c) and one valence (v) band. Both bands are isotropic and parabolic. The (**a**) undoped and (**b**) highly n-doped cases are studied. The free electrons generated by doping are assumed to form a degenerate homogeneous electron gas with uniform density n_e in the conduction band

The gap modification is defined as

$$\Delta E_g(n_e) = \left[\Sigma_{cc}(0, \varepsilon_c^{\mathrm{QP}}(0)/\hbar)|_{n_e} - \Sigma_{cc}(0, \varepsilon_c^{\mathrm{QP}}(0)/\hbar)|_{n_e=0}\right]$$
$$- \left[\Sigma_{vv}(0, \varepsilon_v^{\mathrm{QP}}(0)/\hbar)|_{n_e} - \Sigma_{vv}(0, \varepsilon_v^{\mathrm{QP}}(0)/\hbar)|_{n_e=0}\right]. \quad (15.17)$$

Because of the relatively small perturbation and the homogeneity of the free electron gas, we apply the static COHSEX approximation and neglect LF effects $\mathbf{G} \neq \mathbf{G}'$ in the self-energy differences. Within the effective mass approximation for the two bands [19], the wave functions of the free carriers can be described by plane waves (4.46). The corresponding Bloch matrix elements (13.44) take the form

$$B_{\nu\nu'}^{\mathbf{k}\mathbf{k}'}(\mathbf{Q}) = \delta_{\nu\nu'}\delta_{\mathbf{k}'+\mathbf{Q},\mathbf{k}}, \quad (15.18)$$

where $\mathbf{k}$, $\mathbf{k}'$, and $\mathbf{Q} = \mathbf{q} + \mathbf{G}$ are wave vectors in the entire reciprocal spacc. In the framework of these approximations the COH contributions of conduction and valence bands cancel each other. The gap shift is only influenced by the SEX contribution to the self-energy in (15.11).

As a consequence, the gap shift is solely given by the free-carrier influence on the shift of the valence-band maximum. From (15.17) we find

$$\Delta E_g(n_e) = -\left[\Sigma_{vv}^{\mathrm{SEX}}(0, \varepsilon_v^{\mathrm{QP}}(0)/\hbar)|_{n_e} - \Sigma_{vv}^{\mathrm{SEX}}(0, \varepsilon_v^{\mathrm{QP}}(0)/\hbar)_{n_e=0}\right]$$
$$= \frac{1}{\Omega}\sum_{\mathbf{Q}} \tilde{v}(Q)\left[\epsilon^{-1}(\mathbf{Q}, \mathbf{Q}, 0)|_{n_e} - \epsilon^{-1}(\mathbf{Q}, \mathbf{Q}, 0)|_{n_e=0}\right]. \quad (15.19)$$

This is a pure correlation effect. The gap shift is only driven by free-carrier-induced changes of the screening properties.

We illustrate the result for a pure Thomas-Fermi polarization of the free electrons $(q_{TF}/Q)^2$ [see (13.49)], where the Thomas-Fermi vector q_{TF} is modified by the effective electron mass. Since only small wave vectors contribute to (15.19) the polarizability of the valence electrons of the undoped semiconductor can be nearly described by $\epsilon_\infty - 1$. In the presence of the free carriers the approximate dielectric function is then $\epsilon(\mathbf{Q}, \mathbf{Q}, 0) = \epsilon_\infty + (q_{TF}/Q)^2$. The corresponding screened Coulomb potential is a Yukawa potential. The gap shift is given by

$$\Delta E_g(n_e) = \int \frac{d^3\mathbf{Q}}{(2\pi)^3}\frac{e^2}{\epsilon_\infty\varepsilon_0}\left[\frac{1}{Q^2 + \tilde{q}_{TF}^2} - \frac{1}{Q^2}\right]$$
$$= -\frac{e^2}{4\pi\epsilon_\infty\varepsilon_0}\tilde{q}_{TF} \quad (15.20)$$

with $\tilde{q}_{TF} = q_{TF}/\sqrt{\epsilon_\infty}$. The negative sign indicates a free-carrier-induced shrinkage of the gap. Since $q_{TF} \sim n_e^{1/6}$ the shrinkage increases with rising density of the free carriers. For parameters of ZnO, $m_c^* = 0.3\,m$, and a carrier density of

$n_e = 5 \times 10^{19}\,\text{cm}^{-3}$, a Thomas-Fermi wave vector $\tilde{q}_{TF} = 0.13\,\text{Å}^{-1}$ results. Here ϵ_∞ is chosen to be 6.3, i.e., also lattice polarization has been included in this screening parameter. It results an absolute gap shrinkage of about 0.3 eV.

We have to point out that for the reproduction of the majority of experimental data the model, in particular, the screening model, has to be improved. For instance, for highly n-doped Si and Ge a $n_e^{1/3}$ dependence of $\Delta E_g(n_e)$ has been observed that can be indeed described by a more realistic dielectric function [15]. If a Lindhard-like description (13.47) is employed to the intraband contribution of the electronic polarizability, expression (15.20) is modified to [15]

$$\Delta E_g(n_e) = -\frac{e^2 k_F}{2\pi^2 \varepsilon_0 \epsilon_\infty} \left\{ 1 + \frac{\tilde{q}_{TF}}{k_F} \left[\frac{\pi}{2} - \arctan\left(\frac{k_F}{\tilde{q}_{TF}} \right) \right] \right\}.$$

In the low-density limit $k_F \ll \tilde{q}_{TF}$ expression (15.20) is again valid. In the opposite limit in (15.20) $\tilde{q}_{TF}$ has to be replaced by $4k_F/\pi$. For improved screening descriptions, in order to describe optical high-excitation experiments, the reader is referred to the more specialized literature (see e.g. [20]).

15.2 Direct Modeling of QP Shifts

15.2.1 Approximate Matrix Elements of XC Potential

Many simplifications of the GW schemes are possible and have been tried to implement in several codes in the last two decades. We illustrate some of these approaches for the non-spin-polarized case. The total XC self-energy $\Sigma = \delta_{ss'}\left[\Sigma_{\text{st}} + \Sigma_{\text{dyn}}\right]$ (15.10) in GW approximation is divided into the static contribution

$$\begin{aligned} \Sigma_{\text{st}}(\mathbf{xx}') = {} & \frac{1}{2}\delta(\mathbf{x} - \mathbf{x}')\left[W(\mathbf{xx}', 0) - \text{v}(\mathbf{x} - \mathbf{x}')\right] \\ & - \sum_{\nu,\mathbf{k}}^{(\text{occ})} \varphi_{\nu\mathbf{k}}(\mathbf{x})\varphi^*_{\nu\mathbf{k}}(\mathbf{x}')W(\mathbf{xx}', 0) \end{aligned} \qquad (15.21)$$

according to (15.5) and (15.7) and the effect of the screening dynamics in the spin-diagonal contribution Σ_{dyn}. The summation over occupied Bloch states is indicated by (occ).

To compute the QP shifts (14.45) the XC potential $V_{\text{XC}}(\mathbf{x})\delta(\mathbf{x} - \mathbf{x}')$ has to be subtracted. This is the same as that used to calculate the eigenvalues and eigenfunctions of the starting electronic structure. This local or semilocal spin-independent XC potential $V_{\text{XC}}(\mathbf{x})$ (6.20) occurs in the Kohn-Sham equation (6.22). To compute the QP shifts $\Delta_\nu(\mathbf{k}) = \varepsilon_\nu^{\text{QP}}(\mathbf{k}) - \varepsilon_\nu(\mathbf{k})$ we restrict ourselves to the diagonal approximation

$$\Delta_\nu(\mathbf{k}) = \langle \nu\mathbf{k}|\Sigma_{\mathrm{st}} + \Sigma_{\mathrm{dyn}}|_{\hbar\omega=\varepsilon_\nu^{\mathrm{QP}}(\mathbf{k})} - V_{\mathrm{XC}}|\nu\mathbf{k}\rangle,$$

where it has been taken into account that the frequency in the self-energy must be originally replaced by the QP energy (14.42). After a linearization of the energy dependence of the self-energy in the region of the QP peak in the spectral function [see (14.48)]

$$\begin{aligned}\langle \nu\mathbf{k}|\Sigma_{\mathrm{dyn}}(\omega)|\nu\mathbf{k}\rangle &= \langle \nu\mathbf{k}|\Sigma_{\mathrm{dyn}}(\varepsilon_\nu(\mathbf{k})/\hbar)|\nu\mathbf{k}\rangle - \beta_\nu(\mathbf{k})\left[\hbar\omega - \varepsilon_\nu(\mathbf{k})\right],\\ \beta_\nu(\mathbf{k}) &= -\frac{\partial}{\partial\hbar\omega}\langle \nu\mathbf{k}|\Sigma_{\mathrm{dyn}}(\omega)|\nu\mathbf{k}\rangle|_{\hbar\omega=\varepsilon_\nu(\mathbf{k})},\end{aligned}$$

the QP shift becomes

$$\Delta_\nu(\mathbf{k}) = \langle \nu\mathbf{k}|\Sigma_{\mathrm{st}} + \Sigma_{\mathrm{dyn}}|_{\hbar\omega=\varepsilon_\nu(\mathbf{k})} - V_{\mathrm{XC}}|\nu\mathbf{k}\rangle/[1+\beta_\nu(\mathbf{k})]. \tag{15.22}$$

First, we approximately treat the dynamical contribution [21]. Using the matrix element in (15.11) and neglecting the small influence of the off-diagonal elements of the inverse dielectric function, i.e., local-field effects, the characteristic dynamical quantities may be written as

$$\begin{aligned}\Delta_\nu^{\mathrm{dyn}}(\mathbf{k}) &= \langle \nu\mathbf{k}|\Sigma_{\mathrm{dyn}}\left(\varepsilon_\nu(\mathbf{k})/\hbar\right)|\nu\mathbf{k}\rangle = -\frac{1}{\Omega}\sum_{\mathbf{q},\mathbf{G}}\tilde{v}(|\mathbf{q}+\mathbf{G}|)\sum_{\nu',\mathbf{k}'}\left|B_{\nu\nu'}^{\mathbf{k}\mathbf{k}'}(\mathbf{q}+\mathbf{G})\right|^2\\ &\times\mathcal{P}\int_0^\infty\frac{d\omega}{\pi\omega}\mathrm{Im}\tilde{\epsilon}^{-1}(\mathbf{q}+\mathbf{G},\mathbf{q}+\mathbf{G},\omega)\frac{\varepsilon_{\nu'}(\mathbf{k}')-\varepsilon_\nu(\mathbf{k})}{\mathrm{sgn}\left(\varepsilon_{\nu'}(\mathbf{k}')-\mu\right)\left[\varepsilon_{\nu'}(\mathbf{k}')-\varepsilon_\nu(\mathbf{k})\right]+\hbar\omega}.\end{aligned}$$

The same approximations lead to the positive renormalization coefficient

$$\begin{aligned}\beta_\nu(\mathbf{k}) = &-\frac{1}{\Omega}\sum_{\mathbf{q},\mathbf{G}}\tilde{v}(|\mathbf{q}+\mathbf{G}|)\sum_{\nu',\mathbf{k}'}\left|B_{\nu\nu'}^{\mathbf{k}\mathbf{k}'}(\mathbf{q}+\mathbf{G})\right|^2\\ &\times\mathcal{P}\int_0^\infty\frac{d\hbar\omega}{\pi}\frac{\mathrm{Im}\tilde{\epsilon}^{-1}(\mathbf{q}+\mathbf{G},\mathbf{q}+\mathbf{G},\omega)}{\left[\mathrm{sgn}(\varepsilon_{\nu'}(\mathbf{k}')-\mu)(\varepsilon_{\nu'}(\mathbf{k}')-\varepsilon_\nu(\mathbf{k}))+\hbar\omega\right]^2}.\end{aligned}$$

15.2.2 Consequences of Model Screening

The energy differences $|\varepsilon_{\nu'}(\mathbf{k}')-\varepsilon_\nu(\mathbf{k})|$ are generally small compared to $\hbar\omega$ that is of the order of the plasmon energy $\hbar\Omega(0)$ (13.55). They can be therefore neglected in the denominators of the above expressions. Then, with the closure relation of the Bloch states in (12.69) the renormalization coefficient $\beta_\nu(\mathbf{k})$ becomes a state-independent quantity. The dynamical shift $\Delta_\nu^{\mathrm{dyn}}(\mathbf{k})$ only depends on the average energy distance of the state $|\nu\mathbf{k}\rangle$ to all other single-particle reference states in the system,

$$E_{\nu\mathbf{k}}(\mathbf{Q}) = \sum_{\nu',\mathbf{k}'} \left[\varepsilon_{\nu'}(\mathbf{k}') - \varepsilon_{\nu}(\mathbf{k})\right] \left|B_{\nu\nu'}^{\mathbf{kk}'}(\mathbf{Q})\right|^2 .$$

With the commutator $\left[e^{-i\mathbf{Qx}}, \hat{H}\right]_- = -\frac{\hbar^2}{2m} e^{-i\mathbf{Qx}}(\mathbf{Q}^2 + 2i\mathbf{Q}\nabla_{\mathbf{x}})$, the single-particle reference Hamiltonian $\hat{H} = -\frac{\hbar^2}{2m}\Delta_{\mathbf{x}} + V(\mathbf{x})$ assuming a local potential, and the closure relation, it holds

$$E_{\nu\mathbf{k}}(\mathbf{Q}) = \frac{\hbar^2}{2m}\mathbf{Q}^2 - \mathbf{Q}\nabla_{\mathbf{k}}\varepsilon_{\nu}(\mathbf{k}),$$

where $\nabla_{\mathbf{k}}\varepsilon_{\nu}(\mathbf{k}) = \frac{\hbar}{m}\langle\nu\mathbf{k}|\mathbf{p}|\nu\mathbf{k}\rangle$ is used. In practice, we frequently discuss the QP corrections for high-symmetry points $\mathbf{k}_0$ with $\nabla_{\mathbf{k}}\varepsilon_{\nu}(\mathbf{k})|_{\mathbf{k}_0} = 0$. As a consequence, only state-independent dynamical corrections $\Delta^{\mathrm{dyn}} = \Delta_{\nu}^{\mathrm{dyn}}(\mathbf{k})$ and $\beta = \beta_{\nu}(\mathbf{k})$ with positive values occur in the framework of the described approximations:

$$\Delta_{\nu}^{\mathrm{dyn}}(\mathbf{k}) = -\frac{\hbar e^2}{2m\varepsilon_0}\frac{1}{\Omega}\sum_{\mathbf{q},\mathbf{G}}\mathcal{P}\int_0^{\infty}\frac{d\omega}{\pi\omega^2}\mathrm{Im}\tilde{\epsilon}^{-1}(\mathbf{q}+\mathbf{G},\mathbf{q}+\mathbf{G},\omega),$$

$$\beta_{\nu}(\mathbf{k}) = -\frac{1}{\hbar}\frac{1}{\Omega}\sum_{\mathbf{q},\mathbf{G}}\tilde{v}(\mathbf{q}+\mathbf{G}|)\mathcal{P}\int_0^{\infty}\frac{d\omega}{\pi\omega^2}\mathrm{Im}\tilde{\epsilon}^{-1}(\mathbf{q}+\mathbf{G},\mathbf{q}+\mathbf{G},\omega). \quad (15.23)$$

Using the single-plasmon pole approximation (13.54) the quantities can be even calculated analytically. They only depend on the wave-vector- and frequency-dependent screening function $\tilde{\epsilon}^{-1}(\mathbf{q}+\mathbf{G},\mathbf{q}+\mathbf{G},\omega)$. Merely, if local-field effects in $\tilde{\epsilon}^{-1}(\mathbf{q}+\mathbf{G},\mathbf{q}+\mathbf{G}',\omega)$ with $\mathbf{G} \neq \mathbf{G}'$ are taken into account, a minor variation with the state appears as indicated in Table 15.1. Then the resulting expressions (15.23) can be slightly improved by the replacements of the state-dependent energy terms in the denominators by their average values [21].

The static part of the total QP shift (15.22) also contains nearly state-independent contributions. That is obvious after decomposition of the static XC self-energy in a COH (15.5) and a SEX (15.7) term. According to expression (15.5), without local-field effects, the COH term $\Sigma^{\mathrm{COH}}(\mathbf{xx}',0) = \delta(\mathbf{x}-\mathbf{x}')V_{\mathrm{COH}}(0)$ is given by a local but spatially constant potential

$$V_{\mathrm{COH}}(0) = \frac{1}{2\Omega}\sum_{\mathbf{q},\mathbf{G}}\tilde{v}(|\mathbf{q}+\mathbf{G}|)\left[\tilde{\epsilon}^{-1}(\mathbf{q}+\mathbf{G},\mathbf{q}+\mathbf{G},0) - 1\right].$$

Applying the SPP expression (13.54) with the wave-vector dependence (13.55), i.e., the model dielectric functions as presented in Figs. 13.5 and 13.6, one finds analytically also a constant, state-independent but negative potential energy [21]

$$V_{\mathrm{COH}}(0) = -\frac{e^2}{8\pi\varepsilon_0}q_{TF}\left\{\frac{1}{\alpha}\left(1-\frac{1}{\epsilon_\infty}\right)\frac{1}{1+\frac{q_{TF}}{\alpha k_F}\sqrt{\frac{3\epsilon_\infty}{\epsilon_\infty-1}}}\right\}^{\frac{1}{2}}. \qquad (15.24)$$

Consequently the state dependence mainly arises from the non-local contribution $\Sigma^{\mathrm{SEX}}(\mathbf{x}\mathbf{x}',0) - V_{\mathrm{XC}}(\mathbf{x})\delta(\mathbf{x}-\mathbf{x}')$ in (15.22). The corresponding exchange density and the XC potential have to be taken from the reference KS electronic-structure description.

Improvements are achieved after inclusion of local-field corrections. One possibility is to follow the local-density idea [4] and generalize the screened potential according to (15.8). Then, in all electron gas parameters the homogeneous density n has to be replaced by the local densities $n(\mathbf{x})$ or $n(\mathbf{x}')$ of the inhomogeneous electron gas [22]. This generalization, however, bloats up the numerical calculations and restricts the benefits to use model dielectric functions. A significant reduction of the numerical effort is related to the replacement of $n(\mathbf{x})$ by its average value in the Bloch state $|\nu\mathbf{k}\rangle$, $n_\nu(\mathbf{k}) = \int d^3\mathbf{x} n(\mathbf{x})|\varphi_{\nu\mathbf{k}}(\mathbf{x})|^2$. In materials such as silicon this state-averaged density $n_\nu(\mathbf{k})$ varies in the range from about $0.9n$ for the conduction state X_{1c} to $(1.7\text{-}2.0)n$ for the uppermost valence band $\Gamma_{25'\nu}$ with the average density n. In fact, the use of such average values makes the dynamical corrections (15.23) slightly state-dependent. In the case of the screened-exchange contribution expression (15.21) even suggests the use of effective densities $\sqrt{n_\nu(\mathbf{k}')n_\nu(\mathbf{k})}$ because of the occurrence of wave-function products $\varphi^*_{\nu\mathbf{k}'}(\mathbf{x})\varphi_{\nu\mathbf{k}}(\mathbf{x})$ [21].

15.2.3 QP Shifts for Semiconductors

In the case of the covalently bonded Si and the weakly ionic compounds GaAs and AlAs, respectively, the model QP calculations described above yield shifts listed in Table 15.4 in excellent agreement with the results of more sophisticated one-shot GW computations [6, 23]. The variations are within 0.1 eV except for the direct band gap at Γ of GaAs with a difference of 0.2 eV. Therefore, the physical approximations done in Sect. 15.1.2 shine some light on the physics of the quasiparticle shifts. The dynamical effects provide an upward shift $\Delta_\nu^{\mathrm{dyn}}(\mathbf{k})$ of all bands of about 1.4 eV and a linear energy dependence with a factor $\beta_\nu(\mathbf{k})$ of the order of 0.23. This is in agreement with the renormalization factors $z_\nu(\mathbf{k}) = [1+\beta_\nu(\mathbf{k})]^{-1}$ in Table 15.2. Their average value derived within the model studies is 0.81. On the other hand, the relative shifts of the conduction band states with respect to the vacuum level are mainly driven by the static part $\Delta_\nu^{\mathrm{stat}}(\mathbf{k})$, if the energy dependence of the self-energy, i.e., the renormalization by $\beta_\nu(\mathbf{k})$, is taken into account as in $\Delta_\nu^{\mathrm{stat}}(\mathbf{k})/[1+\beta_\nu(\mathbf{k})]$ according to (15.22). We conclude that this energy dependence cannot be neglected, also not for relatively simple sp-bonded materials.

The picture derived from the GaN values in Table 15.4 is not unique, either due to reduced applicability of the model screening or the much stronger ionic bonds.

Table 15.4 Quasiparticle shifts of KS eigenvalues for Si, GaAs, AlAs, and GaN referred to the value of the valence-band maximum from model calculations [21, 23] using RPA values for the dielectric constants ϵ_∞ and DFT-LDA eigenvalues and eigenfunctions

Semiconductor	Bloch state $\|\nu\mathbf{k}\rangle$	$\Delta_\nu^{\text{stat}}(\mathbf{k})$	$\Delta_\nu^{\text{dyn}}(\mathbf{k})$	$\beta_\nu(\mathbf{k})$	$\Delta_\nu(\mathbf{k})$ (model)	$\Delta_\nu(\mathbf{k})$ (full one-shot GW)
Si	$\Gamma_{25'v}$	−1.19	1.52	0.21	0.00	0.00
	Γ_{15c}	−0.30	1.42	0.24	0.63	0.73
	L_{1c}	−0.24	1.42	0.24	0.68	0.77
	X_{1c}	−0.14	1.36	0.25	0.70	0.72
GaAs	Γ_{15v}	−1.21	1.49	0.21	0.00	0.00
	Γ_{1c}	−0.26	1.42	0.22	0.71	0.91
	L_{1c}	−0.26	1.37	0.24	0.66	0.78
	X_{1c}	−0.08	1.33	0.25	0.76	0.70
AlAs	Γ_{15v}	−1.28	1.48	0.20	0.00	0.00
	Γ_{1c}	+0.03	1.38	0.23	0.98	0.93
	L_{1c}	−0.11	1.34	0.23	0.83	0.90
	X_{1c}	+0.09	1.26	0.26	0.91	0.81
GaN	Γ_{15v}	−2.72	2.26	0.14	0.00	0.00
	Γ_{1c}	−1.33	2.06	0.16	1.03	0.34
	L_{1c}	−2.84	2.14	0.15	−0.20	−0.29
	X_{1c}	+0.29	1.69	0.21	1.57	0.77

In the GaN case the local-field effects are treated according to (13.56) and (15.12). Only the zinc-blende geometry is studied. For comparison one-shot GW shifts [6, 23] are given. The total QP shifts $\Delta_\nu(\mathbf{k})$ are listed with respect to the VBM. All values are in eV

Because of the wider gap of the system the dynamical shift $\Delta_\nu^{\text{dyn}}(\mathbf{k})$ is increased while the $\beta_\nu(\mathbf{k})$ factor is reduced. The agreement with the absolute values of the QP shifts derived in a more sophisticated approach is significantly reduced. However, the shifts of other DFT-LDA+G_0W_0 calculations [24] seem to be closer to the results of the model studies with $\Delta_\nu(\mathbf{k})\big|_{\Gamma_{1c}} = 0.97\,\text{eV}$ and $\Delta_\nu(\mathbf{k})\big|_{X_{1c}} = 0.84\,\text{eV}$. Only the L_{1c} model value seems to be in complete disagreement. Moreover, the QP shifts of GaN exhibit a significant variation with the **k** vector for the lowest conduction band. This variation is much stronger as that found for Si, GaAs, and AlAs (see Table 15.4).

Despite the limited accuracy of the model treatment of the QP corrections of electronic states, it is an interesting method. Because of the significantly reduced numerical effort, its application to systems with many atoms, e.g. 216 in a simple-cubic supercell or more, is suggested in order to treat the electronic structure of nanostructures [25]. Moreover, it gives some insight into the interplay between static and dynamical screening in the corrections for both empty and occupied states.

15.3 Approximate Treatment of XC in Reference System

15.3.1 Self-energy Difference

The most important disadvantage of the approximations described in Sect. 15.2 is the usual treatment of the XC self-energy Σ and the XC potential V_{XC} on different footings. We have seen from Table 15.1 that the matrix elements of Σ and V_{XC} give rise to absolute values of the order of 10 eV. The calculation of QP shifts with an accuracy of 0.1 eV or better therefore asks for a precise calculation of both contributions. Thereby, if possible, a compensation of the errors in both terms of the difference should be requested. For that reason M. Schlüter (see [13]) pointed out the use of an XC potential V_{XC} consistent with the approximation of the XC self-energy Σ.

As a key ingredient for such treatment on equal footings we exploit the observation that for many solids the quasiparticle theory, when carried out by wrongly assuming metallic dielectric screening, approximately reproduces the band structure of the reference DFT calculation, at least, when employing the local density approximation. This has been already observed many years ago [26–28]. One reason is that several descriptions of correlation within LDA are based on GW derivations (see e.g. Sect. 7.2.2 and [14]). As an illustration, Fig. 15.3 shows QP corrections for silicon and solid argon. The standard GW/RPA method yields the well-known openings of the band gap, 0.7 eV for Si (see also Tables 15.1, 15.3 and 15.4) and 6.1 eV for Ar [29]. On the other hand, within the same calculations but employing metallic screening the QP shifts are small, especially for the valence states and the lowest-energy conduction bands. The 'metallic' screening is simulated by a model dielectric function [30] [see also (13.54)]. However, the parameter ϵ_∞, the macroscopic electronic dielectric constant, is set to infinite, $\epsilon_\infty \to \infty$.

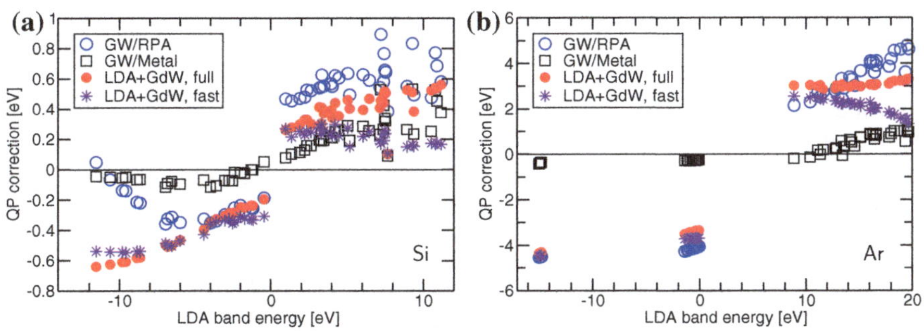

Fig. 15.3 QP corrections $\Delta_\nu(\mathbf{k})$ of (**a**) bulk Si and (**b**) solid Ar. The *open circles* denote standard GW data with RPA screening and DFT-LDA reference electronic structure (GW/RPA). The *squares* are based on the same approximations but a metallic screening function (GW/Metal). The *red filled circles* are obtained from a LDA+GdW approach (full). The *purple asterisks* are due to a reduced number of plane waves (LDA+GdW, fast). The VBM is used as energy zero. Reprinted with permission from [29]. Copyright 2010 by the American Physical Society

If the GW approach to the XC self-energy Σ with metallic screening, $W|_{\epsilon_\infty\to\infty}$, reproduces the DFT-LDA band structure, one can find the true QP band structure approximating the self-energy difference in (14.67) by [29]

$$\Delta\Sigma = \Sigma - V_{XC} \approx i\hbar G\left[W - W_{\epsilon_\infty\to\infty}\right]. \tag{15.25}$$

This approach has been suggested in early studies [27]. Later [29] this procedure has been denoted with 'LDA+GdW'. It nearly replaces the local XC potential V_{XC} by the XC self-energy with a screened potential of a homogeneous electron gas. The approximate expression (15.25) also reflects the observation of small QP corrections computed within the GW framework to DFT-LDA band structures in bulk metals [31, 32]. For non-metals the self-energy difference (15.25) is by one order of magnitude smaller ($\sim$1 eV) than the self-energy itself. All numerical procedures are therefore much more robust to some modifications. Figure 15.3 shows that the LDA + GdW approach leads to QP corrections similar to the full GW/RPA calculations, at least for the band edges around the fundamental gap, of course, only if the basis sets are sufficiently large.

15.3.2 Average Static Result

The observed robustness of the self-energy difference (15.25) suggests further drastic approximations, for instance the neglect of dynamical screening, local-field effects, and off-diagonal elements of the inverse dielectric matrix. The remaining static self-energy difference

$$\Delta\Sigma(\mathbf{x}\mathbf{x}',\omega) = \left[\frac{1}{2}\delta(\mathbf{x}-\mathbf{x}') - n_X(\mathbf{x},\mathbf{x}')\right]\left[W(\mathbf{x}\mathbf{x}',0) - W(\mathbf{x}\mathbf{x}',0)|_{\epsilon_\infty\to\infty}\right] \tag{15.26}$$

with the exchange density distribution $n_X(\mathbf{x},\mathbf{x}')$ (4.20) is dominated by the long-range polarization effects appearing in a non-metal, which are completely missing in the short-sighted LDA of the correlation. The resulting (static) QP shift is given as [see also (15.11)]

$$\Delta_\nu(\mathbf{k}) = \frac{1}{\Omega}\sum_{\mathbf{q},\mathbf{G}}\tilde{v}(|\mathbf{q}+\mathbf{G}|)\left[\frac{1}{2} - \sum_{\nu',\mathbf{k}'}\left|B_{\nu\nu'}^{\mathbf{k}\mathbf{k}'}(\mathbf{q}+\mathbf{G})\right|^2\theta(\mu-\varepsilon_{\nu'}(\mathbf{k}'))\right]$$
$$\times\left[\epsilon^{-1}(\mathbf{q}+\mathbf{G},\mathbf{q}+\mathbf{G},0) - \epsilon^{-1}(\mathbf{q}+\mathbf{G},\mathbf{q}+\mathbf{G},0)|_{\epsilon_\infty\to\infty}\right].$$

With a nearly-free-electron-like approximation (15.18) a symmetric representation for conduction and valence states appear,

$$\Delta_\nu(\mathbf{k}) = -\frac{1}{2}\mathrm{sgn}(\mu - \varepsilon_\nu(\mathbf{k}))\Delta_{\mathrm{scissors}},$$

with a constant widening of all gaps and interband distances between conduction and valence bands, a so-called scissors shift [33]

$$\Delta_{\mathrm{scissors}} = \frac{1}{\Omega}\sum_{\mathbf{q},\mathbf{G}} \tilde{v}(|\mathbf{q}+\mathbf{G}|)\left[\epsilon^{-1}(\mathbf{q}+\mathbf{G},\mathbf{q}+\mathbf{G},0) - \epsilon^{-1}(\mathbf{q}+\mathbf{G},\mathbf{q}+\mathbf{G},0)\Big|_{\epsilon_\infty\to\infty}\right]. \tag{15.27}$$

With the model function (13.54) this quantity can be analytically calculated [34]:

$$\Delta_{\mathrm{scissors}} = \Delta\Big|_{\epsilon_\infty\to\infty} - \Delta,$$

$$\Delta = \frac{e^2}{4\pi\varepsilon_0 R}\left(1-\frac{1}{\epsilon_\infty}\right)^{\frac{3}{4}} \frac{1}{\sqrt{2}\left[1+\frac{\alpha}{2}\left(1-\frac{1}{\epsilon_\infty}\right)^{\frac{1}{2}}\frac{1}{(q_{TF}R)^2}\right]^{\frac{1}{2}}}, \tag{15.28}$$

$$R = \left(\frac{\hbar}{2m\omega_p}\right)^{\frac{1}{2}}.$$

The widening of the average distance between conduction and valence states is dominated by two parameters, ϵ_∞, that characterizes the strength of the static electronic polarizability of the system, and the average density n of the inhomogeneous electron gas, that characterizes the average distance between two electrons. For metallic electron gases with $\epsilon_\infty \to \infty$ this quasiparticle effect disappears. For wide-band-gap materials with $\epsilon_\infty \to 1$, the shift approaches its maximum value $\Delta\big|_{\epsilon_\infty\to\infty}$ for a given electron density. This behavior is clearly illustrated in Fig. 15.4 versus the density

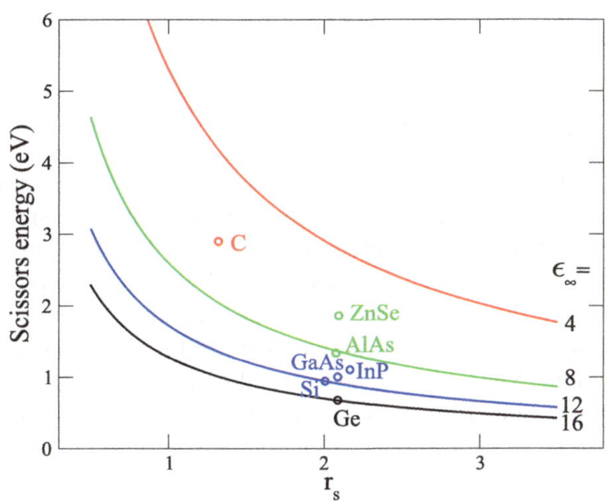

Fig. 15.4 Opening (15.28) of gaps and energy distances between conduction and valence bands $\Delta_{\mathrm{scissors}}$ in static approximation versus the electron-gas parameter r_s. The static electronic dielectric constant ϵ_∞ is varied in the range of values of semiconductors and insulators. The model dielectric function (13.54) with $\alpha = 1.563$ (see Fig. 13.5) has been applied

parameter r_s and some values of ϵ_∞. The scissors operator nearly scales with $1/\epsilon_\infty$. That means that the absolute values of the QP gap openings in free-electron-like materials, such as silicon with $\epsilon_\infty = 11.3$ (see Fig. 13.5) and $r_s = 2$, are smaller in comparison with strongly ionic wide-band-gap systems, such as MgO with $\epsilon_\infty = 2.9$ and $r_s = 1.55$ [35]. According to the discussion of the renormalization of the shifts due to the dynamics of the screening in Sect. 15.1, the resulting gap openings of about $\Delta_{\text{scissors}} = 1.01\,\text{eV}$ (Si) and 4.95 eV (MgO) seem to be slightly overestimated and have to be reduced by a factor $z_\nu(\mathbf{k}) \approx 0.8$. At least, the reduced scissors shift of 0.8 eV resulting for Si is of the order of the values discussed in Tables 14.1, 15.1, 15.3, and 15.4. However, also the value of 4.0 eV for MgO is not too far from that obtained in a more sophisticated calculation [36]. Together with a KS gap of about 4.5 eV [37] a QP gap of 8.5 eV only somewhat above the experimental value 7.7 eV [35] is obtained.

15.3.3 Scissors Operator

For bulk silicon the quasiparticle corrections $\Delta_\nu(\mathbf{k})$ of the KS bands $\varepsilon_\nu(\mathbf{k})$ in Tables 14.1 and 15.4 as well as in Fig. 14.15 vanish for higher valence bands (relative to VBM), while they represent a rigid, almost state (i.e., band and wave-vector)-independent upward shift of the lower conduction bands. This means, that around the fundamental gap the quasiparticle energy shifts can be nearly described by a state-dependent *scissors* operator [33]

$$\Delta_\nu(\mathbf{k}) = \delta_{\nu c}\Delta_{\text{scissors}} \tag{15.29}$$

with a value as illustrated in expression (15.28). It can be interpreted as the consequence of a projection operator

$$\hat{H}_{\text{scissors}} = \Delta_{\text{scissors}} \sum_{c,\mathbf{k},m_s} |c\mathbf{k}m_s\rangle\langle c\mathbf{k}m_s|,$$

which is added to a KS Hamiltonian (6.23). It however leaves the KS wave functions unchanged.

The evaluation of the quality of a scissors operator approximation depends on the precision criterion. For instance, in contrast to silicon it has been found [38] that for other semiconductors such as GaAs, AlAs, and especially diamond quasiparticle band structures cannot be obtained by such an operator starting from DFT-LDA KS eigenvalues. However, taking an uncertainty of about 0.1 eV into consideration, at least the QP conduction band shifts for GaAs and AlAs (see Table I in [38] and Table 15.4) can be approximated by a rigid shift. For systems with larger gaps the uncertainty is increased on an absolute scale [38]. The values presented in Table 15.4 for GaN seem to indicate a complete failure of the scissors operator approach for

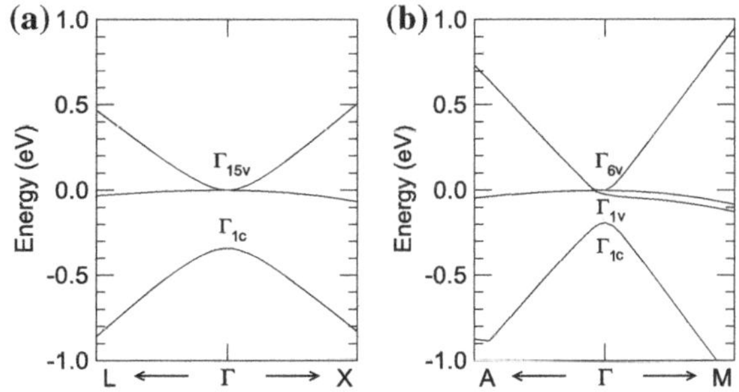

Fig. 15.5 (Inverted) KS band structures of *zb*-InN (**a**) and *wz*-InN (**b**) around the Fermi level (used as energy zero) close to the Γ point. They are computed within DFT-LDA [39] treating the In 4*d* electrons as valence electrons. Spin-orbit interaction is not taken into account

semiconductors with strong ionic bonds such as GaN. On the other hand, Fig. 6.4 suggests that (at least) qualitatively a scissors operator approach is applicable to wide-band-gap compounds like AlN. In any case, one precondition is that the Kohn-Sham theory yields a positive fundamental gap within a local or semilocal approximation to exchange and correlation. Compounds, for which the KS eigenvalues in the framework of DFT-LDA or -GGA lead to negative gaps as wurtzite- and zinc-blende-InN [39] and rocksalt-CdO [37], the scissors operator approach is, of course, not applicable. The inverted KS conduction and valence bands in Fig. 15.5 for InN clearly indicate that a one-shot GW approach does still give rise to a wrong occupation and ordering of bands near the Γ point of the BZ.

An interesting question concerns the applicability of the approach (15.29) to the computation of optical spectra, e.g. the frequency dependence of the macroscopic dielectric function or the absorption coefficient. Because of the many bands ν and $\mathbf{k}$ points, which are needed for a converged spectrum in a wide range of photon energies, the computation of all GW self-energy matrix elements is sometimes prohibited for computer-time reasons. A scissors operator is applied to compute spectra including other many-body effects such as excitonic effects in a wide range of photon energies up to the ultraviolet frequencies for materials without shallow *d* electrons (see e.g. the absorption spectrum of MgO [40]). On such a large energy scale the QP valence (conduction) bands of several materials look as simply shifted by almost constant energy values toward lower (higher) energies compared to the KS reference band structure. As an example the QP band structure of diamond is displayed in Fig. 15.6 together with the reference one computed within DFT-LDA [41]. The fundamental indirect gap between X_{1c} and $\Gamma_{25'v}$ is opened by the QP effects from 4.12 eV (DFT-LDA) to a value of 5.66 eV (GW) close to the experimental value of 5.4 eV [35] measured to the true conduction band minimum Δ_{1c} on the ΓX line near X. The gap opening of 1.54 eV may serve as the value of the scissors operator. However, one has to keep in mind that such a scissors operator remains a good approximation for the description of low-lying excited states. In the case of higher excited states the description of their energy positions may fail [42]. This is not surprising because high-lying conduction states are free-electron-like ones. Such a failure also occurs for the low-energy s-derived valence states (not displayed) because of their stronger localization compared to the near-gap states. Another warning to use

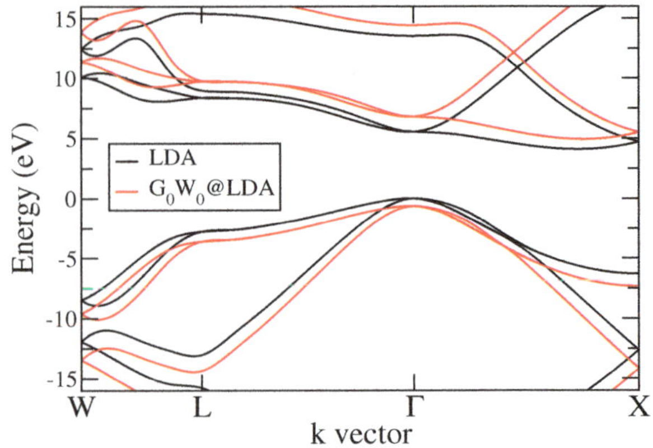

Fig. 15.6 QP band structure of diamond calculated within a one-shot GW (G_0W_0) approximation (*red lines*). The reference KS band structure (*black lines*) is computed using a LDA XC functional. Reprinted with permission from [41]. Copyright 2013 by the American Physical Society

a scissors operator to describe optical properties and, hence, calculate the imaginary part of a dielectric function or its inverse one concerns the consequences for the sum rules of the type (13.27) or (13.38). The blue shift of the spectra with the scissors operator leads to a violation of the f-sum rule. Only with the simultaneous consideration of vertex corrections, resulting in a red shift and a spectral redistribution, the sum rules can be guaranteed [43, 44].

Scissors operators may be also defined for better reference electronic structures than those derived within DFT-LDA. As an example, Fig. 15.7 shows the QP band structure of rocksalt-MgO together with the reference results derived by means of a hybrid XC functional HSE03/06 [see (9.19) and Table 9.1]. The figure demonstrates that a scissors shift of $\Delta_{\text{scissors}} = 1.605\,\text{eV}$ yields conduction bands comparable to the one-shot GW results. The resulting direct fundamental gap of 7.49 eV is in excellent agreement with measured values [35]. Of course the small variations of the valence bands due to QP corrections cannot be reproduced within the scissors operator approximation.

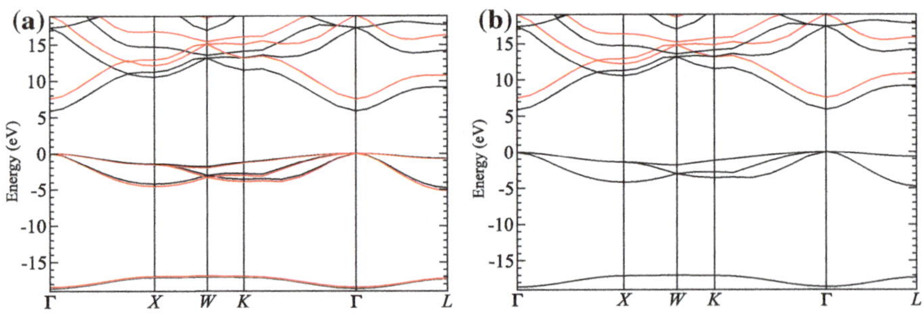

Fig. 15.7 (**a**) Quasiparticle band structure (*red lines*) of *rs*-MgO derived from a one-shot GW calculation on top of a HSE03/06 reference electronic structure (*black lines*) [40]. (**b**) The same electronic structure but with QP effects described by a scissors operator [courtesy of A. Schleife, Friedrich-Schiller Universität Jena]

References

1. L. Hedin, New method for calculating the one-particle Green's function with application to the electron-gas problem. Phys. Rev. **139**, A796–A823 (1965)
2. L. Hedin, S. Lundqvist, Effects of electron-electron and electron-phonon interactions on the one-electron states of solids, in *Solid State Physics*, ed. by F. Seitz, D. Turnbull, H. Ehrenreich (Academic Press, New York, 1969), pp. 1–181
3. F. Bruneval, N. Vast, L. Reining, M. Izquiesdo, F. Sirotti, N. Barrett, Exchange and correlation effects in electronic excitations of Cu_2O. Phys. Rev. Lett. **97**, 267601 (2006)
4. M.S. Hybertsen, S.G. Louie, Model dielectric matrices for quasiparticle self-energy calculations. Phys. Rev. B **37**, 2733–2736 (1988)
5. F. Aryasetiawan, The GW approximation and vertex corrections, in *Strong Coulomb Correlations in Electronic Structure Calculations: Beyond the Local Density Approximation*, ed. by V.I. Anisimov (Gordon and Breach Science Publishers, Amsterdam, 2000), pp. 1–95
6. R.W. Godby, M. Schlüter, L. Sham, Self-energy operators and exchange-correlation potentials in semiconductors. Phys. Rev. B **37**, 10159–10175 (1988)
7. S. Lebègue, B. Arnaud, M. Alouani, P.E. Bloechl, Implementation of an all-electron GW approximation based on the projector augmented wave method without plasmon pole approximation: application to Si, SiC, AlAs, InAs, NaH, and KH. Phys. Rev. B **67**, 155208 (2003)
8. B.I. Lundqvist, Single-particle spectrum of the degenerate electron gas. II. Numerical results for electrons coupled to plasmons. Phys. Kondens. Materie **6**, 206–217 (1967)
9. D.L. Johnson, Local field effects and the dielectric response matrix of insulators: a model. Phys. Rev. B **9**, 4475–4484 (1974)
10. G.E. Engel, B. Farid, C.M.M. Nex, N.H. March, Calculation of the GW self-energy in semiconducting crystals. Phys. Rev. B **44**, 13356–13373 (1991)
11. G.E. Engel, B. Farid, Calculation of the dielectric properties of semiconductors. Phys. Rev. B **46**, 15812–15827 (1992)
12. G.E. Engel, B. Farid, Generalized plasmon-pole model and plasmon band structures of crystals. Phys. Rev. B **47**, 15931–15934 (1993)
13. M.S. Hybertsen, S.G. Louie, Electron correlation in semiconductors and insulators: band gaps and quasiparticle energies. Phys. Rev. B **34**, 5390–5413 (1986)
14. U. von Barth, L. Hedin, A local exchange-correlation potential for the spin polarized case: I. J. Phys. C **5**, 1629–1642 (1972)
15. K.-K. Berggren, B.E. Sernelius, Band-gap narrowing in heavily doped many-valley semiconductors. Phys. Rev. B **24**, 1971–1986 (1981)
16. B.E. Sernelius, K.-K. Berggren, Z.-C. Jin, I. Hamberg, C.G. Granqvist, Band-gap tailoring of ZnO by means of heavy Al doping. Phys. Rev. B **37**, 10244–10248 (1988)
17. M.A.M. Versteegh, T. Kuis, H.T.C. Stoof, J.I. Dijkhuis, Ultrafast screening and carrier dynamics in ZnO: theory and experiment. Phys. Rev. B **84**, 035207 (2011)
18. J.C. Inkson, The effect of electron interaction on the band gap of extrinsic semiconductors. J. Phys. **9**, 1177–1183 (1976)
19. R. Enderlein, J.M. Horing, *Fundamentals of Semiconductor Physics and Devices* (World Scientific, Singapore, 1997)
20. R. Zimmermann, *Many-Body Theory of Highly Excited Semiconductors* (B.G. Teubner Verlagsgesellschaft, Leipzig, 1988)
21. F. Bechstedt, R. Del Sole, G. Cappellini, L. Reining, An efficient method for calculating quasiparticle energies in semiconductors. Solid State Commun. **84**, 765–770 (1992)
22. A.M. Zagoskin, *Quantum Theory of Many-Body Systems: Techniques and Applications* (Springer, New York, 1998)
23. M. Palummo, R. Del Sole, L. Reining, F. Bechstedt, G. Cappellini, Screening models and signified GW approaches: Si & GaN as test cases. Solid State Commun. **95**, 393–398 (1995)
24. M. Palummo, L. Reining, R.W. Godby, C.M. Bertoni, N. Börnsen, Electronic structure of cubic GaN with self-energy corrections. Europhys. Lett. **26**, 607–612 (1994)

25. J. Furthmüller, G. Cappellini, H.-Ch. Weissker, F. Bechstedt, GW self-energy calculations for systems with huge supercells. Phys. Rev. B **66**, 945119 (2002)
26. C.S. Wang, W.E. Pickett, Density-functional theory of excitation spectra of semiconductors: application to Si. Phys. Rev. Lett. **51**, 597–600 (1983)
27. F. Gygi, A. Baldereschi, Quasiparticle energies in semiconductors: self-energy correction to the local-density approximation. Phys. Rev. Lett. **62**, 2160–2163 (1989)
28. V. Fiorentini, A. Baldereschi, Dielectric scaling of the self-energy scissor operator in semiconductors and insulators. Phys. Rev. B **51**, 17196–17198 (1995)
29. M. Rohlfing, Electronic excitations from a perturbative LDA+GdW aproach. Phys. Rev. B **82**, 205127 (2010)
30. F. Bechstedt, R. Enderlein, R. Wischnewski, Binding energies and chemical shifts of least bound core electron excitations in cubic $A_N B_{B-N}$ semiconductors. Phys. Status Solidi B **107**, 637–651 (1981)
31. J.E. Northrup, M.S. Hybertsen, S.G. Louie, Theory of quasiparticle energies in alkali metals. Phys. Rev. Lett. **59**, 819–822 (1987)
32. G.D. Mahan, B.E. Sernelius, Electron-electron interactions and the bandwidth of metals. Phys. Rev. Lett. **62**, 2718–2720 (1989)
33. G. Baraff, M. Schlüter, Migration of interstitials in silicon. Phys. Rev. B **30**, 3460–3469 (1984)
34. F. Bechstedt, Analytical expressions for XC self-energies and quasiparticle shifts in free-electron-like materials. Phys. Status Solidi B **178**, 353–371 (1993)
35. W. Martienssen, H. Warlimont (eds.), *Handbook of Condensed Matter and Materials Data* (Springer, Berlin, 2005)
36. G. Cappellini, S. Boutte-Russo, B. Amadon, C. Noguera, F. Finocchi, Structural properties and quasiparticle energies of cubic SrO, MgO and $SrTiO_3$. J. Phys. Condens. Matter **12**, 3671–3688 (2000)
37. A. Schleife, F. Fuchs, J. Furthmüller, F. Bechstedt, First-principles study of ground- and excited-state properties of MgO, ZnO, and CdO polymorphs. Phys. Rev. B **73**, 245212 (2006)
38. R.W. Godby, M. Schlüter, L.J. Sham, Trends in self-energy operators and their corresponding exchange-correlation potentials. Phys. Rev. B **36**, 6497–6500 (1987)
39. J. Furthmüller, P.H. Hahn, F. Fuchs, F. Bechstedt, Band structures and optical spectra of InN polymorphs: influence of quasiparticle and excitonic effects. Phys. Rev. B **72**, 205106 (2005)
40. A. Schleife, C. Rödl, F. Fuchs, J. Furthmüller, F. Bechstedt, Optical and energy-loss spectra of MgO, ZnO, and CdO from *ab initio* many-body calculations. Phys. Rev. B **80**, 035112 (2009)
41. F. Hüser, T. Olsen, K. Thygesen, Quasiparticle GW calculations for solids, molecules, and two-dimensional materials. Phys. Rev. B **87**, 235132 (2013)
42. G. Cappellini, J. Furthmüller, E. Cadelano, F. Bechstedt, Electronic and optical properties of cadmium fluoride: the role of many-body effects. Phys. Rev. B **87**, 075203 (2013)
43. F. Bechstedt, K. Seino, P.H. Hahn, W.G. Schmidt, Quasiparticle bands and optical spectra of highly ionic crystals: AlN and NaCl. Phys. Rev. B **72**, 245114 (2005)
44. W.G. Aulbur, L. Jönsson, J.W. Wilkins, Quasiparticle calculations in solids, in *Solid State Physics*. Advances in Research and Applications, vol. 54 ed. by H. Ehrenreich, F. Spaepen. (Academic Press, San Diego, 2000), pp. 1–218

Chapter 16
Quasiparticle Electronic Structures

Abstract The energy positions of the main peaks in the spectral functions of inhomogeneous electron gases represent their quasiparticle electronic structures. Discussing numerical results based on *ab initio* Kohn-Sham and generalized Kohn-Sham starting electronic structures the great success of the Hedin GW approximation to the exchange-correlation energy is demonstrated for condensed matter in general. A central result for semiconductors and insulators is the calculation of energy gaps but also bands and their dispersion in reasonable agreement with experimental data. Of course, the obtained quasiparticle energies depend slightly on computational details such as starting point, self-consistency, convergence, and inclusion of vertex corrections beyond the GW approximation. Nevertheless, the resulting global or projected densities of states are applicable to explain spectra measured by means of various spectroscopies. Basically the conclusions derived for non-metals remain valid for bulk metallic and magnetic systems. The great success of the quasiparticle description is also shown for molecules and other low-dimensional systems with confinement of electrons in three, two or one directions.

16.1 Semiconductors and Insulators

16.1.1 Fundamental Energy Gaps

In order to demonstrate the success of the quasiparticle theory (14.42) developed for electronic energy bands and levels within the one-shot GW (i.e., G_0W_0) approximation (14.56) with the frequency-dependent RPA dielectric matrix $\epsilon(\mathbf{q}+\mathbf{G}, \mathbf{q}+\mathbf{G}', z)$ and the corresponding inverse quantity $\epsilon^{-1}(\mathbf{q}+\mathbf{G}, \mathbf{q}+\mathbf{G}', z)$, we list values resulting for the fundamental gaps of 18 semiconductors and insulators in Table 16.1 [1, 2]. Two reference KS or gKS electronic structures are used, DFT-GGA (Sect. 7.3) and HSE03/06 (Sect. 9.2). For the majority of the investigated semiconductors and insulators crystallizing (or assumed to crystallize) in cubic geometries the gap results are plotted against experimental values in Fig. 16.1. The corresponding one-shot or self-consistent GW self-energy is implemented within the scheme described in [1, 3].

F. Bechstedt, *Many-Body Approach to Electronic Excitations*,
Springer Series in Solid-State Sciences 181, DOI 10.1007/978-3-662-44593-8_16

Table 16.1 Quasiparticle gaps (in eV) in one-shot GW quality for two different KS and gKS starting points, GGA and HSE03/06

Material	GGA	GGA + GW	HSE03/06	HSE03/06+GW	Expt.
PbSe	−0.17	0.10	–	–	0.15
PbTe	−0.05	0.20	–	–	0.19
PbS	−0.06	0.28	–	–	0.29
Ge	–	–	0.54	0.79	0.74
Si	0.62	1.12	1.04	1.32	1.17
GaAs	0.49	1.30	1.12	1.66	1.52
SiC	1.35	2.27	2.03	2.60	2.40
CdS	1.14	2.06	1.97	2.55	2.42
AlP	1.57	2.44	2.09	2.69	2.45
GaN	1.62	2.80	2.65	3.29	3.20
ZnO	0.67	2.12	2.11	2.86	3.44
ZnS	2.07	3.29	3.05	3.69	3.91
C	4.12	5.50	5.08	5.84	5.48
BN	4.45	6.10	5.54	6.54	6.1–6.4
MgO	4.76	7.25	6.22	7.94	7.83
LiF	9.20	13.27	11.2	14.1	14.20
Ar	8.69	13.28	10.1	13.7	14.20
Ne	11.61	19.59	14.1	20.2	21.70

The results are compared with experimental values. All values in eV. From [1, 2]

The two different starting points with semilocal or non-local XC potentials derived within GGA [4, 5] and HSE03/06 [6, 7], respectively, lead to different QP gaps in Table 16.1 and Fig. 16.1. The largest discrepancies appear for wide-gap materials or compounds like InN, for which the DFT-GGA gives a negative gap. With the single exception of diamond (C) the GGA reference consistently yields underestimated gap values. The minor overestimation for C is entirely related to the neglect of vibrational effects, not studied in this book. It has been shown that zero-point vibrations reduce the QP gap by about 0.23 eV [8] or even 0.6 eV [9, 10]. In the cases of ZnO, ZnS, CdS, GaAs, and GaN the one-shot GW approach on top of DFT-GGA eigenvalues generates QP gaps which are more than 10 % smaller than the experimental ones. For the examples GaAs and ZnO, this might be explained by rather small DFT-GGA gaps of about or less than 30 % of the measured ones. These findings necessitate a better treatment of the XC effects in the presence of the excitation, e.g. going beyond the one-shot approach and hence the first-order perturbation theory or even including vertex corrections, at least in the screening function, beyond the GW approximation [1, 11].

From Fig. 16.1 and Table 16.1 one recognizes that the HSE03/06 starting point gives rise to much better QP gaps after adding one-shot GW corrections. This holds obviously for (zinc-blende) InN, for which the $\Gamma_{1c} - \Gamma_{15v}$ direct gap changes the

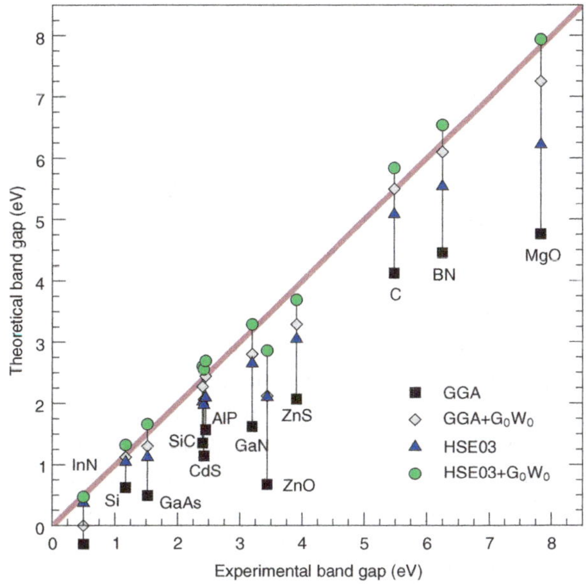

Fig. 16.1 Fundamental gaps of semiconductors and insulators calculated in zeroth-order KS or gKS approximation and corrected by quasiparticle shifts in one-shot GW quality versus experimental values. Starting electronic structures are studied for two different XC potentials: GGA and HSE03/06. The 45° *line* indicates full agreement between QP theory and experimental findings. Drawn using data from [2]

sign from the GGA to the HSE03/06 reference electronic structure. Within the latter approach a QP gap of about 0.5 eV (zinc-blende) [or 0.7 eV for wurtzite] is computed for appropriate lattice parameters in extremely good agreement with measured values (see [12, 13] and references therein). The better performance of the one-shot GW approximation for a reference electronic structure based on a hybrid XC functional such as HSE03/06 for GaAs, CdS, GaN, ZnO, and ZnS may be partially traced back to the better treatment of the binding energies of the shallow *d* electrons. They are also increased with respect to the DFT-GGA values [1, 2]. As a consequence the *pd* repulsion [14] is reduced on the HSE03/06 level, which impacts valence energy levels and wave functions. In particular, the shrinkage of the *sp* gap due to the *pd* repulsion is reduced. For the other materials, usually without shallow *d* electrons, both starting points, GGA and HSE03/06, lead to smaller deviations. Summarizing, Fig. 16.1 and Table 16.1 demonstrate that the modern QP theory may predict fundamental energy gaps with deviations smaller than 0.2 eV compared to experimental results.

Nevertheless, we have to point out that still several issues such as the reference electronic structure, self-consistency, vertex corrections, and also the convergence of the computations with respect to number of bands, **k** points, plane waves, basis sets, treatment of the screening dynamics play an important role for the accuracy of the predictions. The tendency to use better starting points and, hence, energy eigenvalues and wave functions close to the QP results has been realized in various forms, for instance with reference electronic structures based on the static COHSEX approximation for correlated compounds [15], DFT + U approaches for metal oxides with d or f electrons close to band edges [16, 17], and exact exchange DFT calculations in the optimized-effective potential approach for compounds with semicore d states [18].

16.1.2 Challenges and Achievements

The enormous progress made in the last 30 years in predicting quasiparticle gaps of non-metallic solids has been illustrated in Sect. 16.1.1. Nevertheless, there remain some open questions concerning the physics of electronic excitations, for instance, the effect of vertex corrections, but also still numerical problems, among them the convergence with respect to bands, energies and frequencies, basis functions, self-consistency, etc. Moreover, the applicability of the theory on metals, low-dimensional systems and molecules has to be discussed.

A first possible step beyond the GW approximation (12.57) for the XC self-energy is the use of an improved screening description. For details the reader is referred to a review article [19]. The basic idea is the replacement of the functional derivative of the XC self-energy $\delta\Sigma/\delta G$ in the vertex function Γ in (12.55) following the LDA idea by the variation of a local XC potential $K_{\mathrm{XC}} = \delta V_{\mathrm{XC}}/\delta n$ with respect to the electron density. Simultaneously the microscopic polarization function P is replaced by the RPA result L^0 (12.58). Formally this leads again to the GW form of the XC self-energy Σ, however, now with a modified screening of the Coulomb potential [20], for instance with an inverse dielectric function

$$\epsilon^{-1} = 1 + \mathrm{v}L^0\left[1 - (\mathrm{v} + K_{\mathrm{XC}})L^0\right]^{-1}, \tag{16.1}$$

if vertex corrections K_{XC} are only taken into account in W. This expression corresponds to a test charge-test charge dielectric matrix. Including vertex corrections $\Gamma = -[1 + K_{\mathrm{XC}}L^0]^{-1}$ directly in the XC self-energy, the effective screening of the Coulomb interaction of two electrons is replaced by that of a test charge and electrons [19, 21]. The correction given by K_{XC} accounts for XC effects at two levels: (i) An electron of the screening system is surrounded by an XC hole when participating in the dielectric screening. (ii) The potential induced by the excited electron or hole includes XC interactions between the particle and the electronic system.

Indeed, vertex corrections as added to (16.1) have been applied to improve QP calculations. Results for silicon [22] show that the resulting QP energy gaps are close to those without vertex corrections due to the cancellation of many-body corrections. The vertex corrections affect however much stronger the QP binding energies, such as electron affinity or ionization potential [22, 23]. Meanwhile such K_{XC} corrections are implemented in GW codes [1]. Thereby, the kernel K_{XC} is taken from the time-dependent density functional theory within the local framework [19, 24]. Improvements of K_{XC} to non-local kernels [25, 26] are also sometimes used. In many cases the inclusion of vertex corrections is combined with the self-consistent GW approach originally suggested by van Schilfgaarde et al. [27, 28] and called scGW. In the diagonal GW version only the positions of the QP peaks in the Green function G [$\tilde{G}$ in (14.6)] but also in L^0 are updated. However, taking into account off-diagonal elements of the self-energy the wave functions of the starting electronic structure are also updated [1] resulting in $\tilde{\tilde{G}}$ (14.32).

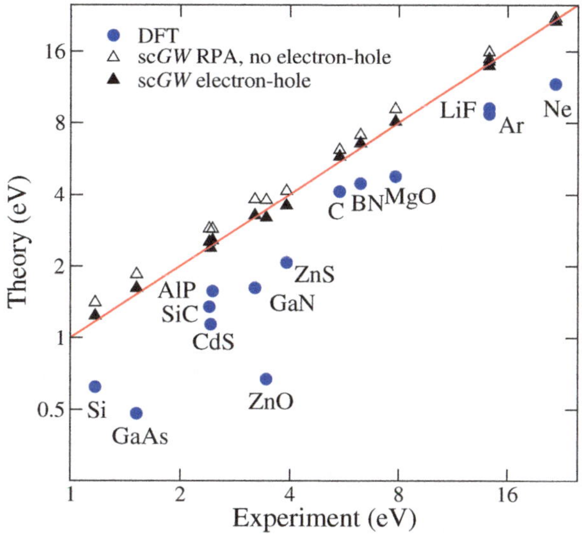

Fig. 16.2 Influence of vertex corrections in W (16.1) on the QP gaps of non-metals. Only cubic polymorphs are investigated, e.g. also for ZnO. Reprinted with permission from [1]. Copyright 2007 by the American Physical Society

Both vertex effects are illustrated in Fig. 16.2 for QP energy gaps starting from a semilocal GGA-PBE XC potential (see Sect. 7.3.2). This starting point is called DFT in the figure. The inclusion (exclusion) of vertex corrections is indicated by "electron-hole" ("no electron-hole"). First of all, the self-consistent QP approach on top of Kohn-Sham eigenfunctions and eigenvalues yields gaps in excellent agreement with experimental values. The inclusion of vertex corrections really improves the agreement. The main reason is that the vertex corrections increase the effective screening and hence reduce the QP corrections (see Sect. 15.3.2). As a consequence the marginal overestimation of the gaps in Fig. 16.2 is almost corrected.

The QP gaps resulting for ZnO and solid rare gases Ar and Ne are extremely promising. However, in the majority of numerical treatments they remain too small, especially for ZnO, within the one-shot GW approximation (see e.g. Figs. 14.15 and 14.17 as well as Table 14.2). The reasons and possible solutions have been recently investigated in a couple of GW studies [29–32]. The wide-band gap semiconductor ZnO which crystallizes in hexagonal wurtzite structure under ambient conditions is used to serve as a prototypical material to shine light especially on the convergence issues. A conventional one-shot GW approach yields a gap of 2.44 eV (using a DFT reference electronic structure) [33] or 2.87 eV (using a hybrid HSE03/06 starting point) [2], much smaller than the experimental gap of about 3.6 eV [34, 35] after correcting the gap for the lattice effects. One aspect is a possible inadequate treatment of the semicore Zn $3d$ electrons, which underbinds the d states and leads to an unphysically strong pd hybridization [14]. This effect seems to be corrected widely within a DFT + U approach (Sect. 9.1). The U parameter shifts the Zn $3d$ states down in energy and hence reduces the gap closing due to the pd repulsion [29, 36].

Another problem is the slow and non-uniform convergence in the calculation of the COH self-energy with the number of plane waves to represent the Bloch band states and the inverse dielectric matrix. Within a plasmon-pole model a number of 3000 plane waves corresponding to a cutoff energy of 67 Ry seems to be sufficient to reproduce the experimental gap [29]. However, taking into account the full frequency dependence of the inverse dielectric function in the GW self-energy, the convergence with the band states is much weaker and the resulting one-shot GW gap value is still below the experimental value [30–32].

Another influence on the one-shot GW approximation is illustrated in Fig. 16.3 together with the self-consistent GW for the fundamental gap of *wz*-ZnO. Starting point of the self-consistent calculations is the hybrid HSE03/06 functional (9.19) with a mixing parameter $\alpha = 0.25$ and a PW91-GGA XC functional (see Sect. 7.3.2). Indeed, as discussed above the self-consistent treatment of the QP eigenvalues leads to a significant enlargement of the gap, but the result is still below the experimental value. The increase of the mixing parameter α results in a gap increase because of the increasing Hartree-Fock character of the XC contribution in the starting electronic structure. Interestingly the one-shot GW gap corrections vanish near $\alpha \approx 0.5$. The two curves HSE03/06 and HSE03/06+G_0W_0 cross. For this mixing parameter the HSE03/06 hybrid DFT computation leads to a gap close to the experimental one. The mixing parameter $\alpha \approx 0.5$ and, hence, the use of half of a Fock operator for the

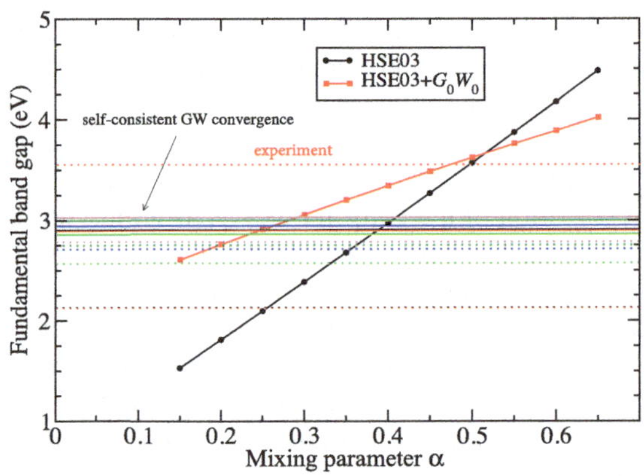

Fig. 16.3 Fundamental quasiparticle gap of *wz*-ZnO starting from a HSE03/06 reference electronic structure with a mixing parameter $\alpha = 0.25$ and using a self-consistent GW procedure. The *solid horizontal lines* represent results for different numbers of self-consistency steps. The small difference to the result of [2] is due to a modified treatment of the core-valence exchange. In addition, results for a self-consistent GW approach based on a DFT-GGA starting point are given by *dotted horizontal lines*. The experimental gap value is given by a *red dotted horizontal line*. The influence of the mixing parameter α, $0 < \alpha < 0.7$, describing the influence of increasing non-local exchange and decreasing XC in DFT-GGA (PW91), is illustrated for the reference value and within the one-shot GW approach by increasing *straight lines* [courtesy of A. Schleife, Friedrich-Schiller-Universität Jena]

exchange supports the idea that the QP effects tend to a stronger localization of the wave functions (see e.g. Fig. 14.12). Its inclusion appears to be important to derive the real gap, especially for wide-gap materials and materials with extremely shallow core levels such as Zn 3*d*.

16.1.3 Bands, Dispersion, and Effective Masses

Besides the fundamental gaps and the energy distances between conduction and valence bands also the wave-vector dispersion of the Bloch bands may be influenced by quasiparticle corrections. This is illustrated in Fig. 16.4 for the one-shot GW approach and two reference electronic structures applied to zinc-blende-AlN [37]. Because of the use of the VBM as energy zero, at first glance, the QP shifts and their wave-vector dispersion are small for the valence-band states. However, it is obvious that they are non-zero. In particular, for the DFT-GGA starting point the QP corrections tend to increase the width of the uppermost valence bands which is obtained by varying the wave vector from the center to the boundaries of the BZ. The QP effects are obviously larger for the conduction band states. Nevertheless, their variation with band index and **k** vector remain small. Rough measures are the differences between the direct Γ-Γ and indirect Γ-X gaps of 0.67 (PBE-GGA), 0.89 (PBE+GW), 0.79 (HSE03/06), and 0.91 (HSE03/06+GW) eV. They indicate that QP-effect-induced variations in the band dispersions are of the order of 0.1–0.2 eV for many not too complex materials. The absolute shift values are, however, generally larger. The absolute one-shot HSE03/06 + GW gap values amount to 5.15 (Γ-X) and 6.06 (Γ-Γ) eV in reasonable agreement with experimental values 5.34/5.3 [38, 39] and 5.93 [39] eV. The small deviations could be a consequence of the used lattice constant that is overestimated within the PBE-GGA framework.

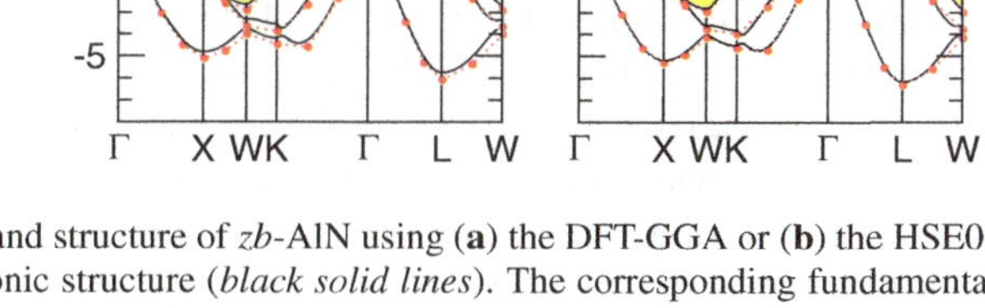

Fig. 16.4 QP band structure of *zb*-AlN using (**a**) the DFT-GGA or (**b**) the HSE03/06 treatment as reference electronic structure (*black solid lines*). The corresponding fundamental gap regions are illustrated by *yellow* areas. The one-shot GW results are displayed as *red dots* and *dotted lines*. The VBM is used as energy zero. From [37]

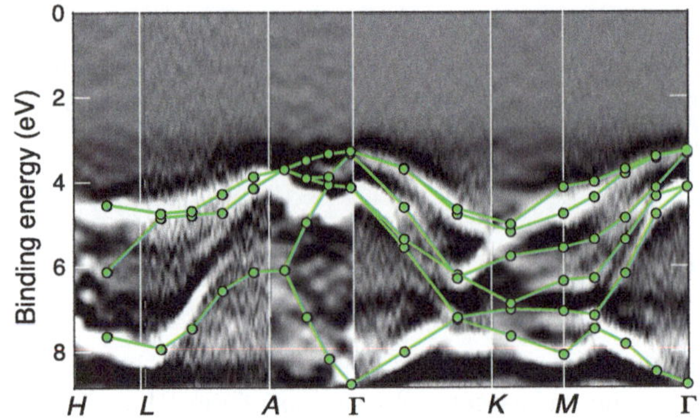

Fig. 16.5 Quasiparticle valence band structure of *wz*-ZnO in one-shot HSE03/06 + G_0W_0 quality [36]. It is compared with results of ARPES measurements [40]. [Courtesy of A. Schleife, Friedrich-Schiller-Universität Jena]

Indeed, the use of a lattice constant closer to the experimental one gives rise to gaps closer to the measured energies [13].

According to the above discussion one expects a reasonable agreement of calculated [36] and measured [40] band dispersions. This is indeed shown in Fig. 16.5 for the valence bands of *wz*-ZnO. Taking the uncertainties due to the photoemission cross sections into consideration, one can state an excellent overall agreement. This holds, in particular, for the high-symmetry directions ΓM and ΓK perpendicular to the c-axis of the wurtzite structure. For the BZ boundaries HL and MK the agreement is satisfying. The largest discrepancies appear in the BZ center parallel to the c-axis along ΓA. Due to matrix-element effects the emission probability seems to be significantly reduced for the uppermost valence band.

Figure 16.4 has clearly shown that the differences between the QP valence bands and the reference bands described by a hybrid XC functional are small, at least, if the VBM is used as joint energy zero, for AlN. This fact suggests the use of the valence bands of the reference electronic structure to derive the corresponding effective hole masses. This procedure has another advantage. In contrast to the GW self-energy calculations, spin-orbit interaction, that significantly influences the uppermost valence bands in tetrahedrally coordinated semiconductors, can be easily included in the computations. As an example, the three uppermost valence bands (without spin) and, hence, six bands including SOC and non-collinear spins of zinc-blende and wurtzite group-III nitrides are displayed in Fig. 16.6 [13]. Interestingly, besides the spin-orbit splittings at the Γ points also **k**-induced spin-orbit splittings of the HH band along ΓL (in the *zb* case) and along ΓM (in the *wz* case) appear for the compounds GaN and InN with heavier cations. They are consequences of the Rashba and Dresselhaus effects [41–43]. In determining the corresponding effective masses we average over the **k**-induced SOC splittings.

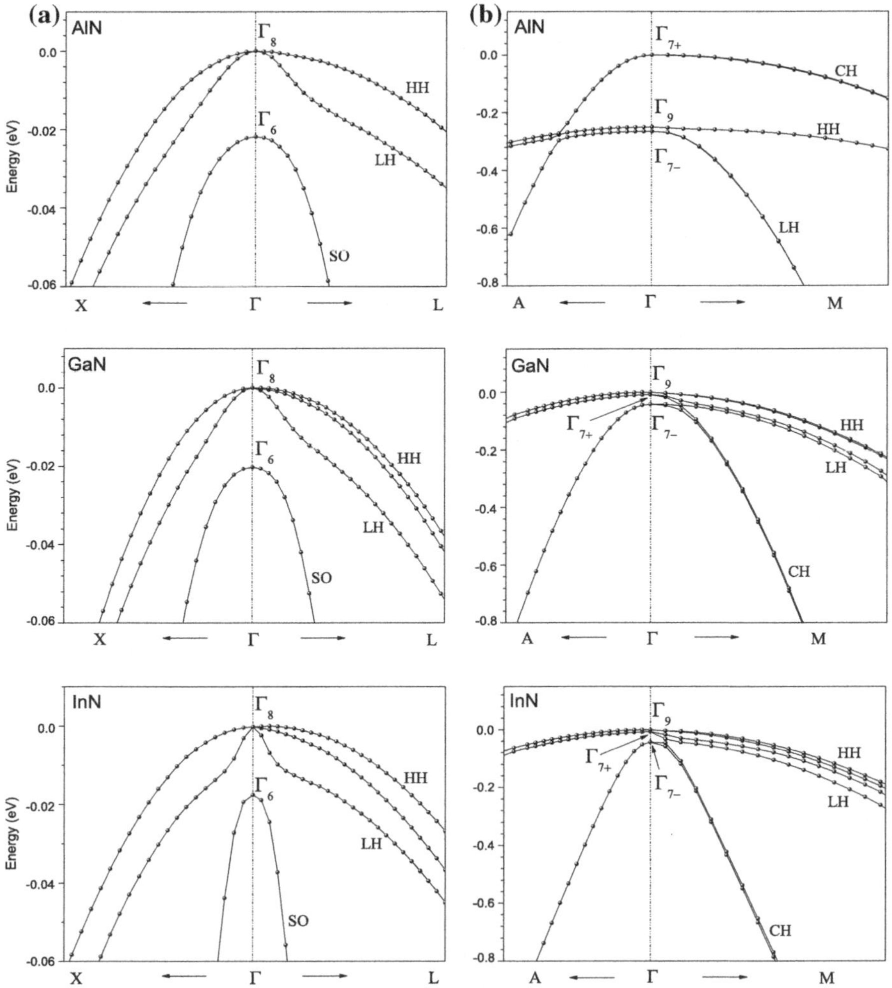

Fig. 16.6 Uppermost valence bands of AlN, GaN, and InN in (**a**) *zb* and (**b**) *wz* structure around Γ along two high-symmetry directions in the BZ. They are derived from the reference electronic structure computed in the HSE03/06+SOC framework. The heavy-hole (HH), light-hole (LH), spin-orbit split-off (SO), and crystal-field split-off hole (CH) bands are labeled. The VBM is used as energy zero. From [13]

Resulting effective hole and electron masses for *wz*-AlN, -GaN, and -InN are compared in Table 16.2 with experimental values. A corresponding comparison is presented for III-V compounds crystallizing in zinc-blende geometry in Table 16.3. Both tables indicate an excellent agreement between theory and experiment taking the uncertainties of the computation, e.g. the choice of **k** points for the parabolic fit, and measurements, e.g. sample quality and free-carrier influence, into consideration.

Table 16.2 Effective hole (hh, lh, ch) and electron (e) masses of III-nitrides crystallizing in wurtzite structure [13] near Γ taken from HSE03/06+SOC computations

Compound	m_{hh}^{A}	m_{lh}^{A}	m_{ch}^{A}	m_{e}^{A}	$m_{hh}^{M,K}$	$m_{lh}^{M,K}$	$m_{ch}^{M,K}$	$m_{e}^{M,K}$
AlN	3.31	3.06	0.26	0.32	6.95	0.35	3.47	0.34
				0.29-0.45				0.29-0.45
GaN	2.00	1.22	0.20	0.21	0.57	0.31	0.92	0.21
	2.20	1.10	0.30	0.20	0.42	0.51	0.68	0.20
InN	1.98	1.02	0.08	0.06	0.44	0.09	0.18	0.06
				0.07				0.07

They are compared with experimental data (second line) mainly collected in [44]. The assumed direction in **k** space is indicated by the corresponding points A or M, K at the BZ boundary. All values are in units of the free-electron mass m

Table 16.3 Effective hole (hh, lh, so) and electron (e) masses of direct III-V compounds crystallizing in zinc-blende geometry taken near Γ along the ΓX direction from HSE03/06+SOC computations [13, 45]

Compound	m_{hh}	m_{lh}	m_{so}	m_{e}
GaN	0.83	0.28	0.34	0.19
				0.16
GaAs	0.31	0.09	0.17	0.07
	0.35	0.09	0.17	0.07
GaSb	0.24	0.05	0.14	0.04
	0.25	0.04	0.12	0.04
InN	0.91	0.08	0.11	0.05
				0.04
InP	0.48	0.12	0.21	0.09
	0.53	0.12	0.21	0.08
InAs	0.34	0.03	0.11	0.03
	0.33	0.03	0.14	0.03
InSb	0.25	0.02	0.13	0.02
	0.26	0.02	0.11	0.02

Experimental values [13, 46] are listed in the second line. All masses are given in units of the free-electron mass m

Indeed, the HSE03/06 starting point combined with the inclusion of spin-orbit interaction seems to give very reasonable band dispersions around the Γ point [45].

16.1.4 Density of States

The quasiparticle band structures also give the global density of states $D(\hbar\omega)$ (11.27). Neglecting the renormalization of the spectral strength of the QP peaks and QP wave functions (see Sect. 14.1.2) it can be directly computed from the QP eigenvalues $\varepsilon^{\mathrm{QP}}_{\nu m_s}(\mathbf{k})$ (14.42), for instance for a system with collinear spins, to

$$D(\hbar\omega) = \sum_{m_s}\sum_{\nu}\sum_{\mathbf{k}} \delta(\hbar\omega - \varepsilon^{\mathrm{QP}}_{\nu m_s}(\mathbf{k})). \tag{16.2}$$

The local density of states (11.26) also allows to restrict the DOS studies to contributions from certain chemical elements, lattice sites, orbital symmetries, etc. by using corresponding projection operators. For instance, using the eigenstates $|n\ell m\rangle$ of a spherical potential located at an atomic position $\mathbf{R}_l$ and explicit spectral functions (12.60) with (12.69), the projected density can be rewritten into

$$D^{n\ell}(\hbar\omega) = \frac{1}{2\ell+1}\sum_{m}\sum_{m_s}\sum_{\nu}\sum_{\mathbf{k}} |\langle \nu\mathbf{k}m_s|n\ell m\rangle|^2\, \delta\left(\hbar\omega - \varepsilon_{\nu m_s}(\mathbf{k})\right). \tag{16.3}$$

As first examples we investigate the valence-band, i.e., occupied, DOS of two crystals in Fig. 16.7. It displays the DOSs of the zinc-blende and the wurtzite polytype of InN together with the corresponding uppermost valence bands [47]. The calculations are performed using HSE03/06 reference electronic structures and quasiparticle shifts derived within a one-shot GW treatment. For the purpose of comparison with X-ray

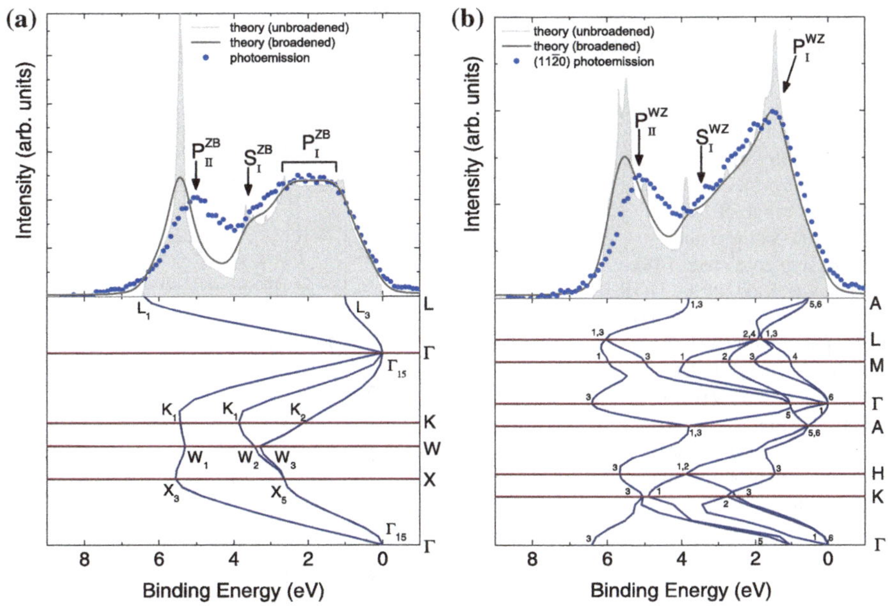

Fig. 16.7 Calculated DOS in comparison with X-ray photoemission spectra (*blue dots*) for (**a**) *zb*- and (**b**) *wz*-InN. The calculated spectra (*grey areas*) are broadened (*solid lines*) with a Lorentzian broadening of 0.2 eV to account for lifetime effects and a 0.45 eV Gaussian broadening to account for instrumental resolution. The photoemission spectra are shifted to align the VBM at 0 eV. From [47]

photoemission data an additional broadening is taken into account and the Shirley background is subtracted from the photoemission spectrum. The theoretical and experimental spectra are normalized to the peak P_{I} intensity, since matrix-element effects do not occur in the calculated curves. The DOS in the energy region from −6.5 to 0 eV of the uppermost valence bands is dominated by p states. Substantial s contributions are only visible for the lower DOS peak. The d contributions due to the pd repulsion are hardly recognizable near the VBM. The shape of the uppermost DOS part depends very much on the interaction of third nearest or more distant neighbors. For zb-InN, due to the relatively flat nature of the Γ_{15v} bands (Fig. 16.7a), the DOS rises rapidly below the VBM, peaking around the critical point L_3. In the wurtzite case (Fig. 16.7b), the onset of the DOS corresponds to Γ_{6v} (or Γ_{5v} depending on the notation) bands followed by the Γ_{1v} one. Instead of rising to a plateau, the DOS continues to rise rapidly due to a number of turning points in the band structure such as Γ_5, A_6, or M_4. The lowest predominant peak in the DOS occurs largely due to the critical points H_3 at H and in the Σ (ΓM) direction between Γ_3 and M_1 (for wurtzite) or the almost degenerate critical points X_3, W_1, and K_1 (for zinc blende). The comparison of the broadened DOS in the HSE03/06 + one-shot GW approximation with results of X-ray photoemission measurements confirms the previous discussion of the peak structure and the differences between the spectra for wz- and zb-InN. The only discrepancy between the calculated DOSs and the experimental spectra is that the low-energy peak occurs at slightly lower binding energies in the experimental spectra for reasons discussed in Sect. 16.1.1.

Using the same quasiparticle approximation as applied to InN several oxides have been investigated [48, 51]. Results for rutile-SnO_2 are displayed in Fig. 16.8.

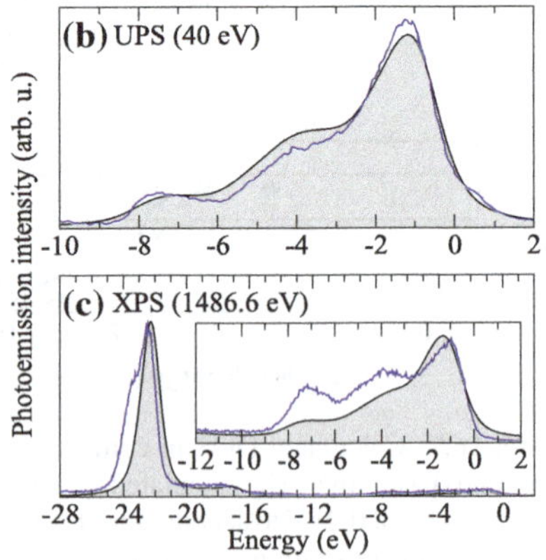

Fig. 16.8 Total density of states of rutile-SnO_2 together with orbital-projected DOS in the energy range of conduction, valence and shallow semicore states [48]. (**a**) The weighted partial DOS (*grey shaded areas*, see text) is compared with (**b**) UPS [49] or (**c**) XPS [50] data (*violet lines*). The VBM is set to zero. From [51]. Copyright Wiley-VCH Verlag GmbH & Co.KGaA. Reproduced with permission

The total density of states is decomposed in Fig. 16.8a into several orbital contributions, which are compared in part in (b) and (c) with photoemission spectra obtained using different incident photon energies [49, 50]. Due to the photoemission cross sections which vary strongly with the incident photon energy, XPS probes mainly d states, whereas ultraviolet photoemission spectroscopy (UPS) is sensitive to both p and d states more or less with equal measure. Taking into account the photoionization cross sections (as tabulated in [52]) as weight factors for the orbital-resolved DOS, the results can be directly compared to experimental data. Thereby, the theoretical spectra are artificially broadened to account for finite lifetime, temperature effects, and instrumental broadening present in the measurements. Altogether, the comparison in Fig. 16.8b and c validates the quasiparticle picture and the GW approximation for occupied valence and semicore states.

The DOS of unoccupied conduction-band states cannot be probed by (one-photon) photoemission spectroscopy. However, there are other spectroscopies with X-rays that allow for the investigation of empty states, e.g. the X-ray absorption spectroscopy (XAS) [12]. Thereby, a direct comparison with the total DOS or some projected DOS is difficult, mainly because of the electron-hole interaction between the electron excited by an X-ray photon into a conduction band state and the hole left back in a core state. Nevertheless, the strong localization of the core states at a certain atom helps to introduce a reasonable approximation, taking the excitation process and its transition matrix elements into account. For core states with vanishing overlap to adjacent atoms and a classification as eigenstates $n\ell$ of a certain atom X, one expects that the transition of the electron can only occur into orbital contributions $n'\ell'$ from the same atom X to the conduction-band state that fulfills the energy conservation. Since such a transition is ruled by the dipole-matrix element of the core and conduction states, the angular-momentum selection rule $\ell' = \ell \pm 1$ [53] should be fulfilled. Excitonic effects which may modify this picture mainly influence the energy conservation law by a rigid red shift that may be interpreted as the exciton binding energy (see Sect. 21.1). In any case, as a consequence of the dipole selection rule the resulting lineshape of the XAS should be determined by that of the orbital-projected DOS.

For illustration, the conduction-band density of states of wz-InN is projected in Fig. 16.9 onto the N $2p$ contributions and compared with X-ray absorption spectra recorded in the total electron yield (TEY) mode using an incident photon energy of 393–423 eV for a wz-InN(000$\bar{1}$) film [54]. The projected DOS is computed within the HSE03/06 + one-shot GW scheme. Two different incident angles 20° and 70° with respect to the c-axis of the hexagonal crystal are investigated [54]. For light polarization perpendicular to the c-axis (the z-axis), transitions from N $1s$ initial states into the N $2p_z$ states are dipole-forbidden. Likewise, for light polarization parallel to the c-axis (z-axis), transitions from N $1s$ initial states into the final N $2p_{x,y}$ states are forbidden. Assuming that excitonic effects do not essentially change the lineshape, the experimental spectra are compared with the partial densities of states projected onto the N $2p_z$ and N $2p_{x,y}$ states. In order to account for the selection rules and the chosen incidence angles (different from 0° and 90°), we compare explicitly to weighted linear combinations of projected densities of states

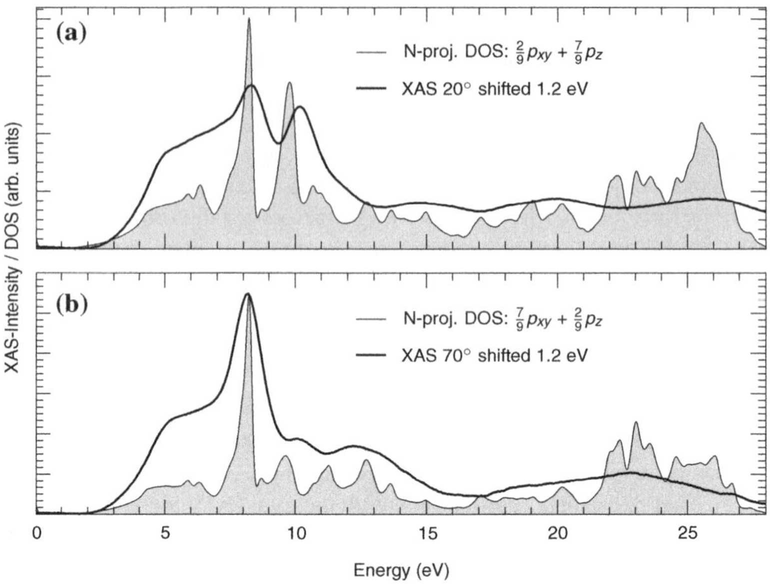

Fig. 16.9 X-ray absorption spectra (*solid line*) taken in the total electron-yield mode for the N K-edge and incidence angles of (**a**) 20° and (**b**) 70° with respect to the c-axis of wz-InN [54]. They are compared with calculated unoccupied partial densities of states (*thin black lines/shaded areas*) projected onto N $2p_z$ or N $2p_{x,y}$ states. Accounting best for the selection rules of N $1s \rightarrow$ N $2p$ transitions for different light polarizations and for the used incidence angles of 20° and 70°, we display calculated spectra corresponding to weighted projections (**a**) 2/9 N $2p_{x,y}$ + 7/9 N $2p_z$ and (**b**) 2/9 N $2p_z$ + 7/9 N $2p_{x,y}$. The theoretical spectra have been reasonably broadened for a comparison with the measurements by a convolution with a Lorentzian (HWHM = 0.3 eV) followed by a Gaussian (FWHM = 0.2 eV) in order to account for lifetime and instrumental broadening. The VBM is used as energy zero. The experimental spectra are shifted by 1.2 eV to higher energies (see text). From [54]

corresponding to 2/9 N $2p_{x,y}$ + 7/9 N $2p_z$ (Fig. 16.9a) and 7/9N $2p_{x,y}$ + 2/9 N$2p_z$ (Fig. 16.9b). For a better comparison the measured spectra are rigidly shifted toward higher energies by about 1.2 eV. This value is interpreted as the binding energy of the N $1s$-N $2p$ core-conduction-band excitons not included in the calculations. The XAS spectrum recorded at an incidence angle of 20° is dominated by two peaks at approximately 8 and 10 eV above the VBM. If the angle of incidence is changed to 70° the intensity of the 8 eV feature increases dramatically while that of the 10 eV peak is reduced. The characteristic peak behavior is also observed in the calculated linear combinations of N $2p_z$ and N $2p_{x,y}$, contributions to the unoccupied partial densities of states shown in Fig. 16.9a, b. The shoulders and the peaks are reproduced by the theoretical spectra for energies up to 14 eV. Especially, the double peak in the 20° spectrum (Fig. 16.9a) and the central peak in the 70° spectrum (Fig. 16.9b) around 8/10 or 8 eV are described correctly.

A site- and symmetry-projected DOS of unoccupied states is also displayed in Fig. 16.10 for rs-CdO. The same QP approximation as for wz-InN has been applied.

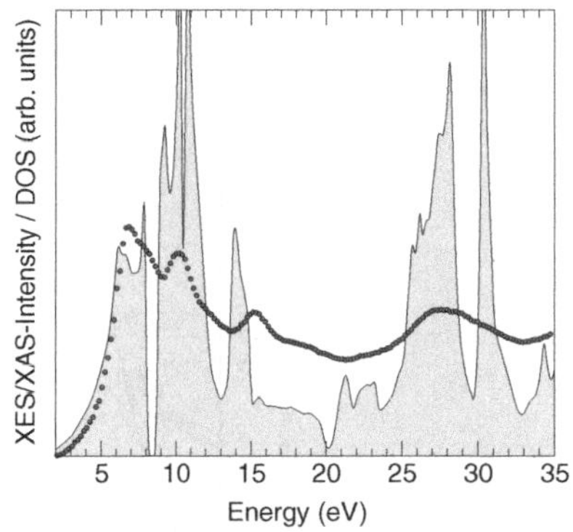

Fig. 16.10 O K-edge XAS spectra (*dots*) of rocksalt-CdO. The calculated unoccupied quasiparticle O 2p partial DOS (with a Gaussian broadening) is plotted together with the respective spectra for direct comparison (*thin black lines/shaded area*). The experimental XAS spectrum is shifted by 1.4 eV to higher energies (see text). From [55]

It is compared with the O K-edge XAS spectrum obtained for photon energies of 528–560 eV. According to the O 1s character of the core states and the dipole-selection rule we have projected the DOS onto the O 2p contributions. For comparison with the XAS spectrum in Fig. 16.10 the unoccupied projected O 2p DOS with a Gaussian broadening of FWHM = 0.2 eV has been applied. The lineshapes of experimental XAS and projected theoretical DOS spectra in Fig. 16.10 agree well, but originally the peak positions are slightly different. For this reason, we have additionally shifted the experimental spectra toward higher energies by 1.4 eV. We note, that since the calculations include already excited-state self-energy corrections, that the additional shift of $\sim$1.4 eV between the experimental and theoretical peaks and the enhancement of the intensity of the first peak should be interpreted to be due to excitonic effects, similarly to the InN case discussed above. Likewise, the shift between experiment and theory in Fig. 16.10 may provide an estimate for the binding energy of the O 1s–2p-like conduction-electron core-hole exciton. The seemingly somewhat reduced agreement between theory and experiment for rs-CdO compared to wz-InN asks for a larger broadening of the theoretical spectrum, probably, because of a reduced quality of the film studied experimentally.

16.2 Metallic and Magnetic Systems

16.2.1 Simple Metals

Simple metals such as alkali metals or aluminum exhibit only weak effective crystal potentials acting on the s and p valence electrons. The weakness of such a potential seems to permit a nearly-free-electron-model description of the conduction electrons in alkali metals [56]. Therefore, they seem to offer experimentally accessible systems for which many-body corrections are weak, allowing a perturbation-theory treatment

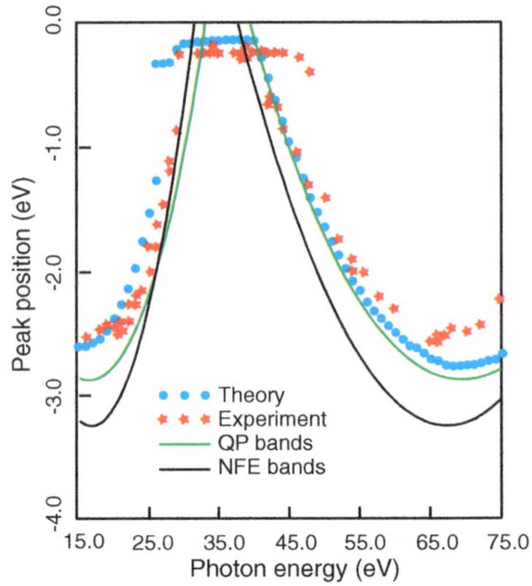

Fig. 16.11 Photoemission peak positions as a function of photon energy for the Na(110) surface. The QP bands include the real and imaginary part of the self-energy and are represented by *green solid curves*. The QP bands are narrower than the nearly-free-electron (NFE) bands (*black solid curves*) by 0.37 eV. The full theory (*blue dots*) also includes surface effects. Experimental data are denoted by *red stars* [57]. Adapted from [58]

of the GW self-energy within a one-shot approach. In principle, this idea is supported by the general findings that the band dispersion is less influenced by many-body effects and metals are gapless systems, so that a scissors operator should tend to be zero. However, already for the sodium metal the experimental determination of the highest occupied s band via photoemission [57] contradicts the nearly-free-electron model (see Fig. 16.11): (i) The measured band width is about 0.6 eV smaller than the nearly-free-electron value of 3.2 eV. (ii) Sharp, non-dispersive peaks in the energy gap can be identified for photon energies of about 35 eV in contrast to the findings within the free-electron-like model.

Indeed, the inclusion of the real and imaginary parts of the XC self-energy improves very much the agreement between the peak positions in photoemission and the quasiparticle energies in Fig. 16.11, although an approach, the Rayleigh-Schrödinger perturbation theory, slightly different from GW has been used [58]. In particular, the Na s band width is reduced by 0.37 eV. An additional improvement is due to the inclusion of surface effects. Surface effects, in particular the finite mean free path of the emitted electrons correct the PES peak positions in the range of photon energies of 35–45 eV. The narrowing of the band width generally happens for simple metals within the GW approximation. This is obvious from Table 16.4, where results for Al, Li, Na, and K are compared with experimental data.

Table 16.4 Width of the s (or sp) band in simple metals

Metal	r_s	DFT-LDA	QP (1)	QP (2)	Expt.
Al	2.1		10.0	10.2	10.6
Li	3.3	3.5	2.9	3.1	3.0
Na	4.0	3.2	2.5	2.7	2.65 ± 0.05
K	4.9	2.3	1.6	1.9	1.4

GW-corrected QP values (1) [20, 59] and (2) [60] are compared with measured values. The values are taken from a collection in [61]

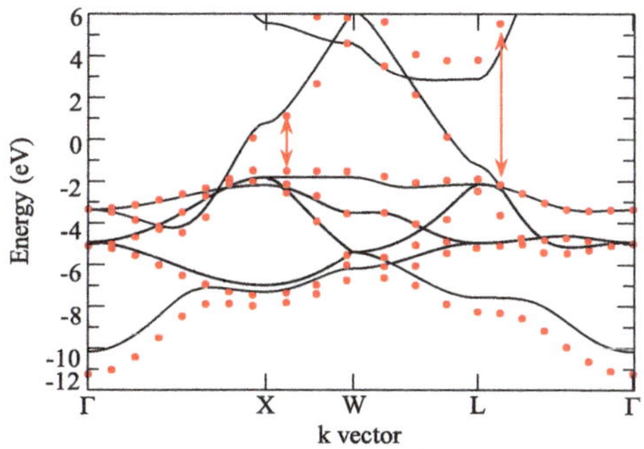

Fig. 16.12 Band structure of fcc gold calculated within DFT-LDA (*black lines*) and one-shot GW approximation (*red dots*). The bands are aligned to the respective Fermi level. Interband transition energies between occupied 5*d* and empty 6*sp* bands are indicated by *red arrows*. Reprinted with permission from [62]. Copyright 2013 by the American Physical Society

16.2.2 d-Electron Metals

Metals with completely filled shallow *d* levels such as gold or partially filled *d* shells such as nickel with *d*-band contributions to the Fermi surface exhibit larger deviations between measured band structures and those derived from Kohn-Sham eigenvalues as illustrated in Figs. 16.12 and 16.13. One striking feature in the Au case is the underestimation of the 5*d*–6*sp* interband gap by ~1.0 eV with respect to available experimental data (see [62, 63]). At least, the *d* band is shifted down by 0.4 eV relative to the DFT-GGA(PBE) values within a self-consistent GW calculation. The dispersion of the bands is well described.

These phenomena seem to be general for *d*-electron metals with occupied *d* bands as illustrated by the band structures of the nobel metals Cu, Ag, and Au [67] in Fig. 16.14 obtained in the framework of the self-consistent GW approximation. However, there is a consistent tendency to slightly underestimate the binding energies

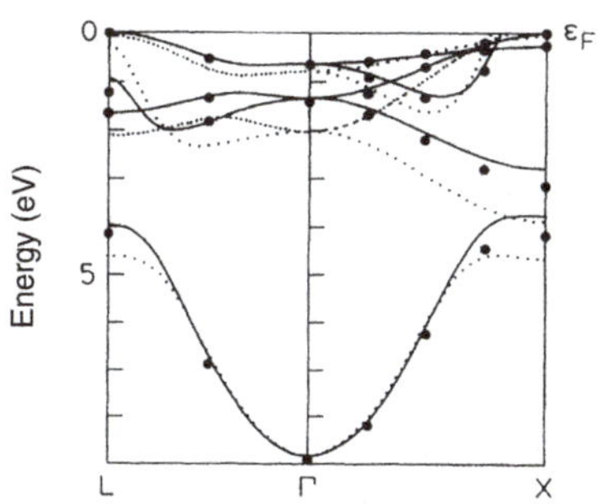

Fig. 16.13 Band structure of Ni along ΓX and ΓL. The *solid curves* represent the experiment and the *dotted curves* are the DFT-LDA results [64]. The *filled circles* are quasiparticle energies in one-shot GW approximation. After [65, 66]. Copyright Royal Swedish Academy of Sciences. Reproduced by permission of IOP Publishing. All rights reserved

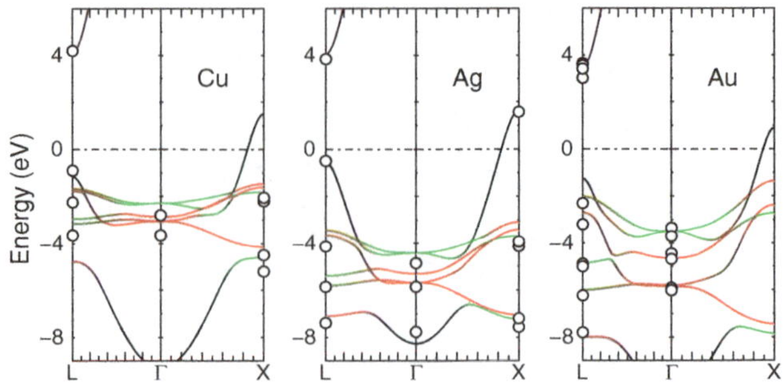

Fig. 16.14 QP band structures of nobel metals Cu, Ag, and Au obtained in the framework of the self-consistent GW approach. The Fermi level defines the energy zero. Energies from ARPES measurements are displayed by open circles. From [67]

of the *d* levels by 0.1 to 0.4 eV. This seems to be a universal feature of the GW approach, probably, because of the missing vertex function in the XC self-energy, for instance, by reduction of the effective screening [67].

Because of the partial filling of the 3*d* electron shell the applicability of the GW approximation, including the neglect of vertex corrections, to the transition metals is under debate for a long time. Indeed, the significant failure of the DFT-LDA and -GGA for TM band energies asks for another handable numerical description of their electronic structure. For instance, in the case of the transition metal Ni four main discrepancies appear [65]: (i) The measured 3*d*-band width of 3.3 eV is about 30 % smaller than its KS value 4.5 eV derived within the DFT-LDA. (ii) The experimental exchange splitting of 0.25–0.30 eV is overestimated by DFT-LDA. (iii) The satellite in the photoemission spectrum at a binding energy of about 6 eV is missing. (iv) The bottom of the 3*d* band cannot be described by sharp excitations. QP peak widths up to 2 eV occur due to the strong interaction between the localized electrons and the other valence electrons.

The first (i) and fourth (iv) discrepancies can be almost lifted within a standard one-shot GW treatment [65, 66]. The DFT-LDA band structure is much improved by QP effects, in particular the 3*d*-band width is narrowed by almost 1 eV, as shown in Fig. 16.13. The QP lifetimes (not shown here) are also described rather well by the GW approximation but the exchange splittings remain essentially unchanged from their KS values and the 6 eV satellite is not reproduced.

The early promising results for Ni have encouraged several authors to apply the many-body perturbation theory also to other metals with partially filled *d* shells. As an example the self-consistent GW band structure of ferromagnetic Fe [67] is displayed in Fig. 16.15. For both the majority- and minority-spin channels the bands near the Fermi energy show nearly ideal agreement with ARPES measurements. Also the *d*-band exchange splitting (in contrast to Ni) and the band widths are well described. The same holds for the magnetic moments of the Fe atoms. The self-consistent GW approximation reproduces the experimental data well, because the

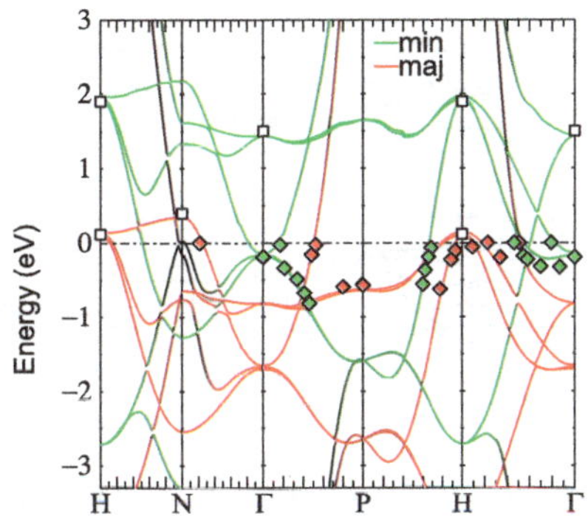

Fig. 16.15 QP band structures of ferromagnetic Fe obtained within the self-consistent GW approach. The bands of the majority (minority) spin channel are described by *red* (*green*) *lines*. Experimental band positions are displayed by *diamonds* and *squares*. The Fermi level is used as energy zero. From [67]

strong non-locality of XC [see (15.6)] is better described as in other approaches, such as DMFT [68].

16.2.3 Antiferromagnetic and Ferromagnetic Insulators

Transition metal oxides such as NiO and MnO have been considered for years as prototypes of Mott-Hubbard insulators. It was pointed out by Mott [69] in the late forties that a system with an on-site Coulomb energy (cf. Sect. 9.1.4) larger than the single-particle band width tends to become an insulator and that the single-particle theory is bound to give a wrong prediction for the state of the system. Indeed, the local or semilocal XC functionals used in DFT predict NiO to be a metal when calculations are performed in the paramagnetic state [70]. In the antiferromagnetic state, more precisely in a (slightly distorted) rocksalt geometry with AF2 magnetic ordering (see Fig. 9.2), these functionals give rise to fundamental (indirect) gaps of about 0.6 eV (NiO) and 0.7 eV (MnO) but much below the experimental values of 3.7–4.3 eV (NiO) and 3.6–4.2 eV (MnO), respectively [16]. With vanishing gaps within the DFT-GGA the situation is worse for the antiferromagnetics FeO and CoO, for reasons which have been discussed in Figs. 9.4 and 9.5.

In the light of these small or vanishing gaps, the applicability of the GW approximation neglecting vertex corrections has been called in question in the past. However, the first QP treatments of the electronic structure of NiO and MnO [71, 72] were rather promising. A tendency to overestimate the fundamental gaps has been traced back to not sufficiently localized wave functions of the 3*d* electrons and, hence, the need for a self-consistent procedure or the use of a better starting point. QP band structures resulting from a hybrid HSE03/06 starting point are presented in Fig. 16.16. Although the gaps are already significantly opened with 2.6 and 4.1 eV within the reference electronic structure the progress due to the one-shot GW corrections is obvious with 3.4 and 4.0 eV (MnO) as well as 4.7 and 5.2 eV (NiO) for the indirect and direct gaps.

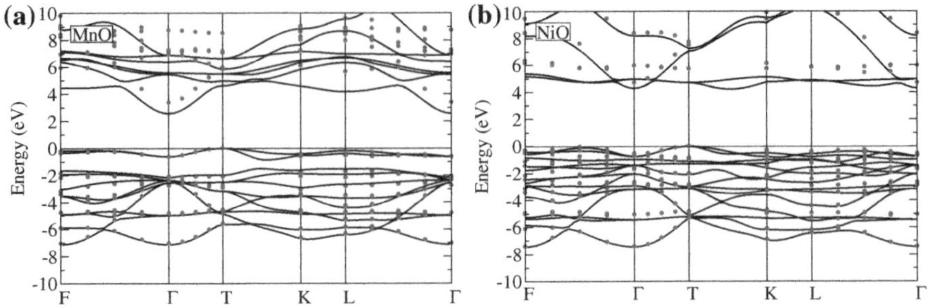

Fig. 16.16 QP band structures for MnO (**a**) and NiO (**b**). *Red dots* denote the QP energies obtained within one-shot GW on top of the HSE03/06 calculation (*black lines*). The VBM is used as energy zero for both descriptions. From [16]

These two antiferromagnetics, whose band structures are displayed in Fig. 16.16, are indirect semiconductors with the VBM located at the T point (using the rhombohedral magnetic unit cell). The highest valence bands are dominated by TM $3d$ states with some admixture of states with O $2p$ character. Consequently, the corresponding band states are strongly localized resulting in a little wave-vector dispersion. The influence of the QP corrections on the dispersion is negligible. They can be mainly represented by a rigid shift. The conduction band minimum is found at Γ. In a small wave-vector region around Γ the parabola-shaped lowest conduction bands exhibit a strong dispersion and feature TM $4s$ character. Above this CBM region, in the entire BZ nearly dispersionless unoccupied bands derived from TM $3d$ states are found.

The drastic differences between the band structures of MnO and NiO are due to the different occupation of the $3d$-derived t_{2g} states in the minority-spin channel. We use the nomenclature $t_{2g}\uparrow$, $t_{2g}\downarrow$, $e_g\uparrow$, and $e_g\downarrow$ of the $3d$-states in an octahedral ligand field (cf. Fig. 9.4). The comparison of their density of states in Fig. 16.17 makes their occupation more visible. While in MnO $e_g\uparrow$ states at the valence band edge are followed by $t_{2g}\uparrow$-derived bands, the uppermost valence bands of NiO are dominated by t_{2g} states. The opposite behavior appears near the conduction band edge with $t_{2g}\downarrow$ states in MnO and $e_g\downarrow$ states in NiO. The density of the s states is too small to be visible in such a figure. The lineshapes and peak positions obtained within HSE03/06 + GW almost match experimental XPS and bremsstrahlung isochromat spectroscopy (BIS) spectra (see [16]). The identification of the orbital symmetry agrees between theory and experiment.

Another interesting class of magnetic systems are ferromagnetically ordered insulators. In the framework of collinear spins their electronic structure can be represented by band structures of the majority- and minority-spin channels. One example is the compound $CrBr_3$ with a critical temperature $T_c = 37$ K [73] for the transition between the ferromagnetic and paramagnetic phases. It crystallizes in a layered hexagonal structure with space group $P3_112$ (No. 151) as illustrated in Fig. 16.18b [73]. The primitive unit cell of $CrBr_3$ is trigonal (rhombohedral) with two formula units in it. The band structures of the ferromagnetic phase in Fig. 16.18a show the pd-derived, uppermost valence bands, the d-derived lowest conduction bands, and also some

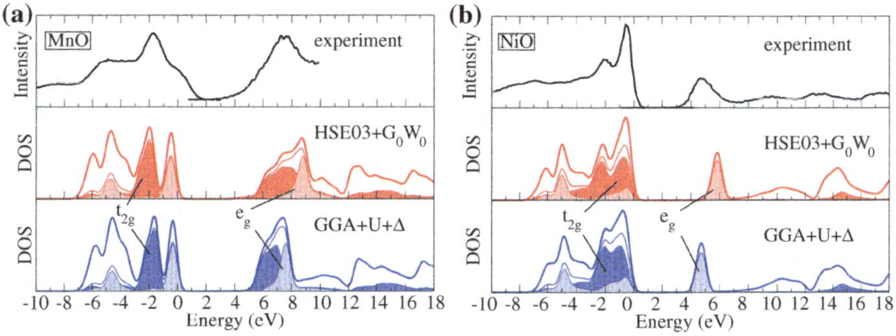

Fig. 16.17 DOS for MnO (**a**) and NiO (**b**) in the HSE03/06 + GW approach (*middle panel*) compared with experimental XPS and BIS spectra (*upper panel*). In the *lowest panel* the DOS obtained by using the GGA + U method with an additional scissors shift Δ_{scissors} is shown. In the calculated spectra the contributions of the TM d states (*thin solid line*) with t_{2g} (*dark shaded*) and e_g (*light shaded*) symmetry to the total DOS (*thick solid line*) are also indicated. For better comparison the computed DOSs are broadened by a Gaussian with 0.6 eV full width at half maximum. The *top* of the valence bands is taken as energy zero. From [16]

higher conduction bands with mixed s, p, and d character. The band structures for both spin channels show the insulating behavior of the ferromagnetic $CrBr_3$ phase. However, in the majority-spin channel six more bands are occupied. This fact is in agreement with a magnetic moment of the Cr atoms of 3 μ_B, in excellent agreement

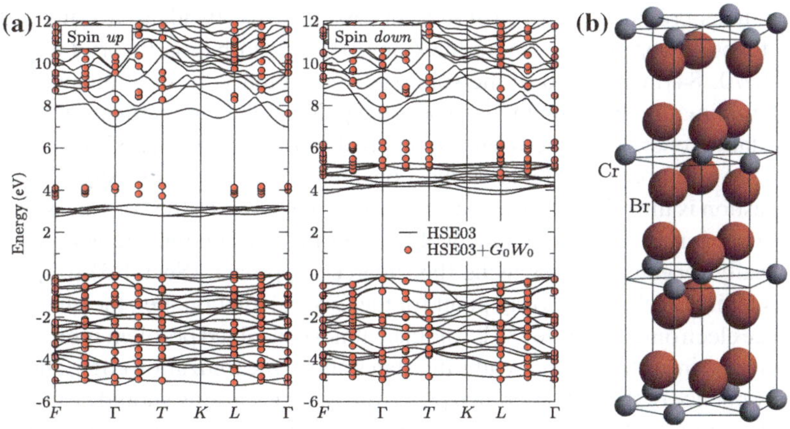

Fig. 16.18 (**a**) Quasiparticle band structure of the ferromagnetic insulator $CrBr_3$ crystallizing in (**b**) a trigonal (rhombohedral) crystal structure with two formula units in a primitive unit cell. (Cr: *grey dots*, Br: *red circles*). The bands of the starting HSE03/06 electronic structure are displayed by *solid black lines*, while the HSE03/06 + G_0W_0 QP band energies are given by *red dots*. The majority (minority)-spin channel is represented in the *left* (*right*) panel of (**a**). The VBM is used as energy zero. From [74]

with the measured values [73]. The QP gap opening with respect to the HSE03/06 value amounts to about 1 eV. The resulting fundamental gap is 3.7 eV (4.7 eV) in the majority (minority)-spin channel. Altogether, from the quasiparticle point of view each spin channel in the ferromagnetic insulator $CrBr_3$ behaves like a non-magnetic insulator.

16.3 Low-dimensional Systems

16.3.1 Molecules

There is a variety of quantum-chemical methods to compute electronic excitations for isolated molecules (for a recent review see, e.g., [75]). Among them are *ab initio* single-configurational methods which use the HF solutions as reference wave functions, e.g., within the so-called configuration interaction-singles (CIS) method [76]. It overestimates the excitation energies due to the absence of correlation effects. The size-extensive coupled-cluster (CC) approach in different versions [77, 78] is also often used for excited-state calculations. Thereby, in general, a zeroth-order HF reference is used. Excited-state calculations carried out using the time-dependent density functional theory [24, 79, 80] are becoming more and more popular because of their simplicity and apparent black-box character. TDDFT approaches seem to have currently no rival when computing excited states in large molecules, e.g., C_{70} [81], at low costs.

A completely different type of methods to calculate excited-state properties is referred to as propagator approach [82, 83] by chemists. Its ideas are similar to the Green function approaches within the many-body perturbation theory (see reviews [19, 58, 70, 84]) employed in the solid-state community. In the framework of the GW approximation these methods, where the poles of the Green function determine the excitation energies, have been also applied to molecules [62, 85–87]. Thereby, a couple of additional problems or questions, that have to be reinvestigated, appear. One question is the relation to the basis sets, e.g. plane waves [85] or Gaussians [86]. Another one is related to the treatment of the long-range Coulomb interaction between periodically repeated images using the supercell approach (Sect. 1.3) [62]. A third point is to find a reliable starting point for the GW calculation, i.e., the quality of the reference electronic structure [87]. Another, almost unsolved problem is related to the empty states, in particular the scattering states above the ionization edge, and, hence, the convergence of the GW approach [85]. At least one warning should be mentioned. The calculations ask for performing sums over all empty states in the correlation contribution (14.59) to the GW self-energy as well as in the "RPA" polarization function (12.70) or (13.43). Whereas for infinite solids the high-energy states can be easily treated, their modeling for finite electronic systems is difficult. There are however suggestions in the literature to solve at least the convergence problem in the self-energy calculations. Recent studies [88] have shown that a modified static COHSEX approach can be used to accurately minimize the empty-state problem in the Coulomb-hole summation in (15.4) and, thus, can be also applied to molecules within a supercell approach [89]. With such tricks it has been demonstrated to improve the self-energy convergence from 1 eV to better than 0.1 eV. For instance, for a molecule containing 46 atoms, bithiophene naphtalene diimide, the polarizability has been calculated with 78 occupied and 875 empty orbitals. These numbers appear to be sufficient also for the COH term.

The convergence problem also exists for the description of electronic excitations in infinite bulk systems. However, the free-electron-like empty states above the vacuum level are usually well described. On the contrary, for molecules in the gas phase and freestanding nanostructures, e.g. colloidal nanocrystals, the continuum of scattering states above the ionization level is still influenced by the interactions in the localized system of interest but should not be modified by the shape and the finite extent of the supercell. Within the modeling the continuum becomes a quasicontinuum of eigenvalues with confined wave functions. The dilemma of finite objects within the numerical description of the scattering states in a supercell approach cannot fully be avoided. It does not only occur using plane-wave basis sets but also when representing scattering state by localized basis functions, e.g. Gaussians. On the contrary, the uncertainties grow since Gaussians do not fulfill the completeness requirement. Because of the reasonable description of occupied states in general, a rewriting of the sum over empty states by using the closure relation and introducing a projection operator onto valence states may give a mathematical answer for the correct treatment of scattering states.

Nevertheless, the GW approach seems to give also reasonable results for molecules, at least for states not to close to the ionization edge, without applying the projection operator. The reason is related to the appearing matrix elements, which reduce the effect of the high-energy states. This is illustrated in Table 16.5 for the lowest eigenvalues of a H_2O molecule. The corresponding self-energies and molecular orbitals are displayed in Fig. 14.8. In general, we can conclude from Table 16.5 that, despite the mentioned difficulties, the GW approach, including a certain self-consistency of the wave functions, gives excellent single-particle excitation energies for the occupied states of molecules such as H_2O. A scaling of the self-energy effects by a factor of 0.5 as observed for less sophisticated GW descriptions [90] is not necessary.

Table 16.5 The lowest vertical ionization energies (binding energies of the highest occupied levels) of the H_2O molecule from QP calculations [85, 91] and measurements [92, 93]

Molecule level	KS	QP [85]	QP [91]	Expt. [92]	Expt. [93]
$1b_1$	7.21	11.94	11.90	12.61	12.78
$3a_1$	9.27	14.42	14.18	14.73	14.83
$1b_2$	13.11	18.79	18.35	18.55	18.72

The KS eigenvalues are computed within a DFT-LDA scheme [85]. All energies are in eV

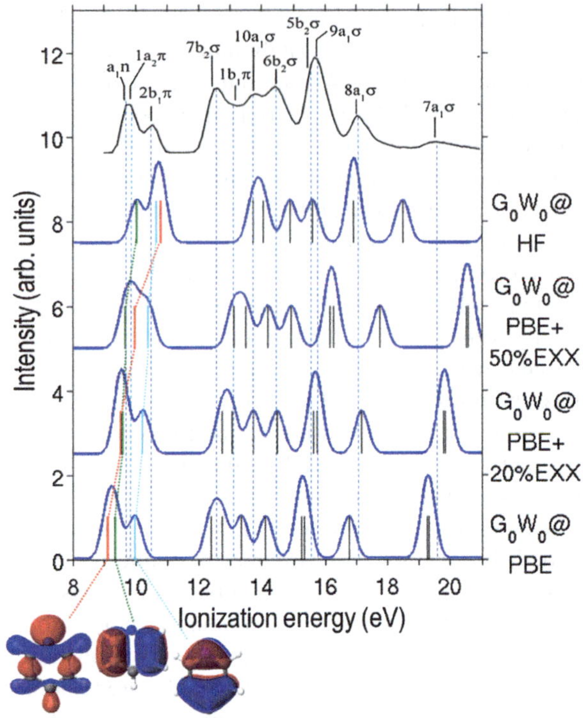

Fig. 16.19 One-shot GW spectra based on HF and various DFT starting points compared to experimental PES [94] (*upper curve*) for pyridine. The Gaussian convolution of the theoretical spectra amounts to 0.3 eV. Reprinted with permission from [87]. Copyright 2012 by the American Physical Society

How to find a reliable starting point for a one-shot GW approach has been recently discussed in terms of XC functionals varying between the pure DFT-PBE one (see Sect. 7.3.2) and the HF approach (Sect. 4.2.3) by mixing DFT exchange with exact (Fock) exchange (EXX) (Sect. 9.2.3) [87]. Results are illustrated in Fig. 16.19 for the pyridine molecule with 11 atoms (see inset in Fig. 16.19). The resulting peak positions and lineshapes vary with the contribution of EXX. That means, there is a significant dependence on the reference electronic structure within the one-shot approach. For the pyridine molecule the DFT-PBE starting point produces electron removal energies and orbital characters in very good agreement with PES data [94, 95]. The entire spectrum is only shifted by 0.4 eV toward too small binding energies. The use of HF eigenvalues and eigenfunctions makes the theory self-interaction-free but also leads to a significant underscreening in GW due to the too large HOMO-LUMO gap. A small amount of EEX (about 20 %) seems to generate the best agreement with the experimental curve. It seems to be essential for the gKS starting point to capture the "right amount" of screening in the molecule. Tests for 11 molecules among them the famous 3,4,9,10-perylene-tetracarboxylic dianhydride (PTCDA) suggest mixing coefficients α (9.21) varying in the interval $0.1 \leq \alpha \leq 0.3$ [86].

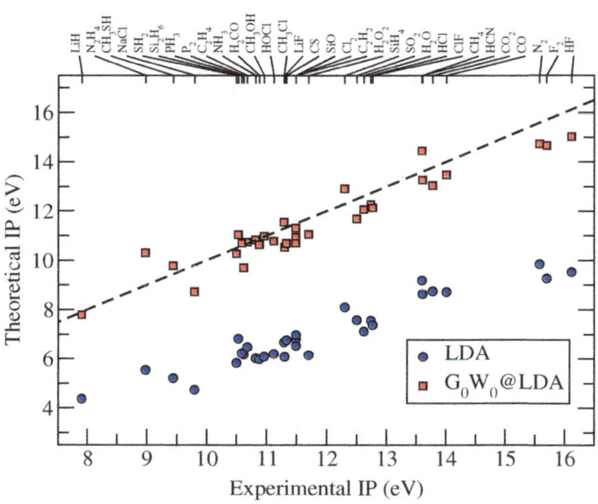

Fig. 16.20 Comparison of theoretical (GW) and experimental ionization potentials of molecules. DFT-LDA is used for the reference electronic structure. Reprinted with permission from [62]. Copyright 2013 by the American Physical Society

Other studies suggest that already a local approach to the XC functional to derive the reference KS electronic structure leads to reasonable excitation energies, in particular, for ionization potentials (IPs), for small molecules. This conclusion is illustrated in Fig. 16.20 for 32 molecules [62]. The convergence problem has been solved by extrapolating to results for an infinite cutoff energy E_{cut} (8.43) of the plane-wave basis set. While KS eigenvalues underestimate the IPs with a mean absolute error (MAE) of 4.8 eV, the QP values obtained within a one-shot GW treatment are typically only around 0.5 eV smaller than the experimental values, although there are a few exceptions where the calculated IP is too large.

Other recent studies with local and semilocal XC starting points [86, 95] confirm these findings despite the use of Gaussian basis sets or using numerical atomic orbitals in an all-electron framework. IP results for the Gaussian basis sets and molecules displayed in Fig. 16.21 are plotted in Fig. 16.22a. The quality of the GW calculations is supported by QP HOMO-LUMO gaps of these molecules in Fig. 16.22b. The IP and gap results in Fig. 16.22 show that the KS eigenvalues computed within DFT-LDA significantly underestimate ionization energies and fundamental gaps also for molecular systems. GW corrections significantly improve the agreement with experimental values. Thereby, a starting-point dependence of around 0.5 eV is visible. The most striking conclusion is that the one-shot values on top of a HF reference electronic structure and the self-consistent GW energies with a DFT-LDA starting point are rather similar. Because of the difficulties to measure the electron affinities only a few experimental HOMO-LUMO gaps are available for comparison with the computed values. They also show good agreement with self-consistent GW results.

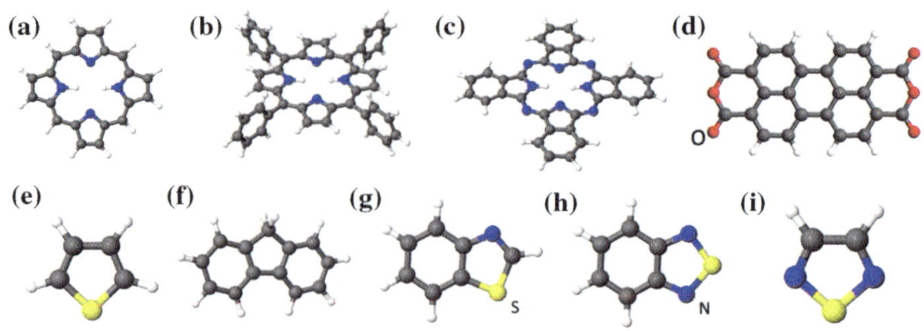

Fig. 16.21 Symbolic representation of (**a**) 21H,23H-porphine (H_2P), (**b**) tetraphenylporphyrin (H_2TPP), (**c**) phtalocyanine (H_2Pc), (**d**) 3,4,9,10-perylene tetracarboxylic acid dianhydride (PTCDA), (**e**) thiophene, (**f**) fluorene, (**g**) benzothiazole, (**h**) 2,1,3-benzothiadiazole, and (**i**) 1,2,5-thiadiazole. *Small white* atoms are hydrogen atoms, *gray* atoms are carbon atoms, while *red*, *blue*, and *yellow* atoms are oxygen, nitrogen, and sulfur atoms, respectively. Reprinted with permission from [86]. Copyright 2011 by the American Physical Society

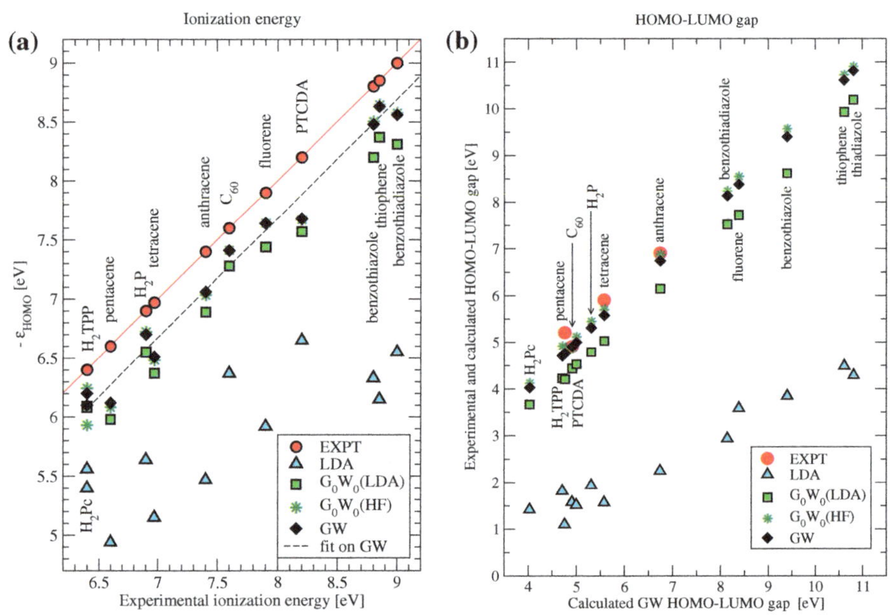

Fig. 16.22 (**a**) Experimental and theoretical ionization energies as well as (**b**) HOMO-LUMO gaps. *Red circles*: experimental values; *light blue triangles up*: LDA Kohn-Sham HOMO energies or HOMO-LUMO gaps; *green squares*: non-self-consistent G_0W_0(LDA) values; *black diamonds*: GW values with self-consistency on the eigenvalues; *green stars*: non-self-consistent G_0W_0(HF) values. The *black dashed line* is a least-squares fit of the GW results. Reprinted with permission from [86]. Copyright 2011 by the American Physical Society

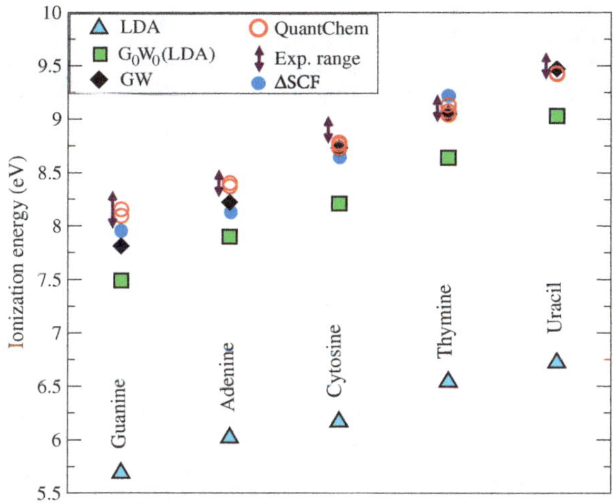

Fig. 16.23 Vertical ionization energies of nucleobases. The vertical (*maroon*) error bars indicate the experimental range. (*Light blue*) *triangles*: LDA values; (*green*) *squares*: G_0W_0(LDA) values; (*black*) *solid diamonds*: GW values; (*red*) *empty circles*: quantum chemistry, namely, CCSD(T), CASPT2, and EOM-IP-CCSD, values. *Blue dots* are from ΔSCF calculations [96] (see also Sect. 6.3.2). Reprinted with permission from [97]. Copyright 2011 by the American Physical Society

Finally, our conclusions are also valid for the basic components of life the DNA and RNA nucleobases adenine, guanine, cytosine, thymine (shown in Fig. 6.3), and uracil. This is illustrated for their vertical ionization potentials in Fig. 16.23.

16.3.2 Clusters and Nanocrystals

One of the first applications of the MBPT on clusters concerned the Na_4 one [98]. Its equilibrium structure turns out to be a planar rhombus. The measurement of characteristic single-particle excitations yields electron addition and removal energies (see Sect. 4.3.2), for instance an adiabatic ionization potential of 4.27 ± 0.05 eV [99]. The KS DFT-LDA value of 2.76 eV is in striking disagreement with the experimental result. Simply adding a one-shot GW correction of 1.53 eV computed for the ground-state geometry of the neutral system yields a vertical ionization potential of 4.29 eV. Inclusion of geometrical relaxations leads to the adiabatic potential of 4.27 eV, in excellent agreement with experiment [100].

For years another model system of the theory was the silane molecule SiH_4, that may be also considered as the smallest hydrogenated Si nanocrystal with tetrahedral symmetry in the series Si_5H_{12}, $Si_{17}H_{36}$, $Si_{41}H_{60}$, ... displayed in Fig. 16.24 [101]. In its ground state SiH_4 possesses the full T_d symmetry. The experimental value for the ionization energy is 12.6 eV [102]. Other studies report a broad peak at

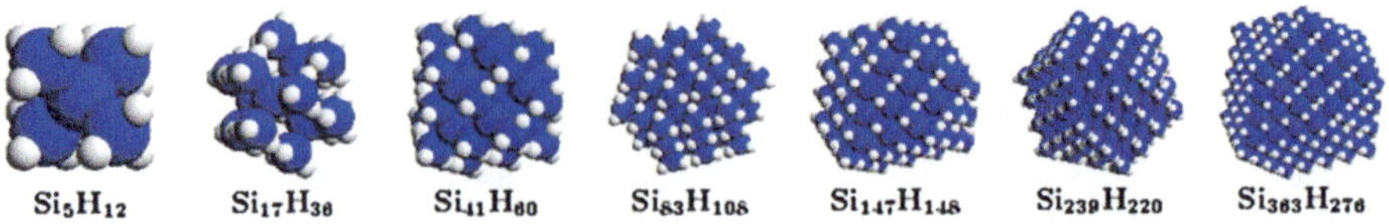

Fig. 16.24 The smallest Si nanocrystals passivated with H atoms (*small bright dots*). A Si atom (*large blue dot*) is assumed in the centre. From [101]. Copyright Wiley-VCH Verlag GmbH & Co.KGa.A. Reproduced with permission

11.5–13.5 eV (cf. discussion in [100]). The highest occupied eigenvalue in the ideal geometry is threefold degenerate and situated at $\varepsilon_{\text{HOMO}} = -8.4$ eV [103, 104] or -8.43 eV [85] in DFT-LDA. The corresponding one-shot GW values are -12.7 or -12.67 eV near the measured value. The measured broad spectrum may be related to a Jahn-Teller distortion of the HOMO level during the ionization process of the cluster [100]. For disilane similar good agreement between theory [85, 103] and experiment [102] can be stated with HOMO values $\varepsilon_{\text{HOMO}} = -7.21/-7.3$ eV in DFT-LDA or $\varepsilon_{\text{HOMO}} = -11.94/-10.6$ eV with GW corrections and $\varepsilon_{\text{HOMO}} = -10.7$ eV taken from the measured ionization energy.

The calculations can be extended to the single-particle excitation energies of larger clusters. The corresponding one-shot self-energy effects are illustrated in Fig. 16.25 for occupied and empty states [105]. They are useful to determine not only the ionization energy but also the electron affinity. Their difference (4.44) yields the corresponding quasiparticle HOMO-LUMO gap. The gap values resulting within

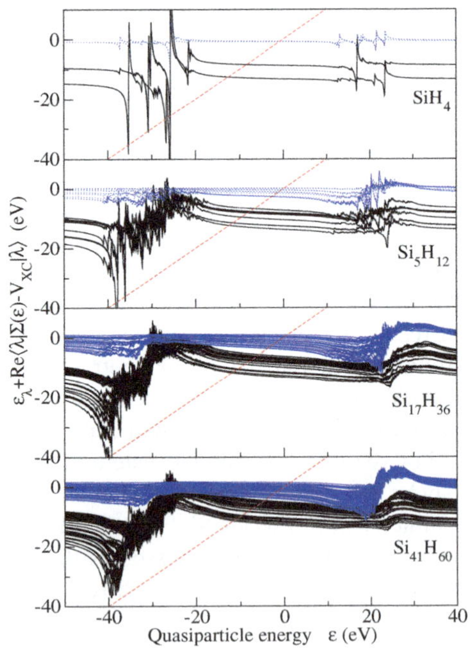

Fig. 16.25 Real part of the diagonal matrix elements of the perturbation operator $\delta\Sigma(\varepsilon/\hbar)$ added to the KS eigenvalues versus single-particle excitation energy. The *black solid* (*blue dotted*) *curves* represent filled (empty) states The *red straight dashed line* is $f(\varepsilon) = \varepsilon$. From [105]

Table 16.6 Quasiparticle gaps (in eV) of small Si nanocrystals with tetrahedral bonds within different approximations: KS (DFT-LDA), gKS (DFT using HSE hybrid functional), and QP (KS eigenvalues with one-shot GW corrections from Fig. 16.25 [105])

Si nanocrystal	KS	gKS	QP	TDDFT
SiH_4	7.95	8.80	12.72	8.2/8.2
Si_5H_{12}	5.64	6.60	10.83	6.6/5.8
$Si_{17}H_{36}$	4.20	5.03	8.09	4.27/4.3
$Si_{41}H_{60}$	3.25	3.96	6.19	3.27

For comparison results of TDDFT using the adiabatic LDA kernel [106, 107] are listed

different approximations are listed in Table 16.6. All approximations confirm the clear trend of gap shrinkage with increasing nanocrystal size. The time-dependent density functional treatment does only slightly open the fundamental gaps with respect to the KS HOMO-LUMO distances. At least, using the adiabatic LDA kernel the TDDFT seems not to be applicable to predict QP gaps for systems with strong three-dimensional confinement. The inclusion of a spatially non-local XC potential in the gKS-HSE framework also opens the gaps. However, the resulting values are still much smaller than the QP gaps computed within a one-shot GW approximation on top of a DFT-LDA reference electronic structure.

16.3.3 Surfaces and Two-dimensional Crystals

The ground-state calculations within the DFT-LDA (or DFT-GGA) allow for an accurate determination of many surface properties, in particular surface geometries, as long as the vdW interaction (see Sect. 9.3) does not play an important role. For these geometries the Kohn-Sham eigenvalues are determined by solving the Kohn-Sham equation (6.22). However, as demonstrated in the last chapters, there is no rigorous justification for the interpretation of the Kohn-Sham eigenvalues as single-particle excitation energies. All spectroscopies discussed, such as STS, PES and IPES, are related to the removal or addition of an electron. The corresponding excitations are described by spectral functions (14.15), the main quasiparticle peaks of which are not located at the KS energies. Rather, the peak positions define quasiparticle energies $\varepsilon^{QP}_{\nu m_s}(\mathbf{k})$ (14.42), which are shifted by $\Delta_{\nu m_s}(\mathbf{k})$ against the KS values.

Meanwhile, there exist many calculations of corresponding QP electronic structures for semiconductor surfaces, including the determination of complete QP band structures [108–110]. Also in the case of surfaces the Kohn-Sham state energies disagree with experimental observations since (i) band gaps between empty and occupied surface-state band energies are underestimated, (ii) the dispersion of the DFT-LDA/GGA surface bands is underestimated in some cases, but too large in others, and (iii) the placement of occupied surface-state energies is in some cases too high by $0.5 - 1.0$ eV relative to the bulk valence band maximum [108]. Three missing physical effects are crucial for the correct energy position of the QP states and, hence, must be considered to remove the DFT-LDA/GGA failures. First, the spatial non-locality of the self-energy operator (14.8) is more sensitive to the localization properties of surface states than just the density- (and gradient-)dependent XC potential (6.20) of the DFT-LDA/GGA. This requires a proper account of the non-locality of the Green function G (14.9). This non-locality leads to a modified dispersion of the quasiparticle energy bands throughout the surface BZ. Second, the inclusion of local fields due to the presence of the surface in the inverse dielectric function ϵ^{-1} (12.46) and, hence, in the screened interaction W (12.47) is crucial for the QP approach, since these local fields describe the strongly inhomogeneous screening (bulk-like versus vacuum, see expression (13.62)) at the surface. Third, an adequate treatment of the dynamical effects in the screening is more important than in the bulk case. This has to do with contributions from both bulk and surface plasmons and the smaller energy distance of bound surface states to the Fermi level. All these effects are important. Focusing on only one effect, e.g., the localization of surface states, cannot give a generally correct answer for the QP shifts [111].

Simple semiconductors with surfaces for which bands of surface bound states appear in the projected fundamental gap are investigated as prototypical electronic systems. Generally these surfaces reconstruct or, at least, relax to passivate the dangling bonds. As an example a 2×1 reconstructed surface geometry is displayed in Fig. 16.26. The six-member bond rings in bulk are replaced by five- and sevenfold rings during reconstruction. Quasiparticle band structures of intrinsic surface states in the prototypical elemental semiconductors silicon and diamond are compared with DFT-LDA electronic structures in Figs. 16.27 and 16.28. The 2×1 reconstructed Si(111) and C(100) surfaces are chosen as examples [114–116]. A buckled π-bonded chain model with a positive buckling and a symmetric-dimer model are applied. In the case of Si(111) 2×1, Fig. 16.27 also shows bound surface-state bands measured by direct and inverse photoemission [112, 113]. In the case of C(100) 2×1 experimental data [117, 118] are not included in Fig. 16.28 because of the presence of hydrogen and contradictory findings. Figure 16.27 illustrates the principal situation. Just as in bulk semiconductors, DFT-LDA is unable to provide an accurate

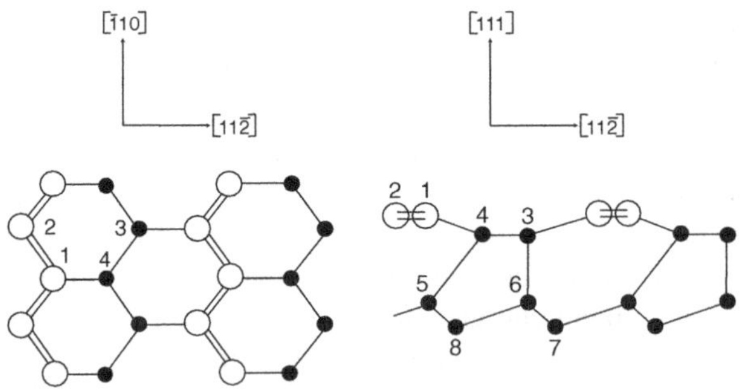

Fig. 16.26 *Top* and *side* view of a group-IV(111) 2×1 surface within the π-bonded chain model. *Large open circles* indicate first-layer atoms

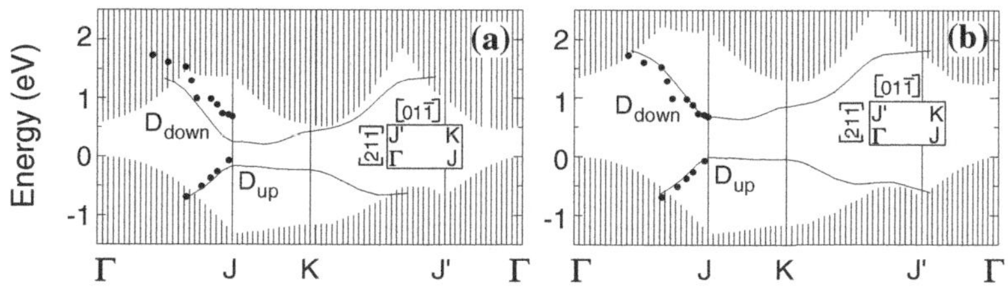

Fig. 16.27 Kohn-Sham (**a**) and quasiparticle (**b**) band structure of the Si(111) 2×1 surface. The hatched areas denote Si bulk states. The dots denote experimental data [112, 113]. $D_{\rm up}$ and $D_{\rm down}$ are π and π^* bands modified by chain buckling. The insets describe the surface BZ. Reprinted with permission from [114]. Copyright 1999 by the American Physical Society

description of the band structure of the surface states. Only the inclusion of a full many-body treatment of the single-particle problem by using the GW approximation allows the reproduction of the experimental electronic structure. In fact, usually QP corrections may be even more important than in bulk semiconductors [119].

The Figs. 16.27 and 16.28, indicate general trends but also specialities for bound surface states of elemental semiconductors. In general, the quasiparticle shifts of empty (filled) surface states are positive (negative). However, with respect to the bulk VBM the surface band positions depend on the difference of their small relative QP shifts. The empty surface states $D_{\rm down}$ and π^* are shifted toward higher energies, while the relative shifts (with respect to the bulk VBM) of the occupied $D_{\rm up}$ and π bands are small. The sign of the net shift for $D_{\rm up}$ (π) is positive (negative). In any case the indirect surface band gaps at $J \to 0.5JK$ and $0.25J'K \to K$ are opened from 0.4 to 0.7 eV (Si(111) 2×1) or from 1.6 to 3.7 eV (C(100) 2×1). The openings of the indirect bulk gaps are somewhat larger (0.6 eV) or smaller (1.7 eV). The dispersion of the surface bands is also influenced by the quasiparticle character. The effect is rather small for Si(111) 2×1 [120] and the π^* band of C(100) 2×1. However, there is a reduction of the dispersion by about 0.2 eV for the π band in the diamond case.

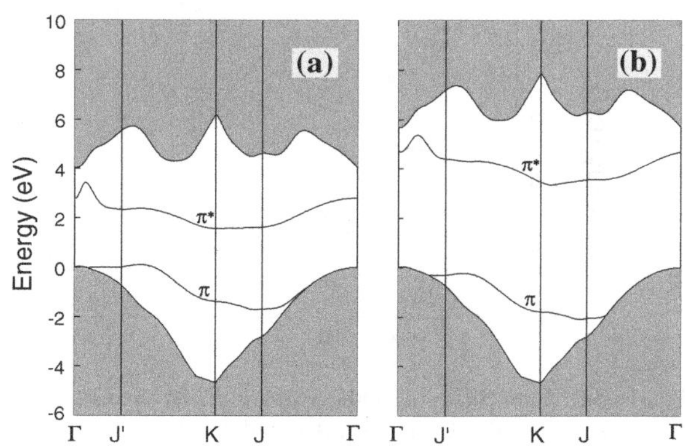

Fig. 16.28 Kohn-Sham (**a**) and quasiparticle (**b**) band structure of the C(100) 2×1 surface. The *shaded regions* indicate the projected bulk band structure. From [115, 116]

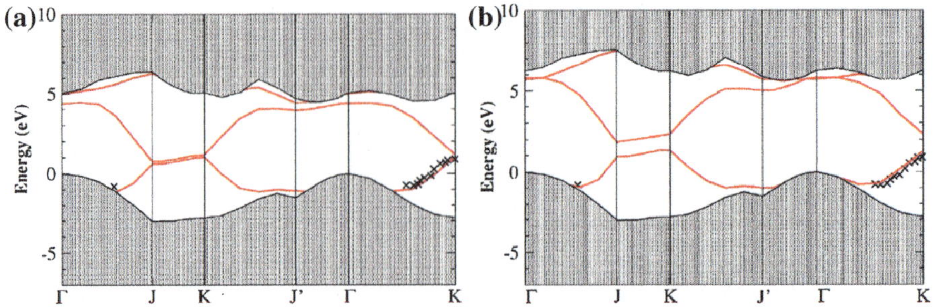

Fig. 16.29 DFT-GGA (**a**) and self-consistent GW (**b**) electronic band structure of the C(111) 2×1 surface. *Crosses* experimental data from ARPES [121, 122]. Reprinted with permission from [123]. Copyright 2005 by the American Physical Society

For the non-buckled π-bonded chains of C(111) 2×1 in Fig. 16.26 the Kohn-Sham and quasiparticle band structures in Fig. 16.29 support the conclusion given above despite the use of a self-consistent QP scheme [123]. A surface gap of about 1 eV is opened along the JK line (in contrast to the vanishing gap in DFT-LDA approach) and the π-band dispersion along ΓK describes the experimental data well [121, 122]. While the order of magnitude of the gap seems to be in agreement with PES findings [122] and the onset of an electron energy loss spectrum [128], it however contradicts recent reflectance anisotropy spectroscopy (RAS) measurements [129], which indicate an optical gap of about 1.5 eV. There is still a debate about the reasons of the discrepancy, e.g. the occurrence of hydrogen at the C(111) surface [130].

The situation is somewhat different for many, in particular cleavage, surfaces of compound semiconductors. As examples the KS and QP band structures are presented for the relaxed InP(110) 1×1 surface in Fig. 16.30. This surface is characterized by nearly resonant C_3 (cation-derived) and A_5 (anion-derived) surface-state bands. The energy overlap between the empty C_3 band and the bulk conduction bands

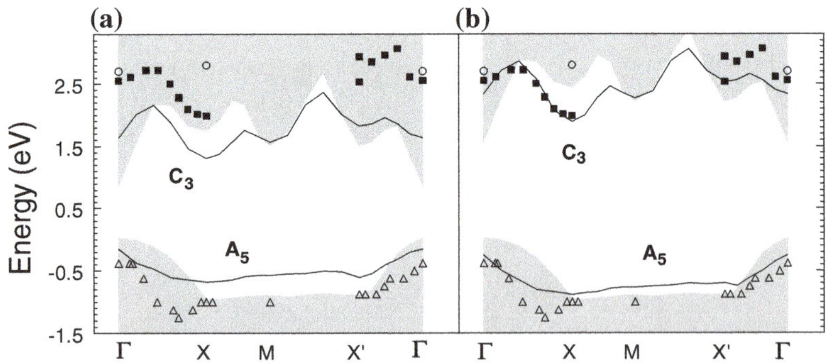

Fig. 16.30 Kohn-Sham (**a**) and quasiparticle (**b**) band structure of the InP(110) 1×1 surface. The *shaded regions* indicate the projected bulk band structure. The *triangles* [124], *squares* [125], and *circles* [126] denote measured surface band energies. From [127]

is somewhat increased within the quasiparticle picture. Effectively the empty C_3 surface-state band is slightly more shifted to higher energies than the bulk conduction-band edge. There is also a small downward shift of the energies of the occupied surface states toward the bulk valence bands. Despite the energy overlap the C_3 and A_5 surface states essentially keep their localization at the surface. The orbital character of the occupied dangling-bond states is p-like (at least at X), and they are localized on the surface anions. The unoccupied dangling-bond state is also primarily p-like and localized on the surface cations. However, in general a hybridization of surface and bulk states due to the off-diagonal elements of the QP self-energy (14.11) cannot be excluded [131].

Characteristic QP gaps between surface states are listed in Table 16.7. They are compared with gaps obtained within the Kohn-Sham theory and gaps measured by combination of PES and IPES or STS. The small direct gaps of bound surface states within the fundamental gap of the bulk elemental semiconductors are considerably opened by the QP corrections. The gap change can be 100 % or more. In most cases the resulting QP gaps are in excellent agreement with measured values. On average the discrepancy is about 0.1 eV or less. The situation is different for the more or less resonant C_3 and A_5 surface states on relaxed III-V(110) 1×1 surfaces. The absolute values of the QP gap openings of about 1 eV are much larger than for the surfaces of elemental semiconductors. However, their relative contributions to the total QP gaps are much smaller than the KS gaps.

The realization of graphene, the 2D sp^2-bonded allotrope of carbon, and its derivates has led to a revolution in the fields of condensed matter physics, basic

Table 16.7 Direct surface-state gaps for selected semiconductor surfaces

Surface	**k** point	KS	QP	Experiment
Si(111) 2×1	J	0.27	0.62 [114, 120]	0.75 [112, 113, 132]
Ge(111) 2×1	JK	0.38	0.66 [133]	0.61 [134–137]
				0.54 [138]
Si(100) 2×1	J	0.20	0.70 [139]	0.9 [140]
Si(100)$c(4 \times 2)$	X	0.39	0.87 [141]	0.9 [140]
Ge(100) 2×1	J	0.40	0.80 [142]	0.9 [143, 144]
				0.9 [145]
GaAs(110) 1×1	Γ	1.8	2.7 [146]	2.4 [147]
	X	1.9	2.9 [146]	3.1 [147]
	M	2.2	3.2 [146]	3.3 [147]
	X'	2.0	2.9 [146]	3.0 [147]
InP(110) 1×1	Γ	1.8	2.5 [146]	2.5 [147]
	X	2.0	2.8 [146]	2.9 [147]
	M	2.2	3.1 [146]	3.2 [147]
	X'	2.4	3.2 [146]	3.1 [147]

A transition energy is characterized by the **k** point in the surface BZ. Three values (all in eV) are given: the difference of the KS eigenvalues, the quasiparticle gap, and the experimental value obtained by combination of ARPES and KRIPES results or from STS

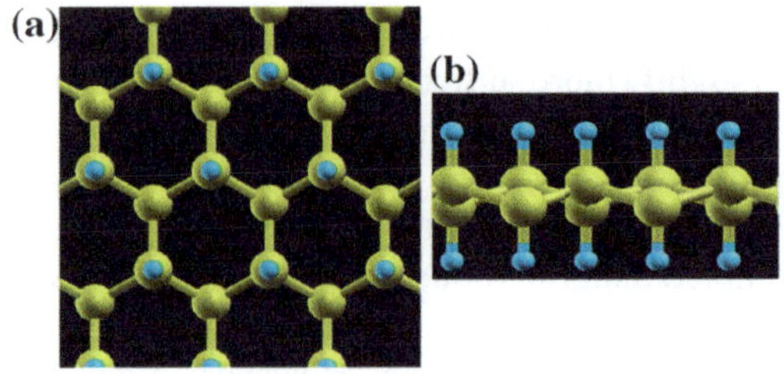

Fig. 16.31 *Top* (**a**) and *side* (**b**) view of a 2D graphane crystal. The carbon (hydrogen) atoms are indicated by *yellow* (*blue*) *dots*

science, and device technology. The understanding and control of their properties are continuously leading to novel effects and device proposals. Structural modifications such as ribbons and chemical treatments, e.g. by hydrogen, make the 2D carbon an ideal material for future nanotechnology. One important example is the fully but alternately hydrogenated and hence sp^3-bonded graphene, the so-called graphane, with still honeycomb symmetry (cf. Fig. 16.31), that has been recently synthesized [148, 149]. The studies of the outstanding properties of graphene and graphane have generated much interest in their silicon and germanium counterparts, silicene/silicane (see Fig. 13.12) and germanene/germanane. Indeed, the preparation of the 2D hydrogenated germanene, the germanane, has been recently reported [150].

The QP band structures of graphane, silicane, and germanane, based on fully relaxed atomic geometries, are depicted in Fig. 16.32. They are completely different from the band structures of the non-hydrogenated sp^2(graphene)- or sp^2/sp^3(silicene, germanene)-bonded group-IV sheets. The Dirac cones at the K points in the 2D hexagonal BZ disappear. Due to the hydrogenation the zero-gap semiconductors transform into insulators with remarkably large gaps almost at the BZ center Γ [151]. The conduction band minimum at M of silicane is however slightly below the

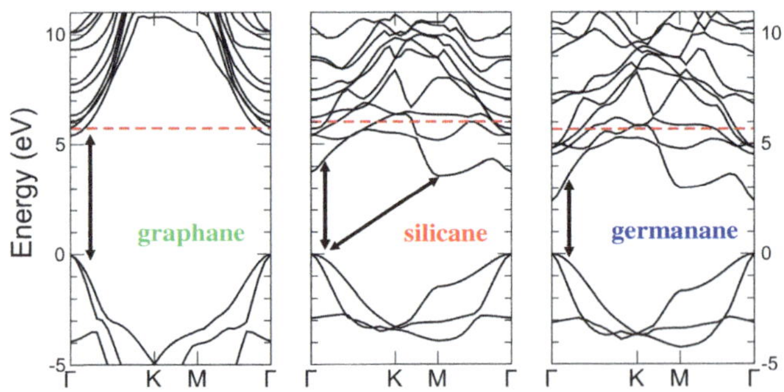

Fig. 16.32 Quasiparticle band structure of the 2D honeycomb crystals graphane, silicane, and germanane. The horizontal *red dashed line* indicates the vacuum level. The *arrows* display the fundamental gaps. The VBM is used as energy zero. From [151] with permission

Γ minimum, while this valley is above the Γ one in germanane. Once many-body excitation effects are included, the opening of the gaps is sizeable, i.e., more than 50 % compared to KS values [152, 153]. Actually the quasiparticle renormalization effects are enhanced in the atomic-layer systems due to the low dimensionality confirming for silicane and germanane what was shown before for graphane [154]. However, there are also strong similarities to the hydrogenated silicongraphane [155], which exhibits significant modifications due to the hydrogenation despite the finite gap in the hydrogen-free silicongraphene (see Fig. 16.33). The resulting $\Gamma \rightarrow \Gamma$ QP gaps amount to 5.4 eV (graphane) and 2.4 eV (germanane). The indirect $\Gamma \rightarrow M$ QP gap of silicane is 3.6 eV, a value only slightly below the direct gap at Γ. The QP effects also increase the effective masses. Their values for electrons (lowest conduction band at Γ) as well as heavy and light holes (highest two valence bands at Γ) in Table 16.8 seem to be interesting for applications in electronics. In particular, the small effective electron mass at Γ of silicane and germanane is promising in terms of a high carrier mobility in two dimensions.

The quasiparticle renormalization significantly modifies electron removal and addition. The absolute position of the lowest empty state in Figs. 16.32 and 16.33 determines the electron affinity A. For graphene one finds a dramatic change of the electron affinity upon hydrogenation from $A = 4.2$ eV (graphene) to the value $A = 0.3$ eV (graphane) [151, 154]. The positive value of A is somewhat in contrast to the negative electron affinity of some hydrogenated diamond surfaces (see for example [156] and references therein). This is due to the presence of compensating group-IV-H dipoles on the top and below the bottom of the sheet. For silicane and germanane A

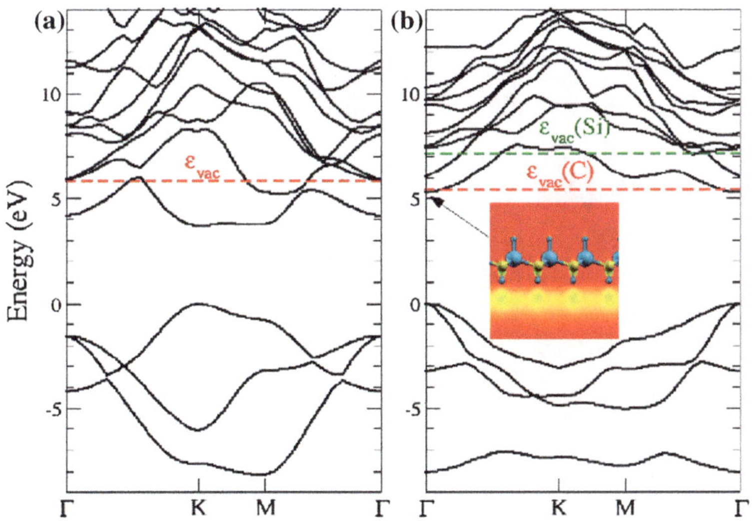

Fig. 16.33 QP band structure of silicongraphene (**a**) and silicongraphane (**b**). The *top* of the valence bands is used as energy zero. The *dashed red/green horizontal lines* indicate the position of the vacuum level(s). The *inset* shows the wave function at Γ for the lowest conduction band. Reprinted with permission [155]. Copyright 2012, AIP Publishing LLC

Table 16.8 Effective masses of electrons and holes (in units of the free electron mass) in hydrogenated group-IV honeycomb crystals at Γ

Mass	Direction	Graphane	Silicane	Germanane
Electron	$\Gamma \rightarrow M$	1.04	0.16 (2.94)	0.08 (4.55)
	$\Gamma \rightarrow K$	1.04	0.16 (0.13)	0.08 (0.11)
Heavy hole	$\Gamma \rightarrow M$	0.68	0.60 (1.47)	0.52 (1.25)
	$\Gamma \rightarrow K$	0.64	0.57 (0.19)	0.49 (0.17)
Light hole	$\Gamma \rightarrow M$	0.27	0.13	0.07
	$\Gamma \rightarrow K$	0.27	0.13	0.07

In the case of silicane and germanane also conduction-band and uppermost valence-band masses at M are given in parenthesis. From [151]

is much larger: $A = 2.5\,\text{eV}$ (4.6 eV in silicene) and 3.3 eV (4.7 eV in germanene), respectively, as for a conventional insulator or semiconductor. The varying values of A for the three materials are mainly a consequence of the different electronegativity of C, Si and Ge atoms with respect to H: while C is more electronegative than H, the opposite holds for Si and Ge [157]. This fact has direct influence also on the character of the indirect/direct gap transition upon dimensionality lowering. For germanane and silicane the transition is due to quantum-confinement effects which shift upward the lowest bulk-unoccupied state (at L and X, respectively), responsible for the bulk indirect gap, while leaving almost unchanged the energy of the lowest unoccupied state at the Γ point. On the contrary, in the case of graphane the presence of H causes the formation of nearly free-electron states (NFES) near the bottom of the conduction band at the Γ point, as shown in Fig. 16.34 [158].

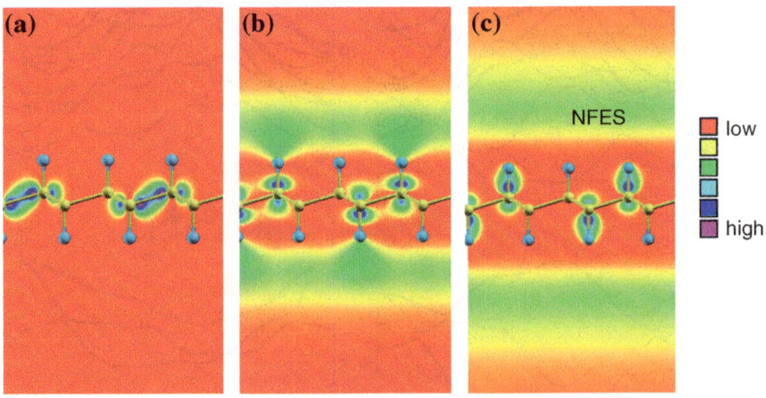

Fig. 16.34 Squares of wave functions at Γ of graphane. (**a**) Uppermost valence band, (**b**) lowest conduction band, and (**c**) second lowest conduction band. *Colors* indicate the strength of the probabilities. The hydrogen contribution to the electron wave function (**b**) and the nearly-free-electron-like character (**b, c**), respectively, are visible. Adopted using data from [151]

References

1. M. Shishkin, M. Marsman, G. Kresse, Accurate quasiparticle spectra from self-consistent *GW* calculations with vertex corrections. Phys. Rev. Lett. **99**, 246403 (2007)
2. F. Fuchs, J. Furthmüller, F. Bechstedt, M. Shishkin, G. Kresse, Quasiparticle band structure based on a generalized Kohn-Sham scheme. Phys. Rev. B **76**, 115109 (2007)
3. M. Shishkin, G. Kresse, Implementation and performance of the frequency-dependent *GW* method within the PAW framework. Phys. Rev. B **74**, 035101 (2006)
4. J.P. Perdew, Unified theory of exchange and correlation beyond the local density approximation, in *Electronic Structure of Solids '91* ed. by P. Ziesche, H. Eschrig (Akademie Verlag, Berlin, 1991), pp. 11–20
5. J.P. Perdew, Y. Wang, Accurate and simple analytic representation of the electron gas correlation energy. Phys. Rev. B **45**, 13244–13249 (1992)
6. J. Heyd, J. Scuseria, M. Ernzerhof, Hybrid functionals based on a screened Coulomb potential. J. Chem. Phys. **118**, 8207–8215 (2003)
7. J. Heyd, G.E. Scuseria, M. Ernzerhof, Erratum: "Hybrid functionals based on a screened Coulomb potential" [J. Chem. Phys. **118**, 8207 (2003)]. J. Chem. Phys. **124**, 219906 (2006)
8. A. Marini, *Ab initio* finite-temperature excitons. Phys. Rev. Lett. **101**, 106405 (2008)
9. F. Giustino, S.G. Louie, M.L. Cohen, Electron-phonon renormalization of the direct band gap of diamond. Phys. Rev. Lett. **105**, 265501 (2010)
10. E. Cannuccia, A. Marini, Effect of the quantum zero-point atomic motion on the optical and electronic properties of diamond and trans-polyacetylene. Phys. Rev. Lett. **107**, 255501 (2011)
11. M. Shishkin, G. Kresse, Self-consistent *GW* calculations for semiconductors and insulators. Phys. Rev. B **75**, 235102 (2007)
12. F. Bechstedt, F. Fuchs, G. Kresse, *Ab-initio* theory of semiconductor band structures: new developments and progress. Phys. Status Solidi B **246**, 1877–1892 (2009)
13. L.C. de Carvalho, A. Schleife, F. Bechstedt, Influence of exchange and correlation on structural and electronic properties of AlN, GaN, and InN polytypes. Phys. Rev. B **84**, 195105 (2011)
14. S.H. Wei, A. Zunger, Role of metal *d* states in II–VI semiconductors. Phys. Rev. B **37**, 8958–8981 (1988)
15. F. Bruneval, N. Vast, L. Reining, Effect of self-consistency on quasiparticles in solids. Phys. Rev. B **74**, 045102 (2006)
16. C. Rödl, F. Fuchs, J. Furthmüller, F. Bechstedt, Quasiparticle band structures of the antiferromagnetic transition-metal oxides MnO, FeO, CoO, and NiO. Phys. Rev. B **79**, 235114 (2009)
17. H. Jiang, R.I. Gomez-Abal, P. Rinke, M. Scheffler, Localized and itinerant states in lanthanide oxides united by GW@LDA $+ U$. Phys. Rev. Lett. **102**, 126403 (2009)
18. P. Rinke, A. Qteish, J. Neugebauer, Ch. Freysoldt, M. Scheffler, Combining GW calculations with exact-exchange density-functional theory: an analysis of valence-band photoemission for compound semiconductors. New J. Phys. **7**, 126 (2005)
19. G. Onida, L. Reining, A. Rubio, Electronic excitations: density-functional versus many-body Green's-function approaches. Rev. Mod. Phys. **74**, 601–659 (2002)
20. J.E. Northrup, M.S. Hybertsen, S.G. Louie, Quasiparticle excitation spectrum for nearly-free-electron metals. Phys. Rev. B **39**, 8198–8208 (1989)
21. M.S. Hybertsen, S.G. Louie, *Ab initio* static dielectric matrices from the density-functional approach. I. Formulation and application to semiconductors and insulators. Phys. Rev. B **35**, 5585–5601 (1987)
22. R. Del Sole, L. Reining, R.W. Godby, GWΓ approximation for electron self-energies in semiconductors and insulators. Phys. Rev. B **49**, 8024–8028 (1994)
23. A. Grüneis, G. Kresse, Y. Hinuma, F. Oba, Ionization potentials in solids: the importance of vertex corrections. Phys. Rev. Lett. **112**, 096401 (2014)
24. E. Runge, E.K.U. Gross, Density-functional theory for time-dependent systems. Phys. Rev. Lett. **52**, 997–1000 (1984)

25. G. Adragna, R. Del Sole, A. Marini, *Ab initio* calculation of the exchange-correlation kernel in extended systems. Phys. Rev. B **68**, 165108 (2008)
26. F. Sottile, V. Olevano, L. Reining, Parameter-free calculation of response functions in time-dependent density-functional theory. Phys. Rev. Lett. **91**, 056402 (2003)
27. S.V. Fateev, M. van Schilfgaarde, T. Kotani, All-electron self-consistent GW approximation: application to Si, MnO, and NiO. Phys. Rev. Lett. **93**, 126406 (2004)
28. M. van Schilfgaarde, T. Kotani, S. Fateev, Quasiparticle self-consistent GW theory. Phys. Rev. Lett. **96**, 226402 (2006)
29. B.-C. Shih, Y. Xue, P. Zhang, M.L. Cohen, S.G. Louie, Quasiparticle band gap of ZnO: high accuracy from the conventional G^0W^0 approach. Phys. Rev. Lett. **105**, 146401 (2010)
30. C. Friedrich, M.C. Müller, S. Blügel, Band convergence and linearization error correction of all-electron GW calculations: the extreme case of zinc oxide. Phys. Rev. B **83**, 081101(R) (2011)
31. C. Friedrich, M.C. Müller, S. Blügel, Erratum: Band convergence and linearization error correction of all-electron GW calculations: the extreme case of zinc oxide. [Phys. Rev. B 83, 081101(R) (2011)]. Phys. Rev. B **84**, 039906 (2011)
32. M. Stankovski, G. Autonins, D. Waroquiers, A. Miglio, H. Dixit, K. Sankaran, M. Giantomassi, X. Gonze, M. Côté, G.-M. Rignanese, G^0W^0 band gap of ZnO: effects of plasmon-pole models. Phys. Rev. B **84**, 241201(R) (2011)
33. M. Usuda, N. Hamada, T. Kotani, M. van Schilfgaarde, All-electron GW calculation based on the LAPW method: application to wurtzite ZnO. Phys. Rev. B **66**, 125101 (2002)
34. S. Tsoi, X. Lu, A.K. Ramdas, H. Alawadhi, M. Grimsditch, M. Cardona, R. Lauck, Isotopic-mass dependence of the A, B, and C excitonic band gaps in ZnO at low temperatures. Phys. Rev. B **74**, 165203 (2006)
35. H. Alawadhi, S. Tsoi, X. Lu, A.K. Ramdas, M. Grimsditch, M. Cardona, R. Lauck, Effect of temperature on isotopic mass dependence of excitonic band gaps in semiconductors: ZnO. Phys. Rev. B **75**, 205207 (2007)
36. A. Schleife, C. Rödl, F. Fuchs, J. Furthmüller, F. Bechstedt, Optical and energy-loss spectra of MgO, ZnO, and CdO from *ab initio* many-body calculations. Phys. Rev. B **80**, 035112 (2009)
37. A. Riefer, F. Fuchs, C. Rödl, A. Schleife, F. Bechstedt, R. Goldhahn, Interplay of excitonic effects and van Hove singularities in optical spectra: CaO and AlN polymorphs. Phys. Rev. B **84**, 075218 (2011)
38. M.P. Thompson, G.W. Auner, T.S. Zheleva, K.A. Jones, S.J. Sinko, J.N. Hilfiker, Deposition factors and band gap of zinc-blende AlN. J. Appl. Phys. **89**, 3331–3336 (2001)
39. M. Röppischer, R. Goldhahn, G. Rossbach, P. Schley, C. Cobet, N. Esser, T. Schupp, K. Lischka, D.J. As, Dielectric function of zinc-blende AlN from 1 to 20 eV: band gap and van Hove singularities. J. Appl. Phys. **106**, 076104 (2009)
40. M. Kobayashi, G.S. Song, T. Kataoka, Y. Sakamoto, A. Fujimori, T. Ohkochi, Y. Takeda, T. Okane, Y. Saitoh, H. Yamagami, H. Yamahara, H. Saeki, T. Kawai, H. Tabata, Experimental observation of bulk band dispersions in the oxide semiconductor ZnO using soft X-ray angle-resolved photoemission spectroscopy. J. Appl. Phys. **105**, 122403 (2009)
41. E.I. Rashba, Properties of semiconductors with an extremum loop. 1. Cyclotron and combinational resonance in a magnetic field perpendicular to the plane of the loop. Fiz. Tver. Tela (Leningrad) **2**, 1224–1238 (1960) [Sov. Phys. Sol. State (English Transl.) **2**, 1109–1122 (1960)]
42. Yu.A. Bychkov, E.I. Rashba, Oscillatory effects and the magnetic susceptibility of carriers in inversion layers. J. Phys. C. Solid State Phys. **17**, 6039–6045 (1984)
43. G. Dresselhaus, Spin-orbit coupling effects in zinc blende structure. Phys. Rev. **100**, 580–586 (1955)
44. P. Rinke, M. Winkelnkemper, A. Qteish, D. Bimberg, J. Neugebauer, M. Scheffler, Consistent set of band parameters for the group-III nitrides AlN, GaN, and InN. Phys. Rev. B **77**, 075202 (2008)

45. Y.-S. Kim, M. Marsman, G. Kresse, F. Tran, P. Blaha, Towards efficient band structure and effective mass calculations for III–V direct band gap semiconductors. Phys. Rev. B **82**, 205212 (2010)
46. I. Vurgaftman, J.R. Meyer, L.R. Ram-Mohan, Band parameters for III–V compound semiconductors and their alloys. J. Appl. Phys. **89**, 5815–5875 (2001)
47. P.D.C. King, T.D. Veal, C.F. McConville, F. Fuchs, J. Furthmüller, F. Bechstedt, J. Schörmann, D.J. As, K. Lischka, H. Lu, W.J. Schaff, Valence band density of states of zinc-blende and wurtzite InN from x-ray photoemission spectroscopy and first-principles calculations. Phys. Rev. B **77**, 115213 (2008)
48. A. Schleife, J.B. Varley, F. Fuchs, C. Rödl, F. Bechstedt, P. Rinke, A. Janotti, C.G. Van de Walle, Tin dioxide from first principles: quasiparticle electronic states and optical properties. Phys. Rev. B **83**, 035116 (2011)
49. J.M. Themlin, R. Sporken, J. Darville, R. Candano, J.M. Gilles, R.L. Johnson, Resonant-photoemission study of SnO_2: cationic origin of the defect band-gap states. Phys. Rev. B **42**, 11914–11925 (1990)
50. T. Nagata, D. Bierwagen, M.E. White, M.Y. Tsai, J.S. Speck, Study of the Au Schottky contact formation on oxygen plasma treated n-type SnO_2(101) thin films. J. Appl. Phys. **107**, 033707 (2010)
51. C. Rödl, A. Schleife, Photoemission spectra and effective masses of *n*- and *p*-type oxide semiconductors from first principles: ZnO, CdO, SnO_2, MnO, and NiO. Phys. Status Solidi A **211**, 74–81 (2014)
52. J.J. Yeh, I. Lindau, Atomic subshell photoionization cross sections and asymmetry parameters: $1 \leq Z \leq 103$. At. Data Nucl. Data Tables **32**, 1–155 (1985)
53. L.D. Landau, E.M. Lifshitz, *Quantum Mechanics* (Academic Press, Oxford, 1959)
54. L.F.J. Piper, L. Colakol, T. Learmonth, P.-A. Glans, K.E. Smith, F. Fuchs, J. Furthmüller, F. Bechstedt, T.-C. Chen, T.D. Moustakas, J.H. Guo, Electronic structure of InN studied using soft X-ray emission, soft x-ray absorption, and quasiparticle band structure calculations. Phys. Rev. B **76**, 245204 (2007)
55. L.F.J. Piper, A. DeMasi, K.E. Smith, A. Schleife, F. Fuchs, F. Bechstedt, J. Zuniga-Pérez, V. Muñoz-Sanjosé, Electronic structure of single-crystal rocksalt CdO studied by soft X-ray spectroscopies and *ab initio* calculations. Phys. Rev. B **77**, 125204 (2008)
56. N.W. Ashcroft, N.D. Mermin, *Solid State Physics* (Sanders College, Philadelphia, 1976)
57. E. Jensen, E.W. Plummer, Experimental band structure of Na. Phys. Rev. Lett. **55**, 1912–1915 (1985)
58. K.W.-K. Shung, B.E. Sernelius, G.D. Mahan, Self-energy corrections in photoemission of Na. Phys. Rev. B **36**, 4499–4502 (1987)
59. M.P. Surh, J.E. Northrup, S.G. Louie, Occupied quasiparticle bandwidth of potassium. Phys. Rev. B **38**, 5976–5980 (1988)
60. E.L. Shirley, Self-consistent GW and higher-order calculations of electron states in metals. Phys. Rev. B **54**, 7758–7764 (1996)
61. W.G. Aulbur, L. Jönsson, J.W. Wilkins, Quasiparticle calculations in solids, in *Solid State Physics*. Advances in Research and Applications vol. 54, ed. by H. Ehrenreich, F. Spaepen (Academic Press, San Diego, 2000), pp. 1–218
62. F. Hüser, T. Olsen, K.S. Thygesen, Quasiparticle GW calculations for solids, molecules, and two-dimensional materials. Phys. Rev. B **87**, 235132 (2013)
63. A. Marini, R. Del Sole, G. Onida, First-principles calculation of the plasmon resonance and of the reflectance spectrum of silver in the GW approximation. Phys. Rev. B **66**, 115101 (2002)
64. H. Mårtensson, P.O. Nilson, Investigation of the electronic structure of Ni by angle-resolved UV photoelectron spectroscopy. Phys. Rev. B **30**, 3047–3054 (1984)
65. F. Aryasetiawan, Self-energy of ferromagnetic nickel in the GW approximation. Phys. Rev. B **46**, 13051–13064 (1992)
66. F. Aryasetiawan, U. von Barth, Dielectric response and quasiparticle energies in nickel. Phys. Scr. T **45**, 270–271 (1992)
67. M. Schilfgaarde, Private communication (2014)

68. A. Georges, G. Kotliar, W. Krauth, M. Rozenberg, Dynamical mean-field theory of strongly correlated fermion systems and the limit of infinite dimensions. Rev. Mod. Phys. **68**, 13–125 (1996)
69. N.F. Mott, The basis of the electron theory of metals, with special reference to the transition metals. Phys. Soc. Lond. Sect. A **62**, 416–422 (1949)
70. F. Aryasetiawan, O. Gunnarsson, The GW method. Rep. Prog. Phys. **61**, 237–312 (1998)
71. F. Aryasetiawan, O. Gunnarsson, Electronic structure of NiO in the *GW* approximation. Phys. Rev. Lett. **74**, 3221–3224 (1995)
72. S. Massidda, A. Continenza, M. Posternak, A. Baldereschi, Band-structure picture for MnO reexplored: a model *GW* calculation. Phys. Rev. Lett. **74**, 2323–2326 (1995)
73. I. Tsubokawa, On the magnetic properties of a $CrBr_3$ single crystal. J. Phys. Soc. Jpn. **15**, 1664–1668 (1960)
74. C. Rödl, Elektronische und exzitonische Anregungen in magnetischen Isolatoren, Ph.D. thesis. Friedrich-Schiller-Universität Jena (2009)
75. L. Serrano-Andrés, M. Mechán, Density functional theory calculations on η^5-monocyclopentadienylnitrilecobalt complexes concerning their second-order nonlinear optical properties. J. Mol. Struct. (Thoechem) **729**, 109–113 (2005)
76. J.B. Foresman, M. Head-Gordon, J.A. Pople, M.J. Frisch, Toward a systematic molecular orbital theory for excited states. J. Phys. Chem. **96**, 135–149 (1992)
77. C.J. Cramer, *Essentials of Computational Chemistry* (Wiley, Chichester, 2002)
78. T. Helgaker, P. Jørgensen, J. Olsen, *Estructure–Structure Theory* (Wiley, Chichester, 2000)
79. M.E. Casida, in *The Parameterized Second-Order Green Function Times Screened Interaction (pGW2) Approximation for Calculation of Outer Valence Ionization Potentials*, ed. by J.M. Seminario. Recent Developments and Applications of Modern Density Functional Theory (Elsevier Science B.V, Amsterdam, 1996), pp. 391–439
80. I. Vasiliev, S. Öğut, J.R. Chelikowsky, First-principles density-functional calculations for optical spectra of clusters and nanocrystals. Phys. Rev. B **65**, 115416 (2002)
81. R.E. Stratmann, G.E. Scuseria, M.J. Frisch, An efficient implementation of time-dependent density-functional theory for the calculation of excitation energies of large molecules. J. Chem. Phys. **109**, 8218–8224 (1998)
82. J. Oddershede, Propagator methods. Adv. Chem. Phys. **69**, 201–239 (1987)
83. J.V. Ortiz, Toward an exact one-electron picture of chemical bonding. Adv. Quantum Chem. **35**, 33–52 (1999)
84. F. Bechstedt, Quasiparticle corrections for energy gaps in semiconductors. Adv. Solid State Phys. **32**, 161–177 (1999)
85. P.H. Hahn, W.G. Schmidt, F. Bechstedt, Molecular electronic excitations calculated from a solid-state approach: methodology and numerics. Phys. Rev. B **72**, 245425 (2005)
86. X. Blase, C. Attaccalite, V. Olevano, First-principles GW calculations for fullerenes, prophyrins, phtalocyanine, and other molecules of interest for organic photovoltaic applications. Phys. Rev. B **83**, 115103 (2011)
87. T. Körzdörfer, N. Marom, Strategy for finding a reliable starting point for G_0W_0 demonstrated for molecules. Phys. Rev. B **86**, 041110 (2012)
88. W. Kang, M.S. Hybertsen, Enhanced static approximation to the electron self-energy operator for efficient calculation of quasiparticle energies. Phys. Rev. B **82**, 195108 (2010)
89. J. Deslippe, S. Samsonidze, M. Jain, M.L. Cohen, S.G. Louie, Coulomb-hole summations or energies for GW calculations with limited number of empty orbitals: a modified static reminder approach. Phys. Rev. B **87**, 165124 (2013)
90. M.E. Casida, D.P. Chong, Physical interpretation and assessment of the Coulomb-hole and screened-exchange approximation for molecules. Phys. Rev. A **40**, 4837–4848 (1989)
91. Y. Shigeta, A.M. Ferreira, V.G. Zakrzewski, J.V. Ortiz, Electron propagator calculations with Kohn-Sham reference states. Int. J. Quantum Chem. **85**, 411–420 (2011)
92. C.R. Brundle, D.W. Turner, High resolution molecular photoelectron spectroscopy. II. Water and deuterium oxide. Proc. R. Soc. Lond. Ser. A **307**, 27–36 (1968)

93. K. Kimura, S. Katsumata, Y. Achiba, T. Yamazaki, S. Iwata, *Handbook of He I Photoelectron Spectra of Fundamental Organic Molecules* (Halsted, New York, 1981)
94. S.Y. Liu, K. Alnama, J. Matsumoto, K. Nishizawa, H. Kohguchi, Y.P. Lee, T. Suzuki, He I ultraviolet photoelectron spectroscopy of benzene and pyridine in supersonic molecular beams using photoelectron imaging. J. Phys. Chem. A **115**, 2953–2965 (2011)
95. F. Bruneval, M.A.L. Marques, Benchmarking the starting points of GW approximation for molecules. J. Chem. Theory Comput. **9**, 324–329 (2013)
96. M. Preuss, W.G. Schmidt, K. Seino, J. Furthmüller, F. Bechstedt, Ground- and excited-state properties of DNA base molecules from planewave calculations using ultrasoft pseudopotentials. J. Comput. Chem. **25**, 112–122 (2004)
97. C. Faber, C. Attaccalite, V. Olevano, E. Runge, X. Blase, First principles GW calculations for DNA and RNA nucleobases. Phys. Rev. B **83**, 115123 (2011)
98. G. Onida, L. Reining, R.W. Godby, R. Del Sole, W. Andreoni, *Ab initio* calculations of the quasiparticle and absorption spectra of clusters: the sodium tetramer. Phys. Rev. Lett. **75**, 818–821 (1995)
99. A. Hermann, E. Schumacher, L.J. Wöste, Preparation and photoionization potentials of molecules of sodium, potassium, and mixed atoms. J. Chem. Phys. **68**, 2327–2336 (1978)
100. L. Reining, O. Pulci, M. Palummo, G. Onida, First-principles calculations of electronic excitations in clusters. Int. J. Quantum Chem. **77**, 951–960 (2000)
101. L.E. Ramos, H.-Ch. Weissker, J. Furthmüller, F. Bechstedt, Optical properties of Si and Ge nanocrystals: parameter-free calculations. Phys. Status Solidi B **242**, 3053–3063 (2005)
102. U. Itoh, Y. Toyoshima, H. Onuki, Vacuum ultraviolet absorption cross sections of SiH_4, GeH_4, Si_2H_6, and Si_3H_8. J. Chem Phys. **85**, 4867–4872 (1986)
103. M. Rohlfing, S.G. Louie, Excitonic effects and the optical absorption spectrum of hydrogenated Si clusters. Phys. Rev. Lett. **80**, 3320–3323 (1998)
104. M. Rohlfing, S.G. Louie, Optical excitations in conjugated polymers. Phys. Rev. Lett. **82**, 1959–1962 (1999)
105. L.E. Ramos, J. Paier, G. Kresse, F. Bechstedt, Optical spectra of Si nanocrystallites: Bethe-Salpeter approach versus time-dependent density functional theory. Phys. Rev. B **78**, 195423 (2008)
106. I. Vasiliev, S. Öğüt, J.R. Chelikowsky, *Ab initio* absorption spectra and optical gaps in nanocrystalline silicon. Phys. Rev. Lett. **86**, 1813–1816 (2001)
107. L.X. Benedict, A. Puzder, A.J. Williamson, J.C. Grossmann, G. Galli, J.E. Klepeis, J.-Y. Raty, O. Pankratov, Calculation of optical absorption spectra of hydrogenated Si clusters: Bethe-Salpeter equation versus time-dependent local-density approximation. Phys. Rev. B **68**, 085310 (2003)
108. M.S. Hybertsen, S.G. Louie, Theory of quasiparticle surface states in semiconductor surfaces. Phys. Rev. B **38**, 4033–4044 (1988)
109. X. Zhu, S.B. Zhang, S.G. Louie, M.L. Cohen, Quasiparticle interpretation of photoemission spectra and optical properties of GaAs(110). Phys. Rev. Lett. **63**, 2112–2115 (1989)
110. F. Bechstedt, *Principles of Surface Physics* (Springer, Berlin, 2003)
111. F. Bechstedt, R. Del Sole, Giant quasiparticle shifts of semiconductor surface states. Solid State Commun. **74**, 41–44 (1990)
112. R.I.G. Uhrberg, G.V. Hansson, J.M. Nicholls, S.A. Flodström, Experimental evidence for one highly dispersive dangling-bond band on Si(111)2 $\times$ 1. Phys. Rev. Lett. **48**, 1032–1035 (1982)
113. P. Perfetti, J.M. Nicholls, B. Reihl, Unoccupied surface-state band on Si(111)2 $\times$ 1. Phys. Rev. B **36**, 6160–6163 (1987)
114. M. Rohlfing, S.G. Louie, Excitons and optical spectrum of the Si(111)-(2$\times$ 1) surface. Phys. Rev. Lett. **83**, 856–859 (1999)
115. C. Kress, M. Fiedler, W.G. Schmidt, F. Bechstedt, Geometrical and electronic structure of the reconstructed diamond (100) surface. Phys. Rev. B **50**, 17697–17700 (1994)
116. C. Kress, M. Fiedler, W.G. Schmidt, F. Bechstedt, Quasi-particle band structure of C(111)2$\times$1 and C(100)2$\times$1 surfaces. Surf. Sci. **331–333**, 1152–1156 (1995)

117. J. Wu, R. Cao, X. Yang, P. Pianetta, I. Lindau, Photoemission study of diamond (100) surface. J. Vac. Sci. Technol. A **11**, 1048–1051 (1983)
118. G. Francz, P. Kania, G. Ganther, H. Stupp, P. Oelhafen, Photoelectron spectroscopy study of natural (100), (110), (111) and CVD diamond surfaces. Phys. Status Solidi A **154**, 91–108 (1996)
119. F. Bechstedt, R. Del Sole, F. Manghi, Giant quasi-particle shifts of semiconductor surface states. J. Phys. Condens. Matter **1**, SB75–SB78 (1989)
120. J.E. Northrup, M.S. Hybertsen, S.G. Louie, Many-body calculation of the surface-state energies for Si(111)2 × 1. Phys. Rev. Lett. **66**, 500–503 (1991)
121. F.J. Himpsel, D.E. Eastmann, P. Heimann, J.F. van der Veen, Surface states on reconstructed diamond (111). Phys. Rev. B **24**, 7270–7274 (1981)
122. R. Graupner, M. Hollering, A. Ziegler, J. Ristein, L. Ley, A. Stampfl, Dispersions of surface states on diamond (100) and (111). Phys. Rev. B **55**, 10841–10847 (1997)
123. M. Marsili, O. Pulci, F. Bechstedt, R. Del Sole, Electronic structure of the C(111) surface: solution by self-consistent many-body calculations. Phys. Rev. B **72**, 115415 (2005)
124. L. Sorba, V. Hinkel, U.U. Middelmann, K. Horn, Bulk and surface electronic bands of InP(110) determined by angle-resolved photoemission. Phys. Rev. B **36**, 8075–8081 (1987)
125. J.M. Nicholls, K.O. Magnusson, B. Reihl, Electronic structure of the InP(110) surface: an inverse photoemission study. Surf. Sci. Lett. **243**, L31–L36 (1991)
126. W. Drube, D. Straub, F.J. Himpsel, Inverse photoemission study of InP, InAs, and InSb. Phys. Rev. B **35**, 5563–5568 (1987)
127. O. Pulci, M. Palummo, R. Del Sole, Private communication (published first as Fig. 5.27 in [111])
128. S.V. Pepper, Diamond (111) studied by electron energy loss spectroscopy in the characteristic loss region. Surf. Sci. **123**, 47–60 (1982)
129. G. Bussetti, C. Goletti, P. Chiradia, T. Derry, Optical gap between dangling-bond states of a single-domain diamond C(111)-2×1 by reflectance anisotropy spectroscopy. Europhys. Lett. **79**, 57002 (2007)
130. A.I. Shkrebtii, M. Marsili, E. Heritage, O. Pulci, R. Del Sole, F. Bechstedt, Defect induced modification of the surface gap and optical properties of C(111)2×1 surface. Phys. Status Solidi A **209**, 669–674 (2012)
131. O. Pulci, F. Bechstedt, G. Onida, R. Del Sole, L. Reining, State mixing for quasiparticles at surfaces: nonperturbative GW approximation. Phys. Rev. B **60**, 16758–16761 (1999)
132. A. Cricenti, S. Selci, K.O. Magnusson, B. Reihl, Position of the empty surface-state band on Si(111)2 × 1. Phys. Rev. B **41**, 12908–12910 (1990)
133. M. Rohlfing, M. Palummo, G. Onida, R. Del Sole, Structural and optical properties of the Ge(111)-(2 × 1) surface. Phys. Rev. Lett. **85**, 5440–5443 (2000)
134. J.M. Nicholls, G.V. Hansson, R.I.G. Uhrberg, S.A. Ladström, Experimental dangling-bond band on the Ge(111)-(2 × 1) surface. Phys. Rev. B **27**, 2594–2597 (1983)
135. J.M. Nicholls, G.V. Hansson, U.O. Karlsson, R.I.G. Uhrberg, Confirmation of a highly dispersive dangling-bond band on Ge(111)-2 × 1. Phys. Rev. Lett. **52**, 1555–1558 (1984)
136. S. Solal, G. Jeszequel, A. Barski, P. Steiner, P. Pinchaux, Y. Petrof, Ge(111)2 × 1: π-bonded chain model or not? Phys. Rev. Lett. **52**, 360–363 (1984)
137. J.M. Nicholls, B. Reihl, Antibonding surface state band of the Ge(111)2×1 surface. Surf. Sci. **218**, 237–245 (1989)
138. R.M. Feenstra, G. Meyer, F. Moresco, K.H. Rieder, Buckling and band gap of the Ge(111)2×1 surface studied by low-temperature scanning tunneling microscopy. Phys. Rev. B **64**, 081306 (2001)
139. M. Rohlfing, P. Krüger, J. Pollmannn, Efficient scheme for GW quasiparticle band-structure calculations with applications to bulk Si and to the Si(001)-(2 × 1) surface. Phys. Rev. B **52**, 1905–1917 (1995)
140. R.J. Hamers, U.K. Köhler, Determination of the local electronic structure of atomic-sized defects on Si(001) by tunneling spectroscopy. J. Vac. Sci. Technol. A **7**, 2854–2859 (1989)

141. J.E. Northrup, Electronic structure of Si(100)c(4 × 2) calculated within the GW approximation. Phys. Rev. B **47**, 10032–10035 (1993)
142. M. Rohlfing, P. Krüger, J. Pollmann, Quasiparticle band structures of clean, hydrogen-, and sulfur-terminated Ge(001) surfaces. Phys. Rev. B **54**, 13759–13766 (1996)
143. M. Skibowski, L. Kipp, Inverse combined with direct photoemission: momentum resolved electronic structure of 2D systems. J. Electron. Spectrosc. Relat. Phenom. **68**, 77–96 (1994)
144. L. Kipp, R. Manzke, M. Skibowski, An intrinsic metallic surface state on Ge(001)2 × 1. Solid State Commun. **93**, 603–607 (1995)
145. J.A. Kubby, J.E. Griffith, R.S. Becker, J.S. Vickers, Tunneling microscopy of Ge(001). Phys. Rev. B **36**, 6079–6093 (1987)
146. O. Pulci, G. Onida, R. Del Sole, L. Reining, *Ab initio* calculation of self-energy effects on optical properties of GaAs(110). Phys. Rev. Lett. **81**, 5374–5377 (1998)
147. H. Carstensen, R. Claessen, R. Manzke, M. Skibowksi, Direct determination of III–V semiconductor surface band gaps. Phys. Rev. B **41**, 9880–9885 (1990)
148. S. Ryo, M.Y. Hahn, J. Maultzsch, T.F. Heinz, P. Kim, M.L. Steigerwald, L.E. Brus, Reversible basal plane hydrogenation of graphene. Nano Lett. **8**, 4597–4602 (2008)
149. D.C. Elias, R.R. Nair, T.M.G. Mohiuddin, S.V. Morozov, P. Blake, M.P. Halsall, A.C. Ferrari, D.W. Boukhvalov, M.I. Katsnelson, A.K. Geim, K.S. Novoselov, Control of graphene's properties by reversible hydrogenation: evidence for graphane. Science **323**, 610–613 (2009)
150. E. Bianco, S. Butler, S. Jiang, O.D. Restrepo, W. Windl, J.E. Goldberger, Stability and exfoliation of germanane: a germanium graphane analogue. ACS Nano **7**, 4414–4421 (2013)
151. O. Pulci, P. Gori, M. Marsili, V. Garbuio, R. Del Sole, F. Bechstedt, Strong excitons in novel two-dimensional crystals: silicane and germanane. Europhys. Lett. **98**, 37004 (2012)
152. J.O. Sofo, A.S. Chaudhari, G.D. Barber, Graphane: a two-dimensional hydrocarbon. Phys. Rev. B **75**, 153401 (2007)
153. L.C. Lew Yan Voon, E. Sandberg, R.S. Aga, A.A. Farajian, Hydrogen compounds of group-IV nanosheets. Appl. Phys. Lett. **97**, 163114 (2010)
154. P. Cudazzo, C. Attaccalite, I.V. Tokatly, A. Rubio, Strong charge-transfer excitonic effects and the Bose-Einstein exciton condensate in graphane. Phys. Rev. Lett. **104**, 226804 (2010)
155. P. Gori, O. Pulci, M. Marsili, F. Bechstedt, Side-dependent electron escape from graphene- and graphene-like SiC layers. Appl. Phys. Lett. **100**, 043110 (2012)
156. F. Bechstedt, A.A. Stekolnikov, J. Furthmüller, P. Käckell, Origin of the different reconstructions of diamond, Si, and Ge(111) surfaces. Phys. Rev. Lett. **87**, 016103 (2001)
157. W. Mönch, *Semiconductor Surfaces and Interfaces* (Springer, Berlin, 2001)
158. O. Pulci, P. Gori, M. Marsili, V. Garbuio, A.P. Seitsonen, F. Bechstedt, A. Cricenti, R. Del Sole, Electronic and optical properties of group IV two-dimensional materials. Phys. Status Solidi A **207**, 291–299 (2010)

Chapter 17
Satellites

Abstract The dynamics of the reaction, in particular, the screening reaction, of the electron gas to the formation of a single-particle electronic excitation, electron or hole, in different spectroscopies gives rise not only to a reduction of spectral weight of the main quasiparticle peak but also to the generation of satellite structures to fulfill the sum rule for the spectral function. The formation of satellite structures is mainly illustrated in terms of intrinsic losses and the sudden limit. The generation of multiple-plasmon losses is discussed in detail. The relation to the GW approximation is described within the Blomberg-Bergersen-Kus method. Its validity is demonstrated in the limit of dispersionless fermions. As examples core-hole and valence-band photoemission spectroscopies are described and discussed in the light of available experiments.

17.1 Facts

17.1.1 Measurement of Spectral Functions

If one illuminates a sample surface with photons, electrons will be emitted, which have an intensity versus energy distribution. Hence, information is obtained directly or indirectly about the bulk and surface electronic states in dependence on the mean-free path of the electrons. The most important and widely used experimental technique to gain information about occupied electronic bulk or surface states is the photoemission spectroscopy (or sometimes called photoelectron spectroscopy) (PES) that is schematically illustrated in Fig. 17.1a. The solid surface is irradiated by monochromatic photons with energy $\hbar\omega$, holes are generated in the sample, and the electrons emitted into the vacuum are analyzed with respect to their kinetic energy $\varepsilon_{\text{kin}} = \varepsilon(\mathbf{k}) = \hbar^2\mathbf{k}^2/2m$. When photons in the ultraviolet (UV) spectral range are used, the technique is called UPS (UV photoemission spectroscopy). Besides the kinetic energy one may also use the emission direction $\mathbf{k}/k = (\cos\phi\sin\theta, \sin\phi\sin\theta, \cos\theta)$ described by the angle θ with the surface normal and the angle ϕ in the surface plane to characterize the geometry of the

F. Bechstedt, *Many-Body Approach to Electronic Excitations*,
Springer Series in Solid-State Sciences 181, DOI 10.1007/978-3-662-44593-8_17

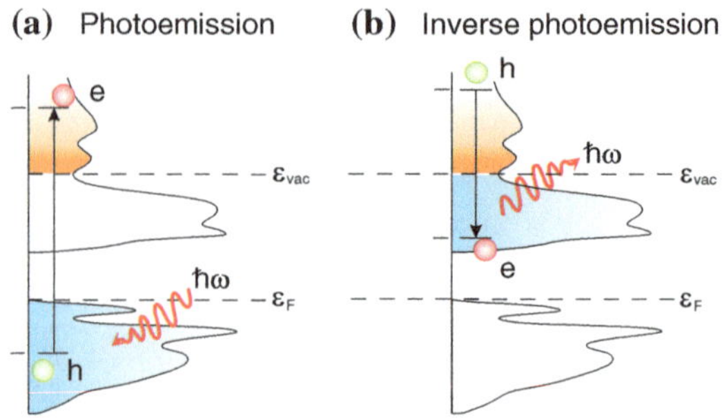

Fig. 17.1 Illustration of the (**a**) PES and (**b**) IPES process using the QP density of states. In (**a**) PES an electron is excited by an incoming photon from an occupied valence state (*lower blue region*) into the continuum (*upper yellow region*) starting above the vacuum level ε_{vac}. In (**b**) IPES, an injected electron with kinetic energy ε_{kin} (*uppermost yellow region*) undergoes a radiative transition into an unoccupied state (*lower blue region*) in-between thereby emitting a photon. The band regions are characterized by the corresponding DOS (ordinate: energy, abscissa: DOS)

experiment. Varying θ and/or ϕ the method is then known as angle-resolved (AR) PES or UPS, ARPES or ARUPS.

Inverse photoemission spectroscopy (IPES) can be regarded as a time-reversed photoemission process as illustrated in Fig. 17.1b. It therefore probes the unoccupied electronic states. In this technique a beam of electrons with energy ε_{kin} and wave vector $\mathbf{k} = k(\cos\phi\sin\theta, \sin\phi\sin\theta, \cos\theta)$ is incident on a surface. The electrons transmitted into the solid decay to empty states with lower energy through the Auger effect or by emitting photons, which are detected. There are two operating modes: either the energy $\hbar\omega$ of the detected photons is held constant and the spectrum is obtained by varying ε_{kin} (bremsstrahlung isochromate spectroscopy (BIS)), or ε_{kin} is kept constant and the spectrum is taken as a function of $\hbar\omega$. If, in addition, one takes advantage of the **k**-vector resolution, one calls the method **k**-resolved (KR) IPES, i.e., KRIPES.

A rigorous theoretical approach of an elementary photoemission (inverse photoemission) process requires a full quantum-mechanical many-body treatment of the complex coherent (and even incoherent) interaction processes starting with a photon (electron) and finishing with an electron (photon) in the detector as well as a hole (electron) in the sample. Theoretical approaches of this kind treat the photoemission quite often as a one-step process, where only the absorption of light is taken into consideration [1–3] (see Fig. 17.1a). A more instructive approach is however the so-called three-step model [4], which includes (i) optical excitation of an electron in the bulk and the surface region, (ii) travel of the electron excited above the vacuum level to the surface, and (iii) escape of this photoelectron into the vacuum. The range of the second step is limited by the mean-free path of the electrons and hence depends on the kinetic energy of the emitted electron.

In this stage of the discussion we will waive the details of the emission or injection processes. In particular, we will not take into account (i) transport through the surface,

(ii) the transition matrix elements, e.g. the dipole matrix elements characterizing the electron-photon interaction, and (iii) the direct interaction between outgoing electron and remaining hole (or vice versa). Thus, the optical excitation of an electron (in a final state) and a hole (in the initial state) can simply be described. Although the attractive interaction between electron and hole may be small, effects of dynamical screening on the formation of quasielectrons and quasiholes as well as their attractive interaction play a role. *Interference effects* [5] might be important resulting in the *adiabatic limit* of the photoionization process instead of the interference-free *sudden limit* [3, 6] (see Sect. 22.2). Then, from the point of view of the many-body interactions a factorization of the two-particle Green function (complete one or irreducible part) in products of single-particle Green functions is possible as shown for the polarization function in (12.58). The spectrum is ruled by the product of the spectral function of the (propagating outward) electron and that of the (leaving) hole. In systems with collinear spins and single-quasiparticle states, which can be classified by the quantum numbers $\nu\mathbf{k}m_s$ and which guarantee almost diagonal spectral functions (see Sect. 14.3), one may write [7]

$$I(\hbar\omega - \varepsilon_{\text{kin}}, \mathbf{k}) \propto \sum_{\nu,\mu} \sum_{m_s} \sum_{\mathbf{k}} \int_{-\infty}^{+\infty} d\omega' f(\hbar(\omega' - \omega))[1 - f(\hbar\omega')] \times A_{\nu\nu}^{m_s m_s}(\mathbf{k}, \omega') A_{\mu\mu}^{m_s m_s}(\mathbf{k}, \omega' - \omega)\Big|_{\varepsilon_{\nu m_s}^{\text{QP}}(\mathbf{k}) = \varepsilon_{\text{kin}} + \varepsilon_{\text{vac}}}, \quad (17.1)$$

where $\varepsilon_{\text{kin}} = \frac{\hbar^2}{2m}\mathbf{k}^2$ is the kinetic energy of the photoelectrons. The vacuum energy ε_{vac} is determined for vanishing kinetic energy. The spectral functions of electron and hole in (17.1) account for the dynamical reaction of the system. The deviations of the hole function from a sharp δ-like quasiparticle peak (see Sect. 14.1) describe *intrinsic* (or shake-up) losses taking place somewhere inside the solid or molecule. On its way out the electron may also lose energy. Such losses are called *extrinsic*. Behind the distinction between extrinsic and intrinsic losses is the three-step model, i.e., photoionization in bulk (in near surface region) plus transport to the surface plus passage through the surface [4]. This model is hard to justify from first principles. Its best defence is that, in general, it seems to give reasonable agreement with experiment and retains a more intuitive physical interpretation.

In the extreme *sudden* limit the interaction of the photoelectron with the sample may be completely neglected, while the reaction of the system to the presence of the photohole is fully taken into account. This may be the case in the limit of high-energy photoelectrons. Then, the electron spectral function can be replaced by a Dirac δ-function. The photoelectron intensity (17.1) reads as [7]

$$I(\hbar\omega - \varepsilon_{\text{kin}}, \mathbf{k}) \propto \sum_{\mu} \sum_{m_s} f(\varepsilon_{\text{kin}} + \varepsilon_{\text{vac}} - \hbar\omega) A_{\mu\mu}^{m_s m_s}(\mathbf{k}, \frac{1}{\hbar}(\varepsilon_{\text{kin}} + \varepsilon_{\text{vac}}) - \omega). \quad (17.2)$$

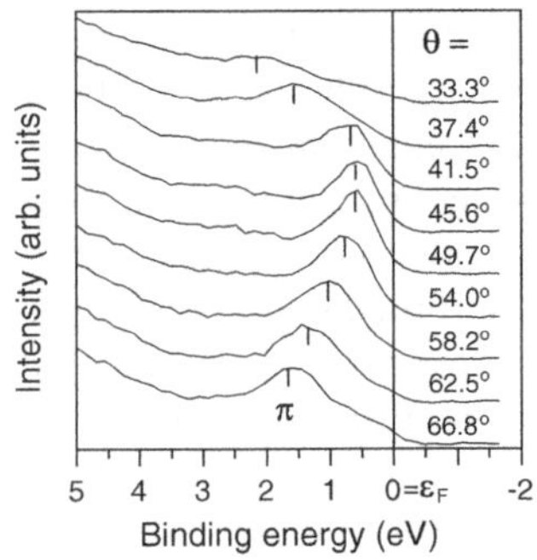

Fig. 17.2 Angle-resolved photoelectron spectra taken from an annealed C(111)2×1 surface. The photon energy is 50 eV at normal incidence of the incoming light and the measured azimuthal direction is $[\bar{1}10]$. Reprinted with permission from [9]. Copyright 1997 by the American Physical Society

Only intrinsic losses are taken into account. If also these losses and the finite lifetime of the quasiparticle are neglected, the hole spectral function becomes a δ-function that indicates the energy conservation during the photoionization process by the Einstein law of the photoelectric effect [8]

$$\varepsilon_{\text{kin}} + \varepsilon_{\text{vac}} - \varepsilon_{\mu m_s}^{\text{QP}}(\mathbf{k}) = \hbar\omega. \tag{17.3}$$

It is illustrated in Fig. 17.2 by photoelectron spectra for the occupied π-band of the C(111)2×1 surface [9], that has been discussed before in Fig. 16.29. The measurements with $\hbar\omega = 50$ eV have been performed under surface-sensitive conditions. The resulting small kinetic energies suggest the validity of the adiabatic limit where all intrinsic, extrinsic and interference losses cancel each other as we will show in Chap. 22. The hole band energies are referred to the Fermi energy, so that $\varepsilon_F - \varepsilon_{\mu m_s}^{\text{QP}}(\mathbf{k})$ corresponds to the binding energy of an electron. The measured azimuthal direction is fixed parallel to $[\bar{1}10]$. Then it holds $k = \sqrt{2m\varepsilon_{\text{kin}}/\hbar^2}\sin\theta$, where the angle of the emitted electrons with respect to the surface normal varies in the interval $\theta = 33.3-66.8°$. Despite the broadening of the spectra, the peak positions should be identified with surface band energies, here with those of the π-band of the π-bonded chain model in Fig. 16.26. The band dispersion visible in Fig. 17.2 is in qualitative agreement with that of the calculated π-band in Fig. 16.29. Near K in the rectangular BZ the occupied π-band comes closest to the Fermi level whereas in the directions toward Γ in the same BZ and Γ in the adjacent BZ a strong dispersion of the π-band toward lower energies is observed.

Expressions similar to (17.1) and (17.2) can be also derived for the time-reversed IPES process. The corresponding spectrum is dominated by the main peak in the spectral function of the electrons in empty final states. Neglecting the influence of the transition matrix elements and the vertex corrections, the spectral variation of KRIPES is governed by the empty **k**-vector-resolved density of states $\sum_\nu A_{\nu\nu}(\mathbf{k}, \varepsilon/\hbar)$. In contrast, in the case of ARPES/ARUPS the spectra are governed by the occupied part of the **k**-vector-resolved density of states of the surface (or bulk) states, $\sum_\mu A_{\mu\mu}(\mathbf{k}, \varepsilon/\hbar)$. Consequently, the combination of ARPES and KRIPES

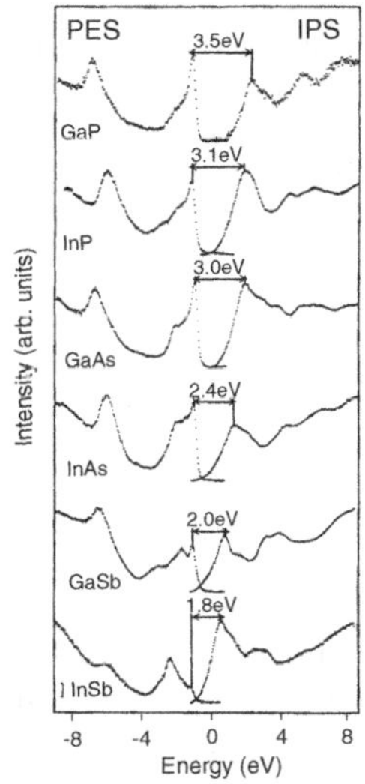

Fig. 17.3 Combined photoemission and inverse photoemission spectra for the determination of surface band gaps at the high-symmetry point X' in the surface BZ of six III-V(110)1×1 surfaces. The energy zero is given by the VBM. The photon energies have been chosen to be $\hbar\omega = 21.2$ eV (PES) and $\hbar\omega = 9.9$ eV (IPES), the energy at which the electron-induced bremsstrahlung is detected. From [10]

allows one to determine the complete **k**-vector-resolved single-particle density of states of a surface (or even bulk) system and hence the quasiparticle band structure.

An example is shown in Fig. 17.3 in which combined photoemission and inverse photoemission spectra [10] are presented for the X' point in the surface BZ in the energy region of the filled anion-derived dangling-bond band A_5 and the empty cation-derived C_3 band (see Table 16.7, Fig. 16.30) for cleaved III-V semiconductor (110)1×1 surfaces. For these surfaces the experiment has the advantage that a common energy reference could be established for the two applied spectroscopies. In all cases in Fig. 17.3, it is evident that the anion-derived surface state gives rise to a ponounced peak, while the cation-derived feature in some cases is a rather broad line with a large slope at lower energies. The surface band gap for the (110)1×1 surface of these materials at the X' point of the surface Brillouin zone is considerably larger than the fundamental bulk gap, but its increase from InSb to GaP has nearly the same slope.

17.1.2 Core-Electron Spectra in Sudden Limit

In accordance to the general discussion of losses in the photoelectron spectra in Sect. 17.1.1, we expect their pronounced evidence in the high-energy limit, the sudden limit, with large kinetic energies of the emitted electrons. For illustration, we investigate high-energy PES spectra of simple sp-bonded semiconductors and metals for situations, where sharp emission peaks are expected for the excitation of holes in deep core levels, whose broadening to bands can be omitted. As examples, the Si $2p$ and Si $2s$ photoelectron spectra of a Si crystal with (111)7×7 surface are presented in Fig. 17.4. They are measured with a photon energy of $\hbar\omega = 1486.7$ eV

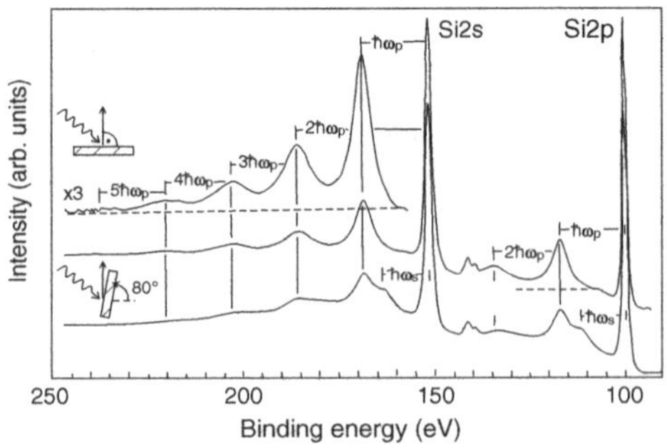

Fig. 17.4 Si $2p$ and Si $2s$ photoelectron spectra of a Si(111)7×7 surface measured for a large escape depth (*upper curves*) but also under surface-sensitive conditions (*lower curve*) relative to the Fermi level. From [11]

(AlK_α radiation) [11]. The large kinetic energy indicates the sudden limit. The escape depth of the photoelectrons is varied by changing the escape direction from normal emission ($\theta = 0°$, maximum depth equal to the mean-free path λ_{mfp} of electrons) to grazing emission ($\theta = 80°$, small depth $\sim \lambda_{\text{mfp}} \cos 80°$). In the first case, the spectra are dominated by bulk losses. In the second case, it is clear that the surface plays an essential role for the PES. Besides the main quasiparticle peak describing the core-hole excitation without losses, the position of which is usually used to define the core electron binding energy, $-\varepsilon_{2p}^{\text{QP}}$ or $-\varepsilon_{2s}^{\text{QP}}$, with respect to a reference level, e.g. the vacuum level ε_{vac} or the Fermi level ε_{F}, each spectrum exhibits a series of incoherent satellite structures. In sp-bonded solids, such as silicon, the satellite structures are mainly due to shake-up of surface and bulk plasmons of the valence electron gas. At the side of higher binding energies, indeed multiple ($n = 1, 2, 3, \ldots$) losses by plasmons are visible. The strongest satellites are due to bulk plasmons at (negative) binding energies of about $\varepsilon_{2\ell} - n\hbar\omega_p$ ($\ell = p, s$) with $\hbar\omega_p \approx 16$ eV (see Fig. 13.7). In the case of a non-metal the plasma frequency defined as the pole in the inverse dielectric function (13.54) may be slightly renormalized to $\Omega(0) = \sqrt{\frac{\epsilon_\infty}{\epsilon_\infty - 1}}\omega_p$ due to the electronic background polarization (see Sect. 13.2). In the lower curves of Fig. 17.4 (measured under surface-sensitive conditions) the broad satellite features exhibit shoulders at the position of the surface plasmon energy $\hbar\omega_s \approx \hbar\omega_p/\sqrt{2}$. Figure 17.5 shows a similar photoelectron spectrum for a simple metal, magnesium [12]. In Sect. 22.2 we will show that such satellite structures cannot be observed in spectroscopies probing the energy region around the fundamental gap as in the case of scanning tunneling spectroscopy or optical absorption.

The satellite structures in photoelectron spectra of core electrons are due to two different mechanisms. The main losses originate from the hole spectral function in (17.2) $A_{n\ell n\ell}(\omega)$ ($\ell = s, p$), where the spin and wave-vector dependences are omitted. They are a consequence of the polarization of the electronic system in the presence of the core hole and are probed by the photoelectron before it leaves the solid. This gives rise to intrinsic losses [3, 6, 13]. On its way to the surface and escape from the system the photoelectron itself polarizes the electronic system. This mechanism, giving rise to extrinsic losses, is of reduced influence because of the

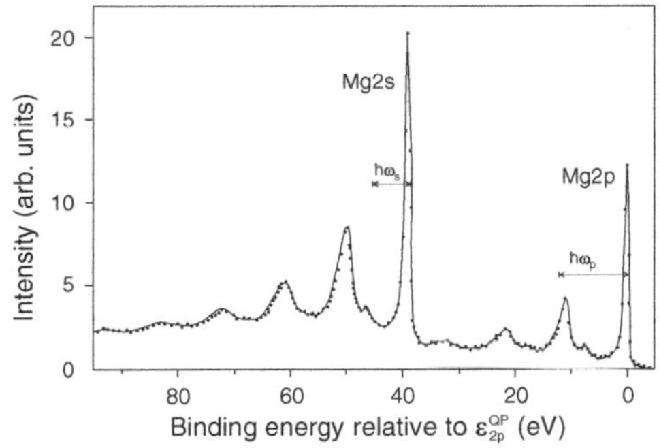

Fig. 17.5 Mg $2p$ and Mg $2s$ photoelectron spectra excited with AlK_α radiation ($\hbar\omega = 1486.7$ eV) relative to the binding energy of Mg $2p$, $-\varepsilon_{2p}^{\mathrm{QP}} = 49.8$ eV. Bulk and surface plasmon losses are indicated. From [12]

large kinetic energies. The same is true for the interference of losses, which however are not considered in the simplified representation (17.1).

Neglecting extrinsic, interference, and surface losses the experimental spectra in Figs. 17.4 and 17.5 may be obviously represented only by a core-hole spectral function in the form of a weighted sum of Dirac δ-functions ($\ell = s, p$) [14]

$$A_{2\ell 2\ell}(\omega) = 2\pi \sum_{n=0}^{\infty} e^{-\beta_{2\ell}} \frac{\beta_{2\ell}^n}{n!} \delta\left(\hbar\omega - \varepsilon_{2\ell}^{\mathrm{QP}} + n\hbar\Omega(0)\right), \qquad (17.4)$$

apart from a broadening of the main spectral line and the satellites due to finite lifetimes and the plasmon dispersion. The position of the main line, $\varepsilon_{2\ell}^{\mathrm{QP}} = \varepsilon_{2\ell} + \Delta_{2\ell}$ (14.42), is renormalized by the same interactions which generate the satellites, hence $\beta_{2\ell} = \Delta_{2\ell}/\hbar\Omega(0)$ holds. The center of gravity of the spectrum (17.4) is identical with the unrenormalized energy $\varepsilon_{2\ell}$ of the reference electronic structure. Interestingly, the spectral function of the type (17.4) is the exact result of a quantum-mechanical many-body problem, where a dispersionless fermion (here: core hole) interacts with bosons (here: plasmons) [13, 15].

17.1.3 Losses and Dynamical Screening

The principal occurrence of incoherent contributions $a_{\nu\nu}^{m_s m_s}(\mathbf{k}, \omega)$ to the single-particle spectral function (14.23) has been discussed in Sects. 14.1.2 and 14.1.3. From these studies the relation of the satellite structures to the frequency dependence of the XC self-energy is obvious. The majority of investigations of the spectral behavior of $A_{\nu\nu}^{m_s m_s}(\mathbf{k}, \omega)$ thereby restrict themselves to (i) the representation (14.19) of diagonal elements of the spectral functions and (ii) a description of the XC self-energy within the GW approximation (12.57). The vertex corrections are neglected in the self-energy but also in the polarization function (12.58) that determines the dynamical screening.

In the last chapters we have however learnt that there are many differences in the explicit treatments of the GW approximation: (i) Treatment of dynamical screening by plasmon-pole approximations or full inclusion of the frequency dependence of the inverse dielectric matrix, (ii) the chosen reference electronic structure, (iii) the kind of self-consistent calculation (mainly with one QP pole in the Green functions with or without update of the wave functions), etc. While the details of these treatments do only quantitatively influence the QP shifts, they may also yield qualitative changes in the spectra including the satellites. We illustrate this problem studying the one-shot GW approximation, the DFT-LDA or -GGA starting point, a certain plasmon-pole approximation for the screening function, and a representation of the spectral functions according to (14.19). Such a treatment is sometimes called ordinary or standard GW approximation. Results are presented in Fig. 14.2 for the valence-band spectral functions of crystalline silicon. For a given Bloch state $|\nu\mathbf{k}\rangle$ they show the main QP peak followed by a broad, structureless satellite feature at lower energies in an energy distance of more than 20 eV, i.e., much larger than the plasmon energy $\hbar\Omega(0)$. Consequently the satellite structure(s) cannot be interpreted in terms of true plasmon losses.

Corresponding earlier applications have been done for a non-spin-polarized homogeneous electron gas [16–18]. The resulting spectral(-weight) function $A(\mathbf{k},\omega)$ exhibits a sharp QP peak for not too large wave vectors due to the neglect of the particle-hole contribution to the dielectric function. In addition, a characteristic satellite structure has been found in the spectra of homogeneous electron gases with varying density, which may be interpreted as the corresponding spectra of simple metals with the same density. Spectral weight is taken from the QP level, and in certain energy regions in a distance different from the plasmon energy a new excitation occurs, called a 'plasmaron' [19]. As an example, a result of a more recent calculation is plotted in Fig. 17.6 for the Na metal and vanishing Bloch wave vector. It is compared with results from a cumulant expansion [20] that almost reproduces the measured XPS spectrum. The spectral weight of the 'plasmaron' satellite structure is much overestimated, no multiple plasmon losses appear, and the satellite peak is shifted toward larger binding energies by about twice of the plasmon energy

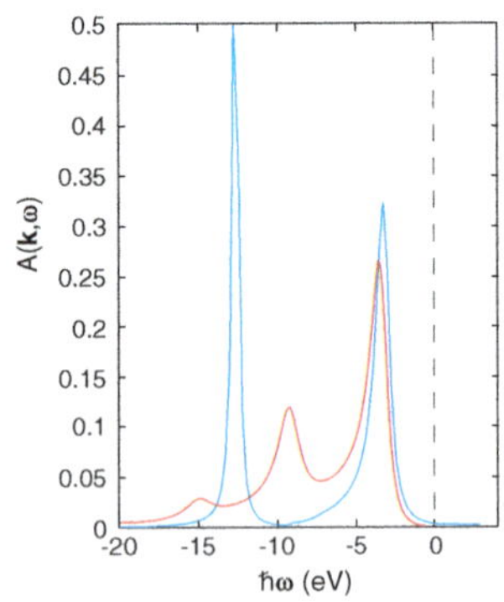

Fig. 17.6 The spectral function of the occupied part of the conduction band in Na at $\mathbf{k} = (0, 0, 0)$. The *red* and *blue lines* correspond to the cumulant expansion and the standard GW approximation, respectively. The Fermi energy represents the energy zero. Reprinted with permission from [20]. Copyright 1996 by the American Physical Society

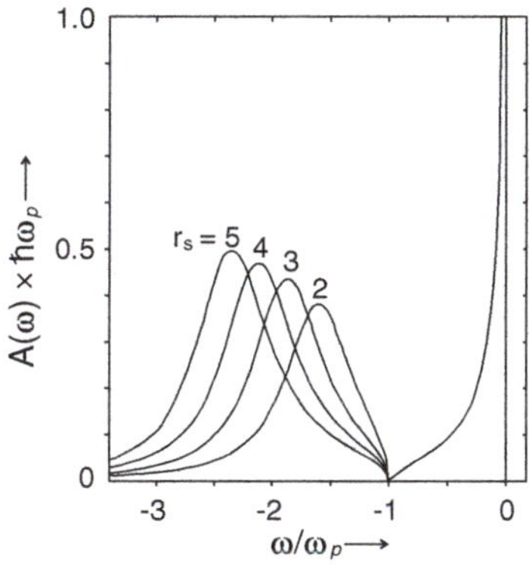

Fig. 17.7 The core electron spectral-weight function in the presence of conduction-electron gases with varying density parameter r_s computed within the standard GW treatment. The position of the QP core electron energy is used as energy zero. From [19]

$\hbar\Omega(0) = \hbar\omega_p = 5.7$ eV. The comparison with the more sophisticated spectrum from the cumulant expansion indicates that a 'plasmaron' excitation does not exist and is a remnant of the approximations done. A similar strange behavior also appears when studying core electron spectral functions within the standard GW approach in the presence of conduction electrons with different densities as displayed in Fig. 17.7 [19]. The spectra in Figs. 17.4 and 17.5 are not qualitatively reproduced. Only one satellite structure appears with a huge spectral weight. Its maximum is shifted by $1.5 - 2.5\hbar\omega_p$ toward larger binding energies. We conclude that standard GW calculations cannot correctly describe satellite structures, neither their position nor their intensity.

One may think that a fully self-consistent calculation of the single-particle Green function G within the GW approximation but still neglecting the vertex corrections may improve the agreement with measured spectra [21]. For a homogeneous electron

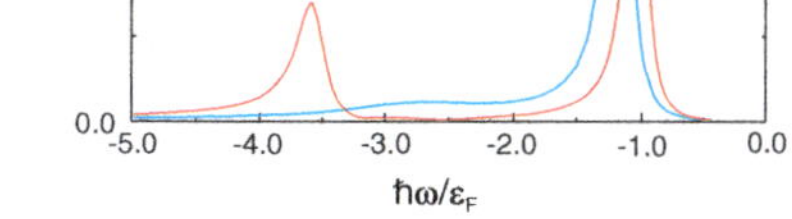

Fig. 17.8 Spectral function $A(\mathbf{k}, \omega)$ from the fully self-consistent GW calculation (*blue*) is compared to the corresponding quantity from the partially self-consistent GW_0 calculation (*red*) for the bottom of the band at $\mathbf{k} = (0, 0, 0)$, and an electron-gas parameter $r_s = 4$. Reprinted with permission from [21]. Copyright 1998 by the American Physical Society

gas a result is illustrated in Fig. 17.8. The fully self-consistent calculation gives rise to a redistribution of lineshape and spectral weight of satellite structures. However, still the description of multiple plasmon losses and their correct positions remains unsatisfactory.

17.2 Reasonable Approaches

17.2.1 Blomberg-Bergersen-Kus Method

Following the ideas of Blomberg, Bergersen, and Kus [22, 23] one obtains a reasonable expression for the (undamped) main QP peak and the first satellite structure in (14.48), in contrast to the standard GW treatment. The most important idea is to start in the self-energy calculation with a Green function that gives the main QP peak at its unknown position $\varepsilon_{\nu m_s}^{\mathrm{QP}}(\mathbf{k})$ and not at the eigenvalue of the reference electronic structure. In the next step, one iterates once the Dyson equation. We have to mention that this procedure is possible without and with neglecting vertex corrections. Nevertheless, we explicitly investigate the result (14.48) in the framework of the GW approach (12.57). To make the underlying physics more clear, we restrict the studies to the diagonal elements of the self-energy in the $T \to 0$ K limit. The Green function G appearing in the XC self-energy (14.58) and (14.59) is replaced by $\tilde{\tilde{G}}$, the Green function with one pole at the quasiparticle energy $\varepsilon_{\nu m_s}^{\mathrm{QP}}(\mathbf{k})$, in agreement with the iterative procedure in (14.48). Since the energy dependence of the self-energy is given by the correlation contribution to $\tilde{\tilde{\Sigma}} = \tilde{\tilde{\Sigma}}_{\mathrm{X}} + \tilde{\tilde{\Sigma}}_{\mathrm{C}}$ (12.13), the correlation part (14.59) of the self-energy fulfills a similar spectral representation (14.16) as the Green function (11.25) [24]

$$\Sigma^{\mathrm{C}\,m_s m_s}_{\;\nu\nu}(\mathbf{k},\omega) = \int_{-\infty}^{+\infty} \frac{d\omega'}{2\pi} \frac{\Gamma_{\nu\nu}^{m_s m_s}(\mathbf{k},\omega')}{\omega - \omega' + i\eta \mathrm{sgn}(\hbar\omega' - \mu)} \tag{17.5}$$

with the positive definite spectral function

$$\begin{aligned}
\Gamma_{\nu\nu}^{m_s m_s}(\mathbf{k},\omega) = &-\frac{2}{\Omega}\sum_{\mathbf{q}}\sum_{\mathbf{G},\mathbf{G}'} \tilde{\mathrm{v}}\left(\sqrt{|\mathbf{q}+\mathbf{G}||\mathbf{q}+\mathbf{G}'|}\right) \\
&\times \sum_{\nu',\mathbf{k}'} B^{\mathbf{k}\mathbf{k}'}_{\substack{m_s m_s\\ \nu\nu'}}(\mathbf{q}+\mathbf{G}) B^{\mathbf{k}\mathbf{k}'*}_{\substack{m_s m_s\\ \nu\nu'}}(\mathbf{q}+\mathbf{G}') \\
&\times \mathrm{Im}\tilde{\epsilon}^{-1}\left(\mathbf{q}+\mathbf{G},\mathbf{q}+\mathbf{G}',\omega - \varepsilon_{\nu' m_s}^{\mathrm{QP}}(\mathbf{k}')/\hbar\right) \\
&\times \left[\theta\left(\hbar\omega - \varepsilon_{\nu' m_s}^{\mathrm{QP}}(\mathbf{k}')\right)\theta\left(\varepsilon_{\nu' m_s}^{\mathrm{QP}}(\mathbf{k}') - \mu\right)\right. \\
&\left. - \theta\left(\varepsilon_{\nu' m_s}^{\mathrm{QP}}(\mathbf{k}') - \hbar\omega\right)\theta\left(\mu - \varepsilon_{\nu' m_s}^{\mathrm{QP}}(\mathbf{k}')\right)\right].
\end{aligned} \tag{17.6}$$

With the dynamical correlation self-energy (17.5) the spectral function within a first iteration (14.48) can be rewritten into

$$
\begin{aligned}
A_{\nu\nu}^{m_s m_s}(\mathbf{k},\omega) &= \left[1 - \mathcal{P}\frac{1}{2\pi\hbar}\int\limits_{-\infty}^{+\infty}\frac{d\omega'}{\omega'^2}\Gamma_{\nu\nu}^{m_s m_s}\left(\mathbf{k},\omega' + \varepsilon_{\nu m_s}^{\mathrm{QP}}(\mathbf{k})/\hbar\right)\right] \\
&\quad \times 2\pi\delta\left(\hbar\omega - \varepsilon_{\nu m_s}^{\mathrm{QP}}(\mathbf{k})\right) \\
&\quad + \mathcal{P}\frac{1}{2\pi\hbar}\int\limits_{-\infty}^{+\infty}\frac{d\omega'}{\omega'^2}\Gamma_{\nu\nu}^{m_s m_s}\left(\mathbf{k},\omega' + \varepsilon_{\nu m_s}^{\mathrm{QP}}(\mathbf{k})/\hbar\right) \\
&\quad \times 2\pi\delta\left(\hbar\omega - \varepsilon_{\nu m_s}^{\mathrm{QP}}(\mathbf{k}) - \hbar\omega'\right).
\end{aligned}
\tag{17.7}
$$

The spectral weight of the main QP peak and that of the first satellite in the spectral function are determined by the imaginary part of the correlation self-energy according to the representation (17.5).

The iterative result (17.7) can be interpreted as the linear expansion of a more general result for the spectral function [24], which also describes higher-order satellite structures,

$$
A_{\nu\nu}^{m_s m_s}(\mathbf{k},\omega) = \frac{1}{\hbar}\int\limits_{-\infty}^{+\infty} dt\, e^{\frac{i}{\hbar}\left[\hbar\omega - \varepsilon_{\nu m_s}^{\mathrm{QP}}(\mathbf{k})\right]t} e^{-C_{\nu m_s}(\mathbf{k},t)}
\tag{17.8}
$$

with a so-called satellite generator [25, 26]

$$
C_{\nu m_s}(\mathbf{k},t) = \mathcal{P}\int\limits_{-\infty}^{+\infty}\frac{d\omega}{2\pi\hbar}\frac{1}{\omega^2}\Gamma_{\nu\nu}^{m_s m_s}\left(\mathbf{k},\omega + \varepsilon_{\nu m_s}^{\mathrm{QP}}(\mathbf{k})/\hbar\right)\left[1 - e^{-i\omega t}\right].
\tag{17.9}
$$

It contains an integral over all linked diagrams of the screened Coulomb interaction.

A rigorous proof of the representation (17.8) of the spectral function for arbitrary band states and screening models cannot be given by the continuation of the iteration procedure in (14.44). This is, in general, only possible in an approximate manner. Following the idea that (17.8) with (17.9) is a result of a cumulant expansion, well known from statistical physics [27], there are three different ways of deriving the cumulant $C_{\nu m_s}(\mathbf{k},t)$: by diagrammatic expansion [28], from the equation of motion of the Green function [3], and by identifying the cumulant expansion with the Green function iteration up to first order [23] as illustrated above.

The satellite generator (17.9), which is only related to the spectral function of the correlation self-energy, describes the deviation of the lineshape of the spectral function from a single Dirac δ-function. Several important properties are obvious: (i) Expression (17.8) fulfills the sum rule

$$1 = \int_{-\infty}^{+\infty} \frac{d\hbar\omega}{2\pi} A_{\nu\nu}^{m_s m_s}(\mathbf{k}, \omega) \tag{17.10}$$

because of $C_{\nu\nu}^{m_s m_s}(\mathbf{k}, 0) = 0$. (ii) The center of gravity of the spectrum is given by

$$\bar{\varepsilon}_{\nu m_s}(\mathbf{k}) = \int_{-\infty}^{+\infty} \frac{d\hbar\omega}{2\pi} \hbar\omega A_{\nu\nu}^{m_s m_s}(\mathbf{k}, \omega) \tag{17.11}$$

with $\bar{\varepsilon}_{\nu m_s}(\mathbf{k}) = \varepsilon_{\nu m_s}^{\mathrm{QP}}(\mathbf{k}) - i\hbar \frac{\partial}{\partial t} C_{\nu m_s}(\mathbf{k}, t)\Big|_{t=0} = \varepsilon_{\nu m_s}^{\mathrm{QP}}(\mathbf{k}) - \mathrm{Re}\,\Sigma^{\mathrm{C}\;m_s m_s}_{\;\nu\nu}\left(\mathbf{k}, \varepsilon_{\nu m_s}^{\mathrm{QP}}(\mathbf{k})/\hbar\right)$. The center of gravity is shifted away from the main QP peak toward the satellite structures, close to a single-particle energy with correction as taken into account in the reference electronic structure. Starting from the HF approach as reference, the center of gravity would be given by the corresponding eigenvalue in agreement with Koopmans theorem (4.41). For Si Fig. 14.13 shows that the correlation self-energy determining the shift is positive (negative) for holes (electrons). (iii) The spectral function (17.8) contains satellite structures in all orders. This can be easily demonstrated by expansion of the integrand in (17.8) with respect to the t-dependent part of the satellite generator. The first order corresponds to the approximation in (17.7).

Very interesting is the spectral weight $z_{\nu m_s}(\mathbf{k})$ of the main QP peak with contributions from all orders of spectral function (17.6). It is modified to

$$\begin{aligned} z_{\nu m_s}(\mathbf{k}) &= \exp\left\{-\beta_{\nu m_s}(\mathbf{k})\right\}, \\ \beta_{\nu m_s}(\mathbf{k}) &= -\frac{\partial}{\partial \hbar\omega} \mathrm{Re}\,\Sigma^{\mathrm{C}\;m_s m_s}_{\;\nu\nu}(\mathbf{k}, \omega)\Big|_{\hbar\omega = \varepsilon_{\nu m_s}^{\mathrm{QP}}(\mathbf{k})} \end{aligned} \tag{17.12}$$

but agrees with the findings (14.27) and (14.47) at first order. At this order all the different definitions of the spectral weight, found in literature, agree.

17.2.2 Excitation of Dispersionless Fermions

In order to make the understanding of the spectral properties clear, we restrict ourselves to the excitation of a photoelectron from a dispersionless core level in a non-spin-polarized material, i.e., with a QP energy $\varepsilon_{\nu m_s}^{\mathrm{QP}}(\mathbf{k}) \rightarrow \varepsilon_{\nu}^{\mathrm{QP}}$. Its Bloch function can be replaced by a Bloch sum of localized orbitals. Since they do not overlap with those localized at adjacent atoms, the Bloch integrals in (17.6) can be approximated by

$$B\,{}^{\mathbf{kk}'}_{\substack{m_s m_s \\ \nu\nu'}}(\mathbf{q}+\mathbf{G}) \approx \delta_{\nu\nu'}\delta_{\mathbf{k},\mathbf{k}'+\mathbf{q}+\mathbf{G}}\rho_\nu(\mathbf{q}+\mathbf{G})$$

with the Fourier transform $\rho_\nu(\mathbf{q}+\mathbf{G})$ of the square of the localized core wave function. In addition, we neglect local-field effects in the screening $\mathbf{G} \neq \mathbf{G}'$. Together with the plasmon-pole model (13.54) and (15.12) it holds

$$-\mathrm{Im}\tilde{\epsilon}^{-1}(\mathbf{q}+\mathbf{G},\mathbf{q}+\mathbf{G},\omega) = \frac{\pi}{2}\frac{\omega_p^2}{\omega}\left[\delta\left(\omega-\Omega(\mathbf{q}+\mathbf{G})\right)+\delta(\omega+\Omega(\mathbf{q}+\mathbf{G}))\right].$$

Since for core levels $\theta\left(\mu-\varepsilon_\nu^{\mathrm{QP}}\right)=1$, the spectral function of the correlation self-energy (17.6) becomes

$$\Gamma_{\nu\nu}(\omega) = 2\pi\sum_{\mathbf{q},\mathbf{G}} g_\nu^2(\mathbf{q}+\mathbf{G})\delta\left(\hbar\omega-\varepsilon_\nu^{\mathrm{QP}}+\hbar\Omega(\mathbf{q}+\mathbf{G})\right) \tag{17.13}$$

with the coupling constant

$$g_\nu(\mathbf{Q}) = \sqrt{\frac{\hbar}{2\Omega}\tilde{\mathrm{v}}(|\mathbf{Q}|)\frac{\omega_p^2}{\Omega(\mathbf{Q})}}\,|\rho_\nu(\mathbf{Q})|$$

of the interaction between plasmons of energy $\hbar\Omega(\mathbf{Q})$ and core holes at level ν with a Fourier-transformed probability $\rho_\nu(\mathbf{Q})$ to find it. The satellite generator (17.9) becomes [13, 15, 26]

$$C_\nu(t) = \sum_{\mathbf{q},\mathbf{G}}\left[\frac{g_\nu(\mathbf{q}+\mathbf{G})}{\hbar\Omega(\mathbf{q}+\mathbf{G})}\right]^2\left[1-e^{i\Omega(\mathbf{q}+\mathbf{G})t}\right]. \tag{17.14}$$

For dispersionless plasmons $\Omega(\mathbf{Q}) \approx \Omega(0)$ the spectral function (17.8) takes the form (17.4) but replacing 2ℓ by ν and setting $\beta_\nu = \sum_{\mathbf{q},\mathbf{G}}[g_\nu(\mathbf{q}+\mathbf{G})/\hbar\Omega(\mathbf{q}+\mathbf{G})]^2$. That means, the lineshape of $A_{\nu\nu}(\omega)$ is replaced by a series of Dirac δ-functions, the intensities of which obey a Poisson distribution. More precisely, the unbroadened primary quasihole peak ($n=0$) is followed by an equidistant family of plasmon replica ($n \geq 1$). The resulting spectral function corresponds to the exact zero-temperature result, if a boson field with quantum energy $\hbar\Omega(\mathbf{Q})$ is coupled with a coupling constant $g_\nu(\mathbf{Q})$ to a hole in the deep core state ν with renormalized energy $\varepsilon_\nu^{\mathrm{QP}}$.

17.2.3 Consequences

For discussion and interpretation we replace the first iteration (17.7) of the Dyson equation for the core-hole excitation in the framework of the above-mentioned approximations (17.13). After some rearrangements it holds for the spectral function

$$A_{\nu\nu}(\omega) = \left\{ 1 - \sum_{\mathbf{q},\mathbf{G}} \left[\frac{g_\nu(\mathbf{q}+\mathbf{G})}{\hbar\Omega(\mathbf{q}+\mathbf{G})} \right]^2 \right\} 2\pi\,\delta\left(\hbar\omega - \varepsilon_\nu^{\mathrm{QP}}\right)$$
$$+ \sum_{\mathbf{q},\mathbf{G}} \left[\frac{g_\nu(\mathbf{q}+\mathbf{G})}{\hbar\Omega(\mathbf{q}+\mathbf{G})} \right]^2 2\pi\,\delta\left(\hbar\omega - \varepsilon_\nu^{\mathrm{QP}} + \hbar\Omega(0)\right) \quad (17.15)$$

or including the plasmon dispersion

$$A_{\nu\nu}(\omega) = \left\{ 1 - \int_0^\infty d\omega' \frac{\alpha_\nu(\omega')}{\omega'} \right\} 2\pi\,\delta\left(\hbar\omega - \varepsilon_\nu^{\mathrm{QP}}\right)$$
$$+ \int_0^\infty d\omega' \frac{\alpha_\nu(\omega')}{\omega'} 2\pi\,\delta\left(\hbar\omega - \varepsilon_\nu^{\mathrm{QP}} + \hbar\omega'\right) \quad (17.16)$$

with

$$\alpha_\nu(\omega) = \frac{1}{\hbar^2\omega} \sum_{\mathbf{q},\mathbf{G}} g_\nu^2(\mathbf{q}+\mathbf{G})\delta(\omega - \Omega(\mathbf{q}+\mathbf{G})).$$

Since in the zero-temperature limit $A_{\nu\nu}(\omega) = 0$ holds for $\hbar\omega > \varepsilon_\nu^{\mathrm{QP}}$, the approximate result (17.16) suggests that the hole spectral function obeys an integral equation

$$\left(\frac{1}{\hbar}\varepsilon_\nu^{\mathrm{QP}} - \omega \right) A_{\nu\nu}(\omega) = \int_0^{\frac{1}{\hbar}\varepsilon_\nu^{\mathrm{QP}} - \omega} d\omega' \alpha_\nu(\omega') A_{\nu\nu}(\omega' + \omega)$$

for $\varepsilon_\nu^{\mathrm{QP}} > \hbar\omega$. This can be shown by partial integration [14]. For a constant quantity $\alpha_\nu(\omega) \approx \alpha_\nu(0)$, in contrast to the result in (17.16), one finds a core-hole spectral function to be

$$A_{\nu\nu}(\omega) \sim \left[\varepsilon_\nu^{\mathrm{QP}} - \hbar\omega \right]^{\alpha_\nu(0)-1}.$$

This lineshape is significantly different from the sum of Poisson-distributed δ-functions (17.4). The drastic modification of the core-hole spectral function corresponds to the Mahan-Noziere-DeDominicis effect [29, 30]. Thus, $\alpha_\nu(0)$ is the asymmetry index of the spectral line.

Expression (17.15) also shines some light on the iterative solution (14.48) of the Dyson equation and the GW approximation used for explicit results:

(i) The energy distance between the main quasiparticle peak and the first plasmon satellite is given by the plasmon frequency $\hbar\Omega(0)$, in agreement with the experimental results depicted in Figs. 17.4 and 17.5. The reason is the use of the QP Green function $\tilde{\tilde{G}}$ as starting point. There is no space for a spurious collective excitation of the electron gas such as the plasmaron [16–19] with a drastically changed loss energy. Such a satellite is a spurious product of the inadequate application of the GW approximation [22, 23]. Its longstanding discussion is amazing. Even a few months after the plasmaron was predicted to appear in core-hole GW spectra [19], Langreth published the exact solution of the dispersionless fermion-boson model (17.4) [13], indicating a plasmon to be responsible for the satellites and not a plasmaron. The reason why the GW approximation seemingly fails is the use of an inappropriate reference Green function in the XC self-energy. The use of a function different from that applied in (14.44), e.g. with a pole at the DFT-LDA or -GGA eigenvalues, automatically modifies the distance between QP peak and satellite.

(ii) The general conclusion that the GW approximation fails in the description of the satellites has to be taken with caution. This certainly holds for the higher-order satellite structures but not for the first one. Expectations to find also correspondence with the plasmon satellites of higher order, if the Dyson equation is solved more exactly with a GW self-energy, cannot be confirmed. Blomberg and Bergersen [22] have shown that the complete iterative solution of the problem of the core-hole spectral function within GW leads to the plasmaron problem. The whole spectral function represents a geometrical series with respect to the coupling strength β_ν, contrary to the exact solution (17.4). The contribution in n-th *order* of β_ν to the satellite structure is proportional to $\beta_\nu^n/n!$, while the continued iteration of the Dyson equation (14.43) would yield contributions proportional to β_ν^n. This means, higher-order contributions are overestimated by iteration of the Dyson equation.

(iii) The two standard statements in the literature that (a) "GW predicts a single satellite instead of a satellite series with decreasing spectral weight and also greatly overestimates the binding energy of the satellite structures" and (b) "The cumulant expansion [3, 13, 20, 28] of the Green function G cures these deficiencies by including significant vertex corrections beyond GW or the $n=1$ iteration" are certainly correct. However, two aspects have to be taken into consideration. The Green function in the GW self-energy and in the computation of W has to be assumed to possess one pole at the QP energy. Satellites and especially the reduction of the spectral weight of the main peak by the factor $z_{\nu m_s}(\mathbf{k})$ should not be taken into account. This approach of the Green function is based on the intuition but also the fact that the vertex corrections stabilize the QP peak. However, in the explicit numerical treatments, even in the so-called self-consistent GW approach, in general, we do not take vertex corrections into account. So, in the sense of perturbation expansions, we are frequently confronted with the problem of not treating many-body effects on an equal footing. In order to treat W correctly in all orders, one has to take into account also vertex corrections or, at least, partly simulate them by stabilizing the QP main peak and restricting the studies to the first satellite.

17.3 Examples

17.3.1 Core-Hole Excitations

In order to verify the relation between the main QP peak and the first plasmon satellite, the expression (17.7) resulting as first iteration has been calculated within the GW approximation and the single-plasmon-pole result (17.13) for Si 2s and Ge 3d core-hole excitations in the corresponding group-IV crystals. The results are summarized in Fig. 17.9 including an additional phenomenological Lorentzian broadening Γ to simulate lifetime and instrumental effects. The resulting agreement in the relative peak positions and intensities between main peak and satellite are satisfying. Deviations in the lineshape of the satellite may be consequences of the used plasmon-pole model. The energies $\hbar\Omega(0) = 16.6$ eV (Si) and 15.6 eV (Ge) define the low-energy onset of the plasmon satellites. The agreement between measured and calculated spectra indicates that the first-iteration procedure (14.48) and the use of the XC self-energy in GW approximation (12.57) gives a reasonable description of the first satellite in core-level PES in the sudden limit.

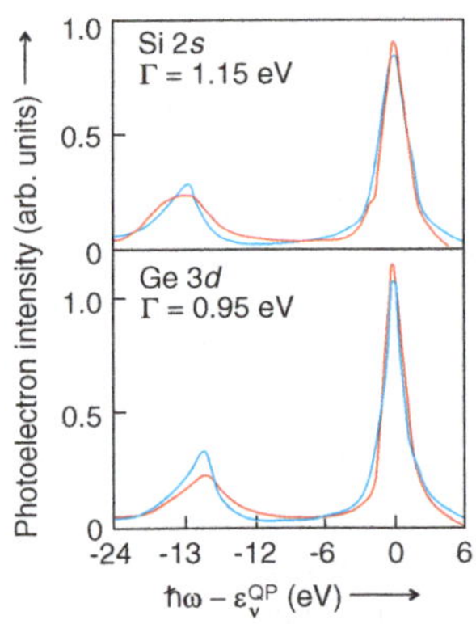

Fig. 17.9 Photoemission spectra for excitation of Si 2s and Ge 3d electrons in crystalline silicon and germanium, respectively, by means of AlK_α radiation. *Red line* – experimental [31], *blue line* – theory. An additional broadening is taken into account. The position of the main QP peak is used as energy zero. The smooth background in the experimental spectra is subtracted. From [26]

17.3.2 Valence-Electron Spectra

The spectral function of valence electrons of silicon has been investigated in a relatively early period of the GW approach [24]. However, only very recently experimental data from angle-integrated XPS for valence-electron excitations of silicon are available also in the energy region of the satellite structures as presented in Fig. 17.10 [32]. The comparison with the trivial form of the standard GW approximation, i.e., the approximation (14.19) for the spectral function and self-energy therein within the GW approximation (14.58) or (14.59), is also depicted. This theory gives rise to a broad satellite structure, the maximum of which is separated by about 23 eV from the QP peak in the corresponding valence-electron spectral function, which is obviously related to the shake-up of fictitious plasmarons. As a consequence a broad structure at about −30 eV, belonging to the entire valence-band spectrum, appears. This satellite feature is much below the experimental low-energy satellite spectrum with a maximum near −24 eV and a second weaker satellite feature near −41 eV.

The disagreement between the trivial application of the GW approach and the experimental spectrum asks for improvements of the description of the emission spectrum along the line discussed in Sects. 17.1 and 17.2. Recently, colleagues [32, 33] followed the idea to solve a Kadanoff-Baym-like equation (see Sect. 12.1.2) for the Green function G in the presence of an effective perturbation potential ϕ_{eff} with $\phi_{\mathrm{eff}} \to 0$. In the time domain they started from a Green function $\tilde{\tilde{G}}$ for a valence electron with one quasiparticle pole at the correct position $\varepsilon^{\mathrm{QP}}_{\nu m_s}(\mathbf{k})$ following the Blomberg-Bergersen-Kus idea. This equation can be formally solved, resulting in a time structure that closely resembles to the cumulant expansion (17.8). Omitting any spin polarization, not indicating the single-particle quantum state, and using only diagonal elements, one finds a solution of such a Kadanoff-Baym-like equation. For the time-dependent QP Green function it can be written as [32, 33]

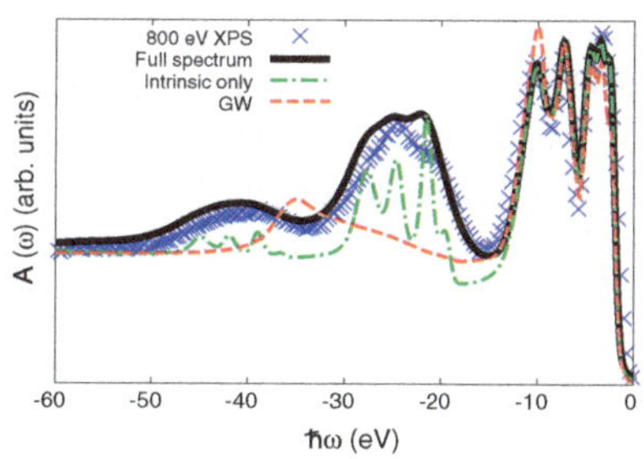

Fig. 17.10 Experimental XPS spectrum of Si at 800 eV photon energy (*blue crosses*), compared to the theoretical intrinsic hole spectral function $A(\omega)$ calculated from standard GW (*red dashed line*), and from (17.17) (*green dot-dashed line*). On top of the latter the improved treatment (*black solid line*) also includes extrinsic and interference effects. All spectra contain photoabsorption cross sections, a calculated secondary electron background, and 0.4 eV Gaussian broadening to account for finite **k**-point sampling and experimental resolution. The Fermi energy is set to 0 eV. Reprinted with permission from [32]. Copyright 2011 by the American Physical Society

$$G(\tau) = \tilde{\tilde{G}}(\tau) e^{\frac{i}{\hbar}\Delta\tau} e^{i\int_0^\tau dt \int_t^\tau dt' W(t-t')} \tag{17.17}$$

with $\tau = t_1 - t_2$ and a quantity $W(t-t')$ that is related to the time-dependent screened Coulomb potential (12.47). The quantity Δ compensates for the self-energy insertion in $\tilde{\tilde{G}}$. The physical meaning of the result (17.17) is obvious within the single-plasmon-pole approximation (13.54) and (15.12). Using similar approximations as applied to find (17.13) the time-dependent quantity is defined as

$$W(\tau) = \frac{-i}{\hbar^2} \sum_{\mathbf{q},\mathbf{G}} g^2(\mathbf{q}+\mathbf{G}) \left[\theta(\tau) e^{-i\Omega(\mathbf{q}+\mathbf{G})\tau} + \theta(-\tau) e^{i\Omega(\mathbf{q}+\mathbf{G})\tau}\right]. \tag{17.18}$$

For generation of photoholes in the valence-band region expression (17.17) with (17.18) leads to a spectral function similar to (17.8), if

$$\Delta = \sum_{\mathbf{q},\mathbf{G}} g^2(\mathbf{q}+\mathbf{G}) / \hbar\Omega(\mathbf{q}+\mathbf{G})$$

is chosen. That means, within the single-plasmon-pole approach the cumulant expansion and the approximate solution (17.17) of the Kadanoff-Baym-like equation lead to the same spectral behavior of valence-hole excitations.

The resulting spectrum is also displayed in Fig. 17.10. It excellently reproduces the integrated emission spectrum in the range from 0 to -15 eV of the valence bands. Also the first and second satellite features appear at correct energy positions in contrast to the standard GW treatment. However, the satellite spectra are still too low in intensity and show some substructures which do not occur in the measured spectrum. For that reason, effects of extrinsic losses and interference terms are additionally taken into account [34] using the formulation of an earlier theoretical work [6]. These effects bring the calculated spectrum in full agreement with the experimental data. Very recently the photoemission spectrum of the Si valence bands measured with a photon energy of 800 eV [32] has been also calculated by another group [35]. A so-called *ab initio* GW plus cumulant theory has been applied. In the light of the above discussions, the results are rather similar to those presented in Fig. 17.10. This also holds for the role of extrinsic and interference effects to reproduce the experimental data.

17.3.3 Conduction Electrons

In Fig. 17.6 the spectral function of the conduction electrons from the BZ center is presented as computed in the framework of the cumulant expansion [20]. The resulting three-peak spectrum with a main QP peak and two plasmon satellites agrees well

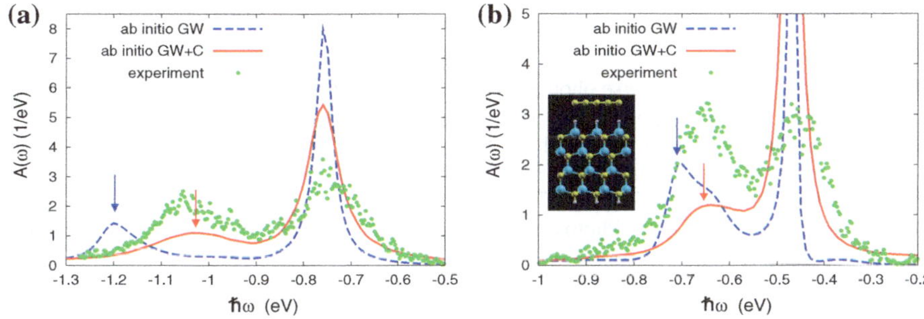

Fig. 17.11 Spectral function of graphene on SiC at the Dirac point calculated within the *ab initio* GW plus cumulant theory as well as the standard GW approach and compared for two carrier densities (**a**) $n_e = 5.9 \times 10^{13}$ cm^{-2} and (**b**) $n_e = 1.5 \times 10^{13}$ cm^{-2}. The *arrows* indicate the positions of the plasmaron and plasmon satellites in the theoretical curves. The *inset* of (**b**) shows the geometry of graphene on a hydrogen-terminated 4H-SiC(0001) surface: carbon atoms are *green*, silicon atoms are *blue* and hydrogens are *white*. Reprinted with permission from [35]. Copyright 2013 by the American Physical Society

with a spectrum [20] extracted from experimental data [36]. That means, the cumulant treatment also works well for conduction bands of three-dimensional metals.

Recently, the spectra of conduction electrons in 2D systems have been studied theoretically and experimentally. Thereby, the plasmaron has been "rediscovered", for instance, for doped graphene. The Green function calculations have been carried out on the Dirac Hamiltonian of graphene with two linear bands within the GW approach [37, 38]. The calculations found strong low-energy plasmaronic structures. The ARPES measurements of doped graphene on a silicon carbide (SiC) substrate [39, 40] do not exhibit simple Dirac cones but indicate a more complex electronic structure. A lucid scaling law as a function of doping is found for the satellites. It has been interpreted in terms of the theoretical predictions as plasmaron excitations caused by strong electron-carrier-plasmon coupling [39, 40]. However, this interpretation is wrong. Also in doped graphene plasmarons do not exist.

This conclusion is clearly demonstrated in Fig. 17.11 which presents measured spectra for two free-carrier concentrations in comparison with results of the standard GW treatment and the *ab initio* GW plus cumulant theory [35]. Standard GW yields plasmarons which accordingly fail with respect to the measured satellite positions. On the contrary, the combination of GW plus cumulant expansion moves the satellite structures at positions in reasonable agreement with experiment. Also the variation with the free-carrier density is in agreement with experiment. These facts clearly demonstrate the non-existence of plasmarons and the importance of an advanced treatment of electron correlation for the description of satellites in photoemission spectra.

References

1. J.B. Pendry, Theory of inverse photoemission. J. Phys. C **14**, 1381–1391 (1981)
2. G.D. Mahan, Theory of photoemission in simple metals. Phys. Rev. B **2**, 4334–4350 (1970)
3. C.-O. Almbladh, L. Hedin, Beyond the one-electron model: many-body effects in atoms, molecules, and solids, in *Handbook of Synchrotron Radiation*, vol. 1, ed. by E.E. Koch (North Holland, Amsterdam, 1983), pp. 607–904
4. C.N. Berglund, W.E. Spicer, Photoemission studies of copper and silver. Theor. Phys. Rev. **136**, A1030–A1044 (1964)
5. F. Bechstedt, K. Tenelsen, B. Adolph, R. del Sole, Compensation of dynamical quasiparticle and vertex corrections in optical spectra. Phys. Rev. Lett. **78**, 1528–1531 (1997)
6. L. Hedin, J. Michiels, J. Inglesfield, Transition from the adiabatic to the sudden limit in core-electron photoemission. Phys. Rev. B **58**, 15565–15582 (1998)
7. A. Damascelli, Z. Hussain, Z.-X. Shen, Angle-resolved photoemission studies of the cuprate superconductors. Rev. Mod. Phys. **75**, 473–541 (2003)
8. A. Einstein, Über einen die Erzeugung und Verwandlung des Lichtes betreffenden heuristischen Gesichtspunkt. Ann. Phys. **322**, 132–148 (1905)
9. R. Graupner, M. Hollering, A. Ziegler, J. Ristein, L. Ley, A. Stampfl, Dispersions of surface states on diamond (100) and (111). Phys. Rev. B **55**, 10841–10847 (1997)
10. H. Carstensen, R. Claessen, R. Manzke, M. Skibowski, Direct determination of III–V semiconductor surface band gaps. Phys. Rev. B **41**, 9880–9885 (1990)
11. U. Höfer, High resolution photoelectron spectroscopy at surfaces: lineshapes, loss processes and applications. Ph.D. thesis. Technical University Munich (1989)
12. P. Steiner, H. Höchst, S. Hüfner, XPS investigation of simple metals. Z. Phys. B **30**, 129–143 (1978)
13. D.C. Langreth, Singularities in the X-ray spectra of metals. Phys. Rev. B **1**, 471–477 (1970)
14. P. Minnhagen, Exact numerical solutions of a Nozieres-de Dominicis-type model problem. Phys. Lett. A **56**, 327–329 (1976)
15. E.P. Gross, Transformation theory, in *Mathematical Methods in Solid State and Superfluid Theory*, ed. by R.C. Clark, G.H. Derrick (Oliver and Boyd Inc, Edinburgh, 1969), pp. 46–120
16. B.I. Lundqvist, Single-particle spectrum of the degenerate electron gas. I. The structure of the spectral weight function. Phys. Kondens. Mater. **6**, 193–205 (1967)
17. B.I. Lundqvist, Single-particle spectrum of the degenerate electron gas. II. Numerical results for electrons coupled to plasmons. Phys. Kondens. Mater. **6**, 206–217 (1967)
18. B.I. Lundqvist, Single-particle spectrum of the degenerate electron gas. III. Numerical results in the random phase approximation. Phys. Kondens. Mater. **7**, 117–123 (1968)
19. B.I. Lundqvist, Characteristic structure in core electron spectra of metals due to the electron-plasmon coupling. Phys. Kondens. Mater. **9**, 236–248 (1969)
20. F. Aryasetiawan, L. Hedin, K. Karlsson, Multiple plasmon satellites in Na and Al spectral functions from *ab initio* cumulant expansion. Phys. Rev. Lett. **77**, 2268–2271 (1996)
21. B. Holm, U. von Barth, Fully self-consistent GW self-energy of the electron gas. Phys. Rev. B **57**, 2108–2117 (1998)
22. C. Blomberg, B. Bergersen, Spurious structure from approximations to the Dyson equation. Can. J. Phys. **50**, 2286–2293 (1972)
23. B. Bergersen, F.W. Kus, C. Blomberg, Single-particle Green's function in the electron-plasmon approximation. Can. J. Phys. **51**, 102–110 (1973)
24. F. Bechstedt, M. Fiedler, C. Kress, R. Del Sole, Dynamical screening and quasiparticle spectral functions for nonmetals. Phys. Rev. B **49**, 7357–7362 (1994)
25. D.C. Langreth, X-ray emission and absorption, in *Interaction of Radiation with Condensed Matter*, vol. 1 (International Atomic Energy Agency, Vienna, 1977), pp. 295–318
26. F. Bechstedt, Electronic relaxation effects in core level spectra of solids. Phys. Status Solidi B **112**, 9–49 (1982)
27. R. Kubo, Generalized cumulant expansion method. J. Phys. Soc. Jpn. **17**, 1100–1120 (1962)

28. L. Hedin, Effects of recoil on shake-up spectra in metals. Phys. Scr. **21**, 477–480 (1980)
29. G.D. Mahan, Excitons in metals: infinite hole mass. Phys. Rev. **163**, 612–617 (1967)
30. P. Noziéres, C.T. De Dominicis, Singularities in the X-ray absorption and emission of metals. III. One-body theory exact solution. Phys. Rev. **178**, 1097–1107 (1969)
31. C.J. Vesely, D.L. Kinston, D.W. Langer, X-ray photoemission studies of silicon and germanium. Phys. Status Solidi B **59**, 121–132 (1973)
32. M. Guzzo, G. Lani, F. Sottile, P. Romaniello, M. Gatti, J.J. Kas, J.J. Rehr, M.G. Silly, F. Sirotti, L. Reining, Valence electron photoemission spectrum of semiconductors: *ab initio* description of multiple satellites. Phys. Rev. Lett. **107**, 166401 (2011)
33. M. Guzzo, Dynamical correlation in solids: a perspective in photoelectron spectroscopy. Ph.D. thesis. Ecole Polytechnique, Palaiseau (2012)
34. M. Guzzo, J.J. Kas, F. Sottile, M.G. Silly, F. Sirotti, J.J. Rehr, L. Reining, Plasmon satellites in valence-band photoemission spectroscopy. Eur. Phys. J. B **85**, 324–330 (2012)
35. J. Lischner, D. Vigil-Fowler, S.G. Louie, Physical origin of satellites in photoemission of doped graphene: an *ab initio* GW plus cumulant study. Phys. Rev. Lett. **110**, 146801 (2013)
36. P. Steiner, H. Höchst, S. Hüfner, Simple metals, in *Photoemission in Solids II*, ed. by L. Ley, M. Cardona. Topics of Applied Physics, vol. 27 (Springer, Heidelberg, 1979), pp. 349–372
37. M. Polini, R. Asgari, G. Borghi, Y. Barlas, T. Pereg-Barnea, A.H. MacDonald, Plasmons and the spectral function of graphene. Phys. Rev. B **77**, 081411 (R) (2008)
38. E.H. Hwang, S. Das Sarma, Quasiparticle spectral function in doped graphene: electron-electron interaction effects in ARPES. Phys. Rev. B **77**, 081412 (R) (2008)
39. E. Rotenberg, A. Bostwick, T. Ohta, J.L. McChesney, T. Seyller, K. Horn, Origin of the energy bandgap in epitaxial graphene. Nat. Mater. **7**, 258–259 (2008)
40. A. Bostwick, F. Speck, T. Seyller, K. Horn, M. Polini, R. Asgar, A.H. MacDonald, E. Rotenberg, Observation of plasmarons in quasi-freestanding doped graphene. Science **328**, 999–1002 (2010)

Part IV
Pair and Collective Excitations

Chapter 18
Bethe-Salpeter Equations for Response Functions

Abstract Based on the Hedin fundamental equations Bethe-Salpeter equations are derived for generalized four-point functions, the polarization function P and the density correlation function L. Their integral kernels are characterized by the effective interaction between two particles or even modified by the Hartree response. Their inhomogeneities are given in random phase and independent-quasiparticle approximation, respectively. The kernels are further approximated within the GW approximation by the screened Coulomb potential W. It leads to a summation over all ladder diagrams. The application of P to frequency-dependent optical properties is described. Only the spin-averaged function is needed. The inclusion of optical local-field effects yields a Bethe-Salpeter equation for the macroscopic polarization function P^M with an integral kernel that contains a short-range Coulomb interaction. The corresponding two-point quantity of P^M determines the macroscopic dielectric function.

18.1 Characteristic Integral Equations

18.1.1 General Four-Point Forms

With the definition of the polarization function P (12.54), the irreducible (or proper) part of the two-particle Green function, the equation for the vertex function (12.55) in the set of Hedin equations yields the BSE [1]

$$P_{s_1s_1',s_2s_2'}(11',22') = L^0_{s_1s_1',s_2s_2'}(11',22') \tag{18.1}$$
$$+ \sum_{s_3,s_4,s_5,s_6} \int d3 \int d4 \int d5 \int d6 L^0_{s_1s_1',s_3s_4}(11',34) \Xi_{s_3s_4,s_5s_6}(34,56) P_{s_5s_6,s_2s_2'}(56,22')$$

F. Bechstedt, *Many-Body Approach to Electronic Excitations*,
Springer Series in Solid-State Sciences 181, DOI 10.1007/978-3-662-44593-8_18

with the kernel

$$\Xi_{s_3s_4,s_5s_6}(34,56) = -\frac{1}{i\hbar}\frac{\delta\Sigma_{s_4s_3}(43)}{\delta G_{s_5s_6}(56)}, \tag{18.2}$$

where the limit of vanishing test potential $\phi \to 0$ can be readily taken in the variational derivative in agreement with the conditions (12.18). The kernel essentially represents the effective interaction of excited electrons and holes modified by the corresponding polarization effects. In contrast to the particle-hole interaction, the particle-particle interaction is of minor interest here. Nevertheless, it is also described by (18.1).

The definition of the density correlation function (12.24) can be rewritten by means of relation (11.44) and the chain rule of the variational derivative as

$$L_{s_1s_1',s_2s_2'}(12) = i\hbar\sum_{s_3,s_4}\int d3\int d4 G_{s_1s_3}(13)\frac{\delta G^{-1}_{s_3s_4}(34|\phi)}{\delta\phi_{s_2's_2}(2)}\bigg|_{\phi=0} G_{s_4s_1'}(41^+).$$

It is sometimes also called two-particle correlation function [1]. The variational derivative of the inverse Green function can be reformulated by means of (11.55) and (12.12). The derivative $\delta\tilde{\Sigma}/\delta\phi$ is replaced by $\delta\tilde{\Sigma}/\delta G$ and the definition (12.24). Together with the four-point generalization we obtain

$$\begin{aligned} L_{s_1s_1',s_2s_2'}(11',22') &= L^0_{s_1s_1',s_2s_2'}(11',22') \\ &+ \sum_{s_3,s_4,s_5,s_6}\int d3\int d4\int d5\int d6 L^0_{s_1s_1',s_3s_4}(11',34)\tilde{\Xi}_{s_3s_4,s_5s_6}(34,56)L_{s_5s_6,s_2s_2'}(56,22') \end{aligned} \tag{18.3}$$

with the kernel

$$\begin{aligned} \tilde{\Xi}_{s_3s_4,s_5s_6}(34,56) &= -\frac{1}{i\hbar}\frac{\delta\tilde{\Sigma}_{s_4s_3}(43)}{\delta G_{s_5s_6}(56)} \\ &= \Xi_{s_3s_4,s_5s_6}(34,56) - \frac{1}{i\hbar}\frac{\delta\Sigma^{\mathrm{H}}_{s_4s_3}(43)}{\delta G_{s_5s_6}(56)}. \end{aligned} \tag{18.4}$$

The integral equation (18.3) is known as the Bethe-Salpeter equation for L (see e.g. [2]). The kernel (18.4) is determined by the total single-particle self-energy $\tilde{\Sigma} = \Sigma^{\mathrm{H}} + \Sigma$ (11.46) that includes, besides exchange and correlation, also the Hartree self-energy (11.47). It contains the full effective two-particle interaction, for instance the electron-hole interaction, including screened and unscreened contributions. The two BSEs are graphically illustrated in Fig. 18.1. In principle, the two BSEs can be used to study an arbitrary two-particle problem in the inhomogeneous electron gas, where the interaction is basically mediated by the longitudinal Coulomb coupling.

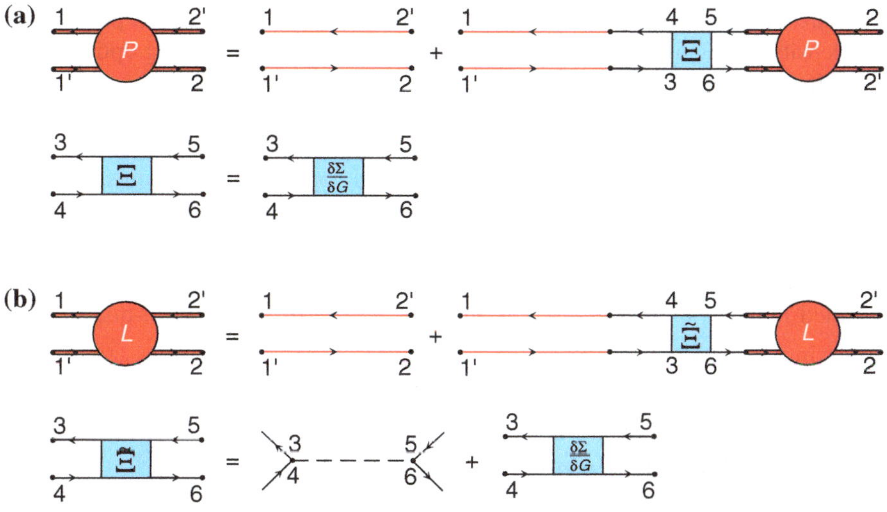

Fig. 18.1 Graphical representation of Bethe-Salpeter equations (18.1) and (18.3) for (**a**) the polarization function P and (**b**) the density correlation function L with the kernels (18.2) and (18.4), respectively. The particle propagation directions have been chosen in accordance with the electron-hole problem studied in the next chapters. Adapted from [1]

18.1.2 Random Phase Approximation

In Sect. 12.3.2 we have characterized the random phase approximation or independent-particle approximation by $\Sigma \equiv 0$. Then, only the variation of the Hartree self-energy (11.47) remains. With

$$\delta \Sigma^{\mathrm{H}}_{s_1 s_2}(12)/\delta G_{s_3 s_4}(34) = -i\hbar \delta_{s_1 s_2}\delta_{s_3 s_4}\delta(1-2)\delta(3^+-4)\mathrm{v}(1-3)$$

the two kernels become

$$\begin{aligned} \Xi_{s_3 s_4, s_5 s_6}(34,56) &\equiv 0, \\ \tilde{\Xi}_{s_3 s_4, s_5 s_6}(34,56) &= \delta_{s_3 s_4}\delta_{s_5 s_6}\delta(3-4)\delta(5^+-6)\mathrm{v}(3-5). \end{aligned} \tag{18.5}$$

Consequently, the BSE (18.1) for the polarization function takes the simplified form (12.58) but, actually with G in Hartree approximation. In this case the RPA can be identified with the Hartree approximation. The BSE for the density correlation function reads

$$\begin{aligned} L_{s_1 s_1', s_2 s_2'}(11',22') &= L^0_{s_1 s_1', s_2 s_2'}(11',22') \\ &+ \sum_{s_3,s_4} \int d3 \int d4 L^0_{s_1 s_1', s_3 s_3}(11',33)\mathrm{v}(3-4) L_{s_4 s_4, s_2 s_2'}(44^+,22'). \end{aligned} \tag{18.6}$$

The bare Coulomb interaction in (18.6) defines the difference between polarization and density correlation function. We come back to this fact discussing optical local-field effects in Sect. 18.3.1.

In the literature, the term RPA is often generalized to the use of $P = L^0 = -i\hbar GG$ with an approximate treatment of XC in the Dyson equation for G, e.g. by $\tilde{G}$. The GW community follows this strategy and computes the screened potential W (12.57) by replacing P by L^0 with single-particle Green functions G including self-energy corrections (14.45) but containing only one undamped pole at the QP energy, i.e., $\tilde{\tilde{G}}$. The application of L^0 instead of P to describe response properties is called independent-particle approximation with $\tilde{G}$ or independent-quasiparticle approximation (IQPA) with $\tilde{\tilde{G}}$.

18.1.3 GW Approximation

The RPA naturally fails, when the density of particles is low [3]. The next step in the hierarchy of approximations for the XC self-energy and the single-particle Green function is the Hedin GW approximation (12.57) for Σ, as illustrated by the magic pentagon in Fig. 12.3. Here, we must point out that, in principle, the screened potential W has to be replaced by that calculated within the RPA. That means in the integral equation (12.49) that P in the kernel has to be replaced by L_0 in the independent-particle approximation $L^0 = -i\hbar GG$ (see discussion in Sect. 12.3.2) with G taken as $\tilde{G}$ of the starting electronic structure for the GW approach. However, within self-consistent schemes of the GW approximation also improved Green functions are applicable. The kernel Ξ (18.2) in the BSE (18.1) for the polarization function P becomes

$$\begin{aligned}\Xi_{s_1's_1,s_2s_2'}(1'1,22') = &-\delta_{s_1s_2}\delta_{s_1's_2'}\delta(1-2)\delta(1'-2')W(1^+1')\\ &-G_{s_1s_1'}(11')\frac{\delta W(1^+1')}{\delta G_{s_2s_2'}(22')}.\end{aligned} \tag{18.7}$$

Contributions to the self-energy and hence the kernel (18.7) that are of higher order in the screened interaction may be obtained, for instance, in a T-matrix approach [4]. The second term in (18.7) is of second order in the screened potential W [5–7]

$$\frac{\delta W(1^+1')}{\delta G_{s_2s_2'}(22')} = -i\hbar\left[W(1^+2)W(2'1') + W(1^+2')W(21')\right]G_{s_2's_2}(2'2). \tag{18.8}$$

One obtains the term related to the variation of the screened potential, using the inversion W^{-1} of W and the chain rule. Using the Dyson equation (12.53) for W within the GW approximation, $P = L^0$, one finds

$$W^{-1}(11') = \mathrm{v}^{-1}(1-1') + i\hbar \sum_{s_1,s'_1} G_{s_1 s_{1'}}(11') G_{s'_1 s_1}(1'1).$$

The variational derivative yields

$$\frac{\delta W^{-1}(11')}{\delta G_{s_2 s'_2}(22')} = i\hbar \sum_{s_1,s'_1} \Big[\delta_{s_1 s_2} \delta_{s'_1 s'_2} \delta(1-2)\delta(1'-2') G_{s'_1 s_1}(1'1)$$
$$+ G_{s_1 s'_1}(11') \delta_{s_1 s'_2} \delta_{s'_1 s_2} \delta(1-2')\delta(1'-2) \Big]$$
$$= i\hbar \Big[\delta(1-2)\delta(1'-2') G_{s'_2 s_2}(1'1) + G_{s'_2 s_2}(11') \delta(1-2')\delta(1'-2) \Big].$$

Plugging this result back in the inversion relation $\delta W/\delta G = W(\delta W^{-1}/\delta G)W$, the validity of expression (18.8) is proven.

The second contribution to the kernel (18.7) of the BSE for P is too complex to take this term routinely into account in the calculations of P. One may argue that it is quadratic in W and therefore negligible compared to the first (linear) contribution. Test calculations of the optical spectrum of bulk silicon showed that the second contribution is indeed negligible [7]. Another argumentation is valid in the limit of vanishing screening, i.e., in systems with low electron density, e.g. molecules. In this limit the neglect of $\delta W/\delta G$ in (18.7) would become exact. For semiconductors, however, the screening of the electron-hole attraction is an essential feature, which cannot be neglected in calculating their optical spectra. Therefore, the role of the second term has to be investigated separately.

Neglecting the second term in (18.7) the kernel Ξ is simply determined by W. The kernel $\tilde{\Xi}$ (18.4), in addition, contains the bare Coulomb potential v as given in (18.5). In the linearized GW approximation the two BSEs (18.1) and (18.3) become

$$P_{s_1 s'_1, s_2 s'_2}(11', 22') \tag{18.9}$$
$$= L^0_{s_1 s'_1, s_2 s'_2}(11', 22') - \sum_{s_3,s_4} \int d3 \int d4 L^0_{s_1 s'_1, s_4 s_3}(11', 4^+3) W(34) P_{s_3 s_4, s_2 s'_2}(34^+, 22')$$

and

$$L_{s_1 s'_1, s_2 s'_2}(11', 22') = L^0_{s_1 s'_1, s_2 s'_2}(11', 22')$$
$$- \sum_{s_3,s_4} \int d3 \int d4 L^0_{s_1 s'_1, s_4 s_3}(11', 4^+3) W(34) L_{s_3 s_4, s_2 s'_2}(34^+, 22')$$
$$+ \sum_{s_3,s_4} \int d3 \int d4 L^0_{s_1 s'_1, s_3 s_3}(11', 33^+) \mathrm{v}(3-4) L_{s_4 s_4, s_2 s'_2}(44^+, 22'). \tag{18.10}$$

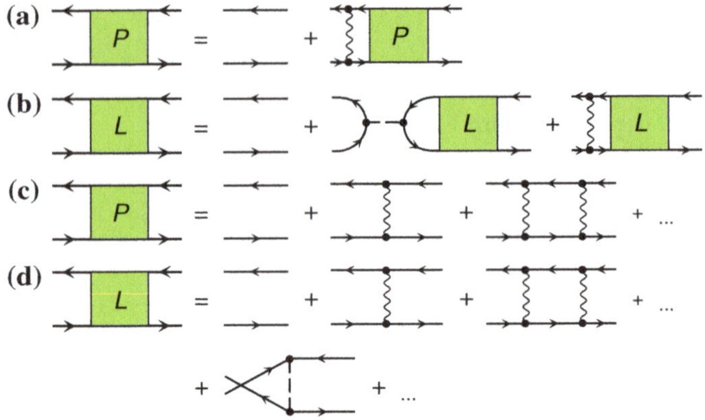

Fig. 18.2 Graphical representation of the Bethe-Salpeter equations (**a**) (18.9) for P and (**b**) (18.10) for L. The *solid lines* represent single-particle *Green* functions. They *wavy* and *dashed lines* describe the screened (W) and unscreened (v) Coulomb interaction, respectively. The iteration of the *right-hand sides* is illustrated in (**c**) and (**d**), in order make the denotation *ladder diagrams* (at least for the *wavy lines*) more obvious

These BSEs in GW approximation go back to Baym and Kadanoff [5]. However, in the early days of many-body theory the underlying GW approximation has been called *shielded interaction approximation*. The two equations are represented by Feynman graphs in Fig. 18.2. The denotation of the approximation as *screened ladder approximation* [4] is obvious for the diagrams including the screened Coulomb interaction W in Fig. 18.2. The rungs of the ladder are clearly visible. Here, we apply screened ladder diagrams as indicated by the wavy lines in Fig. 18.2, instead of unscreened ladder diagrams [8]. The exchange diagrams related to the bare interaction v exhibit a more complex but less clear structure.

18.2 Spin Structure

18.2.1 Singlet and Triplet States

The response functions $L_{s_1s_1',s_2s_2'}(11',22')$ (12.29) and $P_{s_1s_1',s_2s_2'}(11',22')$ (12.31) introduced in Sect. 12.2 are in general four-point functions depending on four sets of coordinates $\mathbf{x}, s, t$. Since they are ruled by two pairs of creation and annihilation operators (see 12.26), an interpretation is possible in terms of electron-hole pairs with a hole as a missing electron. Both particles carry a spin. In the case of collinear systems each of them can be described by the orthonormalized and complete set of single-particle spin functions $\chi_{\frac{1}{2}m_s}(s)$ (4.7). Then, in the spin space of electron and hole a possible basis can be constructed with the basis vectors $\chi_{\frac{1}{2}m_s}(s)\chi^+_{\frac{1}{2}m'_s}(s')$, where the hole spinor is Hermitian conjugated to the electron one.

In the spin space of electron-hole pairs another basis, that of pair states $\zeta_\alpha(ss')$ [9], can be defined. They also form an orthonormalized and complete set with

$$\sum_{s_1,s'_1} \zeta^+_\alpha(s_1s'_1)\zeta_\beta(s_1s'_1) = \delta_{\alpha\beta},$$
$$\sum_\alpha \zeta_\alpha(s_1s'_1)\zeta^+_\alpha(s_2s'_2) = \delta_{s_1s_2}\delta_{s'_1s'_2}. \tag{18.11}$$

The Greek letters $\alpha, \beta = 1, 2, 3, 4$ label the four independent pair states

$$\begin{aligned}
\zeta_1(ss') &= \frac{1}{\sqrt{2}}\left[\chi_{\frac{1}{2}\frac{1}{2}}(s)\chi^+_{\frac{1}{2}\frac{1}{2}}(s') + \chi_{\frac{1}{2}-\frac{1}{2}}(s)\chi^+_{\frac{1}{2}-\frac{1}{2}}(s')\right] && (S=0, M_S=0),\\
\zeta_2(ss') &= \chi_{\frac{1}{2}\frac{1}{2}}(s)\chi^+_{\frac{1}{2}-\frac{1}{2}}(s') && (S=1, M_S=1),\\
\zeta_3(ss') &= \frac{1}{\sqrt{2}}\left[\chi_{\frac{1}{2}\frac{1}{2}}(s)\chi^+_{\frac{1}{2}\frac{1}{2}}(s') - \chi_{\frac{1}{2}-\frac{1}{2}}(s)\chi^+_{\frac{1}{2}-\frac{1}{2}}(s')\right] && (S=1, M_S=0),\\
\zeta_4(ss') &= \chi_{\frac{1}{2}-\frac{1}{2}}(s)\chi^+_{\frac{1}{2}\frac{1}{2}}(s') && (S=1, M_S=-1).
\end{aligned} \tag{18.12}$$

They are eigenstates of the total spin operator of the electron-hole pair with quantum numbers $S = 0, 1$ (quantum number of total spin) and $-S \leq M_S \leq S$ (quantum number of the z-component). The first combination $\alpha = 1$ in (18.12) is symmetric if the two single-particle spin coordinates s and s' are interchanged. It therefore represents a singlet state. The antisymmetric combinations $\alpha = 2, 3, 4$ form a triplet. In the corresponding two-electron states [10] the hole spin in (18.12) has to be replaced by the spin of the other electron.

If the pair states (18.12) are projected onto single-particle states according to

$$\eta_\alpha(m_sm'_s) = \sum_{s,s'} \chi^+_{\frac{1}{2}m_s}(s)\chi_{\frac{1}{2}m'_s}(s')\zeta_\alpha(ss'), \tag{18.13}$$

one finds

$$\begin{aligned}
\eta_1(m_sm'_s) &= \frac{1}{\sqrt{2}}\left(\delta_{\frac{1}{2}m_s}\delta_{\frac{1}{2}m'_s} + \delta_{-\frac{1}{2}m_s}\delta_{-\frac{1}{2}m'_s}\right) = \frac{1}{\sqrt{2}}\begin{pmatrix}1&0\\0&1\end{pmatrix},\\
\eta_2(m_sm'_s) &= \delta_{\frac{1}{2}m_s}\delta_{-\frac{1}{2}m'_s} = \begin{pmatrix}0&1\\0&0\end{pmatrix},\\
\eta_3(m_sm'_s) &= \frac{1}{\sqrt{2}}\left(\delta_{\frac{1}{2}m_s}\delta_{\frac{1}{2}m'_s} - \delta_{-\frac{1}{2}m_s}\delta_{-\frac{1}{2}m'_s}\right) = \frac{1}{\sqrt{2}}\begin{pmatrix}1&0\\0&-1\end{pmatrix},\\
\eta_4(m_sm'_s) &= \delta_{-\frac{1}{2}m_s}\delta_{\frac{1}{2}m'_s} = \begin{pmatrix}0&0\\1&0\end{pmatrix}.
\end{aligned} \tag{18.14}$$

18.2.2 Transformation in Spin Space

For systems without spin polarization the singlet and triplet spin functions of the electron-hole pairs (18.12) are eigenfunctions of the system. In electron gases with spin polarization, regardless of collinear or non-collinear spins, this is not the case. Without spin polarization a representation of the response functions L and P in the four basis states (18.12) leads to significant simplifications [11]. Here, we apply the corresponding transformation to the density correlation function

$$L_{\alpha\beta}(11', 22') = \sum_{s_1,s_1',s_2,s_2'} \zeta_\alpha^+(s_1 s_1')L_{s_1 s_1', s_2 s_2'}(11', 22')\zeta_\beta(s_2' s_2) \qquad (18.15)$$

or to the polarization function of collinear systems with spin polarization. The correct arrangement of the single-particle spin coordinates is obvious taking the definition (12.26) for L into account.

For the purpose of illustration we study first the polarization function $L^0_{s_1 s_1', s_2 s_2'}(11', 22')$ (12.58) of independent quasiparticles. Since this quantity rules the inhomogeneities of the BSEs, its spin-dependent properties determine the spin dependence of the BSEs. The spin dependence of the QP Green function follows that for the spectral function (12.60)

$$G_{ss'}(11') = \sum_{m_s} G_{m_s m_s}(11')\chi_{\frac{1}{2}m_s}(s)\chi^+_{\frac{1}{2}m_s}(s') \qquad (18.16)$$

in the collinear case, in which the matrix elements $G_{m_s m_s}(11')$ of the Green function with the spin functions are diagonal in the quantum number of the z-component of the spin operator. With the definition of L^0 (12.35) and this result, (18.15) can be written as

$$L^0_{\alpha\beta}(11', 22') = -i\hbar \sum_{m_s, m_s'} \eta_\alpha^+(m_s m_s')G_{m_s m_s}(12')G_{m_s' m_s'}(21')\eta_\beta(m_s m_s').$$

With (18.14) the 4×4 matrix in the pair-spin projectors α and β decomposes into a 2×2 matrix for $M_S = 0$ $(\alpha, \beta = 1, 3)$

$$L^0_{\alpha\beta}(11', 22') = -i\hbar\frac{1}{2}\left[G_{\frac{1}{2}\frac{1}{2}}(12')G_{\frac{1}{2}\frac{1}{2}}(21') + (2\delta_{\alpha\beta} - 1)G_{-\frac{1}{2}-\frac{1}{2}}(12')G_{-\frac{1}{2}-\frac{1}{2}}(21')\right]$$

and another 2×2 matrix for $M_S = \pm 1$ $(\alpha, \beta = 2, 4)$

$$L^0_{\alpha\beta}(11', 22') = -i\hbar\delta_{\alpha\beta}\left[\delta_{\alpha 2}G_{\frac{1}{2}\frac{1}{2}}(12')G_{-\frac{1}{2}-\frac{1}{2}}(21') + \delta_{\alpha 4}G_{-\frac{1}{2}-\frac{1}{2}}(12')G_{\frac{1}{2}\frac{1}{2}}(21')\right].$$

In contrast to the non-spin-polarized case [11] the $L^0_{\alpha\beta}$ function is not anymore diagonal, $L^0_{\alpha\beta} = \delta_{\alpha\beta} L^0$, since orbital and spin space are coupled in systems with collinear spin polarization. As a consequence the spatial orbitals of the quasiparticles depend parametrically on the spin quantum number m_s [see (12.69)].

18.2.3 Response Functions in Singlet and Triplet Basis States

In order to transform the Bethe-Salpeter equations (18.9) and (18.10) in ladder approximation to a representation in pair spins, we apply the Dirac formalism to the density correlation function [12]

$$
\begin{aligned}
L_{\alpha\beta}(11', 22') &= \langle\alpha|L(11', 22')|\beta\rangle \\
&= \sum_{s_1, s_1', s_2, s_2'} \langle\alpha|s_1 s_1'\rangle\langle s_1 s_1'|L(11', 22')|s_2' s_2\rangle\langle s_2' s_s|\beta\rangle
\end{aligned}
$$

with (18.14)

$$
\langle ss'|\alpha\rangle \equiv \eta_\alpha(ss').
$$

Equation (18.10) becomes

$$
\begin{aligned}
&\langle\alpha|L(11', 22'|\beta\rangle = \langle\alpha|L^0(11', 22')|\beta\rangle \\
&-\sum_{s_3, s_4}\sum_{\gamma,\kappa}\int d3\int d4 \langle\alpha|L^0(11', 4^+3)|\gamma\rangle\langle\gamma|s_3 s_4\rangle W(34)\langle s_3 s_4|\kappa\rangle\langle\kappa|L(34^+, 22')|\beta\rangle \\
&+\sum_{s_3, s_4}\sum_{\gamma,\kappa}\int d3\int d4 \langle\alpha|L^0(11', 33^+|\gamma\rangle\langle\gamma|s_3 s_3\rangle \mathrm{v}(3-4)\langle s_4 s_4|\kappa\rangle\langle\kappa|L(44^+, 22')|\beta\rangle \\
&= L^0_{\alpha\beta}(11', 22') - \sum_\gamma \int d3\int d4 L^0_{\alpha\gamma}(11', 4^+3) W(34) L_{\gamma\beta}(34^+, 22') \\
&+\sum_{\gamma,\kappa}\int d3\int d4 L^0_{\alpha\gamma}(11', 33^+)\sum_{s_3}\langle\gamma|s_3 s_3\rangle \mathrm{v}(3-4)\sum_{s_4}\langle s_4 s_4|\kappa\rangle L_{\kappa\beta}(44^+, 22').
\end{aligned}
$$

Since only the singlet state in (4.3) possesses a non-vanishing trace, it holds $\langle\gamma|s_3 s_3\rangle = \sqrt{2}\delta_{\gamma 1}$. Finally, one obtains

$$
\begin{aligned}
L_{\alpha\beta}(11', 22') = L^0_{\alpha\beta}(11', 22') &- \sum_\gamma \int d3\int d4 L^0_{\alpha\gamma}(11', 4^+3) W(34) L_{\gamma\beta}(34^+, 22') \\
&+2\int d3\int d4 L^0_{\alpha 1}(11', 33^+) \mathrm{v}(3-4) L_{1\beta}(44^+, 22'). \qquad (18.17)
\end{aligned}
$$

Since a homogeneous integral equation has always the trivial zero solution, we assume that only the pair-spin channels with a finite inhomogeneity $L^0_{\alpha\beta}(11', 22') \neq 0$ lead to solutions $L_{\alpha\beta}(11', 22') \neq 0$. Then, off-diagonal elements only appear due to coupling between $S = 0$, $M_S = 0$ and $S = 1$, $M_S = 0$ states:

$$\begin{aligned} L_{\alpha\beta} &= L^0_{\alpha\beta} - \sum_{\gamma=1,3} L^0_{\alpha\gamma} W L_{\gamma\beta} + 2L^0_{\alpha 1} \mathrm{v} L_{1\alpha} \qquad (\alpha, \beta = 1, 3), \\ L_{\alpha\alpha} &= L^0_{\alpha\alpha} - L^0_{\alpha\alpha} W L_{\alpha\alpha} \qquad (\alpha = 2, 4). \end{aligned} \tag{18.18}$$

In a system without spin polarization all off-diagonal elements vanish. It holds [11]

$$\begin{aligned} L_{\alpha\alpha} &= L^0 - L^0 W L_{\alpha\alpha} + 2L^0 \mathrm{v} L_{\alpha\alpha} \quad \text{for} \quad \alpha = 1 \text{ (singlet)}, \\ L_{\alpha\alpha} &= L^0 - L^0 W L_{\alpha\alpha} \quad \text{for} \quad \alpha = 2, 3, 4 \text{ (triplet)} \end{aligned}$$

with $L^0 = L^0_{\alpha\alpha}$ $(\alpha = 1, 2, 3, 4)$.

For the polarization function P the same (18.17) and (18.18) are valid but with vanishing bare Coulomb potential v. Consequently the relations $P_{11} = P_{33}$ and $P_{13} = P_{31}$ occur, in contrast to the L case, where the bare Coulomb potential, i.e., the electron-hole exchange, lifts this degeneracy. In the non-spin-polarized case there are only two independent diagonal matrix elements of L and P, one, $L_s = L_{11}$ or $P_s = P_{11}$, for singlet states, the other one, $L_t = L_{\alpha\alpha}$ or $P_t = P_{\alpha\alpha}$ $(\alpha = 2, 3, 4)$ for triplet states. Interestingly, in the framework of the GW approximation one obtains

$$\begin{aligned} L_{11} &= L^0 [1 - (W - 2\mathrm{v}) L_{11}] \neq P_{11} = L^0 [1 - W P_{11}] \quad (\alpha = 1), \\ L_{\alpha\alpha} &= L^0 (1 - W L_{\alpha\alpha}) = P_{\alpha\alpha} = L^0 [1 - W P_{\alpha\alpha}] \quad (\alpha = 2, 3, 4). \end{aligned}$$

For singlet excitations the two two-particle functions are different, whereas for triplet excitations polarization and density correlation functions are equal. In addition, it holds $P_s = P_t$.

The results (18.17), (18.18), and the equations below indicate that a transformation of the BSEs for P and L into the singlet-triplet basis is, in general, only advantageous for non-spin-polarized systems. Therefore, we will not make use of this transformation in the following sections. We still work in the product basis of the spin functions of electron and hole.

18.3 Macroscopic Dielectric Function

18.3.1 Relation to Microscopic Dielectric Function

In Sect. 12.2.3 we have seen that the response functions $\epsilon^{-1}(12)$ (12.44) and $\epsilon(12)$ (12.46), which determine the longitudinal and transverse response of a spin-polarized electron gas, are spin-summed quantities. We focus on the dielectric function

$$\epsilon(12) = \delta(1-2) - \sum_{s_2,s_3} \int d3\text{v}(1-3) P_{s_3 s_3, s_2 s_2}(33^+, 22^+). \tag{18.19}$$

Only the two-point polarization function, which is diagonal in pairs of single particle coordinates, is needed to characterize the optical and dielectric properties.

The single-particle spin basis used in (18.19) can be transformed into the singlet-triplet basis according to (18.12),

$$\begin{aligned}\epsilon(12) &= \delta(1-2) - \sum_{s_2,s_3} \sum_{\alpha,\beta} \int d3\text{v}(1-3) \langle s_3 s_3 | \alpha \rangle \langle \alpha | P(33^+, 22^+) | \beta \rangle \langle \beta | s_2 s_2 \rangle \\ &= \delta(1-2) - \sum_{s_2,s_3} \sum_{\alpha,\beta} \int d3\text{v}(1-3) \sqrt{2} \delta_{\alpha 1} P_{\alpha\beta}(33^+, 22^+) \sqrt{2} \delta_{\beta 1} \\ &= \delta(1-2) - 2 \int d3\text{v}(1-3) P_{11}(33^+, 22^+).\end{aligned}$$

That means, only optical transitions from ground state to neutral excited states with electron-hole pairs in a singlet state contribute to optical properties, even in spin-polarized systems though the singlet pairs couple to triplet excitations. Singlet and triplet do not constitute well-defined quantum states anymore. The same holds for the screening. The corresponding transformation yields $\epsilon^{-1} = 1 + 2\text{v}L_{11}$. Only the singlet matrix element L_{11} of the density correlation function contributes.

As a consequence only the spin-averaged polarization function (13.1) rules the dielectric function. All experimental studies of the electronic polarization and the resulting dielectric function are performed in momentum and frequency space. We do this here assuming a certain translational symmetry (13.33), i.e., studying a crystal or an artificial crystal due to a periodic arrangement of some nanoobjects. Together with the time Fourier transformation (13.15) and its generalization to arbitrary frequencies in the complex z-plane, one obtains (13.37)

$$\epsilon(\mathbf{q}+\mathbf{G}, \mathbf{q}+\mathbf{G}', z) = \delta_{\mathbf{GG}'} - 2\tilde{\text{v}}(|\mathbf{q}+\mathbf{G}|) P(\mathbf{q}+\mathbf{G}, \mathbf{q}+\mathbf{G}', z) \tag{18.20}$$

with the spatial Fourier transformations (13.14) and the time-Fourier transformation (13.15) with an analytic continuation $\tilde{z}_m \to z$ into the entire z-plane. The comparison of (18.20) with the above relation to the singlet function shows that the spin-averaged polarization function P and the singlet matrix element of the polarization function P_{11} should be identical in the case of collinear spins.

From the point of view of spatial variations the dielectric function (18.20) is a *microscopic* function. It relates microscopic external and total electric fields to each other [see e.g. relation (12.21) describing these fields by scalar potentials]. These fields exhibit large and irregular variations on the atomic scale in real crystals or on the length scale of the lattice constants in other translationally invariant systems.

Macroscopic electric fields are usually spatially averaged microscopic ones. They do not contain higher Fourier components with $\mathbf{G} \neq 0$. The replacements of fields with Fourier components at $\mathbf{q}+\mathbf{G}$ by fields, which only contain the zero Fourier component $\mathbf{q}$ (with $\mathbf{G} = 0$), occurs, in general, when one moves from the microscopic electrodynamics to macroscopic electrodynamics (see e.g. [13]). The averaging procedure has the result of smoothing out the irregular spatial fluctuations of the microscopic quantities [14]. The differences between microscopic and macroscopic fields are related to local-field effects [1, 15]. They render the response to an electromagnetic field of a crystal different from that of a homogeneous electron gas with the same average density.

Microscopic fields are not the quantities which are dealt with in ordinary electrodynamics, where one usually studies fields that vary on a macroscopic scale and are thus experimentally accessible. To describe such experiments, for instance, optical absorption or reflection, one needs the macroscopic dielectric function $\epsilon^M(\mathbf{q}, z)$ with $\mathbf{q} \to 0$, instead of the microscopic matrix quantity $\epsilon(\mathbf{q}+\mathbf{G}, \mathbf{q}+\mathbf{G}', z)$. The result is in agreement with the limit of vanishing photon wave vectors in the case of optical measurements or, at least, transferred wave vectors $\mathbf{q}$ (in inelastic scattering experiments) which are small compared to the extent of the Brillouin zone.

The relation between the microscopic dielectric tensor ϵ (a second-rank tensor because the elements are labeled by $\mathbf{G}$ and $\mathbf{G}'$) and the macroscopic dielectric function ϵ^M has been derived by several authors [16–18] to be

$$\epsilon^M(\mathbf{q}, z) = \frac{1}{\epsilon^{-1}(\mathbf{q}+\mathbf{G}, \mathbf{q}+\mathbf{G}', z)|_{\mathbf{G}=\mathbf{G}'=0}} \tag{18.21}$$

in the limit of vanishing wave vectors $\mathbf{q}$. An idea to prove this relation can be given starting from the definition (12.22) of the inverse dielectric function. When a crystal or another translationally invariant system is perturbed by an external potential with Fourier components $\phi(\mathbf{q}+\mathbf{G}', \omega)$ at wave vectors $\mathbf{q}+\mathbf{G}'$ and frequency ω, the total (effective) potential seen by a test charge is given, to the first order, by

$$\phi^{\text{eff}}(\mathbf{q}+\mathbf{G}, \omega) = \sum_{\mathbf{G}'} \epsilon^{-1}(\mathbf{q}+\mathbf{G}, \mathbf{q}+\mathbf{G}', \omega)\phi(\mathbf{q}+\mathbf{G}', \omega).$$

The inverse dielectric tensor links the Fourier components of the external perturbation at the reciprocal lattice vectors $\mathbf{G}'$ to the Fourier component of the induced potential at $\mathbf{G}$. We search for a similar relation between Fourier components of the corresponding spatially averaged, i.e., macroscopic, fields, $\langle\phi(\mathbf{q}+\mathbf{G}', \omega)\rangle_{\text{space}}$ and $\langle\phi^{\text{eff}}(\mathbf{q}+\mathbf{G})\rangle_{\text{space}}$, which only contain the $\mathbf{G} = \mathbf{G}' = 0$ components. Assuming that the external perturbation is macroscopic, one has to deal with $\phi(\mathbf{q}+\mathbf{G}', \omega) = \delta_{\mathbf{G}'0}\phi(\mathbf{q}, \omega)$. The spatial average of the total potential leads to $\langle\phi^{\text{eff}}(\mathbf{q}+\mathbf{G}, \omega)\rangle_{\text{space}} = \delta_{\mathbf{G}0}\langle\phi^{\text{eff}}(\mathbf{q}, \omega)\rangle_{\text{space}}$. As a consequence the above equation turns into the form

$$\langle \phi^{\text{eff}}(\mathbf{q}, \omega) \rangle_{\text{space}} = \epsilon^{-1}(\mathbf{q} + \mathbf{G}, \mathbf{q} + \mathbf{G}', \omega) \Big|_{\mathbf{G}=\mathbf{G}'=0} \langle \phi(\mathbf{q}, \omega) \rangle_{\text{space}}.$$

Since $\langle \phi(\mathbf{q}, \omega) \rangle_{\text{space}}$ is related to the dielectric displacement field $\mathbf{D}(\mathbf{q}, \omega)$ of the macroscopic electrodynamics, while $\langle \phi^{\text{eff}}(\mathbf{q}, \omega) \rangle_{\text{space}}$ characterizes the internal electric field $\mathbf{E}(\mathbf{q}, \omega)$ with the relation $\mathbf{D}(\mathbf{q}, \omega) = \epsilon^M(\mathbf{q}, \omega)\mathbf{E}(\mathbf{q}, \omega)$ [13], the definition of the macroscopic dielectric function (18.21) is obtained. It measures the macroscopic response to a macroscopic perturbation. The difference $\epsilon^M(\mathbf{q}, \omega) - \epsilon(\mathbf{q}, \mathbf{q}, \omega)$ between the macroscopic dielectric function and the head element of the microscopic dielectric tensor is due to local-field effects.

Generalizations of the response of the electron gas beyond the density response and scalar potentials require the inclusion of the current density response. This fact asks for the introduction of density-current, current-density, and current-current correlation functions [1] and a subsequent distinction between dielectric and pseudo-dielectric tensors or conductivity and quasi-conductivity tensors [11]. However, these complications will not be discussed here. Moreover, one may argue that the transformation (20.3) and (20.4) between optical transition matrix elements obtained in a longitudinal or transverse gauge corresponds to relation between polarization function and the irreducible part of the current-current correlation function.

18.3.2 Elementary Excitations and Their Measurement

In the optical limit one has to study vanishing wave vectors $\mathbf{q} \to 0$,

$$\epsilon^M(\hat{\mathbf{q}}, z) = \lim_{\mathbf{q}\to 0} \epsilon^M(\mathbf{q}, z), \tag{18.22}$$

with the direction vector $\hat{\mathbf{q}} = \mathbf{q}/|\mathbf{q}|$. Relation (18.22) is valid for crystals of arbitrary space-group symmetry. It may be rewritten in terms of a dielectric tensor with elements $\epsilon_{ij}^M(z)$ of a second-rank tensor according to

$$\epsilon^M(\hat{\mathbf{q}}, z) = \hat{\mathbf{q}}\hat{\epsilon}^M(z)\hat{\mathbf{q}} = \sum_{i,j=1,2,3} \hat{q}_i \epsilon_{ij}^M(z) \hat{q}_j. \tag{18.23}$$

The number of independent components of the second-rank tensor depends on the crystal symmetry. In the limit of vanishing photon wave vectors spatial dispersion [19] and, therefore, polariton effects and the wave vector-induced birefringence are neglected. Usually this limit is possible in the case of optical frequencies, since the corresponding variations of $\epsilon^M(\mathbf{q}, z)$ with $\mathbf{q}$ are rather small, of the order of 10^{-5} in semiconductors.

The derivation of the macroscopic dielectric function via the density response, and not directly taking the current density response into account [11], suggests that relation (18.23) to the second-rank tensor $\hat{\epsilon}^M(z)$ defines the longitudinal dielectric function. However, due to its tensor behavior, at least in the limit of vanishing spatial dispersion $\mathbf{q} \to 0$, also a transverse dielectric function can be defined. It is described by

$$\mathbf{e}\hat{\epsilon}^M(z)\mathbf{e}$$

with a unit vector $\mathbf{e} = \mathbf{e}(\mathbf{q}) \perp \mathbf{q}$. In the optical limit this vector can be interpreted as the vector of light polarization perpendicular to the propagation direction $\hat{\mathbf{q}}$. For cubic systems the second-rank tensor is diagonal with equal elements. Relation (18.22) takes a particularly simple form, $\epsilon^M(\hat{\mathbf{q}}, z) \equiv \epsilon^M(z)$. For uniaxial or even biaxial crystals but especially for crystals with lower symmetry, e.g. in the monoclinic case, the full tensor $\hat{\epsilon}^M(z)$ has to be investigated. In the first cases, one or two optical axes may exist. When plane electromagnetic waves propagate along these axes, they have the same velocity independent of their polarization direction. Along other crystallographic directions, the light velocity varies with the polarization, describing phenomena such as the optical birefringence.

A complete discussion of elementary excitations related to $\epsilon^M(\hat{\mathbf{q}}, z)$ requires an extended discussion including a differentiation between longitudinal and transverse electromagnetic perturbations as well as a study of the current density response [11]. We illustrate the relation to elementary excitations roughly for the cubic case and real frequencies ω shifted by a vanishing imaginary part (see discussion of analytic properties in Sect. 13.2.3). From the homogeneous wave equation one derives the dispersion relation of the transverse excitations, e.g. excitons or exciton polaritons, in such a system to be [20, 21]

$$\mathbf{q}^2 = \frac{\omega^2}{c^2}\epsilon^M(\omega), \tag{18.24}$$

with a wave vector perpendicular to the corresponding electric field in the medium. The frequencies of longitudinal excitations, e.g. plasmons, with an accompanying electric field parallel to the wave vector $\mathbf{q}$ are given by

$$\epsilon^M(\omega) = 0. \tag{18.25}$$

In the transverse case, one usually writes the propagation vector $\mathbf{q}$ in terms of the index of refraction, $n(\omega)$, and of the extinction coefficient, $\kappa(\omega)$,

$$\mathbf{q} = \hat{\mathbf{q}}\frac{\omega}{c}[n(\omega) + i\kappa(\omega)]$$

with

$$\begin{aligned}
\sqrt{\epsilon^M(\omega)} &= n(\omega) + i\kappa(\omega), \\
\mathrm{Re}\epsilon^M(\omega) &= n^2(\omega) - \kappa^2(\omega), \\
\mathrm{Im}\epsilon^M(\omega) &= 2n(\omega)\kappa(\omega)
\end{aligned} \tag{18.26}$$

using relation (18.23). In a medium characterized by $\epsilon^M(\omega)$ an electromagnetic wave propagates with frequency ω and wave number $\mathrm{Re}q = \frac{\omega}{c}n(\omega)$, while its amplitude is weakened by a factor $\exp\left[-\frac{\omega}{c}\kappa(\omega)D\right]$ after propagating a distance D. The power absorbed by the medium is given by the absorption coefficient

$$\alpha(\omega) = 2\frac{\omega}{c}\kappa(\omega) = \frac{\omega}{cn(\omega)}\mathrm{Im}\epsilon^M(\omega) = 2\frac{\omega}{c}\mathrm{Im}\sqrt{\epsilon^M(\omega)}. \tag{18.27}$$

For normal incidence the reflectivity, more precisely the reflectance, of a semi-infinite medium $R(\omega)$ can be characterized by the Fresnel expression of the reflection coefficient

$$\begin{aligned} R(\omega) &= \frac{[n(\omega)-1]^2+\kappa^2(\omega)}{[n(\omega)+1]^2+\kappa^2(\omega)}, \\ &= \left|\frac{\sqrt{\epsilon^M(\omega)}-1}{\sqrt{\epsilon^M(\omega)}+1}\right|^2. \end{aligned} \tag{18.28}$$

The impact of the local-field effects can be illustrated by comparing the optical absorption or, more precisely, the imaginary part of the macroscopic dielectric function $\mathrm{Im}\epsilon^M(\omega)$ to the head element $\mathrm{Im}\epsilon(0, 0, \omega)$ of the microscopic dielectric tensor. The frequency-dependent absorption is displayed for silicon in Fig. 18.3 [22]. The calculations have been performed within a tight-binding scheme applying Gaussian basis functions. The local-field effects tend to a small blue shift and a minor reduction of the spectral strength (not shown) of the most important van Hove singularities E_1 and E_2 [20] but do not significantly modify the lineshape of the absorption spectrum. However, the electron-hole attraction brings the spectrum closer to the measured spectrum [23] but cannot fully bridge the discrepancies between theory and experiment within the used scheme of numerical approximations. The modifications due to local fields are are somewhat more pronounced in more sophisticated computations [24].

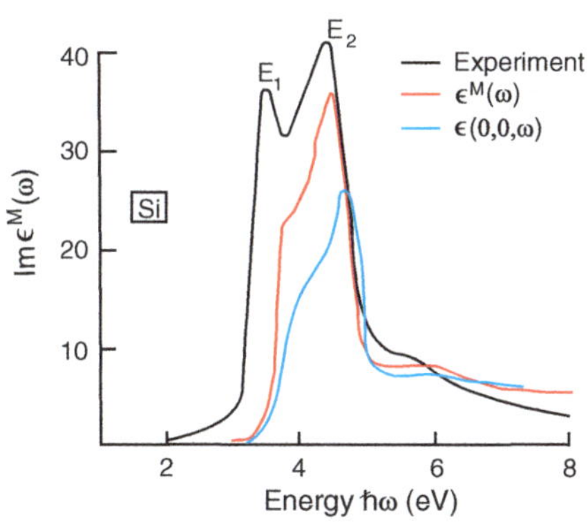

Fig. 18.3 Imaginary part of the dielectric function of Si versus photon energy; $\epsilon(0, 0, \omega)$ describes results of a single-particle calculation within RPA with local-field corrections, while the calculation of $\epsilon^M(\omega)$ includes local-field corrections together with screened electron-hole attraction. From [22]

Another possibility to measure excitation energies is the loss function $L(\omega)$, for instance, probing the inelastic scattering of electrons with velocity v which penetrate the medium characterized by $\epsilon^M(\omega)$. Studying the dissipation rate [11] one finds in the limit of vanishing momentum transfer $\hbar q \to 0$ in the inelastic scattering

$$L(\omega) = \mathrm{Im}\left\{-\frac{1}{\epsilon^M(\omega)}\right\} + (q^2v^2 - \omega^2)\mathrm{Im}\left\{-\frac{1}{\omega^2\epsilon^M(\omega) - c^2q^2}\right\}.$$

The energy loss of the fast electrons is determined by longitudinal elementary excitations in the first contribution and by transverse excitations in the second term. The second term is due to retardation effects. It only significantly contributes for large electron velocities

$$v > c/\sqrt{\mathrm{Re}\epsilon^{\mathrm{M}}(\omega)}.$$

Then, losses are due to the occurrence of Cherenkov radiation [25]. Under common experimental conditions this term does not play a role, and

$$L(\omega) = \mathrm{Im}\left\{-\frac{1}{\epsilon^M(\omega)}\right\} \tag{18.29}$$

describes the loss spectrum for vanishing momentum transfer with poles determined by (18.25) [21]. Such a spectrum may be measurable by means of electron energy loss spectroscopy (EELS) in reflection or transmission as well as inelastic X-ray scattering (IXS).

18.3.3 Macroscopic Polarization Function

In Sect. 18.3.1 we have shown that the transition from the microscopic dielectric tensor (18.20) to the macroscopic dielectric function (18.21) contains a twofold inversion. In any case it requires a computation of the complete microscopic dielectric tensor $\epsilon(\mathbf{q}+\mathbf{G}, \mathbf{q}+\mathbf{G}', z)$ in the limit $\mathbf{q} \to 0$ with its head $\epsilon(\mathbf{q}, \mathbf{q}, z)$, wing $\epsilon(\mathbf{G}, \mathbf{q}, z)$ and $\epsilon(\mathbf{q}, \mathbf{G}', z)$, and body $\epsilon(\mathbf{G}, \mathbf{G}', z)$ matrix elements. The numerical effort to perform the calculations including excitonic effects ruled by Ξ (18.7) for all matrix elements is enormous, in particular, for converged computations with respect to the reciprocal lattice vectors $\mathbf{G}$ and $\mathbf{G}'$. To avoid such heavy computations we formulate the macroscopic dielectric function

$$\epsilon^M(\hat{\mathbf{q}}, z) = 1 - \lim_{\mathbf{q}\to 0} 2\tilde{\mathrm{v}}(|\mathbf{q}|)P^M(\mathbf{q}, \mathbf{q}, z) \tag{18.30}$$

in a similar expression as in (18.20) by introducing a polarization function $P^M(\mathbf{q}, \mathbf{q}, z)$ of the macroscopic response, i.e., including the local-field effects. Thereby, the $\mathbf{q} \to 0$ limit has to be taken with care to keep the correct analytical properties of the macroscopic function and the underlying inverse dielectric matrix [26]. A word of caution is necessary. The factor 2 in (18.30) is a consequence of the fact that in (13.1) spin-averaged instead of spin-summed response functions have been introduced to define the dielectric function (13.32).

To find a BSE for P^M we follow the double inversion procedure of Pick, Cohen, and Martin [26] rewriting (18.21) as

$$\epsilon^M(\hat{\mathbf{q}}, z) = \lim_{\mathbf{q}\to 0}\left\{\epsilon(\mathbf{q}, \mathbf{q}, z) - \sum_{\mathbf{G},\mathbf{G}'(\neq 0)} \epsilon(\mathbf{q}, \mathbf{q}+\mathbf{G}, z) S^{-1}(\mathbf{q}+\mathbf{G}, \mathbf{q}+\mathbf{G}', z)\epsilon(\mathbf{q}+\mathbf{G}', \mathbf{q}, z)\right\},$$

where S^{-1} is the inverse of the lower-right submatrix of $\epsilon(\mathbf{q}+\mathbf{G}, \mathbf{q}+\mathbf{G}', z)$ corresponding to the non-zero reciprocal lattice vectors $\mathbf{G} \neq 0$ and $\mathbf{G}' \neq 0$, the so-called "body" of the dielectric matrix in the limit $\mathbf{q} \to 0$ [27]. Together with the "wing" elements $\epsilon(\mathbf{q}, \mathbf{q}+\mathbf{G}', z)$ and $\epsilon(\mathbf{q}+\mathbf{G}', \mathbf{q}, z)$ it defines the local-field effects $\epsilon^M(\mathbf{q}, z) - \epsilon(\mathbf{q}, \mathbf{q}, z)$. The submatrix S is also defined by (18.20). Its use for the head and wing elements and the comparison with expression (18.30) yields

$$P^M(\mathbf{q}, \mathbf{q}, z) = P(\mathbf{q}, \mathbf{q}, z) + 2\sum_{\mathbf{G},\mathbf{G}'(\neq 0)} P(\mathbf{q}, \mathbf{q}+\mathbf{G}, z) S^{-1}(\mathbf{q}+\mathbf{G}, \mathbf{q}+\mathbf{G}', z)\tilde{\mathrm{v}}(|\mathbf{q}+\mathbf{G}'|) P(\mathbf{q}+\mathbf{G}', \mathbf{q}, z). \quad (18.31)$$

As an intermediate step we introduce a reduced microscopic dielectric function $(\mathbf{G} \neq 0, \mathbf{G}' \neq 0)$

$$\bar{\epsilon}(\mathbf{q}+\mathbf{G}, \mathbf{q}+\mathbf{G}', z) = \delta_{\mathbf{G}\mathbf{G}'} - 2\bar{\tilde{\mathrm{v}}}(|\mathbf{q}+\mathbf{G}|) P(\mathbf{q}+\mathbf{G}, \mathbf{q}+\mathbf{G}', z)$$

with a truncated, short-range Coulomb potential

$$\bar{\tilde{\mathrm{v}}}(|\mathbf{q}+\mathbf{G}|) = \begin{cases} 0 & \text{for } \mathbf{G} = 0 \\ \tilde{\mathrm{v}}(|\mathbf{q}+\mathbf{G}|) & \text{for } \mathbf{G} \neq 0 \end{cases}. \quad (18.32)$$

The result for P^M (18.31) is generalized to a full matrix in $\mathbf{G}$ and $\mathbf{G}'$,

$$\begin{aligned} P^M(\mathbf{q}+\mathbf{G}, \mathbf{q}+\mathbf{G}', z) &= P(\mathbf{q}+\mathbf{G}, \mathbf{q}+\mathbf{G}', z) \\ &+2\sum_{\mathbf{G}'',\mathbf{G}'''} P(\mathbf{q}+\mathbf{G}, \mathbf{q}+\mathbf{G}'', z)\bar{\epsilon}^{-1}(\mathbf{q}+\mathbf{G}'', \mathbf{q}+\mathbf{G}''', z)\bar{\tilde{\mathrm{v}}}(|\mathbf{q}+\mathbf{G}'''|) \\ &\times P(\mathbf{q}+\mathbf{G}''', \mathbf{q}+\mathbf{G}', z). \end{aligned}$$

With the replacement of S by $\bar{\epsilon}$ and $\tilde{\mathrm{v}}$ by $\bar{\tilde{\mathrm{v}}}$ a generalization of the right-hand side also to $\mathbf{G} = 0$ and $\mathbf{G}' = 0$ is possible.

Applying the matrix denotation it holds

$$\hat{P}^M = \hat{P} + 2\hat{P}\hat{\hat{\epsilon}}^{-1}\tilde{\bar{v}}\hat{P}.$$

Due to the introduction of the truncated Coulomb potential also the zero wave vector $\mathbf{G}'' = 0$ or $\mathbf{G}''' = 0$ appears in the summation over all $\mathbf{G}''$ and $\mathbf{G}'''$. The dielectric tensor $\hat{\hat{\epsilon}}$ is related to the microscopic polarization function. With $\hat{\hat{\epsilon}} = 1 - 2\tilde{\bar{v}}\hat{P}$ a Dyson-like equation

$$(\hat{P}^M)^{-1} = \hat{P}^{-1} - 2\tilde{\bar{v}} \tag{18.33}$$

or

$$\hat{P}^M = \hat{P} + \hat{P}2\tilde{\bar{v}}\hat{P}^M$$

is derived for the two different polarization functions. The latter integral equation can be unified with the BSE (18.9) for the spin-singlet matrix elements by

$$\hat{P} = \hat{L}^0 - \hat{L}^0 W \hat{P}$$

or

$$\hat{P}^{-1} = \hat{L}^{0^{-1}}[1 + \hat{L}^0 W]. \tag{18.34}$$

Combining (18.33) with (18.34) and inverting the result, one finds

$$\hat{P}^M = \hat{L}^0 - \hat{L}^0(W - 2\tilde{\bar{v}})\hat{P}^M.$$

Rewriting the spin summations and the Fourier transformations instead of (18.1) or (18.9), a modified BSE results. For the generalization to the macroscopic polarization function with the complete spin dependence, i.e., before spin summation, it holds a BSE for the modified polarization function

$$\begin{aligned} P^M_{s_1s_1',s_2s_2'}(11',22') &= L^0_{s_1s_1',s_2s_2'}(11',22') \\ &+ \sum_{s_3,s_4,s_5,s_6} \int d3 \int d4 \int d5 \int d6 L^0_{s_1s_1',s_3s_4}(11',43)\Xi^M_{s_3s_4,s_5s_6}(34,56)P^M_{s_5s_6,s_2s_2'}(56,22') \end{aligned} \tag{18.35}$$

with the kernel

$$\begin{aligned} \Xi^M_{s_3s_4,s_5s_6}(34,56) &= -\frac{1}{i\hbar}\frac{\delta\Sigma_{s_4s_3}(43)}{\delta G_{s_5s_6}(56)} + \delta_{s_3s_4}\delta_{s_5s_6}\delta(3-4)\delta(5^+-6)\bar{v}(3-5) \\ &\approx -\delta_{s_3s_6}\delta_{s_4s_5}\delta(4-5)\delta(3-6)W(4^+,3) \\ &\quad +\delta_{s_3s_4}\delta_{s_5s_6}\delta(3-4)\delta(5^+-6)\bar{v}(3-5). \end{aligned} \tag{18.36}$$

The BSE (18.35) for the macroscopic polarization function P^M exhibits a great similarity with the BSE (18.10) for the (microscopic) density correlation function L due to the inclusion of optical local-field effects $\sim \bar{\mathrm{v}}$. The only difference is the substitution of the bare Coulomb potential v by the corresponding short-range potential $\bar{\mathrm{v}}$ (without the $\mathbf{G} = 0$ Fourier component) [28]. The appearing dynamically screened potentials W are however the same. Together with the kernel (18.36) the BSE (18.35) derived for P^M is very similar (apart from the difference between v and $\bar{\mathrm{v}}$) to the result of Sham and Rice [29]. Constructing the irreducible part of the effective particle-hole interaction they introduced a repulsive quasielectron-quasihole exchange interaction v, which therefore has to be unscreened, in addition to the screened quasielectron-quasihole attraction. Here we have derived the term $\sim \bar{\mathrm{v}}$ from optical local-field effects. Since exchange interaction is also short-range, the discrepancy between the two descriptions should be however vanishing small. Both interpretations of the contributions $\sim \bar{\mathrm{v}}$ as electron-hole exchange or local-field effects are reasonable.

References

1. G. Strinati, Applications of the Green's functions method to the study of the optical properties of semiconductors. Riv. Nuovo Cimento **11**, 1–86 (1988)
2. C. Csanak, H.S. Taylor, R. Yaris, Green's function technique in atomic and molecular physics. Adv. At. Mol. Phys. **7**, 287–361 (1971)
3. A.M. Zagoskin, *Quantum Theory of Many-Body Systems. Techniques and Applications* (Springer, New York, 1998)
4. H. Stolz, R. Zimmermann, Correlated pairs and mass action law in two-component Fermi systems. Excitons in an electron-hole plasma. Phys. Status Solidi B **94**, 135–146 (1979)
5. G. Baym, L.P. Kadanoff, Conservation laws and correlation functions. Phys. Rev. **124**, 287–299 (1961)
6. A. Schindlmayr, R.W. Godby, Systematic vertex corrections through iterative solution of Hedin's equations beyond the GW approximation. Phys. Rev. Lett. **80**, 1702–1705 (1998)
7. F. Bechstedt, C. Rödl, L.E. Ramos, F. Fuchs, P.H. Hahn, J. Furthmüller, Parameterfree calculations of optical properties for systems with magnetic ordering or three-dimensional confinement, in *Epioptics-9, Proceedings of 39th International School on Solid State Physics*, Erice (Italy), 2006. ed. by A. Cricenti (World Scientific Publishing Co., New Jersey, 2008), pp. 26–40
8. A.L. Fetter, J.D. Walecka, *Quantum Theory of Many-Particle Systems* (Dover Publications Inc, Mineola, 2003)
9. N.E. Bickers, D.J. Scalapino, Conserving approximations for strongly fluctuating electron systems. I. Formalism and calculational approach. Ann. Phys. **193**, 206–251 (1989)
10. C. Cohen-Tannoudji, B. Diu, F. Laloë, *Quantenmechanik*, vol. 2 (Walter de Gruyter, Berlin, 1999)
11. H. Stolz, *Einführung in die Vielelektronentheorie der Kristalle* (Akademie-Verlag, Berlin, 1974)
12. C. Rödl, Spinabhängige GW-approximation. Diploma thesis, Friedrich-Schiller-Universität Jena (2005)
13. J.D. Jackson, *Classical Electrodynamics* (Wiley, New York, 1962)
14. L. Rosenfeld, *Theory of Electrons* (North-Holland, Amsterdam, 1951)

15. R. Del Sole, E. Fiorino, Macroscopic dielectric tensor at crystal surfaces. Phys. Rev. B **29**, 4631–4645 (1984)
16. S.L. Adler, Quantum theory of the dielectric constant in real solids. Phys. Rev. **126**, 413–420 (1962)
17. N. Wiser, Dielectric constant with local field effects included. Phys. Rev. **129**, 62–69 (1963)
18. W. Hanke, Dielectric theory of elementary excitations in crystals. Adv. Phys. **27**, 287–341 (1978)
19. V.M. Agranovich, V. Ginzburg, *Crystal Optics with Spatial Dispersion*, Springer Ser. Solid State Sci. (Springer, Berlin, 1984)
20. P.Y. Yu, M. Cardona, *Fundamentals of Semiconductors* (Springer, Berlin, 1996)
21. Ch. Kittel, *Introduction to Solid State Physics* (Wiley, New York, 2005)
22. W. Hanke, L.J. Sham, Many-particle effects in the optical spectrum of a semiconductor. Phys. Rev. B **21**, 4656–4673 (1980)
23. H.R. Phillip, H. Ehrenreich, Optical properties of semiconductors. Phys. Rev. **129**, 1550–1560 (1963)
24. V.I. Gavrilenko, F. Bechstedt, Local-field and exchange-correlation effects in optical spectra of semiconductors. Phys. Rev. B **54**, 13416–13419 (1996)
25. P.A. Cherenkov, Visible emission of clean liquids by action of γ radiation. Dokl. Akad. Nauk **2**, 451–454 (1934) [English translation: Usp. Fiz. Nauk **93**, 385–388 (1967)]
26. R.M. Pick, M.H. Cohen, R.M. Martin, Microscopic theory of force constants in the adiabatic approximation. Phys. Rev. B **1**, 910–920 (1970)
27. M.S. Hybertsen, S.G. Louie, *Ab initio* static dielectric matrices from the density-functional approach. I. Formulation and application to semiconductors and insulators. Phys. Rev. B **35**, 5585–5601 (1987)
28. W. Hanke, L.J. Sham, Many-particle effects in the optical excitations of a semiconductor. Phys. Rev. Lett. **43**, 387–390 (1970)
29. L.J. Sham, T.M. Rice, Many-particle derivation of the effective-mass equation for the Wannier exciton. Phys. Rev. **144**, 708–714 (1966)

Chapter 19
Electron-Hole Problem

Abstract In order to specify the Bethe-Salpeter equation of the macroscopic polarization function P^M to account for the relevant excitonic effects in optical spectra, spin effects are treated within the collinear approximation. The quasiparticle wave functions are used to represent the space dependence of the polarization function. The resulting integral equation can be formally solved by means of the eigenvectors and eigenvalues of a generalized eigenvalue problem for pairs of particles. Together with the matrix elements of the optical transition operator calculated with the single-particle wave functions, the pair eigenvectors determine the optical oscillator strengths whereas the eigenvalues can be interpreted as the oscillator frequencies. The generalized eigenvalue problem is ruled by a matrix in the single-particle pair states, whose diagonal blocks represent the pairs and antipairs while the off-diagonal blocks describe their coupling. The reduction of the complexity within the Tamm-Dancoff approximation leads to a Hermitian electron-hole-pair problem with screened Coulomb attraction and unscreened electron-hole exchange. The resulting Hamiltonian represents a high-rank matrix which asks for efficient strategies to solve the eigenvalue problem or to compute directly the dielectric function.

19.1 Pair Hamiltonian

19.1.1 Static Screening

For the description of optical and energy loss spectra via the macroscopic dielectric function (18.22), one has to calculate the macroscopic polarization function P^M in (18.30) that depends only on one frequency z. This means that its Fourier transform in time domain should only depend on the difference of two time variables $t_1 - t_2$. However, replacing the time arguments in (18.35) by $t_1' = t_1^+$ and $t_2' = t_2^+$ no closed BSE for such a function $P^M(t_1 - t_2)$ can be derived. The time dependence of the kernel $\Xi^M \sim W(t_4 - t_3)$ [see (18.7) and (18.9)] allows the formulation of a closed BSE only for a polarization function P^M that depends on two time differences. After Fourier transformation one immediately sees that such a closed integral equation is only possible for a function P^M that depends on two frequencies. This can be easily

F. Bechstedt, *Many-Body Approach to Electronic Excitations*,
Springer Series in Solid-State Sciences 181, DOI 10.1007/978-3-662-44593-8_19

seen applying Fourier transformations of types (11.18) and (13.15),

$$P^M_{s_1s_1',s_2s_2'}(\mathbf{x}_1t_1\mathbf{x}_1't_1', \mathbf{x}_2t_2\mathbf{x}_2't_2^+) = \frac{1}{(-i\hbar\beta)^2}\sum_{n,m} e^{-iz_n(t_1-t_1')}e^{-i\tilde{z}_m(t_1-t_2)} \times P^M_{s_1s_1',s_2s_2'}(\mathbf{x}_1\mathbf{x}_1', \mathbf{x}_2\mathbf{x}_2'; z_n\tilde{z}_m). \tag{19.1}$$

The quantity relevant for the optical properties is however given by

$$P^M_{s_1s_1',s_2s_2'}(\mathbf{x}_1\mathbf{x}_1', \mathbf{x}_2\mathbf{x}_2'; \tilde{z}_m) = \frac{1}{-i\hbar\beta}\sum_n P^M_{s_1s_1',s_2s_2'}(\mathbf{x}_1\mathbf{x}_1', \mathbf{x}_2\mathbf{x}_2'; z_n\tilde{z}_m) \tag{19.2}$$

and its analytic continuation $\tilde{z}_m \to z$ into the entire complex z-plane, respectively.

Sometimes one finds the argumentation that only two of the four time variables appearing in general in P^M are independent (see e.g. [1]). Due to the time homogeneity in the absence of external fields, only the difference of these two time variables is finally relevant and allows to carry out a one-dimensional time-frequency Fourier transformation of P^M. This argumentation is however not true because of the dynamics of screening. Correlations between electrons, holes and the entire electron gas evolve during the pair excitation process. This build-up of correlations is nothing but rearrangement of the electrons in the long-range Coulomb field formed by the electron-hole pairs, i.e., the formation of the screening cloud around electron and hole. This process is not instantaneous. Rather it needs some time. In an electron gas the typical time scale for correlation build-up is just the period of the plasma oscillations $\sim 2\pi/\omega_p$. It is extremely short but still influences the formation of an electron-hole pair during its short formation time with an incomplete screening.

To avoid complications due to the small effects of incomplete screening, we neglect the time dependence in the screened potential and write

$$W(34) \approx W(\mathbf{x}_3, \mathbf{x}_4)\delta(t_3 - t_4). \tag{19.3}$$

As a consequence the dynamics of the screening reaction is neglected. We take into consideration that the screening description within this static approximation leads to a minor overestimation of the screening of the quasielectron-quasihole attraction. The direct consequences of the static approximation for other issues, such as the satellite structures in the spectra, will be discussed in Chap. 22. For the time being we follow the standard approach in the literature and neglect the dynamical screening in W.

The time structure of the resulting BSE is

$$P^M(t_1 - t_2) = L^0(t_1 - t_2) + \int dt_3 L^0(t_1 - t_3)\Xi^M P^M(t_3 - t_2)$$

due to the static kernel Ξ^M. In the time-Fourier domain the BSE (18.35) becomes [2]

$$P^M_{s_1s_1',s_2s_2'}(\mathbf{x}_1\mathbf{x}_1', \mathbf{x}_2\mathbf{x}_2'; \tilde{z}_m) = \frac{1}{\beta}\sum_n \Big\{G_{s_1s_2'}(\mathbf{x}_1\mathbf{x}_2', z_n)G_{s_2s_1'}(\mathbf{x}_2\mathbf{x}_1', z_n - \tilde{z}_m)$$
$$+ \sum_{s_3,s_4,s_3',s_4'} \int d^3\mathbf{x}_3 \int d^3\mathbf{x}_4 \int d^3\mathbf{x}_3' \int d^3\mathbf{x}_4' G_{s_1s_4}(\mathbf{x}_1\mathbf{x}_4, z_n)G_{s_3s_1'}(\mathbf{x}_3\mathbf{x}_1', z_n - \tilde{z}_m)$$
$$\times\; \Xi^M_{s_3s_4,s_3's_4'}(\mathbf{x}_3\mathbf{x}_4, \mathbf{x}_3'\mathbf{x}_4')P^M_{s_3's_4',s_2s_2'}(\mathbf{x}_3'\mathbf{x}_4', \mathbf{x}_2\mathbf{x}_2'; \tilde{z}_m)\Big\} \tag{19.4}$$

with a static quasielectron-quasihole interaction kernel Ξ^M.

19.1.2 Spin-Space Representation

For an explicit description of the space and spin dependence of the (macroscopic) polarization function in the case of collinear spin polarization, we apply the factorization (12.69) of the two-component Pauli spinors into

$$\varphi_{\lambda m_s}(\mathbf{x})\chi_{\frac{1}{2}m_s}(s),$$

where spatial orbitals and spinors form complete and orthonormalized basis sets. The orbitals $\varphi_{\lambda m_s}(\mathbf{x})$ are identified as the quasiparticle wave functions discussed in Sect. 14.2. In practice, we are using the wave functions of the KS/gKS/HF reference system, that diagonalize the reference Green function $\tilde{\tilde{G}}$ (14.41). In other words, the single-QP Green function G appearing in the BSE (19.4) is replaced by the reference Green function $\tilde{\tilde{G}}$ within a spin-space representation

$$\tilde{\tilde{G}}_{ss'}(\mathbf{x}\mathbf{x}', z) = \sum_{\lambda,\lambda'}\sum_{m_s,m_s'} \tilde{\tilde{G}}^{m_sm_s'}_{\lambda\lambda'}(z)\varphi_{\lambda m_s}(\mathbf{x})\varphi^*_{\lambda' m_s'}(\mathbf{x}')\chi_{\frac{1}{2}m_s}(s)\chi^+_{\frac{1}{2}m_s'}(s') \tag{19.5}$$

with matrix elements

$$\tilde{\tilde{G}}^{m_sm_s'}_{\lambda\lambda'}(z) = \frac{\delta_{\lambda\lambda'}\delta_{m_sm_s'}}{\hbar z - \varepsilon^{\mathrm{QP}}_{\lambda m_s}}. \tag{19.6}$$

These frequency-dependent functions contain one pole at the quasiparticle energy $\varepsilon^{\mathrm{QP}}_{\lambda m_s}$ that carries the full spectral weight. Satellite structures are neglected because their intensity vanishes in the spectrum of P^M in the limit of optical frequencies [3]. This effect will be discussed in Sect. 22.2. The index λ represents the set of single-particle quantum numbers related to the orbital motion, for example, $\lambda = \nu\mathbf{k}$ in translationally invariant systems.

The corresponding representation of the four-point polarization function is

$$
\begin{aligned}
&P^M_{s_1s_1',s_2s_2'}(\mathbf{x}_1\mathbf{x}_1',\mathbf{x}_2\mathbf{x}_2';\tilde{z}_m) \\
&= \sum_{\lambda_1,\lambda_1',\lambda_2,\lambda_2'}\ \sum_{m_{s_1},m_{s_1}',m_{s_2},m_{s_2}'} \varphi_{\lambda_1 m_{s_1}}(\mathbf{x}_1)\chi_{\frac{1}{2}m_{s_1}}(s_1)\varphi^*_{\lambda_1' m_{s_1'}}(\mathbf{x}_1')\chi^+_{\frac{1}{2}m_{s_1'}}(s_1') \\
&\quad\times \varphi_{\lambda_2 m_{s_2}}(\mathbf{x}_2)\chi_{\frac{1}{2}m_{s_2}}(s_2)\varphi^*_{\lambda_2' m_{s_2'}}(\mathbf{x}_2')\chi^+_{\frac{1}{2}m'_{s_2}}(s_2')P^M_{m_{s_1}m_{s_1'},m_{s_2}m_{s_2'}}(\lambda_1\lambda_1',\lambda_2\lambda_2';\tilde{z}_m).
\end{aligned}
\tag{19.7}
$$

The same representation holds for the inhomogeneity L^0 of the BSE, similar to (12.68). The expansion coefficients are

$$
L^0_{m_{s_1}m_s',m_{s_2}m_{s_2'}}(\lambda_1\lambda_1',\lambda_2\lambda_2';\tilde{z}_m) = \frac{f\left(\varepsilon^{\mathrm{QP}}_{\lambda_1 m_{s_1}}\right) - f\left(\varepsilon^{\mathrm{QP}}_{\lambda_1' m_{s_1'}}\right)}{\varepsilon^{\mathrm{QP}}_{\lambda_1 m_{s_1}} - \varepsilon^{\mathrm{QP}}_{\lambda_1' m_{s_1'}} - \hbar\tilde{z}_m}\delta_{m_{s_1}m_{s_2'}}\delta_{m_{s_1'}m_{s_2}}\delta_{\lambda_1\lambda_2'},\delta_{\lambda_1'\lambda_2}
\tag{19.8}
$$

in accordance with the result (12.70) for the spin-summed case and the diagonal matrix elements of the Green function (19.6). The corresponding representation of the static interaction kernel is

$$
\begin{aligned}
&\Xi^M_{m_4m_3,m_3'm_4'}(\lambda_4\lambda_3,\lambda_{3'}\lambda_{4'}) \\
&= \sum_{s_3,s_4,s_3',s_4'} \int d^3\mathbf{x}_3 \int d^3\mathbf{x}_4 \int d^3\mathbf{x}_3' \int d^3\mathbf{x}_4' \varphi^*_{\lambda_3 m_{s_3}}(\mathbf{x}_3)\chi^+_{\frac{1}{2}m_{s_3}}(s_3)\varphi_{\lambda_4 m_{s_4}}(\mathbf{x}_4)\chi_{\frac{1}{2}m_{s_4}}(s_4) \\
&\quad\times \Xi^M_{s_4s_3,s_3's_4'}(\mathbf{x}_4\mathbf{x}_3,\mathbf{x}_3'\mathbf{x}_4')\varphi_{\lambda_3' m_{s_3'}}(\mathbf{x}_3')\chi_{\frac{1}{2}m_{s_3'}}(s_3')\varphi^*_{\lambda_4' m_{s_4'}}(\mathbf{x}_4')\chi^+_{\frac{1}{2}m_{s_4'}}(s_4').
\end{aligned}
$$

With its explicit form (18.36) one finds

$$
\Xi^M_{m_{s_1'}m_{s_1},m_{s_3}m_{s_4}}(\lambda_1'\lambda_1,\lambda_3\lambda_4) = -\delta_{m_{s_1}m_{s_3}}\delta_{m_{s_1'}m_{s_4}} W^{m_{s_3}m_{s_4}}_{\substack{\lambda_1\lambda_3\\ \lambda_{1'}\lambda_4}} + \delta_{m_{s_1}m_{s_1'}}\delta_{m_{s_3}m_{s_4}} \bar{v}^{m_{s_1}m_{s_3}}_{\substack{\lambda_1\lambda_{1'}\\ \lambda_3\lambda_4}}
\tag{19.9}
$$

with the matrix elements (14.55) of W and the matrix elements of the short-range bare Coulomb potential

$$
\begin{aligned}
W^{m_{s_3}m_{s_4}}_{\substack{\lambda_1\lambda_3\\ \lambda_{1'}\lambda_4}} &= \int d^3\mathbf{x}_3 \int d^3\mathbf{x}_4 \varphi^*_{\lambda_1 m_{s_3}}(\mathbf{x}_3)\varphi_{\lambda_3 m_{s_3}}(\mathbf{x}_3)W(\mathbf{x}_3,\mathbf{x}_4)\varphi_{\lambda_1' m_{s_4}}(\mathbf{x}_4)\varphi^*_{\lambda_{\lambda_4} m_{s_4}}(\mathbf{x}_4), \\
\bar{v}^{m_{s_1}m_{s_3}}_{\substack{\lambda_1\lambda_1'\\ \lambda_3\lambda_4}} &= \int d^3\mathbf{x}_3 \int d^3\mathbf{x}_4 \varphi^*_{\lambda_1 m_{s_1}}(\mathbf{x}_3)\varphi_{\lambda_1' m_{s_1}}(\mathbf{x}_3)\bar{v}(\mathbf{x}_3-\mathbf{x}_4)\varphi_{\lambda_3 m_{s_3}}(\mathbf{x}_4)\varphi^*_{\lambda_4 m_{s_3}}(\mathbf{x}_4).
\end{aligned}
\tag{19.10}
$$

Because of $\bar{v}(\mathbf{x}-\mathbf{x}') = \bar{v}(\mathbf{x}'-\mathbf{x})$ and $W(\mathbf{x},\mathbf{x}') = W(\mathbf{x}',\mathbf{x})$ (12.42), the kernel is symmetric with the matrix elements

$$\Xi^{M*}_{m_{s'_1} m_{s_1}, m_{s_2} m_{s'_2}}(\lambda'_1\lambda_1, \lambda_2\lambda'_2) = \Xi^{M}_{m_{s'_2} m_{s_2}, m_{s_1} m_{s'_1}}(\lambda'_2\lambda_2, \lambda_1\lambda'_1)$$
$$= \Xi^{M}_{m_{s_1} m_{s'_1}, m_{s'_2} m_{s_2}}(\lambda_1\lambda'_1, \lambda'_2\lambda_2). \qquad (19.11)$$

In the representation of the orbitals $\varphi_{\lambda m_s}(\mathbf{x})$ the BSE (19.4) becomes with $\tilde{z}_m \to z$

$$P^{M}_{m_{s_1} m_{s'_1}, m_{s_2} m_{s'_2}}(\lambda_1\lambda'_1, \lambda_2\lambda'_2; z)$$
$$= \frac{f\left(\varepsilon^{\mathrm{QP}}_{\lambda_1 m_{s_1}}\right) - f\left(\varepsilon^{\mathrm{QP}}_{\lambda'_1 m_{s'_1}}\right)}{\varepsilon^{\mathrm{QP}}_{\lambda_1 m_{s_1}} - \varepsilon^{\mathrm{QP}}_{\lambda'_1 m_{s'_1}} - \hbar z} \Bigg\{ \delta_{m_{s_1} m_{s'_2}} \delta_{\lambda_1 \lambda'_2} \delta_{m_{s_2} m_{s'_1}} \delta_{\lambda_2 \lambda'_1} \qquad (19.12)$$
$$+ \sum_{\lambda_3, \lambda_4} \sum_{m_{s_3}, m_{s_4}} \Xi^{M}_{m_{s'_1} m_{s_1}, m_{s_3} m_{s_4}}(\lambda'_1\lambda_1, \lambda_3\lambda_4) P^{M}_{m_{s_3} m_{s_4}, m_{s_2} m_{s'_2}}(\lambda_3\lambda_4, \lambda_2\lambda_{2'}; z) \Bigg\}.$$

19.2 Two-Particle Problem

19.2.1 Effective Hamiltonian

In the next step we reformulate the BSE (19.12) [4, 5] using the abbreviation $\rho \equiv \lambda m_s$. A multiplication with the energy denominator of L^0 yields

$$\left[\varepsilon^{\mathrm{QP}}_{\rho_1} - \varepsilon^{\mathrm{QP}}_{\rho_{1'}} - \hbar z\right] P^M(\rho_1\rho_{1'}, \rho_2\rho_{2'}; z)$$
$$-\left[f\left(\varepsilon^{\mathrm{QP}}_{\rho_1}\right) - f\left(\varepsilon^{\mathrm{QP}}_{\rho_{1'}}\right)\right] \sum_{\rho_3, \rho_4} \Xi^M(\rho_{1'}\rho_1, \rho_3\rho_4) P^M(\rho_3\rho_4, \rho_2\rho_{2'}; z)$$
$$= \left[f\left(\varepsilon^{\mathrm{QP}}_{\rho_1}\right) - f\left(\varepsilon^{\mathrm{QP}}_{\rho_{1'}}\right)\right] \delta_{\rho_1\rho_{2'}} \delta_{\rho_2\rho_{1'}}. \qquad (19.13)$$

Since $\varepsilon^{\mathrm{QP}}_{\rho_1} - \varepsilon^{\mathrm{QP}}_{\rho'_1} - \hbar z \neq 0$ holds for complex frequencies z, one immediately learns that $P^M(\rho_1\rho_{1'}, \rho_2\rho_{2'}; z) \equiv 0$ for $f\left(\varepsilon^{\mathrm{QP}}_{\rho_1}\right) - f\left(\varepsilon^{\mathrm{QP}}_{\rho'_1}\right) = 0$. Therefore, we only have to study the case $f\left(\varepsilon^{\mathrm{QP}}_{\rho_1}\right) - f\left(\varepsilon^{\mathrm{QP}}_{\rho'_1}\right) \neq 0$. For the combination of a fully occupied and an empty state the occupation prefactor is ± 1. In this case the treatment of the inhomogeneous integral equation (19.13) is straightforward. In the case of partially filled bands ($\rho_1 \neq \rho_2$) with

$$F_{\rho_1\rho_2} = -\frac{f\left(\varepsilon^{\mathrm{QP}}_{\rho_1}\right) - f\left(\varepsilon^{\mathrm{QP}}_{\rho_2}\right)}{\left|f\left(\varepsilon^{\mathrm{QP}}_{\rho_1}\right) - f\left(\varepsilon^{\mathrm{QP}}_{\rho_2}\right)\right|} = \mathrm{sgn}\left(\varepsilon^{\mathrm{QP}}_{\rho_1} - \varepsilon^{\mathrm{QP}}_{\rho_2}\right)$$

and after symmetrization with respect to occupation differences

$$P^M(\rho_1\rho_2, \rho_3\rho_4; z) \tag{19.14}$$
$$= \left[\left|f\left(\varepsilon_{\rho_1}^{\mathrm{QP}}\right) - f\left(\varepsilon_{\rho_2}^{\mathrm{QP}}\right)\right|\left|f\left(\varepsilon_{\rho_4}^{\mathrm{QP}}\right) - f\left(\varepsilon_{\rho_3}^{\mathrm{QP}}\right)\right|\right]^{\frac{1}{2}} \mathcal{P}(\rho_1\rho_2, \rho_3\rho_4; z),$$

one finds

$$F_{\rho_1\rho_2}\sum_{\rho_5,\rho_6}\mathscr{H}(\rho_1\rho_2, \rho_5\rho_6)\mathcal{P}(\rho_5\rho_6, \rho_3\rho_4; z) = -F_{\rho_1\rho_2}\delta_{\rho_1\rho_4}\delta_{\rho_2\rho_3} + \hbar z\mathcal{P}(\rho_1\rho_2, \rho_3\rho_4; z) \tag{19.15}$$

with the second-rank tensor

$$\mathscr{H}(\rho_1\rho_2, \rho_3\rho_4) = \left|\varepsilon_{\rho_1}^{\mathrm{QP}} - \varepsilon_{\rho_2}^{\mathrm{QP}}\right|\delta_{\rho_1\rho_3}\delta_{\rho_2\rho_4} \tag{19.16}$$
$$-\left[\left|f\left(\varepsilon_{\rho_1}^{\mathrm{QP}}\right) - f\left(\varepsilon_{\rho_2}^{\mathrm{QP}}\right)\right|\right]^{\frac{1}{2}} \Xi^M(\rho_2\rho_1, \rho_3\rho_4)\left[\left|f\left(\varepsilon_{\rho_3}^{\mathrm{QP}}\right) - f\left(\varepsilon_{\rho_4}^{\mathrm{QP}}\right)\right|\right]^{\frac{1}{2}}.$$

The matrix $\mathscr{H}$ (19.16) in the product basis $|\rho_3\rho_4\rangle$ of two single-particle states with deviating occupation can be interpreted as an effective two-particle or pair Hamiltonian in this basis. As a consequence of the kernel symmetry (19.11) it is indeed Hermitian. We assume that the screened Coulomb interaction is weak enough that $\mathscr{H}$ is a positive definite operator with positive eigenvalues, the pair excitation energies. Otherwise, one has to discuss the instability of the electronic system against the formation of charge or spin density waves or the formation of an excitonic insulator [6, 7]. The positive definiteness is consistent with the assumption of the stability of the statistical operator (10.17). Because of the inclusion of state occupation by the Fermi functions $f\left(\varepsilon_{\rho}^{\mathrm{QP}}\right)$ in the pair Hamiltonian (19.16), the theory can be applied to two-particle and excitonic effects not only in semiconductors, insulators, and molecules but also to highly doped semiconductors, transparent conducting oxides, and metals.

With a generalization of the sign function $F_{\rho_1\rho_2}$ to a diagonal matrix in the product basis of two single-particle states

$$\mathcal{F}(\rho_1\rho_2, \rho_3\rho_4) = F_{\rho_1\rho_2}\delta_{\rho_1\rho_3}\delta_{\rho_2\rho_4}$$

the BSE (19.15) becomes a matrix equation

$$\mathcal{F}\mathscr{H}\mathcal{P}(z) = -\mathcal{F} + \hbar z\mathcal{P}(z). \tag{19.17}$$

The operator $\mathcal{F}\mathscr{H}$, which is a product of two non-commuting Hermitian matrices $\mathcal{F}$ and $\mathscr{H}$ [8], belongs to a class of non-Hermitian operators with a real spectrum, i.e., real eigenvalues.

19.2.2 Generalized Eigenvalue Problem

Following an idea of Sham and Rice [9] the solution of the matrix equation (19.17), which corresponds to the inhomogeneous BSE (19.12), is traced back to a generalized eigenvalue problem, i.e., a 'homogeneous Bethe-Salpeter equation',

$$\mathcal{F}\mathcal{H}|\Phi_\Lambda\rangle = \mathcal{E}_\Lambda|\Phi_\Lambda\rangle, \tag{19.18}$$

with an operator $\mathcal{F}\mathcal{H}$ which is not Hermitian, but possesses real eigenvalues $\mathcal{E}_\Lambda$ because of the positive definiteness of $\mathcal{H}$ and the Hermitian character of both $\mathcal{H}$ and $\mathcal{F}$ [10]. The set of eigenvectors $|\Phi_\Lambda\rangle$ is complete and fulfills the condition

$$\langle\Phi_\Lambda|\mathcal{F}\Phi_{\Lambda'}\rangle = \delta_{\Lambda\Lambda'}\mathrm{sgn}(\mathcal{E}_\Lambda). \tag{19.19}$$

The eigenvectors and eigenvalues of the generalized problem can be interpreted as states and energies of excited pairs of two particles in the interacting inhomogeneous electron gas.

For numerical calculations, instead of the generalized eigenvalue problem (19.18), we study the eigenvalue problem [4]

$$\mathcal{H}\mathcal{F}\mathcal{H}|\Phi_\Lambda\rangle = \mathcal{E}_\Lambda\mathcal{H}|\Phi_\Lambda\rangle \tag{19.20}$$

of the Hermitian operator $\mathcal{H}\mathcal{F}\mathcal{H}$ with the positively definite matrix $\mathcal{H}$. Because of the completeness of the eigenvectors $|\Phi_\Lambda\rangle$, with (19.19) the validity of

$$\langle\Phi_{\Lambda'}|\mathcal{H}|\Phi_\Lambda\rangle = |\mathcal{E}_\Lambda|\delta_{\Lambda\Lambda'} \tag{19.21}$$

is required. Then, the closure relation can be generalized to

$$1 = \sum_\Lambda |\Phi_\Lambda\rangle\frac{1}{|\mathcal{E}_\Lambda|}\langle\Phi_\Lambda|\mathcal{H}.$$

This allows for a spectral representation of the non-Hermitian operator $\mathcal{F}\mathcal{H}$ as

$$\mathcal{F}\mathcal{H} = \sum_\Lambda |\Phi_\Lambda\rangle\frac{\mathcal{E}_\Lambda}{|\mathcal{E}_\Lambda|}\langle\Phi_\Lambda|\mathcal{H}.$$

The solution of the eigenvalue problems of the operators $\mathcal{F}\mathcal{H}$ and $\mathcal{H}\mathcal{F}\mathcal{H}$ can be used to obtain a representation of the polarization function $\mathcal{P}(z)$ (19.4). The solution of the homogeneous problem allows to represent the solution of the inhomogeneous BSE. We obtain the spectral representation

$$\mathcal{P}(z) = -\sum_\Lambda \mathrm{sgn}(\mathcal{E}_\Lambda)\frac{|\Phi_\Lambda\rangle\langle\Phi_\Lambda|}{\mathcal{E}_\Lambda - \hbar z}.$$

An easy proof is possible by inserting this expression in the BSE (19.17) and using (19.18) and (19.19).

19.2.3 Macroscopic Functions

With the transformation (19.14) and the resubstitution $\rho = \lambda m_s$ the macroscopic polarization function in (19.7) reads

$$\begin{aligned} &P^M_{m_{s_1}m_{s'_1},m_{s_2}m_{s'_2}}(\lambda_1\lambda'_1,\lambda_2\lambda'_2;z) \\ &= -\sqrt{\left|f\left(\varepsilon^{\mathrm{QP}}_{\lambda_1 m_{s_1}}\right)-f\left(\varepsilon^{\mathrm{QP}}_{\lambda_{1'} m_{s'_1}}\right)\right|}\sum_{\Lambda}\operatorname{sgn}(\mathcal{E}_\Lambda)\frac{\Phi_\Lambda\left(\lambda_1 m_{s_1}\lambda'_1 m_{s'_1}\right)\Phi^*_\Lambda(\lambda'_2 m_{s'_2}\lambda_2 m_{s_2})}{\mathcal{E}_\Lambda-\hbar z} \\ &\times\sqrt{\left|f\left(\varepsilon^{\mathrm{QP}}_{\lambda_{2'} m_{s'_2}}\right)-f\left(\varepsilon^{\mathrm{QP}}_{\lambda_2 m_{s_2}}\right)\right|}. \end{aligned} \tag{19.22}$$

The matrix elements of the $\mathcal{F}$ operator (19.19) and the antisymmetry of $\mathcal{F}\mathcal{H}$ with respect to some coordinates guarantee that the two-particle eigenvalues $\mathcal{E}_\Lambda$ occur pairwise. With $\mathcal{E}_\Lambda$ also

$$\mathcal{E}_{\Lambda'} = -\mathcal{E}_\Lambda \tag{19.23}$$

is an eigenvalue with the relation of the eigenstates

$$\Phi_{\Lambda'}(\lambda_1 m_{s_1}\lambda_{1'} m_{s'_1}) = \Phi^*_\Lambda(\lambda_{1'} m_{s'_1}\lambda_1 m_{s_1}). \tag{19.24}$$

The representation (19.7) transforms (19.22) back to the space- and spin-dependent four-point (macroscopic) polarization function. With the spin summation (13.1) and the Fourier transformation (13.11) the polarization function $P^M(\mathbf{q},\mathbf{q},z)$ and, thus, the macroscopic dielectric function $\epsilon^M(\mathbf{q},z)$ can be calculated. For collinear spins it obeys the expression

$$\begin{aligned} \epsilon^M(\hat{\mathbf{q}},z) &= 1+\lim_{\mathbf{q}\to 0}\tilde{v}(|\mathbf{q}|)\frac{1}{\Omega}\sum_\Lambda\operatorname{sgn}(\mathcal{E}_\Lambda) \\ &\times\left|\sum_{m_s}\sum_{\lambda,\lambda'}\left[B^{m_s m_s}_{\lambda\lambda'}(\mathbf{q})\right]^*\Phi_\Lambda(\lambda m_s\lambda' m_s)\sqrt{\left|f\left(\varepsilon^{\mathrm{QP}}_{\lambda m_s}\right)-f\left(\varepsilon^{\mathrm{QP}}_{\lambda' m_s}\right)\right|}\right|^2\frac{1}{\mathcal{E}_\Lambda-\hbar z}, \end{aligned}$$

where the definition of the Bloch integrals (13.44) has been used. In this expression the macroscopic dielectric function is simply described by a Λ-sum over harmonic oscillators with resonance frequency $\mathcal{E}_\Lambda/\hbar$ and oscillator strength (here: in units of $(\text{volume})^{-1}$) [5]

$$f_\Lambda(\hat{\mathbf{q}}) = \lim_{\mathbf{q}\to 0} \frac{m|\mathcal{E}_\Lambda|}{\hbar^2|\mathbf{q}|^2\Omega} \left| \sum_{m_s} \sum_{\lambda,\lambda'} [B^{m_s m_s}_{\lambda\lambda'}(\mathbf{q})]^* \, \Phi_\Lambda(\lambda m_s \lambda' m_s) \sqrt{\left| f\left(\varepsilon^{\mathrm{QP}}_{\lambda m_s}\right) - f\left(\varepsilon^{\mathrm{QP}}_{\lambda' m_s}\right)\right|} \right|^2 . \tag{19.25}$$

Because of the resulting (imaginary part of the) macroscopic dielectric function

$$\mathrm{Im}\epsilon^M(\hat{\mathbf{q}}, \omega) = \frac{\hbar^2 e^2 \pi}{\varepsilon_0 m} \sum_\Lambda \frac{f_\Lambda(\hat{\mathbf{q}})}{\mathcal{E}_\Lambda} \delta(\hbar\omega - \mathcal{E}_\Lambda),$$

it fulfills the f-sum rule (13.27) as

$$\sum_\Lambda f_\Lambda(\hat{\mathbf{q}}) = n$$

with the average electron density n of the inhomogeneous electron gas. This can be proven by means of the generalized closure relation for the pair wave functions $|\Phi_\Lambda\rangle$ and the closure relation of the single-quasiparticle wave functions.

With the symmetry relations (19.23) and (19.24) the expression for the macroscopic dielectric function can be restricted to positive eigenvalues. To obtain the retarded response function one has to substitute $z \to \omega + i\eta$

$$\begin{aligned} \epsilon^M(\hat{\mathbf{q}}, \omega) &= 1 + \frac{4\pi}{\Omega} \sum_\Lambda^{(\mathcal{E}_\Lambda > 0)} \left| \sum_{m_s} \sum_{\lambda,\lambda'} [M^{m_s}_{\lambda\lambda'}(\hat{\mathbf{q}})]^* \, \Phi_\Lambda(\lambda m_s \lambda' m_s) \sqrt{\left| f\left(\varepsilon^{\mathrm{QP}}_{\lambda m_s}\right) - f\left(\varepsilon^{\mathrm{QP}}_{\lambda' m_s}\right)\right|} \right|^2 \\ &\times \left[\frac{1}{\mathcal{E}_\Lambda - \hbar(\omega + i\eta)} + \frac{1}{\mathcal{E}_\Lambda + \hbar(\omega + i\eta)} \right] \end{aligned} \tag{19.26}$$

with the optical 'dipole' matrix elements in a certain spin channel m_s

$$M^{m_s}_{\lambda\lambda'}(\hat{\mathbf{q}}) = \lim_{\mathbf{q}\to 0} \frac{e}{\sqrt{4\pi\varepsilon_0}|\mathbf{q}|i} B^{m_s m_s}_{\lambda\lambda'}(\mathbf{q}). \tag{19.27}$$

The phase factor $(-i)$ has been only introduced to define really the matrix element of a dipole operator in the limit of vanishing wave vectors. This general expression to describe linear optical properties via the macroscopic dielectric function takes into account non-trivial occupation factors and the collinear spin polarization of the electronic system. The total transition matrix element in (19.26) for the excitation of a two-particle excitation with energy $\mathcal{E}_\Lambda$ from the ground state contains a sum over the two (minority and majority) spin channels as well as the sum over all single-quasiparticle pairs with different occupation.

In the limit of infinite lifetimes $\eta \to +0$, the retarded version of the macroscopic dielectric function fulfills the Kramers-Kronig relations [see (13.41)]

$$\mathrm{Re}\epsilon^M(\hat{\mathbf{q}},\omega) = 1 + \mathcal{P}\int_{-\infty}^{+\infty}\frac{d\omega'}{\pi}\frac{\mathrm{Im}\epsilon^M(\hat{\mathbf{q}},\omega')}{\omega'-\omega},$$

$$\mathrm{Im}\epsilon^M(\hat{\mathbf{q}},\omega) = -\mathcal{P}\int_{-\infty}^{+\infty}\frac{d\omega'}{\pi}\frac{\mathrm{Re}\epsilon^M(\hat{\mathbf{q}},\omega')-1}{\omega'-\omega} \tag{19.28}$$

with the imaginary part

$$\begin{aligned}\mathrm{Im}\epsilon^M(\hat{\mathbf{q}},\omega) &= \frac{4\pi^2}{\Omega}\sum_{\Lambda}^{(\mathcal{E}_\Lambda>0)}\left|\sum_{m_s}\sum_{\lambda,\lambda'}[M^{m_s}_{\lambda\lambda'}(\hat{\mathbf{q}})]^*\,\Phi_\Lambda(\lambda m_s\lambda' m_s)\sqrt{\left|f\left(\varepsilon^{\mathrm{QP}}_{\lambda m_s}\right)-f\left(\varepsilon^{\mathrm{QP}}_{\lambda' m_s}\right)\right|}\right|^2\\ &\quad\times[\delta(\hbar\omega-\mathcal{E}_\Lambda)-\delta(\hbar\omega+\mathcal{E}_\Lambda)]\end{aligned} \tag{19.29}$$

and a corresponding real part, which fulfills the (anti)symmetry relation

$$\begin{aligned}\mathrm{Re}\epsilon^M(\hat{\mathbf{q}},-\omega) &= \mathrm{Re}\epsilon^M(\hat{\mathbf{q}},\omega),\\ \mathrm{Im}\epsilon^M(\hat{\mathbf{q}},-\omega) &= -\mathrm{Im}\epsilon^M(\hat{\mathbf{q}},\omega).\end{aligned} \tag{19.30}$$

While the real part is an even function of the frequency, the imaginary part is an odd function.

19.3 Electron-Hole-Pair Excitations

19.3.1 Resonant and Antiresonant Pairs

We consider a crystal or another translationally invariant system with band states $\lambda \to \nu\mathbf{k}$, which are either fully occupied (i.e., valence bands $\nu = v$) or completely empty (i.e., conduction bands $\nu = c$). The Fermi functions are Heaviside functions $f(\hbar\omega) = \theta(\mu - \hbar\omega)$ with μ as the chemical potential of the electron gas. Examples are insulators and undoped semiconductors at zero temperature. However, molecules in a supercell description can be treated in the same way. We are interested in real and virtual optical excitations by photons of energy $\hbar\omega$ from valence band states $|v\mathbf{k}m_s\rangle$ into conduction band states $|c\mathbf{k}m_s\rangle$, under conservation of the single-particle spin. Intraband pair excitations are not considered.

In the representation (19.26) in the product basis of two single-(quasi)particle states of the operator $\mathcal{F}\mathcal{H}$ in the homogeneous BSE (19.18) only pairs cv and vc appear. They are coupled by the kernel Ξ^M (19.9) in (19.16). In terms of the band states the operator becomes

$$\mathcal{FH} = \begin{pmatrix} H\left(\mathbf{c}\mathbf{k}m_{s_c}v\mathbf{k}m_{s_v}, c'\mathbf{k}'m_{s_{c'}}v'\mathbf{k}'m_{s_{v'}}\right) & \Xi^M\left(v\mathbf{k}m_{s_v}\mathbf{c}\mathbf{k}m_{s_c}, c'\mathbf{k}'m_{s_{c'}}v'\mathbf{k}'m_{s_{v'}}\right) \\ \Xi^M\left(\mathbf{c}\mathbf{k}m_{s_c}v\mathbf{k}m_{s_v}, v'\mathbf{k}'m_{s_{v'}}c'\mathbf{k}'m_{s_{c'}}\right) & H\left(v\mathbf{k}m_{s_v}\mathbf{c}\mathbf{k}m_{s_c}, v'\mathbf{k}'m_{s_{v'}}c'\mathbf{k}'m_{s_{c'}}\right) \end{pmatrix}. \tag{19.31}$$

Here, we consider only vertical transitions $|v\mathbf{k}m_v\rangle \rightarrow |c\mathbf{k}m_c\rangle$ in the BZ due to the approximation of vanishing photon wave vector $\mathbf{q} \rightarrow 0$ in the long-wavelength limit, as will be demonstrated in Sect. 20.1. In this limit, we completely neglect the influence of spatial dispersion on optical properties as indicated by the definition of $\epsilon^M(\hat{\mathbf{q}}, \omega)$ (19.26). In this way any momentum transfer on the resulting electron-hole pair in conduction and valence bands is neglected [5, 11]. In Chap. 21 we will see that this approach coincides with the neglect of momentum transfer to an electron-hole pair and, thus, the neglect of the center-of-mass motion of a resulting Wannier-Mott exciton [11]. The operator (19.31) is valid for pair excitations, in which electron and hole can carry different spins. This generalization allows to discuss also spin excitations which cannot be directly excited for vanishing photon wave vectors in a system with collinear spins.

The above representation (19.31) of the $\mathcal{FH}$ operator has a block-matrix structure. The diagonal blocks, the resonant and antiresonant electron-hole-pair Hamiltonians, only contain the interaction of pairs cv and $c'v'$, with positive pair energies, and pairs vc and $v'c'$, with negative energies. The off-diagonal blocks represented by matrix elements of the integral kernel (19.9) couple the pairs with positive and negative energies by elements cv and $v'c'$ or vc and $c'v'$. With (19.9), (14.56), and (19.10) the resonant block, the most important part of the electron-hole-pair Hamiltonian, is given by

$$\begin{aligned} &H\left(c\mathbf{k}m_{s_c}v\mathbf{k}m_{s_v}, c'\mathbf{k}'m_{s_{c'}}v'\mathbf{k}'m_{s_{v'}}\right) \\ &= \left[\varepsilon^{\mathrm{QP}}_{cm_{s_c}}(\mathbf{k}) - \varepsilon^{\mathrm{QP}}_{vm_{s_v}}(\mathbf{k})\right]\delta_{cc'}\delta_{vv'}\delta_{\mathbf{k}\mathbf{k}'}\delta_{m_{s_c}m'_{s_c}}\delta_{m_{s_v}m_{s_{v'}}} \\ &\quad - \delta_{m_{s_c}m_{s_{c'}}}\delta_{m_{s_v}m_{s_{v'}}} W^{\mathbf{k}\mathbf{k}'}_{\substack{m_{s_c}m_{s_v}\\ cc'\\ vv'}} + \delta_{m_{s_c}m_{s_v}}\delta_{m_{s_{c'}}m_{s_{v'}}} \bar{\mathrm{v}}^{\mathbf{k}\mathbf{k}'}_{\substack{m_{s_c}m_{s_{c'}}\\ cv\\ v'c'}} \end{aligned} \tag{19.32}$$

with

$$\begin{aligned} W^{\mathbf{k}\mathbf{k}'}_{\substack{m_{s_c}m_{s_v}\\ cc'\\ vv'}} &= \frac{1}{\Omega}\sum_{\mathbf{q}}\sum_{\mathbf{G},\mathbf{G}'} \epsilon^{-1}(\mathbf{q}+\mathbf{G}, \mathbf{q}+\mathbf{G}', 0)\tilde{\mathrm{v}}(|\mathbf{q}+\mathbf{G}'|) \\ &\quad \times B^{\mathbf{k}\mathbf{k}'}_{\substack{m_{s_c}m_{s_c}\\ cc'}}(\mathbf{q}+\mathbf{G})\left[B^{\mathbf{k}\mathbf{k}'}_{\substack{m_{s_v}m_{s_v}\\ vv'}}(\mathbf{q}+\mathbf{G}')\right]^*, \\ \bar{\mathrm{v}}^{\mathbf{k}\mathbf{k}'}_{\substack{m_{s_c}m_{s_{c'}}\\ cv\\ v'c'}} &= \frac{1}{\Omega}\sum_{\mathbf{G}\neq 0}\tilde{\mathrm{v}}(|\mathbf{G}|)B^{\mathbf{k}\mathbf{k}}_{\substack{m_{s_c}m_{s_c}\\ cv}}(\mathbf{G})\left[B^{\mathbf{k}'\mathbf{k}'}_{\substack{m_{s_{c'}}m_{s_{c'}}\\ c'v'}}(\mathbf{G})\right]^*. \end{aligned} \tag{19.33}$$

The electron-hole-pair problem is ruled by two types of interaction (19.33) in (19.32). The terms proportional to $-W$ represents the (statically) screened Coulomb attraction of a negatively charged electron in a conduction band and a positively charged hole in a valence band. In addition, a rather short-range, unscreened Coulomb repulsion,

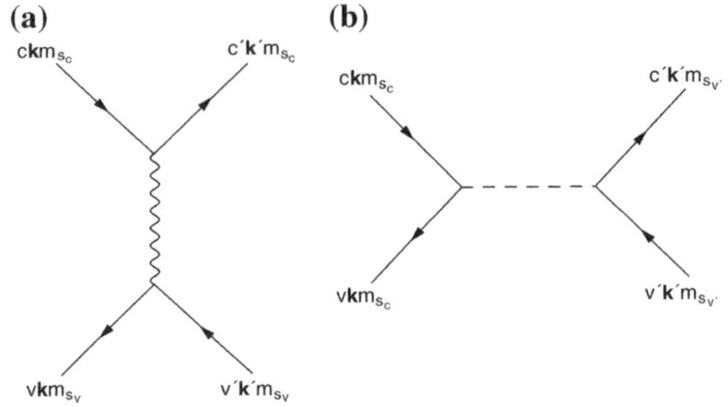

Fig. 19.1 Schematic illustration of the irreducible electron-hole interaction: (**a**) electron-hole attraction mediated by the screened Coulomb interaction (*wavy line*) and (**b**) unscreened electron-hole exchange mediated by the bare Coulomb interaction (*dashed line*)

which may be interpreted as electron-hole exchange, appears. The corresponding Feynman diagrams are displayed in Fig. 19.1. In (19.32) we have assumed that all conduction bands are empty while all valence bands are completely filled. Modifications of this assumption are investigated in Sect. 22.3.

We have to point out that the W matrix elements in (19.33) can be further simplified taking the Bloch character of the QP wave functions

$$\varphi_{\nu\mathbf{k}m_s}(\mathbf{x}) = \frac{1}{\sqrt{\Omega}} e^{i\mathbf{k}\mathbf{x}} u_{\nu\mathbf{k}m_s}(\mathbf{x})$$

with the lattice-periodic Bloch factor

$$u_{\nu\mathbf{k}m_s}(\mathbf{x}+\mathbf{R}) = u_{\nu\mathbf{k}m_s}(\mathbf{x})$$

into account. In the limit $\mathbf{q} \to 0$ it holds

$$B\overset{\mathbf{k}\mathbf{k}'}{\underset{\nu\nu'}{m_s m_s}}(\mathbf{q}+\mathbf{G}) = \delta_{\mathbf{k}'+\mathbf{q},\mathbf{k}} \tilde{B}\overset{\mathbf{k}\mathbf{k}'}{\underset{\nu\nu'}{m_s m_s}}(\mathbf{G})$$

with

$$\tilde{B}\overset{\mathbf{k}\mathbf{k}'}{\underset{\nu\nu'}{m_s m_s}}(\mathbf{G}) = \frac{1}{\Omega_0} \int\limits_{\Omega_0} d^3\mathbf{x} u^*_{\nu\mathbf{k}m_s}(\mathbf{x}) e^{i\mathbf{G}\mathbf{x}} u_{\nu'\mathbf{k}'m_s}(\mathbf{x}),$$

where the integral is restricted to an elementary cell with volume Ω_0.

With the symmetry of Bloch integrals (13.44) and the screened Coulomb interaction, and the bare Coulomb interaction (19.10) with respect to exchange of band indices, the Bloch-matrix structure of the two-particle problem (19.31) can be transformed into

$$\begin{pmatrix} H & \Xi^M \\ -\Xi^{M^+} & -H^+ \end{pmatrix}$$

as a consequence of symmetry of pairs cv and vc. For many systems the off-diagonal blocks $\sim\Xi^M$ have been found to be small and to have nearly no effect on the pair excitations $\mathcal{E}_\Lambda$. Careful studies of the dielectric function of bulk Si and simple molecules [12–14] found practically no impact on the optical absorption spectra. Some other optical properties, however, as the macroscopic electronic dielectric constant ϵ_∞ may be slightly influenced by the off-diagonal elements. Studies of the silane molecule SiH_4 revealed that the excitation energies with and without off-diagonal terms $\sim\Xi^M$ agree with each other within 0.03 eV [1]. The effect of the coupling terms on the energy loss spectrum of Si is slightly stronger [15]. In the case of organic molecules and molecular crystals the neglect of the coupling terms is however not possible [16, 17]. Their importance is reported even for excitons in organic semiconductors [16]. An illustration of the effect of the off-diagonal blocks is given for a nanostructure, a carbon nanotube, in Fig. 19.2. However, in the following we will concentrate on the most important effects and, hence, neglect the coupling terms using only

$$\begin{pmatrix} H & 0 \\ 0 & -H^+ \end{pmatrix}.$$

This is the Tamm-Dancoff approximation (TDA) [18], introduced in literature originally not for excitonic problems. We have to point out that determining optical properties in the framework of the time-dependent density functional theory (TDDFT) [19] the off-diagonal blocks cannot be neglected. Both the diagonal and off-diagonal blocks are of the same order of magnitude [1]. A more qualitative argument for the neglect of the off-diagonal blocks in the BSE treatment is the smallness of the resulting exciton binding energies compared to the excitation energies of electron-hole pairs. The coupling between electron-hole pairs and antipairs should be of minor

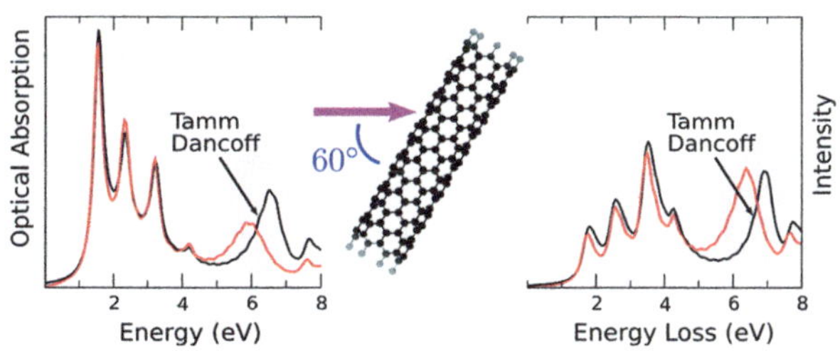

Fig. 19.2 Deviations from the Tamm-Dancoff approximation in the optical absorption and energy loss spectrum of a nanostructure, a carbon nanotube (as illustrated). Reprinted with permission from [17]. Copyright 2009 American Chemical Society

importance. It only gradually modifies the formation and propagation of electron-hole pairs, in particular its internal motion. The most important interactions are included in Ξ^M in H (19.32).

Neglecting the coupling each diagonal block in the two-particle Hamiltonian becomes Hermitian. We can restrict ourselves the studies to the resonant part H (19.32) (in contrast to $\mathcal{F}\mathcal{H}$) which obeys a regular eigenvalue problem with eigenvectors $A_\Lambda(c\mathbf{k}m_{s_c}v\mathbf{k}m_{s_v})$ (instead of $\Phi_\Lambda(c\mathbf{k}m_{s_c}v\mathbf{k}m_{s_v})$) and eigenvalues E_Λ (instead of $\mathcal{E}_\Lambda$),

$$\sum_{c',v'}\sum_{m_{s'_c},m_{s'_v}}\sum_{\mathbf{k}'} H\left(c\mathbf{k}m_{s_c}v\mathbf{k}m_{s_v}, c'\mathbf{k}'m_{s_{c'}}v'\mathbf{k}'m_{s_{v'}}\right) A_\Lambda\left(c'\mathbf{k}'m_{s_{c'}}v'\mathbf{k}'m_{s_{v'}}\right) = E_\Lambda A_\Lambda\left(c\mathbf{k}m_{s_c}v\mathbf{k}m_{s_v}\right), \quad (19.34)$$

with Λ as the set of quantum numbers of the electron-hole pair excitations. The eigenvectors $|A^*_\Lambda\rangle$ and eigenvalues $-E_\Lambda$ of the antiresonant part $-H^+$ can be obtained directly from the solution of the resonant part H of the Hamiltonian without further computational effort. In the Tamm-Dancoff approximation the macroscopic dielectric function (19.26) then reads

$$\epsilon^M(\hat{\mathbf{q}},\omega) = 1 + \frac{4\pi}{\Omega}\sum_\Lambda\left|\sum_{m_s}\sum_{c,v}\sum_{\mathbf{k}} M^{m_s}_{\substack{\mathbf{kk}\\cv}}(\hat{\mathbf{q}})^* A_\Lambda(c\mathbf{k}m_s v\mathbf{k}m_s)\right|^2 \times \sum_{\beta=\pm1}\frac{1}{E_\Lambda - \beta\hbar(\omega+i\eta)}. \quad (19.35)$$

19.3.2 Spin Structure of Pair Hamiltonian

The spin dependence of the response functions, discussed in Sect. 18.2.2, has also consequences for the problem of interacting electron-hole pairs described by the Hamiltonian (19.32). For illustration of the spin structure of the electron-hole-pair Hamiltonian we schematically write down the resonant part of the Hamiltonian in basis functions of spin pairs, i.e., products of single-particle spin functions (see Sect. 18.2.1). With m_{s_c}, m_{s_v}, $m_{s_{c'}}$, and $m_{s_{v'}} = \frac{1}{2}, -\frac{1}{2}$ it schematically results for H [1, 20]

$$\begin{pmatrix} H^{\frac{1}{2}\frac{1}{2}} + \bar{\mathrm{v}}^{\frac{1}{2}\frac{1}{2}} & 0 & 0 & \bar{\mathrm{v}}^{\frac{1}{2}-\frac{1}{2}} \\ 0 & H^{\frac{1}{2}-\frac{1}{2}} & 0 & 0 \\ 0 & 0 & H^{-\frac{1}{2}\frac{1}{2}} & 0 \\ \bar{\mathrm{v}}^{-\frac{1}{2}\frac{1}{2}} & 0 & 0 & H^{-\frac{1}{2}-\frac{1}{2}} + \bar{\mathrm{v}}^{-\frac{1}{2}-\frac{1}{2}} \end{pmatrix}$$

with the abbreviation

$$H^{m_{s_c} m_{s_v}} = \varepsilon^{\text{QP}}_{cm_{s_c}} - \varepsilon^{\text{QP}}_{vm_{s_v}} - W^{m_{s_c} m_{s_v}}$$

for the diagonal elements of the resonant pair Hamiltonian without local-field contributions.

The difference to the non-spin-polarized case is formally described by the dependence of the interaction matrix elements on the same or different spin quantum numbers. It expresses that the Bloch functions (or orbitals) $\varphi_{v\mathbf{k}m_s}(\mathbf{x})$ depend on the spin channel $m_s = \frac{1}{2}$ or $-\frac{1}{2}$ to which the single quasiparticle belongs. It can be rewritten in two block matrices

$$\begin{pmatrix} H^{\frac{1}{2}\frac{1}{2}} + \bar{\mathrm{v}}^{\frac{1}{2}\frac{1}{2}} & \bar{\mathrm{v}}^{\frac{1}{2}-\frac{1}{2}} & 0 & 0 \\ \bar{\mathrm{v}}^{-\frac{1}{2}\frac{1}{2}} & H^{-\frac{1}{2}-\frac{1}{2}} + \bar{\mathrm{v}}^{-\frac{1}{2}-\frac{1}{2}} & 0 & 0 \\ 0 & 0 & H^{\frac{1}{2}-\frac{1}{2}} & 0 \\ 0 & 0 & 0 & H^{-\frac{1}{2}\frac{1}{2}} \end{pmatrix}. \tag{19.36}$$

The exchange matrix elements $\bar{\mathrm{v}}^{\frac{1}{2}-\frac{1}{2}}$ and $\bar{\mathrm{v}}^{-\frac{1}{2}\frac{1}{2}}$ in the first block couple the singlet pair state with $S = 0$, $M_s = 0$ and the triplet pair state with $S = 1$, $M_s = 0$. The second block is diagonal and leads directly to the triplet pair states with $S = 1$ and $M_s = \pm 1$. Even in the collinear case the pair spin is not conserved in spin-polarized systems, in contrast to electron spin in the single-quasiparticle picture. The derived pair Hamiltonian H must therefore be discussed including its full spin structure, in order to solve the homogeneous BSE (19.34).

In the non-spin-polarized case with only two types of interaction matrix elements W and $\bar{\mathrm{v}}$, the Hamiltonian can be diagonalized in spin space by utilizing singlet and triplet pair states, instead of products of one-particle spin functions as basis set. Singlet and triplet states can be treated independently. It is sufficient to solve the pair eigenvalue problems of the Hamiltonians $H = (\varepsilon_c^{\text{QP}} - \varepsilon_v^{\text{QP}}) - W + 2\bar{\mathrm{v}}$ for the singlet excitations and $H = (\varepsilon_c^{\text{QP}} - \varepsilon_v^{\text{QP}}) - W$ for the triplet states.

The optical matrix elements $M^{m_s}_{\substack{\mathbf{kk}\\cv}}(\hat{\mathbf{q}})$ (19.27) in (19.35) are diagonal in the spin quantum numbers of electron and hole. That means, in the collinear limit no spin flip occurs within an optical transition between valence and conduction bands, $|v\mathbf{k}m_s\rangle \rightarrow |c\mathbf{k}m_s\rangle$. The consequence in the non-spin-polarized case is obvious. It turns out that only the singlet contribution is relevant for the calculation of optical spectra (see also discussion in Sect. 18.2.3). If spin polarization is present, still only electron-hole pairs with $m_{s_c} = m_{s_v}$ contribute to the dielectric function in the collinear limit. For the calculation of optical properties the eigenvalue problem (19.34) can be reduced to the case of $m_{s_c} = m_{s_v}$ and $m_{s_{c'}} = m_{s_{v'}}$ and electron-hole-pair amplitudes $A_\Lambda(c\mathbf{k}m_s v\mathbf{k}m_s)$. That means, one must solve only the eigenvalue problem of the upper block in the matrix (19.36) with $H = (\varepsilon_c^{\text{QP}} - \varepsilon_v^{\text{QP}}) - W + 2\bar{\mathrm{v}}$. The rank of the electron-hole-pair matrix can be reduced by a factor 2 to a 2×2 matrix in spin space.

19.3.3 Numerical Methods and Results

The direct solution of the eigenvalue problem (19.34) depends on the rank N of the electron-hole pair Hamiltonian H with N being the number of electron-hole product states. The matrix diagonalization or inversion scales like $O(N^3)$. However, the computation of the Hamiltonian in the framework of the GW approach may be more computationally expensive, since it scales like $O(N^4)$ [1]. For each spin channel N is given by the product $N_v \times N_c \times N_{\mathbf{k}}$ of the number N_v of valence bands, the number of conduction bands N_c, and the number $N_{\mathbf{k}}$ of $\mathbf{k}$ points used to sample the entire BZ. The diagonalization procedure is not only very time consuming, but requires the storage of at least half of the two-particle Hamiltonian matrix (19.32) in the main memory of the computer facility used. Actually $N(N+1)/2$ matrix elements have to be stored to obtain the eigenvectors A_Λ and eigenvalues E_Λ for the computation of the macroscopic dielectric function (19.35). Even for non-spin-polarized bulk systems with two atoms in the unit cell and four valence and four conduction bands N can take large values of $N = 80.000$ with $N_{\mathbf{k}} \approx 5.000$ for converged computations of the optical spectra in a not too wide frequency range. For the optical spectra of bulk CdF_2, a simple three-atomic cubic crystal, an excitonic matrix of rank $N = 210.000$ has been used to account for the needed many band pairs to cover a wide frequency region [21]. For surface optical spectra even more pair states, e.g. $N = 350.000$ [22], have to be used. For reconstructed surfaces or spin-polarized systems with d-states in addition to the sp electrons numbers up to $N \approx 10^6$ can be reached for converged studies [20, 23]. While the matrix diagonalization [1, 12, 13] perfectly works for a relatively small number of electron-hole pair states, direct diagonalization techniques are prohibitively expensive for systems described by too many single-particle states.

For large numbers N iterative methods as the Haydock method [24, 25] or the initial-value/time-evolution method [22, 23] are more efficient. The Haydock method is a generalization of the Lanczos iterative method [26]. Details for its application to the electron-hole pair problem can be found elsewhere [25]. The first computations of the macroscopic dielectric function using the Haydock procedure have been done by Benedict and Shirley [27]. This method to solve the BSE and to compute optical properties is implemented in several codes, e.g. the EXC code [28], ABINIT (optdriver99) [29], and the YAMBO program suite [30].

The main idea of the initial-value or time-evolution method can be easily illustrated for the diagonal components of the dielectric tensor $\epsilon_{jj}^M(\omega) = 1 + 4\pi\beta_{jj}^M(\omega)$ (18.23). According to (19.35) the frequency-dependent macroscopic electronic polarizability can be formally rewritten as [23]

$$\beta_{jj}^M(\omega) = \frac{1}{\Omega}\sum_\Lambda \left|\langle M_j|A_\Lambda\rangle\right|^2 \sum_{\beta=\pm 1} \frac{1}{E_\Lambda - \beta\hbar(\omega + i\eta)}$$

with

$$\langle M_j | A_\Lambda \rangle = \sum_{m_s} \sum_{c,v} \sum_{\mathbf{k}} \left(M \begin{smallmatrix} m_s \\ \mathbf{kk} \\ cv \end{smallmatrix} \right)_j^* A_\Lambda(c\mathbf{k}m_s v\mathbf{k}m_s)$$

and the jth Cartesian component $\left(M \begin{smallmatrix} m_s \\ \mathbf{kk} \\ cv \end{smallmatrix} \right)_j$ of the dipole operator in $\mathbf{M} \begin{smallmatrix} m_s \\ \mathbf{kk} \\ cv \end{smallmatrix} \cdot \hat{\mathbf{q}} = \mathbf{M} \begin{smallmatrix} m_s \\ \mathbf{kk} \\ cv \end{smallmatrix} (\hat{\mathbf{q}})$ in (19.27). This formulation is equivalent to a Fourier representation [4, 23, 31]

$$\beta_{jj}^M(\omega) = \frac{i}{\hbar\Omega} \int_0^\infty dt e^{i(\omega+i\eta)t} \left\{ \langle M_j | \xi_j(t) \rangle - \langle M_j | \xi_j(t) \rangle^* \right\}, \tag{19.37}$$

where the time evolution of the vector $|\xi_j(t)\rangle$ related to the pair states is driven by the quasielectron-quasihole-pair Hamiltonian (19.32)

$$i\hbar \frac{d}{dt} |\xi_j(t)\rangle = H |\xi_j(t)\rangle. \tag{19.38}$$

The initial vector is given by the dipole matrix elements in the jth Cartesian direction

$$|\xi_j(0)\rangle = |M_j\rangle.$$

The equivalence of the two formulations of $\beta_{jj}^M(\omega)$ can be shown by interpreting the solution of the time-dependent 'Schrödinger' equation (19.38) to be $|\xi_j(t)\rangle = e^{-\frac{i}{\hbar}Ht}|M_j\rangle$ [31] and exploiting the spectral representation. In explicit computations the initial-value problem is solved using the central-difference method [32]. Recent versions of the VASP code [33] contain an actual implementation of the time-evolution/initial-value method to compute the frequency-dependent macroscopic dielectric function (19.35). A generalization of (19.38) for the non-Hermitian operator $\mathcal{F}\mathcal{H}$ beyond the Tamm-Dancoff approximation is easily possible [4]. Details of the numerical treatment can be found elsewhere [31, 34].

The upper limit of the Fourier integral (19.37) can be truncated, due to the exponential $e^{-\eta t}$. As a consequence, the number of time steps, i.e., matrix-vector multiplications in the time-dependent 'Schrödinger' equation (19.38) defined for each step, is nearly independent of the system size and essentially governed by the inverse pair lifetime η. The order of 10^3 time steps are typically required applying a broadening parameter of the order of $\hbar\eta = 0.1$ eV. The operation count for this method scales thus as $O(N^2)$, compared to $O(N^3)$ of the matrix diagonalization. In explicit calculations the imaginary part of the dielectric function is computed directly, whereas the real part is found via the Kramers-Kronig transformation (19.28).

The most important advantage of the Haydock and initial-value methods is the reduced quadratic scaling. They are therefore particularly suitable for systems with a large number N of electron-hole-pair basis states. The price to pay is that only the dielectric function with a lifetime broadening η is computed. The eigenvalues E_Λ and eigenvectors A_Λ of the pair Hamiltonian are not calculated. Therefore, a detailed analysis of the optical pair excitations is hampered. For the bound states, the excitons, the exciton binding energy E_B is consequently also not computed. Fortunately, in the limit of strong excitonic effects with large binding energies $E_B > \hbar\eta$, for $\hbar\omega < E_g$, i.e., in the QP gap, isolated peaks may be resolved. Then the relative positions of the bound states to the QP gaps in an absorption spectrum can be identified and the exciton binding energy can be determined approximately.

References

1. M. Rohlfing, S.G. Louie, Electron-hole excitations and optical spectra from first principles. Phys. Rev. B **62**, 4927–4944 (2000)
2. G. Strinati, Applications of the Green's functions method to the study of the optical properties of semiconductors. Riv. Nuovo Cimento **11**, 1–86 (1988)
3. F. Bechstedt, K. Tenelsen, B. Adolph, R. Del Sole, Compensation of dynamical quasiparticle and vertex corrections in optical spectra. Phys. Rev. Lett. **78**, 1528–1531 (1997)
4. C. Rödl, Elektronische und exzitonische Anregungen in magnetischen Isolatoren. Ph.D. thesis. Friedrich-Schiller-Universität, Jena (2009)
5. H. Stolz, *Einführung in die Vielteilchentheorie der Kristalle* (Akademie, Berlin, 1974)
6. D. Jérome, T.M. Rice, W. Kohn, Excitonic insulator. Phys. Rev. **158**, 462–475 (1967)
7. W. Jones, N.H. March, *Theoretical Solid State Physics*, vol. 2 (Dover Publications Inc, New York, 1985)
8. R. Zimmermann, Influence of the non-Hermitian splitting terms on excitonic spectra. Phys. Status Solidi **41**, 23–32 (1970)
9. L.J. Sham, T.M. Rice, Many-particle derivation of the effective-mass equation for the Wannier exciton. Phys. Rev. **144**, 708–714 (1966)
10. A. Mostafazadeh, Pseudo-hermiticity versus PT-symmetry III: equivalence of pseudo-hermiticity and the presence of antilinear symmetries. J. Math. Phys. **43**, 3944–3951 (2002)
11. Ch. Hamaguchi, *Basic Semiconductor Physics* (Springer, Berlin, 2001)
12. S. Albrecht, L. Reining, R. Del Sole, G. Onida, Ab initio calculation of excitonic effects in the optical spectra of semiconductors. Phys. Rev. Lett. **80**, 4510–4513 (1998)
13. S. Albrecht, Optical absorption spectra of semiconductors and insulators: *ab initio* calculation of many-body effects. Ph.D. thesis. Ecole Polytechnique, Palaiseau (1999)
14. P. Hahn, Berechnung von Vielteilcheneffekten in den Anregungsspektren von Kristallen, Oberflächen und Molekülen. Ph.D. thesis. Friedrich-Schiller-Universität, Jena (2004)
15. V. Olevano, L. Reining, Excitonic effects on the silicon plasmon resonance. Phys. Rev. Lett. **86**, 5962–5965 (2001)
16. P. Puschnig, C. Meisenbichler, C. Ambrosch-Draxl, Excited state properties of organic semiconductors: breakdown of the Tamm-Dancoff approximation. arXiv:1306.3790 (2013)
17. M. Grüning, A. Marini, X. Gonze, Exciton-plasmon states in nanoscale materials. Breakdown of the Tamm-Dancoff approximation. Nano Lett. **9**, 2820–2824 (2009)
18. A. Fetter, J.D. Walecka, *Quantum Theory of Many-Particly Systems* (McGraw-Hill, San Francisco, 1971)
19. M.E. Casida, Time-dependent Density Functional Response Theory of Molecular Systems: Theory, Computational Methods, and Functionals, in *Recent Developments and Applications of Modern Density Functional Theory*, ed. by J.M. Seminario (Elsevier Science B.V, Amsterdam, 1996), pp. 391–439

20. C. Rödl, F. Fuchs, J. Furthmüller, F. Bechstedt, *Ab initio* theory of excitons and optical properties for spin-polarized systems. Application to antiferromagnetic MnO. Phys. Rev. B **72**, 184408 (2008)
21. G. Cappellini, J. Furthmüller, E. Cadelano, F. Bechstedt, Electronic and optical properties of cadmium fluoride: the role of many-body effects. Phys. Rev. B **87**, 075203 (2013)
22. P.H. Hahn, W.G. Schmidt, F. Bechstedt, Bulk excitonic effects in surface optical spectra. Phys. Rev. Lett. **88**, 016402 (2002)
23. W.G. Schmidt, S. Glutsch, P.H. Hahn, F. Bechstedt, Efficient $O(N^2)$ method to solve the Bethe-Salpeter equation. Phys. Rev. B **67**, 085307 (2003)
24. R. Haydock, The recursive solution of the Schrödinger equation. Comput. Phys. Comm. **20**, 11–16 (1980)
25. M. Marsili, Electronic and optical properties of the (111)2×1 diamond surface: an *ab-initio* study. Ph.D. thesis. Universitá degli Studi di Roma "Tor Vergata", Rome (2006)
26. E. Dagotto, Correlated electrons in high-temperature superconductors. Rev. Mod. Phys. **66**, 763–840 (1994)
27. L.X. Benedict, E.L. Shirley, Ab initio calculation of $\varepsilon_2(\omega)$ including the electron-hole interaction: application to GaN and CaF_2. Phys. Rev. B **59**, 5441–5451 (1999)
28. http://theory.polytechnique.fr/codes/exc/
29. http://www.abinit.org/helpfiles/
30. A. Marini, C. Hogan, M. Grüning, D. Varsano, Yambo: an *ab initio* tool for excited state calculations. Comput. Phys. Commun. **180**, 1392–1403 (2009)
31. S. Glutsch, D.S. Chemla, F. Bechstedt, Numerical calculation of the optical absorption in semiconductor quantum structures. Phys. Rev. B **54**, 11592–11601 (1996)
32. R. Kosloff, Time-dependent quantum-mechanical methods for molecular dynamics. J. Chem. Phys. **92**, 2087–2100 (1988)
33. http://uni-vienna.at/vasp/
34. S. Glutsch, *Excitons in Low-Dimensional Semiconductors* (Springer, Berlin, 2004)

Chapter 20
Optical Properties

Abstract The macroscopic dielectric function requires the calculation of optical transition matrix elements. The use of wave functions of the single-particle problem with an effective local potential leads to the equivalence of longitudinal and transverse formulations for the optical transition operator. The advantage of the longitudinal approach is the easy inclusion of effects of spin-orbit interaction and non-localities due to exchange and correlation. The resulting matrix elements depend on the symmetry of initial and final state. The scenario of van Hove singularities to interprete the lineshape of optical spectra is significantly modified by the excitonic Coulomb effects. The excitonic redshift of the optical absorption partly compensates the blue shift due to quasiparticle effects. In addition, a redistribution of spectral strength from higher to lower photon energies and the formation of excitonic bound states occur. The combination of quasiparticle and excitonic effects, despite their treatment within the GW approximation, leads to optical and energy-loss spectra in good agreement with experimental findings. This is illustrated for anorganic and organic crystals but also for low-dimensional systems including molecules.

20.1 Transition Matrix Elements

20.1.1 Longitudinal and Transverse Formulation

In order to derive the macroscopic dielectric function $\epsilon^M(\hat{\mathbf{q}}, \omega)$ or tensor $\epsilon_{ij}^M(\omega)$ (18.23) a small scalar perturbative potential (12.3) has been studied in (12.24) and (12.30). The resulting longitudinal dielectric function (19.26) [1] is ruled by transition matrix elements proportional to $B\,{}^{\mathbf{kk}'}_{\substack{m_s m_s \\ \nu\nu'}}(\mathbf{q})$, i.e., Bloch integrals (13.44) of the exponential function $e^{i\mathbf{qx}}$. In Sect. 18.3.2 we have argued that the result can be reformulated in terms of the velocity operator that typically appears when the perturbing electromagnetic field is treated within the Coulomb or transverse gauge with a direct coupling of the paramagnetic current density operator to the vector potential of this field (see e.g. Sect. 2.2). Such a procedure may lead to a transverse dielectric function [1].

F. Bechstedt, *Many-Body Approach to Electronic Excitations*,
Springer Series in Solid-State Sciences 181, DOI 10.1007/978-3-662-44593-8_20

We assume that the effective single-particle Hamiltonian $\hat{H}$, that generates the eigenfunctions $\varphi_{\nu \mathbf{k} m_s}(\mathbf{x})$ and eigenvalues $\varepsilon_{\nu m_s}(\mathbf{k})$ of the electronic reference system with a Green function $\tilde{\tilde{G}}$ (14.41), can be written in the simple form $\hat{H} = \frac{1}{2m}\mathbf{p}^2 + V(\mathbf{x})$ with the momentum operator $\mathbf{p} = -i\hbar\nabla_{\mathbf{x}}$ and an effective potential energy $V(\mathbf{x})$. Then it holds

$$[\hat{H}, e^{i\mathbf{q}\mathbf{x}}]_- = \frac{\hbar}{2m}\left[(\mathbf{q}\mathbf{p}), e^{i\mathbf{q}\mathbf{x}}\right]_+ \tag{20.1}$$

and, consequently,

$$\left\{\varepsilon_{\nu m_s}(\mathbf{k}) - \varepsilon_{\nu' m_s}(\mathbf{k}')\right\} \langle \nu \mathbf{k} m_s | e^{i\mathbf{q}\mathbf{x}} | \nu' \mathbf{k}' m_s \rangle = \frac{\hbar}{2m} \langle \nu \mathbf{k} m_s | \left[(\mathbf{q}\mathbf{p}), e^{i\mathbf{q}\mathbf{x}}\right]_+ | \nu' \mathbf{k}' m_s \rangle. \tag{20.2}$$

This identity shows how the two different couplings of electrons to an electromagnetic field represented by a scalar or a vector potential are related to each other, at least, as long as the solutions of the single-particle problem are explicitly known. We have to mention that the result (20.1) is only correct for local potentials $V(\mathbf{x})$. For non-local ones due to pseudopotentials, self-energy effects, spin-orbit interaction, etc. the momentum operator $\mathbf{p}$ on the right-hand side must be replaced by the velocity operator $\mathbf{p}/m \to \mathbf{v} = \frac{i}{\hbar}[\hat{H}, \mathbf{x}]_-$ (2.6). It is corrected by the commutator of the non-local part of the potential V and the position operator $\mathbf{x}$ [2, 3].

In the limit of vanishing wave vectors $\mathbf{q}$ in the direction $\hat{\mathbf{q}} = \mathbf{q}/|\mathbf{q}|$, the interband matrix element (19.27) becomes

$$M^{m_s}_{\substack{\mathbf{k}\mathbf{k}' \\ cv}}(\hat{\mathbf{q}}) = \lim_{\mathbf{q}\to 0} \frac{e}{\sqrt{4\pi\varepsilon_0}} \frac{1}{i} \frac{1}{|\mathbf{q}|} \langle c\mathbf{k} m_s | e^{i\mathbf{q}\mathbf{x}} | v\mathbf{k}' m_s \rangle = \delta_{\mathbf{k}\mathbf{k}'} \hat{\mathbf{q}} \langle c\mathbf{k} m_s | \frac{e}{\sqrt{4\pi\varepsilon_0}} \mathbf{x} | v\mathbf{k} m_s \rangle. \tag{20.3}$$

Matrix elements of the dipole operator are ill-defined quantities in translationally invariant systems with single-particle eigenstates of Bloch character. This is in contrast to localized systems, such as molecules. For that reason we apply the identity (20.2) and find in the long-wavelength limit

$$M^{m_s}_{\substack{\mathbf{k}\mathbf{k}' \\ cv}}(\hat{\mathbf{q}}) = \frac{e}{\sqrt{4\pi\varepsilon_0}} (-i) \hat{\mathbf{q}} \frac{\hbar}{m} \frac{\mathbf{p}_{cvm_s}(\mathbf{k})}{\varepsilon_{cm_s}(\mathbf{k}) - \varepsilon_{vm_s}(\mathbf{k})} \delta_{\mathbf{k}\mathbf{k}'} \tag{20.4}$$

with the interband Bloch matrix element $\mathbf{p}_{cvm_s}(\mathbf{k}) = \langle c\mathbf{k} m_s | \mathbf{p} | v\mathbf{k} m_s \rangle$ of the momentum operator.

There was a discussion in literature for a long time whether one has to correct the eigenvalues in the energy denominator in (20.4) by QP shifts according to (14.42) or not. As an answer, it has been pointed out that the eigenvalues of the reference system, in which also the wave functions are generated, have to be applied [2, 4]. The influence of the two approaches (20.3) and (20.4) on the optical transition matrix elements in the dielectric function is displayed in Fig. 20.1 within the independent-particle

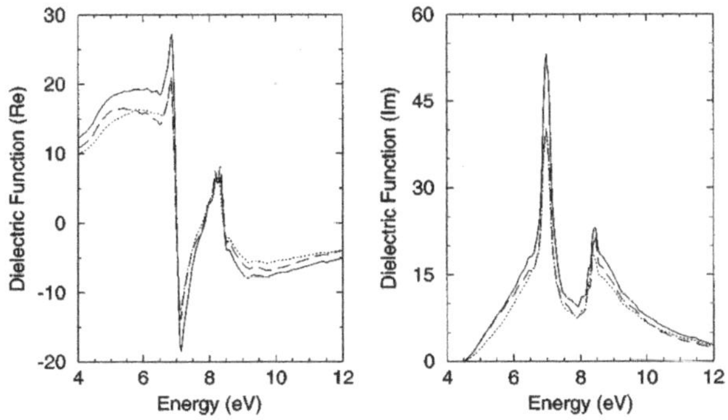

Fig. 20.1 Dielectric function of cubic SiC within the independent-particle approach using pseudo-wave functions. *Solid line*: velocity gauge (20.4) but neglecting the non-local pseudopotential contribution to the transition operator, *dashed line*: velocity gauge (20.4) including non-local contributions, *dotted line*: length gauge (20.3). From [2]

approach [2]. Excitonic $\left(\sim \Xi^M\right)$ and quasiparticle ($\sim \Sigma - V_{XC}$) effects are not taken into account. Single-particle eigenvalues and eigenfunctions are computed within a DFT-LDA scheme (6.22), in which the electron-ion interaction is characterized by non-local pseudopotentials (8.46). From the figure one can conclude that the replacement of the velocity operator by the momentum operator gives rise to a certain error in the 'transverse gauge' (20.4), whereas the 'length gauge' (20.3) is close to the result obtained with the velocity operator **v**.

20.1.2 All-Electron Wave Functions

The problems due to non-local potentials, self-energy contributions or spin-orbit interaction in (20.4) do not appear in the 'length gauge' (20.3). However, there is still a source of uncertainties in the formulation (20.3) if pseudowave functions are used. These functions do not correctly describe the space contributions to the transition matrix elements in the core regions, where the pseudowave functions are too smooth. They do not show the node structure expected for all-electron wave functions. For that reason the use of all-electron wave functions, for instance such generated by means of a self-consistent full-potential linearized augmented plane-wave (FLAPW) method as implemented in the WIEN97 [5, 6] (or a modern version WIEN2k [7]) code, is suggested. Another idea is the use of all-electron wave functions for the valence states derived within a PAW scheme [8, 9].

A comparison of results obtained within a pseudopotential, all-electron and PAW scheme, respectively, is made in Fig. 20.2 [10]. The DFT-LDA approach is used to solve the electronic-structure problem. The dielectric function is computed within the independent-particle approach without taking excitonic and QP effects into account. The joint density of states [1], the spectral density of uncorrelated electron-hole pairs,

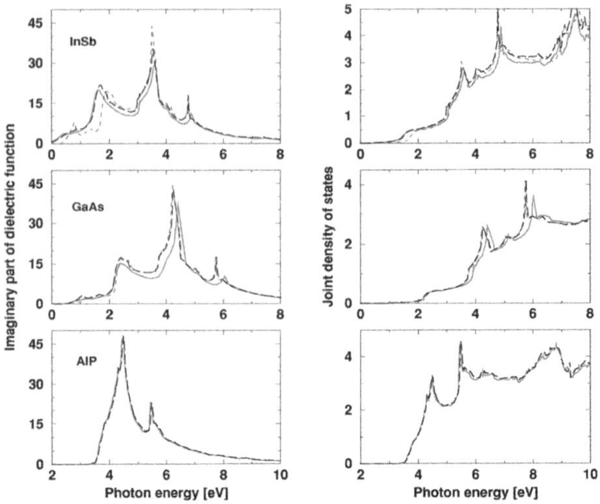

Fig. 20.2 Imaginary part of the dielectric function in independent-particle approximation (*left panels*) and joint density of states (20.5) (*right panels*) for the III-V compounds InSb, GaAs, and AlP. *Solid line*: pseudopotential approach, *thick long-dashed line*: PAW, *thin short-dashed line*: PAW (including shallow d core states). From [10]

$$\mathrm{JDOS}(\hbar\omega) = \sum_{m_s} \sum_{c,v} \sum_{\mathbf{k}} \delta\left(\hbar\omega - \left[\varepsilon^{\mathrm{QP}}_{cm_s}(\mathbf{k}) - \varepsilon^{\mathrm{QP}}_{vm_s}(\mathbf{k})\right]\right) \tag{20.5}$$

is also displayed to distinguish between matrix-element and spectral effects for the studied semiconductors. Apart from the low-energy region of InSb the two all-electron descriptions FLAPW and APW give rather identical results, despite the differences in the numerical treatments. In the case of AlP, a system with solely sp valence electrons, the pseudopotential and all-electron spectra widely agree. In the cases of InSb and GaAs the discrepancies in the dielectric function can be traced back to both matrix-element effects and deviations in the joint density of states (20.5).

A modern implementation of the optical transition matrix elements using the length gauge (20.3) and, in general, linear optical properties are described in detail by Gajdoš et al. [11]. The implementation (20.3) is used in the VASP code [12] and, hence, frequently applied by many theory groups worldwide. Only matrix elements of the exponential function are computed explicitly but in the limit $\mathbf{q} \rightarrow 0$.

20.1.3 Resulting Values and Consequences

The optical matrix elements significantly determine the intensity of the optical spectra versus photon energy. This is clearly shown in Fig. 20.3 for the example of rocksalt-MgO [13]. Excitonic effects or vertex corrections $\sim\Xi^M$ in (19.32) are neglected.

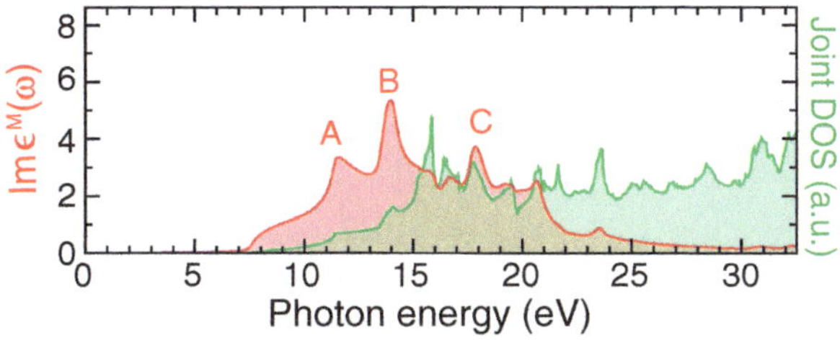

Fig. 20.3 Comparison of the imaginary part of the dielectric function $\epsilon^M(\omega)$ (*red line*) and the joint density of states JDOS($\hbar\omega$) (*green line*) within the independent-quasiparticle picture for rs-MgO. The most important absorption peaks are labeled by A, B, and C. Drawn using data from [13]

The imaginary part of the dielectric function Im$\epsilon^M(\omega)$ is compared with the joint density of states JDOS($\hbar\omega$) (20.5) within the approximation of independent quasiparticles. The matrix elements are computed within the PAW scheme.

The JDOS($\hbar\omega$) is finite above the quasiparticle gap and increases with rising photon energy $\hbar\omega$. It exhibits many spectral features, such as peaks and shoulders, which can be related to critical points in the QP interband structure at high-symmetry points $\mathbf{k}_0$ or high-symmetry lines in the BZ, for which

$$\nabla_{\mathbf{k}} \left[\varepsilon^{\mathrm{QP}}_{cm_s}(\mathbf{k}) - \varepsilon^{\mathrm{QP}}_{vm_s}(\mathbf{k}) \right] \Big|_{\mathbf{k}=\mathbf{k}_0} = 0 \tag{20.6}$$

holds. They give rise to van Hove singularities in the spectrum [1]. The four types of singularities – minimum, two types of saddle points, and maximum – denoted by M_0, M_1, M_2, and M_3, respectively, appear in the joint density of states but should not be analyzed in detail here. The reader is referred to the band structure of rs-MgO displayed in Fig. 15.7 to identify such critical points and van Hove singularities.

The strong influence of the transition matrix elements is evident comparing the JDOS with Im$\epsilon^M(\omega)$. The matrix elements enhance the spectrum for low photon energies $\hbar\omega < 15\,\mathrm{eV}$. They are responsible for the strong slope for energies $\hbar\omega > 21\,\mathrm{eV}$. The latter one guarantees that the f-sum rule is fulfilled and [Re$\epsilon^M(\omega) - 1$] vanishes $\sim\omega^{-2}$ [see (13.38)]. The reason is related to the fact that the strength of optical transitions from sp (especially oxygen) valence states into free-electron-like conduction band states above the vacuum level rapidly vanishes with increasing photon energy.

The actual values of the optical interband matrix elements at certain $\mathbf{k}$ points in the BZ characterize the dipole selection rules for crystalline solids. For systems with spherical symmetry the angular momentum selection rule $\Delta\ell = \pm 1$ is well known. Transitions from p states into s states are allowed, while such between two d states are forbidden. How the $\Delta\ell = \pm 1$ selection rule has to be translated in the case of a crystalline solid is illustrated in Fig. 20.4 for rs-MnO and rs-NiO with antiferromagnetic ordering AF2 (see Fig. 9.2) and partially filled TM 3d shells [14]. Due to the octahedral crystal field the d states split into t_{2g} and e_g ones (see Fig. 9.4) which are differently filled in Mn^{2+} and Ni^{2+} ions. The imaginary part of

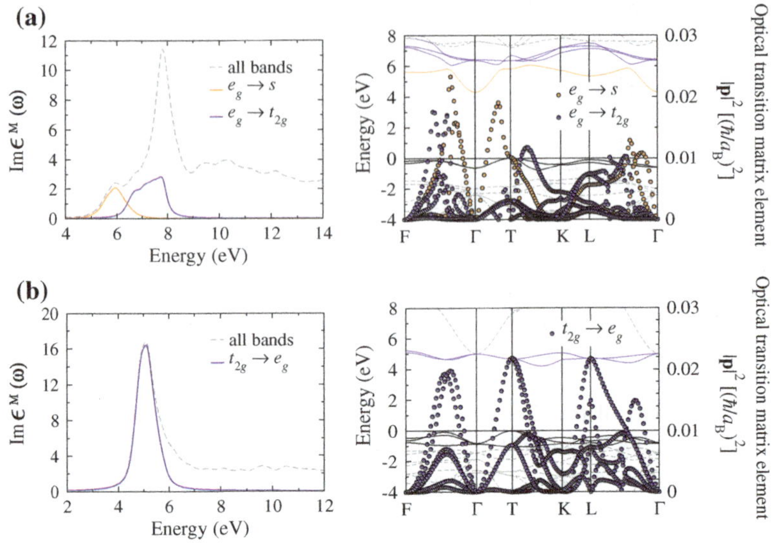

Fig. 20.4 Analysis of optical absorption in MnO (**a**) and NiO (**b**). In the *left panels*, the imaginary parts of the dielectric functions in IQPA are shown together with the contributions of the $e_g \rightarrow s$ and $e_g \rightarrow t_{2g}$ (MnO) or $t_{2g} \rightarrow e_g$ transitions (NiO). In the *right panels*, the corresponding states in the band structures are highlighted. The *black solid lines* indicate the valence-band e_g (MnO) or t_{2g} (NiO) states. The final states of the respective optical transitions are shown in the same *colors* as the partial optical spectra in the *left panels*. The *right panels* also depict the optical transition matrix elements $|\mathbf{p}|^2$ for the interband transitions (*dotted curves*). A GGA+U approach (Sect. 9.1.3) is used. From [14]

the dielectric function is studied in Fig. 20.4 (left panels) within the independent-quasiparticle approximation (IQPA). Special care is taken for the most important contributions due to pure d band – d band transitions.

In the light of the $\Delta\ell = \pm 1$ selection rule the strong $d - d$ contributions to the absorption are somewhat surprising. For that reason the momentum matrix element squares for d-state-derived band transitions are plotted along high-symmetry lines in the BZ of the magnetic rhombohedral structure in Fig. 20.4 (right panels). The full vector $\mathbf{p}_{cvm_s}(\mathbf{k})$ is studied. The corresponding conduction and valence bands are also displayed. The matrix elements are computed within the longitudinal approach (20.3) [11]. Indeed, at Γ whose subgroup of the $\mathbf{k}$ point is equal to the full point group of the crystal, all the $d - d$ transitions are forbidden. However, these interband transitions are dipole-allowed in the rest of the BZ (aside K and F), because the small point groups of these $\mathbf{k}$ points describe a lowered symmetry compared to Γ. Of course, the transition strengths are gradually different for MnO and NiO. Even though the transition matrix elements are one order of magnitude smaller than for sp-bonded semiconductors, nitrides or other oxides (see Table 20.1), they are nonetheless most important for the low-energy absorption spectra of the TMOs. This holds especially for NiO, whose $d - d$ transition matrix elements are larger than those for MnO.

Table 20.1 Squares of the dipole-allowed optical transition matrix elements, more precisely momentum matrix elements $\mathbf{p}_{cvm_s}(\mathbf{k})$, for the lowest-energy interband transitions near the fundamental gap at Γ of hexagonal (wurtzite) and cubic (zinc-blende or rocksalt) semiconductors (III-nitrides and II-oxides) between the uppermost valence bands and the lowest conduction band

Transition	Polarization	AlN	GaN	InN	MgO	ZnO	CdO
$\Gamma_{5v} \rightarrow \Gamma_{1c}$	$\perp$	0.290 (15.78)	0.236 (12.83)	0.172 (9.39)	0.191 (10.39)	0.124 (6.75)	0.099 (5.39)
$\Gamma_{1v} \rightarrow \Gamma_{1c}$	$\parallel$	0.292 (15.92)	0.271 (14.79)	0.193 (10.52)	0.186 (10.12)	0.135 (7.35)	0.103 (5.61)
$\Gamma_{15v} \rightarrow \Gamma_{1c}$	Arbitrary Linear	0.291 (15.86)	0.244 (13.26)	0.175 (9.50)	0.253 (13.74)	0.252 (13.70)	0.196 (10.66)

The orientations $\perp$ ($\parallel$) perpendicular or parallel to the c-axis, i.e., ordinary or extraordinary light polarization, are distinguished in the wurtzite case. The matrix-element squares are given in units of $(\hbar/a_B)^2$. The corresponding energies $E_p^{\perp,\parallel}$ or E_p (20.7) are given in parenthesis in eV. From [16, 17]

The momentum matrix elements of the optical transitions between the highest valence band and the lowest conduction band at 3.65 eV in rutile-SnO_2 as plotted for two projections onto light polarizations perpendicular as well as parallel to the tetragonal axis are displayed in Fig. 20.5. Independent of the polarization direction the optical transition at Γ is dipole-forbidden. This is in agreement with the assignment of the irreducible representations Γ_3^+ and Γ_1^+, which indicate the same parity of initial and final states, and experimental findings. Indeed, in oxides such as rt-SnO_2 but also bixbyite-In_2O_3 the quasiparticle gaps are smaller than the measured absorption edges. However, as shown in Fig. 20.5, the matrix elements for perpendicular polarization grow as soon as the **k** vector becomes finite. Thereby, the matrix elements remain significantly smaller than those associated with typical dipole-allowed transitions for monoxides (see Table 20.1), explaining the small magnitude of the absorption coefficient above the gap [15]. In contrast, the matrix elements for the parallel polarization tend to be very small also away from Γ. There is only one direction ΓR in the BZ along which one observes an increase of the momentum matrix element.

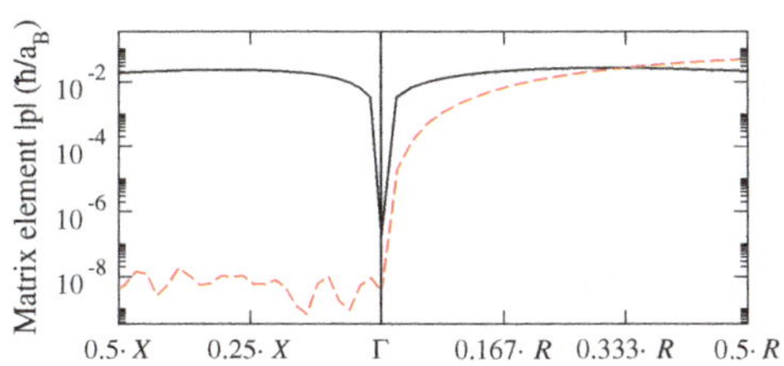

Fig. 20.5 Absolute value of the momentum matrix element (in units of $\hbar/a_B$) for optical transitions between the highest valence band (Γ_3^+) and the lowest conduction band (Γ_1^+) in rutile-SnO_2 for light polarized perpendicular (*solid black line*) and parallel (*dashed red line*) to the tetragonal axis. From [15]

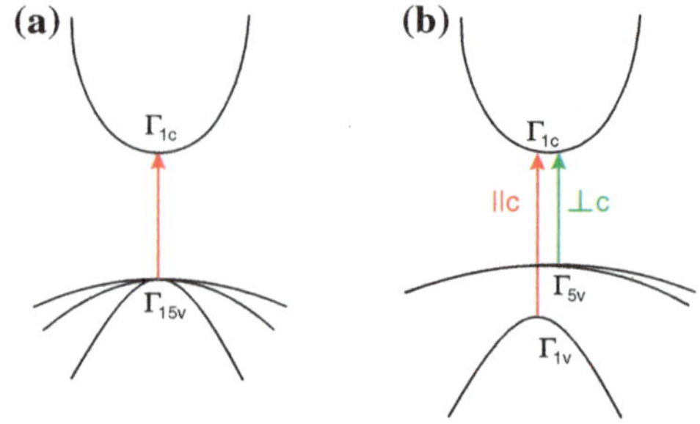

Fig. 20.6 Lowest conduction band and highest three valence bands in (**a**) cubic (*zb*, *rs*) and (**b**) hexagonal (*wz*) nitrides and oxides (schematically). Dipole-allowed optical transitions are characterized by vertical *arrows*. The crystal-field splitting between Γ_{5v} and Γ_{1v} is indicated. The influence of the spin-orbit interaction is not illustrated

Absolute values of momentum matrix elements for $p \rightarrow s$ optical transitions between the highest valence bands Γ_{5v}, Γ_{1v} (wurtzite) or Γ_{15v} (zinc blende, rocksalt) into the lowest conduction band Γ_{1c} are listed in Table 20.1. Here, we use the Bouckaert, Smoluchowski, and Wigner notation [18, 19] Γ_{15v} which leads to Γ_{5v} and Γ_{1v} instead of Γ_{6v} and Γ_{1v} as in the Rashba notation [20] without spin-orbit interaction. The corresponding dipole-allowed transitions are illustrated in Fig. 20.6. In the wurtzite case, ordinary ($\perp$) and extraordinary ($\|$) light polarizations are distinguished. For comparison with literature values also characteristic energies

$$E_p^{\perp,\|} = \frac{2}{m} \left| \sum_{m_s} p_{cvm_s}^{\perp,\|}(0) \right|^2 \tag{20.7}$$

of a $\mathbf{k} \cdot \mathbf{p}$ theory are listed. The values are computed using the HSE03/06 reference electronic structure (see Sect. 9.2.3), which is close to the QP one (see Sect. 16.1 and Fig. 16.4). The listed squares of the momentum matrix elements show clear chemical trends with the reciprocal bond length and the bond ionicity. The matrix elements for *wz*-InN are close to those derived from measured data [21, 22], but seem to underestimate experimental values for *wz*-GaN and *wz*-ZnO [23, 24]. Nonetheless, we will demonstrate in the next two sections that matrix elements of the kind given in Table 20.1 give rise to absolute values of $\mathrm{Im}\epsilon^M(\omega)$ in good agreement with measured spectra in a wide range of photon energies.

20.2 Many-Body Effects

20.2.1 General Trends

Calculating linear optical spectra – the frequency-dependent dielectric function – based on a QP band structure and including screened electron-hole attraction as well

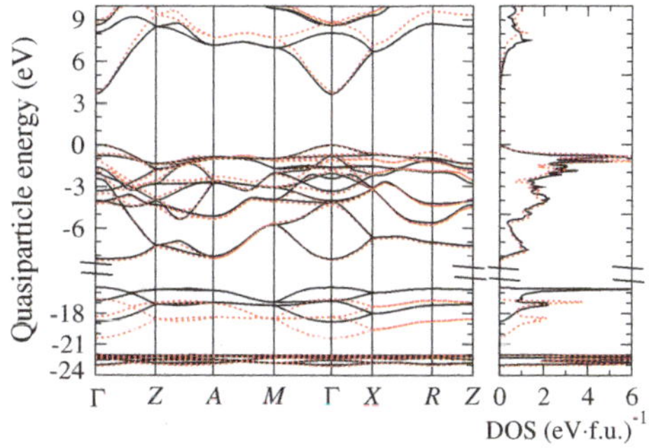

Fig. 20.7 Quasiparticle band structure and density of states of rutile-SnO_2 in HSE03/06 + one-shot GW (*dotted red lines*) and LDA+U+Δ (*solid black line*). The VBM is chosen as common zero of energy. From [15]

as local-field effects by solving the BSE for the macroscopic polarization function is certainly currently state-of-the-art. However, the calculations of QP energies using a HSE03/06 reference electronic structure and a one-shot GW approach or a self-consistent GW starting from another electronic structure are extremely computer-time demanding. This becomes especially obvious remembering that already for bulk systems with two or four atoms in the unit cell converged computations ask for a fine **k**-point sampling of the BZ using up to 5,000 **k** points. Studying bound exciton states at the absorption edge with small exciton binding energies, this number has to be increased to about 100,000 [25]. In the beginning of those calculations therefore the scissors-operator approach (see Sect. 15.3.3) has been applied to the QP band structure [26–28]. Later, in particular in the case of systems with shallow core electrons, a DFT+U+Δ method has been developed [29]. It is based on a LDA+U or GGA+U approach (see Sect. 9.1) with a U that somewhat opens the band gap but keeps the correct (valence) band ordering. To obtain the quasiparticle gap an additional scissors operator Δ (15.27) is applied.

The quality of such an approximate QP description of the electronic structure is illustrated for two approaches to the band structure of rutile-SnO_2 in Fig. 20.7. The HSE03/06+G_0W_0 gap of 3.65 eV [15] is in very good agreement with the experimental one $\sim$3.59 eV. The parameter U is fixed at $U = 4.6$ eV which describes the d-band complex (with respect to the VBM) very well. The scissors operator, chosen as $\Delta = 2.46$ eV to open the gap, rigidly shifts the conduction bands relative to the VBM. However, the O 2s-derived valence bands at approximately 18 eV below the VBM but also the upper valence bands near Γ show deviations from the HSE+G_0W_0 band structure. The conduction band minima in LDA+U+Δ at A, R, and Z are slightly too low in energy. As a consequence, one expects larger deviations in the optical spectra for higher-energy critical points. Nevertheless, the LDA+U+Δ approach, which is by far less computer-time consuming, indeed allows the inclusion of enough **k** points and bands for the computation of optical properties up to high photon energies. We point out that similar findings have been made for wurtzite and rocksalt oxides [13, 14].

For materials with smaller gaps and relatively large dielectric constants a DFT-LDA or -GGA starting point for the GW quasiparticle calculations is good enough.

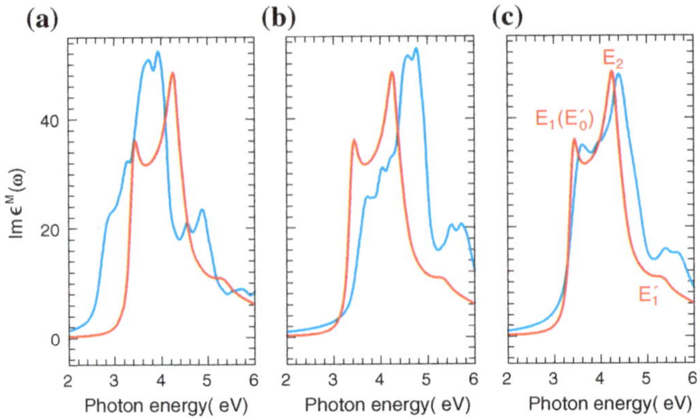

Fig. 20.8 Imaginary part of the dielectric function of Si (*blue solid line*) calculated within the framework of three different approaches: (**a**) Independent-particle approximation using Kohn-Sham eigenvalues of the DFT-LDA, (**b**) independent-quasiparticle approximation including one-shot GW corrections to the KS eigenvalues, and (**c**) correlated quasielectrons and quasiholes including their screened interaction and local-field effects. An artificial broadening of $\hbar\eta = 0.15\,\text{eV}$ is applied. For comparison an experimental spectrum (*red solid line*) [31] is given. The most important critical points [18] are indicated. Adapted from [30]

This is shown in Fig. 20.8 for crystalline Si [30] in comparison to an experimental spectrum [31]. Corresponding KS and QP band structures are displayed in Fig. 14.14. The independent-particle spectrum in Fig. 20.8a exhibits peak positions and a lineshape of the imaginary part of the frequency-dependent dielectric function without similarities with the measured curve. The QP self-energy corrections in Fig. 20.8b lead to a significant blueshift of about 0.8 eV of the characteristic peaks estimated for the E_1 (and E_0') and E_2 critical points [18]. The lineshape is only slightly affected. However, solving the BSE and, hence, taking electron-hole attraction and exchange into account in Fig. 20.8c, a strong redistribution of oscillator strength from the E_2 peak and partially from the E_1' peak to the energy region around E_1 together with a small redshift is observed. The reason for the redistribution is obvious from expression (19.35). Excitonic effects mix various QP interband transitions via the eigenvectors $A_\Lambda(c\mathbf{k}m_s v\mathbf{k}m_s)$. They induce a constructive superposition of the oscillator strengths for transitions at lower energies but lead to a destructive superposition at energies above 5 eV, in order to fulfill the oscillator strength sum rule (13.42). The nature of the sharp peak at 3.4 eV in the experimental spectrum has been under discussion for a long time. The calculations clearly show from *ab initio* that the sharpness and intensity of this peak originate from excitonic effects. While the intensities of the E_1 and E_2 peaks agree well with the measured ones, the calculated positions still occur at energies that are 0.2 eV too high. Reasons for the discrepancy are related to the underestimated theoretical (DFT-LDA) lattice constant, some numerical uncertainties, in particular the **k**-point sampling, and the neglect of temperature effects resulting in a redshift by about 0.05 eV [32]. We have to point out that findings similar to those in Fig. 20.8 have been made in other calculations [3, 27, 32].

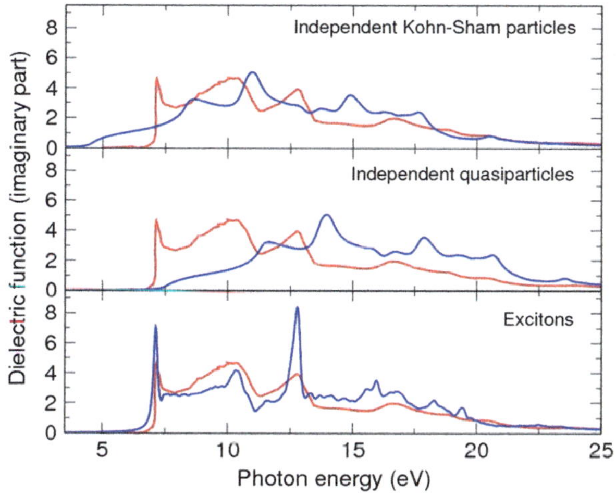

Fig. 20.9 Imaginary part of the dielectric function of *rs*-MgO (*blue solid line*) in three different approximations: Independent KS particles described within DFT-GGA, independent quasiparticles with energies shifted by a scissors energy $\Delta = 2.99\,\mathrm{eV}$, and Coulomb-correlated quasielectron-quasihole pairs including excitonic and local-field effects. A Lorentzian broadening $\hbar\eta = 0.2\,\mathrm{eV}$ is applied. The theoretical spectra are compared with a measured one (*red solid line*) [33]. Adapted from [13]

In wide-band-gap semiconductors and insulators with strong ionic bonds the dielectric screening is much weaker than that in normal semiconductors. The electronic dielectric constants of *rs*-MgO and fluorite-CdF_2 are only $\epsilon_\infty = 3.1$ and 2.5, respectively, compared to 11.7 of Si. Concomitantly, the QP shifts and the electron-hole attraction are much stronger and their effects on the optical spectra are much more pronounced. This is illustrated in Figs. 20.9 and 20.10. Similarly to the

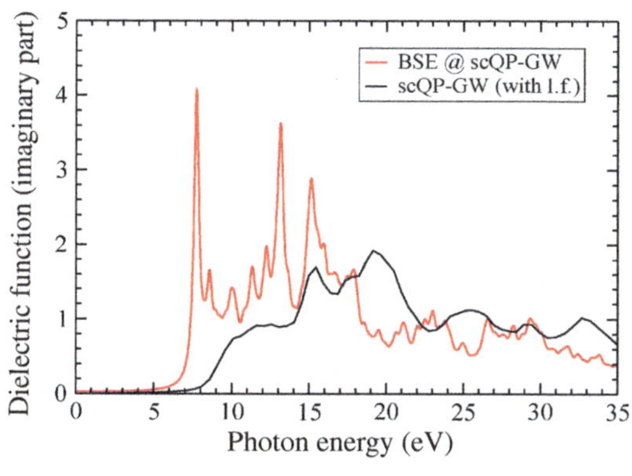

Fig. 20.10 Imaginary part of the dielectric function of CdF_2 within the independent-quasiparticle approach but including local fields (*black solid line*) and after solution of the BSE including excitonic effects (*red solid line*). The HSE03/06 reference electronic structure and a self-consistent GW approximation are used. Adapted from results in [34]

observation for Si, for rs-MgO [13] the QP effects shift the absorption spectrum toward the UV spectral region, here by about 3.0 eV. The position of the calculated and measured absorption edges approach each other. However, the calculated line-shape is still completely different from the measured one. In the entire energy range up to $\hbar\omega = 25$ eV, the spectrum with excitonic effects is completely altered from the non-interacting QP case. Again the absorption spectrum including electron-hole attraction and local-field effects is in much better agreement with experiment [33]. Also the redistribution of spectral strength from higher to lower photon energies due to excitonic effects is clearly visible. The most striking feature in the calculated spectrum is, however, the occurrence of strongly bound singlet excitons. One at about 7.5 eV originates from a bound Wannier-Mott-like excitonic state [25]. In addition, a bound singlet exciton at $\hbar\omega \approx 13$ eV belongs to optical transitions from the highest valence band into the lowest conduction band at critical points X, L, and K at the BZ surface (see band structure in Fig. 15.7). This exciton bound state sits on the continuum of scattering states corresponding to optical transitions around Γ. The other pronounced spectral feature at about $\hbar\omega = 10$ eV originates mainly from transitions between the two highest valence and lowest conduction bands, while that at $\hbar\omega \approx 16$ eV is almost equally composed of transitions from all three O $2p$ valence bands into the conduction bands. The low-energy peaks are almost invisible as distinct peaks in the joint DOS in Fig. 20.3. The peak C in the joint DOS can also hardly be identified in the optical absorption. Within converged calculations the binding energy of the exciton at the absorption edge is evaluated to a value of 0.43 eV. This value significantly overestimates measured binding energies because of neglecting dynamical screening, more precisely the lattice polarizability [25] (see also Sect. 22.1.3). Reports of a better agreement between theory and measurements, despite applying only electronic screening, may be based on less converged computations with a smaller number of **k** points.

The absorption spectrum of a more complex compound, fluorite-CdF_2, is displayed in Fig. 20.10. The independent-quasiparticle approach, but adding local-field effects, is applied starting from the gKS problem based on the HSE03/06 hybrid functional. It is compared to the spectrum of the fully Coulomb-correlated electron-hole pairs [34]. In the studied entire spectral region $0 < \hbar\omega < 35$ eV the two spectra are completely different indicating extremely strong excitonic effects in CdF_2. Again the significant spectral redistribution from higher to lower photon energies is clearly visible. In addition, peaks related to bound excitons belonging to different groups of transitions appear. Summarizing the discussion of Figs. 20.8, 20.9, and 20.10, we have to conclude that optical spectra cannot be calculated without excitonic effects, despite the fascinating and suggestive relation of van Hove singularities in the inter-band electronic structure to absorption features [1, 32].

20.2.2 Validity of Scenario of Van Hove Singularities

The results presented in Sects. 20.1.2 and 20.2.1 have shown the significant influence of the optical transition matrix elements and electron-hole attraction combined

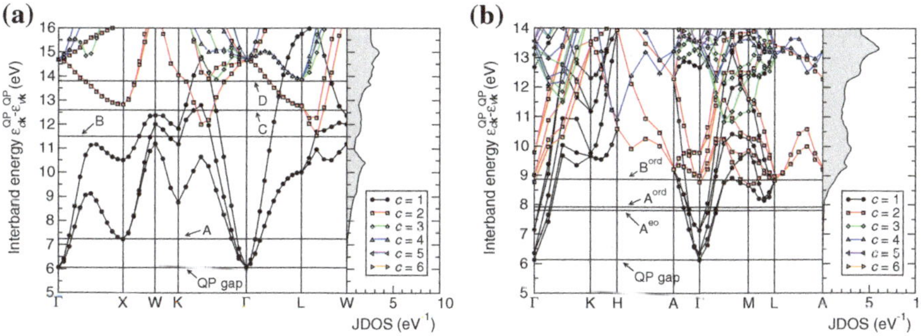

Fig. 20.11 Joint band structure and JDOS (per formula unit) for, (**a**) *zb*-AlN and (**b**) *wz*-AlN in the HSE03/06 + one-shot GW approach. The interband transition energies from all valence bands v to the lowest six conduction bands c are depicted. The calculated positions of the QP gap and the peaks A, B, C, and D (*zb*) or A^{ord} and B^{ord} (*wz* for ordinary light polarization) and A^{eo} (*wz* for extraordinary polarization) are indicated. From [35]

with local-field effects on the frequency dependence of the dielectric function. This holds especially for the comparison of the joint density of states and the imaginary part of the dielectric function in Fig. 20.3 as well as the comparison of the dielectric function computed within the independent-quasiparticle approach and that including additionally excitonic and local-field effects in Figs. 20.8, 20.9, and 20.10. In particular, the discrepancies in the absorption spectra with or without electron-hole interaction suggest a violation of the common picture of the explanation of spectral features in optical absorption spectra by van Hove singularities [1, 18].

In order to investigate the degree of violation we study the optical spectra of zinc-blende- and wurtzite-AlN in detail, based on a starting electronic structure generated within the hybrid HSE03/06 approach. First, we present the joint band structure, the optically relevant interband structure $\varepsilon_c^{\text{QP}}(\mathbf{k}) - \varepsilon_v^{\text{QP}}(\mathbf{k})$, together with the corresponding joint DOS in Fig. 20.11. This figure clearly indicates the peak structures in the JDOS (20.5) and their relation to interband transitions in the QP band structure. The optical absorption spectra based on these QP electronic structures are displayed in Fig. 20.12. Those spectra again show the drastic Coulomb effects related to electron-hole attraction and exchange. The lineshape of the absorption spectra is significantly modified by excitonic effects. In the wurtzite case in Fig. 20.12b the measured most important peaks A^{ord} and B^{ord} (at least for ordinary light polarization) are correctly described with respect to energy position and intensity. In the zinc-blende case in Fig. 20.12a the issue of **k**-point sampling of the BZ is clearly illustrated. An $18 \times 18 \times 18$ MP **k**-point mesh is obviously not sufficient. The redistribution of spectral strength from higher to lower photon energies is still incomplete, as indicated by the underestimation (overestimation) of the intensity of the A (B, C, D) peak(s). An increase of the sampling density to a $40 \times 40 \times 40$ mesh leads, however, to a significant improvement of the spectrum around the A feature. The reason is the interference effect described in (19.35), which sensitively depends on the **k**-point sampling but also on other numerical and physical treatments. For instance, the

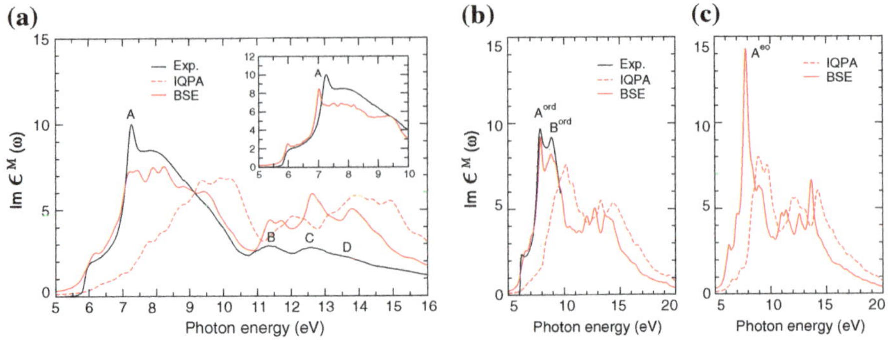

Fig. 20.12 Imaginary part of the dielectric function of, (**a**) *zb*-AlN, (**b**) *wz*-AlN (ordinary polarization), and (**c**) *wz*-AlN (extraordinary polarization). The spectra in IQPA (*red dashed lines*) and the solution of the inhomogeneous BSE including electron-hole attraction and local-field effects (*red solid lines*) are depicted. A Lorentzian broadening of $\hbar\eta = 0.2\,\mathrm{eV}$ has been applied. In the *inset* of (**a**) the **k**-point density is increased by about one order of magnitude, while only quasielectron-quasihole pairs with energies $<10\,\mathrm{eV}$ and a broadening parameter of $\hbar\eta = 0.1\,\mathrm{eV}$ are taken into account. Experimental spectra for *zb*-AlN [36] and *wz*-AlN [37] (*black solid lines*) are given for comparison. From [35]

simplification of the QP treatment, e.g. by a scissors operator, improves the agreement between theory and experiment in the case of AlN (not shown).

In order to answer the question, if the spectral features e.g. in Fig. 20.12a can still be related to the critical points in the quasiparticle band structure, as indicated in Fig. 20.13, and hence to van Hove singularities (20.6), we attempt to visualize the contributions of different **k**-space regions around high-symmetry points in Fig. 20.14. Thereby, we focus on points Γ, X, and L of the highest symmetry, i.e., with the largest subgroups of the **k** point. Spherical regions around these **k** points with radii $0.05 \times 2\pi/\text{Å}$ $(0.1 \times 2\pi/\text{Å})$ are excluded from the **k** summation in (19.35).

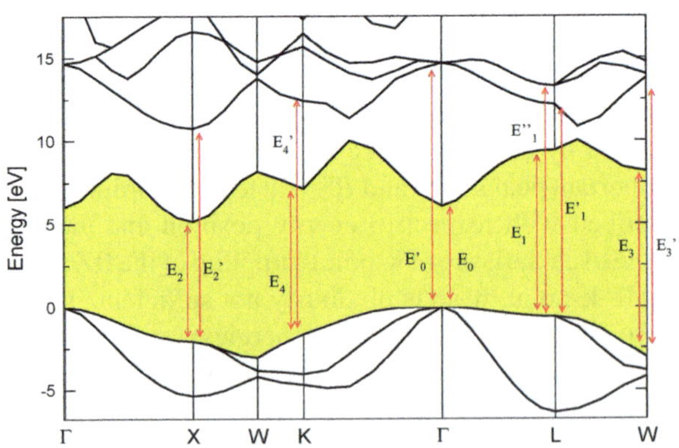

Fig. 20.13 High-symmetry points in the Brillouin zone leading to van Hove singularities in the band structure. As an example *zb*-AlN is chosen. The denotation follows that in [18]. The *yellow* area illustrates the fundamental gap region. Adapted from [35]

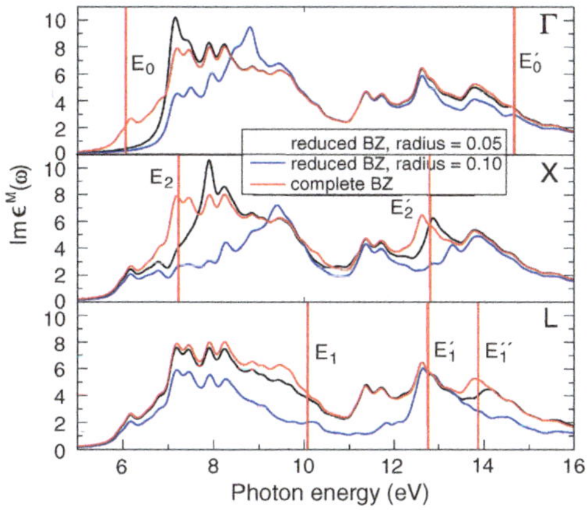

Fig. 20.14 Contribution of Γ-, X-, and L-point regions in the BZ to the optical absorption of *zb*-AlN. The *red line* depicts the solution of the BSE for the entire BZ, i.e., the spectrum shown in Fig. 20.12a. Furthermore, spectra omitting spheres with radii of $0.05 \times 2\pi$ Å^{-1} (*black lines*) or $0.1 \times 2\pi$ Å^{-1} (*blue lines*) around the high-symmetry point(s) are shown. The QP transition energies of important van Hove singularities related to points Γ, X or L are indicated. A Lorentzian broadening of $\hbar\eta = 0.15\,\text{eV}$ has been applied. Adapted from [35]

Contributions of 1 % (9 %) for Γ, 3 % (27 %) for X, or 4 % (36 %) for L of the BZ volume are left out. In a naive picture of contributions of single optical transitions, one would expect a reduction of oscillator strength of a peak which mainly stems from a critical point. However, there are two counteracting effects due to the interference of optical transition-matrix elements $M^{m_s}_{\mathbf{kk}\,cv}(\hat{\mathbf{q}})$ (19.27) weighted by the pair eigenvectors $A_\Lambda(c\mathbf{k}m_s v\mathbf{k}m_s)$: First, there are additional lineshape modifications due to the neglect of the interaction between excluded transitions and the remaining ones in the pair Hamiltonian (19.32). Second, an enhancement of oscillator strength near the energy of the excluded critical point can occur due to constructive superposition [see also (19.35)].

Excluding the surroundings of the Γ point, indeed the contributions to the E_0 structure disappear in Fig. 20.14. For E_0' this effect is significantly reduced due to the comparably strong contributions from other transitions in the relevant spectral region. Similarly pronounced reductions occur near the E_2 (A peak) and the E_2' (C peak) critical-point energies if a vicinity of the X points is omitted. The exclusion of the L points leads to a reduction of the E_1' structure (D peak). However, the C peak in the absorption spectrum near E_1' remains uninfluenced, suggesting that the contribution of optical transitions near the L points is negligible. The absorption near E_1 is also reduced, despite the fact that the corresponding van Hove singularity is only a saddle point (see Fig. 20.11a). Interestingly, in the low-energy region around 7–8 eV one observes artificial peaks with increased intensity compared to a complete BZ sampling despite the reduction of the studied $\mathbf{k}$ space around Γ and X. These peaks are due to enhanced constructive interference effects, overcompensating the reduction of contributing $\mathbf{k}$ points.

We end with somewhat contradictory pictures. One the one hand, the clear discrepancies between the spectra within the independent-quasiparticle approach and those including Coulomb effects via the solution of the BSE in Fig. 20.12 indicate a violation of the picture of van Hove singularities due to excitonic effects. On the other hand, the disappearance or the reduction of peaks close to critical points E_0 and E_0' around Γ, E_2 and E_2' around X, as well as E_1 and E_1'' around L reducing the corresponding $\mathbf{k}$-space contributions suggests that the picture of van Hove singularities and their relation to certain high-symmetry points still makes sense.

20.2.3 Summary and Conclusions

The simultaneous solution of the Dyson and Bethe-Salpeter equations to compute the quasiparticle Green functions and macroscopic polarization functions including quasielectron-quasihole attraction and local-field effects is needed to compute linear optical properties and their frequency dependence in a quality to be comparable with measured optical spectra or to have predicting power for corresponding properties. The most important consequences of the many-body effects are:

i. A significant blueshift of an absorption spectrum toward higher photon energies due to quasiparticle effects.
ii. Excitonic effects give rise to a redshift of higher-energy spectral features accompanied by a drastic redistribution of oscillator or spectral strength from higher photon energies to the low-energy region.
iii. The formation of bound exciton states in the fundamental QP gap or below higher optical interband transitions which are, however, only visible for binding energies larger than the lifetime broadening.

The three consequences are illustrated in Fig. 20.15 for the optical absorption of *rs*-MgO (see also Figs. 15.7 and 20.9). Similar excitonic effects are expected for the spectral variation of non-linear optical properties, e.g. the second-harmonic generation [38].

The blueshift due to quasiparticle effects is obvious from the findings in Chaps. 15, 16, and 17. The redshift of the high-energy part of an absorption spectrum does not result from a negative shift of the interband transition energies. Rather, it is expected within a naive picture of the electron-hole attraction. In such a picture mutual cancellation of the two shifts is expected. However, such a naive picture is only partially valid for low-energy pair excitations in nanocrystals [39]. In the case of crystalline solids the reason of consequence (ii) is more related to the transition strength (19.25) in (19.35). It shows that the oscillator strength of a pair excitation with energy E_Λ is given by a Coulomb-interaction-induced mixture of transitions in the single-particle picture $|v\mathbf{k}m_s\rangle \rightarrow |c\mathbf{k}m_s\rangle$. Instead of one individual band-to-band transition, rather several ones weighted by the eigenvector A_Λ of the pair state

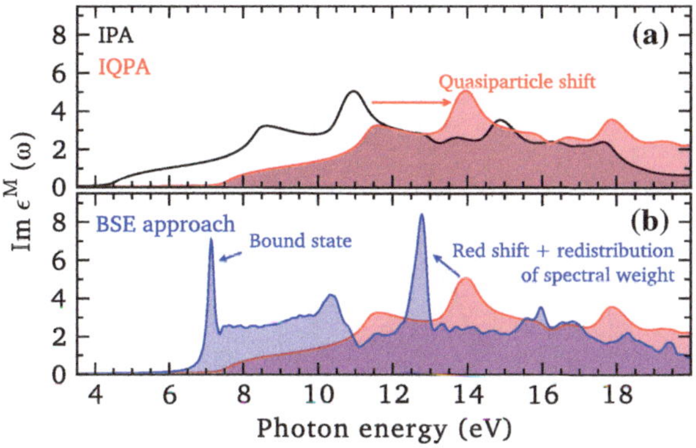

Fig. 20.15 Illustration of (**a**) quasiparticle and (**b**) excitonic effects on an optical absorption spectrum. As an example the wide-band-gap material *rs*-MgO is chosen. Adapted from results in [13]

Λ contribute. The naively expected main effects due to a drastic change in the joint density of states or, more precisely, density of pair states have only a minor influence, at least for materials with weak excitonic effects and large dielectric constants. This is clearly demonstrated in Fig. 20.16 for zinc-blende-GaAs [3]. It is obvious that the joint density of states remains nearly unchanged by the electron-hole interaction, with the exception of bound exciton states below the QP gap, which are, however, not visible in the displayed figure. The drastic modifications of the optical spectrum in independent-quasiparticle approach due to excitonic effects originate from a completely different phenomenon discussed above: They are due to the coherent coupling of the optical transition matrix elements by the electron-hole interaction. It leads to a constructive superposition at lower excitation energies and to a destructive superposition at energies above 5 eV. Such a behavior is not restricted to bulk semiconductors and insulators [3, 13–16, 25–28, 30, 32, 34, 35] but has been found also in low-dimensional systems [3] as we will demonstrate in Sect. 20.3.3.

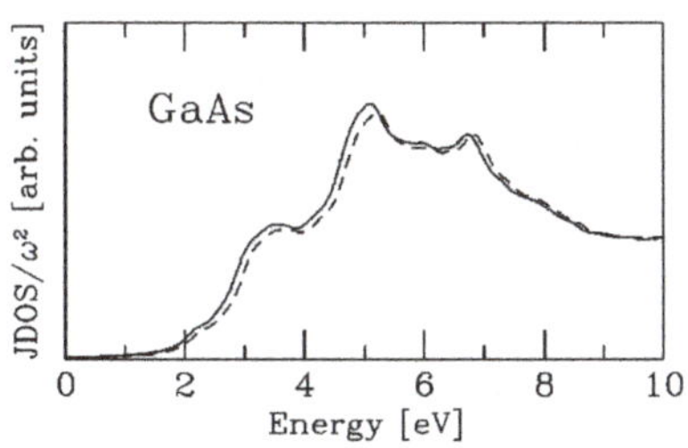

Fig. 20.16 Joint density of quasiparticle states (*dashed line*) and pair density of states including electron-hole attraction and local-field effects (*solid line*) for *zb*-GaAs. For a better comparison with the absorption spectrum the densities are divided by ω^2 and therefore given in arbitrary units. Reprinted with permission from [3]. Copyright 2000 by the American Physical Society

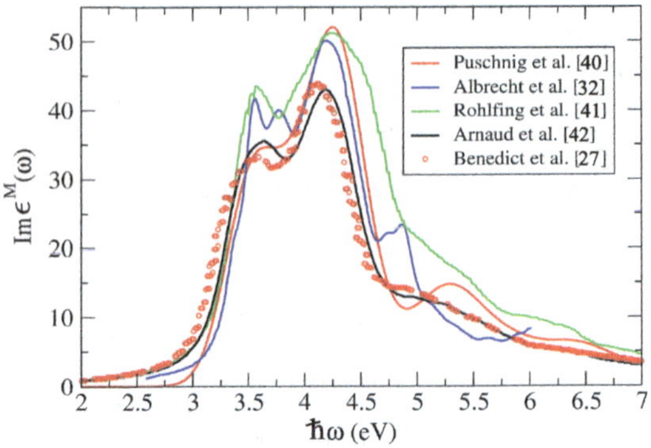

Fig. 20.17 The dielectric function of Si resulting from the solution of the BSE as obtained in various calculations by Puschnig and Ambrosch-Draxl [40], Albrecht et al. [32], Rohlfing and Louie [41], Arnaud and Alouani [42], and Benedict et al. [27]. Reprinted with permission from [40]. Copyright 2002 by the American Physical Society

On the other hand, we have to point out that the correct description of the interference effects depends sensitively on the used numerical procedure to describe the basis set, the wave functions and eigenvalues, the treatment of the GW quasiparticle effects, and the convergence of the BSE concerning the number of bands and **k** vectors in the BZ sampling. This is clearly illustrated in Fig. 20.17 for the imaginary part of the macroscopic dielectric function of Si. Three of the five computations do not correctly describe the redistribution of spectral strength into the low-energy E_1 (E_0') peak (near $\hbar\omega \approx 3.5\,\text{eV}$) whose intensity approaches a value of 40 in experimental spectra. Another discrepancy concerns the description of the E_1' peak (near $\hbar\omega \approx 5.5\,\text{eV}$) which enormously varies with the actual numerical treatment (see also Fig. 20.8).

The exciton bound states (iii), e.g. those occurring below the quasiparticle gap but are not visible in Figs. 20.16 and 20.17, are a clear consequence of the screened Coulomb attraction of electron and hole $\sim W$ (19.33). They will be investigated in more detail in the next chapter.

20.3 Absorption, Refraction, Reflection and Energy Loss Spectra

20.3.1 Bulk Anorganic Crystals

In the last sections we have discussed the influence of many-body effects on the frequency dependence of the dielectric function, mainly for its imaginary part that is directly related to the absorption spectrum (18.27). As model materials bulk crystals of elementary or compound semiconductors and insulators have been chosen. Another example is presented in Fig. 20.18 for the extremely ionic compound LiF with wide gap. This figure clearly proves the superiority of the BSE approach over all other treatments to compute optical properties of matter. It nicely reproduces

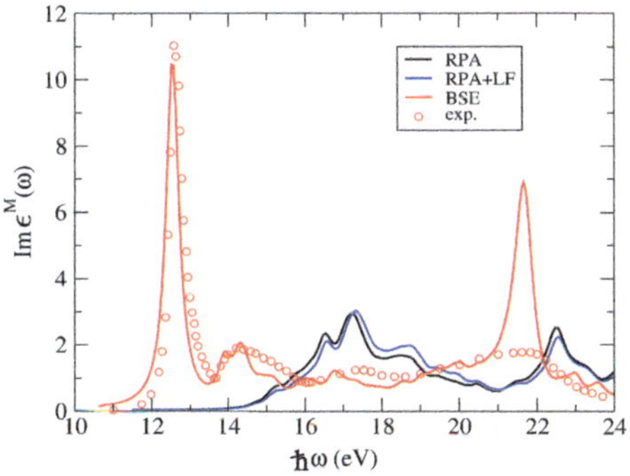

Fig. 20.18 The dielectric function of LiF as obtained from the solution of the BSE (*red solid line*). The *black* (*blue*) *solid line* results for independent quasiparticles without (with) local-field effects. A Lorentzian broadening of $\hbar\eta = 0.2\,\text{eV}$ has been applied. For comparison, experimental data (*circles*) [43] are also shown. Reprinted with permission from [40]. Copyright 2002 by the American Physical Society

the strong bound exciton peak at 12.5 eV. The binding energy of 2.2 eV measured with respect to the QP gap at 14.7 eV is slightly overestimated by 0.2 eV due to the neglect of the screening by the lattice polarization. However, the lattice polarization effects on the exciton binding and QP gap cancel each other in parts, at least near the absorption edge. The intermediate range of photon energies $14 < \hbar\omega < 20$ is well described by the BSE/QP theory [40], despite the extremely small values $\text{Im}\epsilon^M(\omega) \approx 1$. The only major discrepancy remains at roughly 21.5 eV, where the BSE theory predicts a strong peak that does not appear in experimental data [43]. The reason of this discrepancy is still under debate. One reason could be a bound exciton feature belonging to high-energy interband transitions.

Absorption spectra of a completely different class of materials, namely of the metals copper and silver, are displayed in Fig. 20.19 [44] and compared with

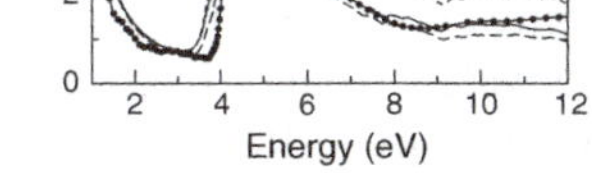

Fig. 20.19 Absorption spectrum of bulk copper and silver. *Dotted line*: BSE. *Dashed line*: BSE including the QP renormalization factors. *Solid line*: result of the Bethe-Salpeter equation including dynamical QP and excitonic effects. *Circles*: experimental data [45]. Reprinted with permission from [44]. Copyright 2002 by the American Physical Society

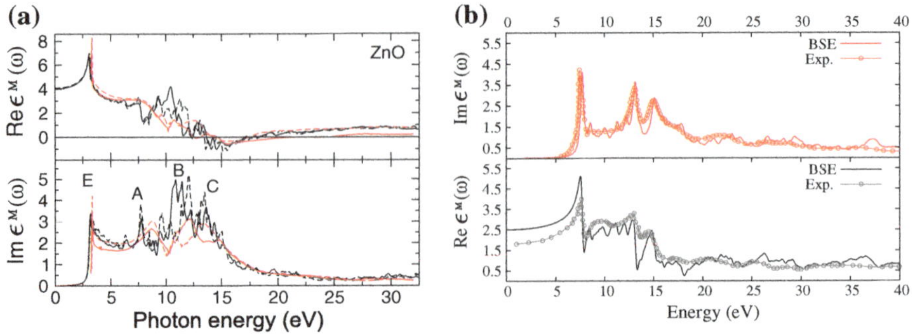

Fig. 20.20 Real and imaginary part of dielectric function calculated within the BSE scheme [13, 34] in comparison with experimental results [46, 47] of, (**a**) wurtzite-ZnO and (**b**) fluorite-CdF_2. In the *wz*-ZnO case theoretical (experimental) curves are displayed as *black* (*red*) *lines*. Ordinary and extraordinary light polarizations are distinguished by *solid* and *dashed lines*, respectively. In the CdF_2 case the theoretical spectra are displayed as *black* and *red solid lines*, whereas the experimental data are given by *dots*. From [13, 34]

experimental data [45]. The metals are examples which indicate a failure of the BSE approach neglecting the dynamics of the screening (19.3), despite the inclusion of effects beyond the Tamm-Dancoff approximation (19.31). This neglect gives rise to an overestimation of the absorption in the range of interband transitions. A rigorous reduction of the quasiparticle peaks by a renormalization factor $z_{\nu m_s}(\mathbf{k})$ (14.47) leads to the opposite tendency. Therefore, dynamical excitonic effects are extremely important to find agreement with experiment. Some details of the inclusion of dynamical screening will be discussed in Sects. 22.1 and 22.2.

The dielectric properties and optical refraction (18.26) are dominated by the real part of the macroscopic dielectric function $\epsilon^M(\omega)$. Since it is frequently computed by means of a Kramers-Kronig transformation, it is displayed in Fig. 20.20 together with the imaginary part for a wide range of photon energies. BSE results [13, 34] are compared with experimental data [46, 47] for wurtzite-ZnO and fluorite-CdF_2. In the case of wurtzite ordinary and extraordinary light polarizations are distinguished. On average the real parts of $\epsilon_\perp^M(\omega)$ and $\epsilon_\parallel^M(\omega)$ are well represented by the solution of the BSE, at least in the low-energy region. Interestingly the first pronounced peak in the absorption spectra gives also rise to a distinct peak in $\mathrm{Re}\epsilon^M(\omega)$. After this peak the spectral intensity decreases with rising photon energy $\hbar\omega$. However, only for ZnO zeros of $\mathrm{Re}\epsilon^M(\omega) = 0$ are observable in the energy regions around $\hbar\omega \approx 14$ and 18 eV indicating some plasmonic excitations. For CdF_2 such zeros are missing due to the decomposition of the lower valence bands in groups of flat bands, which give rise to oscillators with finite contributions to $\mathrm{Re}\epsilon^M(\omega)$ on the low-energy side of each oscillator frequency. The agreement between theory and experiment in the CdF_2 case seems to be better. This is true for the peak positions but also the average intensity. Only for $\omega \to 0$ the computed $\mathrm{Re}\epsilon^M(\omega)$ seems to overestimate the experimental data [47]. However, there are other experimental data [48, 49] which tend to confirm the theoretical value of roughly $\epsilon_\infty = \mathrm{Re}\epsilon^M(0) = 2.5$.

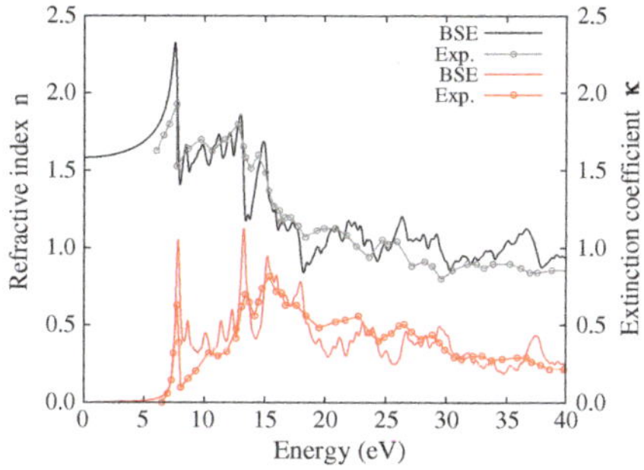

Fig. 20.21 Real and imaginary part of the complex refractive index (n, κ) of fluorite-CdF_2 as a function of photon energy in comparison with measured spectra [47]. From [34]

The reasonable theoretical description of the dielectric function in Fig. 20.20b should lead to a similarly good agreement between BSE theory and experiment for the derived optical quantities, for instance the real and imaginary parts, $n(\omega)$ and $\kappa(\omega)$ (18.26), of the complex index of refraction $\sqrt{\epsilon^M(\omega)}$. This is really demonstrated in Fig. 20.21. The principal spectral features are reproduced by the BSE theory in a quality similar to that seen for the real and imaginary parts of the dielectric function. Due to the **k**-point sampling used in the numerical treatment and the disregard of instrumental and other broadenings the theoretical spectra are somewhat more structured. However, averaged over larger energy intervals the parameterfree theory including many-body effects really gives rise to spectra in good agreement with measurements.

Other quantities derived from the macroscopic dielectric function $\epsilon^M(\omega)$ are the reflectivity $R(\omega)$ (18.28) and the loss function $L(\omega)$ for vanishing momentum transfer (18.29). Examples are again given for group-II oxides crystallizing in rocksalt or wurtzite structure and fluorite-CdF_2 in Fig. 20.22. The low-energy peaks are reasonably described in all reflectance spectra (left panels). In the high-energy region of MgO, ZnO, and CdF_2 the theoretical curves tend to overestimate the reflectivity and show much more spectral features than the measured spectra. Nevertheless, we have to point out that the parameterfree theory applied is able to reproduce the extremely small reflectance values of a few percent in the low-energy spectrum of CdF_2. A similar reasonable agreement is found for the energy loss spectra in Fig. 20.22d (right panel). In the case of rs-MgO the major peak near 23.8 eV and another feature around 22 eV are due to plasmonic losses of the sp valence electrons. Indeed, the plasma frequency of the sp electrons (13.28) derived from the theoretical lattice parameter amounts to $\hbar\omega_p = 23.9\,\text{eV}$. The broad feature around 20 eV in the loss spectra of ZnO may be related to the successive contributions of the p-type, s- and p-type, and s-, p-, and d-type valence electrons to the plasmon resonances with plasma frequencies near 18.1, 21.1, and below 31.6 eV (the d electrons do not contribute fully because of their stronger bonding). For similar reasons a broad three-peak feature appears for CdO in the energy region 15–23 eV. Because of the more localized valence electrons a pronounced plasmon feature is not observed in the loss function of CdF_2. In the low-energy region $\hbar\omega < 15\,\text{eV}$, however, two major peaks appear, which are known

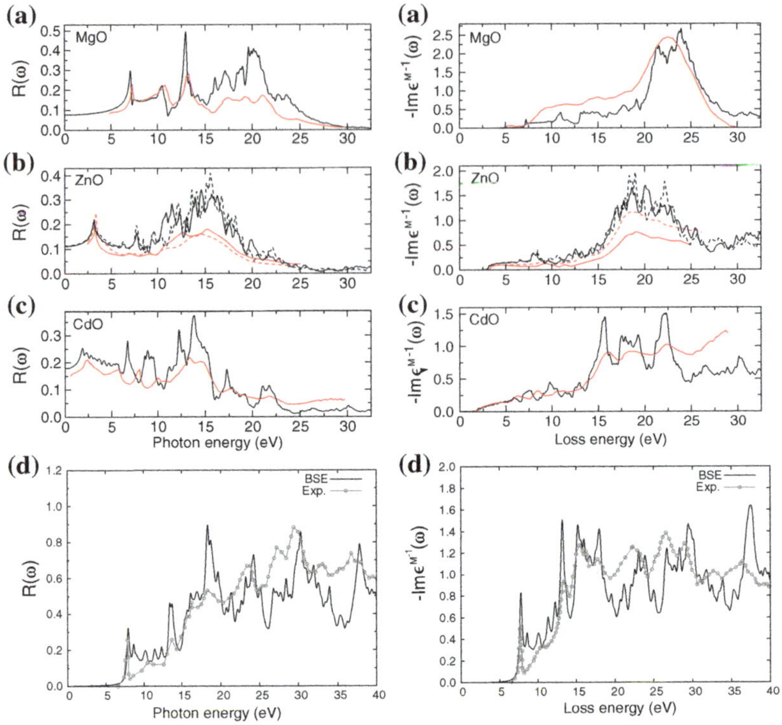

Fig. 20.22 Reflectance (*left panels*) and energy loss (*right panels*) spectra of (**a**) *rs*-MgO, (**b**) *wz*-ZnO, (**c**) *rs*-CdO, and (**d**) fluorite-CdF_2. Results of the solution of the BSE (*black solid lines*) are compared with experimental data (*red solid* or *dotted lines*) [33, 46, 47, 50]. In the *wz*-ZnO case two spectra are shown related to ordinary/extraordinary light polarization or ordinary/extraordinary direction in the measurement geometry. From [13, 34]

from the absorption spectra in Figs. 20.20 and 20.21, in agreement with the fact that the spectral behavior of the loss function is given by $L(\omega) \approx \mathrm{Im}\epsilon^M(\omega)$. The real part $\mathrm{Re}\epsilon^M(\omega)$ mainly determines the absolute values and the shape of the peaks.

20.3.2 Organic, Hydrogen-Bonded, and Magnetic Systems

In addition to the anorganic crystals, organic molecular crystals are of interest for a large number of (opto)electronic applications. They consist of molecular units bonded through weak van der Waals forces. For the examples picene and pentacene the herring-bone monoclinic (picene) and triclinic (pentacene) crystal structures with two molecules in a primitive unit cell are displayed in Fig. 20.23. As a consequence of the weak vdW bonding, their electronic structure is mainly dictated by that of the isolated molecules. Consequently, rather non-dispersive bands with π and σ character appear.

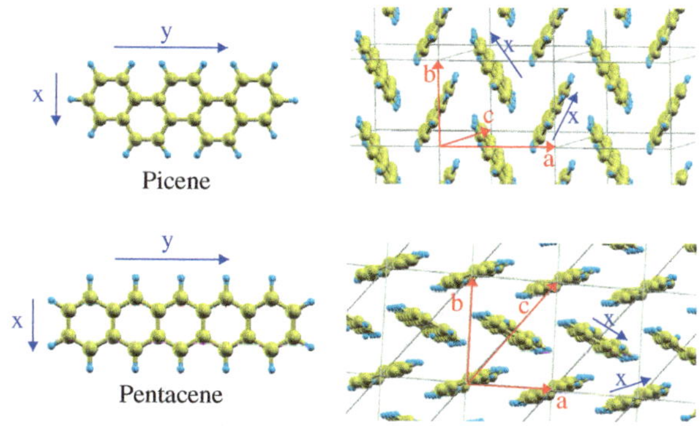

Fig. 20.23 Molecular (*left*) and crystal (*right*) structure of picene and pentacene. The shorter (x) and longer (y) molecular axes and their orientation with respect to the lattice vectors $\boldsymbol{a}$, $\boldsymbol{b}$, and $\boldsymbol{c}$ of the monoclinic (picene) or triclinic (pentacene) crystal are indicated. The *yellow* (*blue*) *dots* denote carbon (hydrogen) atoms. Reprinted with permission from [51]. Copyright 2012 by the American Physical Society

The question arises, whether the optical absorption in such systems is dominated by the direct interband transitions or tightly bound excitons, e.g. Frenkel excitons or even charge-transfer (CT) excitons. These pair excitations will be explained in Chap. 21.

Calculated absorption spectra are displayed in Fig. 20.24. The spectra for the two organic crystals for various directions of the light polarization exhibit two

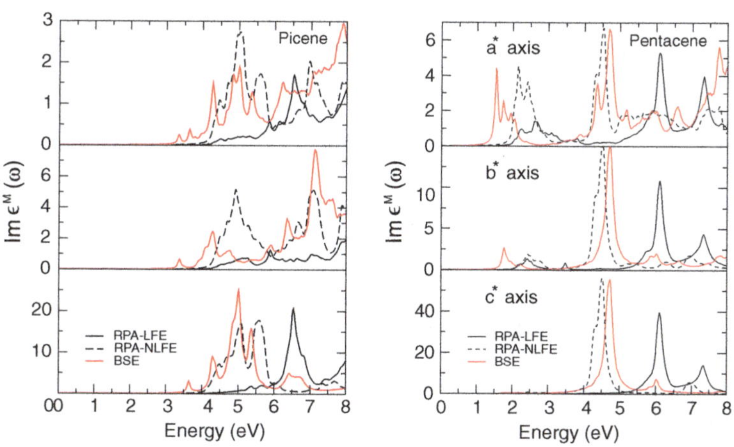

Fig. 20.24 Absorption spectra of picene (*left panels*) and pentacene (*right panels*) molecular crystals for *light* polarization parallel to the reciprocal lattice axes $\boldsymbol{a}^*$, $\boldsymbol{b}^*$, and $\boldsymbol{c}^*$. The spectra are obtained within the independent-quasiparticle approach (RPA but no local-field effects (NLFE), *black solid lines*), IQPA including local-field effects (RPA-LFE, *black dashed lines*), and the full solution of the static Bethe-Salpeter equation (BSE, *red solid lines*). Reprinted with permission from [51]. Copyright 2012 by the American Physical Society

characteristic features, (i) an enormous dependence on the many-body effects and (ii) a pronounced anisotropy:

i. Interestingly, for the organic crystals not only the Coulomb attraction of electrons and holes leads to a drastic modification of the lineshape including the formation of bound exciton states. The corresponding exciton binding energies, defined as the difference between the quasiparticle band gap, 4.08 eV (picene) and 2.02 eV (pentacene), and the position of the first strong peak in the absorption spectrum due to excitonic effects in the fundamental quasiparticle gap, amount to 0.7 eV (picene) and 0.5 eV (pentacene). The position of the first visible bound exciton peak depends on the light polarization. For a polarization parallel to the $\boldsymbol{c}^*$ axis it is 'dark' in pentacene. The corresponding polarization-induced shift may be identified as the Davydov splitting [52]. In addition, in contrast to simple anorganic crystals, the LF effects strongly affect the optical spectra. The reason is related to the strongly localized and polarizable electron density in organic molecular crystals, which gives rise to an important Hartree response [53]. The LF effects lead to a strong redistribution of oscillator strength in picene but mainly to a rigid shift in the case of pentacene.
ii. The optical anisotropy is related to both the arrangement of the organic molecules in the crystal and the symmetry of the molecules themselves. Picene and pentacene molecules possess the C_{2v} and D_{2h} point group, respectively. Therefore, the only dipole-allowed optical transitions are $\pi \rightarrow \pi^*$ and $\sigma \rightarrow \sigma^*$ for light polarization in the molecule plane as well as $\pi \rightarrow \sigma^*$ and $\sigma \rightarrow \pi^*$ for light polarization perpendicular to the molecule plane [51]. Thereby σ and π (σ^* and π^*) denote the most important bonding (antibonding) states of the molecule. The $\boldsymbol{a}^*$ and $\boldsymbol{b}^*$ axes have a large component normal to the molecule plane, while $\boldsymbol{c}^*$ is nearly parallel to the main axis of the molecule. For picene in Fig. 20.24a, roughly speaking, in the low-energy region where mostly $\pi \rightarrow \pi^*$ transitions occur, the oscillator strengths for light polarization in the $\boldsymbol{a}^*\boldsymbol{b}^*$ plane are small and the spectrum is dominated by transitions with polarization along $\boldsymbol{c}^*$. In pentacene in Fig. 20.24b the lowest transitions are forbidden for light polarization parallel to $\boldsymbol{c}^*$. The absorption onset only exhibits contributions from transitions with polarization parallel to the $\boldsymbol{a}^*\boldsymbol{b}^*$ plane.

The strong excitonic effects, i.e., electron-hole attraction and electron-hole exchange, are confirmed by studies of other molecular crystals, e.g. anthracene in Fig. 20.25 [54]. The corresponding molecule structure is also depicted in this figure. According to the monoclinic symmetry of the crystal, the dielectric tensor consists of four independent components ϵ_{xx}^M, ϵ_{yy}^M, ϵ_{zz}^M, and ϵ_{yz}^M. The molecule axes $\boldsymbol{b}$ and $\boldsymbol{c}$ are parallel to the Cartesian axes y and z. The corresponding tensor components are most important. For light polarization parallel to $\boldsymbol{b}$, a strong bound exciton peak appears at $\hbar\omega = 3.11$ eV with a binding energy of 0.64 eV in the calculated spectrum [54]. This is in reasonable agreement with experimental data of 3.1 and 0.8 eV [55]. The quasiparticle gap deduced from the measurements is 3.9 ± 0.1 eV. The main absorption peaks are sensitive to external perturbations. For instance, hydrostatic pressure induces a significant redshift of all absorption peaks in Fig. 20.25.

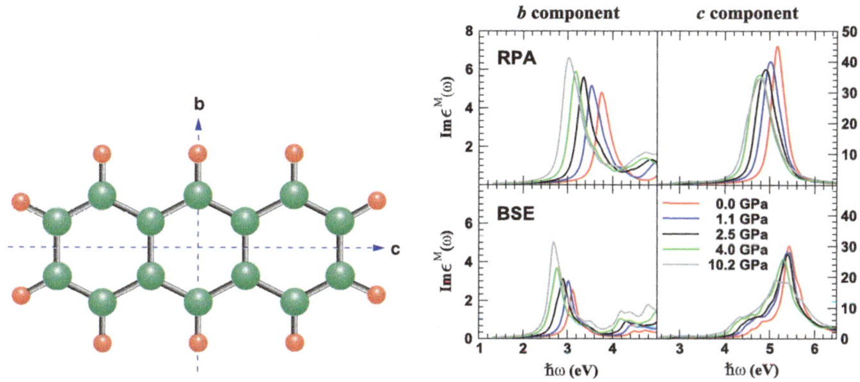

Fig. 20.25 Imaginary part of the dielectric function of an anthracene crystal for light polarization parallel to the short (***b***) or long (***c***) axis of the molecule displayed in the *left panel*. Spectra are shown as obtained within the independent-quasiparticle approximation (RPA, *upper panels*) or solving the BSE (*lower panels*). A lifetime broadening of 0.1 eV is used. The spectra vary with the applied hydrostatic pressure. Reprinted with permission from [54]. Copyright 2004 by the American Physical Society

For light polarization parallel to the ***c*** axis the spectra are drastically changed. Electronic pair excitations appear at higher energies with a much larger oscillator strength. They form a second optically active absorption edge, whereas the excitations at the first one are dipole-forbidden for this polarization direction. Bound states appear only as a shoulder, while the main absorption peak at $\hbar\omega = 5.4\,\text{eV}$ is even above the corresponding quasiparticle transition energy. It seems that it has been observed in measured photocurrent and extinction spectra [56, 57]. An applied hydrostatic pressure again induces a redshift of the spectrum.

Other weakly bonded solids may be characterized by hydrogen(-bridge) [58] or shared-hydrogen [59] bonds. Examples are polymorphs of solid water, ice, e.g. its hexagonal modification (ice-Ih). Absorption spectra of hexagonal ice are depicted in Fig. 20.26. The absorption spectra of cubic and amorphous ice are very similar. This especially holds for the first absorption peak at about 8.7 eV [60]. The independence of its position on the particular ice phase suggests its molecular origin. However, the measured lowest-lying pair excitation of a H_2O molecule occurs at a lower energy of 7.3–7.49 eV in agreement with GW/BSE calculations [61], which yield 7.28 eV. The development of the absorption spectrum with the inclusion of many-body effects and its comparison with a measured spectrum are displayed in Fig. 20.26 for the hexagonal polymorph. We observe the discussed general consequences of the many-body effects, blueshift due to the QP corrections (with a gap opening of about 4.5 eV), redistribution of oscillator strength from higher to lower photon energies together with a redshift, and the formation of a bound exciton peak near 8.7 eV at the absorption edge with a huge exciton binding energy of about 3.2 eV. It is demonstrated in Fig. 20.27 that this peak is related to a bound exciton Λ_0 by the two-particle wave function $\Phi^{m_s}_{\Lambda_0}(\mathbf{x}_e, \mathbf{x}_h)$ with a fixed hole position $\mathbf{x}_h$ at an oxygen atom of a H_2O unit

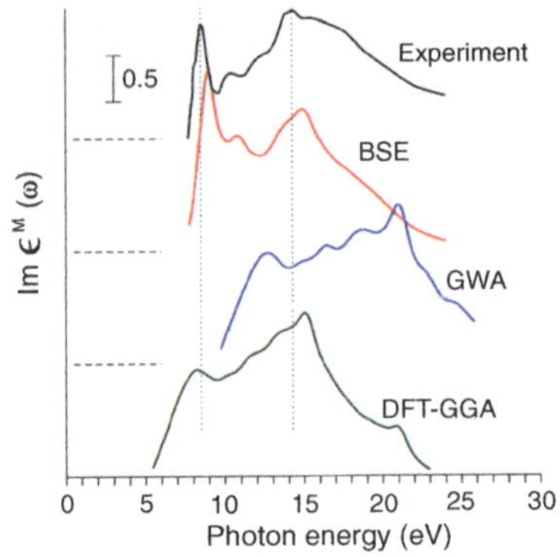

Fig. 20.26 The imaginary part of the macroscopic dielectric function of hexagonal ice as calculated in independent-particle approximation (DFT-GGA), independent-quasiparticle approximation (GWA), and solving the BSE including screened electron-hole attraction and unscreened electron-hole exchange (local-field effects). An average of the polarization directions is presented. For comparison an experimental spectrum [62] is also shown. From [60]

within the hexagonal ice. More strictly speaking, its square, i.e., the probability to find the electron of the pair for a given hole position, is depicted. The pair wave function

$$\Phi^{m_s}_{\Lambda_0}(\mathbf{x}_e, \mathbf{x}_h) = \sum_{c,v}\sum_{\mathbf{k}} \varphi^*_{c\mathbf{k}m_s}(\mathbf{x}_e)\varphi_{v\mathbf{k}m_s}(\mathbf{x}_h)A_{\Lambda_0}(c\mathbf{k}m_s v\mathbf{k}m_s), \tag{20.8}$$

is given by a linear combination of the Bloch functions for conduction and valence states weighted by the exciton eigenvector from (19.34). It is the same for both spin channels m_s since no spin polarization occurs in water and ice. The localization of the wave function with a mean distance of 4.02 Å between electron and hole indicates a Frenkel exciton in solid water [63], although its localization in gas-phase H_2O molecules is larger. The agreement of the GW/BSE spectrum with the experimental absorption [62] in Fig. 20.26 is excellent. This holds for the general lineshape but

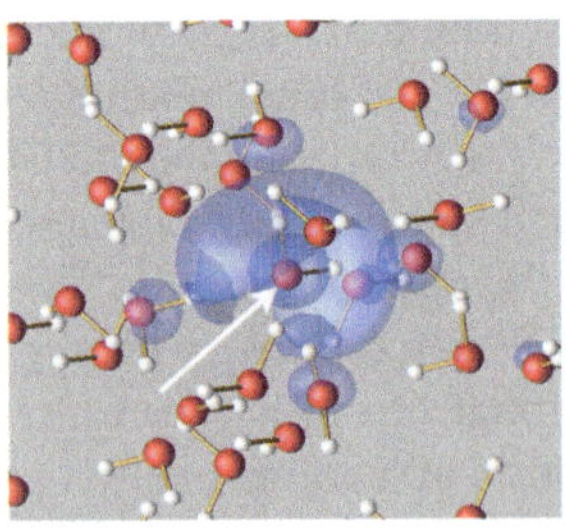

Fig. 20.27 Spatial distribution of the electron in the bound exciton associated with the lowest optical absorption peak of ice-Ih. The hole is located at the oxygen position marked with an *arrow*. Oxygen (hydrogen) atoms are indicated by *red* (*white*) *spheres*. From [60]

also the positions of the peaks and shoulders (within a few tenths of eV) and, more or less, their intensities.

The main spectral features in Fig. 20.26 can be related to exciton-modified optical transitions between molecular electronic states (see for details [61]). However, in contrast to our general expectations, when comparing molecules with sharp energy levels and solids with broad bands and therefore reduced gaps, the lowest absorption peak of a H_2O molecule is redshifted with respect to its position in ice. The HOMO-LUMO peak at about 7.2 eV is strongly modified by excitonic effects. The redshift with respect to the corresponding QP peak of uncorrelated electron-hole pairs is about 5.2 eV. The excitonic effects then largely compensate the QP blueshift of 6.1 eV of the KS position of the HOMO-LUMO transition at 6.3 eV. The redshifted position of the first absorption peak in the molecule is mainly a consequence of the larger exciton binding energy due to the reduced screening in comparison to the case of ice.

Other interesting classes of solids are such with magnetic ordering. The antiferromagnetic transition metal oxides MnO, FeO, CoO, and NiO may serve as examples. Because of their partially filled $3d$ shells, the TM^{2+} ions possess an accompanying local spin polarization and, hence, a magnetic moment. Because of the antiferromagnetic ordering AF2 (see Fig. 9.2) the two spin channels of the oxides contain equal numbers of spin-up and spin-down electrons. Consequently, all states are twofold degenerate. The occupation of the minority-spin t_{2g} subshell with 0 (MnO), 1 (FeO), 2 (CoO), and 3 (NiO) electrons of each TM ion (see Sect. 9.1) leads to drastic differences in the electronic structure, in particular, the ordering of the occupied and empty t_{2g}- and e_g-derived bands. This has been discussed in Fig. 16.16 for MnO and NiO. The question is whether the different electronic structures give rise to different optical properties. The most important optical transitions near the absorption edge and the consequences for the different contributing $3d \rightarrow 3d$ optical transitions have been discussed in Fig. 20.4. Here we study the absorption and energy loss spectra in a wide range of photon and loss energies, respectively, in Figs. 20.28 and 20.29 [14]. The results of the solution of the BSE with a quasiparticle electronic structure based on a GGA+U+Δ approach [29] are compared with available experimental data [64–68].

Almost forbidden bound exciton states [68] are not visible in the fundamental gap region of the spectra in Figs. 20.28 and 20.29. The absorption coefficients in Fig. 20.28 show relatively sharp onsets followed by a broad peak structure whose center changes from about 7 eV to about 5 eV along the row MnO, FeO, CoO, and NiO. It is mainly related to $d \rightarrow d$ transitions outside the Γ point (cf. discussion of Fig. 20.4). The calculated absorption onsets as well as the general lineshape agree well with experimental spectra for MnO and NiO. Only the pronounced $d \rightarrow d$ peak structure is hardly observable in the experimental data. For antiferromagnetic CoO, the computed optical absorption coefficient is higher than the measured spectrum throughout the entire frequency range. However, the qualitative frequency dependence of the absorption is well reproduced. Overall, one has to state that the agreement between experiment and theory is less good compared to other bulk materials. One reason may be the reduced sample quality in the case of the TMOs.

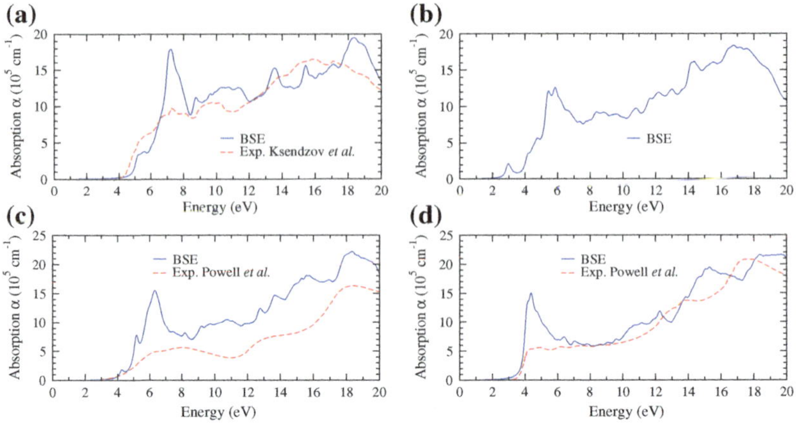

Fig. 20.28 Absorption coefficients for (**a**) MnO, (**b**) FeO, (**c**) CoO, and (**d**) NiO. The spectra have been calculated by solving the Bethe-Salpeter equation (BSE) including excitonic and local-field effects. The experimental *curves* from [64, 65] have been derived from reflectivity data. From [14]

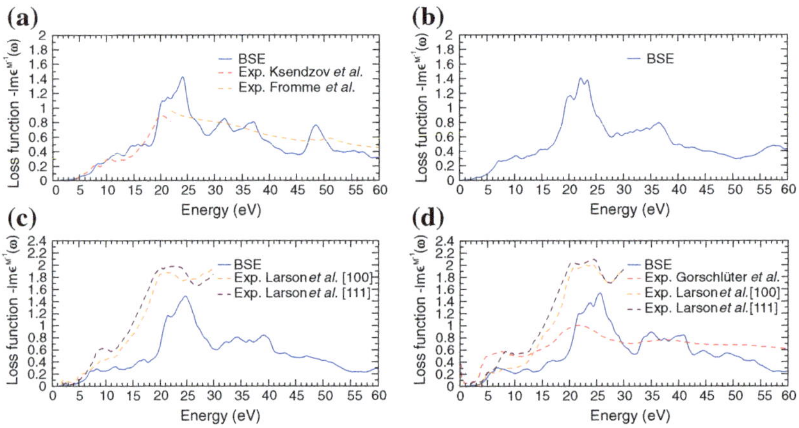

Fig. 20.29 Energy loss spectra for (**a**) MnO, (**b**) FeO, (**c**) CoO, and (**d**) NiO at vanishing momentum transfer, including excitonic and local-field effects (BSE). The experimental reference data have been derived from reflectivity measurements [64], EELS [66, 67] of the MnO and NiO(100) surfaces, and IXS of CoO and NiO along the [100] and [111] directions [68]. The curves from [66, 67] are scaled since they are given in arbitrary units. From [14]

The loss spectra of the TMOs in Fig. 20.29 resemble each other. This is a consequence of their similar electronic structures. Especially the dominant loss peaks in the range between 20 and 25 eV exhibit similar lineshapes, intensities, and peak positions. The steep rise of the loss function at the low-energy edge of this broad feature coincides with a zero in $\mathrm{Re}\epsilon^M(\omega)$ (not shown) at 20.2 (MnO), 19.6 (FeO), 21.5 (CoO), and 22.0 eV (NiO), which corresponds to the plasmon frequency according to (18.25). Consequently, these loss structures are related to the excitation of collective density oscillations of the valence electron gas. The plasma frequency of the O $2p$ and TM $4s$ valence electrons actually amounts to $\hbar\omega_p = 19.5$ (MnO), 20.2 (FeO), 20.8 (CoO), and 21.4 eV (NiO). The contributions of the more localized O $2s$ and TM $3d$ electrons seem to be suppressed or give rise to collective losses at higher energies. If also the TM $3d$ electrons are considered as nearly free electrons one obtains $\hbar\omega_p = 26.4$ (MnO), 28.5 (FeO), 30.6 (CoO), and 32.6 eV (NiO). Including additionally the O $2s$ electrons the plasma frequency further increases to $\hbar\omega_p = 28.7$ (MnO), 30.8 (FeO), 32.8 (CoO), and 34.9 eV (NiO). Indeed, peak structures due to high-energy losses appear in the spectra in the corresponding energy regions. The agreement with experimental data derived from reflectivity spectra or direct measurements for MnO [64, 66] and NiO [67] is reasonable over the entire frequency region but with more broadened features in the experimental spectra. The spectral variation almost agrees with the theoretical predictions, but the intensity of the losses is higher in the experimental spectra. The agreement with the non-resonant inelastic X-ray scattering measurements (not shown) for CoO and NiO [68] only exists for the onset of the losses.

Theoretical absorption spectra of a ferromagnetic insulator are presented in Fig. 20.30 [71]. As an example the trigonal $CrBr_3$ crystal has been chosen. Its atomic structure and resulting quasiparticle bands are discussed in Sect. 16.2.3 and Fig. 16.18. The comparison of the theoretical spectra in different approximations shows a similar influence of the excitonic effects as discussed for non-magnetic semiconductors and insulators. Due to the sum over the two spin channels the differences of the QP band structures in the two spin channels in Fig. 16.18a is hardly resolvable. The principal experimental features [69, 70] are represented by the theory, for instance the absorption edge and the double-peak structure above the absorption edge, though the latter ones are somewhat blueshifted against the measured peak positions. An exciton-mediated coupling between the spin channels (see Sect. 9.3.2) appears but is not visible in the measured optical absorption spectrum.

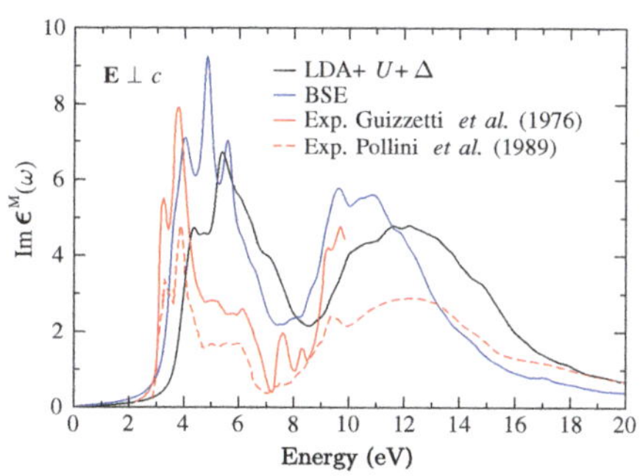

Fig. 20.30 Comparison of theoretical and experimental absorption spectra of $CrBr_3$. An approximate independent-QP (LDA+U+Δ, *black solid line*) spectrum and a spectrum with excitonic effects (BSE, *blue solid line*) are computed with a broadening parameter $\hbar\eta = 0.2$ eV. The two experimental spectra (*red solid* and *dashed lines*) are derived from reflectivity measurements [69, 70]. From [71]

20.3.3 Low-Dimensional Systems

A prototypical two-dimensional system is a graphane sheet as depicted in Fig. 16.31. It represents an sp^3-bonded honeycomb arrangement of carbon atoms. They are alternately bonded to hydrogen atoms in normal or negative normal direction. The resulting quasiparticle band structure is shown in Fig. 16.32. The direct gap of 5.4 eV near Γ is formed by top valence states with A_{1g} symmetry and a conduction band bottom of the A_{2u} symmetry. The VBM corresponds to C$-$C σ bonding states, while the CBM is determined by C$-$H σ^* antibonding states. Therefore, the lowest optical transitions are dipole-forbidden. The optical absorption [72] in Fig. 20.31 without inclusion of excitonic effects, i.e., in independent-quasiparticle approximation, is really zero in the region just above the energy gap.

Independent of the treatment of the QP and excitonic effects, their influences on the spectra in Fig. 20.31 for in-plane light polarization exhibit the same general behavior as summarized in Sect. 20.2.3 for three-dimensional objects. The QP effects treated within the GW approximation shift the entire independent-particle absorption spectrum by 2–4 eV toward higher photon energies. The gap opening with respect to the KS gap of the DFT-LDA reference electronic structure amounts to 2 eV. Excitonic effects lead to a redshift and a significant redistribution of oscillator strength from higher to lower photon energies, thereby compensating the GW shifts. The most prominent physical effect of the electron-hole interactions is the appearance of bound B and C excitons below the QP gap belonging to the (at Γ) dipole-forbidden lowest valence-conduction band transitions. Their binding energies are 1.6 and 0.3 eV, respectively. The A exciton near the position of the B one is dark but its existence is revealed by solving the BSE. Similar qualitative findings have been obtained for the bound excitons in other GW/BSE calculations but with slightly different energy positions and binding energies [73].

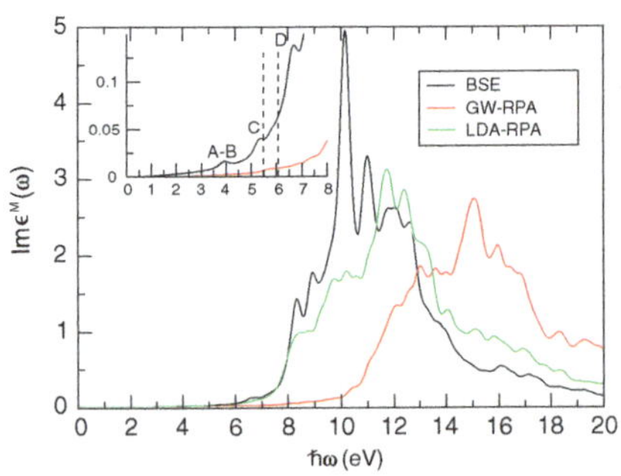

Fig. 20.31 Imaginary part of the macroscopic dielectric function of graphane in independent-particle (LDA-RPA, *green line*), independent-QP (GW-RPA, *red line*) and exciton (BSE, *black line*) approximation. Light of normal incidence and hence polarization in the honeycomb plane is studied. The *vertical dashed lines* in the *inset* indicate the position of the QP gap and the vacuum level, respectively. Reprinted with permission from [72]. Copyright 2010 by the American Physical Society

Other two-dimensional systems are surfaces. Their optical properties can be separately studied if the bulk contribution can be subtracted. There are two possibilities, the *surface differential reflectance* (SDR) *spectroscopy* or the *reflectance anisotropy* (RA) *spectroscopy* [74]. In the first spectroscopy the difference in reflectance due to chemical modification of the surface, for example, by adsorption of oxygen or hydrogen, is measured. In the RA case the relative reflectance difference for two orthogonal light polarizations, x and y, in the surface plane is measured. Since the bulk of cubic crystals is optically isotropic, any RA observed for such crystals must be related to a symmetry-breaking perturbation, e.g. a surface.

A prototypical example is the Si(111)2 $\times$ 1 surface with an atomic structure described by the π-bonded chain model in Fig. 16.26. Its QP band structure can be found in Fig. 16.27. SDR spectra in independent-QP approximation and GW/BSE description [75] are compared with measured spectra [76] in Fig. 20.32. The reference electronic structure is described within the DFT-LDA framework. The full numerical calculations are based on the solution of the homogeneous BSE (19.34) with the two-particle Hamiltonian (19.32) for the electronic subsystem of the two surface bands shown in Fig. 16.27. The forbidden transitions for light polarization perpendicular to the π-bonded chains lead to dark excitons. The spectrum of the allowed optical transitions for light polarization parallel to the chains shows the dramatic influence of the bound exciton states. These generate the broad peak in the differential reflectivity spectrum below the single-quasiparticle absorption edge $E_g^{\mathrm{surf}} = 0.69\,\mathrm{eV}$ (see Sect. 16.3.3). Above the surface QP gap, the differential reflectivity is much reduced. Due to the electron-hole interaction, spectral and oscillator

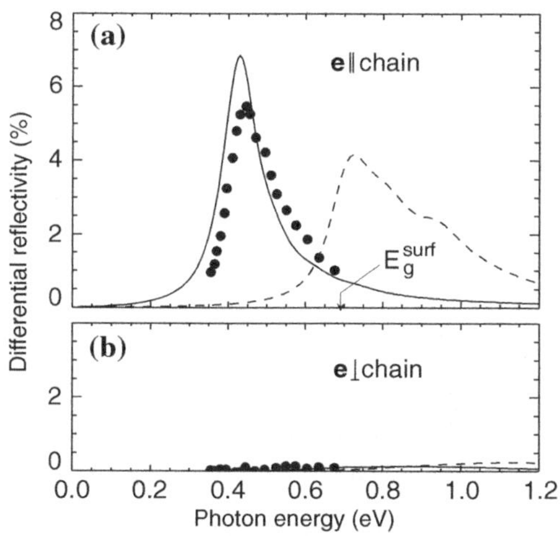

Fig. 20.32 Differential reflectivity spectrum of the Si(111)2 $\times$ 1 surface calculated for normal incidence. The inclusion (neglect) of the electron-hole interaction is indicated by *solid* (*dashed*) *curves*. The two in-plane light polarizations are chosen parallel (**a**) and perpendicular (**b**) to the π-bonded chains. An artificial broadening of 0.05 eV is taken into account. The *dots* denote experimental data [76]. Reprinted with permission from [75]. Copyright 1999 by the American Physical Society

strengths are redistributed to smaller energies. The main reason is the destructive coupling of uncorrelated pair oscillators $c\mathbf{k}v\mathbf{k}$ (without spin) with different wave vectors $\mathbf{k}$ by the eigenvectors $A_\Lambda(c\mathbf{k}v\mathbf{k})$ [cf. (19.34)]. Below the surface QP gap, a number of discrete exciton states are formed. The optical oscillator strength is, however, nearly completely concentrated in the lowest-energy exciton at $\hbar\omega = 0.43\,\text{eV}$.

The dominant spin-singlet exciton at 0.43 eV possesses an exciton binding energy of 0.26 eV. This is more than one order of magnitude larger than the value of about 15 meV in bulk Si [75]. About a factor 4 may be due to the reduced screening in the surface region with an effective dielectric constant of $\epsilon_{\text{surf}} = (\epsilon_\infty + 1)/2$ (cf. the discussion in Sect. 13.3.3). The other main reason for this increased binding is the spatial confinement of both the electron and the hole at the surface. For the direction perpendicular to the surface this is already clear from the atomic geometry. Because of the localization of the surface chain states derived mainly from the D_{down} and D_{up} orbitals (see Fig. 16.27), the bands get a partial 1D character with the strongest dispersion parallel to the π-bonded chains. This 1D character may further increase the exciton binding.

Semiconductor nanowires (NWs) may serve as model systems for electron localization in two directions. As quasi-one-dimensional structures they exhibit extreme quantum confinement effects. The charge carriers are free to move only along the wire axis. Freestanding, infinitely long, and homogeneously hydrogen-passivated Si and Ge NWs with a diameter of about 1 nm and growth orientations along [100], [110], and [111] are prototypical examples. Calculated spectra [77] are presented for a thin Ge NW with a diameter of 0.4 nm in Fig. 20.33. Despite the tetrahedral coordination of the atoms as in the bulk diamond structure, a huge optical anisotropy occurs independent of the approximation used for the many-body effects. It influences both the

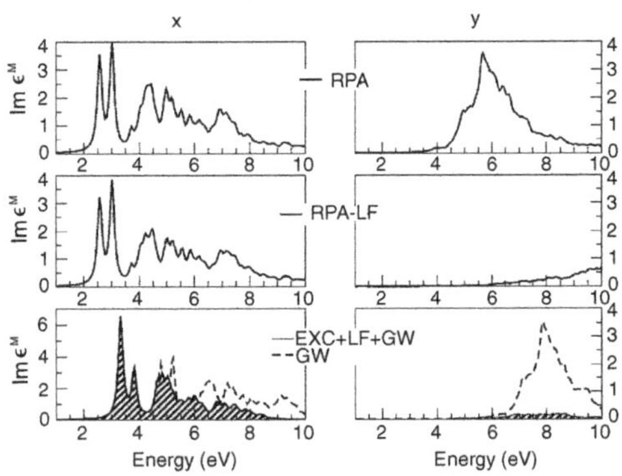

Fig. 20.33 Imaginary part of the dielectric function of a [110] Ge NW with diameter 0.4 nm in four different approximations with respect to the many-body effects: Independent-particle (RPA), independent-particle but including local fields (RPA-LF), independent-quasiparticle (GW), and solution of BSE (EXC+LF+GW). *Left panels*: light polarized along the wire axis (x); *right panels*: light polarized perpendicularly to this axis (y). Reprinted with permission from [77]. Copyright 2005 by the American Physical Society

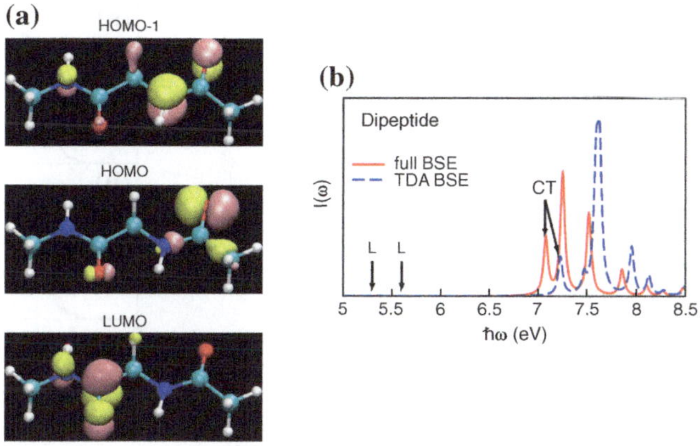

Fig. 20.34 (**a**) Dipeptide molecule computed within DFT-LDA together with the HOMO (n_1) and HOMO-1 (π_1) which are localized on different peptides with respect to the LUMO (π_2^*). *Color code white* for hydrogen, *light blue* for carbon, *blue* for nitrogen, and *red* for oxygen. (**b**) Absorption spectra (in arbitrary units) of a model dipeptide obtained by solving the BSE; with and without the Tamm-Dancoff approximation (TDA). A Lorentzian broadening of about 0.03 eV is used. The local (L) and charge transfer (CT) excitations are highlighted. Reproduced with permission [78]. Copryright 2010, AIP Publishing LLC

lineshape and the energy positions. The interplay of the optical transition strength, the quantization effects, and also the local fields determine the spectra. Apart from the local-field influence, the many-body effects, QP shifts and screened electron-hole attraction tend to partially compensate each other. While LF effects do not influence the absorption for light polarized parallel to the wire axis, they shift the spectrum toward higher energies for the other polarization. Of course, for x polarization bound exciton-like effects remain at the absorption edge and lead to redshifts of the order of 1 eV, which may be identified as exciton binding energies. The drastic differences between x and y polarization underline the importance of local-field effects.

Zero-dimensional electronic systems with completely localized states can be realized in various forms, e.g. as molecules, nanocrystals (NCs) or point defects. The absorption spectrum of a model dipeptide molecule is displayed in Fig. 20.34a together with the molecule geometry [78]. The absorption spectrum in Fig. 20.34b shows pair excitations at energies comparable with results of a precise quantum-chemical method, the complete active space multiconfigurational second-order perturbation theory (CASPT2) [79]. This is especially true for the charge transfer (CT) excitations $\pi_1 \rightarrow \pi_2^*$ and $n_1 \rightarrow \pi_2^*$ between neighboring peptide units with CASPT2 values 7.18 and 8.07 eV, respectively. These excitations correspond to the HOMO-1→LUMO and HOMO→LUMO transitions. Their transition energies are consistent with experimental findings of 7.2–7.6 eV [78]. The quality of the BSE treatment is illustrated by the fact that the lowest-energy absorption peak appears at 7.05 eV only slightly below the CASPT2 and experimental results. The $n_1 \rightarrow \pi_2^*$ excitation is dipole-forbidden. Two local (L), weak intra-peptide excitations, i.e., involving orbitals localized on the same peptide group, from oxygen n (n_1, n_2)

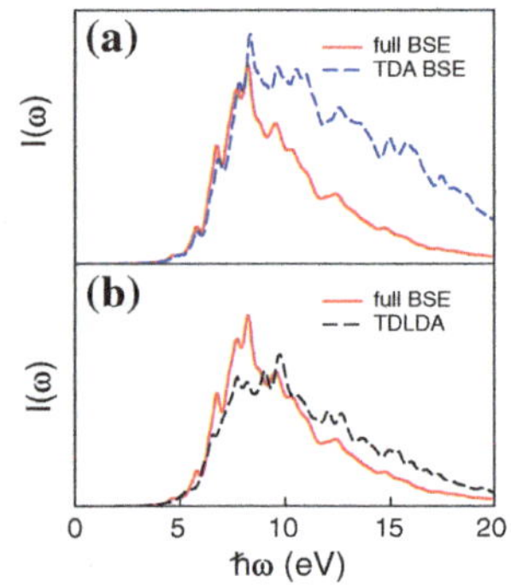

Fig. 20.35 (**a**) Absorption spectra (in arbitrary units) of a 1 nm silicon cluster obtained by solving the BSE with and without the TDA. A Lorentzian broadening of 0.14 eV is used. (**b**) Comparison of the BSE and TDLDA spectra. Reproduced with permission [78]. Copryright 2010, AIP Publishing LLC

lone-pair orbitals to the π^* (π_1^*, π_2^*) orbitals of the amides, are not visible in the spectrum. The model-dipeptide studies indeed yield BSE solutions at 5.30 and 5.60 eV, slightly below the CASPT2 values and the weak band observed at 5.7–5.8 eV in experimental spectra [79]. The influence of the Tamm-Dancoff approximation is weak for the lowest-energy excitations. Beyond TDA only a small (on the large energy scale of the pair excitations) redshift occurs. However, higher-lying excitations are more influenced by the resonant-antiresonant coupling.

The influence of excitonic effects on spectra of a small freestanding nanocrystal is shown in Fig. 20.35 [78]. Resulting absorption spectra are displayed for a hydrogenated Si cluster, $Si_{35}H_{36}$, with a diameter of about 1 nm. The QP approximation is based on results of the DFT-LDA and a scissors operator of 3.48 eV [80]. Since the system only contains 176 electrons, the solution of the BSE in a pair energy range of 20 eV is easily feasible. First, the spectrum in Fig. 20.35a shows that the TDA performs well in the low-energy part of the NC spectrum, up to about 8 eV. However, the agreement with the full calculation beyond TDA worsens significantly at higher energies. Figure 20.34b also shows that for small objects such as $Si_{35}H_{36}$ with strong 3D confinement the difference between full BSE spectra and those obtained within time-dependent DFT within the local density approximation (TDLDA) is reduced. However, the figure still indicates the importance of excitonic effects for the redistribution of oscillator strength from higher to lower photon energies.

Other electronic systems with three-dimensional confinement are the electrons localized on point defects. The influence of the many-body effects on such electrons is illustrated for a defect known for about 150 years, the F center in rocksalt-MgO. It is related to an oxygen vacancy that may occur in the neutral (F^0) but also positively charged (F^+, F^{2+}) states. The main challenge of the many-body theory is to treat the charged defects in supercells with many atoms, which are usually applied to simulate an isolated defect. Cubic 64-atom bulk and 63-atom defect supercells are the smallest possible ones to make the interaction between adjacent vacancies vanishing. Larger supercells, which are desirable for convergence, are hardly tractable within the many-body perturbation theory. In [81] the DFT-LDA reference system is combined with a one-shot GW treatment. The reference wave functions but GW eigenvalues are

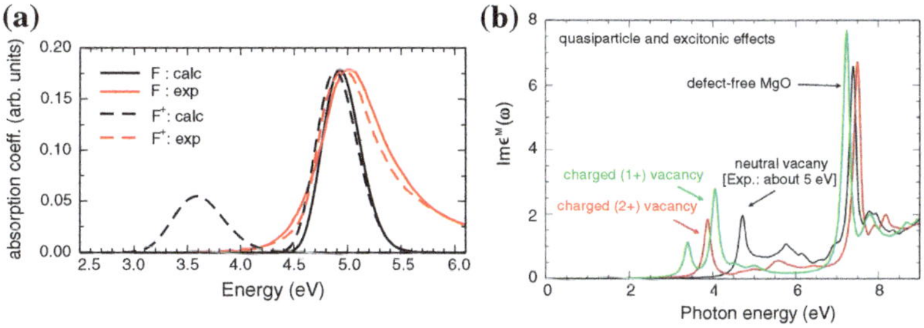

Fig. 20.36 (**a**) Calculated [81] and experimental [82] optical absorption of the F^0 and F^+ centers within the fundamental gap of rs-MgO. The theoretical low-energy peak is due to absorption of an electron from the valence-band top into an F^+ level. (**b**) Comparison of the optical absorption calculated for a defect-free supercell and supercells with a F^0, F^+, and F^{2+} center. (**a**) from [81], (**b**) courtesy of A. Schleife (Friedrich-Schiller-Universität Jena)

used to construct the electron-hole pair Hamiltonian and to solve the BSE for pair energies up to 9 eV. The results are summarized in Fig. 20.36.

Figure 20.36a displays the calculated absorption spectra of the neutral and positively charged F center [81] in comparison with experimental findings [82]. The low-energy peak is associated with the F^+ center, which, since it holds only one electron, offers two spin channels. The electron in the occupied state can be excited at 4.92 eV out of the F^+ center, whereas the empty spin state can be filled at 3.6 eV by excitation of a valence electron. Such light absorption processes are indeed discussed also for the oxygen vacancy in MgO [83]. The low-lying one has, however, to be confirmed experimentally, in contrast to the high-energy absorption line [82]. The observation of such a two-peak structure would be a much more unambiguous fingerprint for the F^+ center than trying to distinguish its peak position at 4.95 eV from the one of the F^0 center at 5.0 eV. Extending to $\hbar\omega = 9$ eV in Fig. 20.36b, i.e., to the absorption edge of the host material, the three absorption spectra of F^0, F^+, and F^{2+} defects show the pronounced bound exciton peak near $\hbar\omega = 7.4$ eV of defect-free MgO (cf. Fig. 20.9). Its position depends slightly on the defect in the supercell but should approach its true bulk value for larger supercells. Its position e.g. in the neutral supercell or that of a higher interband transition should be therefore used to align the defect spectra to that of the defect-free bulk MgO one (see Fig. 20.9). The small energy shifts between the peak positions in Fig. 20.36a, b are due to the inclusion of the coupling of electrons and holes in (a) to longitudinal optical phonons in the framework of a Huang-Rhys-like theory [84]. Nevertheless, the main absorption peak for F^0 without lattice coupling in (b) is also close to the experimental peak position of about 5.0 eV. The phonons mainly affect the line widths.

References

1. G. Grosso, G.P. Parravicini, *Solid State Physics* (Academic Press, Amsterdam, 2000)
2. B. Adolph, V.I. Gavrilenko, K. Tenelsen, F. Bechstedt, R. Del Sole, Nonlocality and many-body effects in the optical properties of semiconductors. Phys. Rev. B **53**, 9797–9808 (1996)
3. M. Rohlfing, S.G. Louie, Electron-hole excitations and optical spectra from first principles. Phys. Rev. B **62**, 4927–4944 (2000)
4. Z.H. Levine, D.C. Allan, Quasiparticle calculation of the dielectric response of silicon and germanium. Phys. Rev. B **43**, 4187–4192 (1991)
5. P. Blaha, K. Schwarz, P. Sorantin, S.B. Trickey, Full-potential, linearized augmented plane wave programs for crystalline systems. Comput. Phys. Commun. **59**, 399–415 (1990)
6. K. Schwarz, P. Blaha, Description of an LAPW DF program (WIEN95). Lect. Notes Chem. **67**, 139–153 (1996)
7. http://www.wien2k.at
8. P.E. Blöchl, Projector augmented-wave method. Phys. Rev. B **50**, 17953–17979 (1994)
9. G. Kresse, D. Joubert, From ultrasoft pseudopotentials to the projector augmented-wave method. Phys. Rev. B **59**, 1758–1775 (1999)
10. B. Adolph, J. Furthmüller, F. Bechstedt, Optical properties of semiconductors using projector augmented waves. Phys. Rev. B **63**, 125108 (2001)
11. M. Gajdoš, K. Hummer, G. Kresse, J. Furthmüller, F. Bechstedt, Linear optical properties in the projector-augmented wave methodology. Phys. Rev. B **73**, 045112 (2006)
12. http://uni-vienna.at/vasp/
13. A. Schleife, C. Rödl, F. Fuchs, J. Furthmüller, F. Bechstedt, Optical and energy-loss spectra of MgO, ZnO, and CdO from *ab-initio* many-body calculations. Phys. Rev. B **80**, 035112 (2009)
14. C. Rödl, F. Bechstedt, Optical and energy-loss spectra of antiferromagnetic transition metal oxides MnO, FeO, CoO, and NiO including quasiparticle and excitonic effects. Phys. Rev. B **86**, 235122 (2012)
15. A. Schleife, J.B. Varley, F. Fuchs, C. Rödl, F. Bechstedt, P. Rinke, A. Janotti, C.G. Van de Walle, Tin dioxide from first principles: quasiparticle electronic states and optical properties. Phys. Rev. B **83**, 035116 (2011)
16. A. Schleife, F. Fuchs, C. Rödl, J. Furthmüller, F. Bechstedt, Band-structure and optical transition parameters of wurtzite MgO, ZnO, and CdO from quasiparticle calculations. Phys. Status Solidi B **246**, 2150–2153 (2009)
17. L.C. de Carvalho, A. Schleife, F. Bechstedt, Influence of exchange and correlation on structural and electronic properties of AlN, GaN, and InN polytypes. Phys. Rev. B **84**, 195105 (2011)
18. P.Y. Yu, M. Cardona, *Fundamentals of Semiconductors* (Springer, Berlin, 1996)
19. L.P. Bouckaert, R. Smoluchowski, E. Wigner, Theory of Brillouin zones and symmetry properties of wave functions in crystals. Phys. Rev. **50**, 58–67 (1936)
20. E.I., Rashba, Properties of semiconductors with an extremum loop. 1. Cyclotron and combinational resonance in a magnetic field perpendicular to the plane of the loop. Sov. Fiz. Tverd. Tela (Leningrad) **2**, 1224–1238 (1960) [Phys. Sol. State (English Transl.) **2**, 1109–1122 (1960)]
21. J. Wu, W. Walukiewicz, W. Shan, K.M. Yu, J.W. Ager, E.E. Haller, H. Lu, W.J. Schaff, Effects of the narrow band gap on the properties of InN. Phys. Rev. B **66**, 201403 (2002)
22. S.P. Fu, Y.F. Chen, Effective mass of InN epilayers. Appl. Phys. Lett. **85**, 1523–1525 (2004)
23. A.V. Rodina, B.K. Meyer, Anisotropy of conduction band g values and interband momentum matrix elements in wurtzite GaN. Phys. Rev. B **64**, 245209 (2001)
24. S. Shokhovets, O. Ambacher, B.K. Meyer, G. Gobsch, Anisotropy of the momentum matrix element, dichroism, and conduction-band dispersion relation of wurtzite semiconductors. Phys. Rev. B **78**, 035207 (2008)
25. F. Fuchs, C. Rödl, A. Schleife, F. Bechstedt, Efficient $O(N^2)$ approach to solve the Bethe-Salpeter equation for excitonic bound states. Phys. Rev. B **78**, 085103 (2008)
26. L.X. Benedict, E.L. Shirley, Ab initio calculation of $\varepsilon_2(\omega)$ including the electron-hole interaction: application to GaN and CaF_2. Phys. Rev. B **59**, 5441–5451 (1999)

27. L.X. Benedict, E.L. Shirley, R.B. Bohn, Theory of optical absorption in diamond, Si, Ge, and GaAs. Phys. Rev. B **57**, R9385–R9387 (1998)
28. B. Arnaud, M. Alouani, Electron-hole excitations in Mg_2Si and Mg_2Ge compounds. Phys. Rev. B **64**, 033202 (2001)
29. C. Rödl, F. Fuchs, J. Furthmüller, F. Bechstedt, *Ab initio* theory of excitons and optical properties for spin-polarized systems. Application to antiferromagnetic MnO. Phys. Rev. B **72**, 184408 (2008)
30. W.G. Schmidt, S. Glutsch, P.H. Hahn, F. Bechstedt, Efficient $O(N^2)$ method to solve the Bethe-Salpeter equation. Phys. Rev. B **67**, 085307 (2003)
31. D.E. Aspnes, A.A. Studna, Dielectric functions and optical parameters of Si, Ge, GaP, GaAs, GaSb, InP, InAs, and InSb from 1.5 to 6.0 eV. Phys. Rev. B **27**, 985–1009 (1983)
32. S. Albrecht, L. Reining, R. Del Sole, G. Onida, Ab initio calculation of excitonic effects in the optical spectra of semiconductors. Phys. Rev. Lett. **80**, 4510–4513 (1998)
33. M.L. Bortz, R.H. French, D.J. Jones, R.V. Kasowski, F.S. Obuchi, Temperature dependence of the electronic structure of oxides: MgO, $MgAl_2O_4$ and Al_2O_3. Phys. Scr. **41**, 537–541 (1990)
34. G. Cappellini, J. Furthmüller, E. Cadelano, F. Bechstedt, Electronic and optical properties of cadmium fluoride: the role of many-body effects. Phys. Rev. B **87**, 075203 (2013)
35. A. Riefer, F. Fuchs, C. Rödl, A. Schleife, F. Bechstedt, R. Goldhahn, Interplay of excitonic effects and van Hove singularities in optical spectra: CaO and AlN polymorphs. Phys. Rev. B **84**, 075218 (2011)
36. M. Röppischer, R. Goldhahn, G. Rossbach, P. Schley, C. Cobet, N. Esser, T. Schupp, K. Lischka, D.J. As, Dielectric function of zinc-blende AlN from 1 to 20 eV: band gap and van Hove singularities. J. Appl. Phys. **106**, 076104 (2009)
37. G. Rossbach, M. Röppischer, P. Schley, G. Gobsch, C. Werner, C. Cobet, N. Esser, A. Dadgar, M. Wieneke, A. Krost, R. Goldhahn, Valence-band splitting and optical anisotropy of AlN. Phys. Status Solidi B **247**, 1679–1682 (2010)
38. R. Leitsmann, W.G. Schmidt, P.H. Hahn, F. Bechstedt, Second-harmonic polarizability including electron-hole attraction from band-struture theory. Phys. Rev. B **71**, 1952 (2005)
39. H-Ch. Weissker, J. Furthmüller, F. Bechstedt, Structure- and spin-dependent excitation energies and lifetimes of Si and Ge nanocrystals from *ab initio* calculations. Phys. Rev. B **69**, 115310 (2004)
40. P. Puschnig, A. Ambrosch-Draxl, Optical absorption spectra of semiconductors and insulators including electron-hole correlations: an *ab initio* study within the LAPW method. Phys. Rev. B **66**, 165105 (2002)
41. M. Rohlfing, S.G. Louie, Electron-hole excitations in semiconductors and insulators. Phys. Rev. Lett. **81**, 2312–2315 (1998)
42. B. Arnaud, M. Alouani, Local-field and excitonic effects in the calculated optical properties of semiconductors from first-principles. Phys. Rev. B **63**, 085208 (2001)
43. D.M. Roessler, W.C. Walker, Optical constants of magnesium oxide and lithium fluoride in the far ultraviolet. J. Opt. Soc. Am. **57**, 835–836 (1967)
44. A. Marini, R. Del Sole, Dynamical excitonic effects in metals and semiconductors. Phys. Rev. Lett. **91**, 176402 (2003)
45. E.D. Palik, *Handbook of Optical Constants of Solids* (Academic Press, New York, 1985)
46. M. Rakel, C. Cobet, N. Esser, P. Gori, O. Pulci, A. Seitsonen, A. Cricenti, N. Nickel, W. Richter, Electronic and optical properties of ZnO between 3 and 32 eV, in *Epioptics-9, Proceedings of the 39th International School on Solid State Physics*, Erice (Italy), 2006. ed. by A. Cricenti (World Scientific Publ. Co, New Jersey, 2008), pp. 115–123
47. A. Bourdillon, J. Beaumont, The reflection spectra of $SrCl_2$ and CdF_2. J. Phys. C **9**, L473–L477 (1976)
48. D.R. Bosomworth, Far-infrared optical properties of CaF_2, SrF_2, BaF_2, and CdF_2. Phys. Rev. **157**, 709–715 (1967)
49. B. Krukowska-Fule, T. Niemyski, Preparation and some properties of CdF2 single crystals. J. Cryst. Growth **1**, 183–186 (1967)

50. J.L. Freeouf, Far-ultraviolet reflectance of II–VI compounds and correlation with the Penn-Phillips gap. Phys. Rev. B **7**, 3810–3830 (1973)
51. P. Cudazzo, M. Gatti, A. Rubio, Excitons in molecular crystals from first-principles many-body perturbation theory: picene versus pentacene. Phys. Rev. B **86**, 195307 (2012)
52. A.S. Davydov, *Theory of Molecular Excitons* (McGraw-Hill, New York, 1962)
53. G. Onida, L. Reining, A. Rubio, Electronic excitations: density-functional versus many-body Green's-function approaches. Rev. Mod. Phys. **74**, 601–659 (2002)
54. K. Hummer, P. Puschnig, C. Ambrosch-Draxl, Lowest optical excitations in molecular crystals: bound exciton versus free electron-hole pairs in anthracene. Phys. Rev. Lett. **92**, 147402 (2004)
55. E.A. Silinsh, *Organic Molecular Crystals* (Springer, Berlin, 1980)
56. M. Pope, C.E. Swenberg, *Electronic Processes in Organic Crystals and Polymers* (Oxford University Press, New York, 1999)
57. L.E. Lyons, G.C. Moris, The intensity of ultraviolet light absorption by monocrystals. Part III. Absorption by anthracene at 295°K, 90°K, and 4°K of plane-polarised light of wavelengths 1600–2750 Å. J. Chem. Soc. **299**, 1551–1558 (1959)
58. Ch. Kittel, *Introduction to Solid State Physics* (Wiley, New York, 2005)
59. W.A. Harrison, *Elementary Electronic Structure* (World Scientific, Singapore, 1999)
60. P.H. Hahn, W.G. Schmidt, K. Seino, M. Preuss, F. Bechstedt, J. Bernholc, Optical absorption of water: Coulomb effects versus hydrogen bonding. Phys. Rev. Lett. **94**, 037404 (2005)
61. P.H. Hahn, W.G. Schmidt, F. Bechstedt, Molecular electronic excitations calculated from a solid-state approach: methodology and numerics. Phys. Rev. B **72**, 245425 (2005)
62. K. Kobayashi, Optical spectra and electronic structure of ice. J. Phys. Chem. **87**, 4317–4321 (1983)
63. R.S. Knox, *Theory of Excitons* (Academic Press, New York, 1963)
64. Y.M. Ksendzov, I.L. Korobova, K.K. Sidorin, G.P. Startsev, Elektronnaya struktura MnO. Fiz. Tverd. Tela (Leningrad) **18**, 173–179 (1976)
65. R.J. Powell, W.E. Spicer, Optical properties of NiO and CoO. Phys. Rev. B **2**, 2182–2193 (1970)
66. B. Fromme, U. Brunokowski, E. Kisker, *d–d* excitations and interband transitions in MnO: a spin-polarized electron-energy-loss study. Phys. Rev. B **58**, 9783–9792 (1998)
67. A. Gorschlüter, H. Merz, EELS study of single crystalline NiO(100). Int. J. Mod. Phys. B **7**, 341–344 (1993)
68. B.C. Larson, W. Ku, J.Z. Tischler, C.-C. Lee, O.D. Restrepo, A.G. Eguiluz, P. Zschakc, K.D. Finkelstein, Nonresonant inelastic X-ray scattering and energy-resolved Wannier function investigation of *d-d* excitations in NiO and CoO. Phys. Rev. Lett. **99**, 026401 (2007)
69. I. Pollini, J. Thomas, B. Carricaburu, R. Mamy, Optical and electron energy loss experiments in ionic $CrBr_3$ crystals. J. Phys. Condens. Mat. **1**, 7695–7704 (1989)
70. G. Guizzetti, L. Nosenzo, I. Pollini, E. Reguzzoni, G. Samoggia, G. Spinolo, Reflectance and thermoreflectance studies of $CrCl_3$, $CrBr_3$, $NiCl_2$, and $NiBr_2$ crystals. Phys. Rev. B **14**, 4622–4629 (1976)
71. C. Rödl, Elektronische und exzitonische Anregungen in magnetischen Isolatoren. Ph.D. thesis. Friedrich-Schiller-Universität Jena (2009)
72. P. Cudazzo, C. Attacalite, I.V. Tokatly, A. Rubio, Strong charge-transfer excitonic effects and Bose-Einstein exciton condensate in graphane. Phys. Rev. Lett. **104**, 226804 (2010)
73. O. Pulci, P. Gori, M. Marsili, V. Garbuio, R. Del Sole, F. Bechstedt, Strong excitons in novel two-dimensional crystals: silicane and germanane. EPL **98**, 37004 (2012)
74. F. Bechstedt, *Principles of Surface Physics* (Springer, Berlin, 2003)
75. M. Rohlfing, S.G. Louie, Excitons and optical spectrum of the Si(111)-(2×1) surface. Phys. Rev. Lett. **83**, 856–859 (1999)
76. P. Chiradia, A. Cricenti, S. Selci, G. Chiarotti, Differential reflectivity of Si(111)2×1 surface with polarized light: a test for surface structure. Phys. Rev. Lett. **52**, 1145–1147 (1984)
77. M. Bruno, M. Palummo, A. Marini, R. Del Sole, V. Olevano, A.N. Kholod, S. Ossicini, Excitons in germanium nanowires: quantum confinement, orientation, and anisotropy effects within first-principles approach. Phys. Rev. B **72**, 153310 (2005)

78. D. Rocca, D. Lu, G. Galli, Ab initio calculations of optical absorption spectra: solution of the Bethe-Salpeter equation within density matrix perturbation theory. J. Chem. Phys. **133**, 164109 (2010)
79. L. Serrano-Andres, M. Fulscher, Theoretical study of the electronic spectroscopy of peptides. III. Charge-transfer transitions in polypeptides. J. Am. Chem. Soc. **120**, 10912–10920 (1998)
80. M.L. Tiago, J.R. Chelikowsky, Optical excitations in organic molecules, clusters, and defects studied by first-principles Green's function methods. Phys. Rev. B **73**, 205334 (2006)
81. P. Rinke, A. Schleife, E. Kioupakis, A. Janotti, C. Rödl, F. Bechstedt, M. Scheffler, C.G. Van de Walle, First-principles optical spectra for F centers in MgO. Phys. Rev. Lett. **108**, 126404 (2012)
82. L.A. Kappers, R.L. Kroes, E.B. Hensley, F^+ and F' centers in magnesium oxide. Phys. Rev. B **1**, 4151–4157 (1970)
83. G.H. Rosenblatt, M.W. Rowe, G.P. Williams, R.T. Williams, Y. Chen, Luminescence of F and F^+ centers in magnesium oxide. Phys. Rev. B **39**, 10309–10318 (1989)
84. B.K. Ridley, *Quantum Processes in Semiconductors* (Clarendon Press, Oxford, 2006)

Chapter 21
Excitons

Abstract The general procedure to account for many-body effects in optical and energy-loss spectra consists of three steps, (i) preparation of a starting electronic structure, (ii) its improvement due to quasiparticle effects, and (iii) the inclusion of electron-hole attraction and exchange. Their combination gives rise to novel quasiparticles, the excitons. Singlet and triplet states differ by twice the electron-hole exchange. Instructive exciton models are derived studying only pairs of conduction and valence bands. For large distances of electron and hole in real space the effective mass approximation and a constant screening can be employed. The resulting pairs are hydrogen-like Wannier-Mott excitons, which give rise to Rydberg series below the ionization edge and a Coulomb enhancement, the Sommerfeld factor, of the pair density of states above the edge. Localized excitons such as Frenkel and charge-transfer excitons possess larger binding energies and electron-hole distances of the order of atomic or molecular distances. Exciton binding is increased by spatial confinement as demonstrated for two-dimensional systems.

21.1 Electron-Hole Pairs: Top-Down Approach

21.1.1 Terminology

In the previous Chaps. 18–20 we have described in detail the response of an electronic system to an external perturbation, which however leaves this system neutral. Especially perturbations which are related to the real or virtual absorption of photons with energy $\hbar\omega$ or the inelastic scattering of high-energy particles with energy losses of energy $\hbar\omega$ have been considered as examples. Thereby we have learnt that the many-electron interactions mediated by the Coulomb potential play a central role. We found that the treatment of their influence can be nearly divided into a three-step procedure that is schematically illustrated in Fig. 21.1 for a non-metal. In a first step an electron-hole pair may be actually or virtually excited by a photon of energy $\hbar\omega$ in the reference electronic structure derived from a Kohn-Sham or generalized Kohn-Sham problem. In the second step the formation of quasiparticles is taken into account. As a consequence an energy shift of the bands is considered resulting in an

F. Bechstedt, *Many-Body Approach to Electronic Excitations*,
Springer Series in Solid-State Sciences 181, DOI 10.1007/978-3-662-44593-8_21

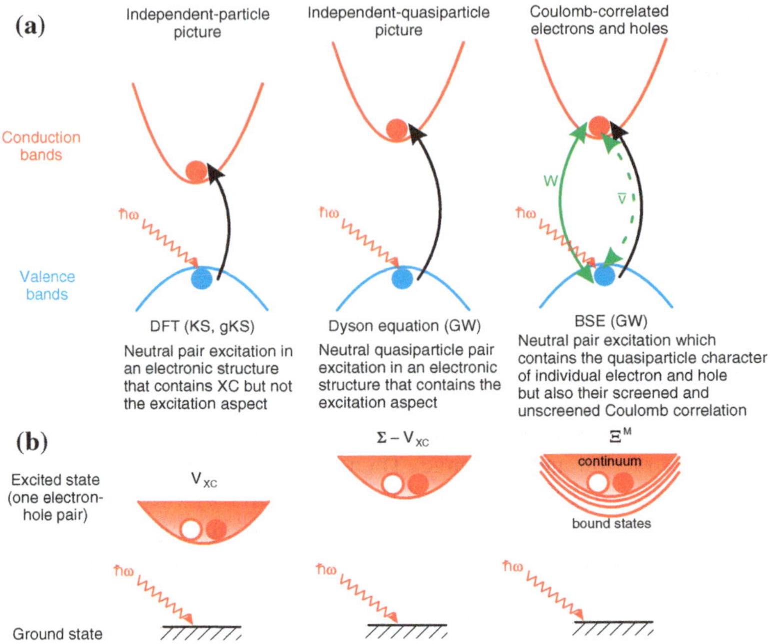

Fig. 21.1 Three-step procedure for the treatment of neutral pair excitations in (**a**) a single-particle picture and (**b**) a two-particle picture (schematically). In the *right panel* of (**a**) the electron-hole interactions and of (**b**) the formation of bound states below the continuum of scattering states are indicated

enlargement of the pair energies. In the last step the attractive screened interaction of quasielectron and quasihole is taken into account together with an electron-hole exchange interaction.

Such a neutral Coulomb-correlated electron-hole pair can form when a photon is actually or virtually absorbed by a non-metal, e.g. semiconductor, insulator, molecule, nanostructure etc. (see Fig. 21.1). Such a photon excites an electron from an occupied state, e.g. in a valence band, into an empty one, e.g. in a conduction band. In turn, this process leaves behind a positively charged hole, a missing electron in the electron density of the ground state from which the electron was removed. The negatively charged electron in the formerly empty single-particle state is then attracted by the hole in the formerly occupied state by Coulomb forces. They are weakened by the screening reaction of the electronic system to the excited electron-hole pair. Apart from a possible electron-hole exchange this attraction is the most important many-body effect. It stabilizes the electron-hole pair, so that one may speak of a new neutral excitation in the electronic system, a new quasiparticle, an *exciton* [1]. If the attractive interaction is strong enough or for a pure Coulomb attraction, even bound states as in an atom with positively charged nucleus and an electron may occur. Consequently, the excitation of such a bound state needs less energy than the unbound electron and hole as indicated in Fig. 21.1b (right panel). The reduction corresponds to the binding energy of the exciton [1, 2].

In the literature the terminology is not unique. Many colleagues call the excited pair only 'exciton' if it is excited in a bound state. Others generally speak about excitons if the Coulomb correlations of electron and hole, screened electron-hole attraction and unscreened electron-hole exchange, are taken into consideration, independent of bound or scattering states. The original concept of excitons was proposed by Frenkel and Peierls [1, 2]. They described the pair excitations as excitation of atoms in an insulator. A more band-like picture of electron-hole pairs with Coulomb attraction has been first formulated by Wannier [3]. Mott [4] then introduced the screening of this interaction, e.g. by the dielectric constant of a semiconductor or insulator. Such a screened attraction W is described in detail in (19.32). The origin of the electron-hole exchange is difficult to determine. However, the physical background of this effect can be partially described by an electric dipole-dipole interaction of electron-hole pairs (see e.g. [5]). The first many-body formulation of this interaction $\bar{\mathrm{v}}$ in (19.33) can be found in a paper of Sham and Rice [6].

21.1.2 Exciton Equation: k-space Formulation

The excitons and, hence, their bound states are described by the homogeneous Bethe-Salpeter equation (19.34). For simplicity we only study a non-spin-polarized electronic system. Then the two-particle states can be characterized by singlet ($S = 0$) and triplet ($S = 1$) states applying the expression (18.12) for their spin part. The remaining part of the electron-hole-pair wave functions in $\mathbf{k}$-space is $A_\Lambda^S(c\mathbf{k}v\mathbf{k})$ ($S = 0, 1$) which obeys a modified equation (19.34)

$$\sum_{c',v'}\sum_{\mathbf{k}'} H^S(c\mathbf{k}v\mathbf{k}, c'\mathbf{k}'v'\mathbf{k}')A_\Lambda^S(c'\mathbf{k}'v'\mathbf{k}') = E_\Lambda^S A_\Lambda^S(c\mathbf{k}v\mathbf{k}) \tag{21.1}$$

with an electron-hole-pair Hamiltonian

$$\begin{aligned} H^S(c\mathbf{k}v\mathbf{k}, c'\mathbf{k}'v'\mathbf{k}') = &\left[\varepsilon_c^{\mathrm{QP}}(\mathbf{k}) - \varepsilon_v^{\mathrm{QP}}(\mathbf{k})\right]\delta_{cc'}\delta_{vv'}\delta_{\mathbf{k}\mathbf{k}'} \\ &-W\begin{smallmatrix}\mathbf{k}\mathbf{k}'\\ cc'\\ vv'\end{smallmatrix} + 2\delta_{S0}\bar{\mathrm{v}}\begin{smallmatrix}\mathbf{k}\mathbf{k}'\\ cv\\ v'c'\end{smallmatrix} \end{aligned} \tag{21.2}$$

and Coulomb matrix elements as given in (19.33).

The eigenvalue problem (21.1) of the homogeneous integral equation can be directly solved in $\mathbf{k}$-space. We do it for true translationally invariant systems such as crystals or artificial periodic systems with a repetition of a nanoobject. The sum over $\mathbf{k}'$, $\sum_{\mathbf{k}'}$, in (21.1) suggests to use a discretized 3D mesh in the BZ, which can be related to a $\mathbf{k}'$-integral $\Omega/(2\pi)^3\int d^3\mathbf{k}'$ (1.16) [7]. For converged computations one uses a fine grid. Usually this grid should be by one or two orders of magnitude more dense than that applied to calculations of the dielectric function (13.43) without excitonic effects. In addition, in general, many band pairs cv have to be taken into account. Of course, by restricting to a certain frequency range $\hbar\omega$ the number of band pairs can be restricted to an appropriate finite number [8]. This number combined with the number of $\mathbf{k}$ points yields the number of $10^4 - 10^6$ basis functions which are necessary to represent the relevant pair excitations in the energy interval investigated. The determination of the $10^8 - 10^{12}$ matrix elements of the electron-hole interaction Ξ^M (18.36), each of which requires a multidimensional integration in (19.33), forms a major bottleneck, because of the numerical expenses and the required computer memory. Therefore, frequently interpolation schemes from a coarse grid to a finer grid are used [7, 8]. Another idea is related to the use of non-uniform, so-called hybrid $\mathbf{k}$-point meshes which only sample the most contributing BZ regions with a finer grid [8]. These ideas are combined with a division of the BZ integration into small subvolumes with a weight related to the square root of the subvolume. The corresponding integrations of the matrix elements over the subvolumes can be easily performed using some subgrids. Only the Coulomb singularity in reciprocal space has to be treated with care [7, 8]. Concomitantly independent of uniform or non-uniform division in subvolumes of the BZ, the homogenous BSE (21.1) no longer represents a standard eigenvalue problem but a generalized eigenvalue problem with an overlap matrix.

The resulting generalized eigenvalue problem can be solved by direct diagonalization. However, this procedure shows a $O(N^3)$ scaling with N as the rank of the actual pair-Hamiltonian matrix (21.2). Often times, not the complete set of pairs is of interest, for instance, restricting only to the bound states with pair excitation energies E_Λ^S below the lowest QP gap. Due to the limited spectral resolution and the possibility of vanishing oscillator strength such excitations cannot be studied in the framework of the dielectric function, which may be computed within the time-evolution method with an $O(N^2)$ scaling (see Sect. 19.3.3). The same holds for the analysis of the excitonic wave functions (20.8), which requires the knowledge of the respective eigenvalues E_Λ^S and eigenvectors $A_\Lambda^S(c\mathbf{k}v\mathbf{k})$.

One possibility to restrict the computations still to an $O(N^2)$ scaling, at least for a finite number of bound states, is the minimization of a Ritz functional based on the Hamiltonian (21.2) [8]. The explicit scheme to do so is the method of Kalkreuter and Simma [9]. It combines consecutive conjugate gradient steps with subspace rotations.

In order to illustrate the bound state problem we focus on electron-hole-pair states with energies near the fundamental quasiparticle gap E_g of a semiconductor or insulator. Thereby, we avoid the label 'QP'. In general, several conduction and valence bands states are mixed in each of these two-particle states (20.8). In real space such an electron-hole pair wave function for the state Λ with total spin S can be formulated as

$$\Phi_\Lambda^S(\mathbf{x}_e, \mathbf{x}_h) = \sum_{c,v} \sum_{\mathbf{k}} A_\Lambda^S(c\mathbf{k}v\mathbf{k}) \varphi_{c\mathbf{k}}^*(\mathbf{x}_e) \varphi_{v\mathbf{k}}(\mathbf{x}_h), \tag{21.3}$$

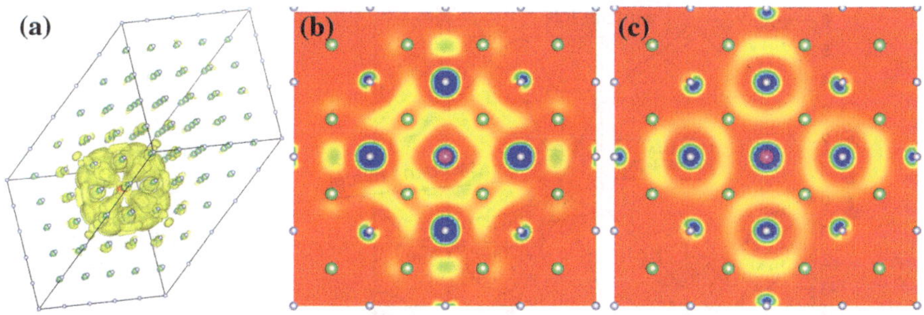

Fig. 21.2 (**a**) Exciton wave function of the lowest energy pair state in LiF for a fixed hole position $\mathbf{x}_h$ at a F^- ion (*violet*, in the center of the cell). (**b**) Cut along the [111] direction for the singlet exciton $S = 0$. (**c**) Same as (**b**), but for triplet excitation $S = 1$. The *green* and *gray circles* indicate atomic positions. Reprinted with permission from [10]. Copyright 2013 by the American Physical Society

where $\mathbf{x}_e$ ($\mathbf{x}_h$) denotes the electron (hole) coordinate. It represents an expansion of the two-particle state in terms of the Bloch functions of conduction and valence bands. Their contributions are weighted by the eigenvector $A_\Lambda^S(c\mathbf{k}v\mathbf{k})$. For illustration, the two-particle wave function of the lowest energy exciton Λ_0 in LiF is plotted in Fig. 21.2 for a fixed hole position at a fluorine atom [10].

In a typical direct semiconductor such as zinc-blende-GaAs the absorption edge is formed by the lowest conduction band c at Γ and three uppermost valence bands v_A, v_B, and v_C which are threefold degenerate at Γ (see e.g. Figs. 16.4, 16.7, and 20.6a). In the wurtzite case these bands split into two upper valence bands v_A and v_B and a lower one v_C by the crystal-field splitting as e.g. in wurtzite-InN (see Figs. 16.6a and 20.6b). Calculations [8] show that only a small region of the BZ around the center Γ contributes to the resulting pair states (20.8) and (21.3), respectively. The valence bands contribute according to their degeneracy at Γ. In the presence of spin-orbit interaction this degeneracy is also lifted (see Fig. 16.6). Therefore, in order to discuss bound exciton states we focus on a two-band model semiconductor as in Fig. 15.2a with bands c and v to represent the most important features of the band structure of a tetrahedrally coordinated direct semiconductor near the absorption edge. An example for a transition from a full band structure to the two-band model is given in Fig. 21.3. The valence band index v may be interpreted as that of the heavy-hole (hh) or light-hole (lh) band. The two-particle Hamiltonian (21.2) becomes diagonal

$$H^S(c\mathbf{k}v\mathbf{k}, c'\mathbf{k}'v'\mathbf{k}') = \delta_{cc'}\delta_{vv'}\left\{\left[\varepsilon_c^{\mathrm{QP}}(\mathbf{k}) - \varepsilon_v^{\mathrm{QP}}(\mathbf{k})\right]\delta_{\mathbf{k}\mathbf{k}'} - W\,{}^{\mathbf{k}\mathbf{k}'}_{\substack{cc\\vv}} + 2\delta_{S0}\bar{\mathrm{v}}\,{}^{\mathbf{k}\mathbf{k}'}_{\substack{cv\\vc}}\right\}. \tag{21.4}$$

When the two bands are isotropic and parabolic, they can be described within the effective mass approximation (EMA) [11] (see also Sect. 16.1.3) as

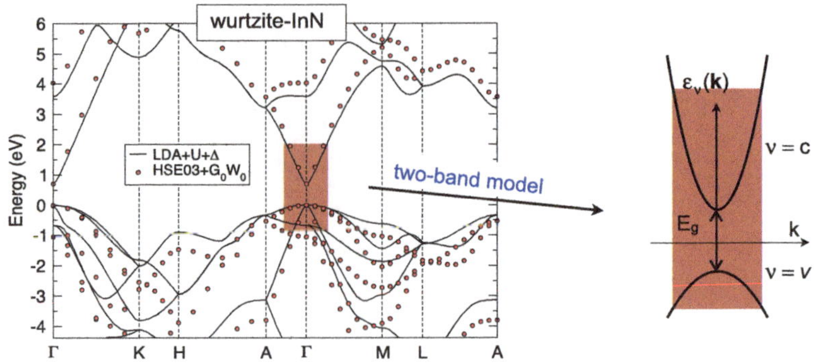

Fig. 21.3 Two-band model of a semiconductor derived from a full QP band structure (here: in two QP approximations for *wz*-InN). Adapted from [8]

$$\varepsilon_c^{\text{QP}}(\mathbf{k}) = E_g + \frac{\hbar^2}{2m_c^*}\mathbf{k}^2,$$
$$\varepsilon_v^{\text{QP}}(\mathbf{k}) = -\frac{\hbar^2}{2m_v^*}\mathbf{k}^2, \tag{21.5}$$

if the band extrema are assumed to be located at the same position for electron and hole in the BZ, e.g. at Γ, i.e., at $\mathbf{k} = 0$. This leads to a QP pair excitation part in (21.2),

$$\varepsilon_c^{\text{QP}}(\mathbf{k}) - \varepsilon_v^{\text{QP}}(\mathbf{k}) = E_g + \frac{\hbar^2}{2\mu_{\text{ex}}}\mathbf{k}^2$$

with $\frac{1}{\mu_{\text{ex}}} = \frac{1}{m_c^*} + \frac{1}{m_v^*}$ as the inverse reduced effective mass of the electron-hole pair. In the EMA the electrons in the conduction band c and the holes in the valence band v nearly behave like free particles (4.46) but with effective masses. The EMA model (21.5) can be easily generalized to band extrema $\mathbf{k}_c$ and $\mathbf{k}_v$ outside the BZ center Γ in order to describe direct Wannier-Mott excitons at high-symmetry points $\mathbf{k}_c = \mathbf{k}_v \neq 0$ or even indirect excitons with $\mathbf{k}_c \neq \mathbf{k}_v$ in indirect multi-valley semiconductors like Ge or Si. Using the Bloch representation of the QP wave functions as introduced below the expressions (19.33), the products of Bloch factors in the Coulomb matrix elements can be nearly replaced by their spatial average [12]. This approximation leads to ($\nu = c, v$)

$$B_{\nu\nu}^{\mathbf{kk}'}(\mathbf{q} + \mathbf{G}) \approx \delta_{\mathbf{k}'+\mathbf{q},\mathbf{k}}\delta_{\mathbf{G}0}$$

for the band-diagonal Bloch matrix elements. A similar argumentation can be applied for the off-diagonal Bloch integrals in the electron-hole exchange interaction [12]. One may argue that only the smallest reciprocal lattice vectors mainly contribute because of the Coulomb singularity. Then, with (20.4) it holds

$$B_{cv}^{\mathbf{kk}}(\mathbf{G}) \approx \frac{\hbar}{m} \frac{\mathbf{G}\mathbf{p}_{cv}(\mathbf{k})}{\varepsilon_c(\mathbf{k}) - \varepsilon_v(\mathbf{k})}.$$

The summation over all $\mathbf{G}$ in the matrix elements (19.33) makes the electron-hole exchange vanish, at least within the discussed approximations for cubic systems and wave vectors $\mathbf{k}$ near the band extrema. Another argumentation is related to the fact that the reciprocal lattice vector $\mathbf{G} = 0$ only appears in the electron-hole attraction but not in the exchange term (18.36). Therefore, in a first approximation the electron-hole exchange is neglected in the following.

Combining these approximations one ends up with the simple two-particle Hamiltonian

$$H^S(c\mathbf{k}v\mathbf{k}, c'\mathbf{k}'v'\mathbf{k}') = \delta_{cc'}\delta_{vv'} H^{\mathrm{WM}}(\mathbf{k}, \mathbf{k}'), \tag{21.6}$$
$$H^{\mathrm{WM}}(\mathbf{k}, \mathbf{k}') = \left[E_g + \frac{\hbar^2}{2\mu_{\mathrm{ex}}}\mathbf{k}^2\right]\delta_{\mathbf{kk}'} - \frac{1}{\Omega}\tilde{\mathrm{v}}\left(\left|\mathbf{k}-\mathbf{k}'\right|\right)\epsilon^{-1}(\mathbf{k}-\mathbf{k}', \mathbf{k}-\mathbf{k}', 0).$$

Our experience with the screened Coulomb attraction $W_{\substack{cc\\vv}}^{\mathbf{kk}'}$ in many semiconductors and insulators at the absorption edge is that mainly small differences $\left|\mathbf{k}-\mathbf{k}'\right|$ contribute due the Coulomb singularity in $\tilde{\mathrm{v}}(|\mathbf{k}-\mathbf{k}'|)$. In such a limit the wave-vector dispersion of the inverse dielectric function can be neglected. With the definition (13.24) of the high-frequency electronic dielectric constant ϵ_∞ of the system, the Hamiltonian $H^{\mathrm{WM}}(\mathbf{k}, \mathbf{k}')$ of a Wannier-Mott (WM) exciton becomes [7, 8, 11, 12]

$$H^{\mathrm{WM}}(\mathbf{k}, \mathbf{k}') = \left[E_g + \frac{\hbar^2\mathbf{k}^2}{2\mu_{\mathrm{ex}}}\right]\delta_{\mathbf{kk}'} - \frac{1}{\Omega\epsilon_\infty}\tilde{\mathrm{v}}(|\mathbf{k}-\mathbf{k}'|) \tag{21.7}$$

in spatial Fourier representation. In this representation the formation of an electron-hole pair is embedded in a continuum described by the dielectric constant ϵ_∞, which reduces the action of electrostatic fields, here the neutral electron-hole attraction in form of a bare Coulomb potential. According to the restriction to the electron-electron interaction this background dielectric constant is ϵ_∞. However, the question arises if the lattice polarization in an ionic system can contribute to an additional screening. The answer will be discussed in Chap. 22. According to the homogeneous BSE (21.1) the Hamiltonian (21.7) fulfills an integral equation

$$\sum_{\mathbf{k}'} H^{\mathrm{WM}}(\mathbf{k}, \mathbf{k}')A_\Lambda(\mathbf{k}') = E_\Lambda A_\Lambda(\mathbf{k}). \tag{21.8}$$

In the described two-band model c, v each stationary (spin-singlet) electron-hole-pair state Λ with energy E_Λ is characterized by a Fourier-transformed envelope function $A_\Lambda(\mathbf{k})$. In the spirit of the EMA the $\mathbf{k}$-summation can be spread out over the entire $\mathbf{k}$-space with a volume $(2\pi)^3/\Omega$ ($\Omega = G^3\Omega_0$). The dimensionless eigenvectors fulfill the orthonormalization and completeness relations

$$\sum_{\mathbf{k}} A^*_\Lambda(\mathbf{k}) A_{\Lambda'}(\mathbf{k}) = \delta_{\Lambda\Lambda'},$$

$$\sum_{\Lambda} A_\Lambda(\mathbf{k}) A^*_\Lambda(\mathbf{k}') = \delta_{\mathbf{k}\mathbf{k}'}.$$

21.1.3 Numerical Studies

In the next section we will study the solution of (21.8) in real space. We will find that in real space (21.8) resembles a hydrogen-atom-like eigenvalue problem with three modifications (i) a shift of the ionization edge by E_g, (ii) a reduced mass μ_{ex} of the particle instead of the free-electron mass m, and (iii) a fictitious proton charge reduced to the value $1/\epsilon_\infty$ by screening. Therefore, the characteristic energy, the Rydberg energy $R_\infty = 13.605\,\mathrm{eV}$, and the characteristic length, the Bohr radius $a_B = 0.529$ Å, of a hydrogen atom have to be replaced by excitonic units

$$R_{\mathrm{ex}} = R_\infty \frac{\mu_{\mathrm{ex}}}{m} \frac{1}{\epsilon_\infty^2},$$
$$a_{\mathrm{ex}} = a_B \frac{m}{\mu_{\mathrm{ex}}} \epsilon_\infty. \tag{21.9}$$

Typically for semiconductors it holds $\epsilon_\infty \approx 10$ and $\mu_{\mathrm{ex}} \approx 0.1\,m$, e.g. for GaAs, and thus the exciton Rydberg is of the order of a few millielectronvolt (meV). For strong insulators such as solid rare gases, e.g. solid neon, we have $\epsilon_\infty \approx 2$ and $\mu_{\mathrm{ex}} \approx m$, and the characteristic energy R_{ex} is of the order of a few electronvolt (eV). The effective distance of electron and hole is thus of the order of $a_{\mathrm{ex}} \approx 50$ Å (semiconductors) or $a_{\mathrm{ex}} \approx 1$ Å (strong insulators). The consequences of the characteristic length to be large compared to the lattice constant/bond length or to be of the order of the lattice constant/bond length of a certain non-metal will be discussed below.

In order to test the numerical procedure for a multiband system, especially the **k**-point convergence, a simple two-band model with $\mu_{\mathrm{ex}} = \frac{1}{3}m$ and $\epsilon_\infty = 4$ is investigated. According to (21.9) the characteristic pair parameters follow as $R_{\mathrm{ex}} = 283.44\,\mathrm{meV}$ and $a_{\mathrm{ex}} = 6.348$ Å. The numerical calculations are performed in a simple cubic **k**-space volume extending over a characteristic edge length of $\frac{2\pi}{3}$ Å^{-1}, which naturally limits the electron-hole pair energy cutoff to roughly 15.5 eV. This value is sufficient to converge exciton binding energies with an underestimation less than 2 % [8]. The 14 bound pair states Λ with the lowest excitation energies $E_\Lambda < E_g$ are studied in Fig. 21.4b versus the density of the **k** points expressed by the minimum distance of two adjacent mesh points in the inner part of the BZ. Two different hybrid meshes of the kind as illustrated in two dimensions in Fig. 21.4a are studied for illustration of the **k**-point sampling on the energy convergence of the excitonic levels. While the density of the outer region of the BZ is kept fixed to a $40 \times 40 \times 40$ Monkhorst-Park sampling [13], the density in the inner region is varied.

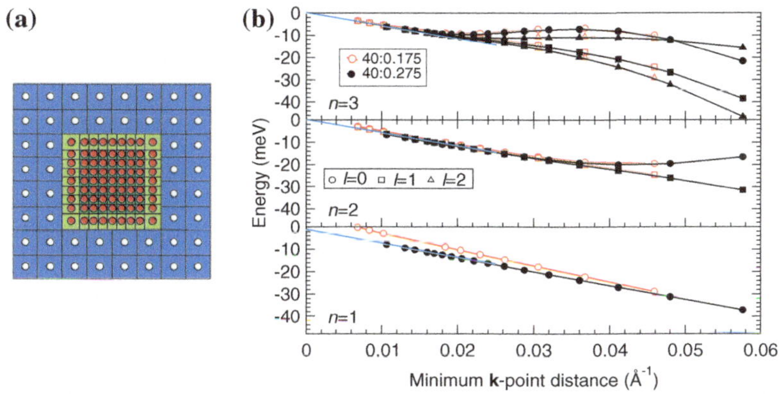

Fig. 21.4 (**a**) Illustration of construction principle of a hybrid **k**-point mesh for a 2D BZ. The **k** points of the (outer) coarse mesh are indicated by *unfilled circles* and those of the (inner) refined mesh by *red dots*. (**b**) Relative binding energies $E_g - E_\Lambda - R_{ex}/n^2$ (principal quantum number $n = 1, 2, 3$) of the excitons with the 14 lowest pair excitation energies E_Λ are plotted versus the minimum distance of the **k** points in the chosen two types of hybrid meshes (*black* and *red* symbols) with fixed outer mesh sampling density but varying inner mesh. Different angular momentum quantum numbers $\ell = 0, 1$, and 2 are encoded by different symbols (*circles*, *squares*, and *triangles*, respectively). The *blue lines* give the interpolation to vanishing **k**-point distances. From [8]

Figure 21.4b shows a clear **k**-point convergence. Within an uncertainty of 0.1 meV the considered 14 lowest eigenvalues of excitons converge to three eigenvalues $n = 1, 2, 3$ which are n^2-fold degenerate according to the spherical symmetry of the bare Coulomb potential. This is in the realm of the two-band model with isotropic and parabolic bands and a Coulombic electron-hole attraction. It should lead to hydrogen-like states $\Lambda = n\ell m$ with the principal quantum number $n = 1, 2, 3, ...$ and the angular momentum quantum numbers $\ell = 0, 1, ..., n-1$ and $-\ell \leq m \leq \ell$ (see Sect. 8.3.1). The lifted degeneracy of the eigenvalues for low **k**-point densities is plausible because of the used lower cubic symmetry. For higher **k**-point densities the resulting energies approach the eigenvalues

$$E_{n\ell m} = E_g - R_{ex}/n^2 \qquad (n = 1, 2, 3, ...) \tag{21.10}$$

of bound states of a hydrogen problem. The numerically computed values are in excellent agreement with analytic results. The hydrogenic results are confirmed by the eigenvectors of (21.7). The square $|A_{\Lambda_0}(\mathbf{k})|^2$ of the eigenvector for the lowest bound state $\Lambda_0 = 100$ in Fig. 21.5a exhibits a variation with **k** that is in agreement with the Fourier transformation of an $1s$ function [14]

$$A_{100}(\mathbf{k}) = 8\sqrt{\frac{\pi a_{ex}^3}{\Omega}} \frac{1}{\left[1 + (\mathbf{k}a_{ex})^2\right]^2}$$

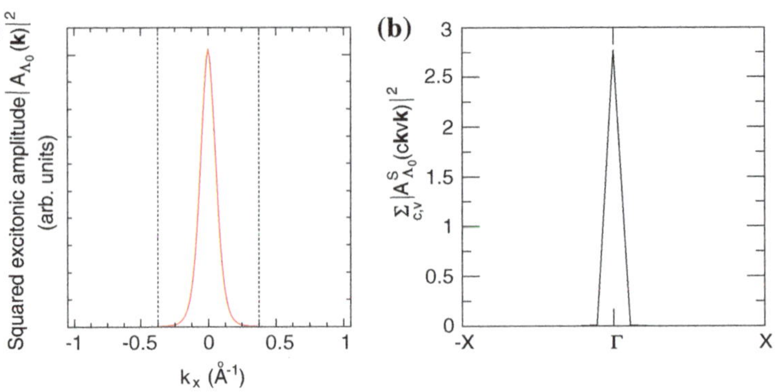

Fig. 21.5 **k**-dependence of band-pair averaged eigenvector square of the lowest bound state in (**a**) a two-band model semiconductor with $\mu_{ex} = \frac{1}{3}m$ and $\epsilon_\infty = 4$ and of (**b**) *zb*-AlN with QP band structure. From [8, 15]

with a width at half maximum of about $|\mathbf{k}| = 2\sqrt{\sqrt{2}-1}/a_{ex} \approx 0.2$ Å^{-1}. This is only a fraction of the BZ extent, in agreement with the results of an all-band calculation of the lowest-energy exciton in *zb*-AlN (see Fig. 21.5b). In other words, the (direct) lowest-energy exciton in *zb*-AlN behaves like a Wannier-Mott exciton. Despite the small deviation of the numerically calculated binding energy from its value in the effective mass theory, the envelope in **k**-space in Fig. 21.5b still indicates the Wannier-Mott character.

The excellent agreement of numerical results in Fig. 21.4b with the analytically calculated values (21.10) suggests the application of the numerically investigated variational approach for a two-band model to the calculation of bound electron-hole pair states, especially excitonic binding energies, for multiband systems with a complete quasiparticle band structure. The cubic semiconductors *rs*-MgO with a wide gap (see QP band structure in Fig. 15.7) via *zb*-AlN (see Fig. 16.4) to the small gap compound *zb*-InN (see valence bands in Fig. 16.6) have been chosen as examples. Due to its rocksalt symmetry, MgO has an *s*-like CBM of Γ_{1c} type and a threefold-degenerate *p*-like VBM of Γ_{15v} type. Two of the uppermost valence bands show almost the same dispersion in the region around the BZ center Γ, while the third one is more dispersive. For this situation three nearly degenerate, *s*-symmetric excitons A, B, and C with the lowest pair excitation energies appear with a splitting between A, B and C of only 0.3 meV. The average binding energy $E_B = E_g - E_{\Lambda_0}$ of the three excitons amounts to the large value of 429 meV for $\epsilon_\infty = 3.0$ [8]. This value drastically overestimates the experimental results of 80 meV [16] and 145 meV [17].

Several bound electron-hole pair excitations below the QP gap of *rs*-MgO are displayed in Fig. 21.6 [18]. The lowest-energy exciton confirms the order of magnitude of the mentioned binding energies. The overestimation of the binding energy by the first-principles-based calculations are attributed to the neglect of dynamical screening [19, 20], in particular the neglect of frequency-dependent lattice screening of the electron-hole attraction [20, 21]. The replacement of the pure electronic screening ϵ_∞ by an effective dielectric constant $\epsilon_{eff} = 6.0$ between ϵ_∞ and the

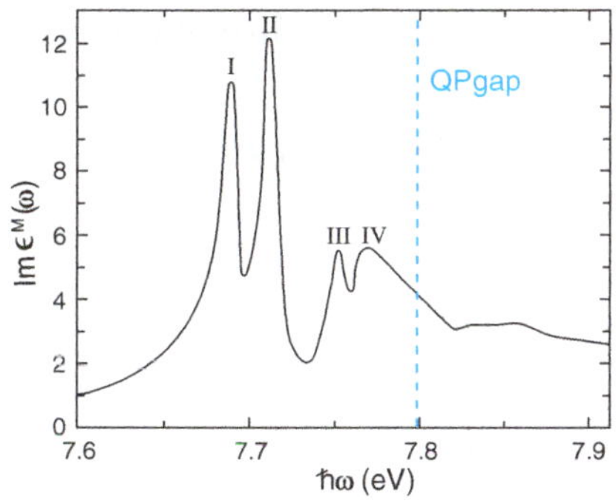

Fig. 21.6 Experimental absorption spectrum of *rs*-MgO near the QP gap. Four bound states I-IV are clearly visible. Reprinted with permission from [18]. Copyright 1969 by the American Physical Society

experimental value for the static dielectric constant $\epsilon_s = 9.8$ including the static lattice polarization [22] would lead to correct binding energies because of the scaling factor $(\epsilon_\infty/\epsilon_{\rm eff})^2 \approx 0.25$. In contrast to the simple WM model (21.7) the binding energy of the lowest singlet exciton in *rs*-MgO is also influenced by the local-field effects $\sim 2\bar{v}$. They decrease the binding energy by about 50 meV compared to that of the triplet exciton [8], i.e., by about 14 %.

The measurements of the exciton fine structure of *wz*-AlN by polarization-resolved photoluminescence spectroscopy indicate a binding energy of 53 meV and an electron-hole exchange interaction value of about 7 meV [23], i.e., a modification also by about 13 % for the wide-gap semiconductor *wz*-AlN. The relative binding energy changes with the local-field effects of this order of magnitude. For ionic compounds one finds their largest possible influence, while they should vanish for more free-electron-like materials. For small-gap semiconductors the situation is more difficult to interprete. Because of the larger dielectric constant the attractive Coulomb interaction is more reduced. However, also the electron-hole exchange effect is decreased in agreement with the more free-electron-like character of the electron and hole wave functions.

The situation for the exciton binding in AlN is rather similar. For the zinc-blende polytype the calculation yields an exciton binding energy of 122 meV [15], which is too large compared with the 58 meV recently derived from ellipsometry measurements on *wz*-AlN [24]. The theoretical investigation of excitonic bound states in the narrow-gap semiconductor InN (see Fig. 21.3) is a challenge because of the small gap, large static electronic dielectric constant, and hence extremely small exciton binding energy. The *ab initio* calculations give 5 meV for the lowest energy A, B excitons of *wz*-InN [8]. Adding the static lattice polarization to the screening the binding energy would be significantly reduced to values below 2 meV. Sometimes better agreement between theoretical and experimental values for exciton binding energies are stated in the literature, e.g. for *wz*-GaN [25]. According to Fig. 21.4 such a better agreement can be however immediately found for less dense sampling of the BZ that tends to decrease the binding. Another (positive) effect of well-converged results concerning the **k**-point density is the avoidance of artificial splittings, which seems to be visible for the excitonic peaks in the absorption spectrum computed for *rs*-MgO, if not sufficiently many **k** points are applied (see e.g. [26]).

21.2 Wannier-Mott Excitons

21.2.1 Hydrogen Problem

The integral equation (21.8) has been formulated in **k**-space with the approximate electron-hole Hamiltonian (21.6) for a non-metallic model crystal with volume $\Omega = G^3\Omega_0$. For many physical discussions a transformation of the electron-hole-pair problem into real space is however more appropriate. One possibility is a Wannier representation of the mixing coefficients $A_\Lambda(\mathbf{k})$ with $\mathbf{R} \in$ Bravais lattice,

$$\Phi_\Lambda(\mathbf{R}) = \frac{1}{\sqrt{\Omega}} \sum_{\mathbf{k}} e^{i\mathbf{kR}} A_\Lambda(\mathbf{k}),$$
$$A_\Lambda(\mathbf{k}) = \frac{\Omega_0}{\sqrt{\Omega}} \sum_{\mathbf{R}} e^{-i\mathbf{kR}} \Phi_\Lambda(\mathbf{R}). \tag{21.11}$$

According to the approximations introduced for this problem, the wave vectors **k** are limited to the BZ. However, in Fig. 21.5 it is demonstrated that the eigenvectors are different from zero only in a small part of the BZ for systems with not too large exciton binding energies. This is the reason why in the EMA, which correctly describes the conduction and valence bands only in a small part of the BZ, the **k**-sum can be extended over the entire **k**-space.

Together with the Fourier transformation of the bare Coulomb potential (13.35) the integral equation with the pair Hamiltonian (21.6) can be approximately transformed into a differential equation

$$\left\{ E_g - \frac{\hbar^2}{2\mu_{\text{ex}}} \nabla_{\mathbf{R}}^2 - \frac{1}{\epsilon_\infty} \text{v}(\mathbf{R}) \right\} \Phi_\Lambda(\mathbf{R}) = E_\Lambda \Phi_\Lambda(\mathbf{R}). \tag{21.12}$$

The vector $\mathbf{R} = \mathbf{R}_e - \mathbf{R}_h$ has to be interpreted as the distance vector between the unit cell $\mathbf{R}_e$, where the electron is situated, and the unit cell $\mathbf{R}_h$, where the hole can be found. Because of the localization of $A_\Lambda(\mathbf{k})$ in the BZ the eigenfunctions $\Phi_\Lambda(\mathbf{R})$ in (21.11) should be extended over several unit cells of the crystal, i.e., distances large compared to the lattice constant. Therefore, the Bravais lattice vector **R** can be replaced by a continuous space coordinate **x**, which can be interpreted as the relative coordinate of the electron-hole motion. The center-of-mass motion of the pair with the total mass $M = m_c^* + m_v^*$ does not appear in our description since the momentum of the translational motion of the electron-hole pair is zero here due to the assumption of vanishing photon wave vectors. We will discuss this problem below. Here, in the spirit of the EMA we assume a continuous variation of $\Phi_\Lambda(\mathbf{x})$ with **x**. In this sense $\Phi_\Lambda(\mathbf{x})$ represents an envelope of the internal motion of the electron-hole pair, which does not care anymore about the atomic nature of matter. The wave functions of the internal motion fulfill the orthonormalization and completeness relations in the form

$$\int d^3\mathbf{x}\Phi_\Lambda^*(\mathbf{x})\Phi_{\Lambda'}(\mathbf{x}) = \delta_{\Lambda\Lambda'},$$
$$\sum_\Lambda \Phi_\Lambda(\mathbf{x})\Phi_\Lambda^*(\mathbf{x}') = \delta(\mathbf{x}-\mathbf{x}').$$

The eigenvalue problem (21.12) is the same as for an electron in a hydrogen atom with the three modifications E_g, μ_{ex}, and $\frac{1}{\epsilon_\infty}$ mentioned in Sect. 21.1.3. Its wave functions can be indexed by three quantum numbers $\Lambda = n$ or k, ℓ, and m: a principal quantum number n (for bound states) or a wave vector k (for scattering states), the angular momentum quantum number ℓ and the magnetic quantum number m. Because of the resulting spherical symmetry the wave functions $\Phi_\Lambda(\mathbf{x})$ can be factorized using polar coordinates r, θ, ϕ as

$$\Phi_{n/k\ell m}(\mathbf{x}) = R_{n/k\ell}(r)Y_{\ell m}(\theta,\phi), \tag{21.13}$$

where $Y_{\ell m}(\theta,\phi)$ are the spherical harmonic functions [see also (8.45)] and $R_{n/k\ell}(r)$ are the corresponding radial parts.

The excitonic bound states $E_{n\ell m} < E_g$ represent a discrete spectrum of eigenvalues,

$$E_{n\ell m} = E_g - \frac{R_{\text{ex}}}{n^2} \qquad (n = 1, 2, 3, \ldots), \tag{21.14}$$

which are n^2-fold degenerate. The binding energy E_B of electron and hole of the lowest exciton with the set of quantum numbers $n\ell m = 100$ amounts to $E_B = R_{\text{ex}}$ in the case of the WM excitons in bulk semiconductors. Their radial functions are

$$R_{n\ell}(r) = -\sqrt{\frac{(n-\ell-1)!}{2n[(n+\ell)!]^3}\left(\frac{2}{na_{\text{ex}}}\right)^3}\, e^{-\frac{r}{na_{\text{ex}}}}\left(\frac{2r}{na_{\text{ex}}}\right)^\ell L_{n+\ell}^{2\ell+1}\left(\frac{2r}{na_{\text{ex}}}\right) \tag{21.15}$$

with the associated Laguerre polynomials $L_{n+\ell}^{2\ell+1}(t)$ [27]. The scattering states with energies above the ionization edge, $E_{k\ell m} > E_g$, possess a continuous energy spectrum (see illustration in Fig. 21.1b)

$$E_{k\ell m} = E_g + \frac{\hbar^2}{2\mu_{\text{ex}}}k^2 \tag{21.16}$$

with radial functions

$$\begin{aligned} R_{k\ell}(r) = {} & \sqrt{\frac{2}{\pi a_{\text{ex}}^3}(a_{\text{ex}}k)e^{\frac{\pi}{2a_{\text{ex}}k}}}\left|\Gamma\left(\ell+1-\frac{i}{a_{\text{ex}}k}\right)\right|\frac{1}{(2\ell+1)!}(2kr)^\ell e^{-ikr} \\ & \times F\left(\frac{i}{a_{\text{ex}}k}+\ell+1, 2\ell+2; 2ikr\right) \end{aligned} \tag{21.17}$$

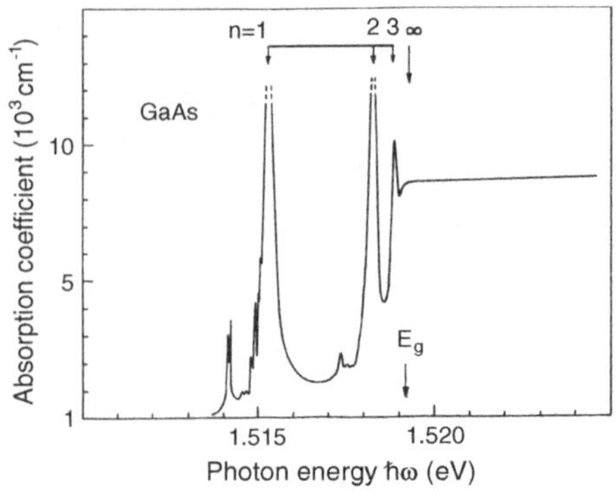

Fig. 21.7 Measured band-edge absorption spectrum of bulk GaAs at $T = 1.2$ K. The $n = 1, 2, 3$ exciton peaks and the extrapolated gap E_g are indicated by *vertical arrows*. The lowest-energy peaks (with reduced intensity) are due to impurity absorption. Adapted from [29]

with the confluent hypergeometric function $F(\alpha, \beta; t)$ [27]. The description of the continuum states is more difficult because of the continuous wave vector k. They cannot be normalized in the usual sense [28]. To circumvent this problem we use later an integral $a_{ex} \int_0^\infty dk$ instead of a sum $\sum_k$ in the analytical expressions for dielectric functions, absorption coefficients or pair densities of states. In the case of GaAs in Fig. 21.7 the measured peak positions can be described by means of $m_c^* = 0.067\,m$, $m_v^* = 0.32\,m$, $\mu_{ex} = 0.055\,m$, $\epsilon_{eff} = 12.5$ (not too far from ϵ_∞), $R_{ex} = 4.82\,\text{meV}$, and $E_g = 1.519\,\text{eV}$.

In the case of the Wannier-Mott excitons one expects a Rydberg series of exciton lines in the absorption spectrum according to (21.14) and measurements displayed in Fig. 21.7 for the semiconductor GaAs [29] with an extremely small binding energy R_{ex} of the $n = 1$ exciton as given in Table 21.1. Values for other semiconductors are also listed in this table together with the ionization edge E_g and the exciton Bohr radius a_{ex}. All these values underline the small binding of electron and hole and,

Table 21.1 Exciton binding energy E_B and Bohr radius a_{ex} of Wannier-Mott excitons in some direct band-gap semiconductors crystallizing in zinc-blende structure

Semiconductor	E_B (meV)	a_{ex} (Å)	R_{ex} (meV)
GaSb	1.7	208	2.11
GaAs	4.9, 4.3	112, 113	4.4, 4.76
InSb	0.4	748	0.52
InAs	1.2	312	1.66
InP	5.1, 3.2	113, 128	5.14, 3.71
CdTe	11	12.2	10.71
ZnTe	13	11.5	11.21
ZnSe	19.9	10.7	22.87
ZnS	29	10.22	38.02

In addition to the experimental values for E_B and a_{ex} also theoretical ones R_{ex} for the exciton Rydberg are listed. From collections in [11, 30]

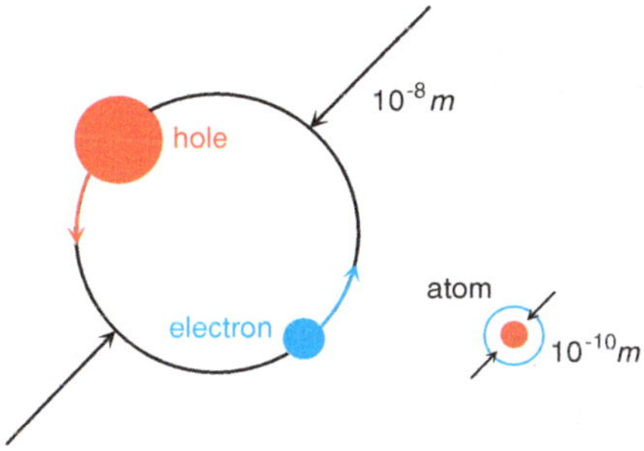

Fig. 21.8 Wannier-Mott exciton (internal motion) as the solid-state analogy of a hydrogen atom. Unlike atoms, the excitons have a finite lifetime. The characteristic distances are by one to two orders of magnitude larger, while the exciton binding is usually two orders of magnitude smaller than that of the electron in the field of the proton

therefore, their large distance in comparison to electron and proton in a hydrogen atom as illustrated in Fig. 21.8. The binding energies of all WM excitons are small compared to the gap energies, while their Bohr radii exceed the lattice constants of the semiconductor crystal by more than one order of magnitude in general.

The redistribution of the pair states due to the Coulomb attraction between electron and hole is described by the modification of the pair or joint density of states (20.5). More interesting is however the weighted pair or joint density of states in units of length$^{-3}\times$ energy^{-1} [28]

$$S(\hbar\omega) = \sum_{\Lambda} |\Phi_{\Lambda}(0)|^2 \delta(E_{\Lambda} - \hbar\omega), \tag{21.18}$$

where the wave-function squares at zero electron-hole distance $|\Phi_{\Lambda}(0)|^2$ illustrate the redistribution of spectral weight due to the excitonic effects. The spin degeneracy is not taken into consideration. The pair energies are given in (21.14) and (21.16). Because of the wave functions (21.15) and (21.17) the weights are

$$|\Phi_{\Lambda}(0)|^2 = \begin{cases} \frac{1}{\pi n^3 a_{\text{ex}}^3} \delta_{\ell 0} \delta_{m0} & \text{for } E_{n\ell m} < E_g \\ \frac{\pi}{a_{\text{ex}} k} \frac{e^{\frac{\pi}{a_{\text{ex}} k}}}{\sinh\left(\frac{\pi}{a_{\text{ex}} k}\right)} \frac{k^2}{a_{\text{ex}} 2\pi^2} \delta_{\ell 0} \delta_{m0} & \text{for } E_{k\ell m} > E_g . \end{cases} \tag{21.19}$$

Only excitons with s symmetry contribute. With the replacement of $\sum_{\Lambda} \rightarrow a_{\text{ex}} \int_0^{\infty} dk \sum_{\ell} \sum_m$ in the case of scattering states it holds

$$
\begin{aligned}
S(\hbar\omega) &= \frac{1}{\pi a_{\text{ex}}^3}\left\{\sum_{n=1}^{\infty}\frac{1}{n^3}\delta\left(E_g - \hbar\omega - \frac{R_{\text{ex}}}{n^2}\right)\theta(E_g - \hbar\omega)\right. \\
&\quad \left. + \frac{1}{2}\int_0^{\infty} dt \cdot t \cdot e^{\pi/t}\frac{1}{\sinh\left(\frac{\pi}{t}\right)}\delta(E_g - \hbar\omega + R_{\text{ex}}t^2)\theta(\hbar\omega - E_g)\right\} \\
&= \frac{1}{4\pi a_{\text{ex}}^3 R_{\text{ex}}}\left\{\sum_{n=1}^{\infty}\frac{4}{n^3}\delta\left(\frac{E_g - \hbar\omega}{R_{\text{ex}}} - \frac{1}{n^2}\right)\theta(E_g - \hbar\omega)\right. \qquad (21.20) \\
&\quad \left. + \left.\frac{e^{\frac{\pi}{t}}}{\sinh\left(\frac{\pi}{t}\right)}\right|_{t=\sqrt{\frac{\hbar\omega - E_g}{R_{\text{ex}}}}}\theta(\hbar\omega - E_g)\right\}.
\end{aligned}
$$

This formula goes back to a derivation of Elliott [31]. In the limit of vanishing excitonic effects $R_{\text{ex}} = \frac{\hbar^2}{2\mu_{\text{ex}}a_{\text{ex}}^2} \to 0$ and $a_{\text{ex}} \to \infty$ the contribution of the bound states vanishes and the pair density of states becomes

$$
S^{\text{free}}(\hbar\omega) = \frac{1}{4\pi^2}\left(\frac{2\mu_{\text{ex}}}{\hbar^2}\right)^{\frac{3}{2}}\sqrt{\hbar\omega - E_g\theta(\hbar\omega - E_g)},
$$

the density of the uncorrelated electron-hole pairs with the relation $2\Omega S^{\text{free}}(\hbar\omega)=$ JDOS$(\hbar\omega)$ to the joint density of states given in (20.5). The modifications of the pair density of states due to the Coulomb attraction are illustrated in Fig. 21.9.

The Elliott formula (21.20) characterizes two important effects due to the Coulomb attraction of electron and hole. For frequencies $\hbar\omega < E_g$ bound states appear. Their discrete spectrum has at limit point at $\hbar\omega = E_g$ and smoothly merges into the continuous spectrum. In contrast to the density of uncorrelated electron-hole pairs, the density (21.20) is non-zero at $\hbar\omega = E_g$. It holds $S(E_g + 0^+) = 1/(2\pi a_{\text{ex}}^3 R_{\text{ex}})$. For high frequencies $\hbar\omega \gg E_g$ one expects that the influence of excitonic effects vanishes. Their effect on the spectrum can be formally expressed by the Sommerfeld factor ($\hbar\omega > E_g$)

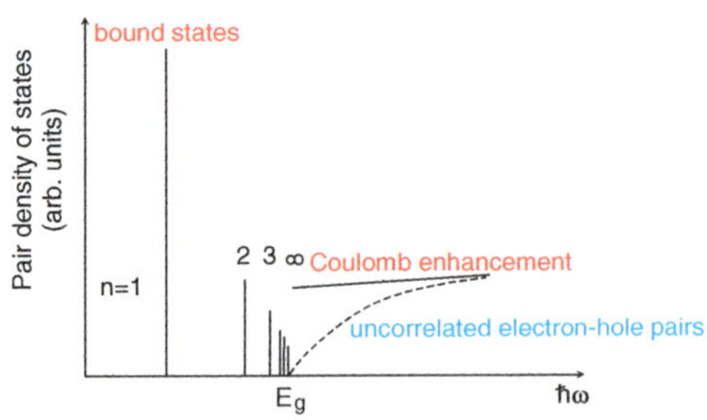

Fig. 21.9 Pair density of states with (*red*) and without (*blue*) Coulomb effects (schematically)

$$S(\hbar\omega)/S^{\text{free}}(\hbar\omega) = \frac{\pi}{t}\frac{e^{\frac{\pi}{t}}}{\sinh\left(\frac{\pi}{t}\right)}\Bigg|_{t=\sqrt{\frac{\hbar\omega-E_g}{R_{\text{ex}}}}},$$

which goes to infinity for $\hbar\omega \to E_g + 0^+$ and to unity for $\omega \to \infty$. More precisely, for $t \to \infty$ the asymptotic behavior is $\left(1+\frac{\pi}{t}\right)$. In other words, the asymptotics of the pair density for large frequencies is $S(\hbar\omega) \sim (\sqrt{\hbar\omega - E_g} + \pi\sqrt{R_{\text{ex}}})$, which is equal to the uncorrelated density plus a constant. It represents a strong Coulomb enhancement of the pair density of states, although only scattering states contribute. The electron-hole attraction has also a pronounced influence near the ionization continuum. While the density of uncorrelated pairs vanishes $\sim\sqrt{\hbar\omega - E_g}$, the Coulomb effects lead to a constant $S(E_g + 0^+) = 1/(2\pi a_{\text{ex}}^3 R_{\text{ex}})$.

The existence of Wannier-Mott excitons with eigenvalues (21.14) and the pair density of states (21.20) is related to two conditions. First, that of the Wannier approximation of large distances

$$a_{\text{ex}} \gg a_0$$

with a_0 as the lattice constant has to be fulfilled. Otherwise, the transition between real and **k** space described in (21.11) and (21.12) cannot be applied. Second, in agreement with the requirement of a positively definite pair Hamiltonian (19.16) it must be valid

$$E_g > R_{\text{ex}}.$$

Otherwise, for $R_{\text{ex}} > E_g$ the normal insulating state with a fully occupied valence band and an empty conduction band would become unstable against the formation of excitons [32]. As a consequence of this proposal, a number of colleagues have discussed a new phase which can be formed at very low temperatures, the so-called *excitonic insulator* [33]. A close formal similarity between the excitonic insulating state and the superconducting state has been predicted. However, physical properties of the two states are different, in particular, there is no Meissner effect in an excitonic insulator. Graphene seems to be a prototypical system because of its character as zero-gap semiconductor. Indeed, there are several papers on the excitonic insulator phase in graphene (see e.g. [34, 35]) but also in other materials.

21.2.2 Allowed and Forbidden Optical Transitions

In order to discuss the excitonic effects at the absorption edge E_g in more detail, we divide the macroscopic dielectric function (19.35) in the part, that is only relevant near $\hbar\omega \approx E_g$, and the rest of exciton-modified higher interband transitions. In addition, we focus the discussion on diagonal elements of the dielectric tensor. We write

$$\epsilon_{jj}^{M}(\omega) = \epsilon_{jj}^{\text{back}} + \chi_{jj}(\omega) \tag{21.21}$$

with an almost frequency-independent background $\epsilon_{jj}^{\text{back}}$ due to the vacuum polarizability and the higher interband transitions. The additional optical susceptibility contribution $\chi_{jj}(\omega)$ will be only correctly described in the vicinity of the quasiparticle gap.

According to (19.35) the susceptibility due to the lowest (spin-singlet) pair excitations can be written in the form

$$\chi_{jj}(\omega) = \frac{8\pi}{\Omega} \sum_{\Lambda} \left| \sum_{\mathbf{k}} (M_{cv}^{\mathbf{kk}})_j^* A_\Lambda(\mathbf{k}) \right|^2 \frac{2E_\Lambda}{E_\Lambda^2 - \hbar^2(\omega + i\eta)^2} \tag{21.22}$$

with the jth component of the dipole operator (matrix element) (20.4)

$$(M_{cv}^{\mathbf{kk}})_j = \frac{e}{\sqrt{4\pi\varepsilon_0}} \frac{\hbar}{im} \frac{[\mathbf{p}_{cv}(\mathbf{k})]_j}{\varepsilon_c(\mathbf{k}) - \varepsilon_v(\mathbf{k})}. \tag{21.23}$$

First, we assume that the optical transition from the valence (v) to the conduction (c) band is dipole-allowed at Γ in the two-band model of Fig. 21.3. This is fulfilled at the fundamental gap of a direct semiconductor such as zb-GaAs (see Fig. 21.7). That means $(M_{cv}^{\mathbf{kk}})_j$ is non-zero at the band extrema at $\mathbf{k} = 0$ with values of the magnitude as given in Table 20.1. In addition, because of the EMA, we assume that the variation of the matrix element (21.23) with the $\mathbf{k}$ vector is weak, i.e.,

$$(M_{cv}^{\mathbf{kk}})_j \approx (M_{cv}^{00})_j.$$

With the envelope function of the internal motion of the electron-hole pairs (21.11) the susceptibility becomes

$$\chi_{jj}(\omega) = 8\pi \left| (M_{cv}^{00})_j \right|^2 \sum_{\Lambda} |\Phi_\Lambda(0)|^2 \frac{2E_\Lambda}{E_\Lambda^2 - \hbar^2(\omega + i\eta)^2}. \tag{21.24}$$

In cubic crystals, which are optically isotropic, the j-dependence disappears. In (21.19) we have seen that the relation $\Phi_\Lambda(0) \neq 0$ is fulfilled only for $\ell = 0, m = 0$, i.e., only excitons with s symmetry can be optically excited. In the limit of vanishing broadening $\eta \to +0$ the resulting imaginary part of the optical susceptibility

$$\text{Im}\chi_{jj}(\omega) = 2(2\pi)^2 \left| \left(M_{cv}^{00}\right)_j \right|^2 [S(\hbar\omega) - S(-\hbar\omega)] \tag{21.25}$$

is directly related to the pair density of states (21.18). The resonant part $\propto S(\hbar\omega)$ of (21.25) corresponds to the Elliott formula (21.20) [31]. Because of the analytic properties and, hence, the Kramers-Kronig relation of the type (19.28), it holds for the real part

$$\mathrm{Re}\chi_{jj}(\omega) = \mathcal{P} \int_{-\infty}^{+\infty} \frac{d\omega'}{\pi} \frac{\mathrm{Im}\chi_{jj}(\omega')}{\omega' - \omega}.$$

The explicit use of the Kramers-Kronig relation can be easily proven by means of the sum (21.24) of harmonic oscillator contributions, at least in the limit of vanishing broadening $\eta \to +0$. However, the use of the representation (21.25) indicates a dilemma of the EMA. For large frequencies the pair density (21.20) increases, so that the Kramers-Kronig integral diverges. To circumvent this problem in many approaches the interband distance $\varepsilon_c(\mathbf{k}) - \varepsilon_v(\mathbf{k})$ in the dipole matrix element (21.23) is replaced by $\hbar\omega$, a procedure which is correct without many-body effects and for the resonant contributions. Since, one is only interested in the frequency region around the fundamental gap, the dipole matrix element can therefore approximately be replaced by $|(M_{cv}^{00})_j|^2 \to |(M_{cv}^{00})_j|^2 \left(\frac{E_g}{\hbar\omega}\right)^2$ [11, 14, 36]. With this replacement the real part can be evaluated by means of the Kramers-Kronig relation.

Another problem concerns the inclusion of the finite lifetime of the electron-hole pairs by the broadening parameter η. With the solutions (21.14)–(21.17) and (21.19) of the eigenvalue problem expression (21.24) becomes

$$\chi_{jj}(\omega) = 8\frac{\left|(M_{cv}^{00})_j\right|^2}{a_{\mathrm{ex}}^3} \left\{ \sum_{n=1}^{\infty} \frac{2}{n^3} \frac{E_g - \frac{R_{\mathrm{ex}}}{n^2}}{\left(E_g - \frac{R_{\mathrm{ex}}}{n^2}\right)^2 - \hbar^2(\omega + i\eta)^2} + \int_0^{\infty} dt \frac{t e^{\frac{\pi}{t}}}{\sinh\left(\frac{\pi}{t}\right)} \frac{E_g + R_{\mathrm{ex}} t^2}{\left(E_g + R_{\mathrm{ex}} t^2\right)^2 - \hbar^2(\omega + i\eta)^2} \right\}. \quad (21.26)$$

For the imaginary part the t-integral can be performed analytically. However, the real part diverges. Only by taking the above-described frequency dependence of the transition matrix element into account, an imaginary part results that generates a finite real part also within the EMA. In this case a corresponding expression in terms of the digamma function is possible [36]. The imaginary part of the excitonic susceptibility is represented in Fig. 21.10 for different excitonic strengths R_{ex} and broadening parameters η. Because of the optical isotropy the dependence on the polarization direction is omitted. The bound states, their broadening, and the Coulomb enhancement of the spectra above the ionization edge $E_g = 1.5\,\mathrm{eV}$, i.e., the Sommerfeld factor, are clearly visible. In contrast to Figs. 21.7 and 21.9 no Rydberg series appears because of the chosen relative large broadening parameter η.

One of the earliest and best examples for an excitonic Rydberg series has been found for the cuprite Cu_2O [37, 38] (see Fig. 21.11). It crystallizes in a cubic (bcc) crystal structure with six atoms in a unit cell and the space group O_h^4 $(Pn\bar{3}m)$. It is a direct semiconductor with the band extrema at the BZ center Γ. Similar to rutile-SnO_2 (see Fig. 20.5) the conduction and valence band extrema in Cu_2O have the same parity under inversion. Optical transitions between these two bands are therefore electric

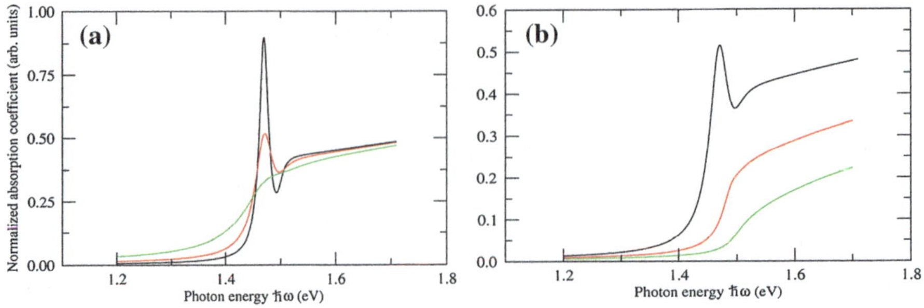

Fig. 21.10 Isotropic excitonic susceptibility $\chi(\omega)$ for a direct bulk semiconductor varying the strength of (**a**) broadening and (**b**) Coulomb interaction. A dimensionless factor $4|(M_{cv}^{00})_j|^2 E_g^{1/2}(2\mu_{\mathrm{ex}}/\hbar^2)^{3/2}$ is not included in the plots. The gap is fixed to be $E_g = 1.5\,\mathrm{eV}$. In (**a**) the *curves* are drawn for $R_{\mathrm{ex}} = 30\,\mathrm{meV}$ and $\hbar\eta = 10$ (*black*), 20 (*red*), and 50 (*green*) meV, while those in (**b**) are for $\hbar\eta = 20\,\mathrm{meV}$ and $R_{\mathrm{ex}} = 30$ (*black*), 10 (*red*), and 0 (*green*) meV

dipole-forbidden but may be observable as magnetic dipole or electric quadrupole transitions, which are however not studied here. Near the extrema the optical matrix element $\mathbf{p}_{cv}(\mathbf{k})$ may be approximated by a linear function as $\mathbf{p}_{cv}(\mathbf{k}) = A\hbar\mathbf{k}$ with the dimension-less prefactor $A \approx 0.15$ [39]. Such a behavior is displayed in Fig. 20.5 for rt-SnO_2.

Correspondingly the dipole moment $(M_{cv}^{\mathbf{kk}})_j$ (21.23) can be also linearized in the $\mathbf{k}$ vector

$$\left(M_{cv}^{\mathbf{kk}}\right)_j = \frac{\partial}{\partial k_j} M_{cv}^{\mathbf{kk}}\bigg|_{\mathbf{k}=0} k_j. \tag{21.27}$$

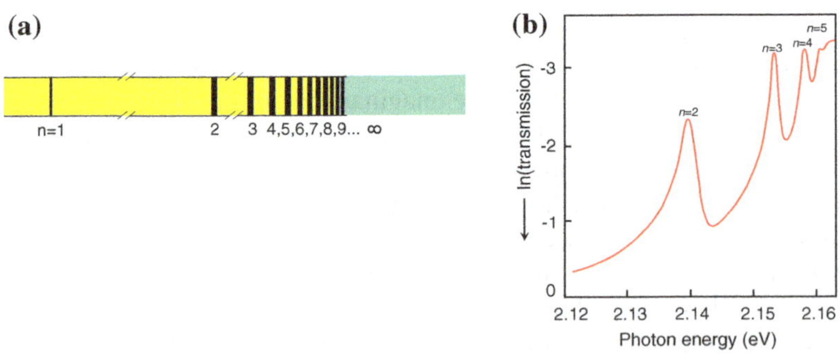

Fig. 21.11 Hydrogen-like yellow excitonic series of Cu_2O observed (**a**) in emission [37] and (**b**) in absorption [38]. The $n = 1$ bound exciton is missing in absorption and extremely weak in emission. The exciton series can be described by $E_g - R_{\mathrm{ex}}/n^2$ ($n = 2, 3, 4, \ldots$) with $E_g = 2.172/2.166\,\mathrm{eV}$ and $R_{\mathrm{ex}} = 97.2/97.0\,\mathrm{meV}$ [37, 38]. Adapted from [37, 38]. The *right panel* is reprinted with permission from [38]. Copyright 1961 by the American Physical Society

Consequently, with (21.10) and (21.12) one finds

$$\frac{1}{\sqrt{\Omega}}\sum_{\mathbf{k}}\left(M_{cv}^{\mathbf{kk}}\right)_j^* A_\Lambda(\mathbf{k}) = -i\frac{\partial}{\partial k_j}M_{cv}^{\mathbf{kk}}\bigg|_{\mathbf{k}=0}\frac{\partial}{\partial x_j}\Phi_\Lambda(\mathbf{x})\bigg|_{\mathbf{x}=0}.$$

The isotropic excitonic susceptibility (21.22) becomes

$$\begin{aligned}\chi(\omega) &= \frac{8\pi}{3}\sum_{j=1}^{3}\sum_{\Lambda}\left|\frac{\partial}{\partial k_j}\left(M_{cv}^{\mathbf{kk}}\right)_j\bigg|_{\mathbf{k}=0}\frac{\partial}{\partial x_j}\Phi_\Lambda(\mathbf{x})\right|^2_{\mathbf{x}=0}\frac{2E_\Lambda}{E_\Lambda^2-\hbar^2(\omega+i\eta)^2}\\ &= \frac{8\pi}{3}\left|\frac{\partial}{\partial k_j}M_{cv}^{\mathbf{kk}}\bigg|_{\mathbf{k}=0}\right|^2\sum_\Lambda\left|\frac{d}{dr}R_{n/k\ell}(r)\bigg|_{r=0}Y_{\ell m}(\theta,\phi)\right|^2\frac{2E_\Lambda}{E_\Lambda^2-\hbar^2(\omega+i\eta)^2},\end{aligned}$$

assuming that the derivatives of the transition matrix element are independent of $j = x, y, z$. Within the hydrogenic model (21.15) and (21.17) it holds [31]

$$\sum_{m=-1}^{+1}\left|\frac{d}{dr}\Phi_\Lambda(\mathbf{x})\bigg|_{r=0}\right|^2 = \begin{cases}\frac{n^2-1}{3\pi n^5 a_{\mathrm{ex}}^5}\delta_{\ell 1} & \text{for } E_{n\ell m} < E_g\\ \frac{\pi}{3a_{\mathrm{ex}}k}\frac{[1+\frac{1}{(a_{\mathrm{ex}}k)^2}]e^{\frac{\pi}{a_{\mathrm{ex}}k}}}{\sinh\left(\frac{\pi}{a_{\mathrm{ex}}k}\right)}\frac{k^4}{2\pi^2 a_{\mathrm{ex}}}\delta_{\ell 1} & \text{for } E_{k\ell m} > E_g\end{cases} \tag{21.28}$$

and consequently

$$\begin{aligned}\chi(\omega) &= \frac{16\left|\frac{\partial}{\partial k_j}\cdot M_{cv}^{\mathbf{kk}}\big|_{\mathbf{k}=0}\right|^2}{9a_{\mathrm{ex}}^5}\left\{\sum_{n=2}^{\infty}\frac{n^2-1}{n^5}\frac{E_g-\frac{R_{\mathrm{ex}}}{n^2}}{\left(E_g-\frac{R_{\mathrm{ex}}}{n^2}\right)^2-\hbar^2(\omega+i\eta)^2}\right.\\ &\quad\left.+\frac{1}{2}\int_0^\infty dt\frac{t(1+t^2)e^{\frac{\pi}{t}}}{\sinh\left(\frac{\pi}{t}\right)}\frac{E_g+R_{\mathrm{ex}}t^2}{(E_g+R_{\mathrm{ex}}t^2)^2-\hbar^2(\omega+i\eta)^2}\right\}\end{aligned} \tag{21.29}$$

omitting the j-dependence. For vanishing damping $\eta \to +0$ its imaginary part is $(\omega > 0)$

$$\begin{aligned}\mathrm{Im}\chi(\omega) &= \frac{8\pi\left|\frac{\partial}{\partial k_j}M_{cv}^{\mathbf{kk}}\big|_{\mathbf{k}=0}\right|^2}{9a_{\mathrm{ex}}^5 R_{\mathrm{ex}}}\left\{\sum_{n=2}^{\infty}\frac{n^2-1}{n^5}\delta\left(\frac{E_g-\hbar\omega}{R_{\mathrm{ex}}}-\frac{1}{n^2}\right)\theta(E_g-\hbar\omega)\right.\\ &\quad\left.+\frac{(1+t^2)e^{\frac{\pi}{t}}}{4\sinh\left(\frac{\pi}{t}\right)}\bigg|_{t=\sqrt{\frac{\hbar\omega-E_g}{R_{\mathrm{ex}}}}}\theta(\hbar\omega-E_g)\right\}.\end{aligned} \tag{21.30}$$

In the limit of vanishing excitonic effects $R_{ex} \to 0$, the excitonic series vanishes and the spectrum of uncorrelated electron-hole pairs $(\hbar\omega > E_g) \sim t^3 \sim (\hbar\omega - E_g)^{3/2}$ known for forbidden optical transitions [11] appears. The Coulomb enhancement is increased by the factor $(1 + 1/t^2)$ compared to the Sommerfeld factor of the allowed optical transitions. The susceptibilities (21.29) and (21.30), respectively, together with the selection rule in (21.28) lead to the conclusion that excitons, bound and scattering states, with p symmetry are weakly dipole-active at a "forbidden" direct band gap. Thereby, the strength of the weak dipole activity is proportional to the excitonic effects. The absorption coefficient (21.30) is identically zero for $n = 1$ in agreement with the fact that $n = 1$ is the only level without a p state. Expression (21.30) is in agreement with Fig. 21.11b, which only shows the weakly allowed excitonic absorption peaks in Cu_2O resulting in the "yellow" exciton series $2p, 3p, 4p$, etc. measured in emission and absorption spectroscopy [37, 38] (see also Fig. 21.11a).

We have to mention that in photoluminescence studies a very sharp but weak exciton peak associated with the $1s$ exciton is observed in Cu_2O at 2.033 eV. The corresponding radiative electron-hole-pair recombination is due to magnetic-dipole and electric-quadrupole $\Gamma_{7v}^+ \to \Gamma_{6c}^+$ transitions [40]. Identifying the ionization edge with $E_g = 2.166$ eV [38], one obtains a binding energy of 133 meV for the $1s$ state. This value is somewhat in contrast to the value $R_{ex} = 97$ meV in Fig. 21.11. The reason for the difference of 36 meV is a central-cell effect influencing the $1s$ exciton very similar to the findings for hydrogenic impurities [11].

21.2.3 Longitudinal-Transverse Splitting

The frequency-dependent, macroscopic dielectric function $\epsilon_{jj}^M(\omega)$ (21.21) near the fundamental absorption edge shows that, in this energy range, the frequency dependence can be mainly represented by a sum over exciton oscillators (21.22). Moreover, spectra for dipole-allowed transitions like those given in Figs. 21.9 and 21.10 indicate that in the limit of strong excitonic effects the spectral variation seems to be reasonably described by a strong oscillator at the lowest exciton frequency $E_{ex} = E_g - R_{ex}$ with a damping η [28]. Only its resonant contribution plays a role. The rest of the oscillators contribute to the background dielectric constant. When one, in addition, assumes an isotropic system the dependence on the direction of the light polarization j can be omitted. The complex dielectric function (21.21) can be written in the simple form [14, 28]

$$\epsilon^M(\omega) = \epsilon^{\text{back}} \left\{ 1 + \frac{\Delta_{LT}}{E_{ex} - \hbar(\omega + i\eta)} \right\} \tag{21.31}$$

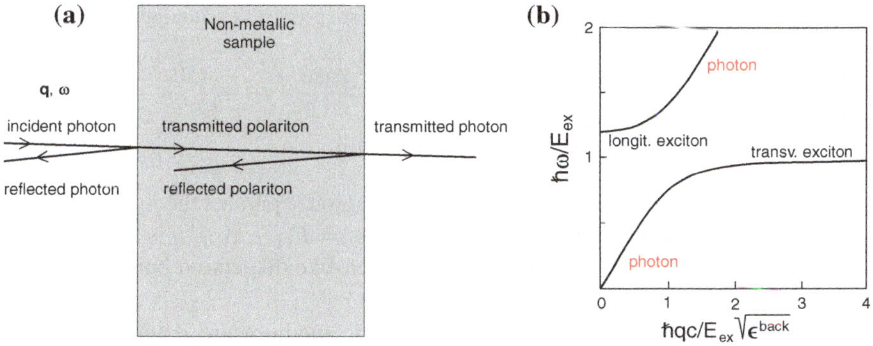

Fig. 21.12 (**a**) Illustration of transition of photons through a materials slab with excitons via propagations of polaritons inside the sample. (**b**) Polariton dispersion without spatial dispersion for $\Delta_{\mathrm{LT}}/R_{\mathrm{ex}} = 0.2$ and $\eta = 0$

with the oscillator frequency $E_{\mathrm{ex}}/\hbar$ and the oscillator strength

$$\begin{aligned}\Delta_{\mathrm{LT}} &= \frac{8\pi}{\epsilon^{\mathrm{back}}}\left|\left(M_{cv}^{00}\right)_j\right|^2 |\Phi_{100}(0)|^2 \\ &= \frac{8}{a_{\mathrm{ex}}^3\epsilon^{\mathrm{back}}}\left|\left(M_{cv}^{00}\right)_j\right|^2 \end{aligned} \tag{21.32}$$

independent of $j = x, y, z$.

The resulting simple form of the lineshape of the complex dielectric function (21.31) and the use of the relations (18.24) and (18.25) make the discussion of the transverse and longitudinal excitations of the coupled light-matter system easy, despite the neglect of spatial dispersion due to the center-of-mass motion of the electron-hole pair. The resulting elementary excitations of the coupled system are exciton-polaritons. They are propagating modes in a dielectric medium, a semiconductor or insulator, in which the electromagnetic wave with frequency ω is coupled with the polarization wave of the excitons as illustrated in Fig. 21.12a.

The coupled excitations follow from the usual electromagnetic wave dispersion $\omega^2 = c^2q^2/\epsilon^M(\omega)$ (18.24), where ω and q are the frequency and wave vector, respectively, of the light. The resulting eigenmodes are schematically represented in Fig. 21.12b for vanishing lifetime broadening. For small frequencies, $\hbar\omega \ll E_{\mathrm{ex}}$, a photon-like dispersion

$$\omega = \frac{cq}{\sqrt{\epsilon^{\mathrm{back}}\left(1 + \frac{\Delta_{\mathrm{LT}}}{E_{\mathrm{ex}}}\right)}}$$

appears with a light velocity $c/\sqrt{\epsilon^{\mathrm{back}}(1 + \Delta_{\mathrm{LT}}/E_{\mathrm{ex}})}$ slightly smaller than $c/\sqrt{\epsilon^{\mathrm{back}}}$. Near the resonance, for $\hbar\omega \to E_{\mathrm{ex}}$, the wave vector diverges, $q \to \infty$. No solution exists for $E_{\mathrm{ex}} < \hbar\omega < E_{\mathrm{ex}} + \Delta_{\mathrm{LT}}$. That means, in this frequency range no propagating

Table 21.2 Longitudinal-transverse splitting of the lowest-energy 1s exciton-polariton measured for a few semiconductors at low temperatures. From [41]

Semiconductor	GaAs	CdTe	ZnSe	CdS	CuCl
Δ_{LT} (meV)	$\approx$0.1	$\approx$0.4	$\approx$1.3	$\approx$2	5.4 – 5.7

modes occur inside the sample. This frequency region separates the upper and lower polariton branches displayed in Fig. 21.12b. At $\hbar\omega = E_{\mathrm{ex}} + \Delta_{\mathrm{LT}}$, q is zero. For large frequencies $\hbar\omega \gg E_{\mathrm{ex}}$ one again obtains a photon-like dispersion but with the light velocity $c/\sqrt{\epsilon^{\mathrm{back}}}$.

The longitudinal eigenmodes of the dielectric medium are determined by the relation $\epsilon^{M}(\omega) = 0$ (18.25). It yields the eigenfrequency of the longitudinal exciton $\hbar\omega_{\mathrm{L}} = E_{\mathrm{ex}} + \Delta_{\mathrm{LT}}$, while the transverse exciton frequency is $\hbar\omega_{\mathrm{T}} = E_{\mathrm{ex}}$. For that reason, the energy Δ_{LT}, representing the oscillator strength in (21.31), is usually called *longitudinal-transverse* (LT) *splitting*. For the longitudinal excitons the vector of the polarization field is parallel to the propagation vector $\mathbf{q}$ of the light. Interestingly the LT splitting (21.32) is proportional to the optical transition matrix element $\left|\left(M_{cv}^{00}\right)_j\right|$ at the BZ center and inversely proportional to the exciton volume $\frac{4\pi}{3}a_{\mathrm{ex}}^3$ and the background dielectric constant ϵ^{back}. The frequency region, in which no transverse wave propagates in matter, is thus particularly large for optical media with small Bohr radii of the excitons a_{ex} and not too large background dielectric constants ϵ^{back}. Experimental Δ_{LT} values are listed in Table 21.2 for a few semiconductors. They are small compared to the exciton binding energies in Table 21.2. Only the wide-gap semiconductor CuCl with a large exciton binding energy of about 190 meV [41] gives rise to a relatively large splitting of about 5 meV.

We have to mention two facts. The measured LT splittings depend on the lifetime broadening η of the excitons. Therefore, sample temperature and quality play a role. In general, i.e., including spatial dispersion, the dielectric functions for longitudinal and transverse excitations are not the same. However, they become degenerate in the long-wavelength limit, at least for cubic crystals, studied here (see [42, 43]). Otherwise, a more careful discussion is needed. There is another problem. The excitation of a singlet Wannier-Mott exciton corresponds to the formation of a longitudinal current density or charge density wave, while the triplet exciton is a spin density wave. The transverse singlet excitons are related to transverse current density waves [42].

21.3 Localized Excitons

21.3.1 Frenkel Excitons

The results for the Wannier-Mott excitons (21.14) and (21.15) clearly indicate a strong tendency for an increase of the electron-hole binding with the average coordinate of the relative motion, i.e., the distance of electrons and holes in space.

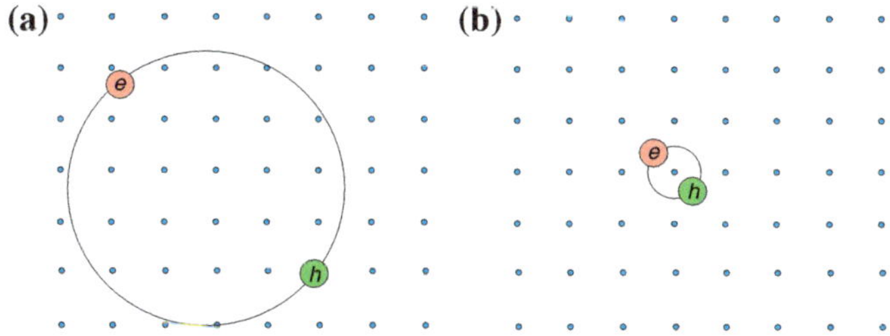

Fig. 21.13 An exciton as a bound electron-hole pair in a crystal described by a 2D square lattice (schematically). (**a**) Wannier-Mott limit with average electron-hole distance large in comparison to the lattice constant. (**b**) Opposite Frenkel limit

This is illustrated in Fig. 21.13. Systems with small electron-hole distances are wide-band-gap insulators such as solid rare gases and alkali halides with narrow bands and small dielectric constants. For instance, LiF and LiCl with the lowest pair excitation energies at $E_{\mathrm{ex}} = 12.8$ and 8.7 eV, respectively, possess dielectric constants of about 1.9 and 2.6, respectively, compared to a value of about 10 for a conventional semiconductor. As a consequence the exciton binding energies derived from experiments in Table 21.3 [44] are large compared to those of Wannier-Mott excitons given in Table 21.1. This picture is also confirmed by the absorption spectra in Fig. 21.14, which display even two series of exciton peaks of solid Ar below the fundamental gap. However, they cannot be interpreted as Rydberg series of hydrogenic excitons despite the labels $n, n' = 1, 2$. Apart from the strong spin-orbit splitting of the valence bands, the exciton radii of the $n, n' = 1$ excitons have been estimated to be near 1.9 Å, i.e., smaller than the nearest-neighbor distance in an Ar crystal of 3.7 Å [45]. In such a case the attractive electron-hole potential will no longer vary as a bare Coulomb potential reduced by a uniform dielectric constant of about 1.66 [46].

Table 21.3 Lowest pair excitation energies E_{ex} of Frenkel excitons in insulating solid rare gases and alkali halide crystals together with the gap energy E_g and the resulting exciton binding energy $E_B = E_g - E_{\mathrm{ex}}$

Crystal	E_g	E_{ex}	E_B
Ne	21.6	17.5	4.1
Ar	14.2	12.1	2.1
Kr	11.7	10.2	1.5
Xe	9.3	8.3	1.0
LiF	13.7	12.8	0.9
KBr	7.4	6.7	0.7
KCl	8.7	7.8	0.9
KF	10.8	9.9	0.9
NaBr	7.1	6.7	0.4
NaCl	8.8	7.9	0.9
NaF	11.5	10.7	0.8

All values are in eV. From [44]

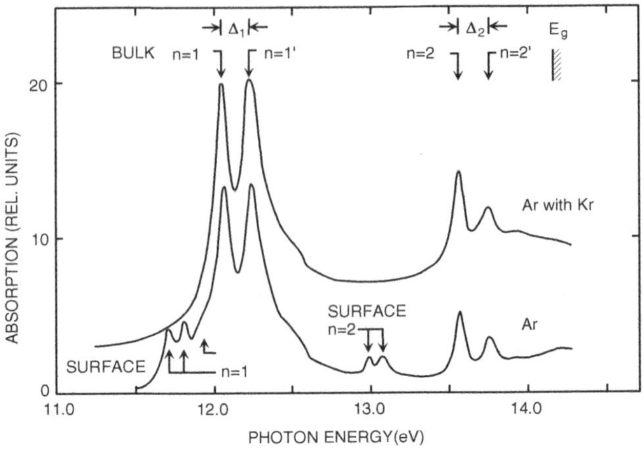

Fig. 21.14 Optical absorption spectrum of a clean Ar film and a Kr-coated Ar film below the fundamental gap. Bulk and surface excitons are labeled by $n, n' = 1, 2$. The spin-orbit splittings of the valence bands Δ_1, Δ_2 are also indicated. The surface excitons, which are not of interest here, are significantly redshifted. Reprinted with permission from [46]. Copyright 1976 by the American Physical Society

The above discussion shows that besides the Wannier-Mott limit of the excitons with large distances $a_{ex} \gg a_0$ between electrons and holes and weak binding energies $R_{ex} \ll E_g$, also excitons in the opposite limit exist. For strong electron-hole interaction, as in ionic crystals or solid rare gases with flat bands and small dielectric constants, the electron and hole of a pair are tightly bound to each other within the same or nearest-neighbor unit cells and, hence, possess large binding energies [1, 2]. In order to illustrate their strong localization we transform the set of algebraic equations (21.1) with the pair Hamiltonian (21.2) and the Coulomb matrix elements (19.33) from the Bloch into the Wannier representation.

For that reason we introduce Wannier functions

$$
\begin{aligned}
a_\nu(\mathbf{x}-\mathbf{R}) &= \frac{1}{G^{3/2}} \sum_{\mathbf{k}} e^{-i\mathbf{k}\mathbf{R}} \varphi_{\nu\mathbf{k}}(\mathbf{k}), \\
\varphi_{\nu\mathbf{k}}(\mathbf{x}) &= \frac{1}{G^{3/2}} \sum_{\mathbf{R}} e^{i\mathbf{k}\mathbf{R}} a_\nu(\mathbf{x}-\mathbf{R}),
\end{aligned}
\tag{21.33}
$$

where $a_\nu(\mathbf{x})$ is a single, localized function implicit in which is a complete characterization of the Bloch band ν. The Wannier function for a given band is however not unique, since any **k**-dependent phase factor can be multiplied to a Bloch function in (21.33). In modern electronic structure codes one uses "maximally localized Wannier functions" [47] as for instance in the WanT package [48]. Because of the corresponding properties of the Bloch functions the Wannier functions are also orthonormalized and complete [49]

$$\int_{\Omega} d^3\mathbf{x} a_\nu^*(\mathbf{x}-\mathbf{R}) a_{\nu'}(\mathbf{x}-\mathbf{R}') = \delta_{\nu\nu'}\delta_{\mathbf{R}\mathbf{R}'},$$

$$\sum_{\nu,\mathbf{R}} a_\nu(\mathbf{x}-\mathbf{R}) a_\nu^*(\mathbf{x}'-\mathbf{R}) = \delta(\mathbf{x}-\mathbf{x}'). \tag{21.34}$$

For the discussion of the Coulomb effects in the Hamiltonian (21.2) we assume strongly localized Wannier functions, so that the overlap between different unit cells is negligible. Since for small distances, i.e., large wave vectors (see Figs. 13.5 and 13.7), screening is less important, with (19.10) or (19.33) we can approximately write

$$\begin{aligned} W\begin{smallmatrix}\mathbf{kk}'\\ cc'\\ vv'\end{smallmatrix} &\approx \int d^3\mathbf{x} \int d^3\mathbf{x}' \varphi_{c\mathbf{k}}^*(\mathbf{x})\varphi_{c'\mathbf{k}'}(\mathbf{x}) W(\mathbf{x},\mathbf{x}')\varphi_{v\mathbf{k}}(\mathbf{x}')\varphi_{v'\mathbf{k}'}^*(\mathbf{x}') \\ &\approx \frac{1}{G^3}\sum_{\mathbf{R}} e^{-i(\mathbf{k}-\mathbf{k}')\mathbf{R}} W_{vv'}^{cc'}(\mathbf{R}), \\ \bar{\mathrm{v}}\begin{smallmatrix}\mathbf{kk}'\\ cv\\ v'c'\end{smallmatrix} &\approx \frac{1}{G^3}\sum_{\mathbf{R}} \bar{\mathrm{v}}_{v'c'}^{cv}(\mathbf{R}). \end{aligned} \tag{21.35}$$

with

$$\begin{aligned} W_{vv'}^{cc'}(\mathbf{R}) &= \int d^3\mathbf{x} \int d^3\mathbf{x}' a_c^*(\mathbf{x}) a_{c'}(\mathbf{x}) W(\mathbf{x}+\mathbf{R},\mathbf{x}') a_v(\mathbf{x}) a_{v'}^*(\mathbf{x}'), \\ \bar{\mathrm{v}}_{v'c'}^{cv}(\mathbf{R}) &= \int d^3\mathbf{x} \int d^3\mathbf{x}' a_c^*(\mathbf{x}) a_v(\mathbf{x}) \bar{\mathrm{v}}(\mathbf{x}-\mathbf{x}'+\mathbf{R}) a_{v'}(\mathbf{x}') a_{c'}^*(\mathbf{x}'). \end{aligned} \tag{21.36}$$

In reality, only a small effective screening remains in $W_{vv'}^{cc'}(\mathbf{R})$ because of the small distance $\mathbf{R}$ between electron and hole.

In agreement with the assumption of the strong localization of the Wannier functions, we may neglect the $\mathbf{k}$ dispersion of the bands and write

$$\varepsilon_c^{\mathrm{QP}}(\mathbf{k}) - \varepsilon_v^{\mathrm{QP}}(\mathbf{k}) \approx E_g. \tag{21.37}$$

The Hamiltonian (21.2) obeys a Fourier representation

$$H^S(c\mathbf{k}v\mathbf{k}, c'\mathbf{k}'v'\mathbf{k}') = \frac{1}{G^3}\sum_{\mathbf{R}} e^{-i(\mathbf{k}-\mathbf{k}')\mathbf{R}} H^S(cv, c'v'|\mathbf{R})$$

with [50]

$$H^S(cv, c'v'|\mathbf{R}) = E_g\delta_{cc'}\delta_{vv'} - W_{vv'}^{cc'}(\mathbf{R}) + 2\delta_{S0}\sum_{\mathbf{R}'} \bar{\mathrm{v}}_{v'c'}^{cv}(\mathbf{R}')\delta_{\mathbf{R}0}. \tag{21.38}$$

The Coulomb contributions can be easily interpreted [50] as the on-site Coulomb attraction $-W^{cc'}_{vv'}(0)$ and the corresponding electron-hole exchange $\delta_{S0}2\bar{\mathrm{v}}^{cv}_{v'c'}(0)$. The long-range Coulomb attraction $-W^{cc'}_{vv'}(\mathbf{R})(1-\delta_{\mathbf{R}0})$ is small. The terms $\delta_{S0}2\sum_{\mathbf{R}\neq 0}\bar{\mathrm{v}}^{cv}_{v'c'}(\mathbf{R})$ can be mainly rewritten in a dipole-dipole interaction with dipoles due to a charge transfer between different sites. It is responsible for a longitudinal-transverse splitting of the localized excitons. The latter fact can be immediately seen when the photon wave vector $\mathbf{q}\to 0$ is taken into account [50]. Effects beyond the Tamm-Dancoff approximation, which may however be important for localized systems, are not considered in (21.38).

The limit of very strong localization can be discussed within a real-space representation. With a Fourier transformation of the exciton eigenvectors (21.11)

$$A^S_\Lambda(c\mathbf{k}v\mathbf{k}) = \frac{\Omega_0}{\sqrt{\Omega}}\sum_{\mathbf{R}} e^{-i\mathbf{k}\mathbf{R}}\Phi^S_\Lambda(cv|\mathbf{R})$$

the excitonic eigenvalue problem (21.1) is rewritten into

$$\sum_{c',v'} H^S(cv,c'v'|\mathbf{R})\Phi^S_\Lambda(c'v'|\mathbf{R}) = E^S_\Lambda\Phi^S_\Lambda(cv|\mathbf{R}), \tag{21.39}$$

where $\mathbf{R}$ can be interpreted as the difference $\mathbf{R}_e-\mathbf{R}_h$ of the Bravais vectors of the unit cells, where one can find electron and hole, respectively.

We study the exciton problem (21.39) for strong ionic model materials, the alkali halides, which crystallize in rocksalt structure with nearest-neighbor separations of $2.0-3.5\,\text{Å}$, extremely small dielectric constants, and large fundamental gaps of 7.3–13.6 eV [51]. The uppermost valence bands are extremely narrow bands based on anion p-states while the lowest, extremely flat conduction band is formed by cation s-state. Therefore, we take three valence bands v_α $(\alpha = x,y,z)$ and one conduction band c into account. The corresponding Wannier functions are assumed to transform as p_x-, p_y-, and p_z-states or as s-states. For $\mathbf{R}\neq 0$ one immediately sees that the Coulomb matrix elements can be approximated by

$$W^{cc}_{v_\alpha v_{\alpha'}}(\mathbf{R}) = \delta_{\alpha\alpha'}W(\mathbf{R}) \tag{21.40}$$

with nearly $W(\mathbf{R})\approx \mathrm{v}(|\mathbf{R}|)$ if the screening vanishes. For $\mathbf{R}=0$ this relation $\sim\delta_{\alpha\alpha'}$ holds approximately.

The electron-hole exchange matrix element in (21.36) can be rewritten to

$$\bar{\mathrm{v}}^{cv_\alpha}_{v_{\alpha'}c}(\mathbf{R}) = \frac{\delta_{\mathbf{R}0}}{\Omega_0}\sum_{\mathbf{G}\neq 0}\tilde{\mathrm{v}}(|\mathbf{G}|)\tilde{B}_{cv_\alpha}(\mathbf{G})\tilde{B}^*_{cv_{\alpha'}}(\mathbf{G})$$

with

$$\tilde{B}_{cv_\alpha}(\mathbf{G}) = \frac{G_\alpha}{|\mathbf{G}|}\tilde{B}_{cv_z}(G_z),$$
$$\tilde{B}_{cv_z}(G_z) = \int d^3\mathbf{x} a_c^*(\mathbf{x}) e^{iG_z z} a_{v_\alpha}(\mathbf{x}).$$

This result easily follows from the symmetry properties of the Wannier functions. Consequently one has

$$\bar{\mathrm{v}}_{v_{\alpha'}c}^{cv_\alpha}(\mathbf{R}) = \delta_{\mathbf{R}0} \sum_{\mathbf{G}\neq 0} \tilde{\mathrm{v}}(|\mathbf{G}|) \frac{G_\alpha G_{\alpha'}}{|\mathbf{G}|^2} \left|\tilde{B}_{cv_z}(G_z)\right|^2. \qquad (21.41)$$

In a cubic crystal this matrix should be also diagonal in α and α', i.e., $\bar{\mathrm{v}}_{v_{\alpha'}c}^{cv_\alpha}(\mathbf{R}) = \delta_{\alpha\alpha'}\bar{\mathrm{v}}(\mathbf{R})$. There is no coupling between excitons belonging to different valence states.

Since the average electron-hole exchange $\frac{1}{G^3}\sum_{\mathbf{R}} \bar{\mathrm{v}}(\mathbf{R})$ is usually smaller than the direct Coulomb interaction $W(\mathbf{R})$ and only appears for singlet excitations, we expect that the excitonic redshift of the eigenvalues E_Λ^S is mainly determined by $W(\mathbf{R})$. For localized excitons the exciton binding energies are strongly influenced by the Coulomb attraction of electron and hole in the same unit cell, $\mathbf{R} = 0$, or between one quasiparticle in the cell $\mathbf{R} = 0$ and another cell with $\mathbf{R}$ = nearest-neighbor vectors. In an on-site approximation, $\mathbf{R} = 0$, and a two-band approximation, $c = c'$ and $v = v'$, the lowest pair energy (21.38) is given by $E_{\Lambda_0}^S = E_{\mathrm{ex}} = E_g - W_{vv}^{cc}(0) + 2\delta_{S0}\sum_{\mathbf{R}} \bar{\mathrm{v}}_{vc}^{cv}(\mathbf{R})$. Then, the eigenvalues of the pair Hamiltonian (21.38) remind us very much of the result (4.43) for pair excitations in the Hartree-Fock theory, at least transforming (4.45) in the singlet-triplet representation by means of the transformation (18.12). The resulting large binding energies of the excitons with the lowest pair excitation energy are illustrated in Table 21.3.

The above described localized character of the excitons is demonstrated in Fig. 21.2 for the ionic crystal LiF by electronic-structure calculations [10]. This result is generalized in a more schematic Fig. 21.15, for another alkali halide crystal, NaCl. Since the upper valence bands are represented by Cl $3p$ states while the lowest

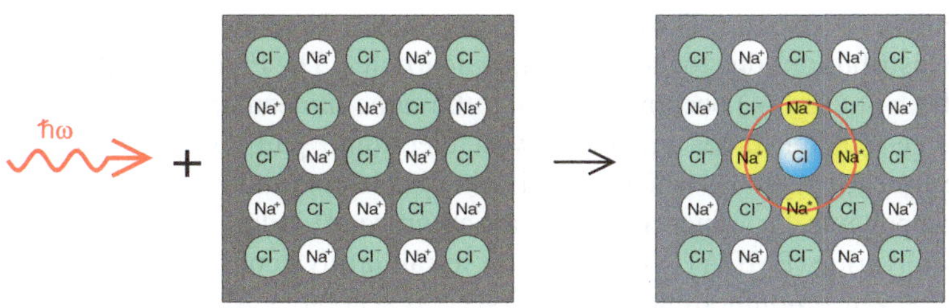

Fig. 21.15 Illustration of the formation of a Frenkel exciton in an ionic crystal NaCl: (**a**) Ground state, (**b**) excited state. Effective ion charges are also displayed. Na* means a partial compensation of the ion charge after excitation of an electron into the conduction band

conduction band mainly possesses Na 3*s* character, the excitation of an exciton can be nearly described as a transition of an electron from a Cl^- ion to the six neighboring Na^+ sites, thereby making these ions more neutral.

21.3.2 Charge-Transfer Excitons

Besides in alkali halides and noble gas crystals localized Frenkel excitons mostly appear in organic materials such as molecular crystals, polymers and many other organic structures [52]. Molecules and crystals of anthracene, tetracene, pentacene, phenanthrene, porphyrin, phenazine or 3, 4, 9, 10-perylene-tetracarboxylic-dianhydride (PTCDA) are favorable materials. In molecular crystals the covalent bonding within a molecule is strong compared to the van der Waals interaction between molecules. Therefore, electronic excitation lines of an individual molecule appear in the corresponding crystalline solids as Frenkel excitons.

Figure 21.15 indicates a certain electron transfer in space during the excitation in an ionic crystal between different ions. Therefore, one may also speak about the formation of a charge-transfer (CT) exciton. Nevertheless, the denotation Frenkel exciton is used in the literature. It is difficult to distinguish between the two types of localized excitations. Sometimes one uses the term "short-range to long-range charge transfer excitations" [53]. In agreement with the different ranges the two types differ with respect to their binding energies. Typically in molecular crystals, the excited lowest-energy pair states are strongly localized Frenkel excitons, where the interacting quasielectron-quasihole pairs are localized on the same molecular unit. CT excitons, in which e-h pairs are delocalized on different units, usually appear at higher energies in the optical spectra. However, when the molecular units are large enough, the effective electron-hole interaction for pairs localized on the same site or on two different sites becomes comparable. Either CT or Frenkel excitons occur. Prototypical molecular crystals are picene and pentacene (see Figs. 20.23 and 20.24). In these systems the localized Frenkel excitons compete with the somewhat more delocalized CT excitons [54], whereas in PTCDA crystals mostly Frenkel excitons are observed [55]. As a somewhat exceptional case a pair excitation is illustrated in Fig. 21.16. In an anthracene crystal with two inequivalent molecules (see Fig. 20.25) per unit cell the lowest exciton state is shown in the ***ab*** plane. It represents a CT exciton. The plus and minus signs refer to the center of gravity of the charge distributions due to the excited quasiparticles. The Frenkel exciton is obtained when both (+) and (–) occupy essentially the same molecular site.

Charge-transfer states are the lowest electronic pair excitations in so-called charge-transfer crystals, which are mixed crystals containing donor- and acceptor-type components. Prominent examples are anthracene (as donor) mixed with pyromellitic dianhydride (PMDA) (as acceptor) or tetrathiafulvalene (TTF) (as donor) mixed with tetracyano-p-quinodimethane (TCNQ) (as acceptor). Many other combinations of organic molecules are possible [56]. As an example the donor/acceptor molecular

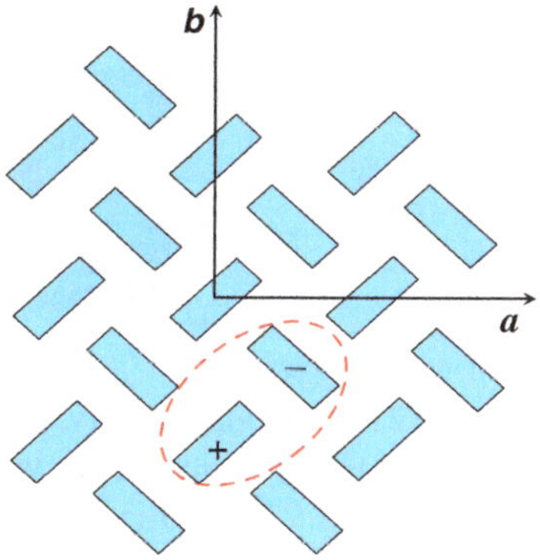

Fig. 21.16 Schematic illustration of a CT exciton in an anthracene crystal with monoclinic herringbone structure. One *rectangle* represents an anthracene molecule

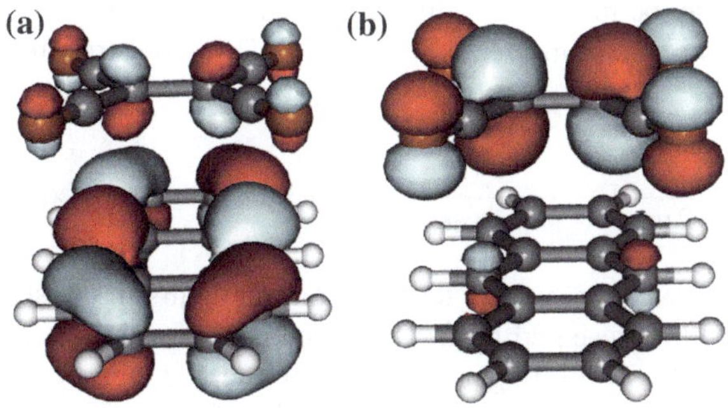

Fig. 21.17 Isocontour representation of the (**a**) HOMO and (**b**) LUMO states of the antracene-TCNE complex. *Gray* and *red colors* indicate different signs of the wave functions. The *gray*, *white*, and *brown* atoms are carbon, hydrogen, and nitrogen, respectively. Reprinted with permission from [56]. Copyright 2011, AIP Publishing LLC

complex of anthracene/tetracyanoethylene (TCNE) is displayed in Fig. 21.17. It shows that the HOMO and LUMO states are localized on different molecules. Consequently, the electron and hole of an excited pair sit on different molecules.

21.3.3 Excitons in Low-dimensional Systems

In order to study the influence of the spatial confinement of electrons and holes we investigate 2D crystals such as graphane, silicane and germanane (see Figs. 13.12 and 20.31) or the extreme limit of thin quantum-well structures based on type-I heterostructures with large potential barriers for electrons and holes [11, 45]. We expect the spatial confinement of the exciton wave function in the direction parallel to the normal of the 2D system, similar to the 3D confinement of such a wave

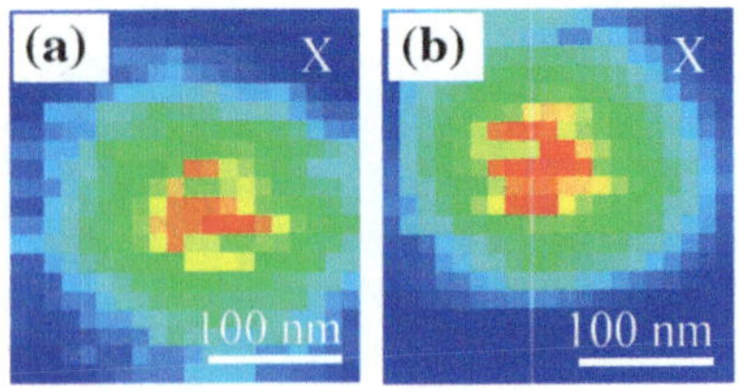

Fig. 21.18 Visualization of an exciton wave function confined in a GaAs quantum dot by photoluminescence spectroscopy using a near-field scanning optical microscope. From [57]

function in a quantum dot in Fig. 21.18. Consequently, the two-dimensional limit of the exciton problem (21.1) with the Hamiltonian (21.2) will be considered. Thereby, as simplest case, a two-band model (Fig. 21.3) with isotropic and parabolic bands with the Hamiltonian $H^{\mathrm{WM}}(\mathbf{k}, \mathbf{k}')$ (20.7) is considered. However, two different screenings of the attractive electron-hole potentials are investigated.

First, a statically screened Coulomb potential in two dimensions is investigated. We still use a dielectric constant ϵ_∞ to describe the screening. Because of the local-field effects, it may be identified with the dielectric constant of the adjacent barrier material in the thin quantum-well case. After a 2D Fourier transformation a differential equation

$$\left\{E_g - \frac{\hbar^2}{2\mu_{\mathrm{ex}}}\nabla^2_{\mathbf{x}_\parallel} - \frac{e^2}{4\pi\varepsilon_0\epsilon_\infty|\mathbf{x}_\parallel|}\right\}\Phi_\Lambda(\mathbf{x}_\parallel) = E_\Lambda\Phi_\Lambda(\mathbf{x}_\parallel) \tag{21.42}$$

for the in-plane relative motion of the electron-hole pairs in the sheet or extremely thin quantum well represented by the xy-plane is obtained. This differential equation is formally the Schrödinger equation of the 2D hydrogen problem. With polar coordinates $\mathbf{x}_\parallel = \rho(\cos\varphi, \sin\varphi, 0)$ the eigenfunctions can be separated into $\Phi_{n/km}(\mathbf{x}_\parallel) = R_{n/km}(\rho)\frac{1}{\sqrt{2\pi}}e^{im\varphi}$ (m - integer). For bound states $E_\Lambda < E_g$, one has $\Lambda = nm$ with $n = 0, 1, 2, \ldots$ as the principal quantum number and $-n \leq m \leq n$ as the quantum number of the normal component of the angular momentum operator. It holds for the $(2n+1)$-fold degenerate eigenenergies [58, 59]

$$E_{nm} = E_g - \frac{R_{\mathrm{ex}}}{\left(n + \frac{1}{2}\right)^2} \tag{21.43}$$

and the $m = 0$ radial parts

$$R_{n0}(\rho) = \frac{\sqrt{2}}{a_{\mathrm{ex}}\left(n + \frac{1}{2}\right)^{3/2}} L_n\left(\frac{2\rho}{\left(n + \frac{1}{2}\right)a_{\mathrm{ex}}}\right) e^{-\frac{\rho}{a_{\mathrm{ex}}\left(n + \frac{1}{2}\right)}}$$

with Laguerre polynomials $L_n(t)$. The scattering states $\Lambda = km$ with $0 < k < \infty$ are described by energies

$$E_{km} = E_g - \frac{\hbar^2}{2\mu_{\mathrm{ex}}} k^2 \tag{21.44}$$

and $m = 0$ radial parts

$$R_{k0}(\rho) = \frac{1}{a_{\mathrm{ex}}} \sqrt{\frac{2}{1 + e^{-\frac{2\pi}{k a_{\mathrm{ex}}}}}} e^{-ik\rho} F\left(\frac{1}{2} + \frac{i}{a_{\mathrm{ex}} k}, 1; 2ik\rho\right)$$

with the confluent hypergeometric function $F(\alpha, \beta; t)$ [27].

With the area element $\frac{k}{2\pi}$ and the replacement of the k sum by an integral the weighted 2D joint density of states in units of $(\text{length})^{-2} \times (\text{energy})^{-1}$ (21.18) becomes [58]

$$\begin{aligned} S(\hbar\omega) = \frac{1}{4\pi a_{\mathrm{ex}}^2 R_{\mathrm{ex}}} & \left\{ \sum_{n=0}^{\infty} \frac{4}{\left(n + \frac{1}{2}\right)^3} \delta\left(\frac{E_g - \hbar\omega}{R_{\mathrm{ex}}} - \frac{1}{\left(n + \frac{1}{2}\right)^2}\right) \theta(E_g - \hbar\omega) \right. \\ & \left. + \frac{e^{\frac{\pi}{t}}}{\cosh\left(\frac{\pi}{t}\right)} \bigg|_{t = \sqrt{\frac{\hbar\omega - E_g}{R_{\mathrm{ex}}}}} \theta(\hbar\omega - E_g) \right\}. \end{aligned} \tag{21.45}$$

The consequences for the contribution of a dipole-allowed optical transition to the 2D frequency-dependent electronic polarizability for $\hbar\omega \approx E_g$ is obvious. With (21.22), Fourier representations (21.11) but reduced to 2D, and a characteristic thickness L of the 2D system (prepared by a 3D one with the dipole matrix element $(M_{cv}^{00})_j$) we find

$$\begin{aligned} L\chi_{jj}(\omega) = 16 \frac{\left|\left(M_{cv}^{00}\right)_j\right|^2}{a_{\mathrm{ex}}^2} & \left\{ \sum_{n=0}^{\infty} \frac{1}{\left(n + \frac{1}{2}\right)^3} \frac{E_g - \frac{R_{\mathrm{ex}}}{\left(n+\frac{1}{2}\right)^2}}{E_g - \frac{R_{\mathrm{ex}}}{\left(n+\frac{1}{2}\right)^2} + \hbar^2(\omega + i\eta)^2} \right. \\ & \left. + \frac{1}{2} \int_0^{\infty} dt \frac{t e^{\frac{\pi}{t}}}{\cosh\left(\frac{\pi}{t}\right)} \frac{E_g + R_{\mathrm{ex}} t^2}{E_g + R_{\mathrm{ex}} t^2 - \hbar(\omega + i\eta)} \right\} \end{aligned} \tag{21.46}$$

for the complex 2D susceptibility in units of $(\text{length})^{-1}$. As in 3D the integral is finite only for the imaginary part.

In Fig. 21.19 the lineshape of its imaginary part is compared with that of the 3D susceptibility for the same finite damping η and bulk exciton Rydberg R_{ex}. It clearly shows the stronger excitonic effects relative to the spectrum of non-Coulomb-correlated electron-hole pairs in 2D in comparison to the 3D case. For the applied finite damping $\hbar\eta = 0.2R_{\mathrm{ex}}$ the absorption of the ionized states and the absorption in the higher bound states join continuously. In 2D the $1s$ exciton is spectrally far better resolved than in 3D as a consequence of the four times larger binding energy $E_B = 4R_{\mathrm{ex}}$ (21.43) of the lowest pair excitation. The oscillator strength of the

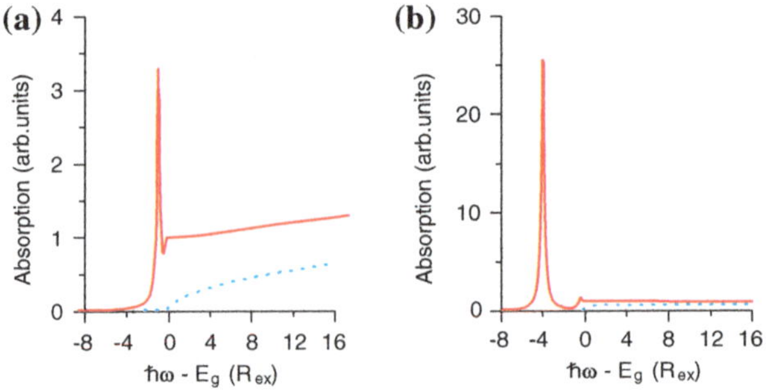

Fig. 21.19 Comparison of exciton-mediated optical absorption (described by the weighted pair density of states) in (**a**) 3D and (**b**) 2D in the framework of the Wannier-Mott approximation (*red lines*). Parameters $R_{\mathrm{ex}} = 4.7\,\mathrm{meV}$ and $\hbar\eta = 0.2R_{\mathrm{ex}}$ of the heavy-hole exciton of GaAs are applied. The *blue dashed lines* represent the spectra without excitonic effects with (**a**) square-root and (**b**) step-like variation with energy. After [60]

ground state $n = 0$ exciton, relative to the continuum spectrum, is 8 times larger, while in the 3D case, an enhancement of only a factor 4 is valid for the lowest-energy $n = 1$ exciton. A similar argument holds for its oscillator strength in comparison to the spectrum of scattering states for $\hbar\omega > E_g$. The continuum absorption of the scattering states again shows a Coulomb enhancement compared to the spectrum of free electron-hole pairs. The enhancement, the so-called Sommerfeld factor, directly follows from the 2D pair density of states (21.45) by $(\hbar\omega > E_g)$

$$S(\hbar\omega)/S^{\mathrm{free}}(\hbar\omega) = \left.\frac{e^{\frac{\pi}{t}}}{\cosh\left(\frac{\pi}{t}\right)}\right|_{t=\sqrt{\frac{\hbar\omega-E_g}{R_{\mathrm{ex}}}}}. \tag{21.47}$$

For $\hbar\omega \to E_g + 0^+$ it approaches an enhancement factor of 2. The absorption at the band edge is thus twice the uncorrelated pair continuum absorption. In the opposite limit $\hbar\omega >> E_g$, the Sommerfeld factor (21.47) goes to unity and, this time, the absorption including Coulomb interaction approaches the uncorrelated-pair absorption.

Second, we study a freestanding 2D system, where the screening of the electron-hole attraction is only determined by the static 2D electronic polarizability α_{2D} of the sheet material itself. The medium embedding the electronic sheet is assumed to be vacuum or air. Therefore, no additional screening due to the environment as studied in (21.42) is taken into account. The stationary 2D Schrödinger equation of the relative motion of electron and hole (21.42) has to be modified to

$$\left\{E_g - \frac{\hbar^2}{2\mu_{\mathrm{ex}}}\nabla^2_{\mathbf{x}_\parallel} - W(\mathbf{x}_\parallel)\right\}\Phi_\Lambda(\mathbf{x}_\parallel) = E_\Lambda \Phi_\Lambda(\mathbf{x}_\parallel) \tag{21.48}$$

with the screened potential (13.70) [61]

$$W(\mathbf{x}_{\parallel}) = \frac{e^2}{16\pi\varepsilon_0\alpha_{2D}}\left[H_0\left(\frac{|\mathbf{x}_{\parallel}|}{2\pi\alpha_{2D}}\right) - N_0\left(\frac{|\mathbf{x}_{\parallel}|}{2\pi\alpha_{2D}}\right)\right], \tag{21.49}$$

where $N_0(s)$ and $H_0(s)$ are the Struve and Neumann functions. Again polar coordinates $\mathbf{x}_{\parallel} = \rho(\cos\varphi, \sin\varphi, 0)$ are introduced.

For large in-plane distances between electron and hole, $\rho/2\pi\alpha_{2D} \gg 1$, the unscreened 2D Coulomb potential

$$W(\rho) = \frac{e^2}{4\pi\varepsilon_0\rho} \tag{21.50}$$

appears. The solutions of (21.48) are well known from the 2D hydrogen problem (21.42). Only the dielectric constant has to be replaced by $\epsilon_\infty = 1$. Therefore, a larger exciton Rydberg R_{ex} and a smaller exciton radius a_{ex}, respectively, have to be considered. In the opposite limit, $\rho/2\pi\alpha_{2D} \ll 1$, i.e., in the limit of small electron-hole distances ρ and/or large sheet polarizabilities α_{2D}, we get a logarithmic behavior for the screened attraction. For vanishing arguments $s \to 0$ it holds $H_0(s) = 2s/\pi^2$ and $N_0(s) = 2[ln(s/2) + C]/\pi$ [62]. The potential becomes

$$W(\rho) = \frac{e^2}{8\pi^2\varepsilon_0\alpha_{2D}}\left[ln\left(\frac{4\pi\alpha_{2D}}{\rho}\right) - C\right], \tag{21.51}$$

where $C = 0.5772$ is the Euler constant. The eigenvalue problem of the 2D hydrogen atom (21.48) with an attractive logarithmic potential energy $-W(\rho) \sim ln(\rho/4\pi\alpha_{2D})$ can be solved numerically [63, 64]. The 2D logarithmic potential has only discrete energy levels since no particle is outside the well, even at infinity $\rho \to \infty$. The true attractive potential is displayed together with the two approximations (21.50) and (21.51) in Fig. 21.20. The characteristic range of average electron-hole distances in some sheet crystals with honeycomb symmetry but finite gap is also indicated.

In order to understand the influence of the screening by the 2D electronic polarizability α_{2D} and the reduced effective mass μ_{ex} of the pairs, we apply a variational approach with an $1s$ trial wave function

$$\Phi_{00}(\rho) = \sqrt{\frac{2}{\pi}}\frac{2\lambda}{a_{ex}}e^{-2\lambda\rho/a_{ex}}, \tag{21.52}$$

where λ is the variational parameter and $a_{ex} = a_B m/\mu_{ex}$ is the corresponding characteristic length parameter. With the characteristic energy parameter $R_{ex} = R_\infty\mu_{ex}/m$ the binding energy of the lowest pair excitation $E_B = E_g - E_{00}$ attracted by the potential (21.49) is then

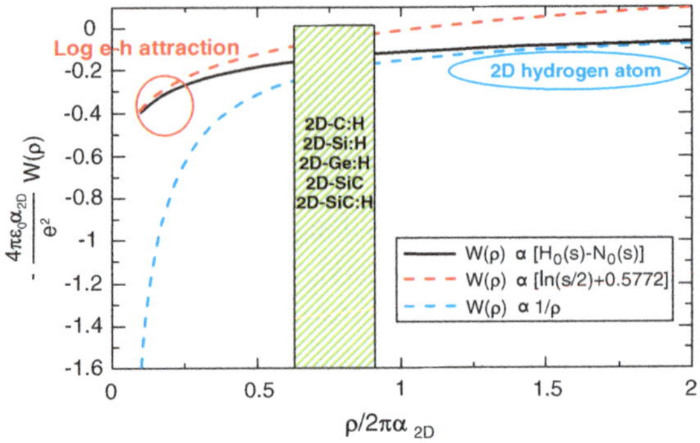

Fig. 21.20 Attractive Coulomb potential (21.49) screened only by the static 2D electronic polarizability α_{2D}. The two limiting cases (21.50) (*blue dashed line*) and (21.51) (*red dashed line*) are indicated together with the range of parameters (*green area*) relevant for true sheet crystals with honeycomb symmetry and finite fundamental gap. Hydrogenated group-IV crystals and SiC with honeycomb symmetry are studied. Prepared using data from [65]

$$E_B(\lambda) = 4R_{ex}\left\{ -\lambda^2 + 2\lambda\beta\left[\frac{ln(\sqrt{1+\beta^2}+\beta)+ln(\sqrt{1+\beta^2}+1)-ln\beta}{(1+\beta^2)^{3/2}} - \frac{1-\beta}{1+\beta^2}\right]\right\}_{\beta=\frac{a_{ex}}{8\pi\lambda\alpha_{2D}}} . \quad (21.53)$$

The maximum binding is found for $\frac{d}{d\lambda}E_B(\lambda)\big|_{\lambda=\lambda_{max}} = 0$. Results for $E_B(\lambda_{max})$ and $r_B = a_{ex}/2\lambda_{max}$ are plotted in Fig. 21.21. The two limits $\beta \ll 1$ and $\beta \gg 1$ give completely different results. In the large polarizability/small excitonic radius limit it holds

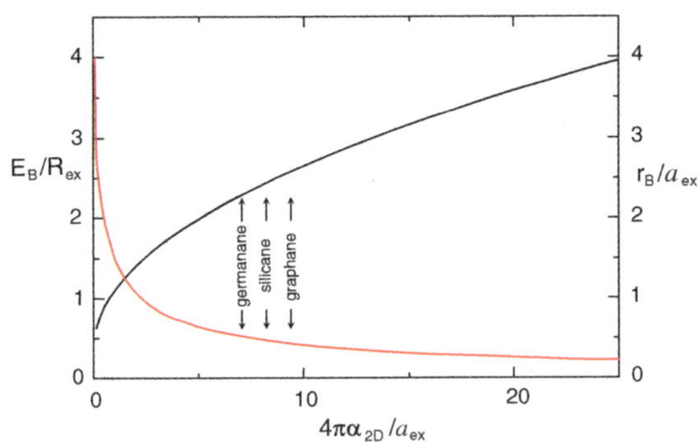

Fig. 21.21 Binding energy (*red line*) and exciton radius (*black line*) of the lowest energy exciton as a function of $4\pi\alpha_{2D}/a_{ex}$ as obtained from the variational procedure (21.53). The parameters a_{ex} and α_{2D} are taken from Table 21.4

Table 21.4 Characteristic parameters, static 2D electronic polarizability α_{2D}, reduced pair mass $\mu_{\rm ex}$, Bohr radius $a_{\rm ex} = a_B m/\mu_{\rm ex}$, fundamental gap E_g, binding energy E_B of the lowest energy exciton, of graphane, silicane, and germanane

Parameter	Graphane	Silicane	Germanane
$\alpha_{\rm 2D}$ (Å)	1.1	3.1	3.5
$\mu_{\rm ex}$ (m)	0.28	0.09	0.05
$a_{\rm ex}$ (Å)	1.9	5.9	10.6
E_g (eV)	3.6	2.7	1.8
E_B (eV)	1.8	0.9	0.6
r_B (Å)	5.8	9.7	16.0

All parameters are taken from *ab initio* calculations including QP effects (E_g) and solving the BSE (E_B, r_B). From [65, 66]

$$E_B(\lambda) = 4R_{\rm ex}\left\{-\lambda^2 + 2\beta\left[ln\left(\frac{\beta}{2\lambda}\right) + 1\right]\right\}.$$

This binding energy is maximized for $\lambda_{\rm max} = \sqrt{\beta}$ with the results

$$E_B = 4R_{\rm ex}\beta\left[-3 + ln\left(\frac{64}{\beta}\right)\right],$$
$$r_B = 2a_{\rm ex}/\sqrt{\beta},$$

if $\beta < 3.18$. Here r_B is the radius of the exciton in the studied sheet. In the opposite limit, i.e., vanishing 2D polarizability/large excitonic radius, one finds

$$E_B(\lambda) = 4R_{\rm ex}(-\lambda^2 + 2\lambda)$$

with $\lambda_{\rm max} = 1$ and

$$E_B = 4R_{\rm ex},$$
$$r_B = a_{\rm ex}/2,$$

hence recovering the exact solution of the 2D hydrogen limit (21.43).

The characteristic input parameters, including the substantial fundamental gap, as well as results for the excitation properties are listed in Table 21.4 for three sheet crystals with honeycomb symmetry, graphane, silicane, and germanane (see Figs. 13.12, 16.31, 16.32 and 20.31). The resulting exciton binding energies E_B and radii r_B derived within the variational procedure (21.53) are close to values obtained within the QP/BSE approach listed in Table 21.4 [66–68].

The full in-plane spectra of the exciton bound states in graphane, silicane and germanane are displayed in Fig. 21.22a. Since the spectral strength of the continuum is small in 2D systems (see Fig. 21.19) we focus the attention on the series of bound states. Their positions and intensities cannot be simply described by a Rydberg series. In any case extremely large binding energies of the lowest-energy pair

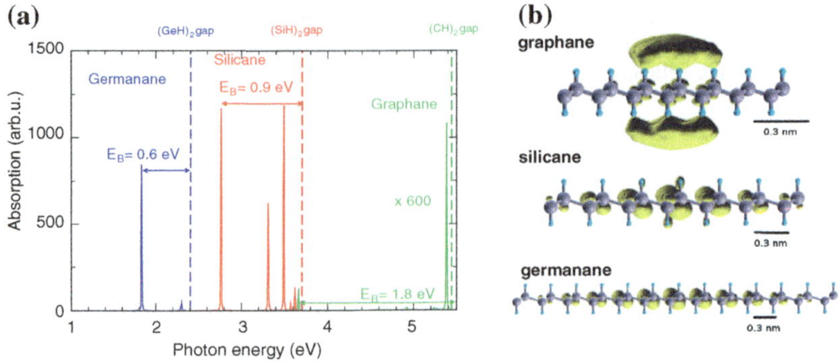

Fig. 21.22 (**a**) Series of bound exciton lines in the absorption of a hydrogenated group-IV sheet crystal with honeycomb symmetry below the fundamental QP gap. In the graphane case the factor 600 indicates that the optical transitions are forbidden. (**b**) Wave function of the lowest-energy exciton for a hole fixed in the center of a IV-IV bond. From [66]

excitations are observed. They are accompanied by strongly localized excitons which are illustrated in Fig. 21.22b. Their extents correlate with the r_B values in Table 21.4. The results shown in Fig. 21.22 suggest two possible interesting applications. The huge exciton binding together with the dipole-forbidden optical transition make graphane a promising candidate to demonstrate the Bose-Einstein condensation of excitons [67]. The giant oscillator strengths and binding energies of silicane and germanane indicate strong polariton effects with a chance for the realization of a polariton laser in the blue spectral region at room temperature [66].

References

1. J.I. Frenkel, On the transformation of light into head in solids. I. Phys. Rev. **37**, 17–44 (1931)
2. R. Peierls, Zur Theorie der Absorptionsspektren fester Körper. Ann. Phys. **13** (5. Folge), 905–952 (1932)
3. G.H. Wannier, The structure and electronic excitation levels in insulating crystals. Phys. Rev. **52**, 191–197 (1937)
4. N.F. Mott, Conduction in polar crystals. II. The conduction band and ultra-violet absorption of alkali-halide crystals. Trans. Faraday Soc. **34**, 500–506 (1938)
5. M.H. Cohen, F. Keffer, Dipolar sums in the primitive cubic lattices. Phys. Rev. **99**, 1128–1134 (1955)
6. L.J. Sham, T.M. Rice, Many-particle derivation of the effective-mass equation for the Wannier exciton. Phys. Rev. **144**, 708–714 (1966)
7. M. Rohlfing, S.G. Louie, Electron-hole excitations and optical spectra from first principles. Phys. Rev. B **62**, 4927–4944 (2000)
8. F. Fuchs, C. Rödl, A. Schleife, F. Bechstedt, Efficient $O(N^2)$ approach to solve the Bethe-Salpeter equation for excitonic bound states. Phys. Rev. B **78**, 085103 (2008)
9. T. Kalkreuter, H. Simma, An accelerated conjugate gradient algorithm to compute low-lying eigenvalues—a study for the Dirac operator in SU(2) lattice QCD. Comput. Phys. Commun. **93**, 33–47 (1996)
10. M. Gatti, F. Sottile, Exciton dispersion from first principles. Phys. Rev. B **88**, 155113 (2013)

11. P.Y. Yu, M. Cardona, *Fundamentals of Semiconductors* (Springer, Berlin, 1996)
12. G. Grosso, G.P. Parravicini, *Solid State Physics* (Academic Press, Amsterdam, 2000)
13. H.J. Monkhorst, J.D. Pack, Special points for Brillouin-zone integrations. Phys. Rev. B **13**, 5188–5192 (1976)
14. Ch. Hamaguchi, *Basic Semiconductor Physics* (Springer, Berlin, 2001)
15. A. Riefer, F. Fuchs, C. Rödl, A. Schleife, F. Bechstedt, R. Goldhahn, Interplay of excitonic effects and van Hove singularities in optical spectra: CaO and AlN polymorphs. Phys. Rev. B **84**, 075218 (2011)
16. D.M. Roessler, W.C. Walker, Optical constants of magnesium oxide and lithium fluoride in the far ultraviolet. J. Opt. Soc. Am. **57**, 835–836 (1967)
17. R.C. Whited, C.J. Flaten, W.C. Walker, Exciton thermoreflectance of MgO and CaO. Solid State Commun. **13**, 1903–1905 (1973)
18. R.C. Whited, W.C. Walker, Exciton spectra of CaO and MgO. Phys. Rev. Lett. **22**, 1428–1430 (1969)
19. K. Shindo, Effective electron-hole interaction in shallow excitons. J. Phys. Soc. Jpn. **29**, 287–295 (1970)
20. R. Zimmermann, Dynamical screening of the Wannier exciton. Phys. Status Solidi B **48**, 603–618 (1971)
21. F. Bechstedt, K. Seino, P.H. Hahn, W.G. Schmidt, Quasiparticle bands and optical spectra of highly ionic crystals: AlN and NaCl. Phys. Rev. B **72**, 245114 (2005)
22. W. Martienssen, H. Warlimont (eds.), *Springer Handbook of Condensed Matter and Materials Data* (Springer, Berlin, 2005)
23. R. Ishii, M. Funato, Y. Kawakami, Huge electron-hole exchange interation in aluminum nitride. Phys. Rev. B **87**, 161204(R) (2013)
24. G. Rossbach, M. Feneberg, M. Röppischer, C. Werner, N. Esser, C. Cobet, T. Meisch, K. Thonke, A. Dadgar, J. Bläsing, A. Krost, R. Goldhahn, Influence of exciton-phonon coupling and strain on the anisotropic optical response of wurtzite AlN around the band edge. Phys. Rev. B **83**, 195202 (2011)
25. R. Laskowski, N.E. Christensen, G. Santi, C. Ambrosch-Draxl, *Ab initio* calculations of excitons in GaN. Phys. Rev. B **72**, 035204 (2005)
26. L.X. Benedict, E.L. Shirley, R.B. Bohn, Optical absorption of insulators and the electron-hole interaction: an *ab initio* calculation. Phys. Rev. Lett. **80**, 4514–4517 (1998)
27. L.D. Landau, E.M. Lifshitz, *Quantum Mechanics*, vol. 3 (Pergamon Press, Oxford, 1959)
28. H. Haug, S.W. Koch, *Quantum Theory of the Optical and Electronic Properties of Semiconductors* (World Scientific, Singapore, 1993)
29. R.G. Ulbrich, Band edge spectra of highly excited gallium arsenide, in *Advances in Solid State Physics*, vol. XXV, ed. by P. Grosse (Vieweg & Sohn, Braunschweig/Wiesbaden, 1985), pp. 301–307
30. J.K. Kübler, The exciton binding energy of III–V semiconductor compounds. Phys. Status Solidi **85**, 189–195 (1969)
31. R.J. Elliott, Intensity of optical absorption by excitons. Phys. Rev. **108**, 1384–1389 (1957)
32. R.S. Knox, Theory of excitons, in *Solid State Physics*, Suppl. 5, ed. by F. Seitz, D.Turnbull (Academic Press, New York, 1963), pp. 100–307
33. D. Jérome, T.M. Rice, W. Kohn, Excitonic insulator. Phys. Rev. **158**, 462–475 (1967)
34. G.-Z. Liu, W. Li, G. Chen, Interaction and excitonic insulating transition in graphene. Phys. Rev. B **79**, 205429 (2009)
35. T. Stroucken, J.H. Grönqvist, S.W. Koch, Optical response and ground state of graphene. Phys. Rev. B **84**, 205445 (2011)
36. C. Tanguy, Optical dispersion by Wannier excitons. Phys. Rev. Lett. **75**, 4090–4093 (1995)
37. E.F. Gross, Spectra of excitons excited in a crystal lattice. Uspekhi Fiz. Nauk **63**, 575–611 (1957)
38. P.W. Baumeister, Optical absorption of cuprous oxide. Phys. Rev. **121**, 359–362 (1961)
39. S. Nikitine, Excitons, in *Optical Properties of Solids*, Chap. 9, ed. by S. Nudelman, S.S. Mitra (Plenum Press, New York, 1969), pp. 197–237

40. Y. Petroff, P.Y. Yu, Y.R. Shen, Study of photoluminescence in Cu_2O. Phys. Rev. B **12**, 2488–2495 (1975)
41. Landolt-Börnstein, in *Numerical Data and Functional Relationships in Science and Technology*, vol. III/17a and III/17b, ed. by O. Madelung, M. Schulz, H. Weiss (Springer, Berlin, 1982)
42. H. Stolz, *Einführung in die Vielteilchentheorie der Kristalle* (Akademie-Verlag, Berlin, 1974)
43. H. Haug, S. Schmitt-Rink, Electron theory of the optical properties of laser-excited semiconductors. Prog. Quant. Electron. **9**, 3–100 (1984)
44. K.S. Song, R.T. Williams, *Self-trapped Excitons* (Springer, Berlin, 1993)
45. Ch. Kittel, *Introduction to Solid State Physics* (Wiley, New York, 2005)
46. V. Saile, M. Skibowski, W. Steinmann, P. Gürtler, E.E. Koch, A. Kozevnikov, Observation of surface excitons in rare-gas solids. Phys. Rev. Lett. **37**, 305–308 (1976)
47. A. Calzolari, N. Marzari, I. Souza, M.B. Nardelli, *Ab initio* transport properties of nanostructures from maximally localized Wannier functions. Phys. Rev. B **69**, 035108 (2004)
48. http://www.wannier-transport.org/
49. W. Jones, N.H. March, *Theoretical Solid State Physics. Perfect Lattices in Equilibrium*, vol. 1 (Dover Publ. Inc, New York, 1985)
50. L. Resca, S. Rodriguez, Exciton states in solid rare gases. Phys. Rev. B **17**, 3334–3340 (1978)
51. W.A. Harrison, *Elementary Electronic Structure* (World Scientific, Singapore, 1999)
52. A. Davydov, *Theory of Molecular Excitons* (McGraw-Hill, New York, 1962)
53. I. Duchemin, T. Deutsch, X. Blase, Short-range to long-range charge-transfer excitations in the zincbacteriochlorin-bacteriochlorin complex: a Bethe-Salpeter study. Phys. Rev. Lett. **109**, 167801 (2012)
54. P. Cudazzo, M. Gatti, A. Rubio, F. Sottile, Frenkel versus charge-transfer exciton dispersion in molecular crystals. Phys. Rev. B **88**, 195152 (2013)
55. I. Vragovič, R. Scholz, Frenkel exciton model of optical absorption and photoluminescence in α-PTCDA. Phys. Rev. B **68**, 155202 (2003)
56. X. Blase, C. Attaccalite, Charge-transfer excitations in molecular donor-acceptor complexes within the many-body Bethe-Salpeter approach. Appl. Phys. Lett. **99**, 171909 (2011)
57. T. Saiki, Visualizing exciton wavefunctions confined in a quantum dot. SPIE Newsroom (June 2011). doi:10.1117/2.1201106.00367823
58. M. Shinada, S. Sugano, Interband optical transitions in extremely anisotropic semiconductors. I. Bound and unbound exciton absorption. J. Phys. Soc. Jpn. **21**, 1936–1946 (1966)
59. S.L. Chuang, *Physics of Optoelectronic Devices* (Wiley, New York, 1995)
60. S. Glutsch, *Excitons in Low-dimensional Semiconductors* (Springer, Berlin, 2004)
61. L.V. Keldysh, Coulomb interaction in thin semiconductor and semimetal films. Pis'ma Zh. Eksp. Teor. Fiz. **29**, 716–719 (1979) [Engl. translation: JETP Lett. **29**, 658–661 (1980)]
62. I.S. Gradstein, I.M. Ryshik, *Sum, Product and Integral Tables*, vol. 1 (Verlag Harri Deutsch, Thun, 1981)
63. F.J. Asturias, S.R. Aragón, The hydrogenic atom and the periodic table of the elements in two spatial dimensions. Am. J. Phys. **53**, 893–899 (1985)
64. K. Eveker, D. Grow, B. Jost, C.E. Monfort III, K.W. Nelson, C. Stroh, R.C. Witt, The two-dimensional hydrogen atom with a logarithmic potential energy function. Am. J. Phys. **58**, 1183–1192 (1990)
65. O. Pulci, M. Marsili, V. Garbuio, P. Gori, I. Kupchak, F. Bechstedt, Excitons in two-dimensional sheets with honeycomb symmetry. Phys. Status Solidi B, 1–6 (2014). doi:10.1002/pssb.201350404
66. O. Pulci, P. Gori, M. Marsili, V. Garbuio, R. Del Sole, F. Bechstedt, Strong excitons in novel two-dimensional crystals: silicane and germanane. EPL **98**, 37004 (2012)
67. P. Cudazzo, C. Attacalite, I.V. Tokatly, A. Rubio, Strong charge-transfer excitonic effects and Bose-Einstein exciton condensate in graphane. Phys. Rev. Lett. **104**, 226804 (2010)
68. W. Wei, Y. Dai, B. Huang, T. Jacob, Many-body effects in silicene, silicane, germanene, and germanane. Phys. Chem. Chem. Phys. **15**, 8789–8794 (2013)

Chapter 22
Beyond Static Screening

Abstract The consideration of the screening dynamics prevents the formulation of a Bethe-Salpeter equation for the macroscopic polarization function that only depends on one frequency. This is only possible within approximate schemes, for instance the Shindo approximation. Due to the incomplete dynamical screening excitonic effects are increased. Including dynamical lattice screening the question which dielectric constant, the static electronic or the total one, has to be used to screen a Wannier-Mott exciton is discussed. Another consequence of the dynamically screened electron-hole attraction is an additional loss mechanism, the interference terms, which modify the spectral strength and the satellite structures over and above the intrinsic and extrinsic losses known from the single-quasiparticle description. It is shown that the combination of intrinsic, interference, and extrinsic effects leads to strong reduction of the satellite strengths, in particular for vanishing energies of an electron emitted in photoemission. The screening in a non-metal is significantly modified in the presence of free carriers. The band filling gives rise to a tendency for a Mott transition, a gap shrinkage, and a Burstein-Moss shift. However, even for large carrier densities Coulomb effects survive resulting e.g. in Mahan excitons.

22.1 Dynamical Effects

22.1.1 Shindo Approximation

The two-particle – in particular the electron-hole and hence optical – properties of condensed matter are determined by the macroscopic polarization function P^M (19.2) which only depends on one frequency ω. The corresponding energy is the excitation energy $\hbar\omega$ of the system. However, there is a dilemma in calculating this function due to the dynamical screening reaction of the electronic system to excitations. Mathematically it is represented by the time or frequency dependence of the screened Coulomb potential W in (18.36), even in the framework of the GW approximation (12.56), in which the screening is only described by that of independent quasiparticles (12.58). So far we have solved the two-particle problem in the framework of the central approximation of static screening of the interaction

F. Bechstedt, *Many-Body Approach to Electronic Excitations*,
Springer Series in Solid-State Sciences 181, DOI 10.1007/978-3-662-44593-8_22

between quasielectrons and quasiholes, $W(t-t) \approx W\delta(t-t')$ (19.3). The neglect of dynamics allows the formulation of a closed BSE (19.4) for the macroscopic polarization function that only depends on one frequency.

Here, we take the screening dynamics into account. For that reason, we formulate and solve a BSE for the macroscopic polarization function (19.1), that depends on two time differences or, after Fourier transformation, on two frequencies. In the time-Fourier domain, instead of (19.4), one has to investigate the generalized BSE

$$\begin{aligned}
P^M_{s_1s_1',s_2s_2'}(\mathbf{x}_1\mathbf{x}_1',\mathbf{x}_2\mathbf{x}_2';z_n\tilde{z}_m) = &-i\hbar\Big\{G_{s_1s_2'}(\mathbf{x}_1\mathbf{x}_2',z_n)G_{s_2s_1'}(\mathbf{x}_2\mathbf{x}_1',z_n-\tilde{z}_m)\\
&+\frac{1}{i\hbar\beta}\sum_{n'}\sum_{s_3,s_4,s_3',s_4'}\int d^3\mathbf{x}_3\int d^3\mathbf{x}_4\int d^3\mathbf{x}_3'\int d^3\mathbf{x}_4'\\
&\times G_{s_1s_3}(\mathbf{x}_1\mathbf{x}_3,z_n)G_{s_4s_1'}(\mathbf{x}_4\mathbf{x}_1',z_n-\tilde{z}_m)\\
&\times\Big[-\delta_{s_4s_4'}\delta_{s_3s_3'}W(\mathbf{x}_3,\mathbf{x}_4;z_n-z_{n'})+\delta_{s_3s_4}\delta_{s_3's_4'}\bar{v}(\mathbf{x}_{3'}-\mathbf{x}_4)\Big]\\
&\times P^M_{s_3's_4',s_2s_2'}(\mathbf{x}_3'\mathbf{x}_4',\mathbf{x}_2\mathbf{x}_2';z_n\tilde{z}_m)\Big\}, \qquad (22.1)
\end{aligned}$$

where the time-Fourier transformed screened potential (14.53) has been introduced. Neglecting its frequency dependence, $W(z_n - z_{n'}) \approx W(0)$, together with the n-summation in (19.2), yields the BSE (19.4) for the static reaction. The small shift (11.41) of the time argument in $W(4^+, 3)$ in (18.36) has consequences for the n'-sum in (22.1), rather similar to that in the exchange self-energy (14.60). Nonetheless, we will not explicitly write down the corresponding exponential in the following mathematical formulation of the dynamical effects.

In a further step we change from the spin-space representation into a representation of two-component Pauli spinors

$$\varphi_{\lambda m_s}(\mathbf{x})\chi_{\frac{1}{2}m_s}(s)$$

with spatial orbitals and eigenfunctions of the spin operator assuming collinear spins. In this way we follow the procedure described in Sect. 19.1.2, in particular, the derivation of expression (19.7). Focussing on the influence of dynamical screening, we disregard a possible spin polarization in the inhomogeneous electron gas, i.e., the orbitals $\varphi_{\lambda m_s}(\mathbf{x}) \equiv \varphi_\lambda(\mathbf{x})$ are independent of the spin quantum number and hence the spin channel. Instead of the BSE (19.13) we derive

$$\begin{aligned}
P^M_{m_{s_1}m_{s_1'},m_{s_2}m_{s_2'}}(\lambda_1\lambda_1',\lambda_2\lambda_2';z_n\tilde{z}_m) &= -i\hbar G_{\lambda_1}(z_n)G_{\lambda_1'}(z_n-\tilde{z}_m)\\
&\times\Bigg\{\delta_{m_{s_1}m_{s_2}'}\delta_{\lambda_1\lambda_2'}\delta_{m_{s_2}m_{s_1}'}\delta_{\lambda_2\lambda_1'}+\frac{1}{-i\hbar\beta}\sum_{n'}\sum_{\lambda_3,\lambda_4}\sum_{m_{s_3},m_{s_4}}\\
&\times\Big[-\delta_{m_{s_1}m_{s_3}}\delta_{m_{s_1'}m_{s_4}}W^{\lambda_1\lambda_3}_{\lambda_{1'}\lambda_4}(z_n-z_{n'})+\delta_{m_{s_1}m_{s_1}'}\delta_{m_{s_3}m_{s_4}}\bar{v}^{\lambda_1\lambda_{1'}}_{\lambda_3\lambda_4}\Big]\\
&\times P^M_{m_{s_3}m_{s_4},m_{s_2}m_{s_2'}}(\lambda_3\lambda_4,\lambda_2\lambda_{2'};z_n\tilde{z}_m)\Bigg\}.
\end{aligned}$$

with

$$G_\lambda(z_n) = \frac{1}{\hbar z_n - \varepsilon_\lambda^{\mathrm{QP}}}.$$

We remember that for the dielectric function (18.19) only a spin-averaged polarization function as in (13.1)

$$P^M(\lambda_1\lambda_1', \lambda_2\lambda_2'; z_n\tilde{z}_m) = \frac{1}{2}\sum_{m_{s_1},m_{s_2}} P^M_{m_{s_1}m_{s_1},m_{s_2}m_{s_2}}(\lambda_1\lambda_1', \lambda_2\lambda_2'; z_n\tilde{z}_m) \tag{22.2}$$

is needed. For singlet excitations the corresponding BSE reads as

$$\begin{aligned} P^M(\lambda_1\lambda_1', \lambda_2\lambda_2'; z_n\tilde{z}_m) = &- i\hbar G_{\lambda_1}(z_n)G_{\lambda_1'}(z_n - \tilde{z}_m)\Bigg\{\delta_{\lambda_1\lambda_2'}\delta_{\lambda_2\lambda_1'} \\ &+ \frac{1}{-i\hbar\beta}\sum_{n'}\sum_{\lambda_3,\lambda_4}\left[-W^{\lambda_1\lambda_3}_{\lambda_{1'}\lambda_4}(z_n - z_{n'}) + 2\bar{v}^{\lambda_1\lambda_{1'}}_{\lambda_3\lambda_4}\right] \\ &P^M(\lambda_3\lambda_4, \lambda_2\lambda_{2'}; z_{n'}\tilde{z}_m)\Bigg\}. \end{aligned} \tag{22.3}$$

In principle, this BSE can be analytically continued into complex planes for z and $\tilde{z}$. To avoid severe numerical problems for $\tilde{z}_m \to \tilde{z} = \tilde{\omega} + i\eta$, the use of generalized spectral representations along two cuts in the complex plane may be helpful [1]. A rather complicated matrix equation in the pair states $\lambda_1\lambda_1'$ and the additional frequency z_n appears, which may be possible to solve with modern computer facilities for optical frequencies $\tilde{\omega}$. However, the analytic continuation $\tilde{z}_m \to \tilde{z} = \tilde{\omega} + i\eta$ poses serious numerical problems. One has to introduce a spectral representation along two cuts in the z-plane [1]. This should lead to two coupled integral equations for the two spectral functions. In principle, it may be solved in a similar way as the gap equation of the superconductivity, which however is simpler because of the replacement of $\lambda_1\lambda_1'$ by one k vector and the limit $\tilde{z} = 0$ [2]. The two facts reduce the numerical efforts tremendously. Therefore, despite the wide availability of modern parallel computer facilities an approximate treatment of the dynamics is certainly helpful. In the limit of weak excitonic effects $\sim[-W + 2\bar{v}]$ or, more precisely, weak dynamical screening influence, the dependence of $P^M(z_n\tilde{z}_m)$ on the frequency z_n can be replaced by that of the non-Coulomb-correlated quasielectrons and quasiholes, i.e., by that of the product $G_{\lambda_1}(z_n)G_{\lambda_1'}(z_n - \tilde{z}_m)$. With (19.2) and (22.3) it holds for the diagonal elements $\lambda_1 = \lambda_2'$ and $\lambda_2 = \lambda_1'$ [1, 3]

$$P^M(\lambda_1\lambda_1', \lambda_2\lambda_2'; z_n\tilde{z}_m) \approx \frac{G_{\lambda_1}(z_n) - G_{\lambda_1'}(z_n - \tilde{z}_m)}{-\frac{1}{i\hbar\beta}\sum_{n'}\left[G_{\lambda_1}(z_{n'}) - G_{\lambda_1'}(z_{n'} - \tilde{z}_m)\right]} P^M(\lambda_1\lambda_1', \lambda_2\lambda_2'; \tilde{z}_m). \tag{22.4}$$

This is the so-called Shindo approximation [4, 5], which directly relates the dependence on two frequencies to that on one frequency, which is relevant for optical studies.

The use of the Shindo approximation (22.4) and the definition (19.2) leads to a new closed BSE for the one-frequency macroscopic polarization function

$$\left[\varepsilon_{\lambda_1}^{\mathrm{QP}} - \varepsilon_{\lambda_1'}^{\mathrm{QP}} - \hbar\tilde{z}_m\right] P^M(\lambda_1\lambda_1', \lambda_2\lambda_2'; \tilde{z}_m) - \sum_{\lambda_3\lambda_4}\left[\tilde{N}_{\lambda_1\lambda_1'}(\tilde{z}_m)\tilde{W}_{\lambda_1'\lambda_4}^{\lambda_1\lambda_3}(\tilde{z}_m) - 2\bar{v}_{\lambda_3\lambda_4}^{\lambda_1\lambda_1'}\right]$$
$$\times P^M(\lambda_3\lambda_4, \lambda_2\lambda_2'; \tilde{z}_m) = -\delta_{\lambda_1\lambda_2'}\delta_{\lambda_2\lambda_2'}\tilde{N}_{\lambda_1\lambda_1'}(\tilde{z}_m) \tag{22.5}$$

with an effective frequency-dependent screened potential

$$\tilde{W}_{\lambda_1'\lambda_4}^{\lambda_1\lambda_3}(\tilde{z}_m) = \frac{1}{\beta^2}\sum_{n,n'}\frac{G_{\lambda_1}(z_n) - G_{\lambda_1'}(z_n - \tilde{z}_m)}{\tilde{N}_{\lambda_1\lambda_1'}(\tilde{z}_m)} W_{\lambda_1'\lambda_4}^{\lambda_1\lambda_3}(z_n - z_{n'})\frac{G_{\lambda_3}(z_{n'}) - G_{\lambda_4}(z_{n'} - \tilde{z}_m)}{\tilde{N}_{\lambda_3\lambda_4}(\tilde{z}_m)} \tag{22.6}$$

and a normalization factor

$$\tilde{N}_{\lambda_1\lambda_1'}(\tilde{z}_m) = -\frac{1}{\beta}\sum_n\left[G_{\lambda_1}(z_n) - G_{\lambda_1'}(z_n - \tilde{z}_m)\right] = f(\varepsilon_{\lambda_1'}^{\mathrm{QP}}) - f(\varepsilon_{\lambda_1}^{\mathrm{QP}}), \tag{22.7}$$

which is frequency-independent in thermal equilibrium. With (12.66) the normalization factor is just the Pauli blocking factor that also has been found in the static case, e.g. in (19.12). The division by the normalization factors in (22.6) is related to obvious shortcomings of the Shindo approximation. The Pauli blocking in the denominators gives rise to divergences for degenerate electron gases where $\tilde{N}_{\lambda\lambda'}$ changes its sign. We circumvent this problem by applying the Shindo approximation only to semiconductors and insulators with filled or empty bands. For the inclusion of dynamical effects in metals the reader is referred to a more specialized literature [6].

Indeed for $W(z_n - z_{n'}) \approx W(0)$ the inhomogeneous BSE (22.5) turns into the known (19.12), if the average over the spin quantum numbers is carried out. Since the normalization factor (22.7) does not depend on the poles $\tilde{z}_m$ of the Bose function, the BSE (22.5) can be easily generalized for arbitrary complex frequencies $\tilde{z}$ by an analytic continuation of $P^M(\tilde{z}_m)$ into the entire complex plane.

The central result of this section is the replacement of the statically screened Coulomb potential by an effective, frequency-dependent one. Since for static screening the BSE (22.5) is exact, the Shindo approximation may be interpreted as a first step in a perturbation treatment with respect to the dynamical nature of W. An explicit formulation of the influence of the dynamical screening requires the spectral representation of the inverse dielectric function of the kind given in (13.19) or (13.41). We again apply the symmetrized inverse dielectric function (14.57). Then, the matrix elements of the dynamically screened Coulomb interaction (14.56) become

$$W_{\lambda_1'\lambda_4}^{\lambda_1\lambda_3}(\tilde{z}_m) = \frac{1}{\Omega}\sum_{\mathbf{q}}\sum_{\mathbf{G},\mathbf{G}'} \tilde{\mathrm{v}}\left(\sqrt{|\mathbf{q}+\mathbf{G}||\mathbf{q}+\mathbf{G}'|}\right) B_{\lambda_1\lambda_3}(\mathbf{q}+\mathbf{G})B^*_{\lambda_1'\lambda_4}(\mathbf{q}+\mathbf{G}')$$
$$\times \left\{\delta_{\mathbf{G}\mathbf{G}'} + \int_{-\infty}^{+\infty} \frac{d\omega}{\pi} \frac{\mathrm{Im}\tilde{\epsilon}^{-1}(\mathbf{q}+\mathbf{G},\mathbf{q}+\mathbf{G}',\omega)}{\omega - \tilde{z}_m}\right\}. \tag{22.8}$$

The n- and n'-summations in (22.6) are carried out by means of the spectral representation (22.8) and the relations to Fermi or Bose functions (12.66) and (14.61). As a result we have

$$\tilde{W}_{\lambda_1'\lambda_4}^{\lambda_1\lambda_3}(\tilde{z}_m) = \frac{1}{\Omega}\sum_{\mathbf{q}}\sum_{\mathbf{G},\mathbf{G}'} \tilde{\mathrm{v}}\left(\sqrt{|\mathbf{q}+\mathbf{G}||\mathbf{q}+\mathbf{G}'}\right) B_{\lambda_1\lambda_3}(\mathbf{q}+\mathbf{G})B^*_{\lambda_1'\lambda_4}(\mathbf{q}+\mathbf{G}')$$
$$\times \left\{\delta_{\mathbf{G}\mathbf{G}'} + \frac{1}{\tilde{N}_{\lambda_1\lambda_1'}\tilde{N}_{\lambda_3\lambda_4}} \int_{-\infty}^{+\infty} \frac{d\hbar\omega}{\pi} \mathrm{Im}\tilde{\epsilon}^{-1}(\mathbf{q}+\mathbf{G},\mathbf{q}+\mathbf{G}',\omega)\right.$$
$$\times \left[\left[f\left(\varepsilon_{\lambda_1}^{\mathrm{QP}}\right) + g(\hbar\omega)\right] \frac{f\left(\varepsilon_{\lambda_1}^{\mathrm{QP}} - \hbar\omega\right) - f\left(\varepsilon_{\lambda_3}^{\mathrm{QP}}\right)}{\varepsilon_{\lambda_1}^{\mathrm{QP}} - \varepsilon_{\lambda_3}^{\mathrm{QP}} - \hbar\omega}\right.$$
$$+ \left[f\left(\varepsilon_{\lambda_1'}^{\mathrm{QP}}\right) + g(\hbar\omega)\right] \frac{f\left(\varepsilon_{\lambda_1'}^{\mathrm{QP}} - \hbar\omega\right) - f\left(\varepsilon_{\lambda_4}^{\mathrm{QP}}\right)}{\varepsilon_{\lambda_1'}^{\mathrm{QP}} - \varepsilon_{\lambda_4}^{\mathrm{QP}} - \hbar\omega}$$
$$+ \left[f\left(\varepsilon_{\lambda_1}^{\mathrm{QP}}\right) + g(\hbar\omega)\right] \frac{f\left(\varepsilon_{\lambda_1}^{\mathrm{QP}} - \hbar\omega\right) - f\left(\varepsilon_{\lambda_4}^{\mathrm{QP}}\right)}{\hbar\omega + \hbar\tilde{z}_m - \varepsilon_{\lambda_1}^{\mathrm{QP}} + \varepsilon_{\lambda_4}^{\mathrm{QP}}}$$
$$\left.\left. + \left[f\left(\varepsilon_{\lambda_1'}^{\mathrm{QP}}\right) + g(\hbar\omega)\right] \frac{f\left(\varepsilon_{\lambda_1'}^{\mathrm{QP}} - \hbar\omega\right) - f\left(\varepsilon_{\lambda_3}^{\mathrm{QP}}\right)}{\hbar\omega - \hbar\tilde{z}_m - \varepsilon_{\lambda_1'}^{\mathrm{QP}} + \varepsilon_{\lambda_3}^{\mathrm{QP}}}\right]\right\}. \tag{22.9}$$

The third and fourth term under the frequency integral in (22.9) can be interpreted as due to virtual transitions from the considered Coulomb-correlated electron-hole pairs with energy $\hbar\tilde{z}_m$ into free quasielectron-quasihole pairs with energies $|\varepsilon_{\lambda_1}^{\mathrm{QP}} - \varepsilon_{\lambda_4}^{\mathrm{QP}}|$ and $|\varepsilon_{\lambda_1'}^{\mathrm{QP}} - \varepsilon_{\lambda_3}^{\mathrm{QP}}|$ during the absorption or emission of collective excitations, e.g. plasmons, with energy $\hbar\omega$. The first two terms lack such a physical interpretation but are – together with the Pauli blocking factors $\tilde{N}_{\lambda_1\lambda_1'}$ and $\tilde{N}_{\lambda_3\lambda_4}$ – important to reach the static limit $\tilde{z}_m = 0$, where $\hbar\omega$ is assumed to exceed all other energy differences. Indeed in this limit it holds $g(\hbar\omega) = -\theta(-\omega)$ and $f(\varepsilon_{\lambda}^{\mathrm{QP}} - \hbar\omega) = \theta(\omega)$ (14.64). The four terms discussed above yield $1/(\hbar\omega)$ when dividing by $\tilde{N}_{\lambda_1\lambda_1'}\tilde{N}_{\lambda_3\lambda_4}$. The sum in the curly brackets becomes $\tilde{\epsilon}^{-1}(\mathbf{q}+\mathbf{G},\mathbf{q}+\mathbf{G}',0)$, i.e., the static screening function.

22.1.2 Dynamically Screened Excitons

The dynamically screened Coulomb attraction (22.9) is now investigated for model systems, crystals or other translationally invariant systems within the Bloch picture $\lambda \to \nu\mathbf{k}$ with the band index ν and the wave vector $\mathbf{k} \in$ BZ. The band states $\nu\mathbf{k}$ are either fully occupied, i.e., valence bands $\nu = v$, or completely empty, i.e., conduction bands $\nu = c$. The Fermi functions can be replaced by Heaviside functions $f(\varepsilon) = \theta(\mu - \varepsilon)$, where the chemical potential of the electrons is located in the fundamental gap. We are again only interested in actual or virtual optical excitations due to photons with energy $\hbar\omega$ from valence into conduction states, $|v\mathbf{k}\rangle \to |c\mathbf{k}\rangle$. A spin polarization is not taken into account. The (interband) normalization factors (22.7) are $\tilde{N}_{cv}(\mathbf{k}, \tilde{z}_m) = -\tilde{N}_{vc}(\mathbf{k}, \tilde{z}_m) = 1$. In the framework of the Tamm-Dancoff approximation (see Sect. 19.3.1) we neglect the coupling between resonant and antiresonant electron-hole pairs. In principle, only electron-hole interactions as taken into account in (19.32) have to be studied with the generalization of $W\,{}^{\mathbf{kk}'}_{\substack{cc'\\vv'}}$ (19.33) to the dynamical screening response.

With the analytic continuation $\tilde{z}_m \to z$ the dynamically screened Coulomb interaction (22.9) becomes [3]

$$\tilde{W}\,{}^{\mathbf{kk}'}_{\substack{cc'\\vv'}}(z) = \frac{1}{\Omega}\sum_{\mathbf{q}}\sum_{\mathbf{G},\mathbf{G}'} \bar{v}\left(\sqrt{|\mathbf{q}+\mathbf{G}||\mathbf{q}+\mathbf{G}'|}\right) B^{\mathbf{kk}'}_{cc'}(\mathbf{q}+\mathbf{G}) B^{\mathbf{kk}'*}_{vv'}(\mathbf{q}+\mathbf{G}')$$
$$\times \left\{\delta_{\mathbf{GG}'} + \int_0^\infty \frac{d\hbar\omega}{\pi} \mathrm{Im}\tilde{\varepsilon}^{-1}(\mathbf{q}+\mathbf{G}, \mathbf{q}+\mathbf{G}', \omega)\right.$$
$$\left.\times \left[\frac{1}{\hbar\omega + \varepsilon_c^{\mathrm{QP}}(\mathbf{k}) - \varepsilon_{v'}^{\mathrm{QP}}(\mathbf{k}') - \hbar z} + \frac{1}{\hbar\omega + \varepsilon_{c'}^{\mathrm{QP}}(\mathbf{k}') - \varepsilon_v^{\mathrm{QP}}(\mathbf{k}) - \hbar z}\right]\right\}. \tag{22.10}$$

Similar expressions have been predicted in literature [4, 7, 8]. The excitations are mainly composed from electron-hole pairs whose transition energies $\varepsilon_c^{\mathrm{QP}}(\mathbf{k}) - \varepsilon_{v'}^{\mathrm{QP}}(\mathbf{k}')$ and $\varepsilon_{c'}^{\mathrm{QP}}(\mathbf{k}') - \varepsilon_v^{\mathrm{QP}}(\mathbf{k})$ are close to the applied excitation energies $\hbar z$, i.e., the relevant differences are much smaller than the plasmon energies which control the dynamics of screening in many solids. For electron-hole pair excitations with energies near $\hbar z$ the static screening appears. In other systems, e.g. molecules, the differences $|\hbar z - (\varepsilon_c^{\mathrm{QP}} - \varepsilon_v^{\mathrm{QP}})|$ may be larger, because of the larger exciton binding energies. In such systems the approximation of static screening may be more violated than in *sp*-bonded solids with large plasmon frequencies. Also for localized electron-hole-pair excitations, e.g. core excitons in solids, the exciton binding energy will be significantly increased by dynamical effects [7, 9].

We will discuss dynamical effects for the example of the inclusion of the lattice polarization below. The formation of polarization clouds around electron-hole pairs has different time constants for the electronic and lattice polarizations. Therefore, they follow the pair excitation to varying degrees. For slow excitations the lattice contribution should be predominant, whereas for fast excitations the electronic screening is more efficient. However, there remains also an influence of the dynamics in the case of pure electronic screening. Intuitively one expects that the dynamics of the screening reaction leads to an effective screening that is reduced compared to the static one. The screening does not fully develop. The resulting attractive potential between electron and hole will be velocity (or energy)-dependent and non-local. The plasma oscillations cannot follow the internal motion of an electron-hole pair at all. During the formation of an exciton the electron gas has not enough time to build up the full, i.e., static, screening. However, the strength of the effect depends on the electron density.

Studying only the coupling between electron-hole pairs, the BSE (22.5) can be reduced to the simplified form

$$\begin{aligned}&\left[\varepsilon_c^{\mathrm{QP}}(\mathbf{k})-\varepsilon_v^{\mathrm{QP}}(\mathbf{k})-\hbar z\right]P^M(c\mathbf{k}v\mathbf{k},v'\mathbf{k}'c'\mathbf{k}';z)\\&-\sum_{c'',v''}\sum_{\mathbf{k}''}\left[\tilde{W}\begin{smallmatrix}\mathbf{kk}''\\cc''\\vv''\end{smallmatrix}(z)-2\bar{\mathrm{v}}\begin{smallmatrix}\mathbf{kk}''\\cv\\c''v''\end{smallmatrix}\right]P^M(c''\mathbf{k}''v''\mathbf{k}'',v'\mathbf{k}'c'\mathbf{k}';z)\\&=-\delta_{cc'}\delta_{vv'}\delta_{\mathbf{kk}'}.\end{aligned}\tag{22.11}$$

This BSE only describes the resonant pair contributions. The contribution of the antiresonant terms to the macroscopic dielectric function (19.35) follows using the symmetry discussed below (19.34). Following the procedure described in Sect. 19.2.2 and (19.22) the solution of this problem for the spin-averaged macroscopic polarization function can be formally traced back to an eigenvalue problem, a homogeneous BSE,

$$\begin{aligned}&\left[\varepsilon_c^{\mathrm{QP}}(\mathbf{k})-\varepsilon_v^{\mathrm{QP}}(\mathbf{k})-E_\Lambda(z)\right]A_\Lambda(c\mathbf{k}v\mathbf{k}|z)\\&-\sum_{c',v'}\sum_{\mathbf{k}'}\left[\tilde{W}\begin{smallmatrix}\mathbf{kk}'\\cc'\\vv'\end{smallmatrix}(z)-2\bar{\mathrm{v}}\begin{smallmatrix}\mathbf{kk}'\\cv\\c'v'\end{smallmatrix}\right]A_\Lambda(c'\mathbf{k}'v'\mathbf{k}'|z)=0,\end{aligned}\tag{22.12}$$

which has to be solved parametrically for each given frequency parameter z. The eigenvectors and eigenvalues are orthornormalized and fulfill a closure relation,

$$\begin{aligned}\sum_{c,v}\sum_{\mathbf{k}}A_\Lambda^*(c\mathbf{k}v\mathbf{k}|z)A_{\Lambda'}(c\mathbf{k}v\mathbf{k}|z)&=\delta_{\Lambda\Lambda'},\\\sum_{\Lambda}A_\Lambda(c\mathbf{k}v\mathbf{k}|z)A_\Lambda^*(c'\mathbf{k}'v'\mathbf{k}'|z)&=\delta_{cc'}\delta_{vv'}\delta_{\mathbf{kk}'}.\end{aligned}$$

They define the macroscopic polarization function as

$$P^M(c\mathbf{k}v\mathbf{k},v'\mathbf{k}'c'\mathbf{k}';z)=\sum_\Lambda\frac{A_\Lambda(c\mathbf{k}v\mathbf{k}|z)A_\Lambda^*(c'\mathbf{k}'v'\mathbf{k}'|z)}{\hbar z-E_\Lambda(z)}.\tag{22.13}$$

The pair resonances of the inhomogeneous electron gas are now not directly given by the eigenvalues $E_\Lambda(z)$, rather by the zeros of the equation

$$\hbar z = E_\Lambda(z). \tag{22.14}$$

The frequency dependence of $\tilde{W}$ in (22.12) generates the z-dependence for E_Λ. In this way the electron-hole-pair and, hence, exciton energies are influenced by the dynamical screening.

22.1.3 Examples

The influence of dynamical screening on the optical absorption and excitons is illustrated in Fig. 22.1. As a system with localized electrons the silane molecule SiH_4 is studied. The three absorption peaks related to singlet excitons are presented based on the quasiparticle approximation, which have been described in Sect. 16.3.2. Apart from the difficulties to describe the unoccupied states [8, 11] with huge QP shifts and due to the large exciton binding energies, the two most pronounced peaks can be mainly identified with electron-hole pairs related to HOMO→LUMO+1 and HOMO→LUMO transitions with redshifts due to the exciton binding energies of about 3.9 eV [11]. The measured positions of the three lowest absorption peaks of 8.8, 9.7, and 10.7 eV [12] are not exactly reproduced but appear in the calculated spectra in Fig. 22.1, although the peak with the lowest intensity is only really visible in the spectrum including dynamical screening. Surprisingly, the two most pronounced peaks in the spectrum in standard approximation, i.e., in the framework of the Tamm-Dancoff approximation and neglecting the screening dynamics, agree well with the peak positions in an experimental extinction spectrum [12]. When the dynamical effects are included, the spectrum is redshifted indicating increased exciton binding energies due to the reduced screening. Interestingly, the inclusion of the

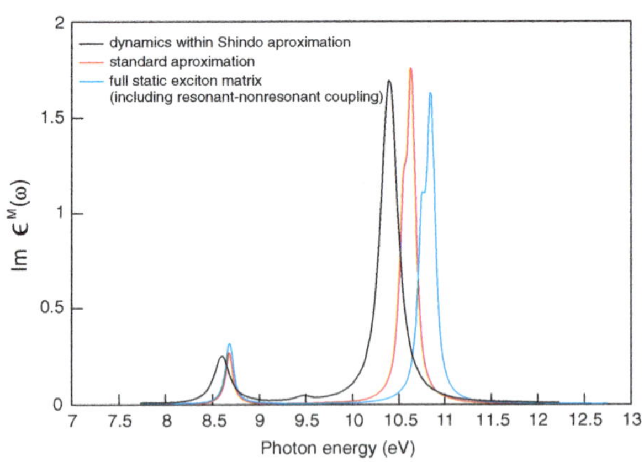

Fig. 22.1 Absorption spectrum of silane SiH_4 as calculated in three different approximations for the electron-hole interaction. From [10]

resonant-antiresonant coupling beyond the TDA gives rise to an opposite blue shift. Therefore, Fig. 22.1 suggests an error compensation, at least for the more pronounced high-energy peak, if dynamical effects and the resonant-antiresonant coupling are simultaneously neglected.

Studying the Wannier-Mott excitons in Sect. 21.2 we asked the question which dielectric constant of a semiconductor, the static electronic dielectric one ϵ_∞ or the static dielectric one ϵ_s, has to be taken into account. In other words, the question if the lattice polarization fully or partially contributes to the screening of the electron-hole attraction has to be answered. For that reason, we modify the static electronic screening $\sim 1/\epsilon_\infty$ by inclusion of dynamical screening of the lattice. We study a polar semiconductor which is characterized by the two dielectric constants, ϵ_∞ and ϵ_s, and the longitudinal optical (LO) and the transverse optical (TO) phonons from the center of the BZ. Their frequencies obey the Lyddane-Sachs-Teller (LST) relation $\epsilon_s/\epsilon_\infty = (\omega_{\rm LO}/\omega_{\rm TO})^2$. In the range of infrared frequencies the resulting dielectric function is given as [13]

$$\epsilon(\omega) = \epsilon_\infty \left[1 + \frac{\omega_{\rm LO}^2 - \omega_{\rm TO}^2}{\omega_{\rm TO}^2 - (\omega + i\eta)^2} \right] \tag{22.15}$$

with $\epsilon(\omega \to \infty) = \epsilon_\infty$ and $\epsilon(0) = \epsilon_s$. That means, in the dynamically screened Coulomb interaction (22.10) the imaginary part of the inverse dielectric function has to be replaced by

$$\mathrm{Im}\tilde{\epsilon}^{-1}(\mathbf{q}+\mathbf{G}, \mathbf{q}+\mathbf{G}', \omega) = -\frac{\pi}{2}\delta_{\mathbf{GG}'}\left(\frac{1}{\epsilon_\infty} - \frac{1}{\epsilon_s}\right)\omega_{\rm LO}\left[\delta(\omega - \omega_{\rm LO}) - \delta(\omega + \omega_{\rm LO})\right] \tag{22.16}$$

but, in addition, a static background screening by ϵ_∞ has to be considered. This static electronic screening governs the system reaction when the contribution of the vibrating ionic lattice is neglected.

In the effective mass approximation the Bloch integrals in (22.10) become Kronecker symbols. It results

$$\begin{aligned} \tilde{W}^{\mathbf{kk}'}_{\substack{cc'\\vv'}}(z) &= \delta_{cc'}\delta_{vv'}\frac{1}{\Omega}\tilde{\mathrm{v}}(|\mathbf{k}-\mathbf{k}'|)\left\{\frac{1}{\epsilon_\infty} - \left(\frac{1}{\epsilon_\infty} - \frac{1}{\epsilon_s}\right)\frac{\hbar\omega_{\rm LO}}{2}\right. \\ &\times \left.\left[\frac{1}{\hbar\omega_{\rm LO} + \varepsilon_c^{\rm QP}(\mathbf{k}) - \varepsilon_v^{\rm QP}(\mathbf{k}') - \hbar z} + \frac{1}{\hbar\omega_{\rm LO} + \varepsilon_c^{\rm QP}(\mathbf{k}') - \varepsilon_v^{\rm QP}(\mathbf{k}) - \hbar z}\right]\right\}. \end{aligned} \tag{22.17}$$

This expression clearly illustrates the dynamics. For vanishing differences $|\varepsilon_c^{\rm QP}(\mathbf{k}) - \varepsilon_v^{\rm QP}(\mathbf{k}') - \hbar z|$, i.e., vanishing exciton binding energies, the bare Coulomb potential is screened by the static dielectric constant ϵ_s that includes lattice screening. In the opposite limit the screening is characterized by the static electronic dielectric constant ϵ_∞. The contribution of the lattice polarizability vanishes.

The effect of the dynamical lattice screening is studied in the framework of the Wannier-Mott model of the excitons (see Sect. 21.2.1). In the EMA we consider only a two-band semiconductor with isotropic and parabolic bands with the reduced mass μ_{ex} (see Fig. 15.2). Only one band pair cv is investigated. The electron-hole exchange contribution vanishes. To solve the eigenvalue problem (22.12) real frequencies $z = \omega$ have to be studied. For the given band pair, instead of (21.8), the homogeneous BSE becomes

$$\left[E_g + \frac{\hbar^2\mathbf{k}^2}{2\mu_{ex}} - E_\Lambda(\omega)\right] A_\Lambda(\mathbf{k}|\omega) - \sum_{\mathbf{k}'} \tilde{W}\,{}^{\mathbf{kk}'}_{\substack{cc\\vv}}(\omega) A_\Lambda(\mathbf{k}'|\omega) = 0. \tag{22.18}$$

We treat the dynamics in the screening of the Coulomb potential by introducing an effective frequency-dependent dielectric constant $\epsilon_{\text{eff}}(\omega)$. Thereby, we focus the attention to the $1s$ exciton ground state $\Lambda = 100$ of the Wannier-Mott exciton. Then a reasonable definition of $\epsilon_{\text{eff}}(\omega)$ is given by

$$\frac{1}{\epsilon_{\text{eff}}(\omega)} \sum_{\mathbf{k},\mathbf{k}'} A^*_{100}(\mathbf{k}|\omega)\tilde{v}(|\mathbf{k}-\mathbf{k}'|) A_{100}(\mathbf{k}'|\omega) = \sum_{\mathbf{k},\mathbf{k}'} A^*_{100}(\mathbf{k}|\omega)\tilde{W}\,{}^{\mathbf{kk}'}_{\substack{cc\\vv}}(\omega) A_{100}(\mathbf{k}',\omega).$$

For the explicit determination of $\epsilon_{\text{eff}}(\omega)$ we use the Fourier transformation of the $1s$ envelope function (21.10)

$$A_{100}(\mathbf{k}) = \sqrt{\frac{\pi a^3}{\Omega}} \frac{8}{\left[1 + (a\mathbf{k})^2\right]^2} \tag{22.19}$$

but with a frequency-dependent exciton radius $a = a(\omega)$ taken from $\hbar^2/[2\mu a^2(\omega)] = E_g - \hbar\omega$. The details of the treatment and the solutions can be found in a paper by Zimmermann [5]. Interestingly, he not only described dynamical corrections but also showed that in real space this treatment leads to the screened Haken potential with exponential modifications of the $1/r$ Coulomb behavior for the electron-hole attraction [14]. A similar potential has been earlier derived by Toyozawa [15] applying the electronic polaron model.

Here, we follow a variational procedure with $a(\omega) = a_{ex}\alpha(\omega)$ and $\alpha(\omega)$ as the variational parameter that for each photon energy $\hbar\omega$ has to be determined by maximizing the resulting exciton binding energy [7]. It follows from (22.18) and $E_B(\omega) = E_g - E_{100}(\omega)$ as

$$E_B(\alpha|\omega) = -\sum_{\mathbf{k},\mathbf{k}'} A^*_{100}(\mathbf{k}|\omega)\left[\frac{\hbar^2 k^2}{2\mu_{ex}}\delta_{\mathbf{kk}'} - \tilde{W}\,{}^{\mathbf{kk}'}_{\substack{cc\\vv}}(\omega)\right] A_{100}(\mathbf{k}',\omega).$$

In order to reduce the complexity of the dynamical screening contribution we investigate the case of heavy holes $m_v^* \gg m_c^*$, i.e., $\mu_{ex} \approx m_c^*$, where the hole is fixed in space and the internal motion of the $1s$ exciton is ruled by the motion of the electron. It results [7]

$$E_B(\alpha|\omega) = R_{\text{ex}}\left\{-\alpha^2 + 2\alpha\left[1 - \left(1 - \frac{\epsilon_\infty}{\epsilon_s}\right)D(x, y)\right]\right\} \tag{22.20}$$

with

$$D(x, y) = \frac{1}{1 + x - y}\left[1 - \frac{4y}{\left(\sqrt{1+x} + \sqrt{y}\right)^2}\right] \tag{22.21}$$

and ($\hbar\omega < E_g$)

$$\begin{aligned} x &= \frac{E_g - \hbar\omega}{\hbar\omega_{\text{LO}}}, \\ y &= \frac{R_{\text{ex}}}{\hbar\omega_{\text{LO}}}\alpha^2. \end{aligned}$$

The function $D(x, y)$ determines the dynamical effects due to the lattice polarization. In the limit of large exciton binding energies E_B compared to the optical phonon frequency $\hbar\omega_{\text{LO}}$, $x, y \gg 1$, it holds $D(x, y) \equiv 0$. The lattice is unable to follow the fast formation of the exciton with the large binding energy. Because of the fast motion of the electron around the localized hole, the lattice vibrations have not enough time for the formation of dynamical dipoles which influence the electron motion. Only the electrons contribute to the screening of the electron-hole attraction with the static electronic dielectric constant ϵ_∞. In this limit, the variational parameter is $\alpha = 1$, and the resulting exciton binding energy is $E_B = R_{\text{ex}}$. In the opposite limit of small binding energies compared to the lattice frequencies, $E_B \ll \hbar\omega_{\text{LO}}$, i.e., $x, y \ll 1$, it holds $D(x, y) \equiv 1$. The formation of the exciton is so slow that the lattice vibrations of the ionic lattice can follow adiabatically. The screening of the electron-hole attraction is replaced by ϵ_s, the static dielectric constant of the semiconductor. The static polarization of the lattice is fully included. The variational parameter becomes $\alpha = \epsilon_\infty/\epsilon_s$. The resulting exciton parameters are $E_B = R_{\text{ex}}(\epsilon_\infty/\epsilon_s)^2$ and $a = a_{\text{ex}}\epsilon_s/\epsilon_\infty$. In between, i.e., for $E_B \approx \hbar\omega_{\text{LO}}$, it holds $0 < D(x, y) < 1$. In any case the screening is enhanced by some lattice contributions, and the exciton binding energy varies between $(\epsilon_\infty/\epsilon_s)^2 R_{\text{ex}} < E_B < R_{\text{ex}}$. This can be qualitatively and quantitatively described by an effective dielectric constant with $\epsilon_\infty < \epsilon_{\text{eff}} < \epsilon_s$. In terms of the materials parameters μ_{ex}, ϵ_∞ and ϵ_s general statements are obvious: For group-IV semiconductors C, Si, and Ge the described dynamical effects do not play a role because of the covalent bonding, i.e., $\epsilon_\infty = \epsilon_s$. For the majority of III-V semiconductors crystallizing in zinc-blende geometry it holds $E_B \ll \hbar\omega_{\text{LO}}$ and the screening is mainly dictated by ϵ_s. In the compounds GaN and AlN, which crystallize in wurtzite structure under ambient conditions, the exciton binding energies are large and dynamical screening has to be taken into account [16, 17]. Because of the higher ionicities of the bonds the situation for many II-VI compound semiconductors is less clear, in particular, how strong the effect of dynamical screening is. However, the band-edge excitons of rs-MgO are certainly influenced by dynamical lattice screening (cf. discussion in Sect. 21.1.3). The pure electronic screening by ϵ_∞ overestimates the exciton binding also for this wide-gap semiconductor [18].

For core excitons the resulting variational problem has been solved numerically. Here, we will only roughly discuss the most important frequency dependence. We follow Zimmermann [5] and set $x = y$ in (22.20) The resulting variational parameter

$$\alpha = 1 - \left(1 - \frac{\epsilon_\infty}{\epsilon_s}\right) D(x, x)$$

is smaller than 1. Consequently the binding energy reads as

$$E_B(\omega) = \left[1 - \left(1 - \frac{\epsilon_\infty}{\epsilon_s}\right) D(x, x)\right]^2 R_{ex}. \qquad (22.22)$$

According to (22.14) the true pair excitation energy of the lowest-energy exciton is the zero $\hbar\omega_0$ of the equation

$$E_B(\omega) = E_g - \hbar\omega \qquad (22.23)$$

with the resulting binding energy $E_B = E_B(\omega_0)$. The variation of (22.22) versus $E_g - \hbar\omega$ together with the graphical solution of (22.23) is illustrated in Fig. 22.2. With rising strength $\left(1 - \frac{\epsilon_\infty}{\epsilon_s}\right)$ of the coupling of the LO phonons to the excited

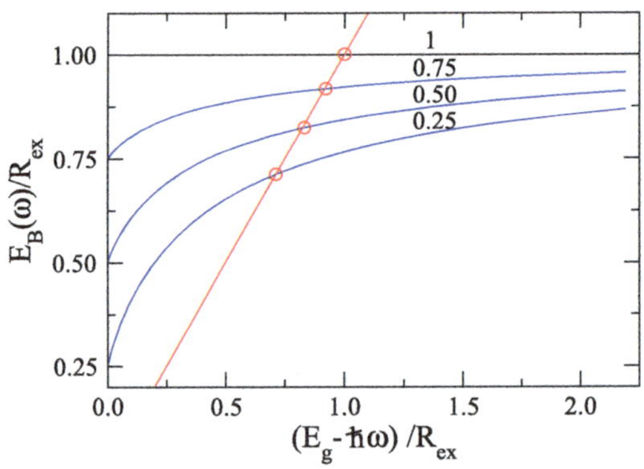

Fig. 22.2 Frequency-dependent binding energy (*blue curves*) of the lowest-energy Wannier-Mott exciton for four different values $1 - \frac{\epsilon_\infty}{\epsilon_s} = 1, 0.75, 0.50$, and 0.25 characterizing the exciton-LO phonon coupling. For comparison also the linear function $E_g - \hbar\omega$ (*red curve*) is shown. The *circles* indicate the influence of the dynamical screening on the resulting binding energies

Table 22.1 Experimental parameters being relevant for dynamical screening and exciton binding of direct III-V and II-VI compound semiconductors crystallizing in zinc-blende, rocksalt or wurtzite geometry

Semi-conductor	Gap (eV)	Phonon energies (meV)		Effective masses (m)			Dielectric constants		Binding energy E_B (meV)		
	E_g	$\hbar\omega_{\mathrm{LO}}$	$\hbar\omega_{\mathrm{TO}}$	m_c^*	m_v^*	μ_{ex}	ϵ_s	ϵ_∞	With ϵ_s	With ϵ_∞	Measured
GaAs	1.43	36.2	33.2	0.068	0.5	0.06	13.1	11.0	4.8	6.7	4.2/4.9
InP	1.27	42.8	37.7	0.067	0.8	0.06	12.4	9.6	5.3	8.9	4.0/5.1
InAs	0.36	30.2	27.1	0.022	0.41	0.02	14.6	11.8	1.3	2.0	
InSb	0.17	24.2	22.6	0.014	0.4	0.01	17.9	15.6	0.4	0.6	(0.4)
MgO	7.67	89	50	0.38	3.2	0.34	9.8	2.9	48.1	550	80/145
ZnO	3.2	72	51	0.27	1.8	0.23	7.9	4.0	50.1	195.6	59
ZnS	3.66	36.9	28.4	0.34		0.26	8.3	4.9	51.3	147.3	29
ZnSe	2.67	31	26	0.17	0.6	0.13	8.1	5.7	27.0	54.4	19.9
ZnTe	2.25	25.5	22.3	0.09	0.68	0.08	9.7	7.4	11.6	19.9	13
CdS	2.42	36.8	32.1	0.20	5	0.19	8.9	6.8	32.6	55.9	29/27
CdSe	1.74	26.2	20.6	0.13	1.2	0.12	10.6	6.6	14.5	37.5	15/15
CdTe	1.5	21.2	17.4	0.11	2.1	0.10	10.9	7.3	11.5	25.5	11

Mainly room-temperature values are collected from [19]. The dielectric constants ϵ_s are taken for III-V compounds from [20]. The experimental binding energies are from [13, 20]. The parameters for MgO are from [21]. The averaged masses of MgO are from [22]. The measured exciton binding energies are taken from [23, 24]. The static electronic dielectric constants are calculated using the LST relation $\epsilon_\infty = \epsilon_s(\omega_{\mathrm{TO}}/\omega_{\mathrm{LO}})^2$. Theoretical binding energies E_B are estimated using both ϵ_s and ϵ_∞

electron-hole pairs the exciton binding is reduced. Indeed the crossing points (22.23) are shifted to smaller binding energies with the lower limit $E_B = R_{\mathrm{ex}}(\epsilon_\infty/\epsilon_s)^2$.

The above qualitative discussion is somewhat more quantitatively underlined by the values for phonon energies, dielectric constants and excitons binding energies listed in Table 22.1 for 12 direct III-V and II-VI compound semiconductors. The picture of the conventional III-V compounds and CdSe, CdTe and ZnTe, i.e., II-VI semiconductors with relatively small exciton binding energies, is rather clear. The static lattice polarizability has to be fully taken into account, i.e., ϵ_s, to screen the electron-hole attraction W. The exciton binding energies for the oxides, MgO and ZnO, clearly indicate that the dynamics of the lattice screening has to be included to understand the measured values. An effective screening constant ϵ_{eff} between ϵ_∞ and ϵ_s has to be chosen. In the case of ZnS and CdS the situation is less clear. The binding energies estimated with ϵ_∞ are only slightly larger than the LO phonon energy. More detailed studies are needed to figure out the effective screening constant ϵ_{eff}.

22.2 Interference Effects

22.2.1 Satellites of Electron-Hole Pairs

In the last sections we have discussed the influence of dynamical screening on optically excited excitons, especially on their binding energies. In Chap. 17 we have studied another dynamical effect, satellite structures in the spectral functions of electrons or holes due to intrinsic or extrinsic losses. Here we investigate if such satellite structures in spectral functions of pairs can be modified by the dynamics of the screening of the electron-hole attraction. We focus on the spectral weights, not on the exact positions of the spectral lines. For that reason, we omit all the effects giving rise to strong static Coulomb correlation of electron-hole pairs or even to bound states. We focus on the dynamical effects on spectral distributions and, hence, the frequency dependence of the screened potential. Therefore, in the BSE (22.3) for the spin-averaged macroscopic polarization function the effect of the vertex corrections is reduced only to the dynamical contributions. On the other hand, we take improved single-particle Green functions with spectral functions (14.50) into account. Consequently, the BSE (22.3) is modified to

$$\begin{aligned} P^M(\lambda_1\lambda_1', \lambda_2\lambda_2'; z_n\tilde{z}_m) = & -i\hbar G_{\lambda_1}(z_n)G_{\lambda_1'}(z_n - \tilde{z}_m)\Bigg\{\delta_{\lambda_1\lambda_2'}\delta_{\lambda_2\lambda_1'} \\ & - \frac{1}{-i\hbar\beta}\sum_{n'}\sum_{\lambda_3,\lambda_4}\Big[W^{\lambda_1\lambda_3}_{\lambda_1'\lambda_4}(z_n - z_{n'}) \\ & - W^{\lambda_1\lambda_3}_{\lambda_1'\lambda_4}(0)\Big]\, P^M(\lambda_3\lambda_4, \lambda_2\lambda_{2'}; z_{n'}\tilde{z}_m)\Bigg\}. \end{aligned} \tag{22.24}$$

Since we are not interested in exciton binding, we only study the first iteration of the polarization function. We neglect multiple electron-hole scattering which is responsible for the formation of excitons. Only the dominant dynamical contribution is taken into consideration. In the language of the ladder approximation in Fig. 18.2c, only the first rung, more precisely its dynamical variation is taken into account. It results

$$\begin{aligned} P^M(\lambda_1\lambda_1', \lambda_2\lambda_2'; \tilde{z}_m) = & \frac{1}{\beta}\sum_n G_{\lambda_1}(z_n)G_{\lambda_1'}(z_n - \tilde{z}_m)\delta_{\lambda_1\lambda_2'}\delta_{\lambda_2\lambda_1'} \\ & - \frac{1}{\beta^2}\sum_{n,n'} G_{\lambda_1}(z_n)G_{\lambda_1'}(z_n - \tilde{z}_m)\left[W^{\lambda_1\lambda_2'}_{\lambda_1'\lambda_2}(z_n - z_{n'}) - W^{\lambda_1\lambda_2'}_{\lambda_1'\lambda_2}(0)\right] \\ & \times G_{\lambda_2'}(z_{n'})G_{\lambda_2}(z_{n'} - \tilde{z}_m) \end{aligned} \tag{22.25}$$

for the polarization function (19.2) that only depends on one frequency. The frequency dependence of the integral kernel is represented by expression (22.8).

We focus the attention on treatment of the modification of the quasielectron-quasihole excitation by dynamical screening effects in the lowest vanishing order in the screened Coulomb potential. To do so, we divide the spin-averaged macroscopic polarization function into two contributions

$$P^M(\lambda_1\lambda_1', \lambda_2\lambda_2'; z_m) = P^M_{\text{Dyson}}(\lambda_1\lambda_1', \lambda_2\lambda_2'; z_m) + P^M_{\text{vertex}}(\lambda_1\lambda_1', \lambda_2\lambda_2'; z_m). \tag{22.26}$$

In the first summand P^M_{Dyson} in (22.26) we take only the dynamical effects into account which are described by the self-energy in the Dyson equation of the single-particle Green function (14.43), in particular the first iteration as expressed by the single-particle spectral function (14.48). The second term P^M_{vertex} contains the (dynamical part of the) vertex corrections in linear order. For this contribution the n- and n'-sums are carried out by means of (12.66) and (14.64). The four single-particle Green functions are described by the zeroth order ones $G_\lambda(z) = 1/(\hbar z - \varepsilon_\lambda^{\text{QP}})$ (14.41), which only contain one quasiparticle peak at the energy $\hbar z = \varepsilon_\lambda^{\text{QP}}$. Here, we only give a simplified expression using a two-band model with Bloch states $|c\mathbf{k}\rangle$ and $|v\mathbf{k}\rangle$ as studied to describe dynamical screening effects on excitons in Sect. 22.1.2. Only the low-temperature limit is investigated, in which the Fermi functions $f(\varepsilon)$ are replaced by Heaviside functions $\theta(\mu - \varepsilon)$, and $g(\hbar\omega) = -\theta(-\omega)$ holds. In addition, the local-field effects $\mathbf{G} \neq \mathbf{G}'$ are disregarded. With (22.10) for the second term in (22.25) one has [25, 26]

$$\begin{aligned}
&P^M_{\text{vertex}}(c\mathbf{k}v\mathbf{k}, v\mathbf{k}'c\mathbf{k}'; \tilde{z}_m) \\
&= -\frac{1}{\left[\varepsilon_c^{\text{QP}}(\mathbf{k}) - \varepsilon_v^{\text{QP}}(\mathbf{k}) - \hbar\tilde{z}_m\right]\left[\varepsilon_c^{\text{QP}}(\mathbf{k}') - \varepsilon_v^{\text{QP}}(\mathbf{k}') - \hbar\tilde{z}_m\right]}\left[\tilde{W}^{\mathbf{k}\mathbf{k}'}_{\substack{cc\\vv}}(\tilde{z}_m) - W^{\mathbf{k}\mathbf{k}'}_{\substack{cc\\vv}}(0)\right] \\
&= \frac{1}{\varepsilon_c^{\text{QP}}(\mathbf{k}) - \varepsilon_v^{\text{QP}}(\mathbf{k}) - \varepsilon_c^{\text{QP}}(\mathbf{k}') + \varepsilon_v^{\text{QP}}(\mathbf{k}')} \\
&\times\left[\frac{1}{\hbar\tilde{z}_m - \varepsilon_c^{\text{QP}}(\mathbf{k}) + \varepsilon_v^{\text{QP}}(\mathbf{k})} - \frac{1}{\hbar\tilde{z}_m - \varepsilon_c^{\text{QP}}(\mathbf{k}') - \varepsilon_v^{\text{QP}}(\mathbf{k}')}\right] \\
&\times \frac{1}{\Omega}\sum_{\mathbf{q}}\sum_{\mathbf{G}} \tilde{v}(|\mathbf{q}+\mathbf{G}|)B^{\mathbf{k}\mathbf{k}'}_{cc}(\mathbf{q}+\mathbf{G})B^{\mathbf{k}\mathbf{k}'*}_{vv}(\mathbf{q}+\mathbf{G})\int_0^\infty \frac{d\hbar\omega}{\pi}\text{Im}\tilde{\varepsilon}^{-1}(\mathbf{q}+\mathbf{G}, \mathbf{q}+\mathbf{G}, \omega) \\
&\times\left[\frac{1}{\hbar\tilde{z}_m - \hbar\omega - \varepsilon_c^{\text{QP}}(\mathbf{k}') + \varepsilon_v^{\text{QP}}(\mathbf{k})} + \frac{1}{\hbar\tilde{z}_m - \hbar\omega - \varepsilon_c^{\text{QP}}(\mathbf{k}) + \varepsilon_v^{\text{QP}}(\mathbf{k}')} + \frac{2}{\hbar\omega}\right].
\end{aligned} \tag{22.27}$$

The result shows some similarities to the treatment of the dynamically screened Coulomb attraction in the Shindo approximation (22.10). We are only interested in the vertex contribution in first order to the two-particle spectral function. To find it, expression (22.27) is analytically continued into the complex $\tilde{z}$-plane. The contribution to the spectral function arises as imaginary part when $\tilde{z}_m \to \omega + i\eta$ and $\eta \to +0$

are investigated. Thereby, the two different types of denominators in (22.27) with $\hbar\tilde{z}_m$ and $\hbar(\tilde{z}_m - \omega)$ indicate that the vertex contribution will modify both the main quasielectron-quasihole peak but also its satellite structures.

In general, the pair polarization function P^M obeys a spectral representation (13.16)

$$P^M(c\mathbf{k}v\mathbf{k}, v\mathbf{k}'c\mathbf{k}'; \tilde{z}_m) = \int_{-\infty}^{+\infty} \frac{d\omega}{2\pi} \frac{\hat{P}^M(c\mathbf{k}v\mathbf{k}, v\mathbf{k}'c\mathbf{k}'; \omega)}{\tilde{z}_m - \omega}.$$

Here we are only interest in the first vertex contribution (22.27). In the limit of small band dispersions it can be represented in a compact form

$$\begin{aligned}\hat{P}^M_{\text{vertex}}(c\mathbf{k}v\mathbf{k}, v\mathbf{k}'c\mathbf{k}'; \omega) = &-\frac{1}{\Omega}\sum_{\mathbf{q}}\sum_{\mathbf{G}} \bar{v}(|\mathbf{q}+\mathbf{G}|) B^{\mathbf{k}\mathbf{k}'}_{cc}(\mathbf{q}+\mathbf{G}) B^{\mathbf{k}\mathbf{k}'*}_{vv}(\mathbf{q}+\mathbf{G}) \\ &\times \int_0^{\infty} \frac{d\hbar\omega'}{\pi} \text{Im}\tilde{\epsilon}^{-1}(\mathbf{q}+\mathbf{G}, \mathbf{q}+\mathbf{G}, \omega') \\ &\times \Big\{ \left[\delta\left(\varepsilon^{\text{QP}}_c(\mathbf{k}) - \varepsilon^{\text{QP}}_v(\mathbf{k}) - \hbar\omega\right) - \delta\left(\hbar\omega' + \varepsilon^{\text{QP}}_c(\mathbf{k}') - \varepsilon^{\text{QP}}_v(\mathbf{k}') - \hbar\omega\right)\right] \\ &\times \frac{1}{\left[\hbar\omega' + \varepsilon^{\text{QP}}_c(\mathbf{k}') - \varepsilon^{\text{QP}}_c(\mathbf{k})\right]\left[\hbar\omega' + \varepsilon^{\text{QP}}_v(\mathbf{k}) - \varepsilon^{\text{QP}}_v(\mathbf{k}')\right]} + \mathbf{k} \leftrightarrow \mathbf{k}' \Big\}\end{aligned}$$

of the vertex contributions to the main spectral line and the first satellite structures. Especially, in the limit of dispersionless valence bands, i.e., core holes, this expression is correct.

For explicit treatments, in particular calculating the absorption coefficient, photoemission cross sections or similar quantities, one has to take into consideration that expression (22.27) appears under a double sum with respect to the $\mathbf{k}$ and $\mathbf{k}'$ vectors weighted by certain transition matrix elements. If their wave vector dependence can be omitted, only sums of $P^M(\mathbf{k}\mathbf{k}')$ have to be considered to determine the contribution of the vertex corrections to the spectral properties.

In order to illustrate the dynamical corrections due to the XC self-energies of an individual electron and hole to the main contribution P^M_{Dyson} to (22.26), we use the single-particle spectral functions in first iteration (17.7), i.e., including satellite structures in first order,

$$\begin{aligned}A_{\nu\nu}(\mathbf{k}, \omega) = &\,2\pi\delta\left(\hbar\omega - \varepsilon^{\text{QP}}_\nu(\mathbf{k})\right) + \frac{1}{\Omega}\sum_{\mathbf{q},\mathbf{G}} \bar{v}\left(|\mathbf{q}+\mathbf{G}|\right) \sum_{\mathbf{k}'} \\ &\times \mathcal{P}\int_0^{\infty} \frac{d\hbar\omega'}{\pi} \text{Im}\tilde{\epsilon}^{-1}(\mathbf{q}+\mathbf{G}, \mathbf{q}+\mathbf{G}, \omega') \left| V^{\mathbf{k}\mathbf{k}'}_{\nu\nu}(\mathbf{q}+\mathbf{G}|\omega)\right|^2 \\ &\times 2\pi\left[\delta\left(\hbar\omega - \varepsilon^{\text{QP}}_\nu(\mathbf{k})\right) - \delta\left(\hbar\omega - \varepsilon^{\text{QP}}_\nu(\mathbf{k}) + (\delta_{\nu v} - \delta_{\nu c})\hbar\omega'\right)\right]\end{aligned}$$

with the reduced weight of the main QP peak and the first satellite structure. The abbreviation

$$V_{\nu\nu}^{\mathbf{kk}'}(\mathbf{q}+\mathbf{G}|\omega) = \frac{B_{\nu\nu}^{\mathbf{kk}'}(\mathbf{q}+\mathbf{G})}{\hbar\omega + (\delta_{\nu v} - \delta_{\nu c})[\varepsilon_{\nu}^{\mathrm{QP}}(\mathbf{k}) - \varepsilon_{\nu}^{\mathrm{QP}}(\mathbf{k}')]} \tag{22.28}$$

has been introduced. With the spectral representation (11.21), a linearization of the result in terms of the screened potential, and the definition (13.16) of the spectral function of a two-particle Green function, one finds for the spectral function containing first-order extrinsic and intrinsic losses

$$\begin{aligned}
\hat{P}_{\mathrm{Dyson}}^{M}(c\mathbf{k}v\mathbf{k}, v\mathbf{k}'c\mathbf{k}', \omega) &= 2\pi\delta_{\mathbf{kk}'}\delta\left(\varepsilon_{c}^{\mathrm{QP}}(\mathbf{k}) - \varepsilon_{v}^{\mathrm{QP}}(\mathbf{k}) - \hbar\omega\right) \\
&\times\left\{1 + \frac{1}{\Omega}\sum_{\mathbf{q},\mathbf{G}} \tilde{v}(|\mathbf{q}+\mathbf{G}|)\mathcal{P}\int_0^{\infty}\frac{d\hbar\omega'}{\pi}\mathrm{Im}\tilde{\epsilon}^{-1}(\mathbf{q}+\mathbf{G},\mathbf{q}+\mathbf{G},\omega')\right. \\
&\times\left.\left[\left|V_{cc}^{\mathbf{kk}'}(\mathbf{q}+\mathbf{G}|\omega')\right|^2 + \left|V_{vv}^{\mathbf{kk}'}(\mathbf{q}+\mathbf{G}|\omega')\right|^2\right]\right\} \\
&- \frac{1}{\Omega}\sum_{\mathbf{q},\mathbf{G}} \tilde{v}(|\mathbf{q}+\mathbf{G}|)\mathcal{P}\int_0^{\infty}\frac{d\hbar\omega'}{\pi}\mathrm{Im}\tilde{\epsilon}^{-1}(\mathbf{q}+\mathbf{G},\mathbf{q}+\mathbf{G},\omega') \\
&\times\left\{\left|V_{cc}^{\mathbf{kk}'}(\mathbf{q}+\mathbf{G}|\omega')\right|^2 2\pi\delta\left(\hbar\omega' + \varepsilon_{c}^{\mathrm{QP}}(\mathbf{k}') - \varepsilon_{v}^{\mathrm{QP}}(\mathbf{k}) - \hbar\omega\right)\right. \\
&+\left.\left|V_{vv}^{\mathbf{kk}'}(\mathbf{q}+\mathbf{G}|\omega')\right|^2 2\pi\delta\left(\hbar\omega' + \varepsilon_{c}^{\mathrm{QP}}(\mathbf{k}) - \varepsilon_{v}^{\mathrm{QP}}(\mathbf{k}') - \hbar\omega\right)\right\}.
\end{aligned} \tag{22.29}$$

The three diagrams counted in the spectral function (22.29) are displayed in Fig. 22.3a–c. The first summand represents the quasielectron-quasihole pair excitation with a significant reduction of its spectral weight due to the generation of satellite structures. The second summand represents the first-order satellites due to

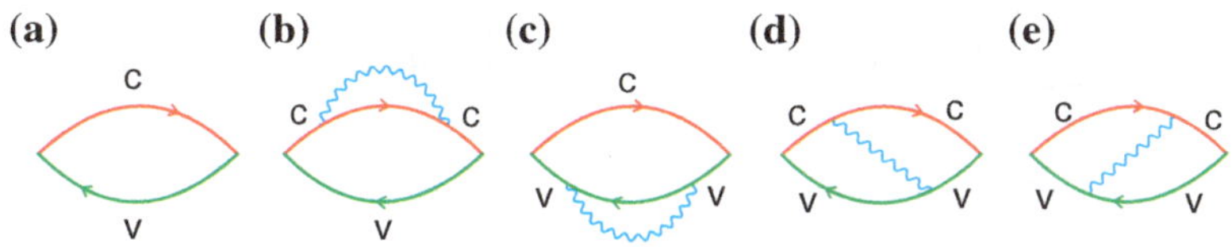

Fig. 22.3 Five diagrams which are included to determine (**a**) the main quasielectron-quasihole excitation and the corresponding dynamical corrections due to (**b, c**) self-energy and (**d, e**) vertex diagrams leading to the first-order satellite structure and the renormalization of the spectral strength of the main pair peak, in addition to the unrenormalized result represented by the *bubble* in (**a**). The diagrams contain extrinsic (**b**), intrinsic (**c**), and interference (**d, e**) losses

extrinsic and intrinsic losses (see Chap. 17). The (dynamical) contributions due to the vertex corrections to the polarization functions (22.27) are illustrated in Fig. 22.3d, e also by the relevant Feynman diagrams.

22.2.2 Spectral Weights

The complexity of expressions (22.27) and (22.29) suggests to focus on the spectral weight of the main peak and not to study explicitly the details of the satellite structures in first order. We know from the single-particle spectral functions, e.g. from (17.15), that the reduction of the spectral weight of the main QP peak is directly related to the strength of the first satellite. There is an additional complication due to the dependence of the vertex contribution on two Bloch wave vectors $\mathbf{k}$ and $\mathbf{k}'$, which can be ignored, recalling the fact that in an observable quantity such as the imaginary part of the dielectric function transition matrix elements $M_{cv}^{\mathbf{kk}}(\hat{\mathbf{q}})$ of the type (20.4) occur. Each $\mathbf{k}$- and $\mathbf{k}'$-sum is weighted by such a matrix element. However, if the $\mathbf{k}$ dependence of the matrix elements can be omitted, e.g. near a critical point in the interband structure in the BZ or studying core level-conduction band transitions, one of the two $\mathbf{k}$-point sums can be carried out, and only a two-particle spectral function depending on one Bloch wave vector $\mathbf{k}$ has to be discussed. Taking from (22.27) only the contribution to the main QP pair peak and the discussed treatment of the wave vector dependence into account, with (2.29) an effectively diagonal spectral function can be introduced [25, 26]

$$\begin{aligned} \hat{P}^M(c\mathbf{k}v\mathbf{k}, v\mathbf{k}'c\mathbf{k}', \omega) &= 2\pi\delta_{\mathbf{kk}'}[1-\beta_{cv}(\mathbf{k})]\delta\left(\varepsilon_c^{\mathrm{QP}}(\mathbf{k})-\varepsilon_v^{\mathrm{QP}}(\mathbf{k})-\hbar\omega\right) \\ &+ \text{ first satellites} \end{aligned} \tag{22.30}$$

with the coefficient

$$\begin{aligned} \beta_{cv}(\mathbf{k}) = &-\frac{1}{\Omega}\sum_{\mathbf{q},\mathbf{G}}\tilde{\mathrm{v}}(|\mathbf{q}+\mathbf{G}|\mathcal{P}\int_0^\infty\frac{d\hbar\omega'}{\pi}\mathrm{Im}\tilde{\epsilon}^{-1}(\mathbf{q}+\mathbf{G},\mathbf{q}+\mathbf{G},\omega') \\ &\times\sum_{\mathbf{k}'}\left[\left|V_{cc}^{\mathbf{kk}'}(\mathbf{q}+\mathbf{G}|\omega')\right|^2-2\mathrm{Re}\left\{V_{cc}^{\mathbf{kk}'}(\mathbf{q}+\mathbf{G}|\omega')V_{vv}^{\mathbf{kk}'*}(\mathbf{q}+\mathbf{G}|\omega')\right\}\right. \\ &\left.+\left|V_{vv}^{\mathbf{kk}'}(\mathbf{q}+\mathbf{G}|\omega')\right|^2\right] \end{aligned} \tag{22.31}$$

of the effective strength of the satellites and, hence, $[1-\beta_{cv}(\mathbf{k})]$ as the strength of the main pair peak. The mixed term in (22.31), containing both electron and hole Bloch matrix elements, is a consequence of the dynamical vertex corrections in $\hat{P}^M_{\text{vertex}}$. It obviously reduces the satellite strength $\beta_{cv}(\mathbf{k})$ compared to the case where only extrinsic and intrinsic satellite losses due to self-energy effects are considered as in (22.29). Assuming that the complex character of $V_{vv}^{\mathbf{kk}'}(\mathbf{q}+\mathbf{G}|\omega)$ can be omitted the expression in the square brackets becomes

$$[...] = \left| V_{cc}^{\mathbf{kk}'}(\mathbf{q}+\mathbf{G}|\omega) - V_{vv}^{\mathbf{kk}'^*}(\mathbf{q}+\mathbf{G}|\omega) \right|^2 . \tag{22.32}$$

This result clearly indicates that the dynamical vertex corrections may lead to interference effects which can significantly reduce the influence of the satellites in electron-hole pair excitations.

22.2.3 Compensation of Dynamical Effects

First, we discuss the influence of the vertex corrections, i.e., the interference terms, on the spectral strength of the electron-hole excitations for photon energies in the range of optical frequencies above the direct fundamental gap. For group-IV semiconductors spectral weights $\beta_{cv}(\mathbf{k})$ (22.31) are depicted in Fig. 22.4. Pseudopotential calculations are combined with a single-plasmon pole model for the screening function. Near the direct fundamental gap the reduction is of the order of $\beta_{cv}(\mathbf{k}) \approx 0.2 - 0.4$ without vertex corrections. This is in agreement with spectral weights discussed for the band edges of group-IV semiconductors in Table 15.2. For higher interband transitions this reduction may increase to values up to 0.5 (diamond) or 0.8 (Si). Such $\beta_{cv}(\mathbf{k})$ values would have a dramatic influence on optical spectra because of the strong reduction of spectral strength of the main peak. However, after inclusion of the vertex corrections a destructive interference is observed. The resulting $\beta_{cv}(\mathbf{k})$ strengths approach values below 0.1 (diamond) or 0.2 (silicon). In other words, with vertex corrections the satellite structures are significantly depressed in optical spectra. This is the reason why plasmon satellites have not been measured in optical

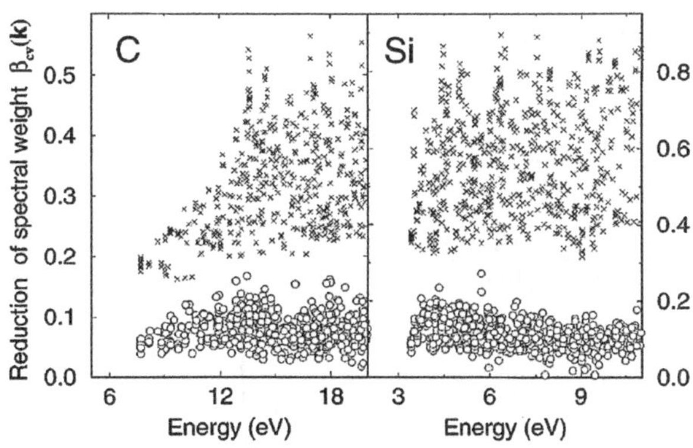

Fig. 22.4 Reduction $\beta_{cv}(\mathbf{k})$ of spectral weight $[1-\beta_{cv}(\mathbf{k})]$ of the main quasielectron-quasihole peak (without static excitonic effects) versus the QP excitation energy $[\varepsilon_c^{\mathrm{QP}}(\mathbf{k}) - \varepsilon_v^{\mathrm{QP}}(\mathbf{k})]$ for diamond and silicon. *Crosses*: without vertex corrections, *open circles*: with vertex corrections. From [26]

spectroscopies in the range of the interband transitions and higher energies, in contrast to core-level photoemission spectra in Figs. 17.4, 17.5, and 17.9 but also the valence-electron photoemission spectra in Figs. 17.10 and 17.11.

This fact is clearly analytically described by the relationship (22.32). For free-electron-like states the Bloch integrals can be expressed by Kronecker symbols describing the momentum conservation. With (22.28) it holds

$$[...] = \delta_{\mathbf{q},\mathbf{k}-\mathbf{k}'} \left[\frac{1}{\hbar\omega - \left[\varepsilon_c^{\mathrm{QP}}(\mathbf{k}) - \varepsilon_c^{\mathrm{QP}}(\mathbf{k}')\right]} - \frac{1}{\hbar\omega + \left[\varepsilon_v^{\mathrm{QP}}(\mathbf{k}) - \varepsilon_v^{\mathrm{QP}}(\mathbf{k}')\right]} \right]^2 .$$

This expression vanishes for small wave vectors $\mathbf{q}+\mathbf{G}$ of the collective excitations characterizing the screening reaction. In a plasmon-pole approximation (13.54) one has an energy $\hbar\omega = \hbar\Omega(\mathbf{q}+\mathbf{G})$ which is much larger than the band width

$$\hbar\Omega(\mathbf{q}+\mathbf{G}) \gg \left|\varepsilon_\nu^{\mathrm{QP}}(\mathbf{k}) - \varepsilon_\nu^{\mathrm{QP}}(\mathbf{k}')\right| \qquad (\nu = c, v),$$

at least for plasmons of the valence electron gas. This inequality also indicates small weights of satellite structures in pair excitations. The (partial) cancellation of quasiparticle and vertex contributions has also been observed in several theoretical papers where the static polarizability of the electrons has been investigated for vanishing momenta [27–29].

Spectral functions of the type (22.30) can also be applied to interpret core-level photoemission spectra. The hole energy $\varepsilon_v^{\mathrm{QP}}(\mathbf{k}) \equiv \varepsilon_{\mathrm{core}}^{\mathrm{QP}}$ has to be identified with the energy of a core level. Core-hole wave functions can be nearly described in terms of atomic ones. The Fourier transform of their square should be $\rho_{\mathrm{core}}(\mathbf{q}+\mathbf{G})$. The conduction band energy has to be approximately replaced by a free-electron parabola $\varepsilon(\mathbf{k}) = \varepsilon_c^{\mathrm{QP}}(\mathbf{k}) = \frac{\hbar^2 k^2}{2m}$, where vanishing momenta and kinetic energy describe the vacuum level. The corresponding Bloch integrals become Kronecker symbols. The reduction of the effective strength (22.31) of the photoemission peak or the strength of the satellite structures can be rewritten to

$$\begin{aligned}\beta_{cv}(\mathbf{k}) = &-\frac{1}{\Omega}\sum_{\mathbf{q},\mathbf{G}} \tilde{v}(|\mathbf{q}+\mathbf{G}|)\mathcal{P}\int_0^\infty \frac{d\hbar\omega}{\pi}\mathrm{Im}\tilde{\varepsilon}^{-1}(\mathbf{q}+\mathbf{G},\mathbf{q}+\mathbf{G},\omega)\\ &\times\left\{\frac{\rho_{\mathrm{core}}^2(\mathbf{q}+\mathbf{G})}{(\hbar\omega)^2} - \frac{2\rho_{\mathrm{core}}(\mathbf{q}+\mathbf{G})}{\hbar\omega\left[\hbar\omega + (\varepsilon(\mathbf{k}+\mathbf{q}) - \varepsilon(\mathbf{k}))\right]}\right.\\ &\left.+\frac{1}{\left[\hbar\omega + (\varepsilon(\mathbf{k}+\mathbf{q}) - \varepsilon(\mathbf{k}))\right]^2}\right\}. \end{aligned} \tag{22.33}$$

With $\mathrm{Im}\tilde{\varepsilon}^{-1}(\mathbf{Q},\mathbf{Q},\omega) \sim -\delta(\omega - \Omega(\mathbf{Q}))$ the three contributions to the strength of the first plasmon satellite have a clear physical meaning. The first term is related to intrinsic plasmons, the third one describes extrinsic plasmons, while the second term is due to their interference, i.e., due to dynamical vertex corrections. In principle, such an argumentation has been first derived by Chang and Langreth [30, 31]. The term 'interference process' has been however introduced by various authors [32–34].

Within the plasmon-pole model and assuming that $\rho_{\text{core}}(\mathbf{Q}) = \rho_{\text{core}}(|\mathbf{Q}|)$ expression (22.33) becomes

$$\beta_{cv}(\mathbf{k}) = \frac{e^2}{4\pi^2\varepsilon_0}\int_0^\infty dQ \frac{\omega_p^2}{\Omega^2(Q)}\frac{1}{\hbar\Omega(Q)} \times \frac{1}{2}\int_{-1}^{+1} dy \left\{ \rho_{\text{core}}^2(Q) - 2\rho_{\text{core}}(Q)\frac{1}{1+\frac{\sqrt{\varepsilon(\mathbf{k})\varepsilon(Q)}}{\hbar\Omega(Q)}y} + \frac{1}{\left[1+\frac{\sqrt{\varepsilon(\mathbf{k})\varepsilon(Q)'}}{\hbar\Omega(Q)}y\right]^2} \right\}$$

with $\varepsilon(\mathbf{k})$ as the kinetic energy of the photoelectron. This result obviously shows that the extrinsic plasmon contribution vanishes as $[\varepsilon(\mathbf{k})]^{-1}$ while the interference term does as $[\varepsilon(\mathbf{k})]^{-\frac{1}{2}}$. The latter prediction has been derived using different approaches [32, 35]. Two limits of photoemission processes are recognizable: (i) For large kinetic energy $\varepsilon(\mathbf{k})$ or high velocity of the emitted electron, the hole or the hole potential is suddenly switched on. In this sudden limit only intrinsic plasmon satellites appear. (ii) In the opposite limit, small kinetic energy and velocity of the emitted electron, the strength reduction $\beta_{cv}(\mathbf{k})$ becomes small. The hole and the hole potential are switched on slowly. The photoelectron and the electron system are able to follow the perturbation adiabatically. In this adiabatic limit only weak satellite structures occur.

This kinetic-energy dependence of the photoelectron spectra for core excitations is experimentally confirmed as demonstrated in Fig. 22.5 for the excitation of Si $2p$ and Al $2p$ levels [36]. In Fig. 22.5a for Si $2p$ the first bulk plasmon loss peak decreases

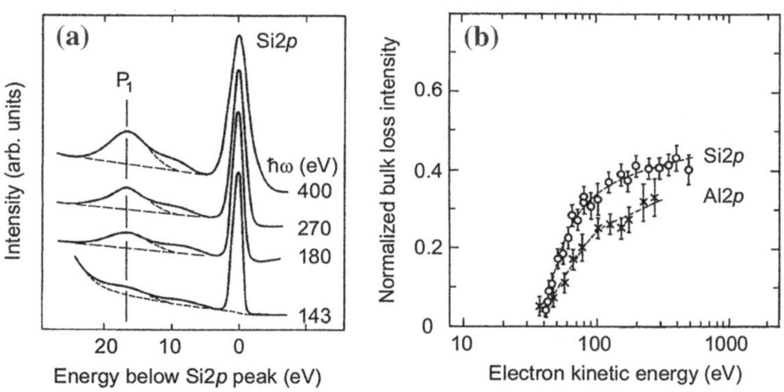

Fig. 22.5 (**a**) Si $2p$ photoemission spectra for different incident photon energies $\hbar\omega$. The first bulk plasmon satellite structure is labeled P_1. The *dashed line* indicates the intensity of the secondary electrons which has to be substracted. (**b**) Intensity of the first bulk plasmon loss peak normalized to the main QP peak for Si $2p$ and Al $2p$, which can be identified with $\beta_{cv}(\mathbf{k})$ (22.33). From [36]

Fig. 22.6 Interplay of extrinsic, intrinsic and interference plasmon losses due to the escaping photoelectron, leaving core hole and their dynamical interactions in an electron gas characterized by the plasmon frequency ω_p

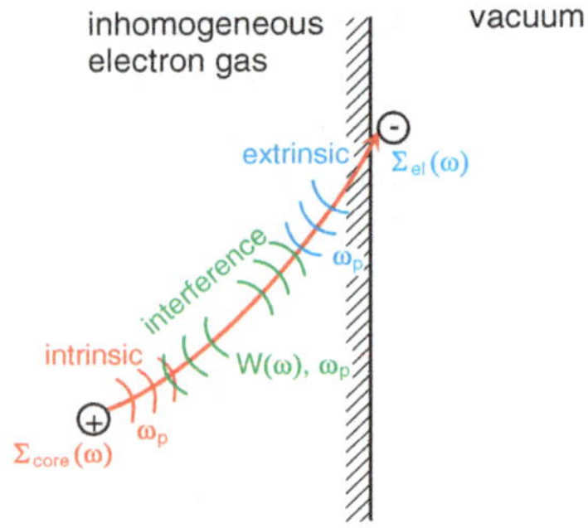

in its intensity relative to the intensity of the main QP peak with decreasing photon energy and, therefore, kinetic energy of the photoelectrons. The same holds for the weaker and closer satellite structure related to surface plasmons. The suppression of the plasmon peaks at decreasing photon energy is apparent. There is an almost complete cancellation of intrinsic, interference and extrinsic losses. Their interplay is schematically summarized in Fig. 22.6. The decrease of the satellite intensity with lowering the kinetic energy of the photoelectrons is illustrated in Fig. 22.5b for both Si $2p$ and Al $2p$ core-level photoemission. The trend of the curves is rather similar, independent of the material. The normalized bulk loss intensity increases rapidly with electron kinetic energy from the onset up to 80–100 eV kinetic energy. The experimental results are in qualitative agreement with the above analytic description of the relative intensity $\beta_{cv}(\mathbf{k})$ of the satellite structures. Recently, similar observations have been made for the first satellite in the valence band photoemission of silicon [37].

22.3 Free-Carrier Screening

22.3.1 Mott Transition and Burstein-Moss Shift

Mobile electrons and holes play a central role in solids. The electrons in a partially filled band of a metal as well as electrons in the conduction band and holes in the valence band of a semiconductor can carry an electrical current. Hence, these free carriers play the central role for transport phenomena such as the conductivity. They also influence the many-body effects on the electronic structure and the spectral properties. One of these effects – the gap shrinkage in heavily doped semiconductors – is discussed in Sect. 15.1.4. The influence of mobile electrons on the band structure of metals is described in Sect. 16.2.

Here we focus the attention on heavily doped semiconductors [38] and insulators with free carriers, e.g. the n-type transparent conducting oxides (TCOs) [39] such as In_2O_3, SnO_2 or ZnO (see Fig. 22.7), and the influence of free electrons.

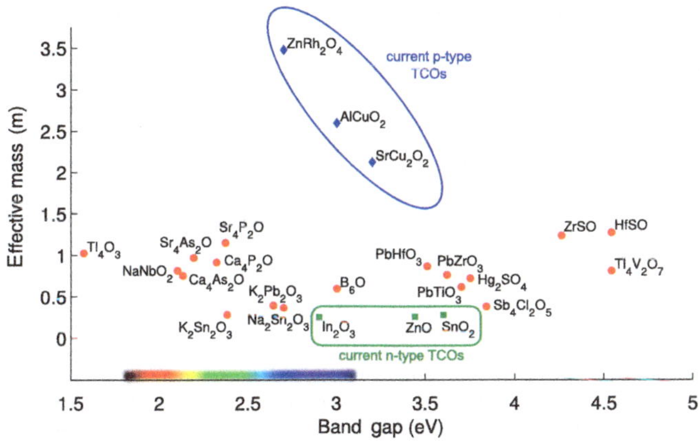

Fig. 22.7 Effective mass versus band gap for TCO candidates. The small electron masses and the wide transparency region of currently known *n*-type TCOs are indicated. From [40]. Permission of Nature Publishing Group is acknowledged

The most important properties studied are optical ones, whose frequency dependence is characterized by the macroscopic polarization function (19.4). The corresponding Bethe-Salpeter (19.12) shows that the free carriers influence an optical spectrum in various parts of the BSE. There is one category which only indirectly influences the screening. It is related to the band occupation in the dielectric function (13.43) in RPA or better in independent-quasiparticle approximation. The interband contributions are modified by the occupation expressed by Fermi functions. In addition, intraband contributions as in (13.46) occur which give rise to the wave vector- and frequency-dependent polarizability of the free carriers. Their contribution modifies the screened Coulomb potential W (12.49) and, consequently, the QP approach to the electronic structure in GW approximation (14.42) as well as the screened potential W in the electron-hole-interaction kernel (19.9). The effect on the QP self-energy and therefore on the band structure and the gaps is illustrated in Sect. 15.1.4. Here we investigate this effect on the formation of exciton bound states and the lineshape of the absorption coefficient.

First, one dominating effect is due to the modified screening. According to the study of (15.19) the attractive electron-hole potential $\frac{1}{\epsilon_\infty}\mathrm{v}(\mathbf{x})$ in the hydrogenic Schrödinger equation (21.12) is replaced by a Yukawa potential

$$W(\mathbf{x}) = \frac{1}{\epsilon_\infty}\mathrm{v}(|\mathbf{x}|) \rightarrow W(\mathbf{x}) = \frac{1}{\epsilon_\infty}\mathrm{v}(|\mathbf{x}|)e^{-\tilde{q}_{TF}|\mathbf{x}|}$$

with the normalized Thomas-Fermi wave number $\tilde{q}_{TF} = q_{TF}/\sqrt{\epsilon_\infty}$ and $q_{TF} \sim n_e^{1/6}$ related to the density n_e of a homogeneous electron gas in the conduction band. In contrast to the bare Coulomb potential, i.e., $\tilde{q}_{TF} = 0$, such a Yukawa potential shows a critical behavior [41, 42]. There is a bound-unbound transition of the exciton binding energy for a critical wave number $\tilde{q}_{TF}^{\mathrm{crit}}$, above which no bound state exists.

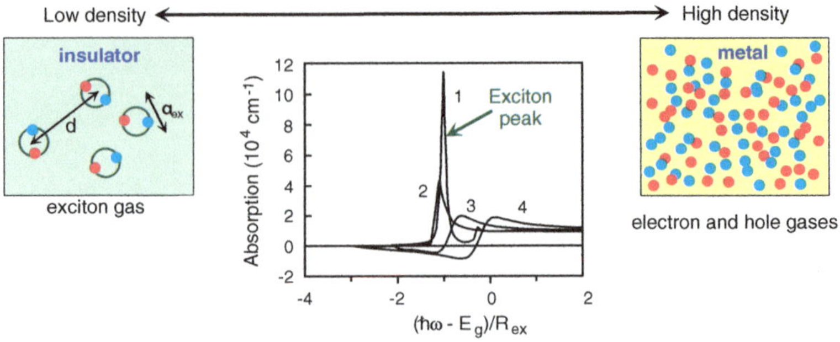

Fig. 22.8 Illustration of the Mott transition of excitons in a semiconductor in the presence of degenerate electron and hole gases with varying density. The absorption spectrum is adapted from a low-temperature computation for GaAs [45]. The labels 1, 2, 3 and 4 indicate increasing densities 0, 5, 30, and 80 $\times$ 10^{15} cm^{-3} of electrons and holes

A variational approach, even with a $1s$ variational function analogous to (21.52) and (22.19), gives such a value. Numerical approaches [41, 42] yield

$$\tilde{q}_{TF}^{\text{crit}} = 1.1906/a_{\text{ex}}. \tag{22.34}$$

with the radius a_{ex} of the corresponding Wannier-Mott exciton characterized by the dielectric constant ϵ_∞, which perhaps has to be replaced by ϵ_s in ionic compounds with weak exciton binding energies (see Sect. 22.1.3).

The disappearance of the excitonic bound states for $\tilde{q}_{TF} > \tilde{q}_{TF}^{\text{crit}}$ and, therefore, the limit of high free carrier densities n_e, may be interpreted in terms of a Mott transition [43]. The original idea by Mott [44] applied to doped semiconductors at zero temperature can be easily extended to excitons. This is illustrated in Fig. 22.8. At low optical excitation density, a gas of excitons with large average distance d is created within the semiconductor. With increasing free-carrier densities of electrons and holes the exciton binding energy reduces due to the more efficient screening. If the first excitons get ionized the screening effect may be enforced, in particular, for a sufficiently large optical excitation density. A phase transition between an insulator and a system with a two-component fermion gas may appear. A similar behavior is expected for low optical excitation but rising density n_e of free electrons in heavily doped samples (see Fig. 22.9). The indicated influence on the lineshape of the absorption however also contains other influences of free carriers. This indicates that the terminology is not well clarified. In the literature 'Mott transition' has no well defined meaning, even not for the exciton gas. Its definition depends on the method of measurement or of calculation. Only the qualitative behavior of bound states is uniquely described.

In the majority of cases a Mott density derived from experimental data n_M is introduced to characterize the phase transition, e.g. for n-doped samples. One possibility is to define it by means of a vanishing density-dependent exciton binding energy

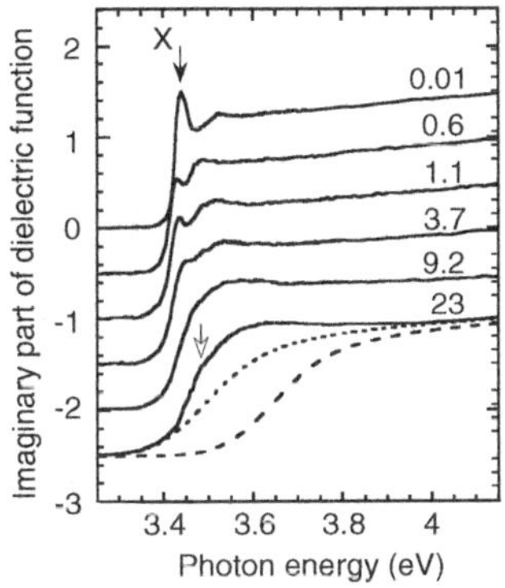

Fig. 22.9 Absorption edge of n-doped wurtzite GaN measured for increasing densities n_e given in units of $10^{18}\,\mathrm{cm}^{-3}$. Reprinted with permission from [50]. Copyright 2009 by the American Physical Society

$$E_B(n_e) = 0 \qquad \text{for} \qquad n_e \geq n_M,$$

in agreement with the definition of the critical Thomas-Fermi wave number (22.34). In this sense the condition (22.34) is frequently used as a rule of thumb [46, 47]

$$n_M a_{\mathrm{ex}}^3 = 1.$$

Actually, it tends to overestimate the Mott density derived from experimental data. For ZnO with $a_{\mathrm{ex}} = 1.8\,\mathrm{nm}$ [48], a value $n_M^{\mathrm{ZnO}} = 1.7 \times 10^{20}\,\mathrm{cm}^{-3}$ about two orders of magnitude too large is derived in comparison with measurements [49]. However, Fig. 22.9 shows the difficulty to derive the Mott density from excitonic absorption spectra with varying n_e, in particular, the density at which the excitons are actually dissociated. Of course, the effective screening due to the valence electron gas and the lattice vibrations has a certain influence. With an effective dielectric constant of about $\epsilon_{\mathrm{eff}} = 4.4$ [50] and masses $m_c^* = 0.3m$ and $m_v^* = 0.5m$ [51] the resulting Bohr radius $a_{\mathrm{ex}} = 1.24\,\mathrm{nm}$ [50] is much smaller. The relation (22.34) leads to a such smaller Mott density $n_M = 6 \times 10^{18}\,\mathrm{cm}^{-3}$ closer to experimental findings [49].

Second, there are also more direct free-carrier effects due to band filling, sometimes called Pauli blocking. It leads to a carrier-dependent renormalization of the electron-hole interaction Ξ^M as shown in (19.16). The interaction of electron-hole pairs in not completely empty and filled bands is reduced. A more drastic effect is due to the blocking of optical transitions in (19.8) or the inhomogeneity of the BSE (19.12). In the case of a degenerate electron gas with the chemical potential in the conduction band as displayed in Fig. 15.2, optical transitions with near-band-gap energies in the center of the Brillouin zone are not possible anymore (see Fig. 22.10). Assuming that the excitons are dissociated, i.e., $n_e > n_M$, the optical absorption edge is shifted from $\hbar\omega = E_g(0)$ to $\hbar\omega = E_g(n_e) + \varepsilon_F$ with $\varepsilon_F = \mu - \varepsilon_c^{\mathrm{QP}}(0) = \Delta E_{\mathrm{BM}}(n_e)$ as the Burstein-Moss (BM) shift. It describes the energy difference between the energy of the highest conduction band state being populated and the conduction band minimum. As a result the 'apparent' optical band gap of a semiconductor, the

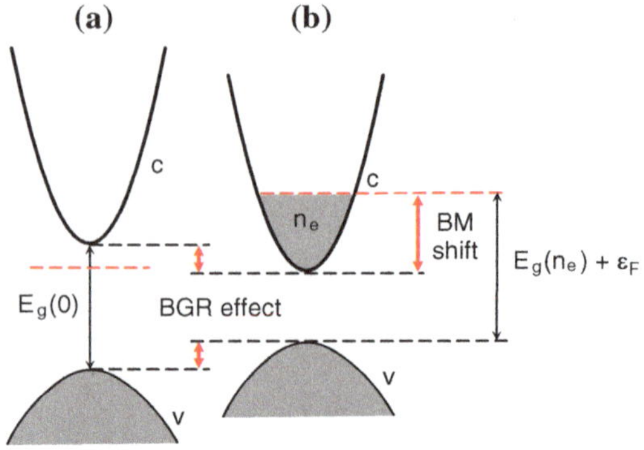

Fig. 22.10 Quasiparticle band structure of a two-band model (**a**) of an undoped semiconductor and (**b**) in the presence of a degenerate electron gas with density n_e. The band filling is indicated by *shaded regions*. The well-known band gap renormalization (BGR) effect resulting in a gap shrinkage $\Delta E_g(n_e) = E_g(0) - E_g(n_e)$ (15.20) together with the Burstein-Moss shift $\Delta E_{\text{BM}}(n_e)$ are displayed

absorption edge as measured using transmission or reflection spectroscopy, is significantly increased by the BM shift. It can be increased to energies of the order of 1 eV for carrier densities $n_e > 10^{20}\,\text{cm}^{-3}$ in many oxides with strongly dispersive and non-parabolic conduction bands. The BM shift in In_2O_3 is illustrated in Fig. 22.11 together with the influence of non-parabolicity on the lowest conduction band [52]. The well-known $\Delta E_{\text{BM}}(n_e) = \varepsilon_F \sim n_e^{2/3}$ dependence (4.47) is clearly visible. The non-parabolicity of the conduction band slightly reduces the increase of ε_F with the density of the free carriers n_e.

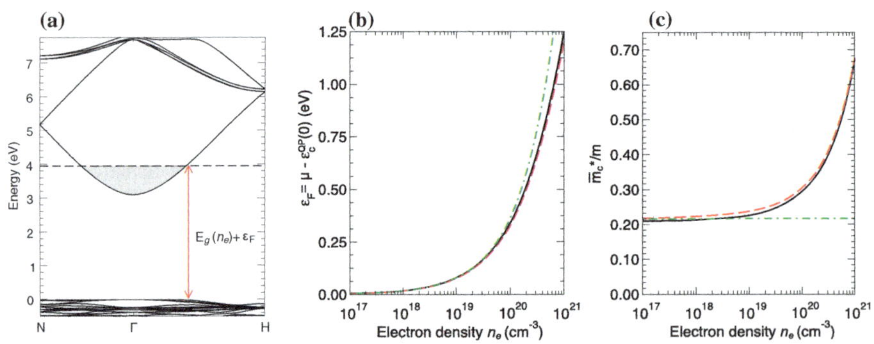

Fig. 22.11 (**a**) QP band structure of rh-In_2O_3 with partial filling of the lowest conduction band, (**b**) Burstein-Moss shift as a function of the free electron density n_e in In_2O_3. The increase of the electron mass averaged over the filled states in the interval ε_F of the non-parabolic conduction band is displayed in (**c**). In (**b**) results for two polymorphs, the bcc bixbyite (*black*) and the rhombohedral (*red dashed*) structures, are compared with the effective mass approximation (*green dot-dashed*) assuming an isotropic and parabolic band. From [52]

22.3.2 Excitons in Transparent Conducting Oxides

In order to appropriately treat the influence of free carriers on an optical spectrum, i.e., on quasiparticle, excitonic and optical local-field effects, within a numerical calculation, we go back to the macroscopic polarization function (19.22). For a translationally invariant system with Bloch conduction and valence bands, $|c\mathbf{k}\rangle$ and $|v\mathbf{k}\rangle$, the (spin-independent) frequency-dependent polarization function is

$$P^M(c\mathbf{k}v\mathbf{k}, v'\mathbf{k}'c'\mathbf{k}'; \omega) = -\sqrt{\left|f\left(\varepsilon_c^{\mathrm{QP}}(\mathbf{k})\right) - f\left(\varepsilon_v^{\mathrm{QP}}(\mathbf{k})\right)\right|} \times \sum_{\Lambda} \frac{A_\Lambda(c\mathbf{k}v\mathbf{k})A_\Lambda^*(c'\mathbf{k}'v'\mathbf{k}')}{E_\Lambda - \hbar\omega - i\eta} \times \sqrt{\left|f\left(\varepsilon_{c'}^{\mathrm{QP}}(\mathbf{k}')\right) - f\left(\varepsilon_{v'}^{\mathrm{QP}}(\mathbf{k}')\right)\right|}. \quad (22.35)$$

Exploiting the Tamm-Dancoff approximation (see Sect. 19.3.1) the effective electron-hole pair Hamiltonian (19.32) reads as

$$H(c\mathbf{k}v\mathbf{k}, c'\mathbf{k}'v'\mathbf{k}') = \left[\varepsilon_c^{\mathrm{QP}}(\mathbf{k}) - \varepsilon_v^{\mathrm{QP}}(\mathbf{k})\right]\delta_{cc'}\delta_{vv'}\delta_{\mathbf{k}\mathbf{k}'} + \sqrt{\left|f\left(\varepsilon_c^{\mathrm{QP}}(\mathbf{k})\right) - f\left(\varepsilon_v^{\mathrm{QP}}(\mathbf{k})\right)\right|}\left[-W\,{}^{\mathbf{k}\mathbf{k}'}_{\substack{cc'\\vv'}} + 2\bar{v}\,{}^{\mathbf{k}\mathbf{k}'}_{\substack{cv\\v'c'}}\right] \times \sqrt{\left|f\left(\varepsilon_{c'}^{\mathrm{QP}}(\mathbf{k}')\right) - f\left(\varepsilon_{v'}^{\mathrm{QP}}(\mathbf{k}')\right)\right|} \quad (22.36)$$

with eigenvalues E_Λ and eigenvectors $A_\Lambda(c\mathbf{k}v\mathbf{k})$. We are interested in the linear optical properties of a crystal described by the frequency-dependent macroscopic dielectric tensor

$$\epsilon_{jj'}^M(\omega) = \delta_{jj'} + \frac{2}{\Omega}\sum_{c,v,\mathbf{k}}\sum_{c',v',\mathbf{k}'}\left(M_{cv}^{\mathbf{k}\mathbf{k}}\right)_j^*\left(M_{c'v'}^{\mathbf{k}'\mathbf{k}'}\right)_{j'}\left[P^M(c\mathbf{k}v\mathbf{k}, v'\mathbf{k}'c'\mathbf{k}'; \omega) + P^{M*}(c'\mathbf{k}'v'\mathbf{k}', v\mathbf{k}c\mathbf{k}; -\omega)\right] \quad (22.37)$$

with $\left(M_{cv}^{\mathbf{k}\mathbf{k}}\right)_j$ as the jth Cartesian component of the dipole matrix element (19.27). The sum of the two polarization functions indicates that resonant and antiresonant contributions are taken into account.

The presence of a degenerate gas of free carriers influences the quasiparticle bands, the screened potential, and the weighting occupation number factors in the Hamiltonian (22.36). Consequently, the polarization function (22.35) is influenced via the eigenvalues and eigenvectors of the pair problem. In addition the polarization function (22.35) is weighted by square roots of differences of Fermi functions, which mainly represent the Pauli blocking of optical transitions and, hence, describe

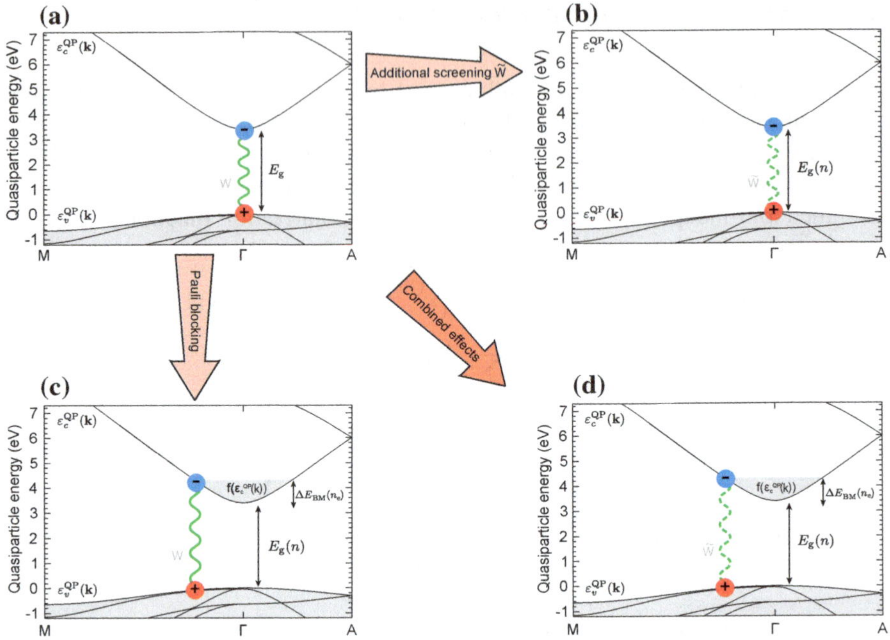

Fig. 22.12 Quasiparticle band structure and electron-hole interaction near the fundamental gap of n-doped wz-ZnO (schematically). (**a**) Electron-hole interaction (*green wavy line*) in undoped material. (**b**) Free-carrier screening affects the electron-hole attraction (*dashed green wavy line*). (**c**) Pauli blocking of the lowest conduction band with Burstein-Moss shift. (**d**) Both the Pauli blocking as well as the modified screening affect the formation of excitons. From [53]

the Burstein-Moss shift. Three of these effects are illustrated in Fig. 22.12 as they influence the allowed pair excitation energies and the screened Coulomb interaction.

The impact of the free-carrier effects on the imaginary part of (22.37), more precisely on the absorption coefficient (18.27), is displayed in Fig. 22.13 [53]. The pure effect of carrier screening on the electron-hole attraction in Fig. 22.13b leads to a weakening of excitonic effects in agreement with experimental findings for GaN, ZnO and In_2O_3 [50, 54–56]. As the striking effect the pronounced excitonic peak below the QP gap disappears with rising carrier concentration. For carrier concentrations of the order of or larger than $1.9 \times 10^{19}\,\mathrm{cm}^{-3}$ a bound exciton is seemingly not anymore visible in the absorption spectra. In Fig. 22.13c, the Pauli blocking of optical transitions shifts the absorption onset toward higher photon energies with increasing density. The exciton peak more or less follows the Burstein-Moss shift with respect to its absolute position, thereby, slightly increases its intensity. This increase of the peak height with increasing n_e is related to the modification of the effective electron-hole interaction by the occupation-number factors in (22.36). Due to the sharpness of the energy variation of the Fermi function for the partially occupied lowest conduction band at low temperatures a pronounced Fermi-edge singularity (FES) [57] occurs

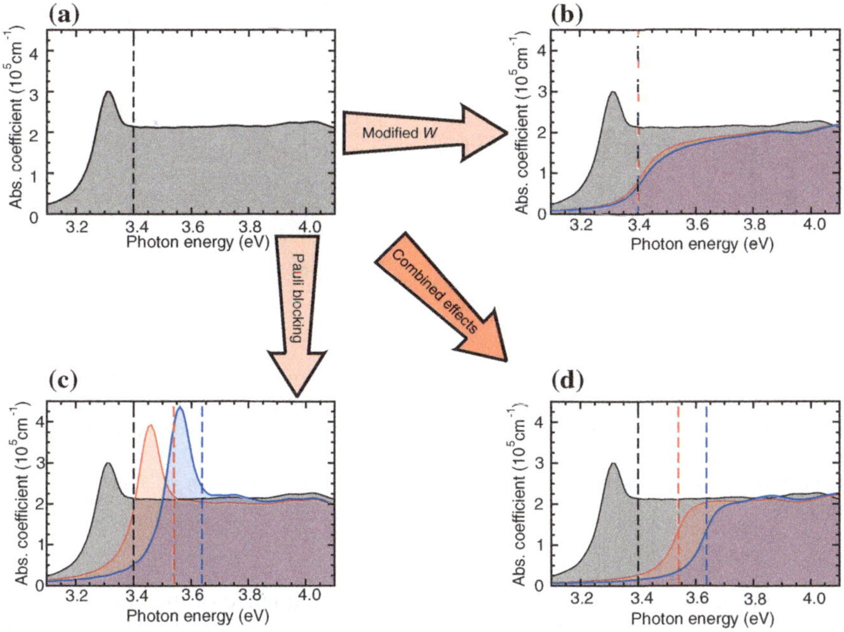

Fig. 22.13 Influence of various free-carrier effects on the frequency-dependent absorption of *wz*-ZnO (ordinary polarization) for densities $n_e = 0$ (*black*), 1.9 (*red*), and 4.8 (*blue*) $\times 10^{19}\,\mathrm{cm}^{-3}$. The *vertical dashed lines* display the position of the fundamental QP gap (without shrinkage) for zero Burstein-Moss shift (*upper panels* (**a**) and (**b**)) and including the BM effect (*lower panels* (**c**) and (**d**)). From [53]

at the absorption onset. This behavior, which contradicts both physical intuition and experimental findings, points out the necessity to simultaneously include the additional screening caused by the free carriers. This is demonstrated in Fig. 22.13d. The curves for finite densities n_e show a much steeper onset than the spectra in Fig. 22.13b. The increase of the oscillator strength in (c) is counteracted by the intraband screening contribution. Still excitonic effects are visible as will be discussed in Sect. 22.3.3.

The inclusion of the BGR effects within a Lindhard description of the intraband polarization as in Sect. 15.1.4 only shows a weak influence of a gap shrinkage of about $0.2 - 0.3\,\mathrm{eV}$ in the range of $n_e = 2{-}5 \times 10^{19}\,\mathrm{cm}^{-3}$ in Fig. 22.14. Nevertheless, the gap shrinkage is important to find agreement with absorption experiments [55] for both the energy position of the absorption edge as well as the steep onset that dominates the lineshape. The simultaneous account of all relevant many-body effects due to the presence of the degenerate electron gas in a parameter free study as illustrated by (22.35), (22.36) and (22.37) yields an unprecedented agreement with measured frequency-dependent absorption coefficients. This especially holds for the lineshape, which is dominated by the formation of Mahan excitons [57, 58] (see following section), but also for the absolute values. The seemingly small underestimation of

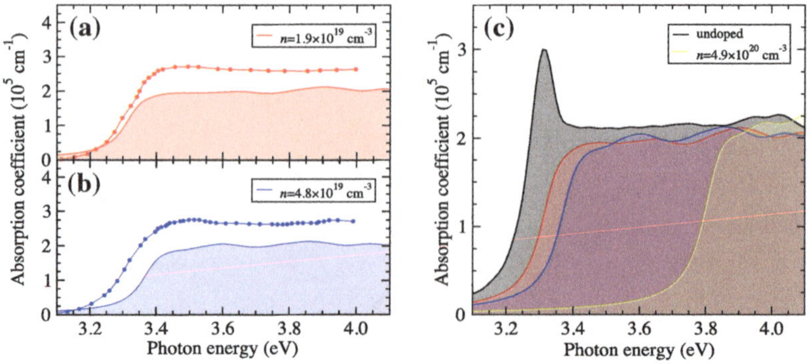

Fig. 22.14 (**a**, **b**) Impact of n-doping on the absorption spectrum of ZnO for ordinary light polarization for two different free-electron concentrations $n_e = 1.9$ (*red*) and 4.8 (*blue*) $\times 10^{19}\,\mathrm{cm}^{-3}$. Calculations: *solid lines*, measurements: *solid lines* with *dots* [55]. In (**c**) the curve for undoped ZnO (*black*) is compared with spectra for $n_e = 1.9$ (*red*), 4.8 (*blue*), and 49.0 (*yellow*) $\times 10^{19}\,\mathrm{cm}^{-3}$. The BGR has been also taken into account. A Lorentzian broadening of $\hbar\eta = 50\,\mathrm{meV}$ is assumed. From [53]

the absorption strength in the theory may be a consequence of the chosen constant lifetime broadening and problems with the sample homogeneity, the determination of free-carrier densities, and the determination of the layer thicknesses in the measurements. While in the studied range of free-carrier densities $n_e = 2\text{–}5 \times 10^{19}\,\mathrm{cm}^{-3}$ the BM and BGR effects tend to cancel each other, for higher densities the Pauli blocking with the resulting Burstein-Moss shift dominates the position of the absorption edge. This is illustrated in Fig. 22.14c for $n_e = 4.9 \times 10^{20}\,\mathrm{cm}^{-3}$. This figure shows another interesting phenomenon. The Sommerfeld factor (see Sect. 21.1.1), i.e., the Coulomb enhancement in the frequency region of the scattering states of the Coulomb-correlated electron-hole pairs, is less influenced by the actual screening, i.e., by the actual value of the free carrier density n_e. The absorption strength remains rather constant independent of the actual carrier density n_e.

22.3.3 Mahan Excitons

For varying free-carrier densities the excitonic effects are characterized in Fig. 22.15. It depicts the 'binding energy E_B' of the lowest pair excitation E_{Λ_0} measured with respect to the QP absorption edge $E_g(n_e) + \Delta E_{\mathrm{BM}}(n_e)$ and the Coulomb enhancement of the corresponding (dimensionless) optical oscillator strength (19.25)

$$f_j^{\Lambda_0} = \frac{8\pi\varepsilon_0 m}{e^2\hbar^2} E_{\Lambda_0} \left| \sum_{c,v} \sum_{\mathbf{k}} \left(M_{cv}^{\mathbf{kk}*} \right)_j A_{\Lambda_0}(cv\mathbf{k}) \sqrt{1 - f(\varepsilon_c^{\mathrm{QP}}(\mathbf{k}))\delta_{cc_0}} \right|^2$$

with c_0 as the lowest conduction band filled partially with free carriers and j as a light polarization direction perpendicular to the c-axis of the wurtzite crystal ZnO.

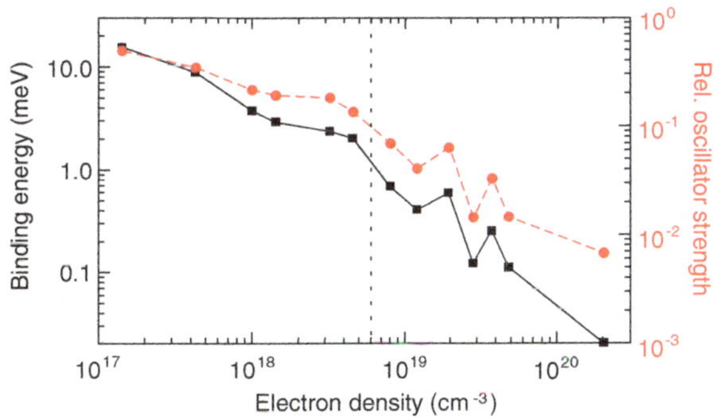

Fig. 22.15 Exciton binding energy (*solid black line*) and relative oscillator strength (*dashed red line*) normalized to the oscillator strength of the A exciton of undoped ZnO versus the free-electron concentration n_e for highly n-doped ZnO. The Mott density n_M estimated in the text is indicated by the *dotted vertical line*. From [53]

For both the binding energy E_B and the oscillator strength $f_j^{\Lambda_0}$ we find a very rapid decrease by orders of magnitude with increasing electron density n_e of the degenerate electron gas. The fluctuations of the values, e.g. in the case of E_B of the order of 1 meV or less for higher carrier densities, illustrate the limited numerical accuracy due to the **k**-point sampling. For densities $n_e < n_M$ the excitonic effects, electron-hole-pair binding and Coulomb enhancement of the oscillator strength, are still visible. Above the estimated Mott density the binding energy is small with $1-2$ meV (but not zero) and the oscillator strength amounts to 7 % of the value found for the corresponding Wannier-Mott exciton in the undoped material. Consequently, the *ab initio* calculations including quasiparticle band structures and the Coulomb matrix elements of the electron-hole interaction do not give rise to a sharp Mott transition, rather, to a continuous transition in another many-body state, the Mahan exciton [57, 58].

An excited electron-hole pair, which interacts with the Fermi sea of the degenerate electron gas forms a new, more collective excitation, the Mahan exciton, i.e., a bound state below the QP absorption edge with only an extremely small binding energy in agreement with experimental observations [50, 55, 59]. The term 'Mahan exciton' dates back to results of Mahan who calculated the influence of doping in semiconductors based on a two-band model [57]. The Mahan exciton can be also interpreted as the optical excitation of an additional electron into the Fermi sea and the strong interaction of the remaining localized hole with the whole Fermi sea [60, 61]. However, still a weak effective binding survives. The effect is especially pronounced for systems with quantum confinement [62, 63]. Therefore, the FES was first observed in the photoluminescence of a two-dimensional quantum well system [64]. The exclusion principle suppresses multiple electron-hole scattering processes as long as the electron energy is below the Fermi energy. At the Fermi edge the scattering rates are strongly enhanced, leading to the singularity and a consistently strong enhancement of the optical absorption compared to the case without interaction. The hole of the Mahan exciton needs to be localized to provide a sufficiently large spread in the Brillouin zone (see Figs. 22.10b and 22.11a), so that the **k**-preserving optical transitions into (absorption) or from (emission) the high-energy (and hence large **k**) electron states are possible. A band-to-band transition with the annihilation of a

Mahan exciton is basically possible but should be of weaker intensity due to the small number of holes in the valence band near the Fermi wave vector k_F.

Because of the sharpness of the Fermi surface at low temperatures and the Pauli exclusion principle, such a Mahan exciton is assumed to cause a singularity in the absorption spectrum at the onset, the FES. It may be described by a power-law divergence of the form [58]

$$\alpha(\omega) \sim [\hbar\omega - E_{\Lambda_0}]^{-\alpha_{\Lambda_0}}$$

with the coupling parameter α_{Λ_0} as a phenomenological measure of the strength of the singularity. In experimental studies, however, the influences of temperature and finite lifetime of the photo-generated electron-hole pairs together with the sample quality broaden and suppress the edge anomalies, thereby make their observation more difficult [50, 55, 59]. The edge character of the absorption coefficients in the theoretical as well as experimental spectra in Fig. 22.14a, b support the interpretation of the formation of Mahan excitons near the absorption onsets. The electric-dipole transitions of Mahan excitons at the Fermi edge should therefore show some resonant behavior in their emission, absorption, and reflection.

References

1. R. Zimmermann, *Many-Particle Theory of Highly Excited Semiconductors* (Teubner-Verlag, Leipzig, 1988)
2. G.M. Éliashberg, Interactions of electrons and lattice vibrations in a superconductor. Zh. Eksp. Teor. Fiz. **38**, 966–976 (1960) [Engl. Transl JETP **11**, 696–702 (1960)]
3. F. Bechstedt, Zur Theorie von Rumpfelektronenanregungen in Halbleitern. Habilitation thesis, Humboldt-Universität, Berlin (1981)
4. K. Shindo, Effective electron-hole interaction in shallow excitons. J. Phys. Soc. Jpn **29**, 287–295 (1970)
5. R. Zimmermann, Dynamical screening of the Wannier exciton. Phys. Status Solidi B **48**, 603–618 (1971)
6. A. Marini, R. Del Sole, Dynamical excitonic effects in metals and semiconductors. Phys. Rev. Lett. **91**, 176402 (2003)
7. F. Bechstedt, R. Enderlein, M. Koch, Theory of core excitons in semiconductors. Phys. Status Solidi B **99**, 61–70 (1980)
8. M. Rohlfing, S.G. Louie, Electron-hole excitations and optical spectra from first principles. Phys. Rev. B **62**, 4927–4944 (2000)
9. G. Strinati, Effects of dynamical screening on resonances at inner-shell thresholds in semiconductors. Phys. Rev. B **29**, 5718–5726 (1984)
10. F. Bechstedt, C. Rödl, L.E. Ramos, F. Fuchs, P.H. Hahn, J. Furthmüller, Parameterfree calculations of optical properties for systems with magnetic ordering or three-dimensional confinement, in *Epioptics-9, Proceedings of 39th Course of the International School on Solid State Physics*, Erice (Italy), ed. by A. Cricenti (World Scientific, New Jersey, 2008), pp. 28–40
11. P.H. Hahn, W.G. Schmidt, F. Bechstedt, Molecular electronic excitations calculated from a solid-state approach: methodology and numerics. Phys. Rev. B **72**, 245425 (2005)
12. U. Itoh, Y. Toyoshima, H. Onuki, Vacuum ultraviolet absorption cross sections of SiH_4, GeH_4, Si_2H_6, and Si_3H_8. J. Chem Phys. **85**, 4867–4872 (1986)

13. P.Y. Yu, M. Cardona, *Fundamentals of Semiconductors* (Springer, Berlin, 1996)
14. H. Haken, Die Theorie des Exzitons im festen Körper. Fortschr. Phys. **6**, 271–334 (1958)
15. Y. Toyozawa, Theory of the electronic polaron and ionization of a trapped electron by an exciton. Prog. Theor. Phys. **12**, 421–442 (1954)
16. F. Bechstedt, K. Seino, P.H. Hahn, W.G. Schmidt, Quasiparticle bands and optical spectra of highly ionic crystals: AlN and NaCl. Phys. Rev. B **72**, 245114 (2005)
17. A. Riefer, F. Fuchs, C. Rödl, A. Schleife, F. Bechstedt, R. Goldhahn, Interplay of excitonic effects and van Hove singularities in optical spectra: CaO and AlN polymorphs. Phys. Rev. B **84**, 075218 (2011)
18. F. Fuchs, C. Rödl, A. Schleife, F. Bechstedt, Efficient $O(N^2)$ approach to solve the Bethe-Salpeter equation for excitonic bound states. Phys. Rev. B **78**, 085103 (2008)
19. M. Bleicher, *Halbleiter-Optoelektronik* (Dr. Alfred Hüthig Verlag, Heidelberg, 1986)
20. Ch. Kittel, *Introduction to Solid State Physics* (Wiley, New York, 2005)
21. W. Martienssen, H. Warlimont (eds.), *Springer Handbook of Condensed Matter and Materials Data* (Springer, Berlin, 2005)
22. A. Schleife, F. Bechstedt, Ab initio description of quasiparticle band structures and optical near-edge absorption of transparent conducting oxides. J. Mater. Res. **27**, 2180–2186 (2012)
23. D.M. Roessler, W.C. Walker, Electronic spectrum and ultraviolet optical properties of crystalline MgO. Phys. Rev. **159**, 733–738 (1967)
24. R.C. Whited, C.J. Flaten, W.C. Walker, Exciton thermoreflectance of MgO and CaO. Solid State Commun. **13**, 1903–1905 (1973)
25. R. Zimmermann, M. Rösler, Theory of electron-hole plasma in CdS. Phys. Status Solidi B **75**, 633–645 (1976)
26. F. Bechstedt, K. Tenelsen, B. Adolph, R. Del Sole, Compensation of dynamical quasiparticle and vertex corrections in optical spectra. Phys. Rev. Lett. **78**, 1528–1531 (1997)
27. D.J.W. Geldart, R. Taylor, Wave-number dependence of the static screening function of an interacting electron gas. I. Lowest-order Hartree-Fock corrections. Can. J. Phys. **48**, 155–165 (1970)
28. D.J.W. Geldart, R. Taylor, Wave-number dependence of the static screening function of an interacting electron gas. II. Higher-order exchange and correlation effects. Can. J. Phys. **48**, 167–181 (1970)
29. S. Hong, G.D. Mahan, Conserving approximations: electron gas with exchange effects. Phys. Rev. B **50**, 8182–8188 (1994)
30. J.J. Chang, D.C. Langreth, Deep-hole excitations in solids. I. Fast-electron-plasmon effects. Phys. Rev. B **5**, 3512–3522 (1972)
31. J.J. Chang, D.C. Langreth, Deep-hole excitations in solids. II. Plasmons and surface effects in X-ray photoemission. Phys. Rev. B **8**, 4638–4654 (1973)
32. J.W. Gadzuk, Plasmon satellites in X-ray photoemission spectra. J. Electron Spectrosc. **11**, 355–361 (1977)
33. D. Chastenet, P. Longe, Intensity of plasmon satellites in ultrasoft-X-ray photoemission spectra. Phys. Rev. Lett. **44**, 91–95 (1980)
34. D. Chastenet, P. Longe, Intensity of plasmon satellites in ultrasoft-X-ray photoemission spectra. Phys. Rev. Lett. **44**, 903 (1980) (Erratum)
35. R. Zimmermann, Plasmon-Satelliten in Röntgen-Photoemissionsspektren. ZIE Preprint **81–3**, 47–60 (1981)
36. L.I. Johansson, I. Lindau, Photoemission studies of the energy dependence of the bulk plasmon loss intensity in Si and Al. Solid State Commun. **29**, 379–382 (1979)
37. M. Guzzo, J.J. Kas, F. Sottile, M.G. Silly, F. Sirotti, J.J. Rehr, L. Reining, Plasmon satellites in valence-band photoemission spectroscopy. Eur. Phys. J. B **85**, 324–330 (2012)
38. V.L. Bonch-Bruevich, *The Electronic Theory of Heavily Doped Semiconductors* (American Elsevier Publishing Company, New York, 1966)
39. T. Minami, Transparent conducting oxide semiconductors for transparent electrodes. Semicond. Sci. Technol. **20**, S35–S44 (2005)

40. G. Hautier, A. Miglio, G. Ceder, J.-M. Rignanese, X. Gonze, Identification and design principles of low effective mass p-type transparent conducting oxides. Nature Commun. **4**, 2292 (2013)
41. O.A. Gomes, H. Chacham, J.R. Mohallem, Variational calculations for the bound-unbound transition of the Yukawa potential. Phys. Rev. A **50**, 228–231 (1994)
42. Y. Li, X. Luo, H. Kröger, Bound states and critical behavior of the Yukawa potential. Sci. China Ser. G **49**, 60–71 (2006)
43. N.F. Mott, *Metal-Insulator Transitions* (Barnes & Noble, New York, 1974)
44. N.F. Mott, The basis of the electron theory of metals, with special reference to the transition metals. Proc. Phys. Soc. A **62**, 416–422 (1949)
45. H. Haug, S.W. Koch, *Quantum Theory of the Optical and Electronic Properties of Semiconductors* (World Scientific, Singapore, 2009)
46. C. Klingshirn, *Semiconductor Optics* (Springer, Berlin, 2007)
47. S.A. Moskalenko, D. Snoke, *Bose-Einstein Condensations of Excitons and Biexcitons* (Cambridge University Press, Cambridge, 2000)
48. Landolt-Börnstein New Series, *Group III, 41B* (Springer, Berlin, 1999)
49. C. Klingshirn, R. Hauschild, J. Fattert, H. Kalt, Room-temperature stimulated emission of ZnO: alternatives to excitonic lasing. Phys. Rev. B **75**, 115203 (2007)
50. S. Shokhovets, K. Köhler, O. Ambacher, G. Gobsch, Observation of Fermi edge excitons and exciton-phonon couplexes in the optical response of heavily doped n-type wurtzite GaN. Phys. Rev. B **79**, 045201 (2009)
51. A. Schleife, F. Fuchs, C. Rödl, J. Furthmüller, F. Bechstedt, Band structure and optical-transition parameters of wurtzite MgO, ZnO, and CdO from quasiparticle calculations. Phys. Status Solidi B **246**, 2150–2153 (2009)
52. F. Fuchs, F. Bechstedt, Indium-oxide polymorphs from first principles: quasiparticle electronic states. Phys. Rev. B **77**, 155107 (2008)
53. A. Schleife, C. Rödl, F. Fuchs, K. Hannewald, F. Bechstedt, Optical absorption in degenerately doped semiconductors: mott transition or Mahan excitons? Phys. Rev. Lett. **107**, 236404 (2011)
54. T. Makino, Y. Segawa, S. Yoshida, A. Tsukasaki, A. Ohtomo, M. Kawasaki, Gallium concentration dependence of room-temperature near-band-edge luminescence in n-type ZnO: Ga. Appl. Phys. Lett. **85**, 759–761 (2004)
55. T. Makino, K. Tamura, C.H. Chia, Y. Segawa, M. Kawasaki, A. Ohtomo, H. Koinuma, Optical properties of ZnO: Al epilayers: observation of room-temperature many-body absorption-edge singularity. Phys. Rev. B **65**, 121201(R) (2002)
56. H. Fujiwara, M. Kondo, Effects of carier concentration on the dielectric function of ZnO: Ga and In_2O_3:Sn studied by spectroscopic ellipsometry: analysis of free-carrier and band-edge absorption. Phys. Rev. B **71**, 075109 (2005)
57. G.D. Mahan, Excitons in degenerate semiconductors. Phys. Rev. **153**, 882–889 (1967)
58. G.D. Mahan, *Many-Particle Physics* (Plenum Press, New York, 1990)
59. M. Feneberg, J. Däubler, K. Thonke, R. Sauer, P. Schley, R. Goldhahn, Mahan excitons in degenerate wurtzite InN: photoluminescence spectroscopy and reflectivity measurements. Phys. Rev. B **77**, 245207 (2008)
60. H. Haug, S. Schmitt-Rink, Electron theory of the optical properties of laser-excited semiconductors. Prog. Quantum Electron. **9**, 3–100 (1984)
61. S. Nojima, Dimensionality of exciton-state renormalization in highly excited semiconductors. Phys. Rev. B **51**, 11124–11127 (1985)
62. P. Hawvylak, Optical properties of a two-dimensional electron gas: evolution of spectra from excitons to Fermi-edge singularities. Phys. Rev. B **44**, 3821–3828 (1991)
63. N.A.J.M. Kleemans, J. van Bree, A.O. Govarov, J.G. Keizer, G.J. Hamhuis, R. Nötzel, P.M. Koenraad, Many-body exciton states in self-assembled quantum dots coupled to a Fermi sea. Nature Phys. **6**, 534–538 (2010)
64. M.S. Skolnick, J.M. Rorison, K.J. Nash, D.J. Mowbray, P.R. Tapster, S.J. Bass, A.D. Pitt, Observation of a many-body edge singularity in quantum-well luminescence spectra. Phys. Rev. Lett. **58**, 2130–2133 (1987)

Index

F. Bechstedt, *Many-Body Approach to Electronic Excitations*,
Springer Series in Solid-State Sciences 181, DOI 10.1007/978-3-662-44593-8

K

L

M

CPSIA information can be obtained at www.ICGtesting.com
Printed in the USA
LVOW01*0140171214

419128LV00001B/9/P